COOL STARS, STELLAR SYSTEMS, AND THE SUN

Seventh Cambridge Workshop

A SERIES OF BOOKS ON RECENT DEVELOPMENTS IN ASTRONOMY AND ASTROPHYSICS

Printed by BookCrafters, Inc.

First published 1992

Library of Congress Catalog Card Number: 92-71694
ISBN 0-937707-45-7

D. Harold McNamara, Managing Editor of Conference Series
408 ESC Brigham Young University
Provo, UT 84602

ASTRONOMICAL SOCIETY OF THE PACIFIC
CONFERENCE SERIES

Volume 26

COOL STARS, STELLAR SYSTEMS, AND THE SUN

Seventh Cambridge Workshop

Edited by
Mark S. Giampapa and Jay A. Bookbinder

A SERIES OF BOOKS ON RECENT DEVELOPMENTS IN ASTRONOMY AND ASTROPHYSICS

Vol. 1-Progress and Opportunities in Southern Hemisphere Optical Astronomy: The CTIO 25th Anniversary Symposium
ed. V. M. Blanco and M. M. Phillips ISBN 0-93707-18-X

Vol. 2-Proceedings of a Workshop on Optical Surveys for Quasars
ed. P. S. Osmer, A. C. Porter, R. F. Green, and C. B. Foltz ISBN 0-937707-19-8

Vol. 3-Fiber Optics in Astronomy
ed. S. C. Barden ISBN 0-937707-20-1

Vol. 4-The Extragalactic Distance Scale: Proceedings of the ASP 100th Anniversary Symposium
ed. S. van den Bergh and C. J. Pritchet ISBN 0-937707-21-X

Vol. 5-The Minnesota Lectures on Clusters of Galaxies and Large-Scale Structure
ed. J. M. Dickey ISBN 0-937707-22-8

Vol. 6-Synthesis Imaging in Radio Astronomy: A Collection of Lectures from the Third NTAO Synthesis Imaging Summer School
ed. R. A. Perley, F. R. Schwab, and A. H. Bridle ISBN 0-937707-23-6

Vol. 7-Properties of Hot Luminous Stars: Boulder-Munich Workshop
ed. C. D. Garmany ISBN 0-937707-24-4

Vol. 8-CCDs in Astronomy
ed. George H. Jacoby ISBN 0-937707-25-2

Vol. 9-Cool Stars, Stellar Systems, and the Sun. Sixth Cambridge Workshop
ed. G. Wallerstein ISBN 0-937707-27-9

Vol. 10-The Evolution of the Universe of Galaxies. The Edwin Hubble Centennial Symposium
ed. Richard G. Kron ISBN 0-937707-28-7

Vol. 11-Confrontation Between Stellar Pulsation and Evolution
ed. C. Cacciari and G. Clementini ISBN 0-937707-30-9

Vol. 12-The Evolution of the Interstellar Medium
ed. L. Blitz ISBN 0-937707-31-7

Vol. 13-The Formation and Evolution of Star Clusters
ed. K. Janes ISBN 0-937707-32-5

Vol. 14-Astrophysics with Infrared Arrays
ed. R. Elston ISBN 0-937707-33-3

Vol. 15-Large-Scale Structures and Peculiar Motions in the Universe
ed. D. W. Latham and L. A. N. da Costa ISBN 0-937707-34-1

Vol. 16-Atoms, Ions and Molecules: New Results in Spectral Line Astrophysics
ed. A. D. Haschick and P. T. P. Ho ISBN 0-937707-35-X

Vol. 17-Light Pollution, Radio Interference, and Space Debris
ed. D. L. Crawfoed. D. L. Crawford ISBN 0-937707-36-8

Vol. 18-The Interpretation of Modern Synthesis Observations of Spiral Galaxies.
ed. M. Duric and P. C. Crane ISBN 0-937707-37-6

Vol. 19-Radio Interferometry: Theory, Techniques, and Applications, IAU Colloquium 131
ed. T. J. Cornwell and R. A. Perley ISBN 0-937707-38-4

Vol. 20-Frontiers of Stellar Evolution, celebrating the 50th Anniversary of McDonald Observatory
ed. D. L. Lambert ISBN 0-937707-39-2

Vol. 21-The Space Distribution of Quasars
ed. D. Crampton ISBN 0-937707-40-6

Vol. 22-Nonisotropic and Variable Outflows from Stars
ed. Laurent Drissen, Claus Leitherer, and Antonella Nota ISBN 0-937707-41-4

Vol. 23-Astronomical CCD Observing and Reduction Techniques
ed. Steve B. Howell ISBN 0-937707-42-4

Vol. 24-Cosmology and Large-Scale Structure in the Universe
ed. Reinaldo R. de Carvalho ISBN 0-937707-43-0

Vol. 25-Astronomical Data Analysis Software and Systems I
ed. Diana M. Worrall, Chris Biemesderfer, and Jeannette Barnes ISBN 0-937707-44-9

Inquiries concerning these volumes should be directed to the:
Astronomical Society of the Pacific
CONFERENCE SERIES
390 Ashton Avenue
San Francisco, CA 94112-1722

TABLE OF CONTENTS

PART I – A BONAS RETROSPECTIVE

PART II – SPACE BASED RESULTS

PART III – CONVECTION

PART IV – MAGNETIC FIELDS

Preface

The Seventh Cambridge Workshop on Cool Stars, Stellar Systems and the Sun was held in Tucson, Arizona, during October 9 - 12, 1991. The National Solar Observatory of the National Optical Astronomy Observatories was the host of the largest Cool Stars Workshop yet with approximately 200 participants. The plethora of exciting new results presented at the Tucson Cool Stars Workshop served to usher in a new decade of discovery in this dynamic area of contemporary astrophysics.

The principal goal of the Seventh in the highly acclaimed series of Cambridge Workshops was to discuss some of the major themes of research that are common to both solar physics and stellar astrophysics. That we achieved a productive synergism between the solar and stellar communities is clearly reflected in these proceedings. We began the meeting with a look back at some of the ideas that prevailed in the 'solar-stellar connection' at the time of the seminal conference on Solar Phenomena in Stars and Stellar Systems which was held at Bonas, France, in 1980. We then focused our attention on fundamental physical processes in the photospheres and outer atmospheres of the Sun and late-type stars.

Convection is a fundamental mode of energy transport in stars that is likely the engine which, together with magnetic fields, gives rise to chromospheres and coronae. The interaction between convection, magnetic fields and energy transport through the outer layers of late-type stars is now being studied through advanced, multi-dimensional computational techniques. Detailed observations of the Sun, along with high spectral resolution stellar observations, are yielding invaluable constraints that the models must satisfy. Synoptic observations, both solar and stellar, are revealing the empirical relationship between magnetic fields and convection over time scales comparable to that of the solar cycle and to the cycle periods of sun-like stars.

The dominant chapter in the proceedings concerns magnetic fields in the Sun and stars. This is indicative of the central role of magnetic fields in solar and stellar physics. The structure seen in the solar atmosphere, and inferred in the case of stellar atmospheres, is defined by magnetic fields arising from an interior dynamo that is not (yet) well understood. A fundamental discovery of the past decade is that magnetic fields modulate the radiative output of the Sun and stars in addition to determining their angular momentum evolution.

Recent observations of the solar corona with novel instrumentation suggest that our early picture of magnetic structures as long-lived, stationary features was incomplete due to the inadequate spatial and spectral resolution of previous work. In fact, the large-scale, seemingly stable, long-lived magnetic configurations are actually composed of short-lived bundles of structures. Parallel progress in the area of stellar magnetic properties indicates that the difference between active and quiet stars probably involves the fractional area coverage of photospheric magnetic fields rather than the actual field strengths. Moreover, the quality of the stellar observations now demands further analytical refinements, such as the consideration of the magnetic field gradient with height, in order to accurately infer stellar magnetic field properties.

Mass loss in the form of winds from stars is also a significant physical process that plays an important role at various stages of stellar evolution. The recent work represented in these proceedings suggests that the usual treatment of mass loss as a steady, spherically symmetric outflow will require revision. In particular, stochastic processes along with a highly localized origin for winds may be a more appropriate description.

The application of sophisticated radiative transfer techniques is becoming a necessary tool to properly and fully interpret stellar spectroscopic data. In view of these developments, the SOC included a session on the advanced computer codes that are becoming available and which incorporate these techniques and atomic data. Ironically, this crucial diagnostic approach is now often ignored or neglected in much of contemporary astrophysics, even to the point where graduate courses in stellar atmospheres are no longer taught in many astronomy departments.

The chapter entitled "Space-Based Results" is dedicated to the memory of Giuseppe S. "Pippo" Vaiana, who so actively contributed to X-ray studies of the Sun and stars. In this chapter the reader is given a glimpse of the latest discoveries arising from the Hubble Space Telescope and the ROSAT X-ray Observatory. These forefront missions have placed us at the threshold of discovery, shedding new light on long-standing problems and stimulating new and ever deeper questions. The data from other operational facilities, such as IUE, along with the archives of past missions, continue to offer new insights on the universe.

Our closing session on "The Future" is in recognition of the fact that the history of astronomy is a chronicle of innovative instruments as much as it is a history of new ideas. Each new instrument produces discoveries that enhance our perception of the universe while constantly challenging our understanding of the laws of physics. Equally challenging is the political process and the role of the citizen-scientist in this arena. We as individuals and as a community must actively participate in the process in order to obtain the precious resources that are required to build advanced facilities. We were indeed fortunate to have in attendance an astronomer who is a former staff member of the National Space Council, Col. Dr. Pete Worden. Col. Worden described to us the challenges that the astronomical community faces, particularly with regard to the conduct of the space science program in the United States. His impromptu remarks given at the end of the workshop are transcribed in these proceedings.

At first sight, the sheer volume of these proceedings may lead the casual observer to conclude that solar and stellar physics are "mature" fields in which most of the current effort is devoted to "detailed refinement." Any such conclusion is a mistake. These proceedings illustrate that now, more than ever, solar-stellar physics is confronting profound questions the answers to which have important implications for our understanding of astrophysical processes in other regimes.

The scientific and logistical organization of a meeting of this scope would not have been possible without the dedicated work of a supporting (and supportive!) cast, beginning with the SOC. The members included: E. Böhm-Vitense, J. Bookbinder, P. B. Byrne, J.-P. Caillault, L. Cram, J. Davila, A. Dupree, R. Gershberg, M. Giampapa (Chairman, SOC and LOC), K. Harvey, J. Linsky, D. Mullan, D. Rabin, M. Rodonò, J. Schmitt, T. Simon, A. Skumanich, J. Zirker, and C. Zwaan. The large SOC was selected to represent the international character of the work in our field. Their ideas were invaluable to the shaping of the scientific program. Many hard choices had to be made, and the SOC made those decisions in a spirit of selflessness and cooperation.

A meeting of this size and scope is characterized by innumerable logistical details. Mary Guerrieri merits special recognition for all her hard work and patient, dedicated efforts that made the workshop into a reality. Ann Barringer was present with Ms. Guerrieri throughout the workshop where, together, they skillfully handled any and all contingencies, both routine and otherwise. La Rae Rothe, Joyce du Hamel, Linda Johnson, and Emmerson Bartlomé also deserve recognition for their contributions. We are especially grateful to Sidney Wolff, Director-NOAO, and John Leibacher, Director-NSO, for their enthusiastic support of both the workshop and the research efforts in solar-stellar physics at our national centers. Doug Kline gave invaluable advice and assistance in our efforts to generate the indexing and formatting software.

We gratefully acknowledge the substantive contributions of the sponsoring organizations and agencies. Money, if not the root of all evil, is certainly at the root of all successful workshops! We acknowledge generous grants from NASA and the NSF to support graduate student and post-doc participation, as well as partial expenses of some invited speakers, other young researchers, and additional secretarial assistance. We were fortunate to receive contributions from corporate sponsors to whom we are most grateful. They included Ball Aerospace Corporation, Hughes-Danbury Optical Systems, Inc., and Photometrics, Ltd. of Tucson, Arizona. We acknowledge the continuing support of this series of workshops by the Smithsonian Astrophysical Observatory. A contribution from Ghost Writers in the Sky of Boulder, Colorado, is also gratefully acknowledged. Finally, we extend our congratulations (and our sympathies) to J.-P. Caillault and the University of Georgia who will host the 8th. Cool Stars Workshop in Athens, GA., in 1993. We wish the organizers of the 8th. Cool Stars Workshop the best of luck!!!!

The editors expended enormous efforts to produce a timely volume that would also continue to serve as a useful reference work for years to come. We trust that you, the reader, will find in these proceedings something of value that will make you, too, want to reach for the stars.

Mark S. Giampapa
Tucson, Arizona

Jay A. Bookbinder
Cambridge, Massachussetts

Workshop Participants

T.B. Ake
NASA/GSFC

Carol Ambruster
Villanova Univ.

B.G. Anandarao
Physical Research Lab
India

L. Anderson
University of Toledo

Philippe Andre
Service d'Astrophysique
France

V. Andretta
Univ. of Naples

J.H. Applegate
Columbia University

P. Avellar
NSO/NOAO

Eugene Avrett
Harvard-Smithsonian C.f.A.

Tom Ayres
Univ. of Colorado

B. Barbuy
Univ. of Sao Paulo
Brazil

G. Basri
University of California

T.S. Bastian
NRAO

Wendy Hagen Bauer
Wellesley College

L. Belkora
CalTech

Erika Bohm-Vitense
Univ. of Washington

J. A. Bookbinder
Harvard-Smithsonian C.f.A.

Arnold I. Boothroyd
University of Toronto

C. Stuart Bowyer
University of California

Gordon Bromage
Rutherford-Appleton Lab
England

Alex Brown
Univ. of Colorado

E.W. Brugel
Univ. of Colorado

J.H.M.J. Bruls
Sterrekundig Instituut
Netherlands

David Bruning
Stellar Research and
Education
Waukesha, WI

R. Burton
Univ. of Colorado

John Butler
Armagh Observatory
Northern Ireland

Derek Buzasi
Johns Hopkins Univ.

P. B. Byrne
Armagh Observatory
Northern Ireland

J.P. Caillault
University of Georgia

Nuria Calvet
Cen. Invest. Astronomica
Venezuela

Mats Carlsson
Institute of Theoretical Astrophysics
Norway

Ken Carpenter
NASA/GSFC

S. Casanova
Penn State University

S. Catalano
Osservatorio Astrofisico di Catania
Italy

F. Cattaneo
Univ. of Chicago

Kwing L. Chan
NASA/GSFC

E.S. Chang
Univ. of Massachusetts

Paul Charbonneau
NCAR

C.-C. Cheng
Naval Research Lab.

K.-P. Cheng
NASA/GSFC

G. Clayton
Univ. of Colorado

W.A. Coles
Univ. of California, La Jolla

Eric R. Craine
Western Research Company, Inc.
Tucson, USA

R. Culver
Colorado State Univ.

Manfred Cuntz
Univ. of Colorado

Giuseppe Cutispoto
Osservatorio Astrofisico di Catania
Italy

C. Deliyannis
Univ. of Hawaii

Andrea K. Dobson
Whitman College

J.F. Donati
Univ. of Western Ontario

J.G. Doyle
Armagh Observatory
N. Ireland

Stephen Drake
NASA/GSFC

Dainis Dravins
Institute for Astronomy
Lund, Sweden

Doug Duncan
S.T.Sc.I.

Andrea Dupree
Harvard-Smithsonian C.f.A.

Bernard R. Durney
National Solar Observatory

Joel A. Eaton
Center of Excellence in Information Systems
Nashville, TN

Robert Erdelyi
Katholiche Universiteit Leuven
Astronomisch Instituut
Belgium

G. Fisher
University of Hawaii

Tom Fleming
MPE
Germany

Bernard Foing
L.P.S.P.
France

D. Fox
Univ. of Colorado

Christ Ftaclas
Hughes Danbury Optical Systems

K. Gayley
Univ. of Colorado

R. E. Gershberg
Crimean Astrophys. Observ.

M.S. Giampapa
NSO/NOAO

David Gibson
MIT Lincoln Lab.

Ron Gilliland
S.T.Sc.I.

L. Golub
Harvard-Smithsonian C.f.A.

David Gray
Univ. of Western Ontario

V. Grinin
Crimean Astrophys. Observ.

M. Guedel
Univ. of Colorado

J.C. Hall
Penn State University

Reiner Hammer
Kiepenhauer-Inst. fur Sonnenphys.
Germany

Arnold Hanslmeier
Karl-Franzens-Universitat
Austria

John W. Harvey
National Solar Observatory

K.L. Harvey
National Solar Observatory

Suzanne Hawley
OCIW

A. Hempelmann
Astrophysikalisches Observatorium Potsdam
und Sternwarte Babelsberg
Germany

Richard N. Herring
Ball Aerospace Systems Group

Kenneth H. Hinkle
KPNO

Keith Horne
S.T.Sc.I.

E.R. Houdebine
Dublin Inst. for Advanced Studies
Ireland

Robert Howard
National Solar Observatory

W. Hubbard
Univ. of Arizona

David Huenemoerder
NASA Headquarters

Shawn M.G. Hughes
Anglo-Australian Obs.
Australia

David G. Iadevaia
Pima College East

C.L. Imhoff
Computer Sciences Corp.

John Jefferies
NOAO

R.D. Jeffries
Univ. of Birmingham
England

L. Jetsu
Univ. of Helsinki
Denmark

C.M. Johns
Univ. of California

Hollis Johnson
Indiana University

H. Jones
NASA/GSFC

J.G. Jorgensen
Niels Bohr Institute
Copenhagen, Denmark

P. Judge
HAO/NCAR

W. Kalkofen
Harvard-Smithsonian C.f.A

Magarita Karovska
Harvard-Smithsonian C.f.A.

Y.-C. Kim
Yale University

R.W. Komm
National Solar Observatory

G. Kopp
National Solar Observatory

J. Krautter
Laudessternwarte
Konigstuhl
Heidelberg, Germany

N. Paul Kuin
7601 Ora Glen Dr., Suite 300
Greenbelt, MD 20770

M. R. Kundu
University of Maryland

Martin F. R. Kurster
MPE, Germany

A. Lanzafame
Osservatorio Astrofisico
di Catania
Italy

Sandy Leggett
U.S. Naval Observatory

John Leibacher
National Solar Observatory

Phillippe Lemaire
LPSP
France

D. Lenz
Univ. of Colorado

E.H. Levy
Univ. of Arizona

J.W. Liebert
Univ. of Arizona

J. Lim
Univ. of Maryland

J.L. Linsky
Univ. of Colorado

Irene R. Little-Marenin
Univ. of Colorado

William Livingston
National Solar Observatory

J.I. Lunine
Univ. of Arizona

Don Luttermoser
Iowa State University

T.J. Lydon
Yale University

K.P. Marshall
Inst. of Astronomy
Cambridge, England

M. Mathioudakis
Armagh Obs.
N. Ireland

D.J. Mullan
Univ. of Delaware

Z. E. Musielak
Univ. of Alabama

R. L. Mutel
University of Iowa

Ana Nash
NRL

J. Neff
Penn. State Univ.

Anastasios Nesis
Kiepenheuer Inst. fur
Sonnenphysik
Germany

J. Newmark
NASA/GSFC

Robert Noyes
Harvard-Smithsonian C.f.A.

Isabella Pagano
Istituto di Astronomia
Universita di Catania

Luca Pasquini
European Southern Obs.

B.M. Patten
Inst. for Astronomy

R.C. Peterson
Lick Obs.

R. Phillips
MIT Haystack Obs.

M. Pinsonneault
Yale Univ.

A.J.M. Piters
Astron. Inst. "Anton Pannekaeh"
Netherlands

Jason Porter
NASA/MSFC

Charles R. Proffitt
S.T.Sc.I.

Charles Prosser
Harvard-Smithsonian C.f.A.

John Pye
Leicester University
England

D. Rabin
NSO/NOAO

Richard R. Radick
AFGL/PHS

Lawrence W. Ramsey
Penn State University

S. Randich
Osservatorio Astrofisico
di Arcetri
Italy

M. Richards
University of Virginia

Richard D. Robinson
NASA/GSFC

P. Rossi
Harvard-Smithsonian C.f.A

Mr. Andrew Rowe
Dept. of Theoretical Physics
Oxford, England

S. Rucinski
York University
Toronto, Canada

G. Rudiger
Astrophysical Obs. of Potsdam

Robert Rutten
Sterrekundig Instituut
The Netherlands

Steve Saar
Harvard-Smithsonian C.f.A

P.V. Sada
New Mexico State Univ.

William H. Sandmann
Harvey Mudd College

Dimitar D. Sasselov
Harvard-Smithsonian C.f.A.

D. Saumon
Univ. of Arizona

J. Schmitt
MPE
Germany

C. J. Schrijver
Astronomical Inst. Utrecht
Netherlands

Salvatore Serio
Osservatorio Astronomico
Palermo, Italy

T.A.A. Sigut
Harvard-Smithsonian C.f.A.

Ted Simon
Institute for Astronomy

Steve Skinner
Univ. of Colorado

R. Smartt
National Solar Observatory

David R. Soderblom
S.T.Sc.I.

Sami K. Solanki
Institut fur Astronomie
ETH-Zentrum, Switzerland

C.P. Sonett
Univ. of Arizona

John R. Stauffer
NASA/Ames

Robert Stein
Michigan State University

Prof. R. Steinitz
Ben-Gurion Univ.

R.E. Stencel
Univ. of Colorado

Robert A. Stern
Lockheed Research Labs

Ms. Karen Strom
Univ. of Massachusetts

G. Tagliaferri
ESTEC (SAC)

R.N. Thomas
Pinebrook Hills
1155 Timberlane Boulder, CO
80302

J.A. Thorburn
Univ. of Chicago

A. Title
Lockheed Research Labs

C. Toner
NSO/NOAO

Han Uitenbroek
Harvard-Smithsonian C.f.A.

P. Ulmschneider
Univ. of Heidelberg
Germany

Roger Ulrich
University of California, L.A.

J.A. Valenti
U. C. Berkeley

R.A. Wade
Pennsylvania State Univ.

G. Wahlgren
NASA/GSFC

Fred Walter
SUNY at Stony Brook

A. Welty
Penn State Univ.

N. White
Lowell Obs.

S. White
Univ. of Maryland

B. Whitney
Harvard-Smithsonian C.f.A.

Peter Wilson
Caltech

S.P. Worden
SDIO/TN
Dept. of Defense

J. P. Zahn
Observatoire du Pic-du-Midi
France

M. Zeilik
Univ. of New Mexico

C. Zwaan
Sterrekundig Instituut
The Netherlands

PART I

A Bonas Retrospective

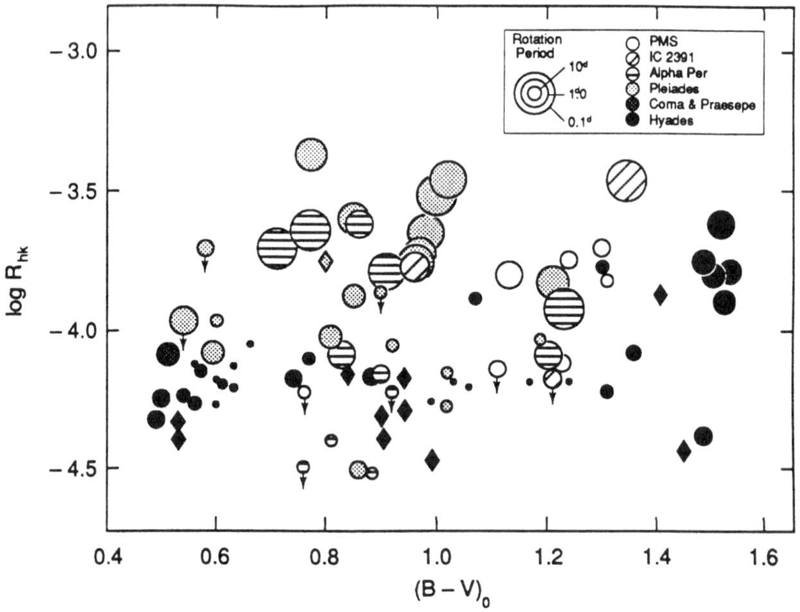

Figure 3 from "A Bonas Retrospective" by T. Simon.

Seventh Cambridge Workshop on Cool Stars, Stellar Systems, and the Sun
ASP Conference Series, Vol. 26, 1992
Mark S. Giampapa and Jay A. Bookbinder (eds.)

BONAS + *11*: THE ACTIVITY-ROTATION-AGE RELATIONS FOR SOLAR-TYPE STARS

THEODORE SIMON
Institute for Astronomy, University of Hawaii, 2680 Woodlawn Drive, Honolulu, HI 96822

ABSTRACT Throughout its life a main-sequence star like the Sun endures an inexorable decline in its rotation and in the magnetic activity generated by its internal dynamo. Others have suggested that this time decay of activity and rotation can be described by an empirical formula, such as the well-known $t^{-1/2}$ relation. However, recently acquired UV and rotational data, as well as theoretical considerations of rotational braking, suggest instead that dynamo activity is first and foremost a function of rotation; that the strength of this activity and its decline, coming as a direct result of magnetic braking, are governed by a complicated feedback process; that the level of activity of individual stars is, as a result, only an implicit function of age; and that the dependence of dynamo activity and rotation upon age may therefore be different from one star to the next. The existing time decay laws for activity and rotation are thus unlikely to be correct, and efforts—past and present—to choose among them can serve no useful purpose. In this talk, by which we commemorate more than a decade of research on the solar-stellar connection since the NATO meeting in Bonas (France), I will state my case for discarding all of these time decay relations and for embracing the new activity-rotation-age paradigm that has been emerging since we met in Bonas.

Keywords: chromospheres; stellar atmospheres; stars: rotations

ORIGINS OF THE PARADIGM

The origin of solar magnetic activity is one of the central problems of modern astrophysics. Astronomers continue to be confounded by the mysterious dynamo within the Sun that appears to be the ultimate cause of this activity. Much of the difficulty we face in unlocking the secrets of the solar interior arises because the Sun is but one star viewed during a brief moment in its evolution. Yet, in the galactic neighborhood of the Sun there exist many examples of both younger Suns and older Suns, stars much like our Sun except for their age. The study of these stars must surely reveal to us both the history of the Sun's past as well as its future. This is the subject of the solar-stellar connection—the study of the Sun as revealed by observations of Sun-like stars.

Work on the solar-stellar connection began more than a quarter of a century ago with the recognition by O.C. Wilson and R.P. Kraft, among others, that the emission reversals in the cores of the Ca II H and K lines could be used to chart out solar-like activity across the cool half of the H-R diagram, and to find out how this activity depends on such fundamental stellar properties as mass, age, and rotation. This connection between solar and stellar activity was based on the knowledge that Ca II emission on the Sun traces out regions of strong magnetic fields and strong mechanical heating. These early studies by Wilson (1963) and Kraft (1967) showed that stellar activity subsides with age, from the strong emission of young stars in clusters like the Pleiades to the weak emission of middle-aged stars like the Sun, and that this decay in activity is accompanied by a decline in rotation, from the rapid rotation of young stars to the barely perceptible rotation of old stars. Encouraged by the success of the Babcock model of the solar dynamo and by studies of the relationship between the solar wind and the spindown of solar rotation (Schatzman 1962, Weber and Davis 1967), astronomers were thus led to the current paradigm of solar-stellar activity, in which the dynamo becomes the driving source of this activity, as well as the ultimate cause of its demise.

With the launch of the *Einstein X-ray Observatory* and the *International Ultraviolet Explorer (IUE)* satellite in the late 1970s, shortly before the conference in Bonas that we commemorate here today, stellar astronomers gained access at last to the wavelength regions where most of the emission from high-temperature coronas, chromospheres, and transition regions can be seen. The extensive UV and X-ray data gathered by those spacecraft have, over the last decade, reaffirmed much that was learned about stellar activity from the H-K lines. But not surprisingly, they have raised many new and profound questions of their own concerning the nature of stellar dynamos. Some of those issues now await study by *ROSAT* and by the *Hubble Space Telescope*. Others are best addressed from the ground, where sensitive electronic detectors and large telescopes now afford us a much better opportunity to explore crucial aspects of this science than ever before.

The purview of my talk, then, is to sum up how our ideas and understanding of the solar-stellar connection, spurred along by these changes in technology, have progressed since the Bonas meeting of scarcely a dozen years ago.

THE TIME DECAY OF ROTATION

In 1972 A. Skumanich published his now famous paper on the square root relations for the time decay of stellar rotation and stellar chromospheric activity. These empirical relations state that the rotational velocity and strength of the chromospheric Ca II H-K emission of a late-type star decline with the square root of the star's age (see Fig. 1). You will recall that at the Bonas meeting Skumanich described in detail how these correlations were derived (Skumanich and Eddy 1981).[1] To my knowledge, it was B. Durney (1972), in remarks published in the

[1] In his derivation, Skumanich implicitly assumed that the time scale for the loss of activity and rotation is independent of mass, *i.e.,* identical for stars more massive or less massive than the Sun. This crucial assumption is now known to be invalid, for the reasons cited in this and later sections of the talk.

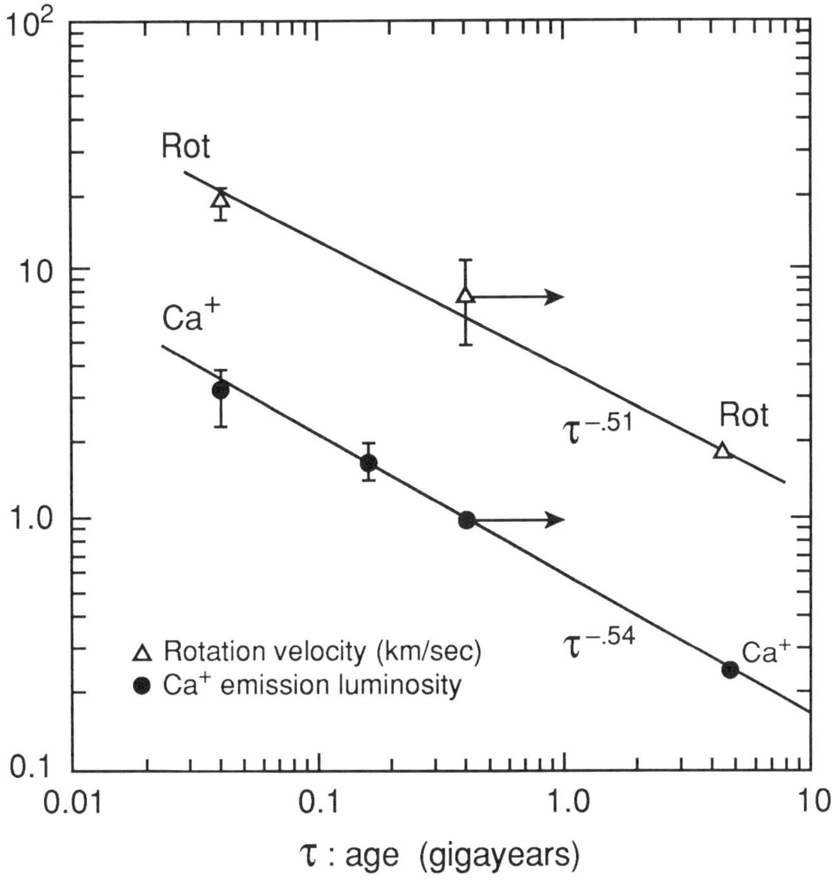

Fig. 1: The time decay of Ca II H-K emission and rotation for main-sequence stars (Skumanich 1972).

proceedings of the 1971 Asilomar conference on the solar wind, who was the first to provide a theoretical framework and derivation of these relations.

Over the years there have been few serious challenges to the Skumanich-Durney square root "laws," and it is no exaggeration to say that these results are by now strongly fixed in the minds of most astronomers. Perhaps the most serious challenge has been raised by two fundamental discoveries concerning stellar rotation, which came not long after the Bonas meeting. In one, Vogel and Kuhi (1981) established from measurements of the rotation rates of a number of T Tauri stars (TTS) that (i) low-mass pre-main-sequence stars on their Hayashi tracks are, without exception, very slow rotators, and (ii) the rapid rotation speeds that everyone expected for such young stars, based on extrapolations (in time and stellar radius) of the square root relations, are characteristic only of the most massive TTS, those with masses above 1.25 $M_\odot$ and located on the radiative portions of their evolutionary tracks. The majority of the low-mass stars surveyed by Vogel and Kuhi proved to be rotating at, or very close to, their photographic measurement limit of 30 km s^{-1}. More recent work using modern detectors has lowered many of those upper limits to $v \sin i < 10$ km s^{-1} (Bouvier 1990), adding further substance to this contradiction with the square root relation (Stauffer and Hartmann 1987).

The second breakthrough came with the equally surprising discovery, first by van Leeuwen and Alphenaar (1982, 1987) and later by J. Stauffer, L. Hartmann, and their collaborators (Stauffer *et al.* 1984, 1985, 1987), that among the low-mass stars in young clusters, there is a class of ultra-rapid rotators having rotation speeds as high as 200 km s^{-1} and rotation periods as short as 6 hours. If these rapid rotators are the descendants of slowly rotating TTS, as current evolutionary models assume they must be, then the process by which they acquire their rapid rotation is quite puzzling. Even if one assumes a complete absence of all rotational braking during the final stages of contraction to the ZAMS, the stellar rotation rates predicted by the most detailed theoretical calculations never get as large as the maximum rates actually observed, regardless of the value of initial angular momentum adopted (Stauffer and Hartmann 1987; Pinsonneault, Kawaler, and Demarque 1990).

Furthermore, the time scale for the loss of angular momentum by these ultra-rapid rotators once they arrive on the ZAMS is in serious conflict with the spindown time predicted by the square root relation. The evidence for this is that while nearly half the G dwarfs in the α Persei cluster rotate with $v \sin i$ values between 25 and 150 km s^{-1}, virtually all of the slightly older G stars in the Pleiades rotate slower than 25 km s^{-1}. Since the turnoff ages of these two clusters differ by no more than 25–50 Myr, the spindown time for such rapidly rotating G stars must be on the order of 50 Myr or less. This is considerably faster than expected from the square root law. It is only much later, after the age of the Hyades according to Pinsonneault *et al.* (1990), that 1 $M_{\odot}$ stars begin a much more gradual loss of angular momentum, on a time scale approximating the decay time predicted by the square root law, that eventually leads to the very slow rotation speeds that typify middle aged and old stars like the Sun.

Rapid rotators less massive than the Sun also appear to undergo strong rotational braking once they reach the ZAMS, but on a time scale that is an order of magnitude longer than for solar-mass stars. Thus, a number of dK stars in the Pleiades are observed to rotate as rapidly as their counterparts in the α Persei cluster, with rotation periods <1 day, while all of the dK stars in the much older Hyades cluster turn out to be relatively slow rotators. The spindown time for M dwarfs must be longer still, since there are dM stars in the Hyades, and even in the field, that rotate appreciably faster than the Sun, with $v \sin i > 5$ km s^{-1} and with rotation periods as short as 3–5 days.

One presumes, and the detailed calculations do show, that the time scale for this spindown of the rapid rotators, as a function of stellar mass along the main sequence, is determined by three main factors: the strength of the external braking torque exerted by the stellar wind, the size of the surface convection zone upon which the torque has to act, and the relative ease with which angular momentum from the rapidly spinning core is shared with more slowly rotating layers at the surface of the star. The most stringent conditions on the efficiency of angular momentum exchange are set by the 50 Myr or less spindown time established for the near-solar-mass G stars (Pinsonneault *et al.* 1989, 1990). It is not certain yet whether the models are also compatible with observed trends

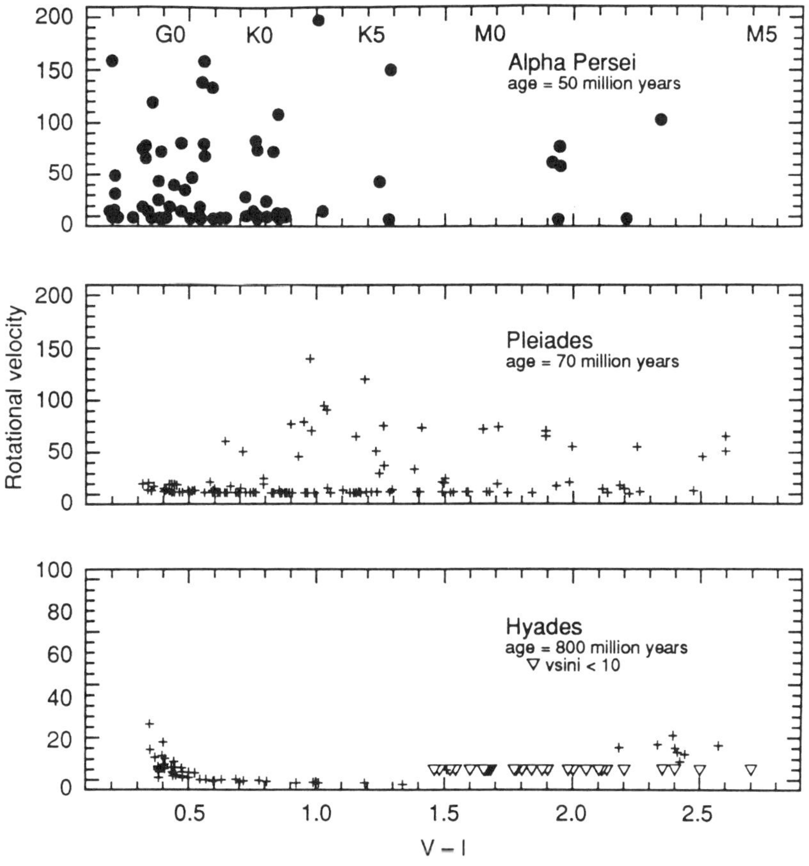

Fig. 2: The rotational velocities of low-mass stars in three open clusters (Stauffer 1987).

in cluster rotation rates higher up the ZAMS, among the very thinly convective F stars, which are known to rotate faster—not slower—than the more deeply convective stars lower down the main sequence.

Still, when one considers the rotation measurements in toto, *e.g.*, in a plot such as Fig. 2, which comes from Stauffer's (1987) invited review at the Fifth Cool Star Workshop, it is apparent that many of the cluster stars rotate rather slowly, with rotation speeds of only 20 km s^{-1} or less. A wide spread in rotation speeds therefore exists at any given mass. The youngest clusters exhibit the widest dispersion over the broadest range in mass, while the oldest clusters exhibit the smallest range confined to the lowest mass. Butler *et al.* (1987) attribute this dispersion to a range in stellar formation times: in a cluster such as the Pleiades, they say, the first stars to form at any given mass have had sufficient time to lose the rapid rotation with which they arrived on the ZAMS, while those that have arrived only recently have not had enough time to spin down yet. This explanation is not entirely convincing for the dG stars in α Persei, however, because the age spread inferred from the rotational braking time for 1 $M_{\odot}$ stars, roughly 50 Myr, is comparable to the age of the cluster itself. A similar argument, if extended to the rapidly rotating dM stars in the Hyades, would imply an age spread of more than 400 Myr, which is far longer than seems reasonable, given our current understanding of star formation. Perhaps most important, among the low-mass stars in the young southern cluster IC 2391, there exists a wide spread in rotation speed, which cannot be due to rotational braking on the ZAMS, since, at an age of 30–50 Myr, none of these stars can have resided there long enough to spin down completely as a result of main-sequence braking.

What seems likely, then, is that stars in any given cluster arrive on the ZAMS with a wide range in their intrinsic rotation rates, from ultra-rapid rotation >100 km s^{-1} to much more modest rotation <20 km s^{-1}. Over time, as stars evolve on the main sequence, those with the fastest rotation suffer the strongest magnetic braking, while those with the slowest rotation suffer the least, so that the distribution of cluster rotation rates is preferentially depleted of stars at its high-velocity end. Eventually, this distribution tends toward the narrow range of nearly uniform rotation exemplified by 1 $M_{\odot}$ stars in the Hyades cluster. This *convergence* toward *equal* rotation rates is what detailed theoretical calculations predict (Pinsonneault *et al.* 1989, 1990), but is totally at odds with the square root relation and with other variations on it, such as the double exponential formula of Barry *et al.* (1981), which are merely scaling laws that preserve the ratio of initial rotation rates of any pair of stars for all time.

The most important conclusion that follows from this discussion is that the spindown of individual stars as a function of main-sequence age cannot be described by a single empirical formula such as the square root law. Because stars on the ZAMS are observed to have a wide range in initial angular momentum, and because theory tells us that rotational spindown is governed by a complicated feedback process that links the instantaneous value of a star's rotation rate with the magnetic torque the dynamo generates via a stellar wind, the dependence of rotation (and therefore dynamo activity) upon age must be far more complex than once believed. Above all, the relationship between rotation and age and between activity and age must be *implicit*, not *explicit* as the scaling formulas assume. Therefore, given differences in the amount of angular momentum stars have, either at birth or when they reach the ZAMS, the history of dynamo activity and rotation must inevitably be different from one star to the next, and no single formula can describe this evolutionary process for each and every star.

THE DEPENDENCE OF ACTIVITY UPON AGE

The point of view I have taken here diverges from the one favored by stellar astronomers for more than 20 years, but additional support for it can be found in measurements of the chromospheric Mg II lines for young cluster stars, which I have been acquiring with *IUE* over the last several years (Simon 1990). The

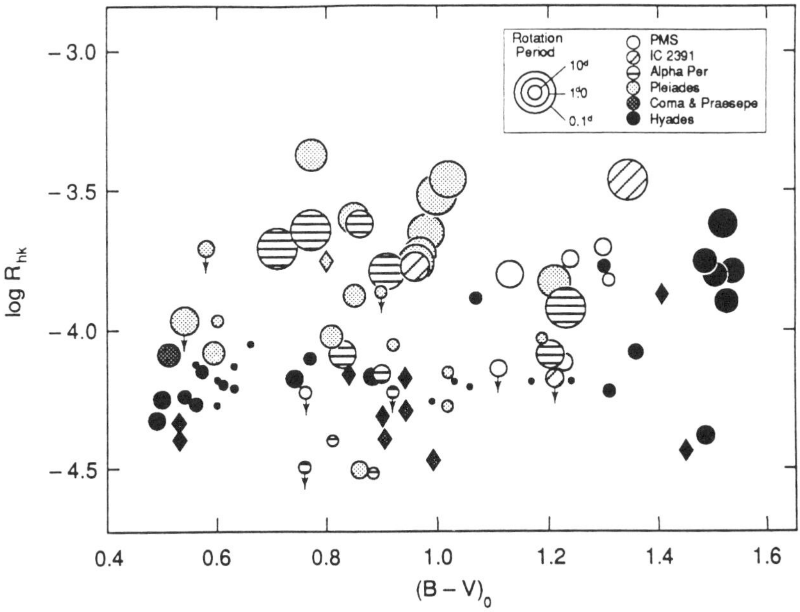

Fig. 3: The normalized chromospheric Mg II flux, corrected for interstellar reddening, versus dereddened $B-V$ color for pre-main-sequence and young cluster stars. The normalized flux is the ratio of the line flux to the bolometric flux at the surface of the star. The diamonds denote stars of unknown rotation period (Simon 1990).

results already published, along with newer spectra for several stars in the Praesepe and Coma clusters, are summarized in Fig. 3. The format I have adopted here, a bubble diagram, is more complicated than the simple line drawing of Ca II H-K strength versus age that Skumanich (1972) devised 20 years ago. However, we need to distinguish the separate influences of mass, age, and rotation, which theory tells us are the major determinants of chromospheric and coronal activity.

From Fig. 3, it is obvious that the sample of early G stars observed at UV wavelengths is too incomplete at this stage to reveal whether 1 $M_{\odot}$ stars experience a sudden decline in activity soon after they reach the ZAMS. Correcting this defect with *IUE* may well be impossible, since the G stars have to be observed at high dispersion and therefore require exposure times of 15 hours or more. This is a task that calls for the light-gathering power of a larger telescope like *HST*. On the other hand, the existing data are now very extensive in the color interval $0.8 < B-V < 1.1$, near spectral type K2, and in this domain there is a spread of more than an order of magnitude in chromospheric emission. The strength of the emission appears to be ordered not by nominal cluster age, but by rotation rate: it is most intense for the fastest rotating stars and weakest for the slowest rotating stars, regardless of cluster membership. Of particular note are four slowly rotating stars from the α Persei and Pleiades clusters near $B - V = 0.8$. their emission is relatively weak, comparable to that of much older stars in the Hyades cluster. While other examples can be pointed out, the importance of these four stars is that they all have very strong Li 6707 Å absorption lines, and so they cannot be shown to be systematically older than the rapid rotators in their respective clusters. That is, there is no evidence these stars have spun down and lost most of their activity (and surface Li) since their arrival on the ZAMS, as suggested by Butler *et al.* (1987); rather, they were slowly rotating and inactive to begin with.

The implication of this result is that stars in young clusters must start out on the ZAMS with a wide range in chromospheric activity, as well as in rotation. By the age of the Hyades, the most rapidly rotating and active stars have lost most of their original angular momentum and activity, while the most slowly rotating and inactive ones have lost far less of what they started with. For the former, the decay of rotation and activity proceeds much more rapidly than the square root law predicts, while for the latter it goes much more slowly. Thus, the time decay of chromospheric activity appears to be as equally at odds with the square root law as is the decay of rotation.

Whether the same conclusions—assuming they are correct—might have been reached in earlier studies of Ca II H-K is an interesting question to ponder. At the time of Wilson's (1963) pioneering studies, photographic observations of faint cluster stars were very difficult and very time consuming. It would have been impractical for him, or anyone else, to obtain spectra of the highest quality for each and every star observed, and so a certain amount of observational scatter in H-K emission would have been considered normal, even among stars in the same cluster, *i.e.*, among stars believed to be the same age. Thus I suspect, given the narrow range in cluster rotation rates then known, that any link between increasing brightness in Ca H-K and increasing rotation speed would have been difficult to find. In fact, Wilson (1963) did call attention to three G8 V stars in the Pleiades, near the faint limit of his survey, with H-K emission that he regarded as exceptionally bright even for the Pleiades. We now know these stars (Hz II 2147, 296, 345) to be rapid rotators; at the time, however, Wilson explained away the anomaly of their activity as the product of some physical interaction with unseen binary companions.

It is also true, as one looks back, that the concept of the dynamo and the relevance of dynamo magnetic fields to chromospheric and coronal heating are ideas that became firmly established only many years later, so there would have been little incentive or reason for observers of the 1960s to look for a correlation between activity and rotation. Thus, for Wilson (1963, 1966) and Kraft (1967), the fundamental relation was always one between activity and age, not rotation. In the years that followed, this remained the prevailing philosophy. By 1972, when the square root relations were being formulated, dynamo and heating theories were still rudimentary, and so the significance of a factor of 5 spread in H-K emission among the group of Pleiades K stars, which Skumanich (1972) used to anchor these curves, might well have been been lost (see Skumanich and Eddy 1981, Fig. 6). It is much harder to defend the same practice by those of us who have worked on this problem in recent times, however, now that the dynamo hypothesis has been so universally accepted; now that the scatter in Ca II strengths within various clusters has been so systematically corroborated (*e.g.*, Barry *et al.* 1987 and references cited therein); and now that the dispersion in cluster rotation rates has been so carefully assayed. It is from within those differences in activity and rotation, I believe, that an understanding of the activity-rotation-age nexus will come. If we sweep aside those differences, then we will sweep aside the science.

THE CONNECTION BETWEEN ACTIVITY AND ROTATION

When stellar astronomers look for predictions about activity that can be tested at the telescope, they usually turn to solar dynamo theories. Those theories, unfortunately, provide very little guidance. Consider, for example, the relationship between the global magnetic activity and rotation rate of a star. As the rotation rate increases, we expect the activity to intensify. Dynamo theories, as far as I can tell, make no specific predictions about the exact form that relationship should take. We are thus left to establish the answer empirically.

What we have learned through observation about stellar activity and its dependence on rotation, starting with what we thought we knew at the time of the Bonas meeting and proceeding on from there, can be summarized as follows. According to Skumanich (1972) and Durney (1972), the strength of the Ca II H-K emission of a star—an index of its overall chromospheric activity—is directly proportional to the velocity at the stellar surface, *i.e.*, to $v \sin i$. Similar kinds of relationships have been proposed in more recent times, for UV emission as well as X-ray emission. In the early work, as well as in the latest, differential rotation has been ignored, for the obvious reason that we know virtually nothing about it except for the Sun. Still, if such activity-rotation correlations are to be believed, differential rotation (for main-sequence stars at least) is either of secondary importance, or it scales directly with the surface rotation rate all along the main sequence, making its observable effects on activity inseparable from those of the average surface rotation rate. Indications are that for stars above the main sequence, for giants and subgiants, differential rotation exerts a more overt influence upon the level of activity we see (Simon and Drake 1989).

In 1984, in what I consider to be a very important paper, Noyes *et al.* (1984) demonstrated that the brightness of the H-K emission of a star located anywhere on the lower half of the main sequence can be related to the Rossby parameter, N_R. As a definition of N_R, Noyes *et al.* used the ratio of the equatorial stellar rotation period (which is observable) to a convective turnover time (which is derivable from models of the stellar interior). While there has been some criticism of this approach—mainly that it introduces an ad hoc scaling in rotation—there is no denying the method works very well with Ca II and with every chromospheric, transition region, and coronal diagnostic measured by *IUE* and *Einstein* (Simon, Herbig, and Boesgaard 1985). The specific advantage of expressing the activity-rotation relations in terms of N_R is that, for each spectroscopic diagnostic, one relation fits all the stars cooler than spectral type F5 V, irrespective of mass (excluding dMe stars during flare outbursts). Furthermore, through a simple change of variables, the Rossby relation can

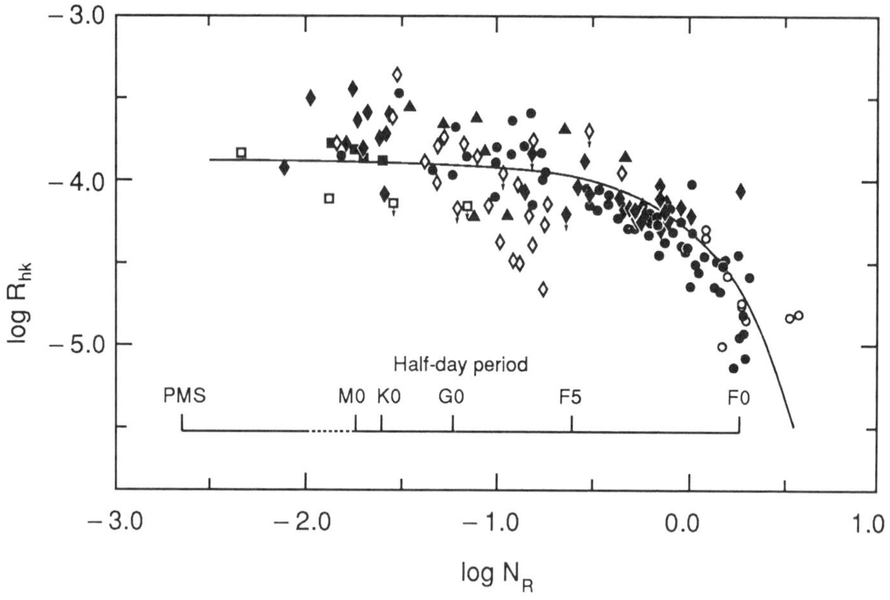

Fig. 4: The normalized Mg II flux versus the Rossby number. The squares denote pre–main-sequence stars, the diamonds represent cluster stars, the circles are late-type field dwarfs, and the triangles are active chromosphere stars, *i.e.,* spotted variables and RS CVn stars. The solid symbols are stars of known rotation period, while the open symbols are those with periods estimated from $v \sin i$. The inset horizontal scale shows the Rossby number for stars with half-day periods, both for the main sequence and for a fully convective pre–main-sequence star (Simon 1990).

be transformed into other activity-rotation relations, so that, in essence, these relationships are all fundamentally equivalent (Simon and Fekel 1987).

An example of such a rotation-activity curve is shown in Fig. 4 for the chromospheric Mg II lines (Simon 1990). Many of the same cluster stars plotted in Fig. 3 are also pictured here, along with a number of field stars for which we now have actual rotation periods (Baliunas *et al.* 1983). The solid line is simply a best-fitting exponential function. Note that at the smallest N_R, among the most rapidly rotating and deeply convective stars, the activity levels off in a plateau, instead of continuing to rise along with the rotation rate as previous correlations with stellar velocity suggested (*e.g.*, Skumanich 1972). This "saturation" mode of the dynamo was discussed by Vilhu (1984), chiefly in the context of short-period contact binaries (W UMa stars), and later by Simon *et al.* (1985) and Simon and Fekel (1987) for normal dwarf stars and detached binaries of the RS CVn class. While we cannot be sure that the same activity mechanism is operating in the W UMa variables as in the normal stars, all groups show the identical roll-off in activity at high rotation speeds, or small N_R. What causes this effect is unknown; it could be a fundamental cutoff on the amount of magnetic flux created by the dynamo, or it could be a more superficial limit on area filling factors, or possibly even a radiative transfer effect. In Mg II, the data are becoming extensive and reliable enough that we can now begin to think about the dispersion in activity about the mean relation at a given N_R. This scatter is not so small that we can simply turn a blind eye to it. Is it observational, or are we seeing the signs of differential rotation, or decade-long activity cycles, or secular variations similar to those that seem to afflict the solar dynamo from time to time? Observations at higher spectral resolution and at better signal-to-noise ratio with the Goddard High Resolution Spectrograph aboard *HST* would be useful here, but only if its original capability for doing high-precision flux-calibrated spectroscopy can be restored.

According to Schrijver (1987), at sufficiently slow rotation rates, *i.e.*, at large values of N_R, the emission in various chromospheric diagnostics levels off to a very weak "basal level," which possibly represents the minimum flux produced by acoustic heating of nonmagnetic regions of the star. Schrijver reached this conclusion by comparing optical and UV emission line fluxes with X-ray fluxes for the same stars. From what I can tell, the results are strongly influenced by the choice of X-ray data; the same data processed with later versions of the *Einstein* reduction software (*e.g.*, Schmitt *et al.* 1990), and stripped of one or two stars whose X-ray data are of questionable quality, show none of the effects claimed by Schrijver, at least none that I can find. Presumably there does exist some minimum or very weak basal activity level component in every chromospheric or transition region spectral line, but whether *IUE* is sensitive enough to measure it in the UV is doubtful, I think. The difficulties of extracting such a tiny emission component from the Mt. Wilson K-line index, given the enormous photospheric contamination of such a broadband measure, have been discussed many times before, without any satisfactory resolution of the issue, and so will not be delved into here. Suffice it to say that the whole question needs to be reexamined with higher-quality data from *HST* and *ROSAT* before we can pass final judgment on it.

THE NEXT SOLAR CYCLE

There is a saying that what seems unimaginable in the future becomes inevitable in the past. That is surely true of astronomy, of our ideas newly conceived as of the old ones we widely believed, of the "new" paradigm of dynamo activity as of the "old." There are some who will claim that what I've said here is wrong, and others will say they've known it all along. Let's see what the next 11 years, the next solar cycle, brings. Our most pressing needs, I submit, are refinements in our theories and computations of magnetic braking, and new classes of experiments, if we are to understand what is happening in this more complicated setting for rotational spindown and evolution of dynamo activity in the Sun and stars that I have described here today.

As we look back over the last 11 years, we realize that the meeting in Bonas came at one of those interesting turning points in astronomy, a dividing line, if you will, between the "old" ways and the "new" ways of looking at the relationship between solar and stellar phenomena. I have not spent time telling you what was said there—you can find that in the proceedings—or what took place there, but rather what ideas were changed there. The true measure of

Bonas is not what happened there, but what happened because of it elsewhere. From that perspective, Bonas was a great success: it helped shape our thinking about the solar-stellar connection. Progress over the last 11 years has been swift. Perhaps, if we are as fortunate, the same will be true of the next 11 years, so that at the end of another solar cycle, we will meet once again to debate the next "new paradigm" of solar-stellar activity.

ACKNOWLEDGEMENTS

I thank E. DeLuca and G. Herbig for their comments on the written text of this talk.

REFERENCES

Baliunas, S.L., Vaughan, A.H., Hartmann, L., Middelkoop, F., Mihalas, D., Noyes, R.W., Preston, G.W., Frazer, J., and Lanning, H. 1983, *Ap. J.*, **275**, 752.
Barry, D.C., Cromwell, R.H., Hege, K., and Schoolman, S.A. 1981, *Ap. J.*, **247**, 210.
Barry, D.C., Cromwell, R.H., and Hege, E.K. 1987, *Ap. J.*, **315**, 264.
Bouvier, J. 1990, *Astr. J.*, **99**, 946.
Butler, R.P., Cohen, R.D., Duncan, D.K., and Marcy, G.W. 1987, *Ap. J. Letters*, **319**, L19.
Durney, B. 1972, NASA SP-308, p. 282.
Kraft, R.P. 1967, *Ap. J.*, **150**, 551.
Noyes, R.W., Hartmann, L.W., Baliunas, S.L., Duncan, D.K., and Vaughan, A.H. 1984, *Ap. J.*, **279**, 763.
Pinsonneault, M.H., Kawaler, S.D., and Demarque, P. 1990, *Ap. J. Suppl.*, **74**, 501.
Pinsonneault, M.H., Kawaler, S.D., Sofia, S., and Demarque, P. 1989, *Ap. J.*, **338**, 424.
Schatzman, E. 1962, *Ann. d'Ap.* **25**, 18.
Schmitt, J.H.M.M., Collura, A., Sciortino, S., Vaiana, G.S., Harnden, F.R., Jr., and Rosner, R. 1990, *Ap. J.*, **365**, 704.
Schrijver, C.J. 1987, *Astr. Ap.*, **172**, 111.
Simon, T. 1990, *Ap. J. Letters*, **359**, L51.
Simon, T., and Drake, S.A. 1989, *Ap. J.*, **346**, 303.
Simon, T., and Fekel, F. 1987, *Ap. J.*, **316**, 434.
Simon, T., Herbig, G.H., and Boesgaard, A.M. 1985, *Ap. J.*, **293**, 551.
Skumanich, A. 1972, *Ap. J.*, **171**, 565.
Skumanich, A., and Eddy, J.A. 1981, in Solar Phenomena in Stars and Stellar Systems, eds. R.M. Bonnet and A.K. Dupree, (Dordrecht: Reidel), p. 349.
Stauffer, J.R. 1987, *Lecture Notes in Phys.*, **291**, 182.
Stauffer, J.R., and Hartmann, L.W. 1987, *Ap. J.*, **318**, 337.
Stauffer, J.R., Hartmann, L., Soderblom, D.R., and Burnham, N. 1984, *Ap. J.*, **280**, 202.
Stauffer, J.R., Hartmann, L.W., Burnham, J.N., and Jones, B.F. 1985, *Ap. J.*, **289**, 247.
Stauffer, J.R., Hartmann, L.W., and Latham, D.W. 1987, *Ap. J. Letters*, **320**, L51.
van Leeuwen, F., and Alphenaar, P. 1982, *ESO Messenger*, No. 28, p. 15.
van Leeuwen, F., Alphenaar, P., and Meys, J.J.M. 1987, *Astr. Ap. Suppl.*, *67*, 483.
Vilhu, O. 1984, *Astr. Ap.*, **133**, 117.
Vogel, S.N., and Kuhi, L.V. 1981, *Ap. J.*, **245**, 960.
Weber, E.J., and Davis, L., Jr. 1967, *Ap. J.*, **148**, 217.
Wilson, O.C. 1963, *Ap. J.*, **138**, 832.
Wilson, O.C. 1966, *Ap. J.*, **144**, 695.

PART II

Space-Based Results

This Part is dedicated to the memory of Giuseppe S. Vaiana in recognition of his pioneering contributions to the study of the Sun and stars at X-ray wavelengths. "Pippo" Vaiana will be remembered by all of us as a tireless worker, an outstanding scientist and a loyal friend.

Seventh Cambridge Workshop on Cool Stars, Stellar Systems, and the Sun
ASP Conference Series, Vol. 26, 1992
Mark S. Giampapa and Jay A. Bookbinder (eds.)

HST OBSERVATIONS OF LATE-TYPE STARS

KENNETH G. CARPENTER
Laboratory for Astronomy and Solar Physics,
Code 681, NASA - Goddard Space Flight Center, Greenbelt, MD 20771

ABSTRACT Observations of cool stars obtained during the first year of HST Science operations are reviewed, and the primary scientific results of each program are summarized.

Keywords: late-type stars; chromospheres; stellar atmospheres; Lyman alpha (H-I, D-I); local interstellar medium

INTRODUCTION

This seems a most appropriate time to look back and review the accomplishments of Hubble Space Telescope (HST), since we are approximately at the one year anniversary of the onset of HST science observations. It is particularly appropriate to be reviewing the results of cool star investigations since they have been a very significant portion of the scientific output of the HST, and in particular of the Goddard High Resolution Spectrograph (GHRS). I would like to point out that, although I am a member of the GHRS Investigation Definition Team, I am *not* intentionally biasing the contents of this talk toward observations taken with the GHRS. However, to the best of my knowledge, with the exception of an FOC image of the dusty symbiotic star R Aqr, all observations of cool stars to date have been obtained with the GHRS. These observations have been acquired as parts of the Science Verification (SV), Science Assessment Observation (SAO), Guaranteed Time Observation (GTO), Director's Discretionary (DD) and Guest Observer (GO) Programs. In this paper, I review only the scientific results of these programs. Performance information can be found in the Proceedings of the HST Workshop, held in May 1991 at STScI.

In contrast to the limited variety in instrumentation utilized, there is an impressive variety in objects observed. Targets observed during the previous year include the interacting binary systems W Ser and R Aqr, the dwarf flare stars AD Leo and AU Mic, the red giants Capella (α Aur), γ Dra, and α Tau (Aldebaran), and the red supergiant α Ori (Betelgeuse). The red giants include examples of an active star, a hybrid chromosphere star, and a non-coronal object. The observations of Capella include exciting high-resolution spectra of the Ly-α line, which will allow a determination of the D/H ratio in the interstellar medium along the line of sight - not strictly speaking a result on 'cool stars', but one of such obvious general interest that it must be included. Very recent preliminary results on observations of UMa cluster stars and on boron abundances in metal-poor subdwarfs will also be noted. It is impossible to do justice to all of these observations in a single review, but I will attempt to highlight the primary scientific results to date from each of these programs. Analysis of much of these data continues and further fascinating results are to be expected.

INTERACTING BINARIES

Ed Beaver, Janet Weiland, and Ed Rosenblatt have recently acquired GHRS spectra (medium resolution at 1400 Å) of the close, interacting binary W Ser. This system is thought to have a giant secondary, with a filled Roche Lobe, which suffers mass loss via an accretion disk around the dwarf primary. In this model, the low-ionization emission arises from the wind and bipolar flow, while the high-ionization emission originates in the accretion disk. This GTO program is designed to test this picture by obtaining high signal/noise line profiles of Si IV at both primary and secondary eclipse. The theoretical picture suggests that these lines should be asymmetric (excess red flux) and blue-shifted at primary eclipse, but more symmetric at secondary eclipse, since neither red or blue flux should be blocked from our line of sight at that time. The observations have been successfully obtained and the answer is ... to be revealed at the Atlanta AAS meeting (January 1992), after the investigators have had the proper amount of time to digest the information.

The FOC observations of R Aqr, acquired by Francesco Paresce *et al.* (1991) during the SAO Program, are the sole non-GHRS observations of cool stars obtained to date. The images are of the core of R Aqr, which is a dusty-type symbiotic (0.1" separation, 250 pc distant) containing a 387 day period M7e Mira variable and surrounded by nebulosity and a bright jet. The pictures were taken in the F/96 mode through the F501N and F372M filters to image in the light of [O III] 5007 Å and [O II] 3727 Å, respectively. The highlights of the observations include the: 1) resolution of bright arcs, knots, and filaments extending SW-NE from Mira, 2) resolution of the core into 2 knots with a position angle of 50 degrees, 3) detection of a stream of material seen extending out 3-4" (1000 AU) to the north which, as described by the investigators, 'seems to bend around an opaque obstacle, and form a spiral, before breaking up into discrete wisps and knots, deviating toward the north', and 4) the resolution of a smaller stream containing discrete parcels of gas and curving to the SE. Paresce *et al.* state that the observed ratio of [O III]/[O II] $\gg$ 1, indicating that the features are either density-bounded or that the ionizing radiation follows a non-thermal power law.

DWARF FLARE STARS

Three observing programs have successfully obtained data on flare stars: a DD program led by Jay Bookbinder to observe AD Leo, and two programs to observe AU Mic, a GTO program led by Steve Maran and a Cycle 1 GO program led by Bruce Woodgate. All of them sought high time resolution spectra of stellar flares and simultaneous multi-wavelength coverage, but each had different specific goals. I shall briefly review them in the order of execution. More detailed discussions are presented elsewhere in this volume by Richard Robinson (for AU Mic) and Jay Bookbinder (AD Leo).

The Bookbinder AD Leo observations, obtained in May 1991, utilized the Rapid Readout mode of the GHRS with 1 second integrations, the G140L grating (R=2,000 design resolution), and 30 minute sequences (containing approximately 1800 spectra each) which were alternately centered at 1300 and 1500 Å. The observations were coordinated with ROSAT, VLA, GINGA, and various ground-based optical observatories. The investigators successfully utilized a rabbit's foot (it is widely believed) to ensure the detection of a stellar flare. In fact, they saw two flares (gotta watch out for those rabbits!). The stronger flare displayed a 1-minute impulsive phase during which the Si IV flux increased ≈ 11x, the C IV flux ≈ 6x, and the He II flux ≈ 3x. The rise time of the flare was less than 20 seconds and all three lines were substantially red-shifted during the impulsive phase of the flare, indicating an ≈ **2000 km** $\mathbf{s^{-1}}$ **downflow**. The red-shifted emission is gone by the end of the impulsive phase. A surface plot showing a time series of spectra (from lower left to upper right) containing the Si IV, C IV, and He II lines during the period around the flare is shown in Figure 1.

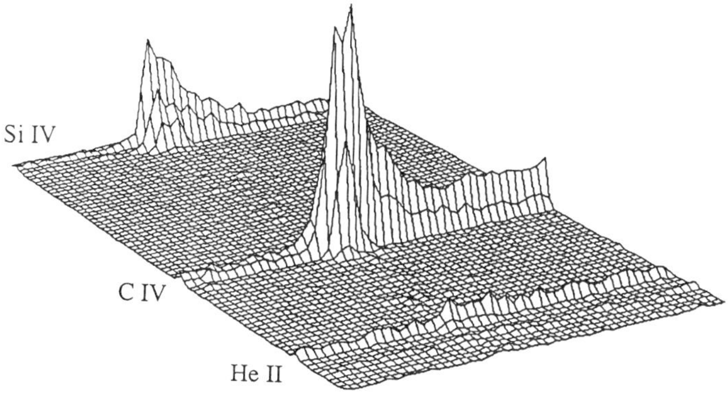

Fig. 1: A surface plot showing a time series of GHRS G140L spectra of AD Leo, with pre-flare data to the lower left and time increasing to the upper right. The Si IV, C IV, and He II lines are clearly seen before, during and after the flare. (Courtesy Jay Bookbinder).

The AU Mic GO program, lead by Bruce Woodgate, obtained Rapid Readout GHRS observations of the dMe 1.6 star using the G160M medium resolution (R=20,000) centered at 1216 Å and 0.4 second integrations. A prime goal was to check for evidence of proton downflow during the impulsive stage of a flare, as might be evidenced by excess emission in the red-wing of Lα . Four 30-minute sequences, containing more than 4500 spectra each, were successfully obtained and a modest-size flare was detected. We have not been able to determine whether rabbits were involved. The Si III (UV 2) line at 1206 Å increased in flux by ≈ 7x. An apparent simultaneous increase in Lα flux is attributed primarily to a coincidental increase in sky (geocoronal) Lα flux rather than true flare emission, although a modest increase in the blue-wing of the line is thought to be real. Figure 2 shows the excellent quiescent spectrum obtained by co-adding the three 30-minute sequences during which no large flares were seen. It is shown on a logarithmic scale to illustrate clearly the large dynamic range of the GHRS and the successful detection of the O V line (UV 0.01) near

1218 Å which is nearly 50x weaker at its peak than the Lα line itself. Real signal is clearly seen at 100x smaller levels than the peak Lα signal.

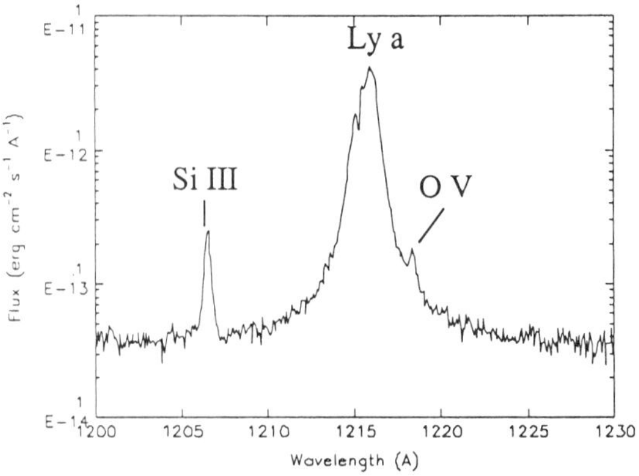

Fig. 2: The quiescent spectrum of AU Mic in the Lα region, as seen in a 90-minute integration with the GHRS.

The AU Mic GTO program led by Steve Maran also utilized 0.4 second integrations with G160M, but was centered at 1360 Å instead of at Lα, to allow the simultaneous monitoring of chromospheric, transition region, and coronal lines. Seven 30-minute sequences (> 4500 spectra each) were acquired, but no flares were seen (forgot the rabbit). However, a good quiescent spectrum (shown in Robinson *et al.* , this volume) does result when the 3.5 hours of integration are co-added. In this spectrum, Fe XXI and O V (UV 7) are resolved for the first time in a star other than the sun and lines of Cl I, C I, and O I are also seen.

RED GIANTS

Red giant stars are well-represented in early GHRS/HST observations, with three considerably different types observed: the rapidly-rotating **active** giant Capella Ab, the hybrid chromosphere giant γ Dra, and the non-coronal giant α Tau.

Capella is a 104-day binary system consisting of G9 III (Capella Aa) and G0 III (Capella Ab) stars. IUE observations have shown that Ab dominates the UV emission spectrum. It was the target of a GTO program led by Jeff Linsky and Alex Brown. The spectra have been discussed previously in Linsky, Brown, and Carpenter (1991) and details of the Lα observations are presented elsewhere in this volume by Gayley. I will summarize the major points here. The observations included G140L (low resolution) data covering the 1150 - 1700 Å region, medium resolution spectra at 1300, 1400, 1550, 1655, and 1900 Å, and echelle (R=100,000) spectra at Lα and Mg II. Highlights from these observations include the resolution of numerous density-sensitive lines such as C II, N II, O III, O IV, Si III, and S IV, and the determination that the surface fluxes of the high-temperature lines are only about a factor of 3x below the saturation limit defined by Vilhu (1987), at which the star is covered with active regions and the maximum possible radiative emission is seen. Perhaps most exciting of all, the echelle spectra obtained of H and D Lα, Fe II, and Mg II will allow the determination of the D/H ratio in the ISM along the line-of-sight to Capella to a much higher precision than heretofore possible. Data analysis and calibration to support the latter are still in progress, and the investigators promise a result by the Atlanta AAS meeting (January 1992). Figure 3 shows a medium resolution spectrum centered near 1400 Å. Figure 4 compares the interstellar absorption lines of D I, Mg II, and Fe II - note the different widths of the lines, which will allow the separation of thermal and turbulent broadening in the ISM and facilitate the D/H ratio determination.

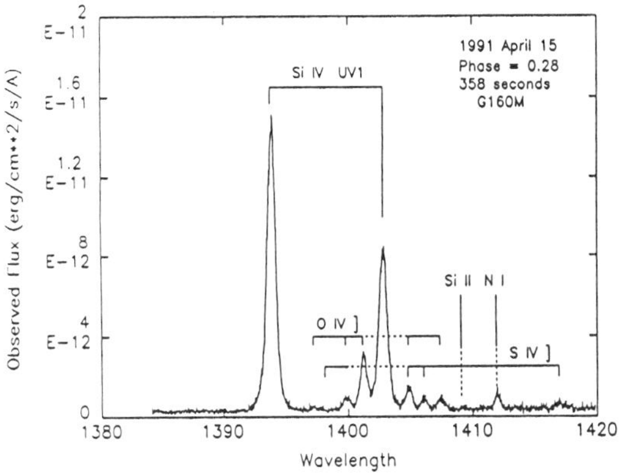

Fig. 3: A G160M GHRS small aperture spectrum of Capella, showing several density-sensitive intercombination lines, including the S IV] lines seen for the first time in a a star other than the sun. (from Linsky, Brown, and Carpenter 1991).

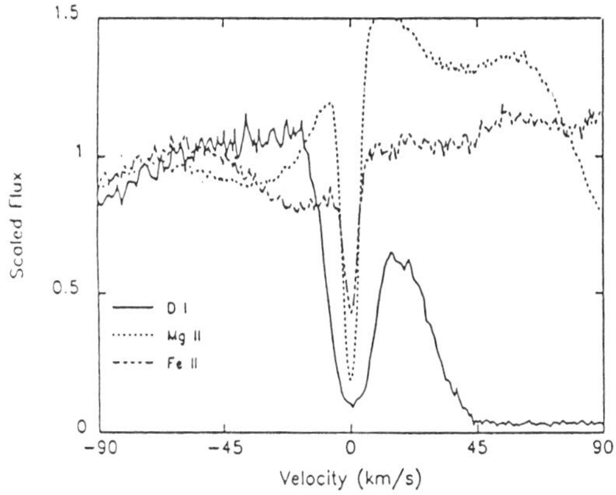

Fig. 4: Comparison of the interstellar absorption lines of D I, Mg II, and Fe II on a common velocity scale. (from Linsky, Brown, and Carpenter 1991).

γ Dra was observed as part of the GHRS Science Verification Program by Carpenter, Linsky, *et al.* , to demonstrate the degree (very) to which the GHRS detectors are solar-blind, while at the same time studying the wind and high-temperature plasma of the star. The observations of this hybrid chromosphere star thus included low resolution G140L spectra (1150 - 1700 Å), echelle spectra of Mg II, and spectra of the O I (UV 2) region at low, medium, and echelle resolution. The surface fluxes measured for the high-temperature lines in this star (*e.g.* N V and Si IV in Fig. 5) are the smallest ever measured ($\approx$ 10,000x below the saturation limit) and suggest that the amount of hot plasma and/or active regions in its outer atmosphere are very small indeed. Figure 6 shows the 1300 - 1308 Å region of low, medium, and high resolution spectra of γ Dra and the value of increasing resolution in resolving first the individual lines, then the detailed line profiles of the O I triplet.

The echelle observations fully resolved the O I (UV 2) triplet near 1304 Å and illustrate the **wealth** of information contained in high signal/noise line profiles which can be obtained with the GHRS. Figure 7 shows an overplot of the three O I lines in velocity space. The interstellar absorption can be seen in the 1302 Å line, along with a strong absorption due to O I in the stellar wind out to as far as 70 km s^{-1} . The wind absorption is seen in the intrinsically weaker lines as well, but they become optically thin sooner, and the emission lines themselves are less broad, so the terminal wind velocity is not seen. The detection of this wind absorption confirms the hybrid nature of the star inferred from the Mg II observed by IUE (Drake *et al.* 1984). S I (UV 9) absorption is seen near the center of the 1306 Å line, thus confirming the Brown and Jordan (1980) O I - S I fluorescence mechanism. Finally, the narrow absorption feature seen in the red wing of all three lines *could* be due to infalling O I, but it may very well be telluric O I - we still need to check the velocity vector of the spacecraft at the time of the observations to determine which is the case.

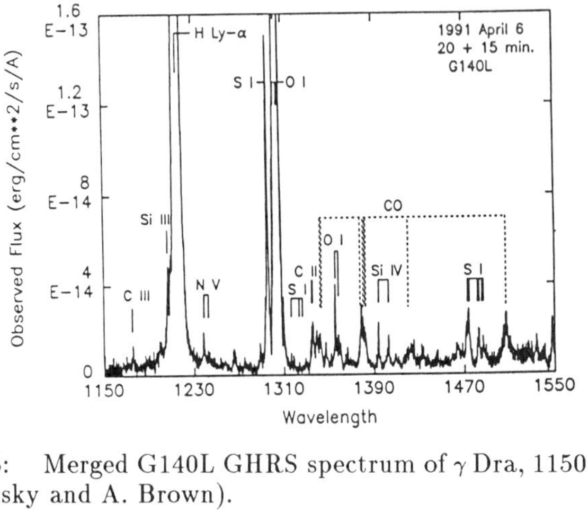

Fig. 5: Merged G140L GHRS spectrum of γ Dra, 1150 - 1550 Å(Courtesy J. Linsky and A. Brown).

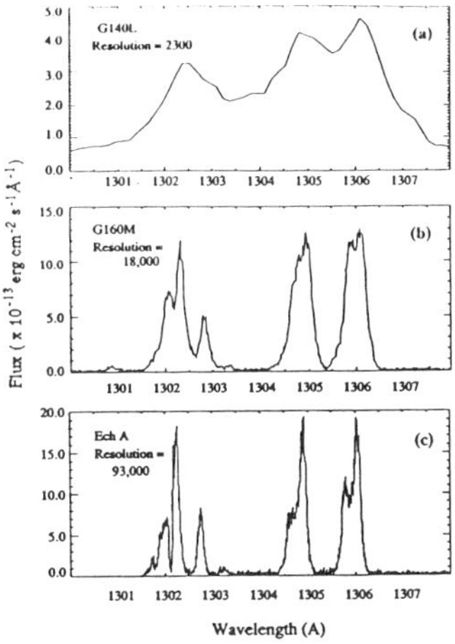

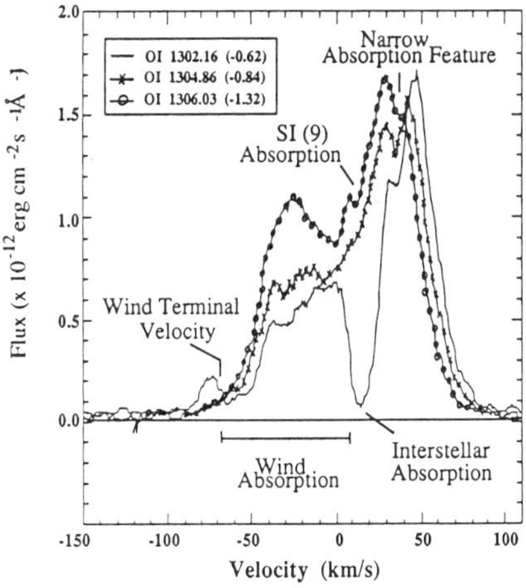

Fig. 6 (left): The O I (UV 2) lines near 1304 Å in γ Dra, as seen at 3 different resolutions with the GHRS. Fig. 7 (right): GHRS Echelle profiles of the three O I (UV 2) lines in γ Dra, overplotted in velocity space.

The non-coronal giant α Tau was observed by Carpenter *et al.* (1991a) during the GHRS SAO program, using both the Large (LSA) and Small (SSA) Science Apertures and the G270M (R=20,000) and Echelle-B (R=100,000) gratings centered at 2345 and 2327 Å, respectively. The observations were designed to assess the relative utility of the 4 instrument modes for studying narrow emission-line sources and to measure flow and turbulent velocities in the stellar chromosphere. The C II lines were resolved by the echelle observations, allowing a direct measurement of the mean turbulence at about 20.5 km s^{-1} . A careful study of the line shapes shows that the C II lines, unlike the nearby Co II and Si II lines, cannot be fit by a single gaussian. A dual-gaussian fit, which combines gaussians with the same central wavelength but different widths and maxima, does fit well. The full-disk integration of an anisotropic velocity field can generate such profiles (Gray 1988), although more complicated causes are possible, such as formation of C II in regions where the turbulent velocity increases with effective temperature as in the hybrid chromosphere models of Harper (1991). Both the G270M and Echelle observations confirm a downflow of the C II plasma of about 4.0 km s^{-1} , as well as a possible first detection of a chromospheric continuum in a red giant (see Fig. 8). Twenty-five new emission lines were detected in the 50 Å of medium-resolution spectra, including the collisionally-excited members of Co II (UV 8), thereby confirming the presence of the ion in the chromosphere and the Carpenter, Wing, and Stencel (1982) identification of the 2330 Å feature as fluorescently-excited Co II (UV 8). Additional presentations of the spectra can be found in Carpenter *et al.* (1991a and 1991b).

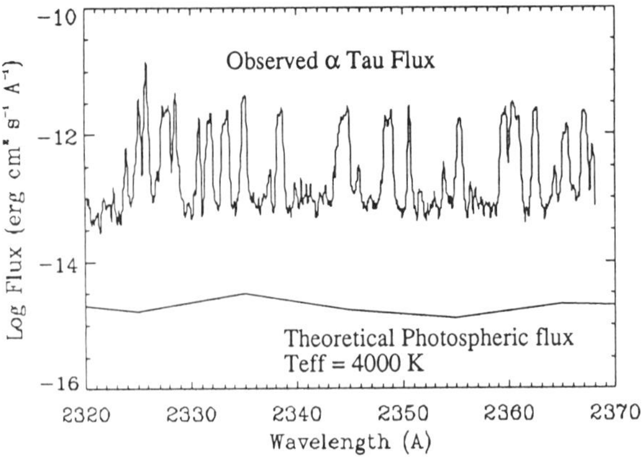

Fig. 8: A comparison of the GHRS G270M SSA spectrum of α Tau and the theoretical photospheric flux, indicating the presence of a chromospheric continuum.

RED SUPERGIANT

The non-coronal supergiant α Ori was observed as part of the Carpenter GTO program in collaboration with Robinson, Wahlgren, and Linsky. The initial observation set includes low-resolution GHRS spectra of the 1150 - 1700 Å region, medium resolution spectra centered at 1655 and 1994 Å, and echelle observations at 2799 Å. The spectra illustrate the value of serendipity in science. IUE had already shown us (Basri *et al.* 1981, Carpenter *et al.* 1990) that the far-UV spectrum of α Ori was very atypical of late-type stars in that it had a very low O I (UV 2) flux, but excess flux longward of those lines. It was generally anticipated that the latter was due to a forest of weak emission lines throughout the far-UV and that is what we expected to see in these deep short-wavelength exposures. However, we found instead a very **strong** far-UV continuum, upon which was superposed a series of very strong, regularly spaced absorption features from 1300 to 1550 Å (see Fig. 9). These absorptions are clearly due to the 4th positive A-X system of the CO molecule and likely originate in the circumstellar (CS) shell. They will provide an unexpected opportunity to probe conditions in the circumstellar shell, as the paper in this volume by Wahlgren, Robinson and Carpenter (1991) illustrates. Preliminary results indicate that a temperature $\tilde{5}00$ K, a column density of $\tilde{1}0^{18}$, and a turbulence of $\tilde{5}$ km s^{-1} are needed to fit the (v',v")=(1,0) band near 1500 Å. The presence of the CO, and in particular the (9,0) band which is coincident in wavelength with O I (UV 2), may also provide clues to the deficiency of the O I flux in supergiants relative to other cool stars, although simple absorption by a single component CO shell does not appear capable of removing all O I photons from the line of sight. The circumstellar shell can also be examined with the echelle observations of Mg II near 2800 Å (Fig. 10), which contain CS Mn I and Fe I absorptions superposed over the chromospheric Mg II emission. These absorptions have been discussed by Bernat and Lambert (1976) as a possible explanation of the k-line asymmetry seen in Copernicus spectra and seen by Dupree *et al.* (1984) with IUE, but not with the same combination of high signal/noise and resolution obtained with the GHRS.

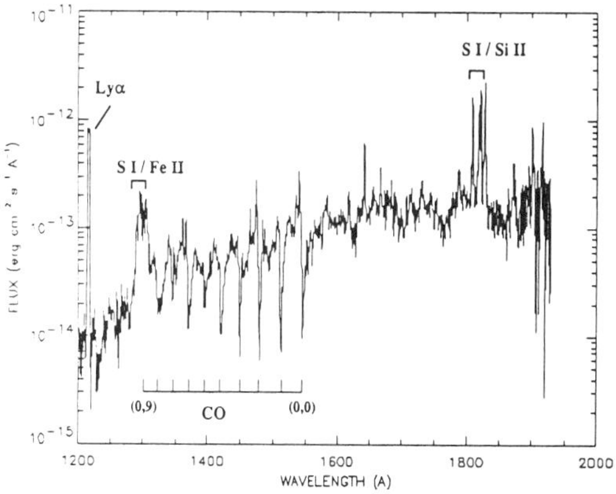

Fig. 9: Merged G140L LSA spectra of α Ori, for 1200 - 1940 Å. Note the strong far-UV continuum and CO molecular absorption features.

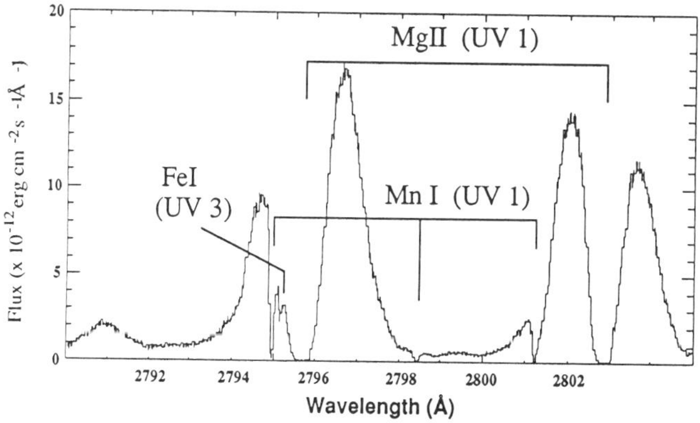

Fig. 10: GHRS Echelle observation of the Mg II resonance lines in α Ori, resolving circumstellar absorptions from Mn I and Fe I.

LATE-BREAKING ADDITIONS

During the early days of this meeting I was briefed on two additional cool star programs for which preliminary results are now available. The first is a program by Fred Walter to obtain spectra of stars in the UMa cluster. The GHRS data for HD 109011 indicate that the trend, identified on the basis of IUE data, for surface flux to decrease with decreasing mass is continued when this K2 V star is included in the sample. The second program by Dave Lambert and Doug Duncan is designed to measure Boron abundances in metal poor subdwarfs and combine these GHRS results with ground-based estimates of the Berylium abundance to compare the observed B/Be ratio to theoretical values based on formation by cosmic-ray spallation and by non-standard cosmologies (the big-bang does not produce B and Be). The preliminary result of the observations is a ratio of 6 ±3. This compares with an expected value of 12 for the usual cosmic-ray spallation theory and would suggest formation by a non-standard cosmology, though it may be possible to explain the lower value via CR-spallation by changing the distribution of energetic particles at the time of formation of the galaxy (Schramm, private communication). Thus a non-standard cosmology may not be required. This story is still developing.

SUMMARY

I have summarized here the observations and major scientific results which have been obtained on cool stars with the HST during its first year of science operations. These data and results, although still limited in quantity, are impressive in their quality and give us reason to be hopeful that HST will be extremely useful and productive as it leaves the calibration era and enters the era of dedicated science operations.

ACKNOWLEDGEMENTS

Thanks to the PI's of the various programs for permission to quote and display the results of their observations, frequently in advance of formal publication, and to R. Robinson, G. Wahlgren, and J. Linsky for comments on the manuscript.

REFERENCES

1 Basri, G. S., Linsky, J. L., and Eriksson, K. 1981, *Ap. J.*, **251**, 162.
2 Bernat, A. P., and Lambert, D. L. 1976, *Ap. J.*, **204**, 830.
3 Brown, A. and Jordan, C. 1980, *M.N.R.A.S.*, **191**, 37P.
4 Carpenter, K. G., Norman D., Robinson, R., Fernandez-Villacanas, J. L., Jordan, C., and Judge, P. 1990, "On the Disappearance of OI in Some High- Luminosity Cool Stars," in Evolution in Astrophysics: IUE Astronomy in the Era of New Space Missions. ed. E. J. Rolfe, ESA SP310, p. 307.
5 Carpenter, K. G., Robinson, R. D., Wahlgren, G. M., Ake, T. B., Ebbets, D. C., Linsky, J.L., Brown, A. and Walter, F. M. 1991a, *Ap. J. Letters*, **377**, L45.
6 Carpenter, K. G., Robinson, R. D., Linsky, J. L., and Brown, A. 1991b, in Proceedings of HST Workshop: The Year of First Light, in press.
7 Carpenter, K. G., Wing, R. F., and Stencel, R. E. 1982, BAAS, **14**, 614.
8 Drake, S. A., Brown, A., and Linsky, J. L. 1984, *Ap. J.*, **284**, 774.
9 Dupree, A., Sonneborn, G., Baliunas, S. L., Guinan, E. F., Hartmann, L., and Hayes, D. P. 1984, in The Future of Ultraviolet Astronomy Based on Six Years of IUE Research, NASA SP-2349, p. 462.
10 Gray, D. F. 1988, chap. 1 in Lectures on spectral Line Analysis: F, G., and K Stars, (The Publisher: Arva, Ontario).
11 Harper, G. 1991, *M.N.R.A.S.*, , submitted.
12 Linsky, J. L., Brown, A., and Carpenter, K. G. 1991, Proceedings of HST Workshop: The Year of First Light, in press.
13 Paresce, F. *et al.* 1991, *Ap. J. Letters*, **369**, L67.
14 Vilhu, O. 1987, in Cool Stars, Stellar Systems, and the Sun, ed. J.L. Linsky and R.E. Stencel (Berlin: Springer-Verlag), p. 110.

Seventh Cambridge Workshop on Cool Stars, Stellar Systems, and the Sun
ASP Conference Series, Vol. 26, 1992
Mark S. Giampapa and Jay A. Bookbinder (eds.)

HST OBSERVATIONS OF AD LEO

JAY A. BOOKBINDER Smithsonian Astrophysical Observatory, Cambridge, MA 02138

FREDERICK M. WALTER SUNY, Earth and Space Sciences, Stony Brook, NY 11794-2100

ALEXANDER BROWN University of Colorado, JILA, Campus Box 440, Boulder, CO 80309

ABSTRACT We present results from a multiwavelength campaign to study the flares and quiescent emission from AD Leo. High time resolution UV observations provide the first evidence for rapid chromospheric heating, as well as extremely high bulk velocities in the transition region.

Keywords: Stellar Flares; Stellar Atmospheres; Flare Stars

INTRODUCTION

In May 1991 we undertook a coordinated campaign to study flares on the dM4e flare star AD Leonis. This campaign included ROSAT and GINGA X–ray observations, VLA and Arecibo observations at centimeter wavelengths, the Caltech Sub-millimeter Observatory, ultraviolet spectroscopy with IUE and HST, and ground-based spectroscopy and photometry, as well as magnetic field observations. In total, over 25 observatories participated in this campaign.

OBSERVATIONS

The HST observations were made with the photon-counting Goddard High Resolution Spectrograph (GHRS). The grating we used for these observations (G140L) has a spectral resolution of $\lambda/\Delta\lambda = 2000$. We utilized the large aperture (2") and read out the data once every second using the rapid readout mode; there is no deadtime between readouts. The wavelength coverage of the grating was alternated every 5 minutes between 1170 and 1450 Å and 1390 and 1670 Å, the overlap region includes the Si IV lines. Gaps in the coverage of the Si IV line occur only for earth occultations and guide star re-acquisitions. Approximately 20 minutes of data was collected during each spacecraft orbit. In all, 16 sets of data were collected, 8 for each of the grating settings, before a hardware error on HST terminated the observations prematurely.

The calibration we employed was based both on the GHRS team software and the Space Telescope Institute's software. No significant differences were found in the processing methods. No background subtraction has been employed; the mean observed background rate was equivalent to $5.45 \pm 1.8 \times 10^{-3}$

counts diode^{-1} sec^{-1}. We have not yet attempted to deconvolve the mirror abberation from the line profiles.

There are two significant results from this series of observations. First, a set of very deep quiescent spectrum was generated by co-adding all of the available HST data for our two spectral regions. These quiescent spectra shows a number of features that are not visible in IUE spectra, including the coronal Fe complex near 1354 Å, and a large number of as-yet unidentified weak emission features. A total exposure time for the two spectra are 2686 and 2193 seconds, respectively.

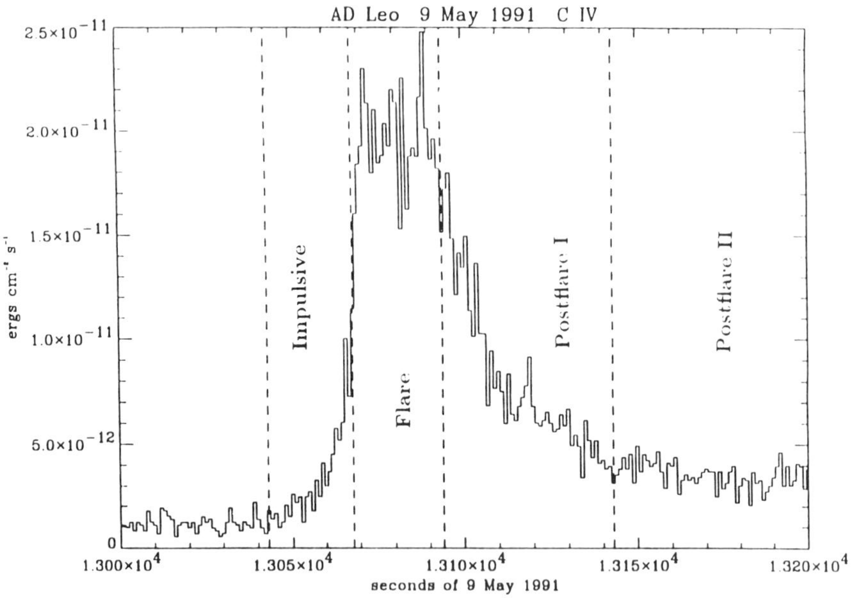

Figure 1: The light curve for C IV for the larger flare, at 1 second time resolution. The impulsive, flare, and post-flare phases of the flare are labelled for comparison with Figure 3. The vertical dashed lines indicate elapsed time from the start of the observation: t = 100 sec, 125 sec, 161 sec, and 200 sec respectively. The spectrum was integrated over 13.6 Å.

The primary goal of the observations was realized when HST caught at least two UV flares on AD Leo. A small flare near the beginning of the observation is probably the smallest stellar UV flare ever detected. A larger flare was seen at 13100 seconds, showing a peak enhancement of the Si IV line of about a factor of 60 with a duration of about 15 seconds. At C IV, both the enhancement and duration are larger: a factor of 90 for 25 seconds (see Figure 1). Note the flat peak to the C IV intensity, and the extended tail on the emission that has not completely decayed at the end of this segment of the observation, but which has decayed before the next segment's observations were begun. A smaller enhancement of about a factor of two is seen in He II (1640 Å), and there is no appreciable enhancements of the low temperature C I lines at 1561 and 1656 Å. The total energy in the UV spectrum for this flare is of order 4.5×10^{30} ergs. If this flare had been seen by IUE, it would have resulted in about a 40% increase in line strength if integrated over 1 hour in the SWP. Hawley and Pettersen (1991) discuss a flare on AD Leo with UV line (e.g., C IV, N V, etc) enhancements of factors of 10 – 20 when integrated over 30 minutes. In fact, the time-averaged "flare flux" of C IV in their event ($\approx 2 \times 10^{-11}$ ergs cm^{-2} sec^{-1}), is comparable to our peak flux. Clearly the GHRS is capable of studying extremely small flares! A dynamic spectrum of this event is shown on the cover of this volume.

For a very short interval during the impulsive phase (as delineated in Figure 1), a number of lines were seen in the spectrum (Figure 2) that do not otherwise appear – even in the very deep exposures. We are in the process of identifying these lines, which remain in the spectrum for about 5 seconds. A few of the lines have been tentatively identified as C I and O I.

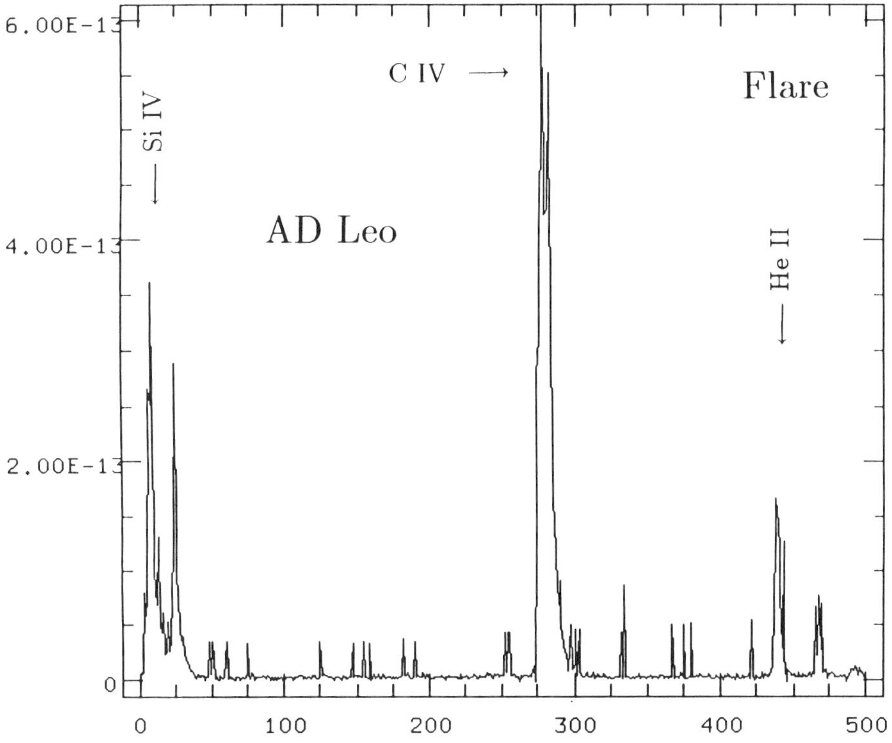

Figure 2: A short interval of the impulsive phase of the flare showing the emergence of a number of lines that are not otherwise seen – even in the very deep exposures. The scale is in detector pixels rather than wavelength units. The prominent UV lines of C IV, Si IV and He II are labelled.

The UV line profiles for this event are of particular interest. The primary lines in the region of the spectrum we were observing, Si IV and C IV, generally behave in a similar fashion throughout the flare. Near the flare maximum the line profiles exhibit tremendous redshifts (see Figure 3) with a maximum redshift of about 1800 km sec^{-1}, corresponding to mass flows towards the stellar surface. There is no evidence for any change is the blue side of the lines. The He II (1640 Å) also shows a similar redshifted component.

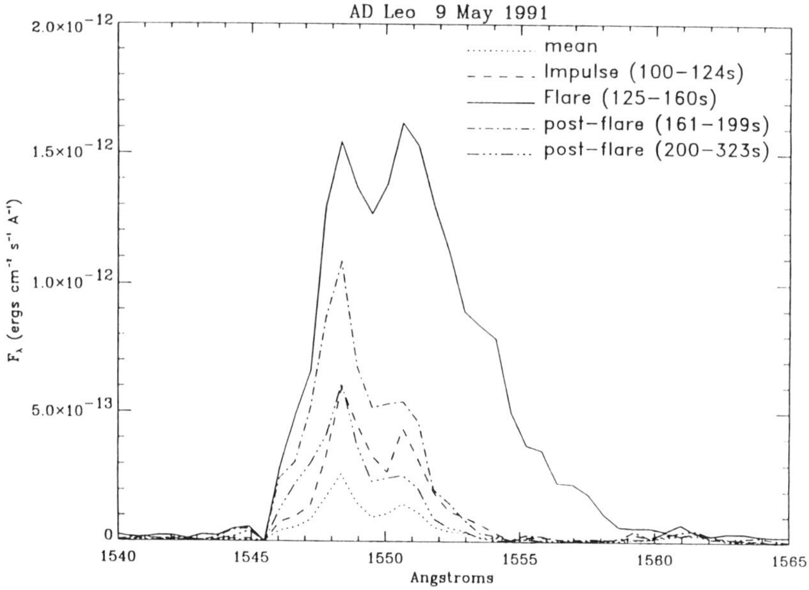

Figure 3: Time resolved spectra of the C IV lines. The feature to note is that during the peak of the flare (15 seconds at Si IV and 25 seconds for C IV), the line wings exhibit tremendous redshifts. The maximum redshift is ≈ 1800 km sec^{-1}; the velocity of the half-maximum point is 650 km sec^{-1}. There is no evidence for any change in the blue side of the lines.

SUMMARY

The time-resolved spectroscopy shows that the flare we observed on AD Leo is an extremely dynamic event, accompanied by mass motions with velocities up to 1800 km sec^{-1}. Unfortunately, these are likely to be the only such data to be obtained with HST, as the failure of the Side 1 electronics has removed the low-resolution grating from service. Future flare observations will be limited to use of the medium resolution gratings, which are capable of – at most – about 35 Å of instantaneous spectral coverage. The FOS cannot be used for such studies because of the red-leak and its lower sensitivity below 1500 Å.

REFERENCES

Hawley, S., and Pettersen, B. 1991 *Ap. J.*, **378**, 725.

Seventh Cambridge Workshop on Cool Stars, Stellar Systems, and the Sun
ASP Conference Series, Vol. 26, 1992
Mark S. Giampapa and Jay A. Bookbinder (eds.)

HST OBSERVATIONS OF THE FLARE STAR AU MIC

RICHARD D. ROBINSON, STEVEN N. SHORE Computer Sciences Corporation, Goddard Space Flight Center, Code 681/CSC, Greenbelt, MD 20771

KENNETH G. CARPENTER, BRUCE E. WOODGATE, STEPHEN P. MARAN NASA/Goddard Space Flight Center, Code 681, Greenbelt, MD 20771

JOHN C. BRANDT Univ. of Colorado, LASP , Campus box 392, Boulder, CO 80302

MUKUL R. KUNDU, STEPHEN M. WHITE Univ. of Maryland, Astronomy Program, Space Science Bldg., College Park, MD, 20742

JEFFREY L. LINSKY Univ. of Colorado, JILA, Campus box 440, Boulder, CO 80302-0440

FREDERICK M. WALTER SUNY , Earth and Space Sciences, Stony Brook, NY 11794-2100

ABSTRACT We report the results of two observations of the flare star AU Mic taken with the Goddard High Resolution Spectrograph as part of a multi-wavelength observing program. The first observation shows unambiguous evidence of Fe XXI (1354.1 Å), formed at 10 million degrees, in the quiescent spectrum. The second showed a transient event in which Si III (1206.5 Å) increased by a factor of 6.6 in integrated flux.

Keywords: Stellar atmospheres; stellar activity; flare stars; HST.

INTRODUCTION

In early Sept, 1991 the flare star AU Mic (Gl 803) was the subject of an intense observing campaign involving optical, UV and radio wavelengths. This campaign was in support of two observing programs carried out by the Goddard High Resolution spectrograph aboard the *Hubble Space Telescope* (HST). The first of these (GTO #1158) examined the UV spectrum between 1345 Å and 1375 Å. This region contains emission lines formed over a wide range of temperatures, going from Fe XXI 1354, formed at 10^7 K, to C I and O I lines formed near 10^4 K. By observing flux variations and Doppler shifts in these lines during the impulsive and decay phases of a stellar flare it was hoped to gain information concerning the initial site of energy release as well as energy balance and transport mechanisms.

The second program (GO #2321) examined the Ly-α region in an attempt to detect evidence for the acceleration of high energy protons during the impulsive phase of a flare. According to a theory first proposed by Orrall and Zirker (1976) protons in the 10-300 keV energy range are accelerated downward during the flare and change exchange with ambient hydrogen atoms, becoming energetic, highly excited neutral hydrogen. These atoms then emit redshifted Ly-α radiation up to 20 Å from line center.

In this paper we describe some of the major results of these observations.

OBSERVATIONS

Both HST programs were performed using the Goddard High Resolution Spectrograph (GHRS) in rapid readout mode. The star was observed through the 2 arcsec aperture with a medium resolution grating (G160M), giving a wavelength coverage of 32 Å and a spectral resolution ($\lambda/\Delta\lambda$) of 10,000. Observations were grouped into a series of orbits with one 30 minute time sequence for each 90 minute orbit. Individual spectra had an integration time of 0.4 s.

RESULTS

There was no evidence for flaring during the 3.5 hours of on-source observations taken in program GTO #1158. The integrated spectrum, shown in Figure 1, should therefore be an accurate representation of the "quiescent" phase of this star. Of most interest is the clear detection of the Fe XXI line. Preliminary analysis suggests a differential velocity of 12 ± 5 km s^{-1} between the Fe XXI and the C I / O I lines, with the neutral atoms red-shifted with respect to Fe XXI. There is also weak evidence for an enhancement in the red wing of the the Fe XXI line, possibly indicating a downflow of as much as 200 km s^{-1}.

A comparison of the spectrum in Figure 1 with Skylab spectra (see Cohen, 1981) shows a closer agreement with a solar flare than a solar active region.

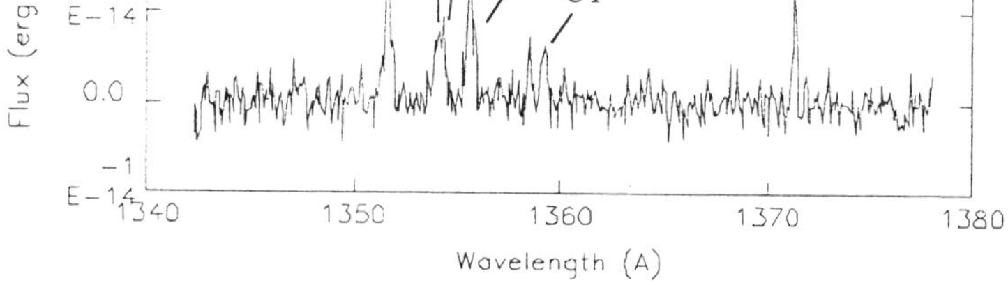

Fig 1: Flux calibrated spectrum after summing the entire 3.5 hour time sequence of program GO #1158. The background, before subtraction, had 135 counts. The Cl I line has 230 counts above background.

Of the four successful orbits in program GO #2321 only the last, starting at 13:09 UT on 1991 Sept 3, showed any transient activity. The integrated counts in the Si III (1206.5 Å) line showed an enhanced level at the onset of observations and peaked 20s later at approximately 6.6 times the quiescent level. The flux then decayed exponentially with an e-folding time of 160 s. Figure 2a shows the Si III profile during various phases of the event. Aside from a slight asymmetry near the peak of the event, the profile shape remained remarkably constant throughout, with no indications of the large Doppler shifts often associated with flares.

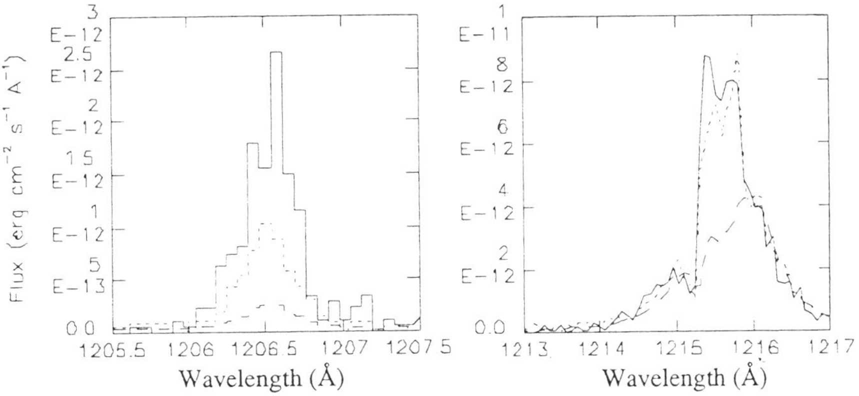

Fig 2: (a) Profiles of the Si III (1206.5 Å) line during the transient event on 1991 Sept 03. The solid line is an integration over the first 50 seconds, the short dashed line integrations from 100 to 200 s, the long-dashed line is the quiescent star. (b) Profiles of the Ly-α line integrated over the same time periods as the Si III in (a).

Integrating over the shortward (1216-1217 Å) and longward (1217-1227 Å) part of the red wing of Ly-α shows no indications of changes correlated with the Si III transient. Thus, there is no evidence for energetic protons associated with the event. There is however a detectable enhancement in the blue wing of Ly-α (see Figure 2b) which may be a signature of chromospheric evaporation. The large changes in the core of Ly-α between 1215.2 and 1215.9 Å ¡can be accounted for by orbit-modulated variations in the geocoronal emission.

ACKNOWLEDGEMENTS

The authors would like to acknowledge the efforts of the staff of the STScI and the BASG engineers in the complex process of scheduling and acquiring these observations. We also wish to thank B. Carter, A. Page and K. Thompson for photometric monitoring of AU Mic.

REFERENCES

Cohen, L. 1981, NASA Reference Publication 1069.
Orrall, F.Q. and Zirker J.B. 1976, *Ap. J.*, **208**, 618.

Seventh Cambridge Workshop on Cool Stars, Stellar Systems, and the Sun
ASP Conference Series, Vol. 26, 1992
Mark S. Giampapa and Jay A. Bookbinder (eds.)

THE CAPELLA SUBCORONAL HEATING EFFICIENCY

KENNETH G. GAYLEY   Joint Institute for Laboratory Astrophysics, Campus Box 440, Boulder, CO 80309-0440

ABSTRACT Using transition region fluxes and high resolution spectra of the Lyα and Mg II h and k lines observed by the GHRS spectrograph, I determine the total flux radiated from each temperature regime in the Capella upper chromosphere and transition region. This contains essentially the same information as the emission measure distribution, but to better suggest physical interpretations of the heating mechanisms, it is expressed instead in terms of the efficiency for extracting convection zone energy flux and depositing it over the temperature ranges considered.

Keywords: emission measures; chromospheres; stellar atmospheres

INTRODUCTION

The binary star Capella consists of two "yellow" giants, an active G0 "secondary" of 2.5 solar masses and 9 solar radii, and a less active G8 "primary" of 2.6 solar masses and 12 solar radii (Batten 1991). The active secondary is also a rapid rotator, with a $v \sin i$ of 36 km/s, while the less active primary is a slow rotator, with a $v \sin i$ of 5 km/s (Ayres 1988). These stars are of significance in the H-R diagram because the active star is a rare Hertsprung gap star, and the less active star is a recent addition to the core Helium burning (CHeB) clump (Ayres 1988). Thus these stars provide a good example of the behavior of both rapid and slow rotators near the Simon-Drake (1989) dividing line between stars that exhibit a clear connection between rotation and activity, and those that do not. Since it seems likely that the atmospheric heating may undergo an important transition at this boundary, it is important to understand this heating in Capella.

THE DIFFERENTIAL HEATING EFFICIENCY

In this preliminary effort to understand the heating of the chromosphere and transition region of Capella, I introduce a more physical way of expressing the information contained in an emission measure distribution. By multiplying the usual differential emission measure distribution by the radiative loss function (Cox and Tucker 1969) and dividing by the total bolometric stellar luminosity, I obtain the fractional distribution of emergent flux with temperature over the atmosphere. This will be termed the differential heating efficiency, defined as

$$\epsilon = \frac{R(T)n_e^2}{L_{total}} \frac{dV}{d\log T}$$

where $R(T)$ is taken to be the radiative loss function of Raymond *et al.* (1976).

If the efficiency distribution is integrated over a given range in $\log T$, the result is the fraction of the total bolometric surface flux that emerges from this atmospheric region. Assuming these losses are balanced by heating, and assuming the source of this heating is free energy extracted from the convection zone, the result can be interpreted as the thermodynamic efficiency for removing energy from the convective heat engine and depositing it in a given atmospheric temperature range. In the transition region and corona, this energy is further redistributed by thermal conduction prior to being radiated, and for the present purpose this conduction is considered to be part of the overall energy deposition process.

The efficiency distribution for the Capella GHRS data, using fluxes from Lyα and the multiplets of C II (1335 Å), C III] (1176 Å), C IV (1549 Å), N V (1240 Å), Si II (1812 Å), Si III] (1892 Å), Si IV (1400 Å), and the method of Brown *et al.* (1984) to find the emission measure, is shown in Figure 1. Here I have assumed that the majority of the flux originates from the rapidly rotating active star, based on Ayres (1988). A comparison with theoretical results for a classical transition region at a pressure of 5 dyne cm^{-2} (also shown in Fig. 1) suggests that the primary energy deposition mechanism above about 40,000 K may be thermal conduction of free energy that has been dissipated as heat in the corona, although a pressure of 5 dyne cm^{-2} is too high to be consistent with chromospheric observations if it is uniform over the surface (Kelch *et al.* 1978).

From the figure, it is clear that the total efficiency of the transition region heating is about $1-2\times10^{-4}$. It is also clear that a separate heating mechanism is required below 40,000 K, having a total efficiency that is also $1-2\times10^{-4}$ above about 10,000 K. For comparison, the solar efficiency distribution (multiplied by a factor of 100) is also shown, and it can be seen that the active Capella secondary is at least 50-100 times more efficient at heating its outer atmosphere.

It is interesting to note that if the convective turnover time scales inversely with gravity, then the inverse Rossby number of the Capella rapid rotator is about 50 times greater than the Sun, in rough proportionality to the enhanced upper-atmospheric heating. However, as we move to lower temperatures, the Capella heating efficiency enhancements over the solar values are not as great. The ratio of the Capella Lyα flux (approximately corrected for interstellar absorption) to the solar value is only about 20 times larger than the bolometric flux ratio, and the Mg II flux ratio is only about 5 times greater than the bolometric ratio. It appears that the Capella heating mechanism is relatively more efficient at higher temperatures than the solar heating mechanism. This could indicate that the higher temperatures are heated by mechanisms that are sensitive to the Rossby number (such as magnetic dynamo activity), and this sensitivity to rotation drops with temperature in the chromosphere (such as would occur if an acoustic heating component became increasingly important deeper in the chromosphere).

ACKNOWLEDGEMENTS

The author would like to thank A. Brown for determining the transition region emission measure distribution. This work was supported in part by NASA Grant S-56500-D to the National Institute of Standards and Technology.

REFERENCES

Ayres, T. R. 1988, *Ap. J.*, **331**, 467

Batten, A. H., Hill, G., and Lu, W. 1991, *Pub. A.S.P.*, **103**, 623

Brown, A., Jordan, C., Stencel, R. E., Linsky, J. L., and Ayres, T. R. 1984, *Ap. J.*, **283**, 731

Cox, D. P. and Tucker, W. H. 1969, *Ap. J.*, **159**, 1159

Kelch, W. L., Linsky, J. L., Basri, G. S., Chiu, H.-Y., Chang, S.-H., Maran, S. P., and Furenlid, I. 1978, *Ap. J.*, **220**, 962

Simon, T. and Drake, S. A. 1989, *Ap. J.*, **346**, 303

Raymond, J. C., Cox, J. P., and Smith, B. W. 1976, *Ap. J.*, **204**, 290

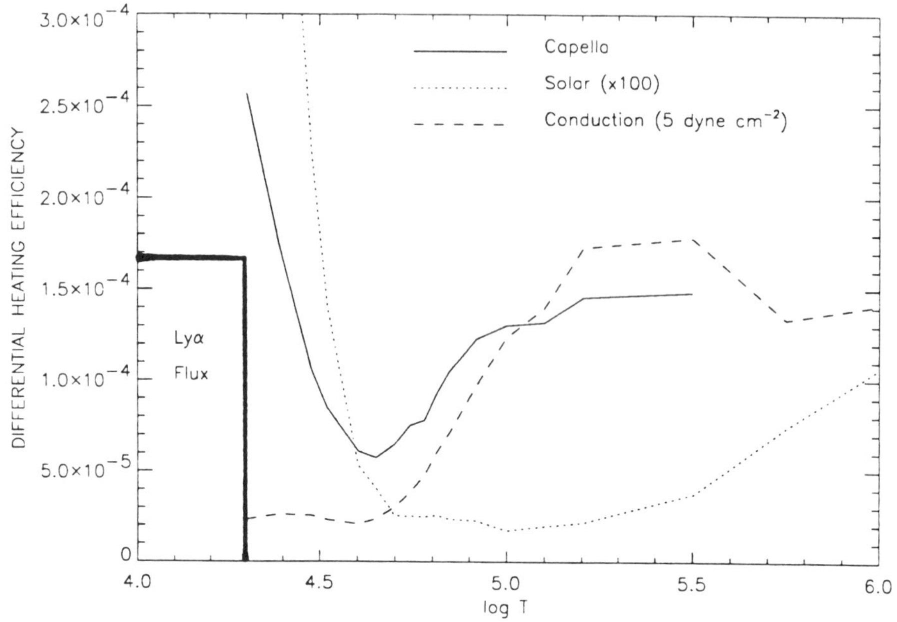

Figure 1. Capella differential heating efficiency in the upper chromosphere and transition region, compared to the solar case (multiplied by 100) and a classical transition region for Capella with a pressure of 5 dyne cm^{-2}. The Lyα flux, divided by bolometric surface flux, is given by the area of the rectangle. The temperature resolution is crude and has not been smoothed.

Seventh Cambridge Workshop on Cool Stars, Stellar Systems, and the Sun
ASP Conference Series, Vol. 26, 1992
Mark S. Giampapa and Jay A. Bookbinder (eds.)

MOLECULAR ABSORPTION IN THE UV SPECTRUM OF α ORI

GLENN M. WAHLGREN, RICHARD D. ROBINSON
Computer Sciences Corporation/GHRS Science Team, GSFC

KENNETH G. CARPENTER
Laboratory for Astronomy and Solar Physics, NASA/GSFC

ABSTRACT Spectra obtained with the GHRS of the cool supergiant α Ori display absorption by CO, in addition to atomic and molecular emissions and a strong continuum in the far-UV. An isothermal slab model is used to obtain the physical parameters characterizing the circumstellar CO which is absorbing the far-UV radiation.

Keywords: Ultraviolet:spectra–stars:circumstellar shells

INTRODUCTION

The presence of CO in the circumstellar (CS) environment of α Ori has been known from spectra obtained at IR and cm wavelengths. From the vibration-rotation spectra at 4.6μm, Bernat *et al.* (1979) identified two CS shells which were characterised by excitation temperatures of 200 and 70 K, CO column densities of 4.7×10^{17} and 1.2×10^{16} cm^{-2}, and expansion velocities of 10 and 17 km/sec, respectively. Knapp, Phillips, and Huggins (1980) noted CS material at the single expansion velocity of 15 km/sec from the emission of the J = 2-1 rotational transition at 230 GHz. Studies of the CO photospheric spectrum have also been conducted, particularly addressing the carbon isotope ratio, $^{12}C/^{13}C$.

Ultraviolet spectra taken with the *IUE* satellite have also displayed the spectral signature of CO. Ayres, Moos, and Linsky (1981) identified CO emission lines in α Boo; Sitko (1983) observed CO absorption in the CS environment of the Red Rectangle central star HD 44179; and Basri, Linsky, and Eriksson (1981) suspected CO emission in α Ori. In this paper we present an ultraviolet spectrum of α Ori and show that the CO absorption originates in a CS environment.

OBSERVATIONAL DATA

The Goddard High-Resolution Spectrograph (GHRS) aboard the *Hubble Space Telescope* was used to obtain low-resolution (R=1000) spectra with the G140L grating and the Large Science Aperture. Due to the primary mirror spherical aberration the spectral resolution is approximately half of that which may be obtained by using the Small Science Aperture. Three wavelength settings were necessary to cover the 1200 - 2000 Å region. At each setting a total exposure time of 1040 seconds provided a signal-to-noise ratio of 20 in the 'continuum' and nearly 40 for the strongest emission lines. Figure 1 displays the spectral region containing obvious absorption arising from the first nine resonance bands of the CO (A $^1\Pi$ - X $^1\Sigma^+$) 4th positive system. The CO gas has an expansion velocity of approximately 20 km/sec, relative to the mean stellar radial velocity. A more rigorous value requires a higher spectral resolution.

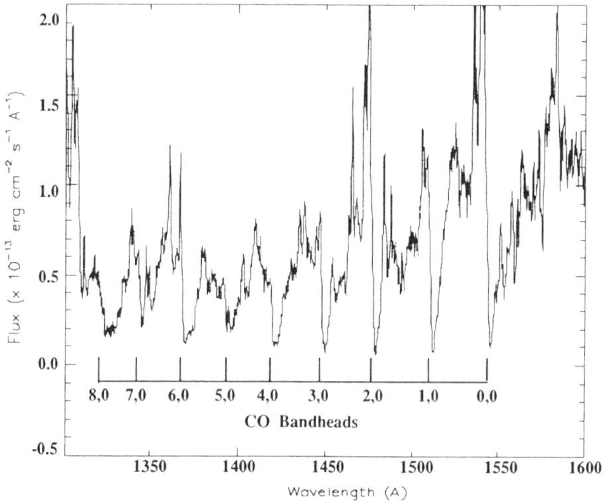

Fig. 1. CO absorption bands dominate the far-UV spectrum of α Ori. Emission at the (0,0), (2,0) and (6,0) bandheads is noticeable and other CO emissions may be present.

SPECTRUM SYNTHESIS

The continuum flux can be shown to originate from a non-photospheric source by modeling a photospheric spectrum. The program SYNTHE was run with an ATLAS9 model atmosphere for T_{eff} = 4000K, log g = 1.0, turbulent velocity = 2 km/sec and solar elemental abundances. The resultant spectrum, comprised of CO and atomic line data from Kurucz, displays a noticeable slope over even a limited wavelength range, whereas the observed spectrum is considerably flatter. The observed flux is orders of magnitude stronger and does not show obvious absorption from higher excited CO levels as does the computed spectrum.

On the assumption that CO absorption in the ultraviolet originates in a layer of material found above the photosphere and continuum forming region, we have employed a single isothermal slab model to describe the physical characteristics of this CO gas. The slab is parameterized by its temperature, turbulent velocity, CO column density, and carbon isotope ratio $^{12}C/^{13}C$. The (v',v")=(1,0) band was chosen for modeling as it appears to be uncontaminated by emission lines. A grid of models was computed in LTE, with the best fit defined by the parameter values T = 500K, N(CO) = $1.\text{x}10^{18}$ cm^{-2}, and V(turb) = 5 km/sec. The carbon isotope ratio could not be unambiguously determined from low-resolution data; however, it is clear that a solar value provides too little absorption while a value close to 20 produces a better overall fit (Figure 2).

The best fit model provides good fits to other resonance bands. As a result, weaker absorptions are revealed, perhaps arising from CO excited states or other molecules. Many coincidences are found between observed absorption and excited CO band structures. For example, models with temperatures between 1000 -1500 K predict the onset of absorption due to the (2,1) and (5,3) bands. However, such temperatures do not allow a fit to the resonance bands. Also, the best fit model does not fully explain the observed O I flux deficiency (Carpenter, this volume). A higher temperature component to the CS CO gas may exist. In the future we will explore the possibility of higher temperature CO gas, the existence of SiO in the CS environment, and the O I fluorescence processes.

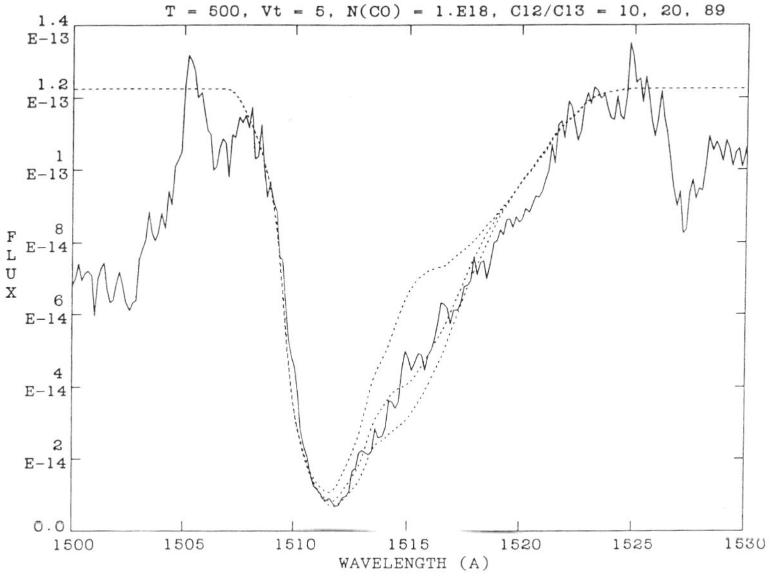

Fig. 2. The observed spectrum (solid) is compared against the best fit slab model (dashed) for, top to bottom, $^{12}C/^{13}C$ = 89, 20, and 10.

REFERENCES

Ayres, T.R., Moos, H.W., and Linsky, J.L., 1981, *Ap. J. Letters*, **248**, L137.

Basri, G.S., Linsky, J.L., and Eriksson, K. 1981, *Ap. J.*, **251**, 162.

Bernat, A.P., Hall, D.N.B., Hinkle, K.H., and Ridgway, S.T. 1979, *Ap. J. Letters*, **233**, L135.

Knapp, G.R., Phillips, T.G., and Huggins, P.J. 1980, *Ap. J. Letters*, **242**, L25.

Sitko, M.L., 1983, *Ap. J.*, **265**, 848.

Seventh Cambridge Workshop on Cool Stars, Stellar Systems, and the Sun
ASP Conference Series, Vol. 26, 1992
Mark S. Giampapa and Jay A. Bookbinder (eds.)

HUT OBSERVATIONS OF THE RS CVN SYSTEM HR 1099

DEREK L. BUZASI & THE HUT SCIENCE TEAM

Johns Hopkins University, Baltimore, MD

ABSTRACT The Hopkins Ultraviolet Telescope, part of the Astro observatory, observed the RS CVn system HR 1099 (V711 Tau) for some 1800 seconds during its flight. We carried out emission measure modeling of the star, using newly available UV lines to determine electron densities. The models are carried to greater heights in the stellar atmosphere than has been previously possible, due to the detection of the high-temperature O VI line, and are more concordant with the coronal X-ray emission than earlier IUE-based models. A two-component continuum was detected; we show that this radiation must originate from below the $T = 2 \times 10^4$K level of the atmosphere, although its exact origin remains unknown.

Keywords: Variable stars; chromospheres; stellar atmospheres

INTRODUCTION

HR 1099 (V711 Tau) is the brightest RS CVn system in the sky, and hence the most studied. The system contains a K0 IV primary and a G5 V secondary, and has a period of 2.822 days (Bopp and Fekel 1976).

Observations in the ultraviolet provide a natural diagnostic for these solar-like active stars, thanks to the high contrast between the emission-line emitting chromosphere and transition region and the dark photosphere at these wavelengths. Simon & Linsky (1980) and Byrne *et al.* (1987), among others, carried out IUE observations of HR 1099 and derived atmospheric models based on these observations using the techniques of density-sensitive line ratios and emission measure analysis (see, *e.g.* Brown & Jordan 1981). Walter *et al.* (1980) used soft X-ray observations to describe the nature of the stellar corona. However, modeling has been hampered by the large gap in temperature between the hottest line diagnostic available to IUE (C IV; $T_{max} = 2 \times 10^5$) and the temperature represented by the soft X-ray data ($\sim 10^6$). Data reported here from the Hopkins Ultraviolet Telescope (HUT) go a significant distance toward filling this gap.

OBSERVATIONS

Observations were carried out with HUT as part of the Astro-1 space shuttle mission in December 1990. In first order, the free spectral range of the instrument is the region from 850 to 1850 Å, at a plate scale of 0.5 Å/pixel. Resolution is approximately 3 Å.

During the flight of Astro-1, a single, 30 minute observation of HR 1099 was made on 3 December 1990 at 22:06 UT. The total integration time was 1572 s, with all of the observation but 60 s taking place at night.

Figure 1 shows the observed first-order spectrum, smoothed to the spectrograph resolution (3 Å). Traditional IUE diagnostics such as N V λ1237, C IV λ1550, and Si II λ1808 are visible. In addition, a number of new lines are visible, such as C III λ977, N II $\lambda\lambda$990, 992, O VI $\lambda\lambda$1032, 1038, N II λ1086, and C II λ1176. The HUT spectrum, with lower noise levels than any IUE spectrum of HR 1099, also clearly shows the presence of an underlying stellar continuum.

DISCUSSION

The emission measure formalism is outlined in Brown & Jordan (1981), and will not be reviewed here. Models were calculated both with and without lines below Ly-α, in order to illustrate the usefulness of lines at short wavelengths. The modeling procedure requires as a starting point a known value of (T_e, N_e), and this was provided by the density diagnostic [C III λ 1176]/[C III λ 977]. This ratio in HR 1099 is 0.52, which corresponds to a density of $n_e = 1.4 \times 10^{10}$ from the graph in Cook and Nicholas (1979). Both of these lines are formed at approximately the same temperature, $T = 9 \times 10^4$ K, and this provides us with a model starting point.

The models derived for the atmosphere of HR 1099 are shown in Figure 2. Height is given in scale heights at $T = 4 \times 10^4$. A point corresponding to a simple corona is also shown. Note that including those lines below the Lyman limit, particularly O VI with its 3×10^5 temperature of formation, permits the extension of the model to a greater height. In this model, the turnover of the transition region en route to the corona is much more clearly visible. The so-called "top pressure," or pressure at the top of the transition region, is approximately 0.17 dynes cm^{-2}, which is somewhat lower than the pressure range of 0.2 - 1.0 dynes cm^2 derived by Simon & Linsky (1980) based on IUE data.

As noted above, a continuum is clearly visible in the HUT spectrum of HR 1099. The continuum appears to consist of at least two components. A component with a thermal shape corresponding to a brightness temperature of 5500K is responsible for continuum emission longward of 1450 Å, while a second component dominates the shorter wavelengths and shows a gradual increase to a brightness temperature of 7500K at 950 Å.

Examination of Ultraviolet Solar Photometer (UVSP) data near the C IV λ1550 line kindly supplied by Dr. A. Poland shows the solar continuum in this region to be more than a factor of ten fainter relative to the bright line flux than is the case in HR 1099. This continuum apparently arises from the photosphere and low chromosphere. However, there does not seem to be a solar analog to the high brightness temperature emission seen in HR 1099 below 1450 Å. The emission measure models described above were used to calculate the continuum emission arising from heights above $T = 2 \times 10^4$K, considering the hydrogen bound-free and free-free opacity; these calculations indicate that all of the continuum visible in HR 1099, including the short wavelength, high brightness temperature component, must arise from below $T = 2 \times 10^4$K.

While the long-wavelength continuum has a readily explicable origin which is based on a solar paradigm, the short wavelength continuum is less readily explained. It must be caused either by "excess" opacity (compared to the Sun) high above the temperature minimum, or, perhaps, by the extended emitting structures found in HR 1099 by Buzasi (1989) and recently confirmed by Hall (1991). Further calculations may reveal the solution to this problem.

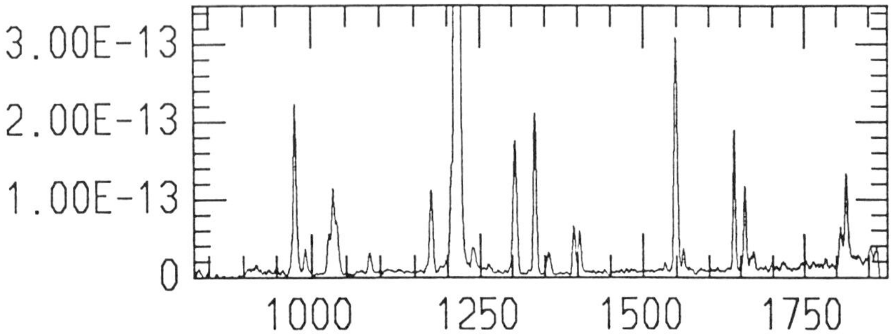

Fig. 1: The first-order HUT spectrum of HR 1099

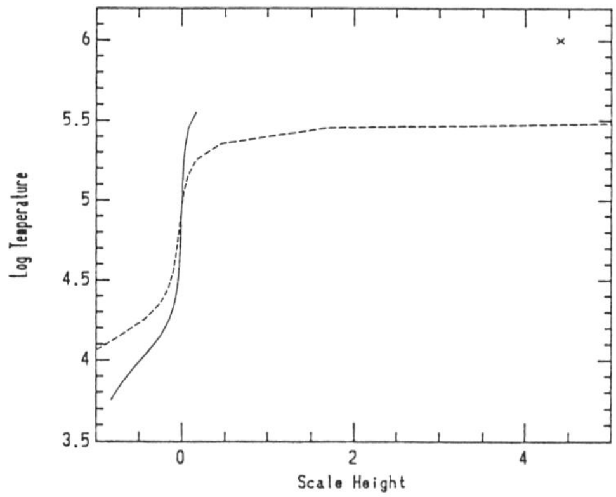

Fig. 2: Emission Measure models of HR 1099. The solid line includes all strong HUT lines, while the dashed line includes only those lines visible to IUE. The cross represents a typical coronal point.

REFERENCES

Bopp, B.W. and Fekel, F.C. 1976 *Ap. J.*, **81**, 771.
Brown, A. & Jordan, C. 1981 *M.N.R.A.S.*, **196**, 757.
Buzasi, D.L. 1989, Ph.D. thesis, The Pennsylvania State University
Cook, J.W. & Nicholas, K.R. 1979 *Ap. J.*, **229**, 1163.
Hall, J.C. 1991, Ph.D. thesis, The Pennsylvania State University
Murthy, J., Henry, R.C., Moos, H.W., Vidal-Madjar, A., Linsky, J.L., & Gry, C. 1990 *Ap. J.*, **356**, 223.
Walter, F.M., Cash, W., Charles, P.A., & Bowyer, C.S. 1980 *Ap. J.*, **236**, 216.
Simon, T. & Linsky, J.L. 1980 *Ap. J.*, **241**, 759.

Seventh Cambridge Workshop on Cool Stars, Stellar Systems, and the Sun
ASP Conference Series, Vol. 26, 1992
Mark S. Giampapa and Jay A. Bookbinder (eds.)

A SEARCH FOR SHORT-TIMESCALE VARIATIONS IN THE CHROMOSPHERE AND UV SPECTRUM OF THE M-GIANT GAMMA CRUCIS

K.-P. CHENG and KENNETH G. CARPENTER
Code 684, NASA/GSFC/LASP, Greenbelt, MD 20771

ABSTRACT The results of a search for variations in the UV spectrum of γ Cru on timescales comparable to a chromospheric crossing time are presented and compared with longer-term variations previously reported (Carpenter 1988).

Keywords: stars: chromospheres; ultraviolet: spectra

SUMMARY

IUE observations of γ Cru taken in 1978 shows dramatically different Fe II line profiles than spectra taken in 1984 and later (see Fig. 1). We therefore obtained a series of observations, in IUE program LGLKC, to check if similar variations could be seen on a timescale comparable to the chromospheric crossing time (1-10 days). We have examined a subset of Fe II, Mg II, and C II lines in the LGLKC spectra for such variations and present preliminary results here. Figures 2 and 3 show measurements of the total flux and the ratio of red/violet peak flux in Fe II and Mg II lines in these spectra which were taken over a 10-day period in 1989. We find no convincing evidence of real variations on timescales of < 10 days, with the possible exception of a minor variation on the violet edge of the Mg II h-line. This suggests that the changes seen in the earlier spectra are probably not caused by shocks propagating outward in the chromosphere, but rather are more likely due to an increase in the star's mass-loss rate or a one-time ejection of mass sometime in the period between the 1978 and 1984 IUE observations.

TABLE I
IUE DATA USED FOR THE LONG-TIMESCALE VARIABILITY

Date (year)	Date (day)	Camera Id	Image No.	Disp Mode	Program Id	Exp Time (sec)
78	111	LWR	1356	H	LTRFW	4320.
84	95	LWR	17340	H	CSGJL	4860.
86	175	LWP	8472	H	OD90K	3600.
87	217	LWP	11329	H	MGJKC	4680.
89	176	LWP	15785	H	LGLKC	3240.

TABLE II IUE DATA USED FOR THE SHORT-TIMESCALE VARIABILITY

Date (year)	Date (day)	Camera Id	Image No.	Disp Mode	Program Id	Exp Time (sec)
89	172	LWP	15763	H	LGLKC	3240.
89	172	LWP	15764	H	LGLKC	80.
89	173	LWP	15769	H	LGLKC	3240.
89	173	LWP	15770	H	LGLKC	110.
89	174	LWP	15775	H	LGLKC	3240.
89	174	LWP	15776	H	LGLKC	110.
89	175	LWP	15780	H	LGLKC	3240.
89	175	LWP	15781	H	LGLKC	83.
89*	176	LWP	15785	H	LGLKC	3240.
89	176	LWP	15786	H	LGLKC	110.
89	177	LWP	15791	H	LGLKC	3240.
89	177	LWP	15792	H	LGLKC	110.
89	178	LWP	15803	H	LGLKC	110.
89	178	LWP	15804	H	LGLKC	3240.
89	179	LWP	15810	H	LGLKC	110.
89	179	LWP	15811	H	LGLKC	2160.
89	180	LWP	15813	H	LGLKC	2880.
89	180	LWP	15814	H	LGLKC	110.
89	181	LWP	15816	H	LGLKC	2700.
89	181	LWP	15817	H	LGLKC	110.

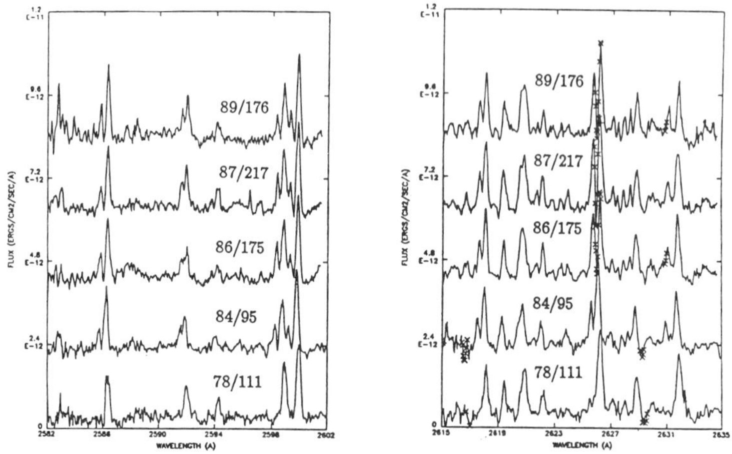

Fig. 1. Portions of 5 IUE high-resolution spectra, taken over the period 1978 - 1989 (see Table I), showing emission lines of Fe II UV multiplets 1 and 64. The intrinsically stronger lines near 2586, 2592, 2618, and 2626 Å are relatively narrow and single-peaked in the initial spectrum, but have broadened and developed self-reversals in the latter spectra. This change in line profiles persists in all later spectra and there is a suggestion that the effect may increase slightly with time.

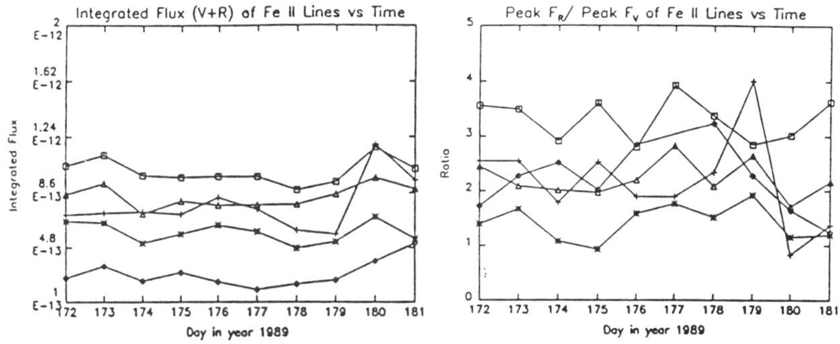

Fig. 2. The total flux and the ratio of the peak flux of the red/violet components of 5 Fe II lines in the spectra (2586, 2592, 2594, 2598, and 2599 Å) from 1989 (see Table II) versus time. No correlations are seen in the behavior of these lines and no convincing evidence has been found in these spectra for real variations on timescales comparable to the chromospheric crossing time (1-10 days).

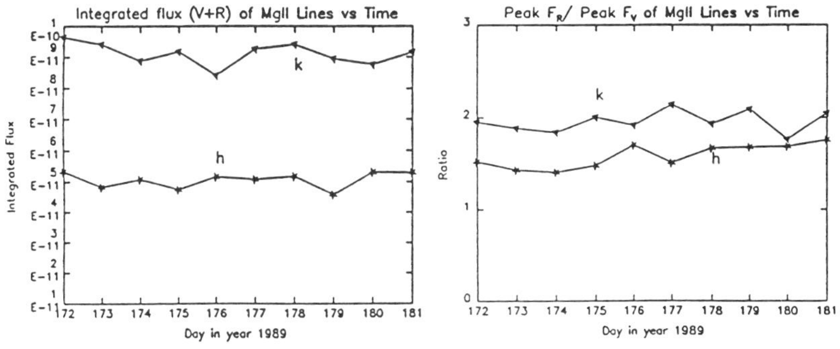

Fig. 3. The total flux and the ratio of the peak flux of the red/violet components of the Mg II lines (2799.5 and 2802.7 Å) in the spectra from 1989 versus time. As is the case for the Fe II lines, no convincing evidence (such as simultaneous similar changes in both the h and k lines) for real variations on timescales comparable to the chromospheric crossing time is seen. The presence of a reseau mark in the violet edge of the k-line does however limit our ability to verify small changes in the violet edge of the h-line.

REFERENCES

Carpenter, K.G. 1988, in IAU colloquium 94, Physics of Formation of Fe II Lines Outside LTE, ed. R. Viotti, A. Vittone, and M. Friedjung (Dordrecht: Reidel), p. 95.

Seventh Cambridge Workshop on Cool Stars, Stellar Systems, and the Sun
ASP Conference Series, Vol. 26, 1992
Mark S. Giampapa and Jay A. Bookbinder (eds.)

IUE AND ROSAT SURVEY OBSERVATIONS OF SYMBIOTIC STARS

ROBERT E. STENCEL and EDWARD W. BRUGEL APAS/CASA, University of Colorado at Boulder

SCOTT J. KENYON Harvard-Smithsonian Center for Astrophysics

K.F. BICKERT, T.A. FLEMING AND J.H.M.M.S. SCHMITT Max Planck Institute fur Physik und Astrophysik, Garching

and the IUE ROSAT ALL-SKY SURVEY TEAM

ABSTRACT The set of long-period, interacting binary stars, known as Symbiotics, offer an excellent opportunity to examine the *atmospheric structure and state of evolution of the red giant and AGB stars* in such systems. Mass transfer, leading to UV and X-ray enhancements, is modulated as a function of mass loss rate, stellar wind acceleration and homogeneity, and degree of Roche lobe filling by the red giant. In addition, many of these sources exhibit radio lobes indicative of jets, suggesting the formation of large scale jets and shocks. We report here some preliminary results of coordinated IUE and ROSAT all sky survey (RIASS) observations of selected Symbiotics, including AG Dra and CH Cyg. While the line fluxes in AG Dra have declined since the early 1980s, the X-ray emission measure appears to have increased. The upper limit on X-ray output of CH Cyg provides new constraints for the dust formation and starspot models in that system.

Keywords: ultraviolet spectra; IUE; X-ray observations; ROSAT; interacting binaries; symbiotic stars

INTRODUCTION

For several decades, cool evolved stars in binary systems have been studied by use of the hot companion as a probe of the atmospheric structure. The yellow and red supergiants in widely separated ζ Aur and VV Cep binaries (see Hack and Strickland, 1987) have been successfully explored in this way. The analysis of red giants in binaries has been complicated by stronger effects of interaction (see Stencel 1987), although a promising attempt to map the velocity profile in EG And has been made by Vogel (1991), using some of the original eclipse discovery IUE data of Stencel (1984). Further progress should be possible using eclipsing symbiotics containing either Mira/AGB stars or first ascent red giant branch primary stars (see Judge, these Proceedings).

The study of the class of long period interacting binaries, known as symbiotic stars, underwent a renaissance as a result of early *Einstein* X–ray detections and the first few years of IUE observations of these sources. Known optically for their combination of red giant (cool, molecular absorption bands) and nebular (semi-forbidden emission lines) spectral features, IUE spectra have clarified in many cases the nature of the hot companion by revealing the far UV continuum characteristic of 50,000K sources. *Einstein* and EXOSAT observations have placed constraints on high temperature emission measures in the presumed disks in these interacting binaries. In addition, access to the UV transitions of abundant ions has made it possible to derive more precise motions and abundances in these systems (*cf.* Nussbaumer and Stencel 1978). The analogy between symbiotics and active galactic nuclei, in terms of morphology (radio and optical jets) and spectra (emission line profiles) has begun to receive increasing attention (*cf.* Penston 1987; Judge 1991).

Nevertheless, symbiotics exhibit a bewildering array of observed phenomena that defy simple interpretation. For example:

– UV emission lines of AG Peg and RX Pup for example, are broad and complex, and differ from ion to ion. These differences could arise from formation in an extended accretion or wind-wind shock structure;

– R Aqr is enveloped in a vast, arc minute sized bipolar nebula, with jets. This is thought to be the residue of a hundred year old thermal pulse of the Mira, modulated by the binary interaction;

– thermal emission at radio wavelengths has a spectral index average of about 0.9 that is not readily explained;

– Mira symbiotics show greatly reduced maser emission in OH, H_2O and SiO compared to normal Miras, etc. This could depend on the far UV ionizing flux of the hot companion.

In each case, the "normal" behavior of a red giant or AGB star is changed by the intervention of the companion's mass, radiation field and stellar wind. Therefore, the extent of change as a function of companion type, could provide a way to reveal the underlying red giant characteristics.

Prior to ROSAT, only a handful of the symbiotics were observed by X–ray telescopes. The majority of these appear to be very soft X ray sources. Analysis of the limited spectral information suggests the signal originates in a moderate density region at temperatures of several times 10^5K. Theory argues that such heated regions would occur in the wind-wind shock interface between the components of the binary. Thus, it is plausible that the majority of all known symbiotics are X–ray sources *at some level.*

CASE STUDIES: AG Dra and CH Cyg

AG Dra is (so far) the brightest X–ray symbiotic. It is an essentially unreddened, dust-free (s-type) object with a K3 spectrum plus emission lines of H I and He II (Kenyon 1986). Kenyon and Webbink (1984) argued that the 1981 outburst was powered by thermonuclear runaway processes. Reportedly, the ROSAT WFC also detected AG Dra (G. Bromage, private communications).

Viotti *et al.* (1984) report on the IUE spectral changes during the 1980-83 outburst. Comparing recent spectra with these (Figure 2), it is clear that AG Dra is relatively quiescent compared with the outburst epoch.

Anderson, *et al.* (1981) reported the *Einstein* detection of AG Dra and derived a log column density (N_H) of 20.5 and log emission measure [EM] of 55.4 and a log Temperature of 6.04 [K]. Preliminary ROSAT survey observation reductions find log N_H = 20.1, log EM = 55.8 and log T = 5.8. Therefore, one has the impression that the system has cooled slightly since the flare event in the early 1980s.

For the case of AG Dra, the observed He II 4686 flux indicates approximately 2 x 10^{44} He+ ionizing photons are absorbed every second into the AG Dra nebula. The luminosity in absorbed He+ photons, about 2 x 10^{34} ergs/sec, exceeds the luminosity of the source deduced from the X–ray flux by a factor of two! The degradation of the X–rays by the surrounding nebula thus causes an underestimate in the actual source luminosity, which can be recovered only by an accurate estimate of the soft X–rays absorbed locally. Since V1016 Cyg and HM Sge also have strong and variable He II 4686 emission lines, the X–rays emitted by these objects are poorly determined.

A second example of how symbiotics might contribute to understanding of red giant star mass loss behavior is given by CH Cyg (HD 182917). In addition to showing an M7 III (SRb) spectrum, there is evidence for a magnetic white dwarf companion and associated multi-periodicities of 500 sec, 100 days and 770 days (Mikolajewski et al. 1990) in addition to the arc-second size bipolar nebula observed optically and at radio wavelengths. Taranova (1990) had proposed that some of the long term variations in the Mira-like behavior could be explained in terms of dust condensation in the outer atmosphere, whereas Mikolajewski *et al.* (1991) argue for the presence of large, long-lived starspots.

Maggio, *et al.* (1990) report an Einstein X–ray flux upper limit of 1.5 x 10^{-13} ergs/cm2/s. Preliminary analysis of ROSAT data suggests an upper limit of this order, as well. AO1 pointed observations (Kafatos) may show differently.

The IUE observations suggest continued decline since the flare-like events of the early 1980s, when compared with Mikolajewska, Hack and Selvilli (1988) who analyzed the 1979-1986 spectra.

ACKNOWLEDGEMENTS

We are pleased to acknowledge partial support for these efforts under NASA grants NAG5-816, NAG5-1456, NAS5-29337 to the University of Colorado. This research has made use of the Colorado Regional Data Analysis Facility and the Simbad database, operated at CDS (Strasbourg, France). I also thank Dan Potter for help in preparing some of the figures.

We are delighted to acknowledge, as well, the heroic efforts of the NASA and ESA IUE and ROSAT Observatory staffs, in obtaining the observations, including: J. Truemper (ROSAT PI), W. Wamstecker (RIASS Coordinator), C. Gonzales (ESA IUE), J. Pye (ROSAT WFC), M. Mas-Hesse (ESA IUE and ROSAT), J. Bonnell (NASA IUE Designated Observer) plus D. deMartino (ESA Vilspa Designated Observer). These observations were made in further collaboration with H. Nussbaumer (ETH), R. Viotti (Frascati) and R. Cardini (Frascati). More complete reports will appear elsewhere.

Seventh Cambridge Workshop on Cool Stars, Stellar Systems, and the Sun
ASP Conference Series, Vol. 26, 1992
Mark S. Giampapa and Jay A. Bookbinder (eds.)

HIGH SPECTRAL RESOLUTION SOLAR DATA DEGRADED TO LOW DISPERSION IUE RESOLUTION

J.G. DOYLE Armagh Observatory, Armagh BT61 9DG, N.Ireland

J.W. COOK Code 4163, E.O. Hulburt Center for Space Research, Naval Research Laboratory, Washington, DC 20375, USA

ABSTRACT We present high spectral resolution (~ 0.06Å) solar data for an active region and a large two-ribbon flare degraded to the typical resolution of low dispersion (~ 5Å) spectra of the International Ultraviolet Explorer (IUE) satellite. This clearly shows the amount of detail yet to be acquired in stellar spectra and indicates the main spectral regions where line blending is a problem.

Keywords: Sun:flare — Sun:plage — ultraviolet:spectra — spectra:line blending

INTRODUCTION

The short wavelength region covered by the SWP camera onboard the International Ultraviolet Explorer (IUE) satellite (*i.e.* 1150 - 1950Å) is diagnostically important as it can give valuable information on the density and the volume of the emitting structure in the mid-transition region. Numerous papers dealing with spectra of solar and late-type stars has appeared since the lunch of IUE in 1978. Some of these papers have either quoted or in some instances produced plots of degraded solar spectra in their discussion. However, none have re produced the original high spectral resolution data, which is vital if one is to comprehend the actual number of spectral lines present.

DATA REDUCTION & RESULTS

The spectra presented here were acquired in September 1973 by the Naval Research Laboratory's S082B ultraviolet spectrograph onboard Skylab. These data were originally used in a study of the enhancement of the ultraviolet continuum in flares by Cook & Brueckner (1979). That paper gives a fuller description of the instrument, the observations, and the data reduction. The S082B instrument was a double dispersion, normal incidence grating spectrograph. It covered the 960 – 1960Å wavelength range with a spectral resolution of 0.06Å, the projected slit dimensions on the solar disk were 2 x 60 arc sec (1450 × 43500 km), with no spatial resolution along the slit. The spectra were recorded on photographic film using Kodak 104 emulsion.

The spectra were scanned using the NRL Grant microdensitometer. The S082B spectra are slightly curved in the dispersion direction, and so 16 individual scans of approximately 65Å length, with some overlap at each end, were made to cover the complete 960 – 1960Å range in order to keep the densitometer slit always in the central area of the dispersion direction.

The densitometer scans were converted from film densities to absolute intensities using the film characteristic (H&D) curve, which had a standard deviation in relative intensity of 7% above 1400Å but around three times greater below 1400Å, and the instrumental efficiency calibration of Kjeldseth-Moe & Nicolas (1977). The error in the final absolute intensities is estimated as $\sim$ 50%.

DISCUSSION

The flare spectrum comes from a large double ribbon flare (Hα class 2B, X–ray class X1) on 1973 September 7. (A discussion of the morphology of this flare can be found in Doyle & Widing, 1990). Spectrograph observations began after the soft X–ray maximum observed by the Solrad 9 satellite and continued for 26 min of the extended ($\sim$ 1 hr) decay phase. We present the spectrum of the earliest 20 s exposure (the longest exposure time available), taken at 12:22 UT, 17 min after the soft X–ray peak. The active region spectrum comes from a 40 s plage exposure on September 11 2:20 UT.

In order to compare directly with IUE-type data, we degraded the solar spectrum to the spectral resolution of the SWP camera, $\sim$5Å. This was done using a simple 'running gaussian' function from the STARLINK program DIPSO (Howarth & Murray 1987). This degraded spectrum is super-imposed on the original data. Line identifications are provided for many of the stronger lines.

As can be clearly seen from the above figures the degraded spectrum hides a wealth of detail. In some instances a great deal of line blending occurs. In the solar flare spectrum, there are very few regions which are free of emission regions. Hence, care should be exercised when extracting continuum fluxes from low resolution spectra. (*Copies of both the high resolution and degraded data may be obtained on either magnetic tape or floppy disk from JGD upon request — email SPAN 19884::JGD*).

These figures can also aid greatly in the identification of weak emission features. For example, we illustrate in the opposite figure a section of a flare spectrum of the active RS CVn star II Peg, observed in September 1990 by JGD with IUE (unpublished). Several weak emission features can now be identified, *e.g.* Si II 1264Å, the Si III 1298Å multiplet and O V 1370Å, which have not previously been noted in late-type stars. The feature at 1356Å is due to a blend of O I, C I and the high temperature coronal line Fe XXI.

Acknowledgments Research at Armagh Observatory is grant-aided by the Department of Education for N. Ireland. This work was supported in part by a grant for International Collaboration in Research from the NATO Scientific Affairs Division (grant number 0469/87). We also acknowledge the support provided in terms of both hardware and software by the STARLINK Project which is funded by the UK SERC. The work of JWC was supported by NASA under DPR W-14,541 and by the Office of Naval Research.

REFERENCES

Cook, J.W. and Brueckner, G.E. 1979, *Ap. J.*, **227**, 645.
Doyle, J.G. and Widing, K.G. 1990, *Ap. J.*, **352**, 754.

Howarth, I.D. and Murray, M.J. 1987, Starlink User Node 50, Rutherford Appleton Laboratory.

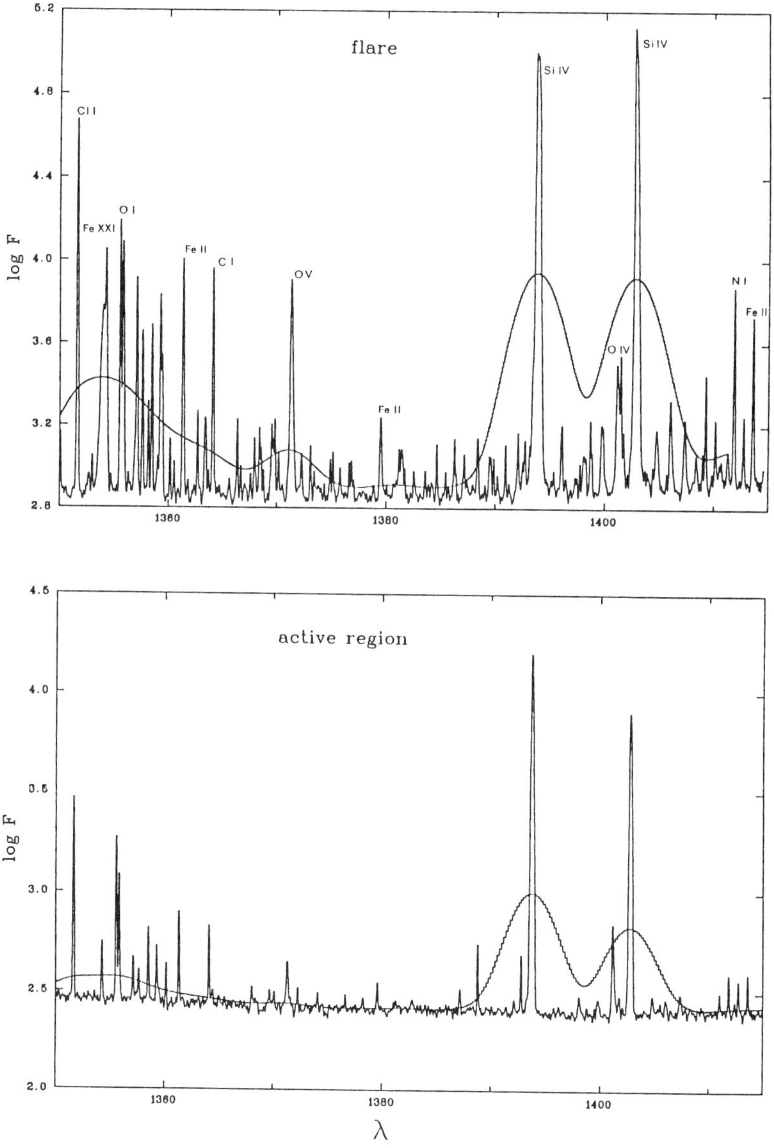

Fig. 1: A sample of the high spectral (~ 0.06Å) resolution solar data for (a) a two-ribbon flare and (b) a plage. Superimposed is the same spectra degraded to ~ 5Å, typical of the resolution of low dispersion IUE spectra. Intensity units are log $erg\ cm^{-2}\ s^{-1}\ \AA^{-1}\ sr^{-1}$ at the solar surface

Seventh Cambridge Workshop on Cool Stars, Stellar Systems, and the Sun
ASP Conference Series, Vol. 26, 1992
Mark S. Giampapa and Jay A. Bookbinder (eds.)

CONTINUOUS MONITORING OF TY PYX WITH IUE DURING THE ROSAT ALL-SKY SURVEY

JAMES E. NEFF
Astronomy Department, Penn State University, University Park, PA

ISABELLA PAGANO AND MARCELLO RODONÒ,
Istituto di Astronomia, Universitá di Catania, and
Osservatorio Astrofisico di Catania, Catania, Italy

JERRY BONNELL
CSC & IUE Observatory, NASA/GSFC, Greenbelt, MD

ABSTRACT We observed the eclipsing-binary system TY Pyx with the IUE spacecraft simultaneously with the ROSAT all-sky survey. The observations continuously spanned the 3.2 day period of this system. We are comparing the rotational modulation of the ultraviolet line fluxes and line profiles with the rotational and eclipse modulation of the X–ray flux. These data will permit us to spatially correlate the magnetic activity in the chromosphere and transition region with the coronal structure.

Keywords: TY Pyx, Chromospheres, Coronae, Stellar Surface Structure

INTRODUCTION

The ROSAT All-Sky Survey provided an excellent opportunity to correlate the temporal or phase-locked variability of stellar coronae with that of their chromospheres and photospheres. For rapidly-rotating stars and for eclipsing binaries, these temporal variations can be used to map the spatial structure of these different atmospheric levels. The procedures for inferring the spatial structure of the photosphere was described by Vogt, Penrod, and Hatzes (1987), who exploit the correspondence between wavelength on a rotationally-broadened line profile and location on the stellar surface. We have developed a related procedure to map the structure of stellar chromospheres using ultraviolet emission line profiles (Neff *et al.* 1989). In an eclipsing binary system, a detailed analysis of the light curve can provide some useful constraints on the spatial structure of the system. Walter, Gibson, and Basri (1983) demonstrated the application of eclipse mapping to stellar coronae, using EINSTEIN observations. Culhane *et al.* (1990) mapped TY Pyx using full-orbit, LE and ME light curves from EXOSAT. The structure of the corona, chromosphere, and photosphere likely are all induced by magnetic activity. By conducting spectroscopic observations in the visible (photosphere) and ultraviolet (chromosphere) *simultaneously* with broad-band observations in the X–ray (corona), we can effectively map the magnetic active regions in three-dimensions. We report herein preliminary results from such an observing campaign.

OBSERVATIONS AND DATA ANALYSIS

We observed the RS CVn-type binary system TY Pyx 10–13 November 1990 with IUE continuously throughout the system's 3.2 day orbital cycle. The observing dates were chosen to fall within the ROSAT all-sky survey window. TY Pyx is not an optimal candidate for such a campaign from any single consideration, but it might be the target that fulfills the largest number of requirements. The system is composed of two nearly identical, slightly-evolved G5 stars. These stars rotate synchronously with their orbital period, and their mutual partial eclipses permit an accurate determination of the system parameters and enable eclipse-mapping of the X-ray coronae.

We obtained 33 LWP-HI and 11 SWP-LO spectra. We interleaved the observations to acquire maximum time resolution of LWP spectra, and we obtained accurate FES magnitudes between each exposure.

The data were reduced at Penn State using the current RDAF software available from the IUE project at NASA/GSFC. We measured the low-dispersion line fluxes using gaussian fits to the emission lines plus a quadratic fit to the background. Our fitting routines allow multiple gaussians, so blended lines can be fit accurately by constraining the line width to be the instrumental resolution. The high-resolution Mg II k line profiles were analyzed using the procedures described by Neff *et al.* (1989). Gaussian profiles are fit to the two stellar emission lines and to the interstellar absorption line. The interstellar line is constrained to have instrumental width. Because the instrumental width is known, and because the equivalent width of the interstellar absorption should remain constant, the interstellar line merely serves to provide a velocity reference on the spectrum. Unfortunately, we do not know *a priori* the interstellar velocity, but using the formula of Crutcher (1982) we predict $v_{ism} = 12.52$ km s^{-1}. Unfortunately, the Crutcher velocity law is not accurate enough to use to rectify the systematic offset of the wavelength scale, which can be incorrect by as much as 20 km s^{-1}.

The FES Light Curve

We acquired 73 FES measurements during the period of the IUE observations. Because these were obtained uniformly in phase, with no gaps due to clouds or day/night cycles, the resulting light-curve is extremely valuable. The combined effect of the photometric "wave" and coarse time sampling conspired to produce a secondary eclipse that was apparently fainter than the primary eclipse. The minimum of the photometric wave occurred slightly after secondary eclipse.

Variability of Emission Line Fluxes

The integrated fluxes of the chromospheric and transition-region emission lines clearly showed variation due to the eclipses, and the flux during phase 0.6–0.9 was slightly higher than at 0.1–0.4. We were not able to continue observing into the next orbital cycle, so we cannot distinguish time and phase.

Variation of Mg II K Line Profiles

To first order, at least, the Mg II emission lines from the two stars were constant in flux outside of eclipse. They were of constant and equal FWHM

(~0.86 Å). They varied primarily in wavelength, blending near the eclipse phases. There is no pronounced difference in emission line flux or width between opposite hemispheres, as is always seen on the similar system AR Lac. We are continuing our analysis, though, because the measured centroid velocities of the two components do not match the expected (*i.e.*, photospheric) velocity curve. This could be due to inaccurate photospheric parameters, fixed-pattern noise in the IUE images, systematic flows in the chromospheres, or to large-scale structures in the stellar atmospheres. We must systematically work through the less appealing explanations before we can attribute the result to the more physical explanations.

Other Data

Results from the supporting VLA observations, obtained over a total of 22 hours on 8-11 November, are described by Linsky *et al.* at this meeting. TY Pyx should have been observed over a 2.26 day period by the ROSAT X-ray telescope, and over 7 days with the ROSAT wide-field camera. Those observations consist of a few minutes during each ~90 minute orbit of the satellite. The ROSAT results are presented elsewhere at this meeting.

Using the stellar spectrograph at the McMath telescope (National Solar Observatory/Kitt Peak), we obtained two spectra per night 10–15 November 1990. These spectra of the Ca II H and K lines are comparable in resolution to the IUE spectra of the Mg II h and k lines, but they are of much higher S/N. We solicited, by various means, other ground-based observations, and we would be interested to hear whether these were successful.

ACKNOWLEDGEMENTS We are grateful to the IUE project for providing us observing time under the RIASS program and to Ron Pitts for assisting with the observations.

REFERENCES

Crutcher, R.M. 1982, *Ap. J.*, **254**, 82.
Culhane, J.L., White, N.E., Shafer, R.A., Parmer, A.N. 1990, *M.N.R.A.S.*, **243**, 424.
Neff, J.E., Walter, F.M., Rodonò, M., and Linsky, J.L. 1989, *Astr. Ap.*, **215**, 79.
Vogt, S.S., Penrod, G.D., Hatzes, A.P. 1987, *Ap. J.*, **321**, 496.
Walter, F.M., Gibson, D.M., Basri, G.S. 1983, *Ap. J.*, **267**, 665.

Seventh Cambridge Workshop on Cool Stars, Stellar Systems, and the Sun
ASP Conference Series, Vol. 26, 1992
Mark S. Giampapa and Jay A. Bookbinder (eds.)

CHROMOSPHERIC ACTIVITY AND STELLAR EVOLUTION: CLUES FROM IUE DATA

LUCA PASQUINI European Southern Observatory, Casilla 19001, Santiago 19, Chile

ENZO BROCATO Osservatorio di Collurania, 64100 Teramo, Italy

ABSTRACT We investigate the dependence of chromospheric activity on stellar physical parameters in a sample of cool giants and supergiants. Ca II surface fluxes for stars of known masses and evolutionary status are derived from published IUE Mg II observations. Adding these data to the sample of Pasquini *et al.* (1990, hereafter PBP) we find that activity well correlates with T_{eff} and *Mass*. We suggest that the stellar evolutionary status plays a fundamental role in determining the level of chromospheric activity.

Keywords: Chromospheres, Stars: Giants, Stars: Evolution

Mg II and Ca II Fluxes

In a recent series of papers Dupree *et al.* (1990 and references therein) have analysed Mg II observations in evolved stars. These authors have observed a significant sample of stars in two well studied open clusters (Hyades and M67), as well as globular clusters and field metal deficient giants. All these objects have masses, metallicities and ages rather well defined. In this study we link this sample to the work of PBP who studied field stars, covering the upper part of the H-R diagram.

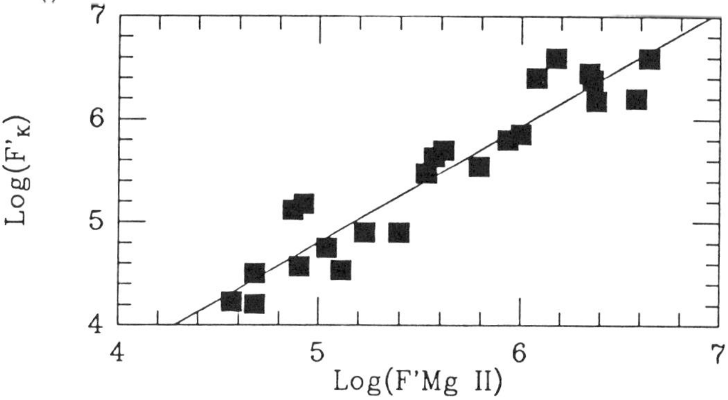

Fig. 1: Log(F'_k) vs. Log(F'Mg II) for the stars common to the PBP and Dupree *et al.* (1990) samples. The best fit relationship is shown.

In order to properly merge IUE data with the Ca II observations analysed by PBP, Mg II fluxes must be transformed in Ca II fluxes; a comparison between the stars in common to the two data sets will provide the required transformation. The resulting flux *vs* flux diagram is shown in Figure 1 for the 23 stars in common: the data are well fitted by a linear relationship giving:

$$Log(F'_k) = 1.125 \times Log(F'MgII) - 0.82 \tag{1}$$

Where F'_k is the net chromospheric flux in the K line of Ca II and F′(Mg II) the equivalent in the *h* and *k* Mg II lines. Equation 1 was applied to the Hyades, M67 and Pop II stars having fairly good Mg II determination, giving a total of 19 new Ca II entries, which, added to the 65 of PBP give a total sample containing 84 objects.

Analysis of the total sample

In Figure 2 an H-R diagram, containing the sample stars is shown. Different symbols indicates different levels of chromospheric activity and the new entries are marked with open circles. The Mg II converted data give Ca II fluxes in perfect agreement with those of field stars in the same position of the H-R diagram.

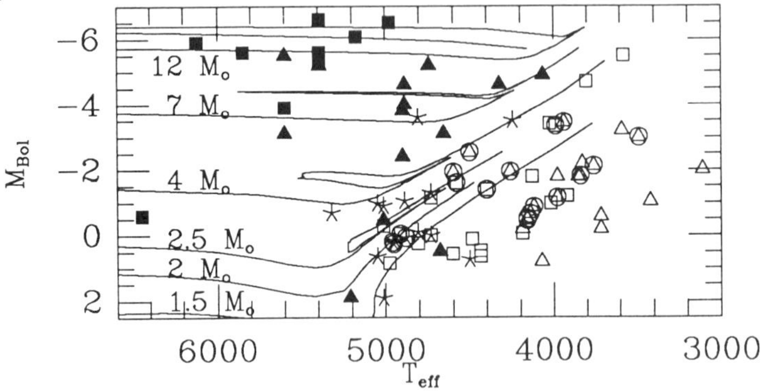

Fig. 2: H-R diagram containing the stars of the final sample; new entries are marked with circles. Different symbols are used according to the stellar activity level: Log(F'_k) < 4.6 Open Triangles; 4.6 < Log(F'_k) < 5.2 Open Squares; 5.2 < Log(F'_k) < 5.8 Asterisks; 5.8 < Log(F'_k) < 6.4 Filled Triangles; 6.4 < Log(F'_k) Filled Squares.

In order to proceed in the analysis, we have obtained a table containing Log F'_k, M_{Bol}, Log T_{eff} Log($M/M_\odot$) and Log(g) for all the objects. Masses have been derived in the following way:

- For the IUE stars masses were taken from detailed evolutionary studies obtaining $M/M_\odot = 2.05 \pm 0.15$, 1.2 ± 0.15 , 0.85 ± 0.2 for Hyades, M67 and Pop II objects respectively.
- For the field stars we were forced to adopt rather coarse divisions, due to the relatively small number of objects and to the errors involved in the determination of the stellar physical quantities. We have divided the field stars in 4 subgroups, according to Figure 2:

$M/M_\odot = 0.85 \pm 0.15$ for the 5 objects found to probably be Pop II stars by PBP.

$M/M_\odot = 1.15 \pm 0.35$ for stars lying below the 1.5 $M/M_\odot$ evolutionary track.

$M/M_\odot = 2.75 \pm 1.25$ for stars comprised between the 1.5 and 4 $M/M_\odot$ tracks.

$M/M_\odot = 8 \pm 4$ for stars over the 4 $M/M_\odot$ track.

We have finally performed a series of one and two dimensional best fits, whose results are given in Table 1 (Log(g) was computed using M_{Bol}, T_{eff} and Mass).

TABLE I : Best fit to Log(F'_k). In first column the dispersion around the best fit equation is given; in columns 2-5 the angular coefficients of the relationships.

σ	Log(T_{eff}	(Log($M/M_\odot$)	Log(M_{Bol})	Log(g)
0.43	11.34			
0.50		1.69		
0.75			-0.145	
0.8				0.227
0.27	8.09	1.01		
0.31	11.2		-0.13	
0.36	13.7			-0.35

From Table I clearly emerges that activity strongly depends on T_{eff}; this is a well known result (Linsky *et al.* 1979, Pasquini *et al.* 1988). From Table I also emerges that a biparametric fit much better represents the data; *at least two parameters are necessary to describe chromospheric activity of evolved stars.* In particular the minimum scatter is obtained when T_{eff} and Mass are considered

$$F'_k \sim T^8_{eff} \times (M/M_\odot) \qquad (2)$$

It is worth to note that this scatter (0.27 in Log F'_k) is comparable with the uncertainty with which absolute chromospheric fluxes can be determined in evolved stars (Pasquini *et al.* 1988); the activity level of evolved stars is *at least in first approximation* determined when knowing the stellar T_{eff} and *Mass.*

Equation (2) supports the conclusions derived by PBP: for a given temperature, more massive stars appear to be more active; *viceversa*, for a fixed mass , activity is decreasing with temperature.

We finally note that the evolutionary status of the more active stars is the core He-burning phase and that less active stars are in the shell H-burning phase. Since the evolutionary phase and the mass of a star determine its position in the HR diagram, relation (2) may suggest that chromospheric activity is determined by stellar evolutionary status.

REFERENCES

Dupree, A.K., Hartmann, L., Smith, G.H. 1990: *Ap.J.* **353**, 623.

Linsky, J.L. *et al.* 1979: *Ap.J.Suppl.* **41**, 47.

Pasquini, L., Pallavicini, R., Pakull, M. 1988: *Astron. Astrophys.* **191**, 266

Pasquini, L., Brocato, E., Pallavicini, R. 1990: *Astron. Astrophys.* **234**, 277.

Seventh Cambridge Workshop on Cool Stars, Stellar Systems, and the Sun
ASP Conference Series, Vol. 26, 1992
Mark S. Giampapa and Jay A. Bookbinder (eds.)

UV VARIABILITY OF PRE-MAIN SEQUENCE STARS

BRIAN M. PATTEN AND THEODORE SIMON University of Hawaii, Institute for Astronomy 2680 Woodlawn Drive, Honolulu, Hawaii 96822

ABSTRACT There are some 633 large aperture, low dispersion, long wavelength spectra of PMS stars in the *IUE* archives. We examine the time variability of the integrated Mg II h and *k* flux and its underlying continuum. Spectra for some 15 PMS stars, which have been observed at least 8 times by *IUE* on timescales ranging from hours to days, have been analyzed. We find that many of the PMS stars show substantial variability in both the Mg II line and the continuum.

Keywords: pre-main sequence stars; T Tauristars; variable stars

INTRODUCTION

There are some 633 large aperture, low dispersion spectra of pre-main sequence stars in the *IUE* archives. We have conducted a survey of these UV spectra in order to determine which stars have sufficient coverage to allow an analysis of variability on various timescales. We present an analysis of variability of the integrated Mg II *h* and *k* line flux and its underlying continuum on timescales of hours, days, and many days for 15 stars from the archives. Through this analysis we try to differentiate between two accretion models, an axisymmetric boundary layer (Kenyon and Hartmann 1987, BBB 1988) and a hot spot via magnetic channeling (Uchida and Shibata 1984, Königl 1991). The former model is expected to introduce variability in the line and continuum fluxes on a timescale of principally half a day, while the latter is expected to produce variability on a timescale of days. In both cases, non-steady accretion rates may also produce variability in the observed UV fluxes due to "flaring" events (Gahm, 1990) on timescales of many hours to a couple of days.

ANALYSIS

We analyzed the variability of the 15 PMS stars for which the *IUE* archives contain 8 or more low dispersion, large aperture observations with the long wavelength cameras. The spectra were calibrated, corrected for camera blemishes, and were measured using the portable Unix RDAF software provided by the *IUE* RDAF at Goddard. We used the routine FEATURE to measure the integrated Mg II *h* and *k* flux and the underlying continuum above the zero flux level. The line and continuum fluxes measured in this way are generally accurate to the 5-10% level. The Fine Error Sensor data associated with these UV spectra were also obtained and reduced to absolute fluxes using the IUE calibration procedures and the absolute flux calibration of Johnson (1966).

The measured fluxes were searched for variability on timescales of less than 12 hours, 12 hours to 2 days, and a few days. These timescales were chosen to allow us to differentiate between an axisymmetric boundary layer, non-steady accretion, and a hot spot on the stellar surface, respectively. We characterize a star as variable if, on the examined timescale, the amplitude of variation exceeds our flux measurement error level of 5-10%. The amplitude of variability was determined by inspection due to the lack of data points for a Fourier analysis. The results of this analysis are summarized in Table I. RY Tau, a particularly well sampled star by *IUE*, is shown as an example in Figure 1.

ACKNOWLEDGEMENTS

This research was supported by NASA grant NAG 5-146 to the University of Hawaii. We thank Jerry Bonnell for providing the Unix-based RDAF data reduction package, and Lyla Taylor, Jen Gagliardi, and Ruth Bradley of the Goddard RDAF staff for their assistance in the reduction of these spectra.

REFERENCES

Bertout, C., Basri, G. and Bouvier, J. 1988, *Ap.J.* **330**, 350. (BBB)

Gahm, G.F. 1990, in IAU Symposium 137 Flare Stars in Star Clusters, Associations, and Solar Vicinity eds. E.R. Priest and V. Krishan (Dordrecht: Kluwer).

Johnson, H.L. 1966 *Ann.Rev.A.A.* **4**, 193.

Kenyon, S.J. and Hartmann, L. 1987 *Ap.J.* **323**, 714.

Königl, A. 1991 *Ap.J.* **370**, L39.

Uchida, Y. and Shibata, K. 1984 *P.A.S.J.* **36**, 105.

TABLE I Amplitude of Variability Summary

Star	# Obs	% variability UV continuum flux[1] ≤12h	12h-2d	>2d	% variability Mg II h and k flux ≤12h	12h-2d	>2d	IUE[2] coverage	variability[3] mechanism
T Tau	20	30 (9)	27 (4)	...	≤5 (4)	30 (1)	...	i	-----
RY Tau	59	13 (26)	22 (14)	44 (10)	9 (21)	10 (14)	21 (10)	e	H ?
BP Tau	25	26 (10)	75 (2)	96 (10)	14 (10)	17 (2)	41 (10)	e	H
DE Tau	7	6 (1)	...	...	≤5 (1)	...	...	i	-----
DF Tau	16	12 (2)	119 (8)	31 (5)	8 (2)	138 (8)	25 (5)	g	-----
DG Tau	18	12 (3)	28 (2)	12 (4)	17 (2)	≤5 (2)	47 (4)	g	-----
DN Tau	14	...	28 (7)	64 (6)	...	18 (7)	45 (6)	i	-----
DR Tau	31	24 (10)	28 (10)	38 (9)	20 (10)	22 (10)	67 (9)	e	H
V410 Tau	8	...	196 (3)	172 (3)	...	36 (3)	44 (3)	i	-----
RW Aur	25	17 (8)	30 (4)	50 (2)	20 (5)	≤5 (1)	...	i	B
SU Aur	14	17 (3)	12 (6)	19 (3)	18 (3)	19 (6)	23 (3)	g	B
FU Ori	7	24 (3)	...	...	8 (3)	...	...	i	-----
GW Ori	9	7 (2)	19 (3)	7 (2)	≤5 (1)	16 (2)	7 (2)	i	-----
TW Hya	8	24 (3)	132 (2)	42 (2)	26 (3)	34 (2)	32 (2)	i	-----
RU Lup	14	12 (2)	27 (1)	...	...	...	...	i	-----

[1] Numbers in parenthesis indicate the number of data pairs used to calculate amplitude
[2] "e" = excellent, "g" = good, and "i" = insufficient designate levels of coverage in terms of both the number of data points available and the temporal distribution of those points
[3] "H" = hot spot and "B" = boundary layer

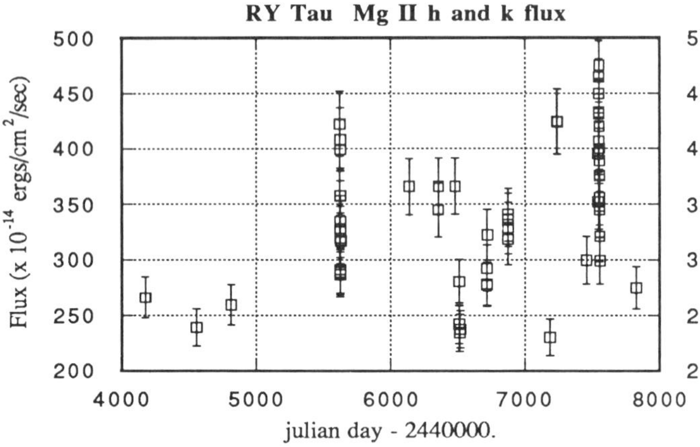

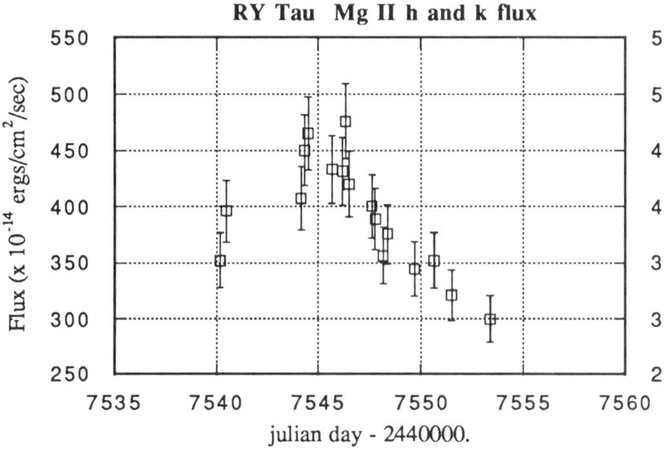

Fig. 1: Line Fluxes for RY Tau. Above is the total line flux for all observations in the archives while below is the line flux for the Julian dates 7535-7560. RY Tau, a particularly well sampled star by *IUE*, shows variations on timescales of weeks.

Seventh Cambridge Workshop on Cool Stars, Stellar Systems, and the Sun
ASP Conference Series, Vol. 26, 1992
Mark S. Giampapa and Jay A. Bookbinder (eds.)

ROSAT/WFC EUV OBSERVATIONS OF COOL STARS

GORDON E. BROMAGE Space Astronomy Group, Rutherford Appleton Laboratory, Didcot, Oxon OX11 0QX, U.K.

ABSTRACT The UK Wide Field Camera on ROSAT has carried out the first-ever all-sky survey in the EUV region. A large number of active cool stars have been detected. The WFC instrument and preliminary overall survey results are briefly described. Examples of the wide variety of active cool star survey observations are presented, including large flares on known dMe flare-stars and the discovery of activity on new dMe, RS CVn and fast-rotating single stars in the solar neighbourhood.

Keywords: Stellar coronae; EUV; flares; sky surveys

INTRODUCTION

The Extreme Ultraviolet (EUV) region has often been referred to as the last great window in the electromagnetic spectrum waiting to be opened on the universe. In the subject area of this workshop, the great importance of observation of EUV emission from plasma in the outer atmospheres of late-type stars has been known for years, from solar observations and modelling (*e.g.* the coronal resonance Fe IX - Fe XIV line emission, temperature and density diagnostics, emission measure distribution studies, continuum emission mechanisms). We strongly expect that radiative energy losses in the EUV region are a very important component of the overall energy balance. Knowledge of the emission measure distribution enables us to start to understand the structure of the atmosphere and consider heating mechanisms – but until ROSAT/WFC there was a gap between $\sim 2\ x\ 10^5$ and $\sim 2\ x\ 10^6\ {}^\circ K$ in stellar data, between the highest-temperature IUE lines and the temperatures probed by previous X-ray astronomy missions.

Whilst we have long known that this temperature range was crucial, it has been essentially inaccessible to detailed observation, due to a combination of two historical factors: technological barriers and a "fear of the interstellar medium" (because of course the photoelectric absorption by the ISM is crippling in the EUV in high-density directions). Prior to ROSAT we had a closed, shuttered EUV window – with just one or two cracks that had been looked through to give a tantalising glimpse of the EUV universe. Figure 1(a) shows the known EUV sources before ROSAT: a handful of hot white dwarfs and cataclysmic variables, and one active cool star – Proxima Centauri. With the all-sky survey completed by the UK Wide Field Camera on ROSAT, we have opened the window at last. Figure 1(b) shows the resulting EUV sky, with a dramatic increase in numbers of sources detected. These two figures encapsulate the impact of the ROSAT WFC on our knowledge of the EUV sky and stellar EUV sources.

In this review I will briefly describe the WFC instrument, and in particular the wavelength regions covered and the thin-film filters used to select them. Then, concentrating on the all-sky survey results, I will give examples of the EUV sources and their variability, including stars previously known to be active and also some of the exciting new discoveries. Further results from the WFC, including statistical results for cool stars, will be presented by my WFC Consortium colleagues John Pye (Pye *et al.* , Hodgkin & Pye, these proceedings) and Robin Jeffries (Jeffries & Jewel, these proceedings). See also Bromage *et al.* (these proceedings).

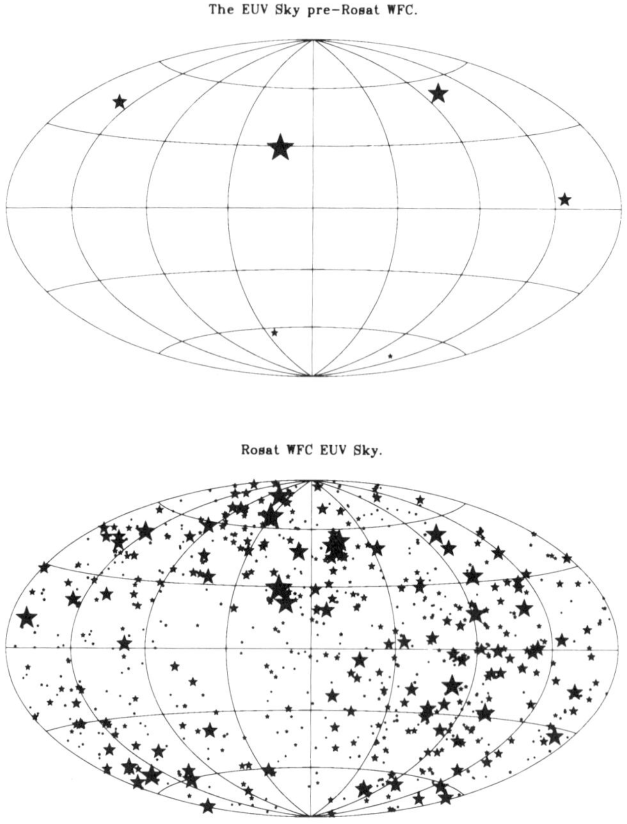

Fig. 1: The known sources in the EUV sky before and after the ROSAT WFC all-sky survey. Note that the sizes of symbols represent the significance of detections in the WFC survey: these are roughly correlated with fluxes, depending on exposure times.

THE WIDE FIELD CAMERA ON ROSAT

Schmitt (these proceedings) has described the ROSAT mission and the soft-X-ray telescope on board. The complementary instrument, performing co-aligned observations during both the survey and pointed phases, is the UK Wide Field Camera. Because of the wider field of view, most areas of the sky were observed in the WFC survey for 5 days, with typical exposure times of 50 seconds every 96-minute orbit. High-ecliptic-latitude areas were covered more deeply, as with the XRT. The WFC (Wells *et al.* , 1990) was built and is operated by a consortium of four UK University groups and the Rutherford Appleton Laboratory; the PI is K.A. Pounds (Leicester). The instrument is shown schematically in Figure 2.

The field of view is 5 degrees diameter, and for the all-sky survey results presented here the spatial resolution was ~ 3 arcmin and absolute positional accuracy of sources is generally between 20 arcsec and 1 arcmin. The instrument consists of a grazing incidence telescope with a set of three nested (co-axial, confocal) Wolter-Schwarzschild Type I gold-coated aluminium mirrors. A focal-plane turret is used to select one of two identical detector assemblies. The detectors are microchannel plates curved to match the optimum focal surface, and with CsI photocathode deposited on the front face to enhance the EUV response. Eight filters are mounted on a wheel in front of the detector.

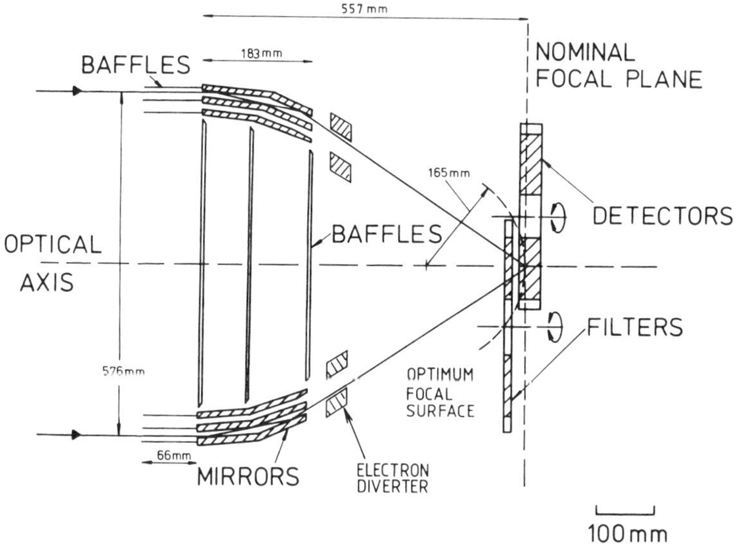

Fig. 2: Wide Field Camera optical path schematic.

The filters (one of the hardware contributions from the Rutherford Appleton Laboratory; see Kent *et al.* 1990) allow broad band photometry to be performed. They are large-area (7 cm diameter) thin-film filters typically 0.5 μm thick, with transmission centred on ~ $\lambda\lambda$ 100, 140, 180 and 600 Å respectively: the first two were used alternately during the all-sky survey, swapping normally once per day.

In addition to selecting bandpasses, the filters are required to suppress the intense geocoronal radiation which would otherwise saturate the detectors, and to suppress UV radiation from hot stars which would otherwise be counted due to the residual UV detector sensitivity. The last requirement implies that the transmission of the survey filters between Ly-α and 2000Å is less than 10^{-8}, a condition that has been met successfully. EUV passband filters are necessarily very thin and therefore fragile, and the design and fabrication of the filters was governed by severe mechanical and optical constraints. The filters are performing exceptionally well in orbit. Details of fabrication, qualification and calibration are described in Kent *et al.* (1990). The main astronomical filters are coated polycarbonate (lexan). Two (S1) lexan/carbon filters (one with a layer of boron carbide for protection against atmospheric oxygen erosion) and two redundant (S2) lexan/beryllium filters were designed for both survey and pointed-phase use: S1a and S2a were used throughout the all-sky survey. In addition, lexan/aluminium (P1) and tin/aluminium (P2) filters provide two further wavelength bands for pointed phase observations. The overall WFC effective-area curves including each filter in turn are shown in Figure 3.

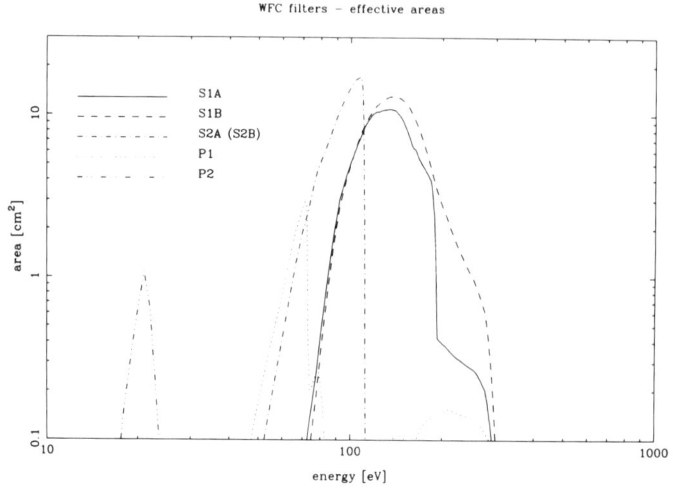

Fig. 3: WFC transmission curves, labelled by filter number.

THE WFC ALL-SKY SURVEY RESULTS

Early in the ROSAT mission, a five-day mini-survey was performed (Pounds *et al.* 1991) and of the 23 identified EUV detections, 15 are coronal sources. This was followed by the all-sky survey, the vast majority of which was completed in the 6-month period from August 1990 onwards. Provisional analysis of all the data is complete and the following summary refers to secure detections from the current stage of analysis: further processing is in hand and is expected to significantly increase the totals.

There are some 700 secure EUV sources so far, of which 250 remain unidentified at present. Identifications via catalogue correlations with known X-ray sources and nearby active stars and hot white dwarfs have been greatly supplemented by an intensive series of optical identification observing campaigns (Mason *et al.* 1991) organised by the WFC Consortium. For coronal sources, an indication of the detection levels is that a 6^m main-sequence active G star with barely detectable Ca II K line emission core at 1 Å resolution is typically also just detected in the WFC survey, whilst active dMe stars are detected down to $\tilde{1}3^m$ in V.

Of the ~ 450 identified objects, ~ 10 are Active Galactic Nuclei seen through low-density directions of the interstellar medium, ~ 20 are cataclysmic variables including some exciting new AM Her objects, and ~ 120 are hot white dwarfs — including many of the brightest WFC objects and of which half are new discoveries. Nearly 300 are late type stars, including some 50 identified as RS CVn (evolved) binaries, and 50 dMe stars (of which about half are known to be flare stars).

An example of a pair of fairly close WFC sources is shown in Figure 4. their separation is 8 arcmin. The softer source (brighter in S2a filter) has now been identified as a new white dwarf, and this filter ratio is typical of such nearby WDs. The somewhat harder source is coincident with a bright double G6 main sequence star, of which one component is optically variable and may be a new RS CVn system.

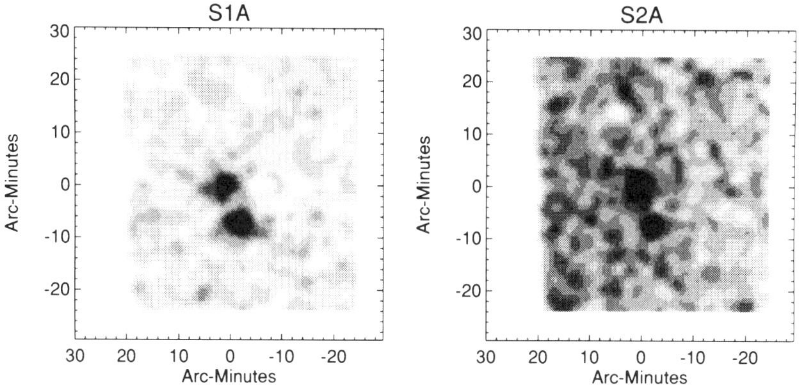

Fig. 4: Example of a pair of close WFC sources illustrating the spatial resolution. The northerly object is a new WD and the other is a coronal source.

For quiescent coronal sources, the WFC filter ratio can be used in conjunction with plasma models to constrain the effective temperature, provided that a reasonably accurate value for the interstellar absorption along the line of sight is known. (*e.g.* Figure 5). More information on the emission measure distribution will be obtainable by combining WFC and XRT data, at least for the ranges of temperature with large emission measure that are tuned to the ROSAT instrument sensitivities; information on low emission-measure ranges will necessarily be more difficult to recover from these broad-band observations. It can be seen from Figure 5 and bearing in mind that most coronal sources with high interstellar columns are too far away to be detectable by the WFC, that coronal sources are expected to have S2/S1 filter ratios between 1 and 2. Finally, it should be noted that it is normally impossible to derive temperature information solely from WFC data on *transient events* during the survey, since the filters were swapped only daily.

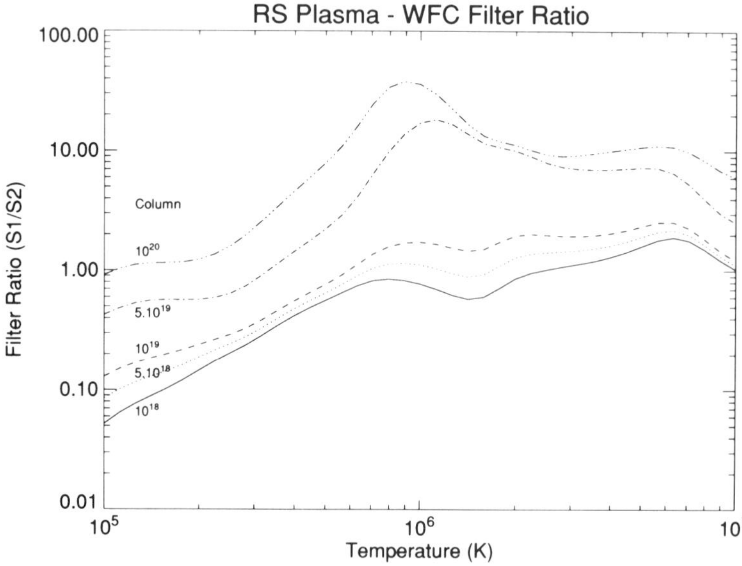

Fig. 5: WFC filter ratio as a function of plasma temperature for single-temperature Raymond-Smith plasma models, and for various values of N_H interstellar hydrogen column density (cm^{-2}).

PREVIOUSLY-KNOWN CORONAL SOURCES: EUV FLARES

The first unambiguous detection of a stellar flare in the EUV — also seen in the ultraviolet (IUE), X-ray (XRT) and optical regions — was recorded by the WFC on 1990 October 1, from the famous double flare star BY Dra. We have already reported the first results from this data set (Barstow *et al.* , 1991). The S2/S1 filter ratio for BY Dra during *quiescence* and a single-temperature model fit yielded $\approx 3x10^6$ ^oK. The quiescent luminosity and emission measure were $4x10^{28}$ erg s^{-1} and $2x10^{52}$ cm^{-3} in the energy range 0.08 - 0.18 keV. The total energy radiated in the EUV *flare* was $\approx 10^{33}$ erg in the S2a filter — a substantial fraction of the total energy released. The EUV and UV (IUE) flare apparently lasted for several hours, and the corresponding XRT flare decay was detectable for nearly 10 hours, making this a particularly long-lasting event for dMe stars.

With only part of the full survey time-series analysed so far, we have also found large EUV flares that occurred on the dMe stars EV Lac (December 17 at ~15:30 UT and December 20 at ~18:00), YY Gem (October 3 at ~10:30) and AU Mic (October 18 at ~19:00). The last is illustrated in Figure 6. The presence of flaring activity in the S1a filter explains the (at first puzzling) anomalous average filter ratio for AU Mic, which was outside the allowable range for coronal models for this star. The first EV Lac flare showed even larger enhancements over quiescent levels, peaking at 1.5 count s^{-1} in the S2a filter, a factor of 20 enhancement. Many of the known shorter-period RS CVn binaries were clearly detected in the survey, and their time variability is now being investigated. For example, an EUV flare has been detected (Barstow & Guinan, private communication) in HR 1099 near the maximum brightness phase in the optical photometric rotational-modulation light curve.

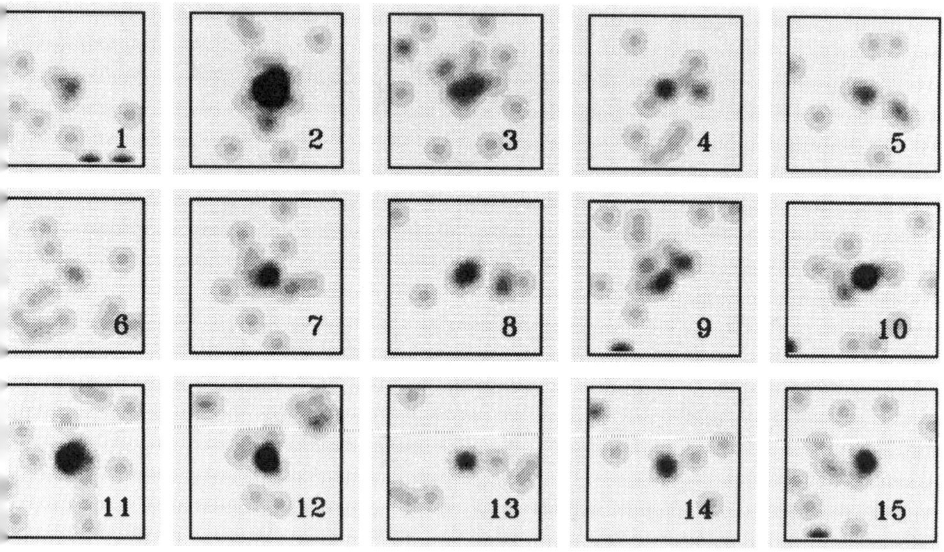

Fig. 6: Time-resolved images of WFC data of the dMe flare star AU Mic, showing EUV (S1 filter) flaring activity. Each frame corresponds to one "snapshot" of ~50 seconds integration, and the frames are for consecutive orbits (separated by 96 minutes) totalling one day's data.

DISCOVERY OF NEW ACTIVE STARS

Many main-sequence F-G-K stars in the solar neighbourhood appear in the WFC survey source list, and a large fraction of these have no previously published record of activity (as evidenced normally by Ca II H and K line emission cores). In our optical identification and follow-up observations, we have confirmed that many of these are indeed chromospherically active, but some reveal nearby dMe companions or nearby WD stars that are the real EUV emitters. The most exciting categories of new discoveries are (a) new dMe flare stars, (b) new RS CVn binaries, and (c) new rapidly-rotating single dwarf stars. In all these cases it is as usual the objects' youth-linked or binary-locked rapid stellar rotation that is assumed to be the key contributing factor (via differential rotation and magnetic fields) to give optical activity and high EUV emission.

An example discovery of a new dMe flare star is shown in Figure 7. This is the WFC source RE0604-34, which had no catalogued counterpart. In the S2a filter observations, a three-orbit flare event occurred on September 12 (labelled section B). Images integrated over the other three labelled time periods show the source was below the WFC sensitivity level at those times, although there is a small positive detection in the averaged S1a filter data. The optical identification search in the WFC error circle yielded a new dMe star with V magnitude 13. A similar object with no catalogued counterpart, RE0447-27, and detected only in the S2a filter, proved to be another new dMe, which actually flared serendipitously during optical follow-up spectroscopy, showing enhanced Balmer and Ca II emission including downward-moving (redshifted) chromospheric material in the early stages of the flare.

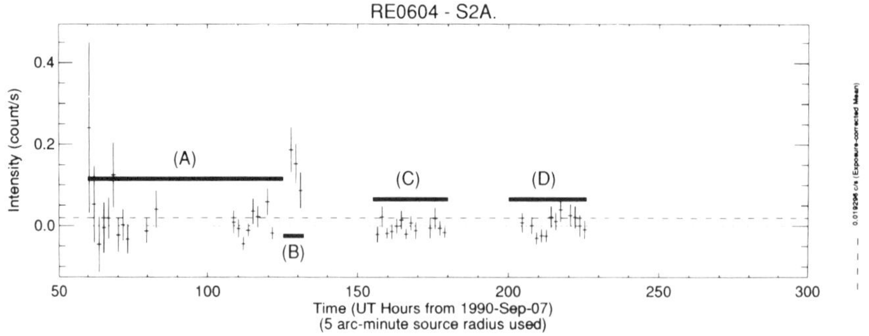

Fig. 7: Time series of WFC survey (S2a filter) observations of the new dMe flare star RE0604-34.

Newly discovered RS CVns include RE0418+23, a double-lined binary with a period of slightly less than 2 days, and RE1603-57, also double-lined with a period of 6 days. The latter is particularly interesting in that it is a G-type 8^m star very close to Iota Normae, a tight cluster of 4^m mid A-type stars in the WFC error circle. We have found on a number of occasions after careful investigation that apparent coincidences of EUV sources with A-type main-sequence stars have proved illusory. In some cases, a faint but very hot white dwarf close-by in the error circle proves to be the EUV-emitter, as with the famous case of Sirius B. In the case of RE1603-57 the EUV source proved to be a coronal late-type active binary rather than the much (optically) brighter A stars.

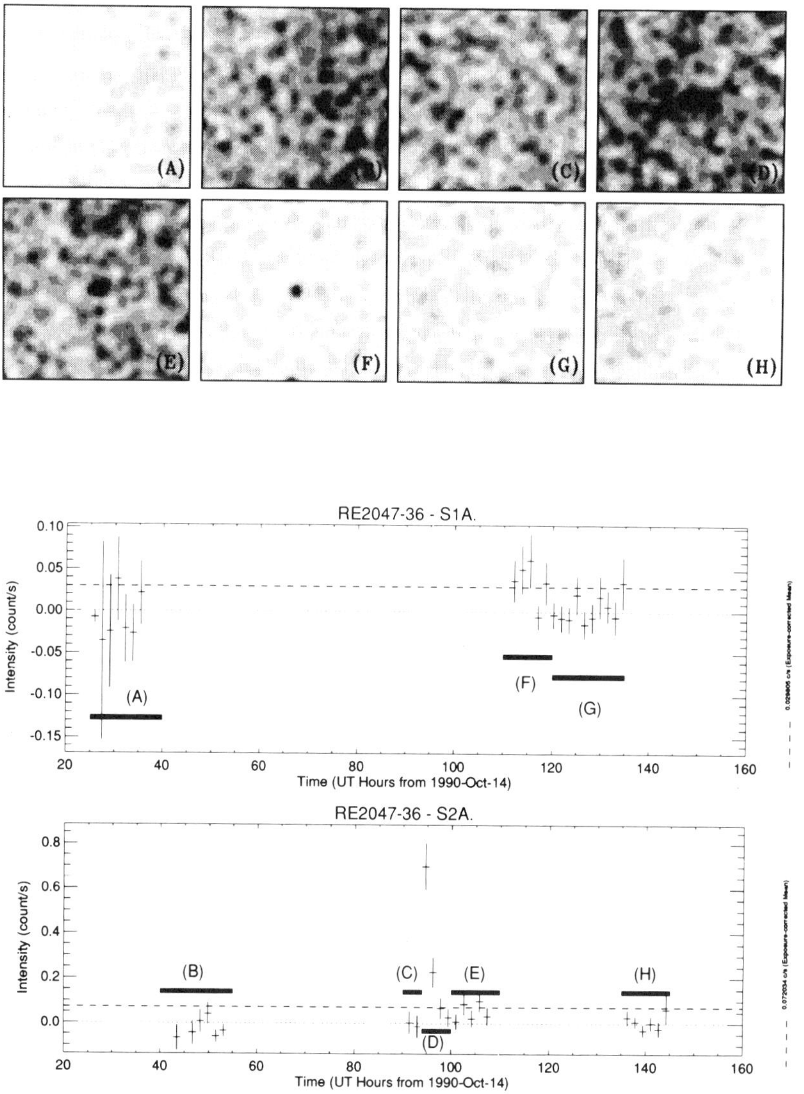

Fig. 8: Time series and time-resolved image sequence for "Speedy" Mic WFC EUV observations, showing a large flare on this K0V fast rotator.

These results strengthen the existing evidence that these A stars do not possess significant coronae.

Finally, a small number of WFC sources have been identified with young, very fast rotating single K dwarfs, including the well-known example AB Dor.

The most spectacular discovery so far is HD 197890 or "Speedy" Mic, in the constellation of Microscopium (Bromage *et al.* , these proceedings). Despite the apparent positional correlation with the HD star, this was at first thought to be a new white dwarf, because of the large average S2a/S1a filter ratio, compared with a normal one of approximately unity for the nearby dMe flare star AT Mic. However, as with AU Mic (see above) which had an apparently irreconcilably *low* S2a/S1a ratio due to flaring in S1a observations, this third Mic source also showed a large flare, this time in S2a. The time series and time-resolved images are shown in Figure 8. Detailed optical follow-up observations proved that this is the fastest single-dwarf rotator known amongst the solar-neighbourhood late-type stars, and is a member of the young Pleiades moving group.

CONCLUDING REMARKS

The WFC instrument is performing excellently and during the all-sky ROSAT survey has produced a wealth of EUV observations of coronal sources. It has opened wide the EUV window on the (relatively nearby) universe. The examples of active star observations presented here are but a small part of this wealth, but have already demonstrated a range of exciting discoveries in this area of astrophysics. We can confidently expect that further analysis of survey data, especially in conjunction with simultaneous XRT observations, further optical and IUE follow-up of the new EUV sources, and indeed the more recent pointed-phase ROSAT WFC observations, will yield more insights into the outer atmospheres of active cool stars.

ACKNOWLEDGEMENTS

The WFC project is supported by the UK Science and Engineering Research Council. The WFC instrument and associated Ground System were developed by a consortium of five UK institutes: the University of Leicester, SERC's Rutherford Appleton Laboratory, the University of Birmingham, the Mullard Space Science Laboratory of University College London, and Imperial College STM London. I would particularly like to thank Barry Kellett (RAL), Robin Jeffries (Birmingham) and Martin Barstow and John Pye (Leicester) for many aspects of help, discussion and collaboration in the analysis and interpretation of data on WFC coronal sources.

REFERENCES

Barstow, M. A., *et al.* 1991, *Nature*, **353**, 635.
Kent, B. J., *et al.* 1990, *Proc. S.P.I.E.*, **1344**, 255.
Mason, K. O., *et al.* 1991, *Vistas in Astronomy*, **34**, 343.
Pounds, K. A., *et al.* 1991, *M.N.R.A.S.*, **253**, 364.
Wells, A., *et al.* 1990, *Proc. S.P.I.E.* **1344**, 230.

Seventh Cambridge Workshop on Cool Stars, Stellar Systems, and the Sun
ASP Conference Series, Vol. 26, 1992
Mark S. Giampapa and Jay A. Bookbinder (eds.)

ROSAT EUV LUMINOSITY FUNCTIONS OF NEARBY F & G STARS

S.T. HODGKIN & J.P. PYE
Department of Physics & Astronomy, Leicester University, Leicester, LE1 7RH, UK.

ABSTRACT Using observations from the ROSAT Wide Field Camera all-sky survey, we have constructed the first EUV (100–150 Å) luminosity functions, for nearby (distance $\leq$ 22 pc) stars of spectral types F and G.

Keywords: EUV; luminosity functions; nearby stars

1. THE ROSAT ALL-SKY EUV SURVEY

The ROSAT EUV all-sky survey was conducted with the UK Wide Field Camera (WFC; Wells *et al.* 1990; Bromage 1991) over the six month period July 1990 to January 1991, in two 'colours' using broadband filters to define wavebands centred at about 100 Å (120 eV; S1A) and 140 Å (90 eV; S2A). The survey was fully imaging, with effective spatial resolution of $\sim$ 3 arcminutes (FWHM) and point-source location accuracy of typically $\lesssim$ 1 arcminute. We present here a preliminary investigation of the F and G star EUV luminosity functions.

2. THE STELLAR SAMPLE AND EUV LUMINOSITIES

In order to construct the luminosity functions, we need to have a (complete) volume limited sample of stars. Our sample is defined as those stars in the Gliese (1969) catalogue classified as spectral types F or G (but excluding white dwarfs) and with distances $\leq$ 22 pc. We have not, in the present preliminary analysis, separated dwarfs and giants, or singles and binaries.

We have uniformly analysed the EUV data to yield 'detected' count rates or upper limits, for each star. The detection threshold was set such that we expect $\lesssim$ 1 false detection due to Poisson noise on the background counts. Similarly, the upper limit level was set such that we expect the true (and hence unknown) count rates to exceed the derived upper limit values for $\lesssim$ 1 object in each of the F and G star samples.

We have converted the WFC count rates into EUV luminosities (in the photon energy bands 90–190, 60–100 eV for the S1A and S2A filters respectively; these are roughly the points at which the effective area in each band falls to 10% of maximum), using constant conversion factors appropriate for a cosmic-abundance, optically thin, hot plasma spectrum with T$\sim 10^6$ K, and line-of-sight photoelectric absorption $n_H \lesssim 10^{18.5}\ cm^{-2}$.

Out of a total of 132 (229) F (G) stars in our Gliese sample (corresponding to mean space densities of $3.0 \times 10^{-3}\ pc^{-3}$ and $5.2 \times 10^{-3}\ pc^{-3}$ respectively), 20 (18) were detected in the S1A filter, and 22 (19) in the S2A filter. Hence we detect $\sim$ 20% of the F stars and $\sim$ 10% of the G stars. Fig. 1 shows the detected and upper limit luminosities (L_{EUV}) plotted against distance, for the S2A waveband. It can be seen that the luminosity upper limits are highly significant in the sense that they are of the same order, or in many cases (especially for the G stars) lower, than the detected levels (*i.e.,* the non-detections are not due to poor coverage). The results for the S1A waveband are similar to those for S2A.

3. THE EUV LUMINOSITY FUNCTIONS

We have constructed and compared the EUV luminosity functions using 'survival analysis' methods to take account of the upper limits (Feigelson and Nelson 1985). The resulting cumulative luminosity distribution functions in the S2A waveband are shown in Fig. 2; again the results for S1A are similar. The mean (and median) values of $\log(L_{EUV}\ erg\ s^{-1})$ are: F-stars/S1A 27.6 ± 0.2 (27.7); G-stars/S1A 27.2 ± 0.1 (27.3); F-stars/S2A 27.6 ± 0.1 (27.8); G-stars/S2A 26.9 ± 0.1 (27.2). Comparison of the F and G star luminosity functions indicates that the probability that both are drawn from the same parent population is $\sim 5\%$ and $\lesssim 0.5\%$ for the S1A and S2A wavebands respectively. (These probabilities should be considered as provisional in that we have yet to investigate the reliability of the statistical tests for such heavily censored data.) Hence, the G star distribution appears to lie below that for the F stars. A similar effect has previously been noted in comparisons of the soft X–ray (energy range $\sim 0.2 - 4$ keV) luminosity functions of dF and dG stars based on EINSTEIN Observatory measurements (*e.g.* Vaiana 1990 and references therein).

ACKNOWLEDGEMENTS

The WFC project is supported by the UK Science and Engineering Research Council (SERC). The WFC instrument and associated Ground System for mission operations and data processing were developed by a consortium of five UK institutes: Leicester University, SERC's Rutherford Appleton Laboratory, Mullard Space Science Laboratory – University College London, Birmingham University, and Imperial College of Science, Technology and Medicine – London. The authors acknowledge the financial support of the SERC.

REFERENCES

Bromage, G.E. 1991 in Cool Stars, Stellar Systems, and the Sun, ed. J. Bookbinder & M. Giampapa, *Pub. A.S.P.*, Conference Series, this volume.

Feigelson, E.D., and Nelson, P.I. 1985, *Ap. J.*, **293**, 192.

Gliese, W. 1969, *Veroff Astron. Inst. Heidelberg*, No. 22.

Vaiana, G.S. 1990 in Imaging X–ray Astronomy, ed. M. Elvis (Cambridge University Press), p. 61.

Wells, A.A., *et al.* 1990, *Proc. S.P.I.E.*, **1344**, 230.

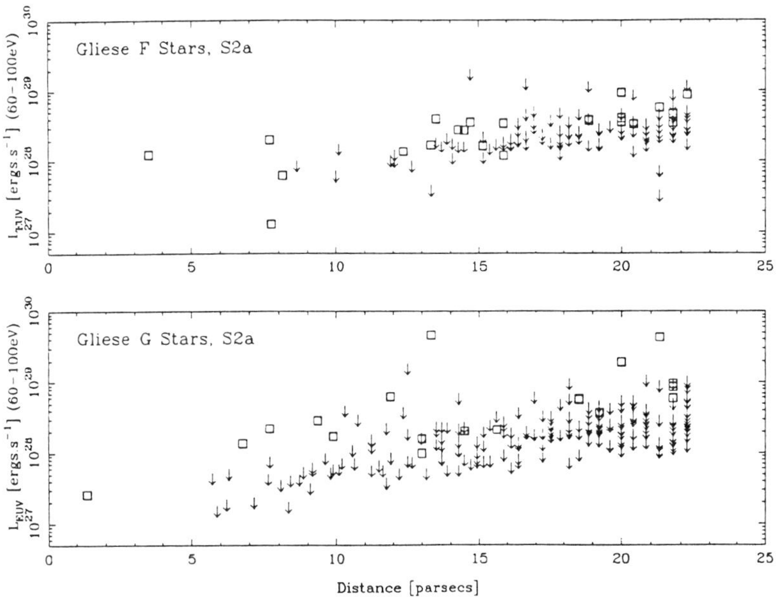

Fig. 1: EUV luminosity (in the S2A waveband) versus distance for the sample stars. Symbols: squares – detections; arrows – non-detections (upper-limit luminosities).

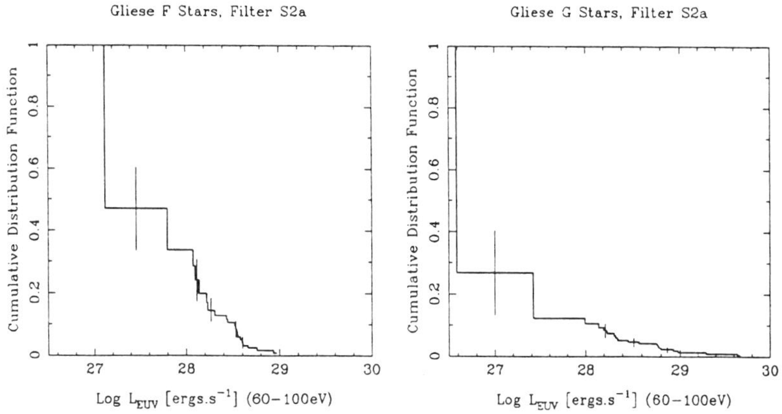

Fig. 2: Cumulative EUV luminosity distribution function for F and G stars, in the S2A waveband. The error bars were computed according to the method of Feigelson and Nelson (1985).

Seventh Cambridge Workshop on Cool Stars, Stellar Systems, and the Sun
ASP Conference Series, Vol. 26, 1992
Mark S. Giampapa and Jay A. Bookbinder (eds.)

THE KINEMATICS OF ACTIVE STARS

R. D. JEFFRIES & S. J. JEWELL
School of Physics and Space Research, University of Birmingham, Edgbaston, Birmingham B15 2TT, U.K.

ABSTRACT Late-type stars detected by the ROSAT Wide Field Camera (WFC) are used to investigate the kinematics of active stars in the solar neighborhood. The most active single stars are found to be kinematically young and there is significant subclustering around the space motions of well known groups of young stars. Many RS CVn and BY Dra close binary systems were also detected. The RS CVn systems are kinematically older than the main sequence active stars, as expected, but the youthful kinematics of the BY Dra systems is puzzling and poses the question - where are the old BY Dra systems?

Keywords: late-type stars; activity; age; kinematics

SINGLE STARS

Data from the WFC have been used to study the kinematics of nearby active late-type stars. The all-sky nature of the ROSAT survey and comparative sensitivity of EUV emission to activity make this a useful tool to study the magnetic activity, age relationship. Our initial sample of stars consisted of those WFC sources that are closely correlated (within 2.5$'$) with a single main-sequence star in the Woolley or Gliese catalogues of nearby stars. Our detection threshold was such that no spurious correlations are expected. To measure absolute activity and discriminate against those stars which have been detected by virtue of their proximity to the sun rather than their activity, an EUV flux at the surface of each star was calculated in the waveband 0.05-0.20 keV. The details of this are given by Jeffries and Jewell (1991). For some stars the calculation was impossible due to the lack of an accurate spectral type or in the cases of close visual binaries.

U, V and W space motions of these objects were obtained from the Woolley or Gliese catalogues. The accuracy of these is dependent on the accuracy of the distance determination and proper motion measurements and is typically a few km s^{-1} . Figure 1 shows the U,V plot for our sample. The size of circle represents the EUV surface flux, and ranges from $6 \times 10^4 - 2 \times 10^7$ ergs cm^{-2} s^{-1} . There are two major features of this plot. Firstly, the most active stars are kinematically very young. Comparison with distributions for A stars in the Woolley catalogue suggest an age less than 10^9 years for most of the sample. Secondly, there appears to be significant clustering in the U,V plane and this is confirmed by cluster analysis techniques, which indicate three significant velocity groups. The centroid of the first group is at (-9, 2, -9), which compares favourably with the space motions of the Sirius Supercluster (*e.g.* Eggen 1984). Several of these stars are known members of this group (*e.g.* π^1 UMa). The second cluster is at (-22, -12, -6), and is probably a combination of young disk stars around the local standard of rest at (-12, -12, -6), and a few Hyades Supercluster members at (-41, -18, -2). The third grouping is at (-7, -28, -13), in excellent agreement with the Local Association (*e.g.* Eggen 1983) at (-7, -25, -15). This is an exciting find; there are only about half a dozen suspected members of this very young ($5 - 20 \times 10^7$ yrs) kinematic group outside of open clusters such as the Pleiades (see Bromage *et al.* 1991). Presumably their EUV activity arises as a result of rapid rotation, and nearby examples are essential to study in detail the rapid angular momentum loss as stars approach the main-sequence. This Local Association grouping is contradictory to the findings of Soderblom (1990), based upon chromospheric activity, although only two of our suspected Association members are included in his sample.

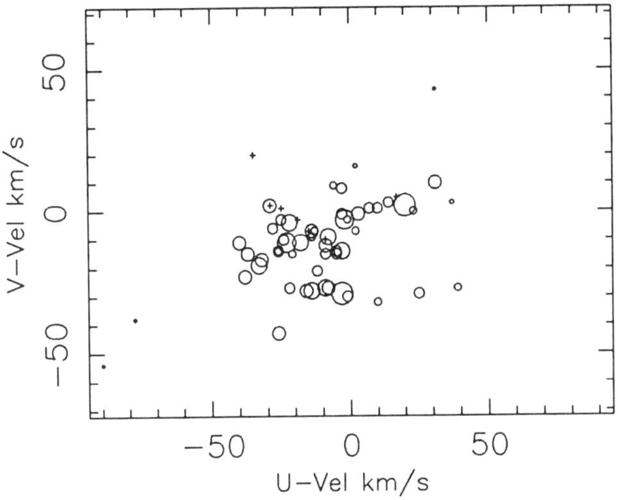

Fig. 1: U,V (heliocentric) space motions for nearby single stars. Size of circle indicates magnitude of EUV surface flux, crosses indicate an uncertain EUV flux.

BINARY STARS

Similar plots have been constructed for the BY Dra and RS CVn active late-type binary systems detected by the WFC. Here the distinction between these types is that RS CVn systems contain an evolved component, whereas both stars in a BY Dra binary are on the main-sequence. U,V,W velocities were taken from Strassmeier *et al.* (1988) and are considerably more uncertain than for the Woolley and Gliese samples. Figure 2 shows the U,V space motion plot for BY Dra systems, which is quite remarkable. Except for one or two stragglers, the BY Dra systems appear kinematically similar to the single main-sequence stars, even though their binarity might be expected to keep activity high even for older systems. Soderblom (1990) suggested that angular momentum loss and coalescence to W UMa systems might explain the lack of old BY Dra binaries. Another possibility might be that tidal damping of internal differential rotation may eventually turn off the dynamo in the shorter period systems. Again, structure is present in the diagram with six systems clustered around the Local Association space motion, and a bar stretching between the Hyades and Sirius superclusters. The RS CVn systems have rather more dispersed space motions, which might be expected if they contain at least one evolved component, but still appear kinematically younger than field stars in the solar neighborhood.

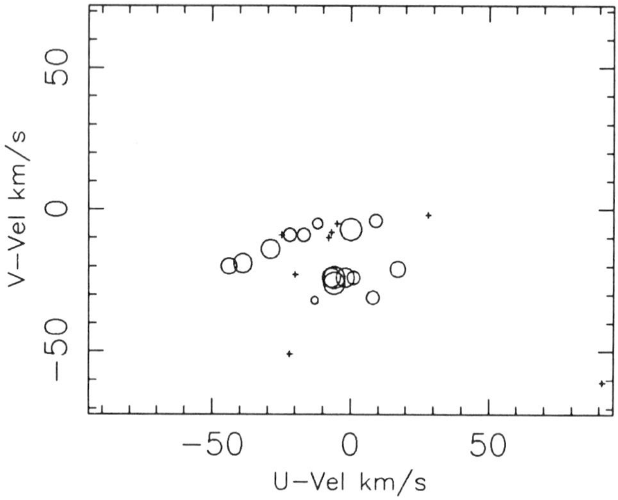

Fig. 2: U,V (heliocentric) space motions for BY Dra systems. Size of circle indicates magnitude of EUV surface flux, crosses indicate an uncertain EUV flux.

ACKNOWLEDGEMENTS

The WFC project is supported by the UK Science and Engineering Research Council (SERC). The WFC instrument and data analysis were the responsibility of the WFC consortium institutes (Leicester University, SERC's Rutherford Appleton Laboratory, Mullard Space Science Laboratory - University College London, Birmingham University and Imperial College of Science, Technology and Medicine - London). The authors acknowledge funding from SERC.

REFERENCES

Bromage, G.E., Kellett, B.J., Jeffries, R.D., Innis, J.L., Matthews, L.M., Anders, G.J. and Coates, D.W. 1991, in Cool Stars, Stellar Systems, and the Sun, eds. J. Bookbinder, M. Giampapa, *Pub. A.S.P.*, Conference Series, this volume.

Eggen, O.J. 1983, *M.N.R.A.S.*, **204**, 377.

Eggen, O.J. 1984, *A. J.*, **89**, 1350.

Jeffries, R.D. and Jewell, S.J. 1991, in Cool Stars, Stellar Systems, and the Sun, eds. J. Bookbinder, M. Giampapa, *Pub. A.S.P.*, Conference Series, this volume.

Soderblom, D.R. 1990, *A. J.*, **100**, 204.

Strassmeier, K.G., Hall, D.S., Zeilik, M., Nelson, E., Eker, Z. and Fekel, F.C. 1988, *Astr. Ap. Suppl.*, **72**, 291.

Seventh Cambridge Workshop on Cool Stars, Stellar Systems, and the Sun
ASP Conference Series, Vol. 26, 1992
Mark S. Giampapa and Jay A. Bookbinder (eds.)

EUV EMISSION FROM LATE TYPE STARS

R. D. JEFFRIES & S. J. JEWELL
School of Physics and Space Research, University of Birmingham, Edgbaston, Birmingham B15 2TT, U.K.

ABSTRACT Late-type stars detected by the ROSAT Wide Field Camera (WFC) are used to investigate the contribution of EUV radiative losses to the energy budget of their outer atmospheres. Evidence is found for an upper limit to the EUV losses as a fraction of the bolometric luminosity that is independent of spectral type. General correlations with coronal and transition region diagnostics are presented, but investigation of how these correlations vary with spectral type must wait until a better knowledge of the source plasma is obtained.

Keywords: late-type stars; EUV; activity; stellar coronae

INTRODUCTION

The EUV wavelength range is one of the last unobserved regions in which a major proportion of the radiative energy losses from magnetically heated late-type stellar atmospheres is thought to emerge (Doyle 1989). As such, measurement of these losses will prove fundamental in the testing of heating mechanisms and models for the generation of magnetic fields. The WFC all-sky survey instrumentation is sensitive to photons in the range 60-250Å (0.05-0.20 keV), and for the optically thin, thermal plasmas producing this emission, implies a peak count rate from plasmas at $\sim 10^6$ K.

A sample of stars observed during the WFC all-sky survey were chosen based upon the criteria that the WFC detection was close ($< 2.5'$ from) to a bright catalogued star, that there was a significant detection in at least one of the two survey filters used (S1 - 0.08-0.20 keV, S2 - 0.05-0.12 keV) and that the catalogued star has an accurate spectral type and visual magnitude. Of this sample 70 are in the Woolley catalogue of nearby stars and 51 are in the Strassmeier *et al.* (1988) catalogue of chromospherically active binary systems. Most of the remaining 49 stars in the sample have been previously established as active, by observation of Ca II H and K emission cores.

The count rate in the WFC filters was converted to a flux at the earth (in the range 0.05-0.20 keV) by making an assumption concerning the temperature distribution of the plasma responsible for the EUV emission. Where an S1/S2 count rate ratio was available, (approximately half the sample) it was used to calculate an *effective* EUV plasma temperature. The mean of this temperature determination was $log\,T = 6.5$ with a standard deviation of 0.1. Otherwise, a temperature of $log\,T = 6.5$ was assumed. In all cases a value of the interstellar absorbing hydrogen column was needed. This was estimated from values given in Frisch and York (1983), but because these estimates are only accurate to $\sim 50\%$, those stars with an estimated column $> 2 \times 10^{19}$ cm^{-2} were not used in the subsequent analysis as the error could be of the order of the mean free path of an EUV photon. Conversion to a flux at the surface of the star was achieved using the apparent visual magnitude and empirical relationships between colour indices ($B-V$ for $B-V < 1.3$ and $R-I$ for $B-V \geq 1.3$), photospheric temperature and bolometric correction. In the case of binary stars count rates were split between the two components unless Strassmeier *et al.* (1988) indicated that only one of the components was active, or that one component was not a late-type star. Errors in this overall process are dominated by uncertainties in the temperature distribution of the EUV emitting plasma and amount to between ±30% where a filter ratio is available and a factor of two where it is not.

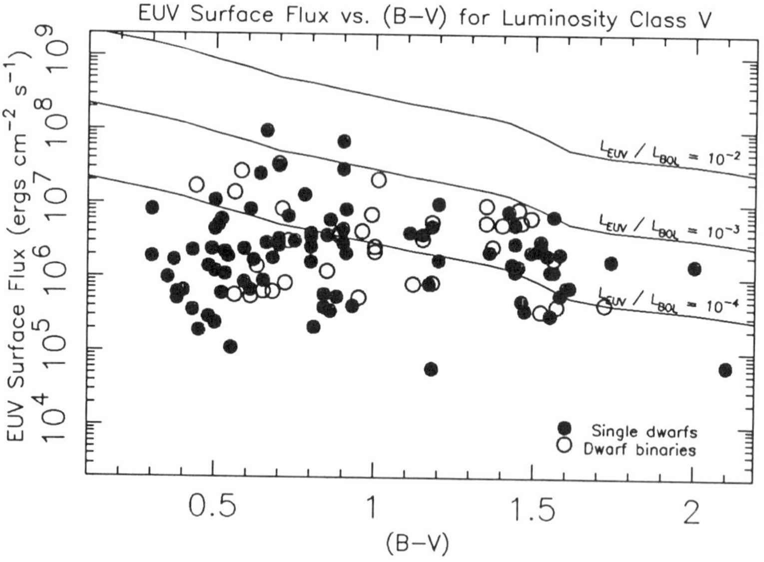

Fig. 1: Surface EUV flux (0.05-0.20 keV) versus $B-V$ index for a sample of late-type dwarfs. Lines of constant L_{euv}/L_{bol} are drawn.

RESULTS

Figure 1 illustrates a compilation of surface EUV fluxes for main sequence stars (both single and binary) as a function of $B-V$. Superposed are lines of constant EUV to bolometric radiative loss ratio. The main feature of this diagram is the empirical upper limit of $L_{euv}/L_{bol} \sim 10^{-3}$ which is defined by the most rapidly rotating stars such as AB Dor, HD 197890 (Bromage *et al.* 1991) and short period BY Dra systems (here any close binary with two main sequence components is given that name). The limit appears to be more or less constant between spectral types G0 and M5. This is close to the observed behaviour in the X-ray band (0.2-4 keV) where an upper limit of $L_x/L_{bol} \sim 2 \times 10^{-3}$ was found, and has been attributed to the maximum amount of mechanical flux available in subphotospheric regions (Vilhu 1987). The main significance of our result is that the EUV region (of which only the short wavelength part is sampled) is at least as important as X-rays in determining the total radiative losses from active stellar atmospheres. There are also considerably more single stars close to the saturation boundary than shown by Vilhu (1987), many of which were not previously known to be active.

Correlations were performed between the EUV surface flux and those in the X-ray waveband (0.2-4 keV) (from the *Einstein IPC* pointed and slew survey catalogues) and the C IV line (λ1550) (from Rutten *et al.* and the IUE Uniform Low Dispersion Archive, G.E. Bromage and D.J. Stickland personal communication). The flux at the earth in the X-ray waveband was calculated assuming a uniform coronal temperature. Fluxes at the earth in these two wavebands are then converted to surface fluxes in an identical manner to the EUV values before correlation. Best fit power laws to these correlations are $F_{euv} \propto F_x^{1.03\pm0.12}$ and $F_{euv} \propto F_{C\,IV}^{1.30\pm0.22}$ (90% confidence limits). The scatter about these relations is up to an order of magnitude, which is larger than we would expect from our estimated errors, although non-simultaneity of the observations would account for some of this scatter. The one to one relationship between EUV and X-ray flux might lead to speculation that the coronal temperature distribution is similar in all stars. However, the assumption of a uniform coronal temperature in calculating the X-ray flux and scatter in the EUV fluxes will mask variations. Investigation of any detailed dependence of EUV to X-ray flux ratio with $B-V$ or rotation, for instance, must wait until more accurate source spectra are defined by combining ROSAT XRT and WFC results.

ACKNOWLEDGEMENTS

The WFC project is supported by the UK Science and Engineering Research Council (SERC). The WFC instrument and data analysis were the responsibility of the WFC consortium institutes (Leicester University, SERC's Rutherford Appleton Laboratory, Mullard Space Science Laboratory - University College London, Birmingham University and Imperial College of Science, Technology and Medicine - London). The authors acknowledge funding from SERC.

REFERENCES

Bromage, G.E., Kellett, B.J., Jeffries, R.D., Innis, J.L., Matthews, L.M., Anders, G.J. and Coates, D.W. 1991, in Cool Stars, Stellar Systems, and the Sun, eds. J. Bookbinder, M. Giampapa, *Pub. A.S.P.*, Conference Series, this volume.

Doyle, J.G. 1989, *Astr. Ap.*, **214**, 258.

Frisch, P.C. and York, D.G. 1983, *Ap. J. Letters*, **271**, L59.

Rutten, R.G.M., Lemmens, A.F.P. and Zwaan, C. 1987 in Magnetic Activity of Cool Stars and its Dependence on Rotation and Evolution, Utrecht University Publication, p. 99.

Strassmeier, K.G., Hall, D.S., Zeilik, M., Nelson, E., Eker, Z. and Fekel, F.C. 1988, *Astr. Ap. Suppl.*, **72**, 291.

Vilhu, O. 1987, in Cool Stars, Stellar Systems, and the Sun, eds. J.L. Linsky, R.E. Stencel, (Springer: Verlag), p. 110.

Seventh Cambridge Workshop on Cool Stars, Stellar Systems, and the Sun
ASP Conference Series, Vol. 26, 1992
Mark S. Giampapa and Jay A. Bookbinder (eds.)

SPEEDY MIC: A YOUNG, NEARBY, EXTREMELY FAST ROTATING K STAR DISCOVERED BY THE ROSAT/WFC EUV SURVEY.

G. E. BROMAGE, B. J. KELLETT Space Astronomy Group, Rutherford Appleton Laboratory, Didcot, Oxon OX11 0QX, U.K.

R. D. JEFFRIES, J. L. INNIS School of Physics and Space Research, University of Birmingham, U.K.

L. MATTHEWS Imperial College of Science Technology & Medicine, University of London, U.K.

G. J. ANDERS, D. W. COATES Department of Physics, Monash University, Australia.

ABSTRACT In October 1990, the ROSAT/WFC all-sky survey detected strong EUV emission from the 9th magnitude K0 star HD197890 in the constellation Microscopium. Prior to ROSAT, this object was not known to be an active-chromosphere star. Most of the EUV emission was contained in a large flare event together with continued activity for about 20 hours. Subsequent optical spectroscopy and photometry have revealed that the star is a single, extremely fast rotator with a photometric period of 8.1 hours. We have nicknamed the star *Speedy Mic*. Kinematic evidence and lithium abundance are consistent with its membership of the very young disk population Local Association (Pleiades group).

Keywords: Stellar coronae; EUV; flares; Local Association

INTRODUCTION

The U.K. Wide Field Camera on board *ROSAT* carried out the first-ever all-sky EUV survey in 1990/91 (Wells *et al.* 1990, Bromage 1991). During this survey, several hundred active late-type stars were detected, including dMe and earlier type active-chromosphere stars and BY Dra, RS CVn and W UMa type binaries.

One of the important classes of EUV emitter, though its members are very rare, is the group of nearby G-K type active stars belonging to the very young disk population Local Association or Pleiades moving group (Innis *et al.* 1986, Anders *et al.* 1991). The stars are most likely to be in a short-lived evolutionary phase following core contraction to the main sequence. In order to study this important phase, it is crucial to discover further nearby members of this group in order to be able to (*inter alia*) investigate rotationally modulated activity in as wide a wavelength range as possible. We report here the discovery of a new member with the fastest rotation found so far.

EUV OBSERVATIONS

The WFC observed the Speedy Mic area from 1990 October 15 to 19. As a result of the survey strategy, the data on a particular source consist of a series of 'snapshots' of about 50 seconds duration separated in time by one ROSAT orbit (1.6 hours) and with the two survey filters (S1 and S2 with energy ranges approximately 0.08 - 0.18 and 0.06 - 0.11 keV respectively) being swapped each day. For the particular coverage including Speedy Mic, for technical reasons a gap of about 35 hours exists on Oct 16-17. The background-subtracted time series of EUV observations is illustrated in the WFC review paper in these proceedings (Bromage 1991).

Outside of the time Oct 17.9 - 19.0, the flux in both filters was largely below the sensitivity of the WFC. However, a large flare occurred in S2 on Oct 17.9 lasting at least 2 orbit-samples and with peak flux at least 20 times the quiescent level. In addition, there was clearly enhanced activity for the next day, up to Oct 19.0 (first in S2 and then in S1 filter observations). Using a distance of 40 parsec (see below) combined with interstellar hydrogen density estimates for this direction, and assuming also that a single smoothly decaying flare was sampled by the WFC snapshots, the total energy emitted was 10^{33} erg within the S2 wavelength range (approximately $\lambda\lambda$ 115 - 180 Å).

The detected source had a positional error circle of 40 arcsec radius and was 10 arcsec from the V = 9.3 mag K0 V star HD197890. We are not aware of any previously published record of activity in this star. If this was indeed the the EUV source that flared, it must be a very active star indeed, and must be either a close binary of the RS CVn type or a fast rotating single star.

OPTICAL SPECTROSCOPY AND PHOTOMETRY

HD 197890 was observed with high dispersion optical spectroscopy covering both Ca H and K and Hα and surrounding spectral regions, at the South African Astronomical Observatory in July and September 1991, and in the red at Mount Stromlo Observatory in July/August 1991. In both cases these observations formed part of a series of optical identifications and follow-ups of a large number of WFC sources, performed by some of the present authors.

The SAAO observations used the 1.9m telescope with image tube spectrograph and RPCS detector, with a spectral resolution of 1 Å at Ca HK. HD 197890 showed by far the broadest K emission line of all stars observed, with a FWHM of 240 ± 40 km s^{-1} after deconvolution from the instrumental profile. Sequences of observations showed no detectable wavelength shifts or other evidence for binarity. A measurement of the Li I λ6708 Å absorption line equivalent width of 220 ± 50 mÅ added to the evidence that this star is a single, young, extremely fast rotating dwarf similar to the well-known AB Dor but even faster.

The Mount Stromlo spectra were taken in the Hα and Na D line region using the echelle spectrograph and an uncoated GEC CCD at the coude focus of the 1.9m telescope. The resolution was 0.2Å . Radial velocities were determined from the Na D profiles using cross-correlation and templates from standard stars. The radial velocity was measured on 11 occasions over 6 days: the average value was -6.5 km s^{-1} with an rms deviation of 2.0 km s^{-1}, and no non-random variations were apparent. We estimate a $v_e \sin i$ value of 120 ± 20 km s^{-1} from these spectra, compared with 70 for AB Dor. The large value of $v_e \sin i$ together with no significant radial velocity variations constitute further strong evidence that the star is single. The Hα profile showed very clear variability from night to night.

Finally, the star was observed on 6 nights during August/September 1991 using the Monash 45 cm telescope for V-band photometry. The differential V magnitude (relative to the average of two comparison stars, whose internal differential was better than 0.01 mag) showed 0.2 mag periodic variations which we attribute to rotational modulation due to large cool spots on the photosphere. A period-search Fourier analysis yielded 8.103 hours.

DISCUSSION AND CONCLUSIONS

The rapid rotation, activity and apparent lithium abundance of Speedy Mic are strongly reminiscent of young active stars like AB Dor. Membership of the kinematic group known as the Local Association (the Pleiades group) has been demonstrated for AB Dor, PZ Tel, V838 Cen and V343 Nor (Innis *et al.* 1986, Anders *et al.* 1991). The total space motion of a Local Association star, V_T, is ~ 30 km s^{-1}. The expected radial velocity of a group member, $V_{radial} = V_T \cos\lambda$, where λ is the angular distance from the convergent point. For Speedy Mic, $V_{radial} = -6$ km s^{-1}, in excellent agreement with the measured value. In addition, limited supporting evidence comes from the proper motion: this is directed to within 25° of the convergent point with a rather large uncertainty. The observed $v_e \sin i$ and photometric period give $R_\star \sin i \sim (0.8\pm0.1)R_\odot$, and $\sin i$ must be close to unity or no photometric modulation would be seen. Thus $R_\star$ is consistent with the catalogued spectral type of K0 V and the distance is $\sim$ 40pc.

We conclude that the evidence for membership of Speedy Mic in the Local Association is very strong. The age of the association members is of order 50 to 70 million years. The rapid rotation, and hence activity, is a consequence of the star's youth, as seen in stars like AB Dor, and the Pleiades cluster K dwarfs. The ROSAT/WFC has apparently discovered the fastest known rotator amongst the nearby members identified so far.

ACKNOWLEDGEMENTS

The WFC project is supported by the UK Science and Engineering Research Council. The WFC instrument and associated Ground System were developed by a consortium of five UK institutes: the University of Leicester, SERC's Rutherford Appleton Laboratory, the University of Birmingham, the Mullard Space Science Laboratory of University College London, and Imperial College STM London.

REFERENCES

Anders, G. J., *et al.* 1991, *M.N.R.A.S.*, **252** 408.

Bromage, G. E. 1991, in Cool Stars, Stellar Systems, and the Sun, ed. J. Bookbinder & M. Giampapa, *Pub. A.S.P.*, Conf Series, this volume.

Innis, J. L., *et al.* 1986, *M.N.R.A.S.*, **223** 183.

Wells, A., *et al.* 1990, *Proc. S.P.I.E.* **1344**, 230.

Seventh Cambridge Workshop on Cool Stars, Stellar Systems, and the Sun
ASP Conference Series, Vol. 26, 1992
Mark S. Giampapa and Jay A. Bookbinder (eds.)

ROSAT OBSERVATIONS OF LATE-TYPE STARS

J.H.M.M. SCHMITT
Max-Planck-Institut für extraterrestrische Physik, 8046 Garching, FRG

ABSTRACT I will review the first results from observations of late-type stars in the ROSAT all-sky survey. The various background components and the resulting sensitivity of the all-sky survey will be discussed; further, the stellar content of the overall X–ray source population will be described. Among the scientific issues covered are coronal dividing lines and hybrid stars, the problem of X–ray emission from A-type stars, and X–ray variability studies from the all-sky survey data.

Keywords: late-type stars, stellar coronae, X–rays

INTRODUCTION

Observations with the *Einstein Observatory* and EXOSAT have shown the ubiquity of stellar X–ray emission (*cf.*, Vaiana *et al.*, 1981). While the data collected with these observatories were extremely useful in acquiring information on individual sources, a major disadvantage resulted from the fact that only a very limited fraction of the sky was viewed (less than 10 percent of the sky was covered by pointed *Einstein* IPC observations). Consequently, the construction of large unbiased samples of objects was difficult if not impossible. A major goal of the ROSAT mission was to carry out the first all-sky survey with an imaging telescope which - in particular - allows the construction of such large and unbiased samples of various classes of X–ray sources. This first major goal of the ROSAT mission has been accomplished: The ROSAT all-sky survey has been completed and almost the entire sky has been scanned down to a limiting flux of approximately 2×10^{-13} erg/s/cm^2 in the pass band 0.1–2.0 keV; in regions of deeper exposures near the poles of the ecliptic considerably fainter flux limits have been achieved. In this article I will try to summarise the main properties of the ROSAT all-sky survey, to characterise the data resulting from the ROSAT all-sky survey, and highlight some of the new scientific results obtained by the stellar working group at MPE.

THE ROSAT SATELLITE AND ITS HARDWARE

For the purposes of this article it is unnecessary to provide an in depth description of the hardware onboard ROSAT; the essential information has been provided by Trümper *et al.* (1991) and Pfeffermann *et al.* (1986). Here it suffices to point out that the ROSAT all-sky survey has been carried out with the ROSAT X–ray telescope (XRT) and a position sensitive proportional counter (PSPC) in the focal plane. This detector has an energy resolution of $\Delta E/E \sim 0.43$ at 1 keV and thus allows the definition of $\sim$ 3 more or less independent energy bands. However, limitations in photon count usually limit the available bands to two, a soft band between 0.1 - 0.28 keV, and a hard band between 0.5 - 1.5 keV. The instrumental properties of the PSPC are therefore in many ways similar to the well known UBV system of optical astronomy.

THE PSPC BACKGROUND

A full understanding of the results of the ROSAT all-sky survey requires an understanding of the PSPC background (*cf.*, Schmitt 1991). Point sources such as stars must be detected on a background typically composed of three components: Particle-induced background, scattered solar X-rays, diffuse cosmic X-ray background. Particle-induced background is produced by cosmic rays penetrating the PSPC without being rejected by the counter's veto electronics. Extensive measurements of this background have been carried out under a variety of orbital conditions (*cf.*, Snowden *et al.*, 1991); typical particle-induced background rates are $\sim 1 - 2 \times 10^{-5}$ cts/s/arcmin^2. The pulse-height spectrum of this component is essentially flat with a noticeable increase towards the very lowest energies. The second background component are solar X-rays scattered off ionospheric atoms into the beam of the XRT. The magnitude of this background depends very sensitively on the viewing geometry. It is significant only at large zenith angles during day-side viewing when count rates of up to 10^{-3} cts/s/arcmin^2 are encountered, while typical rates are between zero (night-side viewing) and 10^{-4} cts/s/arcmin^2. Obviously, this background component essentially reflects the incoming solar X-ray spectrum and hence is very soft, with the majority of the events usually appearing at energies below 0.28 keV. However, quite frequently a strong oxygen K_α line is also observed in the scattered solar X-ray spectrum. Finally, sources must be detected against the general cosmic diffuse X-ray background; this background varies over the sky with rates of $\sim 5 \times 10^{-4}$ cts/s/arcmin^2 encountered a low galactic latitudes up to $\sim 20 \times 10^{-4}$ cts/s/arcmin^2 found at higher galactic latitudes. The spectrum of the diffuse X-ray background is also rather soft, with approximately 70 percent of its intensity appearing in the C-band below 0.28 keV. I would like to emphasise that the ROSAT PSPC's sensitivity is – at least for normal viewing conditions – determined by the diffuse X-ray background which represents a major advantage over all previously flown imaging X-ray instruments.

SENSITIVITY OF THE ALL SKY SURVEY

During the all sky survey the XRT scanned the sky once per orbit along great circles containing the north and south ecliptic poles. The scanned longitudes slowly moved with the apparent motion of the Sun along the ecliptic, *i.e.*, at a rate of $\sim$ 1 degree/day. Since the field of view of the XRT/PSPC combination is 2 degrees, sources at low ecliptic latitudes could be viewed for 2 days; for sources at higher latitudes the viewing time increased roughly as $1/cos(\beta)$, with β being the ecliptic latitude. An observation of a source consists of a sequence of individual scans, each lasting between 20 to 30 seconds. During each scan the XRT/PSPC point response function varies rapidly. On-axis 50 percent of the photons are contained within a circle with 15 - 25 arcsec (dependent on energy) while at 50 arcmin off-axis an energy-independent circle with 160 arcsec radius is required. In the survey mode one finds about 80 percent of the photons in a 150 arcsec circle. Within such a detect cell the background rate (assuming "nominal" conditions of 10^{-3} cts/s/arcmin^2) becomes 2×10^{-2} cts/s/detect cell. The minimum detectable count rates are typically of the same order as this mean background rate. Putting these numbers differently, an average time of 51 sec elapses between arrival of background events in such a detect cell, and if only the hard band above 0.5 keV is considered, this average time increases to more than 150 sec. Consequently the Poissonian nature of the photon arrival times must be explicitly considered when dealing with source detection or source light curves in the ROSAT all-sky survey. Highly significant sources may be detected with only a few counts and correspondingly large count rate errors.

THE STELLAR CONTENT OF THE ALL-SKY SURVEY

The processing of the all-sky survey data has been largely (albeit not totally) completed and therefore the total number of sources detected is not known yet. However, extrapolation of the presently available information leads to an expected overall source count of $\sim$ 60000 X–ray sources. An optical identification program with the goal to reach a complete optical identification of selected areas on the sky has been initiated within the context of an ESO key program.

The results so far obtained are consistent with the results previously obtained with the identifications obtained from the *Einstein* Medium Sensitivity Survey (*cf.*, Stocke *et al.* 1991), *i.e.*, about a quarter of the high latitude sources are of coronal origin. At lower latitudes the fraction of coronal sources increases to 50 percent (*cf.*, Motch *et al.* 1991), again consistent with the results from the *Einstein* Galactic Plane Survey (*cf.*, Hertz and Grindlay 1988). In special regions such as open clusters (for example, Pleiades or α Per cluster) or star forming regions (for example, Orion, Chamaeleon etc.) we find an even higher density of coronal X–ray sources (*cf.*, Krautter *et al.* 1992). In total we may then expect about a third of the source content of the all-sky survey to be of coronal origin.

The overwhelming majority of these stellar X–ray sources are late-type stars. Early-type stars (such as O stars or Wolf-Rayet stars) and white dwarfs are of course also found, however, in far smaller numbers than late-type stars. It is important to realise that the overwhelming majority of the late-type stars found in the all-sky survey do not comply with the solar paradigm usually employed to interpret stellar X–ray emission. Given a typical sensitivity limit of the all-sky survey ($\sim 2 \times 10^{-13}$ erg/s/cm^2), an object emitting X–rays at typical solar levels ($L_X \sim 2 \times 10^{27}$ erg/s) can be detected out to a distance of $\sim$ 9 pc only. The typical late-type star detected in the all sky survey is much more distant and must hence emit X–rays at levels significantly higher than those found for the Sun. Therefore the stars discovered in the ROSAT all-sky survey X–ray sources constitute the most extensive sample of active stars known.

Obviously, the identification of the active stars found in the all-sky is a rather time-consuming effort requiring the labor of many astronomers in the years to come. Quicker results can be obtained by studying "known" stars selected from various stellar catalogs. Most efforts in working with such optically selected stellar samples has so far been undertaken with stars selected from the

Bright Star Catalog (Hoffleit and Jaschek 1982) and the Gliese catalog (Gliese 1969). From entries in these catalogs a large number of X-ray detections has been obtained. In figure 1 I plot an HR diagram with X-ray detections marked; only those detections with reliable distances were used. As is clear from figure 1, stellar X-ray emission is ubiquitous for (almost) all types of stars and thus figure 1 confirms a result first obtained from the *Einstein Observatory* stellar surveys.

Three types of stars have traditionally been considered to show a deficiency of X-ray emission: First, very red M dwarf stars of spectral type M5 or later, whose interior structure is presumably fully convective; second, A dwarf stars, whose interior structure is fully radiative; and third, red giants and supergiants. The ROSAT results on very late-type M dwarf stars are discussed by Fleming *et al.* (1992); I will discuss in the following two sections first ROSAT results on A-type stars and the coronal dividing line.

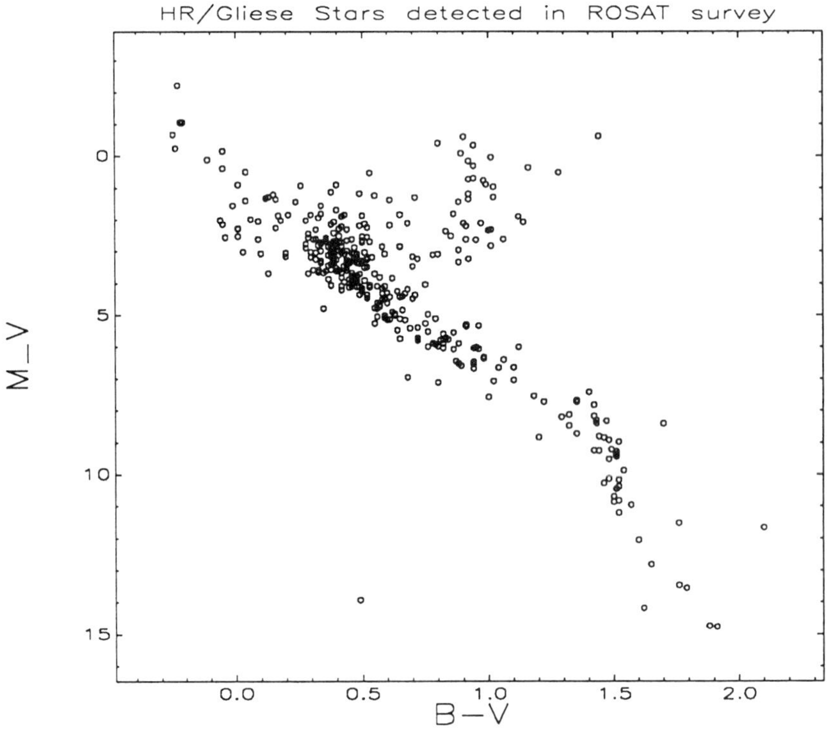

Fig. 1: HR diagram of bright and/or nearby stars detected in the ROSAT all-sky survey. Only about two thirds of the survey data have been utilised, stars are plotted regardless of their binarity nature. Note that all types main sequence stars are detected, while there are no detections for red giants and supergiants.

THE CORONAL DIVIDING LINE

A so-called coronal dividing line was discovered by Ayres *et al.* (1982) and Haisch and Simon (1982) as a line separating X–ray emitting yellow giants and supergiants from non X–ray emitting red giants and supergiants using X–ray data obtained with the *Einstein Observatory*; the discovery of this X–ray dividing line was preceded by the discovery of a so-called wind dividing line with the IUE satellite (Linsky and Haisch 1979), *i.e.*, a line separating giants with evidence for transition region temperature material in their ultraviolet spectra from stars showing evidence for massive cool winds (but no warm or hot material). The data collected in the ROSAT all-sky survey offers the possibility to test for the presence of a dividing line using large data samples.

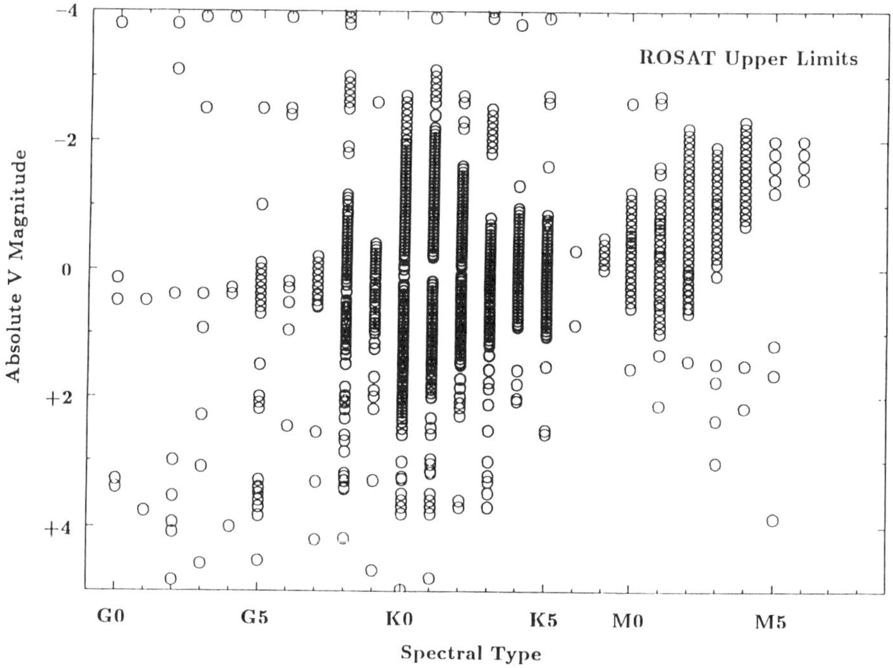

Fig. 2: HR diagram of bright giants covering the spectral range between G0 to M8 and luminosity class II to IV. Shown are 803 (single) giants scanned but not detected in the ROSAT all-sky survey (*cf.*, Haisch, Schmitt and Rosso 1991); only about two thirds of the survey data has been utilised for this figure. Note that all types of giants have been sampled in the all-sky survey.

Haisch, Schmitt and Rosso (1991) constructed a sample of 868 giant stars for which X–ray data was available and which could be considered as single stars. Figure 2 shows an HR-diagram for 803 stars in this sample which could not be detected in the ROSAT all-sky survey; figure 2 makes it clear that this large sample constructed from the all-sky survey data samples all regions of the upper cool half of the HR-diagram quite extensively.

If one considers then the locations of the detected stars in the HR diagram (*cf.*, figure 3), quite a different picture emerges: Almost all the detections are of spectral type K3 or earlier, only one star, HR 4289, is of spectral type K4 or later. To be specific, for giants of spectral type K3 or earlier, 64 detections and 540 upper limits were obtained, *i.e.*, a detection rate of 10.6 percent, for stars of spectral type K4 or later, 1 detection and 263 upper limits, *i.e.*, a detection rate of 0.4 percent. The very large data sample makes the coronal dividing an extremely significant result: Making the assumption that a detection rate of 10.6 percent applies for both yellow and red giants, the probability of obtaining on detection in 264 trials is less than 10^{-11} and hence the null hypothesis (that there is no dividing line) can be safely rejected. It is finally worthwhile noting that the star HR 4289, the only detection later than K3, is very little studied and it is unclear whether its X-ray emission might be due to an hitherto unknown companion; we encourage observers at other wavelengths to carry out follow-up observations of this star.

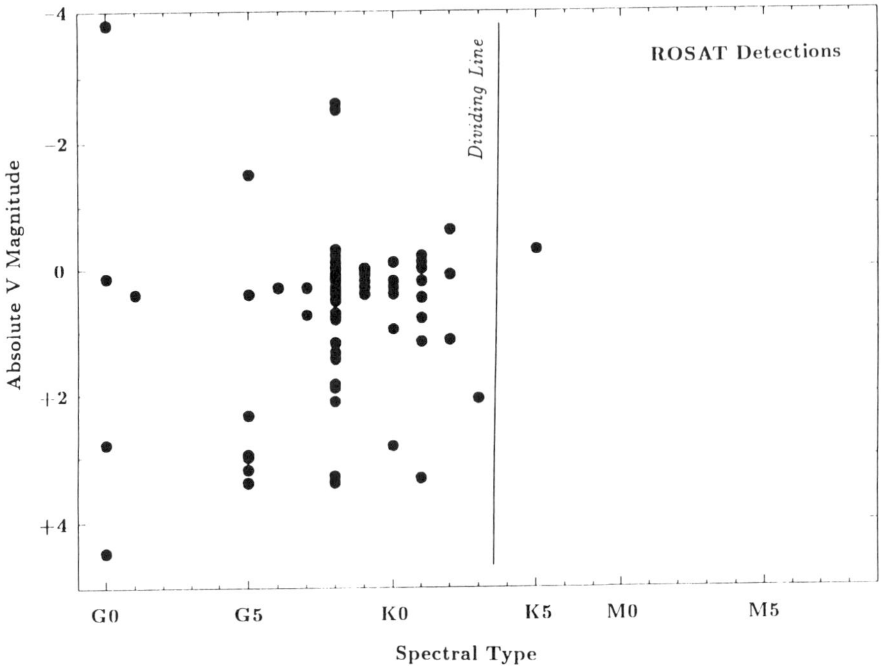

Fig. 3: Same as figure 2, but now only detections are shown. 64 out of 65 detections are of spectral type earlier than K4 (*cf.*, Haisch, Schmitt and Rosso 1991). While the exact location of the dividing line is debatable, the existence of such a line is beyond any doubt.

A particularly interesting class of stars are hybrid stars; these stars are located in the vicinity of the dividing line, yet their ultraviolet spectra show evidence both for cool wind material as found in the red giants to the right of the dividing line as well as warm material at transition region temperatures. The X-ray properties of hybrid stars are unclear. Haisch (1987) reviewed the *Einstein Observatory* data on hybrid stars and concluded that no convincing detections of hybrid stars were obtained at X–ray wavelengths. Brown *et al.* (1990) present EXOSAT observations of the hybrid star α TrA, demonstrating that (at least some) hybrid stars also sustain coronae. Haisch, Schmitt and Rosso (1991) checked the ROSAT all-sky survey data for detections of hybrid stars; the detection of α TrA by Brown *et al.* (1990) could be confirmed, another hybrid star, δ And was detected, while all other hybrids listed by Haisch (1987) were not detected in the ROSAT all-sky survey. Unfortunately, δ And is a rather complex binary system and the observed X–ray emission cannot be unambiguously attributed to the giant star; however, Haisch, Schmitt and Rosso (1991) present arguments that the bulk of the X–ray emission does indeed come from the hybrid star. Finally, both the detections for α TrA and δ And were obtained in both the soft and hard energy band of the ROSAT PSPC, possibly indicating X–ray temperatures of $\sim 10^7$ K; however, the detections are quite weak, and the spectral results definitely require confirmation through pointing data.

X–RAY EMISSION FROM A-TYPE STARS

The detection of X–ray emission from A-type stars is of particular interest. "Canonically" such stars should be devoid of any X–ray emission; they do not possess significant radiatively driven winds which are thought to be responsible for the X–rays observed from O-type stars, nor do they possess deep convection zones harboring magnetic dynamos, which are thought to be the ultimate root of the activity phenomena observed in late-type stars. X–ray studies on the basis of data collected with the *Einstein Observatory* confirmed the validity of this picture at large. Golub *et al.* (1982) argued that almost all their X–ray detections obtained for A-type main sequence stars could be attributed to late-type companions; the only single star detections obtained, *i.e.,* those of Sirius and Vega, were made with the *Einstein Observatory* HRI, and later Schmitt *et al.* (1985) argued that these detections were most likely due to UV contamination of the *Einstein Observatory* HRI. In their study of the onset of convection, Schmitt *et al.* (1985) found a rapid increase of X–ray detections for stars between spectral types A7 and F0; for stars of earlier spectral types they obtained a large number of upper limits only, and all their detections except for two, *i.e.,* HR 1014 and HR 7160, were binaries.

However, some "non-canonical" measurements of A-type stars have also been reported; for example, Pallavicini *et al.* (1990) found quiescent and flaring X–ray emission from the system Castor A/B (a result incidentally confirmed by extensive pointed ROSAT observations, *cf.*, Schmitt *et al.* 1992) and Caillault and Zoonematkermani (1989) and Micela *et al.* (1990) found X–ray emitting A-type stars in the Orion region and Pleiades cluster respectively.

At MPE we decided to take a dual approach to pursue the problem of X–ray emission from A-type stars; on the one hand we are systematically searching the ROSAT all-sky survey data base for X–ray sources matched with A-type stars, on the other hand we are carrying out a pointing program of selected nearby A dwarf stars for which hitherto no X–ray emission has been found. Here I will present a status report of these studies.

As far as the ROSAT all-sky survey is concerned, we found more than 140 X–ray sources associated with bright (*i.e.* $m_v < 6.5$) A-type stars (utilising two thirds of the all sky survey data). We classified the stars into various categories, such as dwarf A stars, giant A stars, A stars with early-type, late-tape dwarf or late-type giant companion, A stars with white dwarf companions etc.; the most interesting category in our context is of course the class of apparently single A-type stars. Removing out of this category those stars which may have a companion, whose colors are inconsistent with spectral type, and those with no color measurements at all, we are left with - so far - 12 apparently single A-type stars with X-ray emission detected. This "black list" includes the stars HR 943, HR 1014, HR 1042, HR 3761, HR 4502, HR 4855, HR 5216, HR 5511, HR 5729, HR 5870, HR 6436, HR 6618, and we are in the process of collecting information on these stars. Again we would like to encourage observations at other wavelength which may serve to indicate either the presence of chromospheres and coronae or the presence of fainter late-type companions which could then be responsible for the observed X-ray emission. Incidentally we note that the star HR 1014 was already reported by Schmitt *et al.* (1985) as an X-ray source.

As far as nearby single A-type stars are concerned, a deep (10 ksec) pointing on Vega, the nearest single A star, has been performed. A preliminary analysis showed no obvious X-ray source present at the position of Vega. It may be the case that a weak source ($\sim 10^{-3}$ cts/s) is present in the very softest channels, consistent with the pre-launch predictions of the PSPC UV count rate expected from a bright UV source such as Vega (Schmitt *et al.* 1992). At any rate, one can safely assume that any X-ray count rate from Vega must be less than 10^{-3} cts/s; in particular, if the *Einstein* HRI detection had been caused by X-rays, a PSPC count rate of $\sim$ 0.1 cts/s would have been expected, more than two orders of magnitude above the upper limit obtained. I therefore conclude that the detections of Vega reported by Topka *et al.* (1979) and Golub *et al.* (1982) are caused by UV radiation and not X-ray radiation.

X-RAY LIGHT CURVES IN THE ROSAT ALL-SKY SURVEY

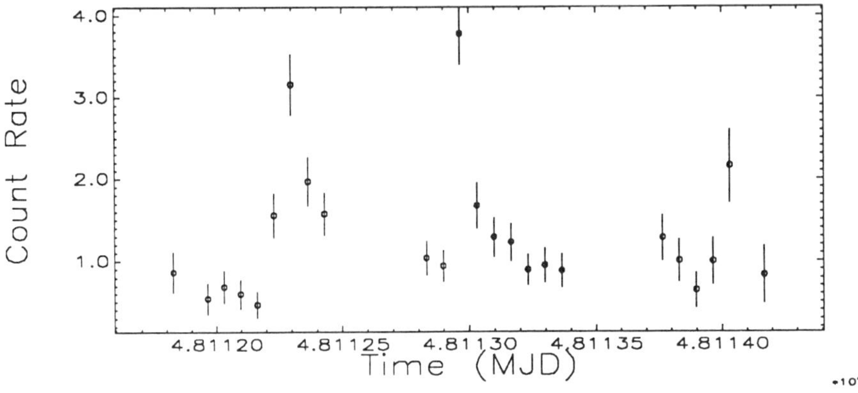

Fig. 4: ROSAT all-sky survey light curve of Proxima Centauri; plotted is PSPC total count rate vs. time given in MJD. Note the occurrence of three major flares during the two days Proxima Centauri was scanned by the XRT.

The ROSAT all-sky survey offers a new, unprecedented way to study variability of coronal X-ray emission. As pointed out above, the data consist of snapshots (~ 20 sec long), regularly obtained every 96 minutes (barring any data dropouts), for two days or longer. Thus, for the stronger sources, one can construct light curves covering variability time scales of 5000 seconds and more. Obviously, for the study of individual (short duration) flares, the all-sky survey sampling is rather sparse; however, flares with decay time scales of 3000 seconds or more can be easily detected. As an example I show (*cf.*, figure 4) the light curve of Proxima Centauri, a well known flare star at optical and X-ray wavelengths. Obviously, during the two day coverage in the all-sky survey, three major flares similar to the one observed by Haisch *et al.* (1981) occurred.

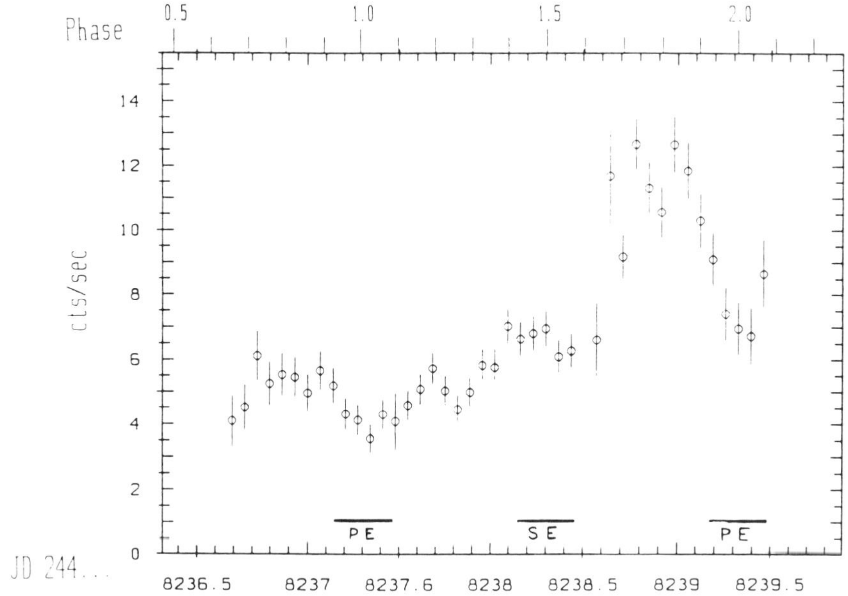

Fig. 5: ROSAT all-sky survey light curve of the eclipsing RS CVn system AR Lac (courtesy M. Kürster); plotted is the total PSPC count rate vs. phase (top label) and time (in JD: bottom label), also shown are the times of primary and secondary minimum during the all-sky survey scans. Note that both primary minima are observed, while the secondary minimum is marginal; the general level of emission is enhanced after some flaring activity between phases 1.6 - 1.8.

In addition to the frequency of flares, the time scales of decay times of long duration flares as well as those of stellar rotation and binary orbits become accessible to study in the all-sky survey data. As an example I show the all-sky survey light curve (in the total PSPC band) of the best studied (at X-ray wave lengths) RS CVn system AR Lac (*cf.*, figure 5). The binary period of this system is 1.98 days, and hence the system was covered in the all-sky survey for a little more than one binary cycle. Luckily, the primary minimum could be detected in two consecutive cycles, while a secondary minimum is marginal at best; a more detailed discussion of the survey AR Lac light curve is given by Kürster *et al.* (1992).

ACKNOWLEDGEMENTS

I would like to acknowledge the contributions of the MPE stellar working group (T. Fleming, B. Haisch, M. Kürster, R. Ottmann, C. Rosso) without whose efforts this presentation would have been impossible.

REFERENCES

Ayres, T., 1981, *et al.*, *Ap. J.*, **250**, 293.
Brown, A., Drake, S.A., van Steenberg, M.E., and Linsky, J.L., 1991, *Ap. J.*, **373**, 614.
Caillault, J.-P. and Zoonematkermani, S., 1989, *Ap. J. Letters*, 338, L57.
Fleming, T.A., 1992, *et al.*, these proceedings.
Gliese, W., 1969, Catalogue of Nearby Stars, Veröff. d. Astron. Recheninst., Heidelberg, 22.
Golub, L. *et al.*, 1983, *Ap. J.*, **271**, 264.
Haisch, B.M., 1987, in *Lecture Notes in Physics*, 291, 269.
Haisch, B.M., *et al.*, 1981, **245**, 1009.
Haisch, B.M. and Simon, T., 1982, *Ap. J.*, **263**, 253.
Haisch, B.M., Schmitt, J.H.M.M. and Rosso, C., 1991, *Ap. J. Letters*, in press.
Haisch, B.M., Schmitt, J.H.H.M. and Rosso, C., 1992, *Ap. J. Letters*, submitted.
Hertz, P. and Grindlay, J.E., 1988, *A. J.*, **96**, 233.
Hoffleit, D. and Jaschek, C., 1982, The Bright Star Catalogue, (New Haven: Yale Univ. Obs.).
Krautter, J., *et al.*, 1992, these proceedings.
Kürster, M., Fleming, T.A., Schmitt, J.H.M.M., 1992, these proceedings.
Linsky, J.L. and Haisch, B.M., 1979, *Ap. J. Letters*, **229**, L27.
Micela, G., Sciortino, S., Vaiana, G.S., Schmitt, J.H.M.M., Stern, R.A., Harnden, F.R., Jr., and Rosner, R., 1990, *Ap. J.*, **290**, 307.
Motch, C., *et al.*, 1991, *Astr. Ap. Letters*, **246**, L24.
Pallavicini, R., Tagliaferri, G., Pollock, A.M.T., Schmitt, J.H.M.M., Rosso, C., 1990, *Astr. Ap.*, **277**, 483.
Pfeffermann, E., *et al.*, 1986, SPIE, **733**, 519.
Schmitt, J.H.M.M., Golub, L., Harnden, F.R., Jr., Maxson, C.W., Rosner, R., and Vaiana, G.S., 1985, *Ap. J.*, **290**, 307.
Schmitt, J.H.M.M., 1991, *Adv. Space Res.*, in press.
Schmitt, J.H.M.M., *et al.*, 1992, in preparation.
Snowden, S.L., Plucinsky, P.P., Briel, U., Hasinger G., and Pfeffermann, E., 1992, *Ap. J.*, submitted.
Stocke, J.T., *et al.*, 1991, *Ap. J. Suppl.*, **76**, 813.
Topka, K., Fabricant, D., Harnden, F.R., Jr., Gorenstein, P., and Rosner, R., 1979, *Ap. J.*, **229**, 661.
Trümper, J., *et al.*, 1991, *Nature*, **349**, 579.
Vaiana, G.S., *et al.*, 1981, *Ap. J.*, **244**, 163.

Seventh Cambridge Workshop on Cool Stars, Stellar Systems, and the Sun
ASP Conference Series, Vol. 26, 1992
Mark S. Giampapa and Jay A. Bookbinder (eds.)

THE NATURE OF THE DYNAMO AT THE END OF THE MAIN SEQUENCE: A ROSAT SURVEY OF THE LATE M DWARFS

THOMAS A. FLEMING Max Planck Institute for Extraterrestrial Physics, 8046 Garching bei Muenchen, Germany

MARK S. GIAMPAPA National Solar Observatory, Tucson, AZ 85726

J. SCHMITT Max Planck Institute for Extraterrestrial Physics, 8046 Garching bei Muenchen, Germany

JAY A. BOOKBINDER Center for Astrophysics, Cambridge, MA 02138

ABSTRACT We present X-ray data, both detections and upper limits, from the ROSAT all-sky survey for most known M dwarfs later than type M5, as well as from selected ROSAT pointed observations of some of these stars. We compare these data to similar data for early M dwarfs in an attempt to probe the nature of the magnetic dynamo and coronal heating mechanism for the very late M dwarfs, which are presumably totally convective. Our preliminary results indicate that the "saturation" boundary for the late M dwarfs is an extrapolation of that for the early M dwarfs and that coronal heating efficiency (as evidenced from L_X/L_{bol}) remains constant beyond spectral type M5.

Keywords: stellar coronae; ROSAT; very low mass stars

INTRODUCTION

One of the questions regarding stellar X-ray emission left unsolved by the *Einstein* Observatory was whether the very late M dwarfs (also known as Very Low Mass Stars; VLMS) do indeed have coronae (cf. reviews by Rosner, Golub, and Vaiana 1985; Liebert and Probst 1987). Bookbinder's (1985) study of Gliese Catalogue stars either detected or not by the *Einstein* Observatory seemed to indicate a deficiency of X-ray emission among dwarfs of spectral type later than M5. Furthermore, no dwarfs later than type M5 were detected in either the *Einstein* Extended Medium Sensitivity Survey or the EXOSAT High Galactic Latitude Survey (Fleming *et al.* 1988; Giommi *et al.* 1991). These results begged the question whether there existed yet another dividing line in X-ray H-R Diagram.

A plausible explanation for why VLMS could be deficient in X-ray emission stems from the current view that stars with masses $\lesssim 0.1\ M_{\odot}$ are fully convective. Rosner (1980) has discussed how a magnetic dynamo could operate in such a star without the differential rotation which occurs at the base of the convection zone. The question is if such a "distributive" dynamo could be as efficient as a "shell" dynamo in creating magnetic flux. Separate from this issue is that of coronal heating efficiencies, *i.e.*, whether the mechanism which actually heats the corona is dependent upon the internal stellar structure. We already know that the VLMS can create chromospheres efficiently (Giampapa and Liebert 1986; Liebert *et al.* these proceedings) and the VLMS VB 8 was detected by both *Einstein* and EXOSAT (Johnson 1987; Tagliaferri *et al.* 1990). But the paucity of X-ray data on VLMS has prevented us from drawing any conclusions regarding the strength of their coronae.

With the completion of the ROSAT all-sky survey, we have a uniform set of soft X-ray observations for the VLMS. To answer the question posed above, we have initiated a two-pronged observation program. First, we have searched the ROSAT all-sky survey data at the positions of known VLMS for either X-ray detections or upper limits. Second, we supplement these data with a program of ROSAT pointed observations of selected VLMS in order to model in detail the coronae of these stars. To date, pointed observations of VB 8, VB 10, and LHS 2924 have been carried out. In this paper, we present preliminary results with emphasis on the nature of the "saturation" boundary for VLMS.

OBSERVATIONS

The data presented in this paper were taken with ROSAT which, from August 1990 to January 1991, carried out the first all-sky survey in X-rays with an imaging telescope (cf. the review by Schmitt in these proceedings). We have attempted to compile a catalogue of all known M dwarfs having $M_V > 16.0$. We began with the list of Bessell (1991) and supplemented it with other stars found in the literature. Stars discovered or brought to our attention since the submission of this catalogue to the ROSAT Standard Analysis Software System (SASS) in Spring 1990 will be appended to the catalogue for remission to SASS before the next processing.

During SASS processing, the position of each star in the catalogue was checked against the master list of ROSAT survey source positions. If an X-ray source was found within 2 arcmin of the star, it was flagged as a probable detection. The large search radius was necessary due to initial errors in the attitude solution. This has since been empirically corrected and all detections have been confirmed to be within 30 arcsec of the optical positions, which is the typical 90% confidence error circle for survey positions. When no source was found, a maximum likelihood source detection algorithm was run on the survey data at the position of the star to find the most likely "bump" in the photon distribution. The number of counts 2-σ out on the likelihood curve was divided by the exposure time to get a 95%-confidence upper limit in count rate. Three stars in the catalogue: VB 8, VB 10, and LHS 2924, were targets of ROSAT guest investigator observations. Although not detected in the survey, VB 8 was detected in the pointed observation. This detection, as well as the upper limits on the other two, are used in the subsequent analysis.

The results are displayed in Fig. 1, where we plot L_X in the 0.1 to 2.4 keV band versus M_V for the VLMS as well as for the M dwarfs from the *Einstein* EMSS (Fleming *et al.* 1988). A conversion factor of 7 x 10^{-12} ergs cm^{-2} cnt^{-1}, which is appropriate for flare stars, was multiplied by the count rate to obtain

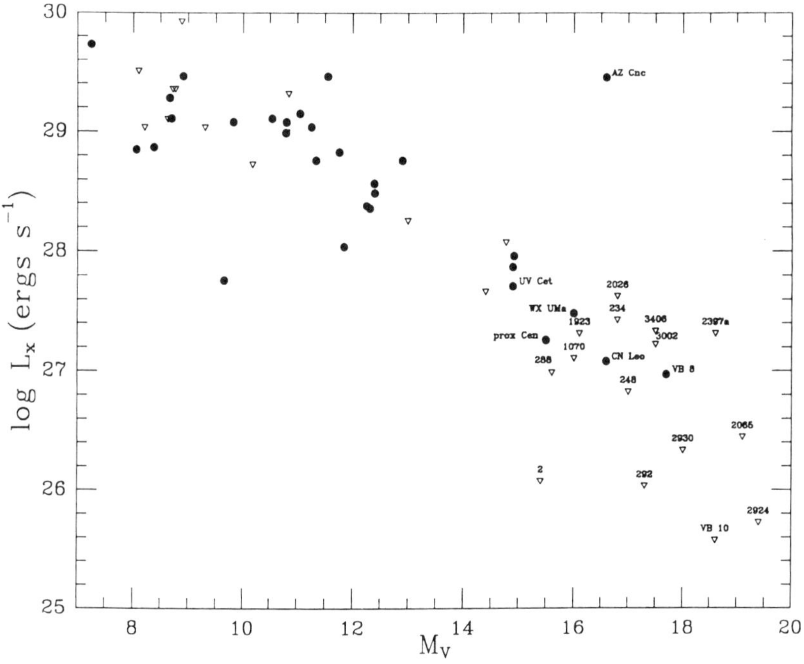

Fig. 1 X–ray luminosity (0.1 to 2.4 keV) plotted versus absolute visual magnitude for most known M dwarfs later than type M5 (labelled points) and for the EMSS M-dwarf sample. Solid circles indicate detections by ROSAT, while open triangles indicate 95% confidence upper limits. The labelled points are designated by either their LHS number or by their more commonly recognized name.

fluxes. Luminosities were calculated from trigonometric parallaxes for the VLMS and mostly photometric parallaxes for the EMSS stars. One can see the continuation of the "saturation boundary" (cf. Vilhu and Walter 1987) from early M to later M types. An inspection of the survey light curve of AZ Cnc shows that all of the photons came within the first six orbits of the two-day (32 orbits) observation window. This means that the quiescent flux of AZ Cnc is below the survey sensitivity, but that it was detected only in flare.

DISCUSSION

In order to test whether coronal heating efficiency decreases among the VLMS, it is easiest to consider only stars at the "saturation boundary", *i.e.*, the maximum activity level, since it is more difficult to identify M dwarfs of all subtypes at the same moderate or weak activity levels for comparison. We use the EMSS M dwarfs since, being an X–ray selected sample, it defines the "saturation boundary" at early M types. As for the VLMS, we consider only the dMe stars (except AZ Cnc, for which we have only a flare value) in the belief that they have the highest probability of being saturated. In Fig. 2, we plot L_X/L_{bol} versus

M_V for the dMe stars only. We use L_X/L_{bol} as our proxy indicator for coronal heating efficiency. Use of survival statistics analysis on these data yield a 5% probability that they are correlated (Cox-Hazard test) and the best regression fit gives a slope of zero with constant value of $L_X/L_{bol} = 0.0016 \pm 0.0007$.

These results indicate that coronal heating efficiencies do not change as one goes to totally convective stars along the end of the main sequence. For "saturated" M dwarfs, 0.1% of all energy is typically radiated in X-rays, while this number can increase during flaring (*e.g.* 8% for AZ Cnc). There is no evidence so far to suggest that VLMS are less efficient than more massive dwarfs in creating coronae. This may indicate that the distributive dynamo is just as efficient in producing magnetic flux as the shell dynamo. However, one can not yet rule out the possibility that the VLMS do indeed still have radiative cores. Or it could be that the coronal heating mechanism depends purely on the amount of mechanical energy available in the photosphere, making the question of whether a shell or distributive dynamo is at work irrelevant.

REFERENCES

Bessell, M.S. 1991, *A. J.*, **101**, 662.
Bookbinder, J.A. 1985, Ph.D. Thesis, Harvard University.
Fleming, T.A., Liebert, J., Gioia, I.M., and Maccacaro, T. 1988, *Ap. J.*, **331**, 958.
Giampapa, M.S. and Liebert, J. 1986, *Ap. J.*, **305**, 784.
Giommi, P. *et al.* 1991, preprint.
Johnson, H.M. 1987, *Ap. J.*, **316**, 458.
Liebert, J. and Probst, R. 1987, *Ann. Rev. Astron. Ap.*, **25**, 473.
Rosner, R. 1980, in Cool Stars, Stellar Systems, and the Suned. A.K. Dupree (SAO Rep. No. 389), p. 79.
Rosner, R., Golub, L., and Vaiana, G.S. 1985, *Ann. Rev. Astron. Ap.*, **23**, 413.
Tagliaferri, G., Doyle, J.G., and Giommi, P. 1990, *Astr. Ap.*, **231**, 131.
Vilhu, O. and Walter, F.M. 1987, *Ap. J.*, **321**, 958.

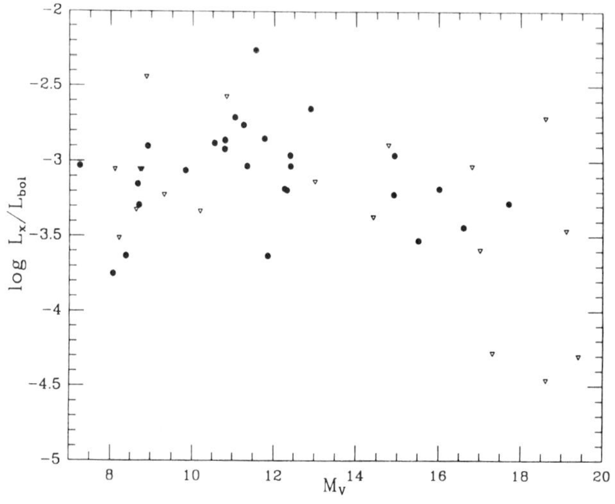

Fig. 2 Ratio of X-ray (0.1 to 2.4 keV) to bolometric luminosity plotted versus absolute visual magnitude for the dMe stars only. The symbols are the same as in Fig. 1.

Seventh Cambridge Workshop on Cool Stars, Stellar Systems, and the Sun
ASP Conference Series, Vol. 26, 1992
Mark S. Giampapa and Jay A. Bookbinder (eds.)

ROSAT OBSERVATIONS OF THE PLEIADES CLUSTER

JEAN-PIERRE CAILLAULT AND MARC GAGNÉ Department of Physics and Astronomy, University of Georgia, Athens, GA 30602

JOHN STAUFFER Harvard-Smithsonian Center for Astrophysics, 60 Garden St., Cambridge, MA 02138

ABSTRACT We have used the ROSAT PSPC to observe the Pleiades Cluster for 4,000 seconds. The observation was centered about 7.5' west of the bright star, Alcyone. Forty-seven X-ray sources have been detected, all but one of which have been identified with optical candidates; 36 of these are cluster members. The spectral types of those detected range from B stars (two of the Seven Sisters) to M dwarfs. Roughly half of these ROSAT sources were not detected in the EINSTEIN Observatory survey of the Pleiades, but half of the EINSTEIN sources have not been detected with this short ROSAT exposure, either; the implication of this is that the X-ray emission from these stars is variable on long timescales (about 10 years). The completion of our survey (50,000 seconds for this field and two 25,000 second adjacent fields) will provide us with an order of magnitude improvement in sensitivity over the EINSTEIN data.

Keywords: Young clusters, X-ray emission, ROSAT

INTRODUCTION

ROSAT presents us with the opportunity to observe the Pleiades cluster with a full order of magnitude improvement in sensitivity over that which the EINSTEIN Observatory could provide (Caillault and Helfand 1985; Micela *et al* 1990). This allows us to address the issue of the evolutionary dependence of X-ray emission by establishing the complete stellar X-ray luminosity functions of the late-type stars in this cluster. Also, the Pleiades is the best open cluster to study the dependence of X-ray emission on rotation because it is nearby, well-studied, and is young enough that its low mass members still show a wide range of rotational velocities (Stauffer and Hartmann 1987). Finally, the short, medium, and long-term variability characteristics of its young stars will provide us with information about the spatial and temporal distribution of their magnetic loop structures.

OBSERVATIONS

We observed the Pleiades Cluster, centered on RA(2000) = 3H 47M, DEC(2000) = 24° 09', with the ROSAT PSPC during three intervals in early 1991: on 9 February for 2,000 seconds, on 26 Feb. for 2,000 seconds, and on 4 March for 15,000 seconds. However, the latter segment was taken during AMCS standby, making the data useless in its current form. The data from the first two segments is shown in Figure 1. Our proposal was approved for 50,000 seconds, though; the remaining time was made up in August (for which we have not yet received the data; in addition, a second round proposal for two additional 25,000 second exposures adjacent to this central field were performed in September and will be presented elsewhere). The implications of this are that we will ultimately obtain a sensitivity about 3 times better than what we present here in this preliminary report, *i.e.*, that 4,000 seconds allows us to detect sources down to about 7E-14ergs cm^{-2} s^{-1} , already better than EINSTEIN. This sensitivity, of course, degrades rapidly with off-axis distance.

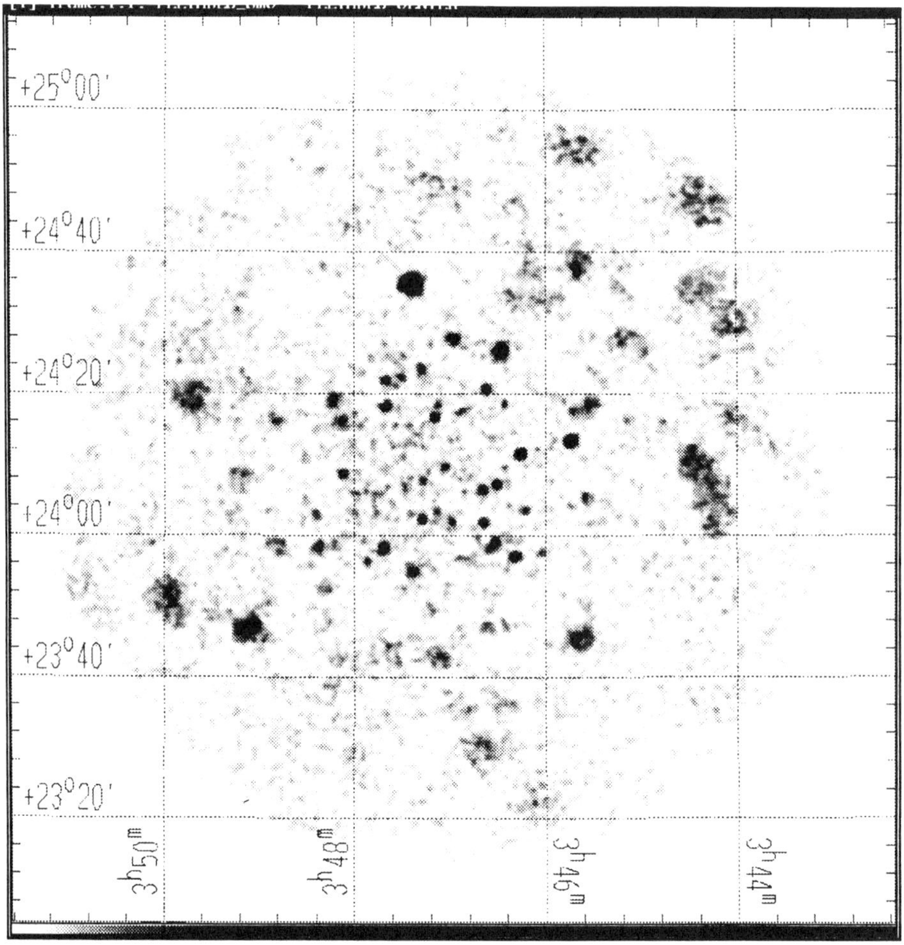

Fig. 1: A 4,000 second ROSAT PSPC exposure of the Pleiades Cluster.

RESULTS

Forty-seven X-ray sources were detected in the 4,000 second ROSAT observation. Of these 47, 36 have been identified with cluster members and the remainder (except for one) have non-member stellar candidates within 1' of the X-ray position. The stars are well distributed in spectral type, with 2 of the Seven Sisters (B stars), 3 A stars, 10 F stars, 5 G stars, 11 K stars, and 1 M dwarf detected among those with known spectral type. Of the 23 late-type (F6 and later) stars detected, 19 have known V sin i values; these range from 7.5 km s^{-1} to 85 km s^{-1} (Stauffer *et al.* 1984; Stauffer and Hartmann 1987; Soderblom *et al.* 1992), implying that both rapid and slow rotators are X-ray sources. We can summarize our results as a function of off-axis distance. Splitting our image into concentric rings of radius 15', *i.e.,* 0-15', 15-30', and >30', allows us to see the effect of vignetting on sensitivity. The sources detected within the first ring had count rates as low as .007 c/s. The lowest in the second ring was .015 c/s, while the outermost ring's sources were all brighter than .030 c/s. The number of sources did not vary much: 18 in the innermost ring, 16 in the intermediate ring, and 13 in the outermost. The number of cluster member identifications, though, did: 15 (out of 18 sources) in the first ring, 13 (out of 16) in the second ring, and only 8 (out of 13) in the outermost ring. All 28 members detected within 30' of the center of the pointing had X-ray/optical position offsets less than 30". The eight beyond 30' ranged in offset from 12" to 87". Finally, roughly half of these sources were not detected in the earlier survey, however, it is also true that about half of the original EINSTEIN survey's sources fail to show up in this short ROSAT image. Therefore, we conclude that this result implies long-term X-ray variability for these stellar sources.

CONCLUSION

Upon receipt of the full 100,000 seconds of ROSAT time on the Pleiades, we will easily be able to accomplish all of the goals set out in the Introduction to this paper. We are waiting with great anticipation for mail from GSFC.

ACKNOWLEDGEMENTS

This research was supported by NASA Grant NAG5-1608 to the University of Georgia.

REFERENCES

Caillault, J.-P. and Helfand, D. J. 1985, *Ap. J.*, **289**, 279.
Micela, G., *et al.* 1990, *Ap. J.*, **348**, 557.
Soderblom, D., *et al.* 1992, in preparation.
Stauffer, J., *et al.* 1984, *Ap. J.*, **280**, 202.
Stauffer, J. and Hartmann, L. 1987, *Ap. J.*, **318**, 337.

Seventh Cambridge Workshop on Cool Stars, Stellar Systems, and the Sun
ASP Conference Series, Vol. 26, 1992
Mark S. Giampapa and Jay A. Bookbinder (eds.)

ROSAT ALL-SKY SURVEY OBSERVATIONS OF THE HYADES CLUSTER

ROBERT A. STERN
Lockheed Palo Alto Research Laboratory, O/91-30, Bldg. 252, 3251 Hanover St., Palo Alto, CA, 94304, USA.

JÜRGEN H.M.M. SCHMITT, CHRISTINA ROSSO
Max-Planck-Insitut für Extraterrestriche Physik, D-8046 Garching, FRG.

JOHN P. PYE, SIMON T. HODGKIN
Physics and Astronomy Department, University of Leicester, University Road, Leicester LE1 7RH, UK.

JOHN R. STAUFFER
Harvard-Smithsonian Center for Astrophysics, 60 Garden Street, Cambridge, MA 02138, USA.

ABSTRACT We present preliminary ROSAT All Sky X-ray Survey results for the Hyades cluster. Over 120 Hyades cluster members were detected at a S/N > 10. Seven cluster members were also seen in the WFC EUV all-sky survey. We are currently updating our survey database, which should produce a significant increase in the number of detected Hyades members.

Keywords: X-rays; galactic clusters; stellar atmospheres

INTRODUCTION

Over a decade ago, at the first Cambridge Cool Star Workshop, Stern *et al.* (1980) reported the detection of X-ray emission from Hyades stars. Results from the completed *Einstein* Observatory cluster center survey (Stern *et al.* 1981) demonstrated that coronal X-ray emission was a common feature throughout the Hyades H-R diagram. The more extensive Hyades region survey of Micela *et al.* (1988) confirmed the earlier cluster center results, while significantly increasing the number of cluster members detected, and accurately determining the X-ray luminosity function for the brighter (F-G) main sequence stars.

A complete survey of all known or suspected cluster members, covering over 40° by 40° in the sky, was not feasible with *Einstein* pointings. Such a flux-limited survey is, however, ideally suited to the recently completed all-sky X-ray survey, carried out with the ROSAT PSPC (Position Sensitive Proportional Counter). The typical limiting X-ray sensitivity for the survey is ≈ 0.02 ct s^{-1} or $\approx 1.2 \times 10^{-13}$ erg cm^{-2} s^{-1} in the 0.1–2.4 keV band. This corresponds to a limiting L_x of $\approx 3 \times 10^{28}$ erg s^{-1} at the Hyades cluster distance, comparable to the sensitivity limit for a typical 2000 sec *Einstein* IPC pointing. The all-sky

EUV survey conducted with the co-aligned WFC (Wide Field Camera) has a limiting sensitivity of $L_{EUV} \approx 3 \times 10^{29}$ erg s^{-1} in the 0.09–0.19 keV band.

Source detections in the Hyades region are obtained by processing a list of catalogued star positions. Our Hyades input catalog consists of ≈ 500 confirmed or suspected cluster members down to V ≈ 16, taken largely from the open literature. We were somewhat liberal in including suspected members, given the requirement for an input catalog. We estimate that no more than 5-10% of the catalogued stars are likely to be non-members. The source detection threshold corresponds to a signal-to-noise level of ≈ 10 over adjacent sky background.

RESULTS

Preliminary X–ray detections in our Hyades survey are shown in Figure 1, plotted on an H-R diagram of the input catalog cluster members. *We note that, due to problems with a PSPC gas valve, 2 days of survey data scanning the cluster center are presently missing from our database.* Hence the number of detected cluster members should increase significantly when this data is incorporated into the survey. Despite the temporary loss of data, 88 of the 120 ROSAT sources were previously unobserved or undetected in earlier X–ray studies of the cluster.

We estimate stellar X–ray luminosities using the known cluster distance (45 pc) and the conversion from PSPC count rate to X–ray flux (this is temperature dependent, and is also still being refined). Note that cluster depth effects are not included in this L_x estimate. In Figure 2 we plot L_x vs. B-V for the detected stars only (no upper limits are available at this time). Six of seven stars detected with the WFC during the survey are also indicated in the figure. The brightest X–ray emitters are generally associated with spectroscopic binary systems. V471 Tau is an anomaly because of its white dwarf, which produces photospheric X–ray and EUV emission. VB141 (=71 Tau) remains the brightest object in the cluster after V471 Tau, as was the case in the *Einstein* surveys. In the WFC survey, the brightest two objects are also V471 Tau and VB141, the latter with $L(0.09-0.19\ keV) \approx 4 \times 10^{29}$ erg s^{-1}. Two more spectroscopic binary systems, Leiden 20 and BD +22 669, are near the top of the X–ray luminosity distribution. Finally, we have unambiguously detected the white dwarf EG37 (=VA490) in the PSPC, with a marginal detection in the WFC survey.

ACKNOWLEDGEMENTS

R.A.S. acknowledges support from the Lockheed Independent Research Program.

REFERENCES

Micela, G., Sciortino, S., Vaiana, G.S., Schmitt, J.H.M.M., Stern, R.A, Harnden, F.R., Jr., and Rosner, R., 1988, *Ap. J.*, **325**, 798.

Stern, R.A., Underwood, J.H., Zolcinski, M.-C., and Antiochos, S.K., 1980, in Cool Stars, Stellar Systems, and the Sun, ed. A.K. Dupree, SAO Special Report 389, p.127.

Stern, R.A., Zolcinski, M.-C., Antiochos, S.K., and Underwood, J.H., 1981, *Ap. J.*, **249**, 647.

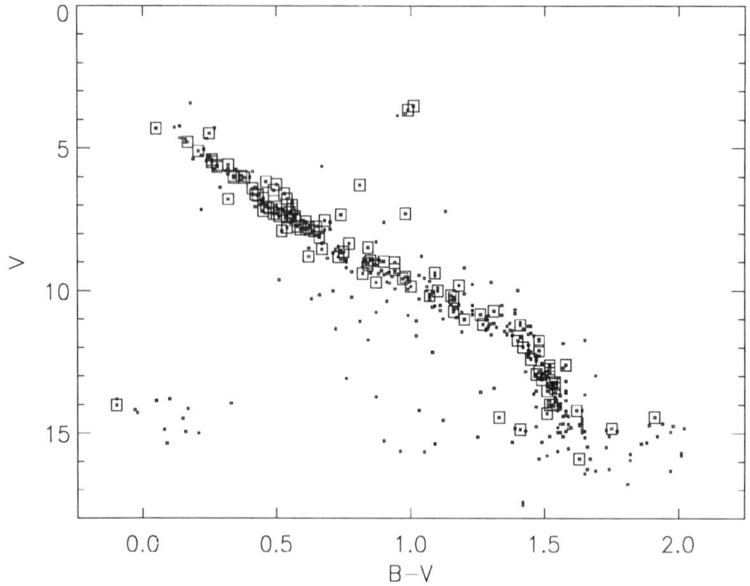

Figure 1. Hyades H-R Diagram showing detected X–ray sources (boxes).

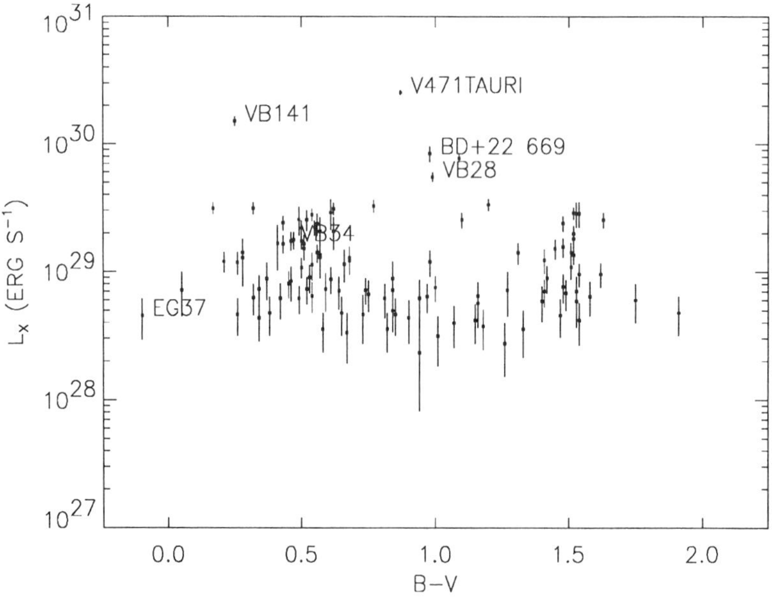

Figure 2. X–ray luminosities of detected stars. WFC detections are labeled.

Seventh Cambridge Workshop on Cool Stars, Stellar Systems, and the Sun
ASP Conference Series, Vol. 26, 1992
Mark S. Giampapa and Jay A. Bookbinder (eds.)

A DEEP ROSAT OBSERVATION OF THE HYADES

ROBERT A. STERN
Lockheed Palo Alto Research Laboratory, O/91-30, Bldg. 252, 3251 Hanover St., Palo Alto, CA, 94304, USA.

JÜRGEN H.M.M. SCHMITT
Max-Planck-Insitut für Extraterrestriche Physik, D-8046 Garching, FRG.

JOHN P. PYE
Physics and Astronomy Department, University of Leicester, University Road, Leicester LE1 7RH, UK.

JOHN R. STAUFFER
Harvard-Smithsonian Center for Astrophysics, 60 Garden Street, Cambridge, MA 02138, USA.

THEODORE SIMON
Institute for Astronomy, University of Hawaii, 2680 Woodlawn Dr., Honolulu, HI 96822, USA.

ABSTRACT We present preliminary results from a 40,000 sec ROSAT pointed observation of the Hyades cluster. At field center, the limiting Hyades L_x is $\approx 2 \times 10^{27}$ erg s^{-1}, increasing to $\approx 2 \times 10^{28}$ erg s^{-1} at 40′ off-axis. At this early stage in the analysis, over 30 sources have been detected in the PSPC data, 17 of which are associated with known Hyades members. We also discuss preliminary WFC EUV results for the same pointing.

Keywords: X–rays; galactic clusters; stellar atmospheres

INTRODUCTION

From previous *Einstein* X–ray observations of the Hyades (Stern *et al.* 1981, Micela *et al.* 1988), and the current ROSAT All-Sky Survey data (Stern *et al.* 1991), we have a fairly accurate picture of the X–ray luminosity function for F and G main sequence cluster members. However, the typical limiting sensitivity of both the *Einstein* pointings and the ROSAT sky survey ($L_x \approx 3 \times 10^{28}$ erg s^{-1}) is sufficient to detect only the X–ray brightest of the K and M dwarfs. Deep ROSAT PSPC exposures are therefore required to probe the Hyades lower main sequence. Such deep exposures also produce many high quality proportional counter pulse-height spectra, which are essential in determining coronal temperatures and emission measures.

OBSERVATIONS AND RESULTS

As part of the U.S. AO-1 ROSAT pointed program, we obtained a single long PSPC exposure in the Hyades at $\alpha(2000)= 4^h24^m50.4^s$, $\delta(2000)=+15°06'00''$. The observation extended over $2\frac{1}{2}$ days from 27 Feb 1991 0610 UT to 1 Mar 1991 1859 UT, with a net exposure time of 4.04×10^4 sec.

A gray-scale image of the 2° PSPC field-of-view is shown in Figure 1. Also shown (boxes) are the source detection locations from the standard ROSAT processing (SASS MPLSX list, for ROSAT aficionados). In all, 38 sources are reported, 27 of which have optical identifications within 3'. In general, the ROSAT PSPC positions should be much better than an arc-minute; unfortunately, the processed data for this field has known systematic position offsets which will be corrected in later reprocessing.

There are 24 Hyades stars from our ≈500-member catalog in this PSPC field: 17 of these have been detected, including several new M dwarf detections. The estimated X-ray luminosities for the Hyades sources are shown in Figure 2, on the same scale as presented for the all-sky survey results in the accompanying paper by Stern *et al.* Note that the pointed observation has a much greater sensitivity than the survey: the X-ray luminosity of the faintest detected object, H366=VA388, is $\approx 6 \times 10^{27}$ erg s^{-1}, roughly 5 times below the survey limit. Upper limits have not yet been determined, but they should range from $L_x \approx 2 \times 10^{27}$ erg s^{-1} at field center, increasing to $\approx 2 \times 10^{28}$ erg s^{-1} at 40' off-axis.

High quality proportional counter pulse-height spectra are available for 21 sources in the field. Of these, 14 are from Hyades stars ranging in spectral type from F to M, including one (VB 141=71 Tau) with 20,000 counts. Two stars, VB 141 and VB 71=77 θ^1 Tau, were also detected in the S2 (Be/Lexan) filter band ($\overline{\lambda} \approx 140$ Å) of the co-aligned Wide Field Camera (WFC). Combined with the PSPC pulse-height spectra, the WFC filter photometry will provide significant *simultaneous* constraints on the coronal emission measure distribution in the $10^{5.5}$-10^7 K range, and on the line-of-sight absorption column to the cluster. The remaining PSPC data constitute an excellent sample with which to study the relationship between X-ray luminosity, coronal temperature, and other stellar parameters.

ACKNOWLEDGEMENTS

R.A.S. acknowledges support from NASA Contract NAS5-31212 under the U.S. ROSAT Guest Investigator Program.

REFERENCES

Micela, G., Sciortino, S., Vaiana, G.S., Schmitt, J.H.M.M., Stern, R.A, Harnden, F.R., Jr., and Rosner, R., 1988, *Ap. J.*, **325**, 798.

Stern, R.A., Schmitt, J.H.M.M., Rosso, C., Pye, J.P, Hodgkin, S., and Stauffer, J.R., 1991, these proceedings.

Stern, R.A., Zolcinski, M.-C., Antiochos, S.K., and Underwood, J.H., 1981, *Ap. J.*, **249**, 647.

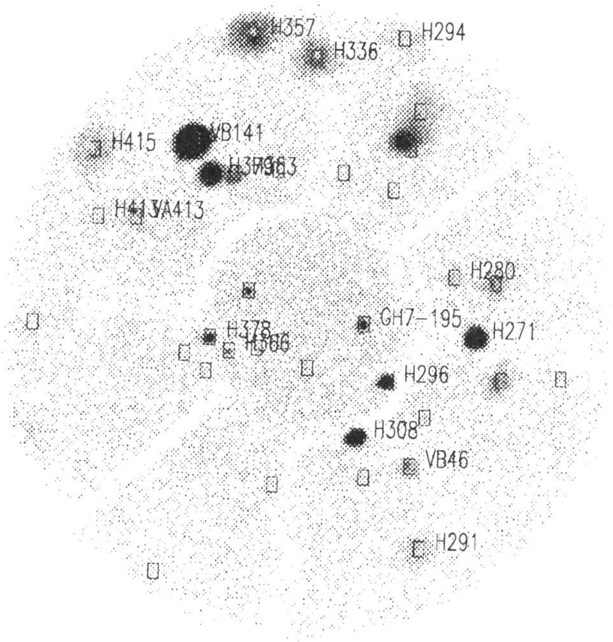

Figure 1. Gray-scale image of ROSAT field showing detections (boxes) and Hyades identifications.

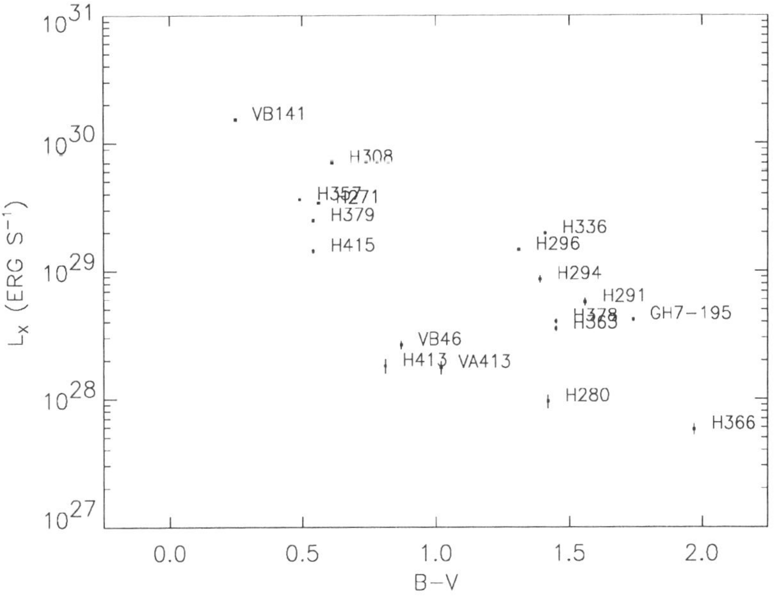

Figure 2. X–ray Luminosities of detected stars

Seventh Cambridge Workshop on Cool Stars, Stellar Systems, and the Sun
ASP Conference Series, Vol. 26, 1992
Mark S. Giampapa and Jay A. Bookbinder (eds.)

FIRST RESULTS FROM A COORDINATED ROSAT, IUE, AND VLA STUDY OF RS CVN SYSTEMS

J. L. LINSKY,[1] D. FOX, A. BROWN, R. DEMPSEY, and C. SCHMITT
Joint Institute for Laboratory Astrophysics, NIST and Univ. of Colorado, Boulder, CO 80309-0440.

J.H.M.M. SCHMITT and T. FLEMING Max-Planck-Institut für Extraterrestrische Physik, Giessenbachstrasse, D-8046 Garching bei München, Germany

M. RODONÒ and I. PAGANO Osservatorio Astrofisico di Catania, Viale Andrea Dorio 6, I-95125 Catania, Italy

J. E. NEFF Department of Astronomy and Astrophysics, The Pennsylvania State University, University Park, PA 16802

G. BROMAGE Rutherford Appelton Lab., Chilton, Didcot, England

ABSTRACT This is a preliminary report on our extensive set of coordinated observations of RS CVn systems obtained during the ROSAT all-sky survey. Simultaneous with the PSPC and WFC observations, we observed five RS CVn systems with IUE and 15 systems with the VLA, typically at three wavelengths. We list the times of the coordinated observations and show our derived fluxes for one system - TY Pyx.

Keywords: RS CVn binaries; TY Pyx; X-ray emission; radio emission; ultraviolet spectra

We report here on the first results from our extensive set of coordinated observations of RS CVn systems obtained during the ROSAT all-sky survey. Under the RIASS (ROSAT-IUE All Sky Survey) program we obtained IUE observations of five of these systems simultaneous with the ROSAT observations. A total of 15 systems were observed by the VLA simultaneous with ROSAT, including three systems (TY Pyx, Z Her, and SV Cam) also observed simultaneously by IUE. Nearly all of the VLA observations were at three wavelengths (3.6, 6, and 20 cm) with the array typically divided between two wavelengths for simultaneous coverage. The total amount of on-target subarray observing time was about 77 hours. Table 1 summarizes the 20 RS CVn systems for which we have VLA and/or IUE observations. The last column indicates PSPC, WFC, and IUE detections. We would appreciate learning of other coordinated observations of these systems.

[1] Staff member, Quantum Physics Division, National Institute of Standards and Technology

Three stars were extensively monitored by the VLA – HD39576, SZ Pic, and TY Pyx. During the period 9-12 November 1990, IUE obtained 40 spectra nearly continuously of the 3.20 day period TY Pyx (G5 IV + G5 IV) system, ROSAT observed Ty Pyx every 90 minutes for more than 2 days, and the VLA obtained 140 individual observations over the course of 21 hours. Figure 1 summarizes the C IV, PSPC, and 3.6-cm fluxes with 1σ error bars for the PSPC and 3.6-cm data. Although TY Pyx did not flare in this period, the fluxes varied at all wavelengths. In particular, the 3.6-cm flux increased by a factor of 2 on November 12 with no corresponding changes at the other wavelengths, indicating that at this time the radio radiation from nonthermal electrons and the soft X-ray emission from the thermal electrons was not correlated.

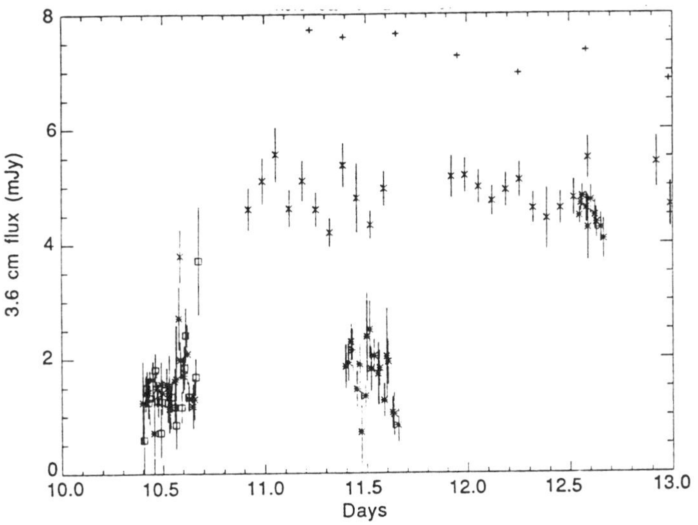

Fig. 1: Coordinated observations of TY Pyx in November 10–13, 1990. The C IV fluxes (+), in units of 10^{-13} erg cm^2 s^{-1}, are displaced up by 2 in the ordinate. The PSPC X-ray counts s^{-1} (x) are displaced up by 3. The 3.6-cm fluxes ($*$ and $\square$) are in mJy.

The correlations of the C IV, X-ray, and 6-cm radio fluxes shown by Drake, Simon, and Linsky (*Ap. J. Suppl.*, **71**, 905 (1989)) for a large sample of RS CVn systems are for data sets obtained years apart in time. We note that the fluxes obtained during the coordinated observations of TY Pyx fit very close to these regression curves, confirming that the statistical correlations are real even though the fluxes change independent of each other at specific times.

ACKNOWLEDGMENTS

We thank NASA for partial support of this work through grant NAG5-82 to the University of Colorado, and Interagency Transfer W-17,772 to the N.I.S.T.

Table 1. Times of VLA Observations during the ROSAT survey

Target	RS CVn Name	Day 1990 (UT)	Time (hours) (UT)	ROSAT & IUE Detections
HD4502	ζ And	31 Dec	None	PSPC, WFC,IUE
HD19845	LX Per	09 Aug	9–19	PSPC
HD21242	UX Ari	09 Aug	9–18	PSPC, WFC
HD22403	–	09 Aug	9–18	PSPC
HD26337	EI Eri	09 Aug	10–19	PSPC, WFC
HD37824	–	06 Sept	12–16	PSPC
		07 Sept	12–16	PSPC
HD37847	TW Lep	06 Sept	12–16	PSPC, WFC?
HD39576	–	07 Sept	12–16	PSPC
		08 Sept	12–16	PSPC
		09 Sept	12–16	PSPC
		10 Sept	12–16	PSPC
HD39917	SZ Pic	06 Sept	12–16	PSPC
		07 Sept	12–16	PSPC
		08 Sept	12–16	PSPC
		09 Sept	12–16	PSPC
		10 Sept	12–16	PSPC
		11 Sept	12–16	PSPC
HD44982	SV Cam	12 Sept	6–7	PSPC
		13 Sept	None	PSPC, IUE
HD77137	Ty Pyx	09 Nov	13–14	PSPC, WFC
		10 Nov	9–17	PSPC, WFC, IUE
		11 Nov	9–17	PSPC, WFC, IUE
		12 Nov	13–16	PSPC, WFC, IUE
		13 Nov	None	PSPC, WFC, IUE
HD106677	DK Dra	20 Oct	None	PSPC, IUE
HD108102	IL Com	09 Dec	None	PSPC, WFC, IUE
HD150708	WW Dra	08 Jan 91	None	PSPC, IUE
HD163930	Z Her	11 Sept	None	PSPC, IUE
		12 Sept	7–8	PSPC
HD165590	V772 Her	12 Sept	6–7	PSPC, WFC
HD210334	AR Lac	12 Dec		PSPC, WFC, IUE
		13 Dec		PSPC, WFC, IUE
HDE250810	CQ Aur	11 Sept	12–16	PSPC
		12 Sept	7–8	PSPC
HDE341475	MM Her	12 Sept	6–7	PSPC
–	RW UMa	10 Nov	9–16	PSPC<
		11 Nov	10–18	PSPC<
		12 Nov	13–17	PSPC<

Seventh Cambridge Workshop on Cool Stars, Stellar Systems, and the Sun
ASP Conference Series, Vol. 26, 1992
Mark S. Giampapa and Jay A. Bookbinder (eds.)

FIRST ROSAT ALL SKY SURVEY X-RAY LIGHT CURVES OF ACTIVE STARS.

MARTIN KÜRSTER, JÜRGEN H.M.M. SCHMITT, THOMAS A. FLEMING
Max–Planck–Institut für extraterrestrische Physik, D–8046 Garching bei München, Federal Republic of Germany.

ABSTRACT ROSAT all sky survey X–ray light curves are presented for three active stars. We find primary eclipses for AR Lac as well as indications for rotational modulation of the X–ray flux of AB Dor. In the case of TY Pyx data quality prohibits any such interpretation.

Keywords: Coronal X–ray emission; rotational modulation; eclipses

INTRODUCTION

First X–ray light curves extracted from the ROSAT all sky survey are presented for the active stars AB Dor, AR Lac, and TY Pyx. While AB Dor is a single pre–main sequence K0–star the other two stars are RS CVn–type close binaries.

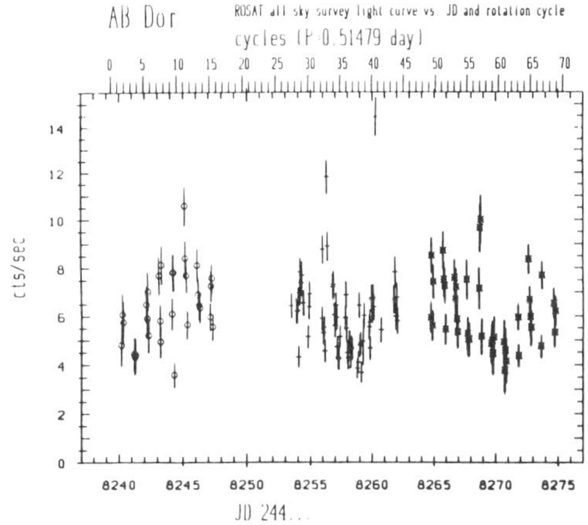

Fig. 1: ROSAT all sky survey light curve of AB Dor in the total PSPC bandpass plotted vs. time (Julian day) and rotational cycle (indicated at top).

RESULTS ON INDIVIDUAL STARS

AB Dor = HD 36705: AB Dor was visible for more than a month in the ROSAT survey. Short term emission variation by about a factor of two occurred plus some flaring events (fig. 1). In order to look for rotational modulation of the non–flaring emission the whole data set was subdivided into three groups of data (different symbols in fig. 1) and then folded with the rotational period (ephemeris by Innis *et al.* 1988; fig. 2).

Quite erratic flux variations occur, only some fraction of AB Dor's X–ray emission may be rotationally modulated. This modulation is perhaps best seen in fig. 2b where most data points are grouped around a sinusoidal band (maximum near phase 0.25, minimum near phase 0.75), but deviations to both sides of the band occur. While positive deviations can be attributed to flares we do not yet have an explanation for the negative ones at hand. A similar sinusoidal band may be seen in fig. 2a with positive deviations only. There is also some slight indication of such a grouping of data in fig. 2c, but with the maximum shifted towards a higher phase.

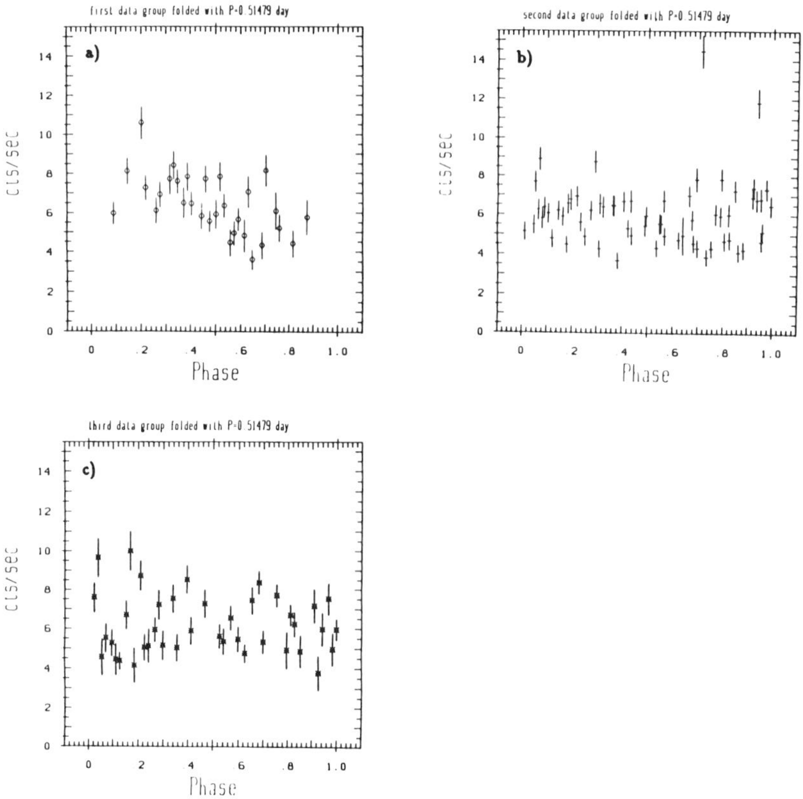

Fig. 2: Three groups of AB Dor data folded with the rotational period.

AR Lac = HD 210334: The ROSAT survey light curve of this eclipsing RS CVn system is shown in fig. 5 of the review article by *Schmitt* (this volume). It spans 1.5 binary cycles including two primary eclipses ('PE') and one secondary eclipse ('SE'). Phases are calculated from the ephemeris by Rodonó *et al.* (1986). A drop in X–ray count rate is seen around the first primary eclipse which seems to occur during a time of quiescent emission. A similar drop at the second primary eclipse follows an episoide of multiple flaring and reaches a much higher minimum level. There is no indication of secondary eclipse in X–rays.

These results — primary eclipses, but no secondary eclipse — are consistent with previous X–ray observations of AR Lac by *Einstein* (Walter *et al.* 1983) and EXOSAT (White *et al.* 1990). The same behaviour has been found during a ROSAT pointed observation of AR Lac carried out during the calibration phase of the ROSAT mission (Ottmann, Schmitt, Kürster, *in prep.*).

While at primary eclipse the G2IV–star is totally occulted by the larger K0IV–star, secondary eclipse is annular covering less than one third of the stellar disk. This suggests a complete X–ray eclipse of the G–star and a lack of emission from the eclipsed equatorial region of the K–star.

TY Pyx = HD 77137: The ROSAT survey light curve for the partially eclipsing system TY Pyx spans ≈ 85% of the binary cycle covering one secondary eclipse (fig. 3). Phases are calculated from the ephemeris by Andersen *et al.* (1981). Count rate statistics are a too poor for the dip around phase 0.5 to be considered as conclusive evidence for an X–ray eclipse. A previous X–ray observation of TY Pyx with EXOSAT (Culhane *et al.* 1990) also yielded just marginal detections of the eclipses. Other than with EXOSAT no large flares were detected with ROSAT.

REFERENCES

Andersen, J., Clausen, J.V., Nordstrom, B., Reipurth, B. 1981, *Astr. Ap.*, **101,** 7.
Culhane, J.L., White, N.E., Shafer, R.A., Parmar, A.N. 1990, *M.N.R.A.S.*, **243,** 424.
Innis, J.L., Thompson, K., Coates, D.W., Lloyd Evans, T. 1988, *M.N.R.A.S.*, **235,** 1411.
Rodonó, M. *et al.* : 1986, *Astr. Ap.*, **165,** 135.
Walter, F.M., Gibson, D.M., Basri, G.S. 1983, *Ap. J.*, **267,** 665.
White, N., Shafer, R., Horne, K., Parmar, A., Culhane, J. 1990, *Ap. J.*, **350,** 776.

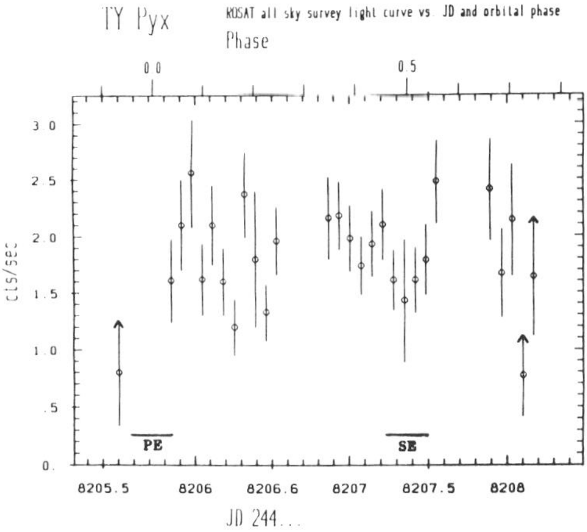

Fig. 3: ROSAT all sky survey light curve of TY Pyx in the total PSPC bandpass. Orbital phase is indicated at the top. Horizontal bars mark the times of (optical) primary ('PE') and secondary ('SE') eclipse.

Seventh Cambridge Workshop on Cool Stars, Stellar Systems, and the Sun
ASP Conference Series, Vol. 26, 1992
Mark S. Giampapa and Jay A. Bookbinder (eds.)

DETECTION OF STRONG CHROMOSPHERIC AND CORONAL ACTIVITY IN POP II BINARIES

LUCA PASQUINI European Southern Observatory, Casilla 19001, Santiago 19, Chile

T.A. FLEMING Max Planck Für Extraterrestrische Physik, Garching, BRD

FRANCOIS SPITE and MONIQUE SPITE Observatoire de Paris, 92195 Meudon Cedex, France

ABSTRACT We present Ca II and Hα observations of 4 short period Pop II binaries. All the sample stars show evidence for rather strong activity. The two brightest and nearest were detected in X–rays by the ROSAT satellite, being the first Pop II coronal source ever detected and we suggest Pop II short period binaries as a new class of soft X–ray emitters.

Keywords: Chromospheres, Coronae, Stars: Pop II, Stars: Binaries.

INTRODUCTION

According to our knowledge of stellar activity in single main sequence stars, activity in Pop II objects is expected to be rather low, or not existing. It is in fact generally accepted that activity in main sequence stars is related to the presence of surface magnetic fields, maintained during the stellar lifetime by the so called stellar dynamo (Parker 1970); a powerful dynamo needs the presence of subphotospheric convective zones and high stellar rotational velocity. Since Pop II stars are old and slow rotators, a vigorous dynamo is not supposed to be present.

So far activity in Pop II stars has been detected at very low levels in several giants (Dupree *et al.* 1990) and in HD 89499 (Hartman & Gehren 1988), one of the binaries of our sample. Very recently, Murphy *et al.* (1991) observed active stars in the core of the globular cluster M 15, suggesting that this activity is due to their binarity.

SAMPLE AND OPTICAL OBSERVATIONS

Our sample consists of 4 Pop II spectroscopic binaries, having well established orbital periods between $\sim$ 3.4 and $\sim$ 10 days. Assuming metallicity as the primary parameter for Pop II classification, out of the 4 binaries, one (HD 89499) can be considered as an extreme halo star (Ardeberg and Lindgren 1991) while the other three are intermediate Pop II (Latham *et. al.* 1988). The sample stars, together with their physical parameters, are described in table 1. Effective temperatures are computed using the available photometry and the calibrations of Laird *et al.* (1988) and Magain (1987) . Periods, metallicities and distances are taken from Ardeberg and Lindgren (1991), Laird *et al.* (1988), Latham *et al.* (1988).

TABLE I : Parameters of the observed binaries: Distances are expressed in parsecs, Periods in days.

Identifier	V	B-V	M_v	T_{eff}	dist.	Period	Fe/H
HD85091	7.62	0.605	5.81	5600	23	3.39	-1.09
HD89499	8.66	0.72	7	5000	21	5.57	-2.3
BD+30 2130	9.29	0.62	5.37	5500	61	6.57	-0.67
BD+05 3080	9.14	0.835	6.98	4750	27	9.94	-1.11

Optical observations were obtained at ESO, La Silla using the CASPEC spectrograph attached at the Cassegrain focus of the 3.6 m telescope and the CES spectrograph, fed by the 1.4m CAT telescope. The resolving powers, as deduced from the FWHM of the lines of a comparison lamp, were ~ 15000 and ~ 60000 respectively.

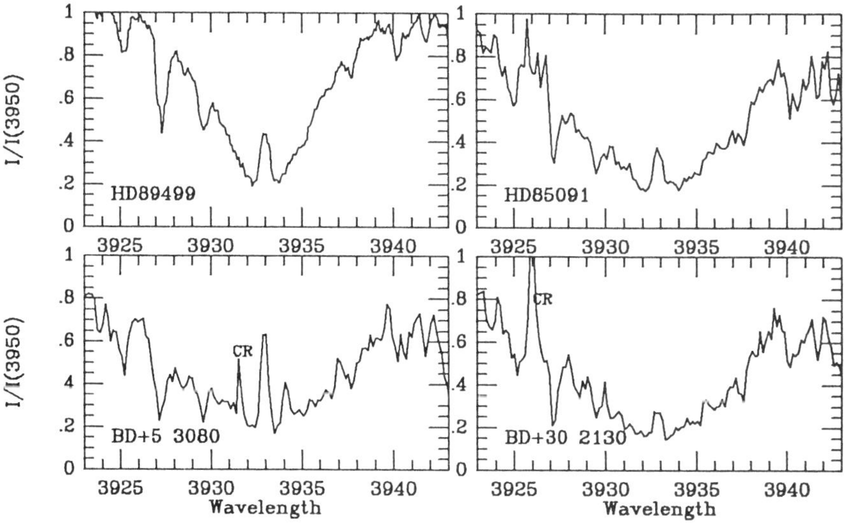

Fig. 1: CASPEC spectra of 4 Pop II binaries in the K line of Ca II. CR labels indicate the detection of high energy particles.

In Figure 1, we present the CASPEC Ca II K line spectra obtained for the sample stars, normalized to the 3950 Å pseudocontinuum (Pasquini *et al.* 1988). All the targets show a central emission, indicating the presence of a rather strong chromosphere.

In order to further investigate and to confirm the nature of the observed chromospheres, we have obtained Hα spectra for 3 of the sample stars. Since Hα even in quite active stars does not appear in emission but only shows a filling of the central core (Pasquini and Pallavicini 1991), we observed three Pop II single stars having parameters as close as possible to the binaries in order to facilitate a comparison. CES spectra of HD 140283 ($T_{eff} \sim$ 5600 K , [Fe/H] $\sim$ -2.5) HD 189558 ($T_{eff} \sim$ 5660 K , [Fe/H] $\sim$ -1.3) and HD 102365 ($T_{eff} \sim$ 5420 K , [Fe/H] $\sim$ -0.70) (Cayrel de Strobel *et al.* 1985) were also obtained. Differences in metallicities and effective temperatures between the binaries and the single stars allow only a qualitative comparison: the residual intensity of the $H\alpha$ core in single stars is only 0.2 or less, while for the binaries it is between 0.4 and 0.8, indicating the presence of an active chromosphere.

X-RAY OBSERVATIONS

As mentioned in the introduction, strong activity is not expected to be present in single Pop II stars and up to now no evidence of X-ray emission from Pop II coronal sources has been found. The ROSAT X-ray satellite (Pfeffermann *et al.* 1990) has performed a soft X-ray survey of almost the entire sky which, for its sensitivity and spatial resolution, is much more accurate than any previous soft X-ray survey. All the four sources of our sample happened to be in parts of the sky already processed at MPE (Garching). Two of them, the brightest and nearest, were detected by ROSAT. The field of HD 85091 was observed by ROSAT for $\sim$ 600 seconds. An X-ray source with a count-rate of 0.12 $\pm$ 0.02 counts/sec was detected by the ROSAT PSPC only 12 arcsec away from the expected stellar position. Since the 3σ error in position for this source is 15 arcsec, the identification of the X-ray source with HD 85091 is virtually certain.

The field of HD 89499 was observed for 301 seconds; an X-ray source in the proximity of the optical position of HD 89499 was detected, for a total of 49 photons, giving a countrate (after vignetting correction) of 0.233 $\pm$ 0.048 counts/sec.

Because of the limited spectral resolution and coverage of X-ray instruments like the PSPC on board ROSAT, the conversion between observed count-rates and X-ray luminosities is not straightforward and requires an assumption regarding the spectral distribution of the photons (see *e.g.* Pallavicini *et al.* 1988). The total number of counts recorded for each source ($\sim$ 50) does not allow us to perform a reasonable fit of the X-ray spectrum and to put firm constraints on the temperature and the emission measure of the emitting plasma. Typically, RS CVn binaries exhibit a two-component temperature distribution in their coronal plasma, one at T = a few $\times 10^6$ K and the other at $10^7 <$ T $< 2 \times 10^7$ K (*e.g.* Mewe *et al.* 1986). PSPC calibration data on AR Lac show components at 3×10^6 and 1.4×10^7 K (Ottmann 1991). Such an X-ray spectrum would imply a PSPC conversion factor of 7.5×10^{-12} ergs cm^{-2} cnt^{-1}. However, it must be said that the PSPC is very sensitive to plasmas around 10^6 K. The conversion factor changes by a factor of three for plasmas between 1 and 5 $\times 10^6$ K. Since HD 89499 and HD 85091 are Pop II analogs of RS CVn binaries, we can not be certain that their coronal temperatures will be as hot as their Pop I cousins. We hope to find out with ROSAT pointed observations of these two stars which are scheduled for the second observing program. Until then, we adopt a conversion factor, as for all ROSAT survey cool stars, of $5 \pm 2 \times 10^{-12}$ ergs cm^{-2} cnt^{-1}, which covers the range of probable conversion factors. Combined with the distance given in Table 1 and the observed count-rate, this implies an X-ray luminosity of $\sim 4 \times 10^{28}$ and $\sim 6 \times 10^{28}$ ergs/sec respectively in the 0.1 - 2.4 Kev ROSAT passband. These luminosities are remarkably similar and they are typical for active, single main sequence Pop I stars, but a couple of orders of magnitude lower than what is found in RS CVn binaries (Majer *et al.* 1986). We caution, however, that this X-ray luminosities are rather uncertain.

DISCUSSION AND CONCLUSIONS

The detection of activity in 4 Pop II binaries opens several interesting topics, not only related to activity, but also to the determination of physical parameters of Pop II stars and the dynamics of globular clusters (Murphy *et al.* 1991).

With respect to stellar activity, the data are not extensive enough to fully understand the nature of the observed chromospheres and coronae, as well as their relation with Pop I analogs. However the presence of rather strong coronae and chromospheres would suggest that we are dealing with magnetic phenomena, in analogy to the Sun. Spectroscopic and photometric studies searching for variability in these objects have not yet been carried out; Ardeberg and Lindgren (1991), however, find evidence for photometric variability in HD 89499. Since the star appears to be very active, we may expect it as due to the transit of inhomogeneities on the stellar surface, like active regions or spots.

The fact that single Pop II stars do not show strong activity explicitly suggests that binarity is a key parameter and, on a reduced scale, the systems presented in this paper recall Pop I active binaries like RS CVns.

Our observations are limited to short period systems; the behaviour of longer period binaries is presently unknown; further observations will show if activity depends on orbital periods, metallicity and other stellar parameters. They may also give some additional constraints on dynamo theories.

The coronae of Pop II binaries, if confirmed, could represent natural laboratories for the study of hot plasma at low metallicities. New observational and theoretical efforts are necessary to better investigate these exciting astrophysical problems.

REFERENCES

Ardeberg A., Lindgren H. 1991, *Astr. Ap.*, **244**, 310.

Cayrel R., Cayrel de Strobel G., Campbell B. 1985, *Astr. Ap.*, **146**, 249.

Dupree A.K., Hartmann L., Smith G.H. 1990, *Ap. J.*, **353**, 623.

Hartman K. & Gehren T. 1988, *Ap. J.*, **199**, 269.

Laird J.B., Carney B.W., Latham D.W. 1988, *A. J.*, **95**, 143.

Latham D.W. *et al.* 1988, *A. J.*, **96**, 567.

Magain P. 1987, *Astr. Ap.*, **181**, 323.

Majer P. *et al.* 1986, *Ap. J.*, **300**, 360.

Mewe R. *et al.* 1986, *Adv. Space Res.*, **8**, No. 6, 133.

Murphy W.B. *et al.* 1991, *Nature* **351**, 130.

Ottmann R. 1991, in preparation.

Pallavicini R. *et al.* 1988, *Astr. Ap.*, **191**, 109.

Parker E.N. 1970, *Ann. Rev. Astr. Ap.*, **10**, 1.

Pasquini L., Pallavicini R., Pakull M. 1988, *Astr. Ap.*, **191**, 266.

Pasquini L., Pallavicini R. 1991, *Astr. Ap.*, in press.

Pfeffermann E. *et al.* 1990, *Proc. SPIE*, **733**, 519.

Seventh Cambridge Workshop on Cool Stars, Stellar Systems, and the Sun
ASP Conference Series, Vol. 26, 1992
Mark S. Giampapa and Jay A. Bookbinder (eds.)

OPTICAL SPECTROSCOPY OF CHROMOSPHERICALLY ACTIVE BINARY SYSTEMS IN CONJUNCTION WITH THE ROSAT ALL-SKY SURVEY

ALAN D. WELTY, JEFFREY C. HALL, ROBERT J. PATTERER, LAWRENCE W. RAMSEY
The Pennsylvania State University, Department of Astronomy and Astrophysics, University Park, PA. 16802

DAVID P. HUENEMOERDER
Jet Propulsion Laboratory. NASA Headquarters, Astrophysics Division, Code SZ, Washington, DC 20546

ABSTRACT We compare optical indicators of chromospheric activity with X–ray luminosities for a sample of active binary systems. Excluding known mass transfer systems, a correlation between Hα and X–ray luminosity exists. Hα/Hβ ratios suggest we are seeing solar quiescent prominence-like material off-limb.

Keywords: Chromospherically active stars; RS CVn systems

INTRODUCTION

In RS CVn binaries, excess emission in the Ca II resonance (H and K) and infrared triplet (IRT) lines and in Hα is generally cited as evidence for chromospheric activity. The activity is usually modeled as scaled up solar-type activity. RS CVn systems are also sources of X–ray emission, which is believed to arise from coronal loop structures (see Rosner, Tucker, & Vaiana 1978). We investigate the connection between processes that produce optical activity indicators and X–rays.

We present results for 39 RS CVn systems whose spectra were obtained with the 1.6 m telescope at Penn State's Black Moshannon Observatory, and with the Coudé Feed telescope of the Kitt Peak National Observatory. The former has $R = 5000$, and the latter $R = 12000$. Gaps in wavelength coverage, and poor signal-to-noise ratio at short wavelengths limit us to Hα and the λ8542Å IRT line, and Hβ for objects observed at KPNO.

All spectra were obtained during the *ROSAT* All-Sky Survey, although few are simultaneous with *ROSAT* observations. These data have not yet been analyzed with the *ROSAT* data; here we compare them with *Einstein* IPC data. While the properties of individual systems may have changed during the decade since *Einstein*, we expect results for the sample as a whole to remain constant.

We have used our technique of spectral subtraction to determine the non-photospheric contributions to activity sensitive lines: Spectra of non-active stars are used to fit the "quiet" regions of an object spectrum. The photospheric component to activity sensitive lines is calculated with the best fit parameters (radial and rotational velocities and relative weights) and subtracted from the object spectrum, yielding the spectrum of the active region(s).

RESULTS

In Figure 1 we show excess Hα luminosity *vs. Einstein* IPC X-ray luminosity. We use X-ray fluxes from Schmitt *et al.* (1991) when available, X-ray luminosities from Strassmeier *et al.* (1988) otherwise. Distances, V magnitudes, and colors (used to convert flux to luminosity) are those reported in Strassmeier *et al.* (1988). Each point represents a single observation of a program object. We estimate the uncertainties in the excess line luminosities to be $\sim 20\%$. Triangles are for mass transfer systems known to us; pentagons are for AR Mon.

Excluding points with $L_x < 2 \times 10^{31}$ erg s^{-1} and for AR Mon (mostly mass transfer systems), a correlation (slope 0.59 ± 0.07) is present, suggesting physical connection between X-ray and excess Hα emission. However, a situation with no underlying correlation, punctuated by chance Hα-bright events is consistent with these data.

Complicating interpretation of the data is the apparent presence of circumstellar material that produces absorption in some subtracted spectra (*e.g.* Hall & Ramsey, this volume; other scenarios are possible). Also, the excess Hα luminosity can be highly variable in a single system, as evident in AR Mon.

Figure 2 shows excess Hα *vs.* excess Hβ luminosity. Solid circles are for objects with $L_x > 2.0 \times 10^{31}$ erg s^{-1}, asterisks for objects with L_x less than that value, and stars represent AR Mon. The presence of extended or distributed material causes some "excess absorption."

Two trends are apparent for points in the first quadrant, least affected by this absorption. First, the segregation of symbol types suggests a correlation between excess Balmer emission and X-ray emission (see Figure 1). Second, the slope, 4.42 ± 0.60 from a χ^2 minimization straight line fit (Press *et al.* 1986), gives a clue to the origin of excess Balmer emission. Buzasi (1989) found that excess Hα/Hβ ratios greater than ~ 3 can be obtained only when viewing material similar to solar quiescent prominences *off-limb*.

Our excess Hα *vs.* excess λ8542Å luminosity data is consistent with that seen in solar active regions and a small sample of the most active RS CVn systems (Chester 1991).

CONCLUSIONS

1.) Based on our $L_{H\alpha}$ *vs.* L_x and $L_{H\alpha}$ *vs.* $L_{H\beta}$ results, it appears that excess Balmer and X-ray emission in RS CVn binaries arises in extended, prominence-like structures. Other types of activity likely are present, and probably are responsible for "deviant" points.

2.) The non-photospheric Hα profiles of $\sim$ 40% of our objects reveal excess absorption (some of these also have emission, *e.g.* P Cygni-like profiles), suggesting extended material (possibilities include: prominence [occulting other star; see Hall & Ramsey, this volume], accretion stream, distributed circumstellar material, *etc.*).

ACKNOWLEDGEMENTS

The authors acknowledge support from NASA grant NAG 5-1435.

REFERENCES

Buzasi, D. L. 1989, Ph. D. Thesis, Pennsylvania State University.
Chester, M. M. 1991, Ph. D. Thesis, Pennsylvania State University.
Schmitt, J.H.M.M. *et al.* 1990, *Ap. J.*, **365**, 704
Strassmeier, K.G. *et al.* 1988, A&ASS, **72**, 291
Press, W.H. *et al.* 1986, "Numerical Recipes," (Cambridge Univ. Press: New York)
Rosner, R., Tucker, W. H., and Vaiana, G. S. 1978, *Ap. J.*, **220**, 643

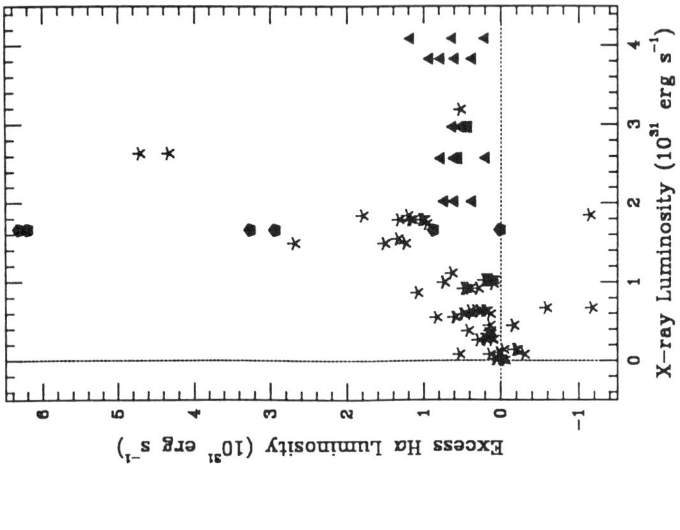

Figure 1. Pentagons = AR Mon.
Triangles = other mass-transfer systems.

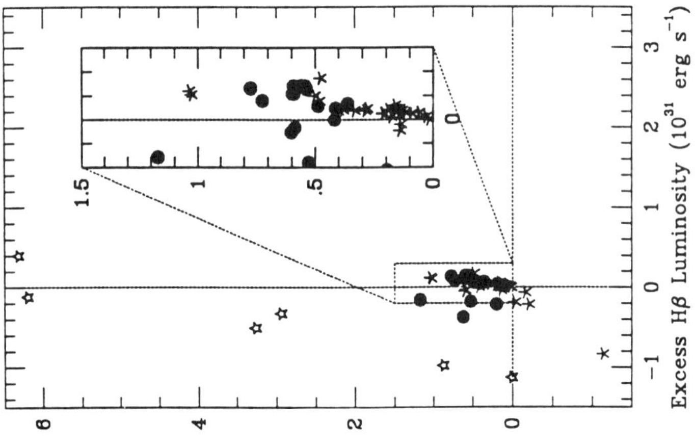

Figure 2. Stars = AR Mon.
Circles = systems with $L_x > 2 \times 10^{31}$ erg s^{-1}.

Seventh Cambridge Workshop on Cool Stars, Stellar Systems, and the Sun
ASP Conference Series, Vol. 26, 1992
Mark S. Giampapa and Jay A. Bookbinder (eds.)

PHOTOMETRY OF SERENDIPITOUS X-ray SOURCES DETECTED BY EXOSAT

GIUSEPPE CUTISPOTO Catania Astrophysical Observatory, viale A.Doria 6, I-95125 Catania, Italy

ROBERTO PALLAVICINI Arcetri Astrophysical Observatory, Largo E.Fermi 5, I-50125 Firenze, Italy

LUCA PASQUINI ESO, La Silla, Casilla 19001, Santiago 19, Chile

MARCELLO RODONÓ Institute of Astronomy, Catania University, viale A.Doria 6, I-95125 Catania, Italy

GIANPIERO TAGLIAFERRI EXOSAT Observatory, SSD-ESTEC, 2200 AG Noordwijk, The Netherlands

ABSTRACT More than one hundred stars have been detected serendipitously as X-ray sources in the energy range 0.05-2 keV by the Low Energy Telescope and Channel Multiplier Array detector on board the EXOSAT satellite (Giommi *et al.* 1991). We discuss the photometry of some of these sources.

Keywords: Cool stars, Photometry, X-ray, RS CVn variables, BY Dra variables.

OBSERVING PROGRAM Cool stars are strong X-ray emitters, their X-ray luminosity increases with rotation velocity, and appears to be largely independent of the bolometric luminosity, thus supporting the hypothesis that dynamo-generated magnetic fields are responsible for coronal heating in stars with sub-photospheric convective zones (Pallavicini *et al.* 1981). Cool stars include about 40% of the full sample of sources detected by EXOSAT, but owing to UV contamination some B and A-type stars have also been found. Bright early type stars cannot be regarded as true serendipitous sources; but for B and A stars fainter than $V \simeq 7$ little or no UV contamination is expected. Although these stars should not be strong X-ray emitters (*e.g.* Pallavicini A&A Rev. 1990), they were included in our list to ascertain whether the X-ray emission is due to a late-type companion. In order to study in detail the nature of EXOSAT serendipitous X-ray sources, we started a campaign of spectroscopic and photometric observations using telescopes at ESO, La Silla (Chile), Catania Observatory (Italy) and San Pedro Martir (Mexico). The first results of this program have been presented by Cutispoto *et al.* (1991) and by Tagliaferri *et al.* (1991). In Table 1 we summarize some results of $UBV(RI)_c$ photometry carried out with the 50cm and 1.0m telescopes at ESO in 1989 and 1990 for 45 of the 60 sources observable from La Silla. Twenty variables and 11 suspected variable were found. In Figure 1 some typical V-band light curves are shown. By means of photometric and spectroscopic criteria we classified most of the detected variables as RS CVn or BY Dra-type systems.

SAO 202618, one of the sources we classified as an RS CVn-type system, showed also eclipses (Figure 1, top panel). Besides RS CVn and BY Dra-type systems, an Ap star (HD 20629) and a PMS object were found to be variable. The variables we observed over different seasons showed light curves with variable amplitude, shape and mean brightness, as is typical for spotted stars. None of the sources was found tr have spectral type later than M5V.

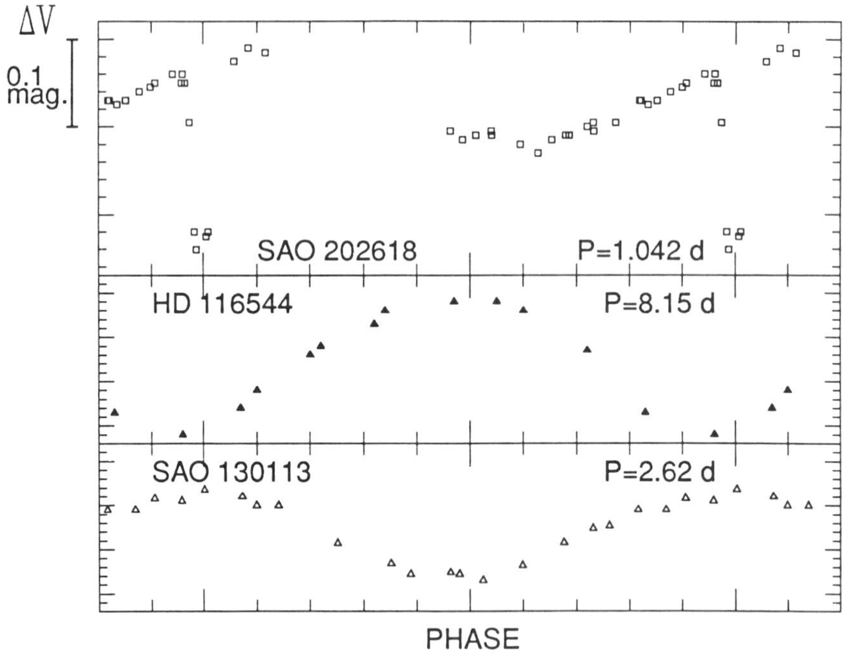

Figure 1: *V*-band light curves for 3 EXOSAT X–ray serendipitous sources found to be optically variable.

REFERENCES

Cutispoto G., Tagliaferri G., Giommi P., Gouiffes C., Pallavicini R., Pasquini L., Rodonó M.: 1991, *A&ASS* **87**, 233.

Giommi P., *et al.* : 1991, *Ap. J.*, **378**, 77.

Pallavicini R., Golub L., Rosner R., Vaiana G.S., Ayres T., Linsky J.L.: 1981, *Ap. J.*, **248**, 279.

Pallavicini R.: 1990, *A&A Rev.*

Tagliaferri G. Cutispoto G., Giommi P., Pallavicini R., Pasquini L., Rodonó M.: 1992, these proceedings.

TABLE 1: EXOSAT serendipitous sources photometrically observed

EXOSAT name	Other name	Sp Type	V	Var	P days	DV	H&K	Type
0012-7307		M2 Ve	11.6	S		0.03		
0109-2613	HD 7172	G3 V?	9.7	Y		0.06	E	
0134+2026	HD 9902	G8IV + F6V	8.69	Y	7.9	0.06	E	RS CVn
0146+0608		M1V + M1V	12.2	Y		0.05		BY Dra
0235-5216		M2 Ve	13.0	Y		0.05		BY Dra
0244-0024	SAO 130113	G9V + K4V	9.63	Y	2.62	0.10	E	RS CVn
0316-1853	HD 20629	A0 SiSrCr	7.4	Y	4.7	0.09		Ap
0336+0929	HD 22670	G4 V	9.1	S		0.02	E	
0408-7134		K7/M0 Ve	12.5	Y	4.5	0.06		BYDra
0425+1735		M3 Ve	14.4					
0508-0414	HD 293857	K1IV + F6V	9.33	Y		0.04	E	RSCVn
0512+0000	HD 34179	B8 V	8.03	N				
0519-4544		K5 V	11.8	Y	6.95	0.10		BYDra
0524-1156	HD 35850	F7 V	6.26	S		0.02	E	
0527-3329		G+K IV/V	10.6	Y		0.08	E	RSCVn
0532-0510		K ?	10.7	Y	5.6	0.20		PMS
0554-3806	HD 40216	F7 V	7.46	N			E	
0556-3804		G8IV + F7V	9.57	Y	3.72	0.09	E	RSCVn
0635-7535		M2V + M3V	10.09	N				
0740+0336	HD 62323	F8V + GV	7.0	N			E	
0848+1228		KV + KV	11.8	Y	2.69	0.08		BYDra
0851+2025	SAO 80493	F6 V	8.6	N			E	
0905-0938	HD 78423	A0/1 V	7.85	N			N	
1122-6122	HD 99279	K5V + M0V	7.23	N				
1133-3745	SAO 202618	K IV + G	10.58	Y	1.042	0.12	E	RSCVn
1219+0213		K7/M0 Ve	12.9	Y		0.03		BYDra
1229+2039		M3 Ve ?	12.9	S		0.02	N	
1259+1238		M3 Ve	12.2	S		0.02	N	
1321-0203	HD 116544	K5V + KV	9.24	Y	8.15	0.15	E	RSCVn
1330-0811	HD 117860	G0 V	7.36	N			E	
1615+0624		M3 Ve	12.7	S		0.02	N	
1621-4901	HD 147633	G2IV + G0V	7.38	S		0.01	E	
1703-3745	HD 154338	G5/6 V ?	9.7	Y	3.27	0.09	E	RSCVn
1727-3430	HD 158394	G8IV + A4	8.2	Y	≥16	0.07	E	?
1735-4404	HD 159808	F6 V	7.9	N				
1808-6456		O5 + G3 V	12.5	S		0.01	N	
1818-1601		B7 II	9.2	N				
1818-1602		B	9.8	N				
1827-6517	HD 170196	B9 III	7.8	N				
1854+0116	BD+013828	G8IV+K2/7V	9.4	S			E	
2041-3129		M4 V	15.3	Y		0.20	N	BYDra
2205-4725		K3/5 IV/V	12.5	S		0.01		
2214-0903	GL 852AB	M4Ve+M5Ve	13.1	Y		0.13		BYDra
2306+0424	HD 218515	A0V	7.6	N				

Legenda Var: optical variability, Y=Yes, S=Suspected, N=No
DV : Maximum amplitude observed in the V band
H&K: Ca II H and K lines, E=Emission, N=Not observed spectroscopically

Seventh Cambridge Workshop on Cool Stars, Stellar Systems, and the Sun
ASP Conference Series, Vol. 26, 1992
Mark S. Giampapa and Jay A. Bookbinder (eds.)

OPTICAL SPECTROSCOPY OF COOL STARS DETECTED BY EXOSAT

GIANPIERO TAGLIAFERRI EXOSAT Observatory, SSD-ESTEC, 2200 AG Noordwijk, The Netherlands

GIUSEPPE CUTISPOTO Catania Astrophysical Observatory, v.le A.Doria,6 I-95125 Catania, Italy

PAOLO GIOMMI European Space Information System, ESA/ESRIN, Frascati, Italy

ROBERTO PALLAVICINI Osservatorio Astrofisico di Arcetri, Largo E.Fermi,5 I-50125 Firenze, Italy

LUCA PASQUINI ESO, La Silla, Casilla 19001, Santiago 19, Chile

MARCELLO RODONÒ Institute of Astronomy, Catania University, v.le A.Doria,6 I-95125 Catania, Italy

ABSTRACT Optical spectroscopy of southern serendipitous X–ray stars detected by EXOSAT are presented. We detected the Hα line in emission or filled-in in more than 50% of our stars. Strong Ca II H&K emission lines were detected in almost all sources. We also detected the 6707 Å Lithium line in 17 out of the 26 observed. In particular for 10 sources this line was stronger than the nearby Ca I 6717 Å line; a clear indication of youth.

Keywords: Cool stars; X–ray; Spectroscopy; Lithium

INTRODUCTION

X–ray observations are a powerful tool for studying stellar surface activity, its variation with stellar evolution and the determination of the stellar contribution to the diffuse soft X–ray background (*e.g.* Rosner *et al.* 1985; Pallavicini 1989). By studying samples of sources serendipitously detected in the X–ray band, it is possible to investigate the high tail of the stellar X–ray luminosity function and the highest levels of activity among stars (Caillault *et al.* 1986; Favata *et al.* 1988; Fleming *et al.* 1988, 1989). The ROSAT full sky survey, in conjunction with optical studies, will fully exploit the potentials of this approach (*e.g.* Schmitt 1991).

About one hundred late-type stars have been detected with the Low Energy Telescope and CMA detector on board the EXOSAT satellite (Giommi *et al.* 1991). This sample includes many variable stars; to date, three new CVs (Beuermann *et al.* 1987, 1989; Osborne *et al.* 1988), a flaring PMS star

(Tagliaferri *et al.* 1988) and a few flaring M dwarfs have been identified. Among the latter is the faint M dwarf VB 8 (=GL 644C, $m_v \sim 17$), one of the coolest stars (Sp=dM7) so far detected at X–ray wavelengths (Tagliaferri *et al.* 1989). The nature of most stellar EXOSAT sources remains, however, to be determined.

RESULTS

In order to study the nature of X–ray selected cool stars in the EXOSAT sample, we started a campaign of optical observations. In March 1990 we obtained, at the European Southern Observatory at La Silla, Chile, moderate-resolution spectra both in Hα and in the Ca II H&K region, and high-resolution spectra of the Ca II H&K lines and of the Li I (6707 Å) line. UBVRI photometric observations were also obtained (Cutispoto *et al.* 1991). The Hα line was detected in emission or filled-in in more than 50% of our stars (here we exclude the dMe stars which by definition have Hα in emission and constitute $\sim$ 40% of our sample), providing a sure indication of extremely high chromospheric activity. Strong Ca II H&K emission was also detected in almost all our sources.

Among PMS the Post T-Tauri stars (PTTS) are particularly interesting. The T-Tauri phase occupies only a small fraction ($\approx 5 - 10\%$) of the contraction time of a 1 $M_\odot$ star towards the main sequence Herbig (1978). As a consequence,

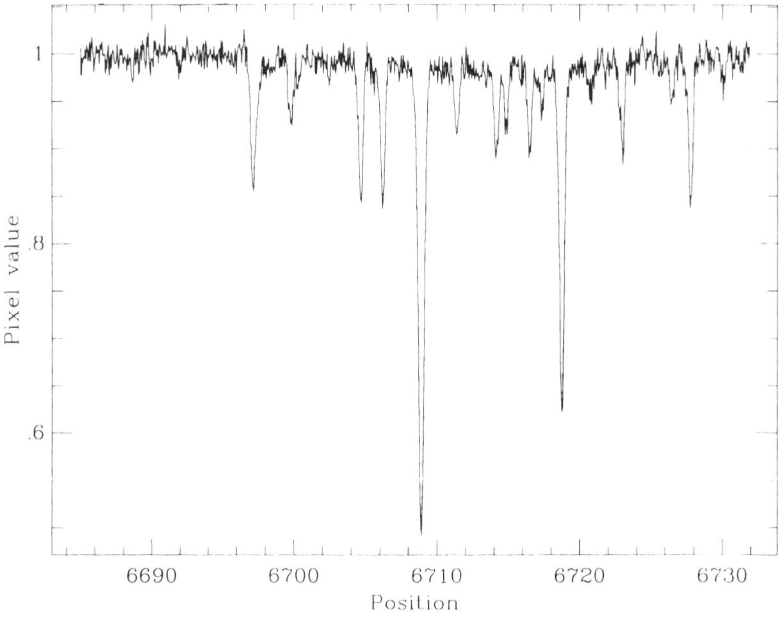

Fig. 1: High resolution spectra of an EXOSAT serendipitous star in the Li I 6707 Å line region. Note that the Li line is stronger than the nearby Ca I 6717 Å line.

many more PTTS should exist than classical T-Tauri stars. But so far, only a handful of *bona fide* PTTS have been identified outside star-forming regions (see reviews by Herbig (1978) and Haro (1983)). However PTTS are expected to be vigorous X-ray coronal sources and X-ray surveys are the best way to identify them. One PTTS (HD 560B) has already been detected in the EXOSAT sample (Tagliaferri *et al.* 1988). Some of these elusive objects, whose study can provide crucial information on the evolution of stellar coronal activity during the pre-main sequence phase, could be present in our sample. To search for PTTS we took high resolution spectra of the 6707 Å lithium line region. The line was clearly detected in 17 stars out of the 26 observed. All these stars are of spectral type later than G. In particular for 10 sources this line was stronger than the nearby CaI 6717 Å line (fig. 1); a clear indication of youth. Unless extremely strong, the Li line, in fact, is not sufficient *per se* to classify a star as a young object, since this line has recently been detected in some RS CVn binaries (Pallavicini *et al.* 1990), which are evolved systems.

The high resolution spectra in the lithium region also revealed that many stars in our sample are very fast rotator, with a v*sini* between 15 and 50 Km s^{-1} (these are preliminary estimates from comparison with spectra of template stars obtained with the same telescope configuration).

In summary, our optical data show that a large fraction of the EXOSAT serendipitous sources are very active stars, a number of which are especially interesting objects such as pre-main sequence stars, RS CVn systems and BY Draconis type stars. Further study of this sample will provide important information on stellar coronal activity and its evolution.

REFERENCES

Beuermann, K., Thomas, H., Giommi, P., Tagliaferri, G.: 1987, *Astr. Ap.* **175**, L75

Beuermann, K., Thomas, H., Giommi, P., Tagliaferri, G., Schwope, A.: 1989, *Astr. Ap.* **219**, L7

Caillault, J.-P., Helfand, D.J., Nousek, J.A., Takalo, L.O.: 1986, *Ap.J.* **304**, 318

Cutispoto G., Tagliaferri G., Rodonò M., Giommi P., Pallavicini R., Pasquini L.: 1991, this conference

Favata, F., Rosner, R., Sciortino, S., Vaiana, G.S.: 1988, *Ap.J.* **324**, 1010

Fleming, T.A., Gioia, I.M., Maccacaro, T.: 1989, *Ap. J.*, **340**, 1011

Fleming, T.A., Liebert, J., Gioia, I.M., Maccacaro, T.: 1988, *Ap.J.* **331**,958

Giommi, P., Tagliaferri, G., *et al.* : 1991 *Ap.J.* **378**, 77

Haro, G.: 1983, . *Rev. Mexicana Astr. Ap.* **7**, 183

Herbig, G.H.: 1978, in *Problems of Physics and the Evolution of the Universe*, ed. L.V. Mirzoyan (Yerevan: Armenia Academy of Science), p.171

Osborne, J., Giommi, P., Angelini, L., Tagliaferri, G., Stella, L.: 1988, *Ap.J.* **328**, L45

Pallavicini, R.: 1989, *Astr. Ap. Rev.* **1**, 177

Pallavicini, R., Randich, S., Giampapa, M., Cutispoto, G.: 1990, *The Messenger*, n. **62**, p. 51

Rosner, R., Golub, L., Vaiana, G.S.: 1985, *Ann. Rev. Astr. Ap.* **23**, 413

Schmitt, J.H.H.M.: 1991, this conference

Tagliaferri, G., Giommi, P., Angelini, L., Osborne, J.P., Pallavicini, R.: 1988, *Ap.J.* **331**, L113

Tagliaferri, G., Doyle, J.G., Giommi, P.: 1990, *Astr. Ap.* **231**, 131

PART III

Convection

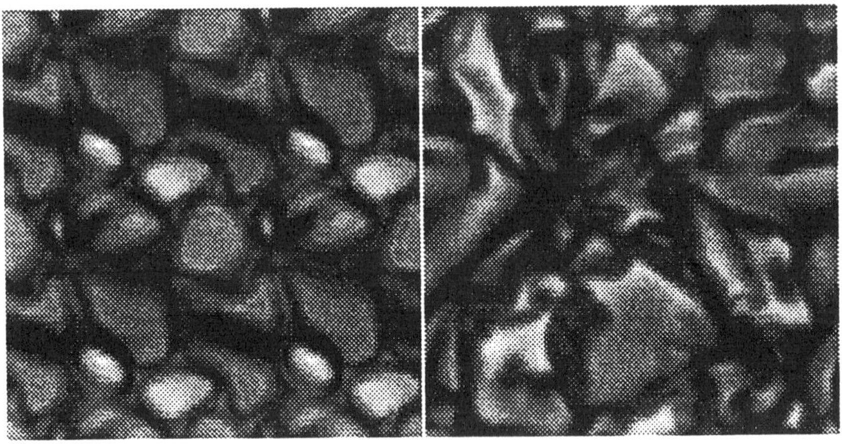

Intensity (MHD) Intensity (HD)

Figure 15 from "Magneto-Convection" by R.F. Stein, A. Brandenburg, and Å. Nordlund.

Seventh Cambridge Workshop on Cool Stars, Stellar Systems, and the Sun
ASP Conference Series, Vol. 26, 1992
Mark S. Giampapa and Jay A. Bookbinder (eds.)

STELLAR CONVECTION: THE OBSERVATIONS

DAVID F. GRAY Department of Astronomy, University of Western Ontario, London, Ontario N6A 3K7, Canada

ABSTRACT Convection is a major factor in the life of stars. We can observe the top of the convective envelope through the granulation in the photosphere and with the macroturbulence dispersion. Both show increasing strength with increasing effective temperature and luminosity. It is not known if the large velocities inferred for stars on the hot side of the granulation boundary arise from convection or other wise. Temperature inversions are seen to grow rapidly in strength as stars evolve across the H-R diagram between the granulation boundary and the rotation boundary. Possibly convection alone drives this activity. Changes in temperature and granulation are seen on time scales of stellar cycles.

Keywords: Convection; Stellar Cycles

WHY ARE WE INTERESTED IN STELLAR CONVECTION?

Convection affects or controls a remarkably wide range of stellar phenomena. Numerous observations enter the picture, but none give definitive information on the characteristics of convection. Consider the following 'chores' performed by convection.

- convection carries thermal power through the convection zone.
- convection mixes the material, countering settling and diffusion processes.
- convective envelopes cycle material down into hot layers resulting in the destruction of Li and Be.
- convection affects, probably controls, the internal distribution of angular momentum within the envelope.
- convection combines with rotation resulting in dynamo action.
- convection probably winds up magnetic fields above the photo-sphere, resulting in dissipation of magnetic energy into heat.
- convection generates acoustic waves.
- convection generates Alfvén waves.

These last four points are intimately connected with the existence of temperature inversions seen as chromospheres and coronae. In addition,

- convection dominates the shaping of spectral-lines in cool stars.

Further, inhibited convection makes starspots dark, and enhanced convection may be the cause of starpatches. Convection-induced motions at the surface may "comb" the magnetic field into network structures.

No surprise then that we need to know everything we can about convection. And indeed each of the phenomena on the list potentially supplies us with information about convection.

HOW CAN WE FIND OUT ABOUT STELLAR CONVECTION?

We can do hydrodynamical calculations, pushing the best physics we can through the fastest computers. Examples of this kind of work can be seen in papers by Nordlund (1985), Fox & Van der Borght (1985), Stein *et al.* (1989), Steffen *et al.* (1989), Spruit *et al.* (1990), Nordlund & Dravins (1990), and others, especially in the proceedings of the granulation workshop (Rutten and Severino 1989). And some real progress is being made on this front. In particular, we gain insight into what happens below the surface where we cannot see. These topics are the subject of the paper given next by Cattaneo and Stein. The other thing we can do is look at real stars. Observations come in from a number of different directions, but I will concentrate on the surface phenomenon of granulation because in granulation we see directly the top-most layers of the convection zone. Other lines of evidence will be apparent as they come up in this talk. So how can we measure stellar granulation? The most effective way has been to look at the asymmetries of spectral lines. (Systematic differences in radial velocity with excitation potential has proved to be less useful, ref. Glebocki & Stawikowski 1980, Dravins *et al.* 1981). Since the discovery of asymmetries of spectral lines (Gray 1981b, 1982a), several methods of specifying the asymmetry have been tried, but the best one is the line bisector. Line bisectors are constructed by connecting the midpoints of horizontal line segments running between the sides of the line profile. Figure 1 shows this process. A symmetrical profile has a straight vertical line as its bisector, and we see in the example of Figure 1, in the left-hand portion, a bisector that appears to be nearly straight and vertical. Asymmetry of lines are small and requires careful measurement. Signal-to-noise ratios of several hundred are needed along with spectral resolving power $\lambda/\Delta\lambda \approx$ 100,000 or more. Specifically, resolving power $\lesssim$ 50,000 is quite inadequate and results in the asymmetry being lost in the smearing of the instrumental profile.

It is necessary to expand the velocity (wavelength) scale in plots like Figure 1 if we are to see what bisectors have to show us. Expansions of 20-50 times the scale used for line profiles themselves are typical, as shown in Figure 1, on the right. The C shape we see in this example is characteristic of stars on the cool side of the H-R diagram, a point we will return to in the next section. How does granulation cause asymmetries in lines? It does so quite directly because the distribution of its Doppler shifts is the dominant broadener of the spectral lines we are dealing with. That distribution is not symmetric because more photons come from the hotter rising convection cells than come from the cooler material falling back down between the cells. That brings us to the next step.

WHAT DO LINE BISECTORS LOOK LIKE?

Most spectral lines are blended with others to a lesser or greater degree, and so it is important to compare the results of several lines, weeding out those that are strongly blended and therefore unsuited for the task at hand. For those that

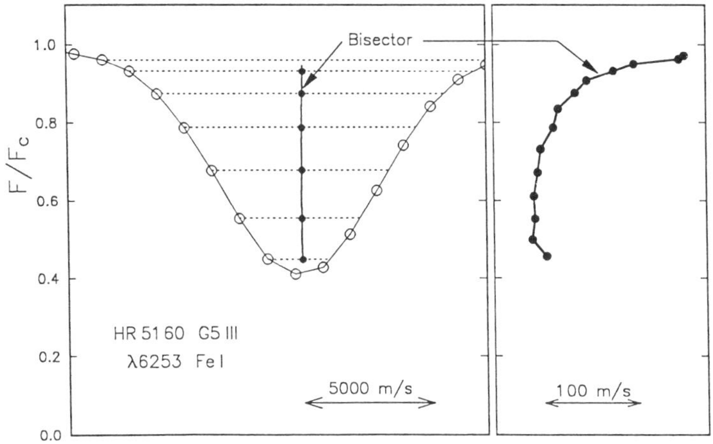

Fig. 1: The bisector is constructed by connecting the midpoints (solid dots) of the dashed horizontal lines. A typical spectral line is very nearly symmetric, and the wavelength (velocity) scale has to be expanded by a factor of 10-50 (note the velocity scales indicated) to see the shape of the bisector. The profile has been interpolated to give the extra points on the expanded version of the bisector on the right.

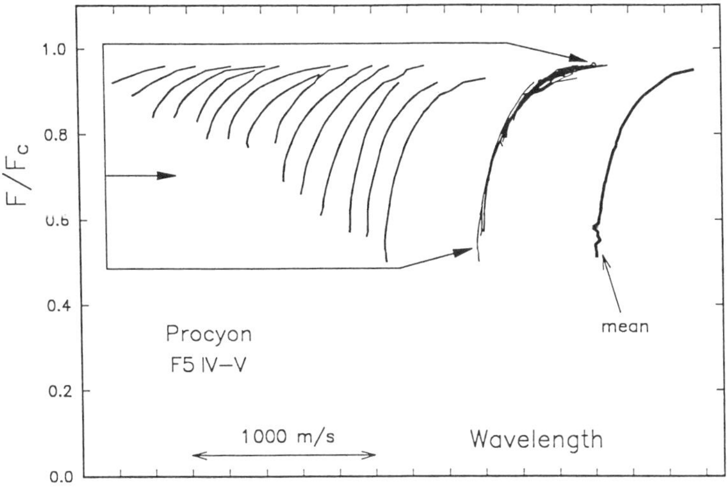

Fig. 2: Line bisectors for lines of different strength are shown on the left with those from the weakest lines left-most. Telescoping these together produces the composite shown in the middle, and the mean of the composite is shown on the right.

are usable, one finds that lines of different strength map out parts of the same bisector within the errors of observation. Consequently, several lines can be put together to form a mean bisector for a star. An example is reproduced here in Figure 2. This makes sense when we reflect on the point that the wings of strong lines and all of weak lines are formed at essentially the same physical depth and should experience the same velocities. Bear in mind, lines formed at distinctly different depths can be expected to have different bisector, *i.e.*, lines having large differences in excitation potential, or those formed in different spectral regions where the continuous opacity is significantly different, because the granulation contrast and velocities vary with depth. Figure 3 shows some typical examples of average line bisectors. A useful way to understand the general "C" shape and the variations they illustrate is as follows. The larger the velocities of granulation and the larger the brightness difference between rising and falling material, the larger the line asymmetry and the larger span in velocity over which the bisector extends. So the velocity span is a measure of how vigorous the granulation is. Because the larger portion of photons comes from the rising granules, the characteristic mean or centroid of the distribution of Doppler shifts is shifted toward shorter wavelengths (which I will refer to as "blue" shifts), and consequently, so is the spectral line it shapes. These displacements can amount to up to 500 m/s, and could be quite significant as systematic errors in precision measurements of radial velocities. We can measure these net line shifts directly and unambiguously in the solar spectrum, but they must be assumed or measured indirectly in stellar spectra because the absolute zero is lost owing to the motion of the star in space. Next consider the displacement of the bottom of the bisector, that part that comes from the core of the line. The core of a strong line is formed higher in the photosphere than the rest of the line. If the granulation penetrates to the height where the core is formed, the bottom of the bisector is blue shifted. If the granulation dies off below that height, the bottom of the bisector is near zero velocity. In Figure 3, Procyon shows strong granulation high up in its photosphere. In contrast, the cooler dwarfs show none. The maximum velocity span goes hand-in-hand with the behavior of the bottom of the line bisector, both features telling different aspects of the same story. Rough estimates of the mean granule velocities are $\approx$ 1.5 km/s (Gray & Toner 1985, Gray 1986) for star similar to the sun. But the strength of convection does depends on the position of the star in the H-R diagram.

HOW DOES CONVECTION CHANGE ACROSS THE H-R DIAGRAM?

We have relatively little information so far. Generally the strength or "vigor" of the granulation increases with effective temperature and with luminosity on the cool half of the H-R diagram. Some general scaling laws are available (Gray 1982), but we urgently need a more complete and a more detailed grid of "standard" bisectors for theoretical interpretation and against which deviant behavior can be compared. Macroturbulence dispersion ζ_{RT} also increases with effective temperature and luminosity (ref. Gray 1988). It would be hard not to suspect that the main component of macroturbulence is granulation. Look at Figure 4, for example. Here the bisector velocity span characterizing line asymmetries of dwarfs is plotted as a function of the macroturbulence dispersion.

Going right along with this pattern of variation are the velocities found in mixing-length calculation (*e.g.*, de Loore 1970, Renzini *et al.* 1977).

A remarkable discovery (Gray & Toner 1986, Gray and Nagel 1989) showed that the nature of photospheric velocity fields changed rather dramatically across a nearly vertical line in the H-R diagram called the "granulation boundary."

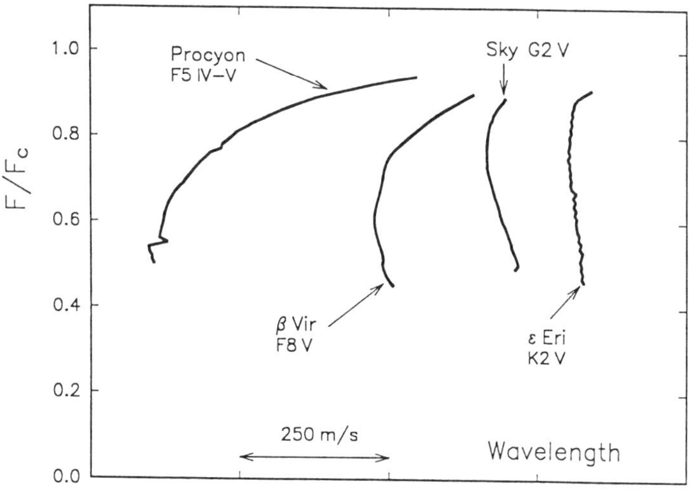

Fig. 3: Mean bisectors are shown for dwarfs having a range in spectral type. Notice the increase in velocity span and the increasing blueward displacement of the lower portion of the bisector with increasing effective temperature.

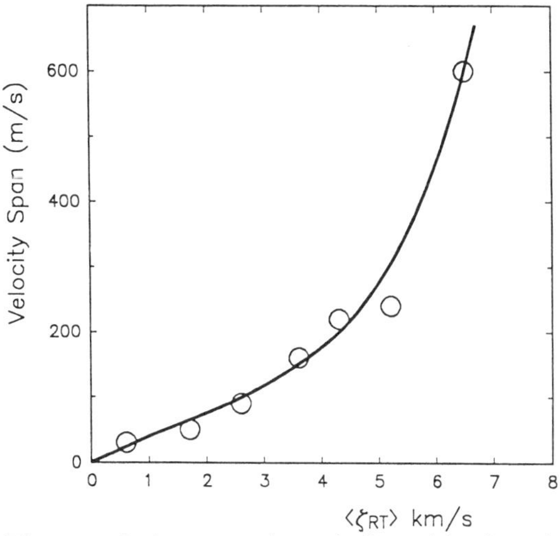

Fig. 4: The granulation strength, as indicated in the velocity span, is closely related to the Doppler broadening of macroturbulence parameterized here by the radial-tangential dispersion. These data are for dwarfs.

Granulation is expected on the cool side of this boundary since these stars have convective envelopes. But "hot" stars are not generally thought of as having convective envelopes. With no convection, we expected little Doppler broadening, and yet velocities inferred from the spectral-line asymmetries are several times larger than for the cool stars. This boundary is truly the separator of "hot" stars from "cool" ones. It runs from G1 Ib down to about F0 on the main sequence. It may be that stars on the hot side of the granulation boundary do have shallow convection zones, but even so, the velocity fields are very different from granulation as we see it in the cooler stars. Not only are the apparent velocities roughly one order of magnitude larger, but the fraction of photons coming from rising material is tiny, running at about 10%. Equally significant is the fact that each line, no matter what its strength, shows the same shape bisector, unlike the behavior in Figure 2 where the bisectors could be telescoped together. The implication is that there is no visible depth dependence in the velocity field. An explanation invoking gravity waves has been offered (de Jager 1990). So, is convection involved here, or should reversed bisectors be discussed in some other talk?

HOW INTIMATE ARE CONVECTION AND MAGNETIC FIELDS?

We all know that rotation and convection are supposed to get together in a process called a dynamo to generate magnetic fields. So perhaps we are not surprised to find that chromospheric activity rises sharply across the granulation boundary. An example of this is shown in Figure 5. Here the carbon emission for class III giants is shown as a function of spectral type (which is essentially a non-linear but monotonic time coordinate since these stars are evolving across the H-R diagram). The position of the granulation boundary is shown by the "GB" at the top of the graphs. The depth of the convection zone grows monotonically after a giant crosses the granulation boundary; one might naturally attribute the growth of the carbon emission to this factor. A short interval later, this chromospheric emission nearly dies out because the rotation drops to a fraction of its previous value as it crosses the rotation boundary (Gray 1981, 1982, 1988, 1989, 1991), labelled "RB" in Figure 5. In a nutshell, add convection to the envelope of a rotating star and you get a chromosphere; add deeper convection, and the chromosphere gets stronger. But take away rotation, and the chromosphere all but disappears. So it behaves just as we would expect for a dynamo. But what is actually happening? I have argued in the references cited in the last paragraph that the rotation boundary is the result of a turn-on of a magnetic brake; the brake resulting from the convection zone having grown deep enough to combine with the rotation to turn on the dynamo. Calculated Rossby/dynamo numbers support this explanation. So also does the observed change in distribution of rotation rates and the rate of decline in rotation seen on the cool side of the rotation boundary. But if dynamos do not turn on until stars reach the rotation boundary, then why do we see magnetic activity between the granulation boundary and the rotation boundary? Although we might think of alternative explanations, might it be that the magnetic field between the boundaries is not a result of dynamo action at all. Could the very appearance of granulation give structure to the photosphere and "comb" an already existing field into something coherent and capable of driving a temperature inversion?

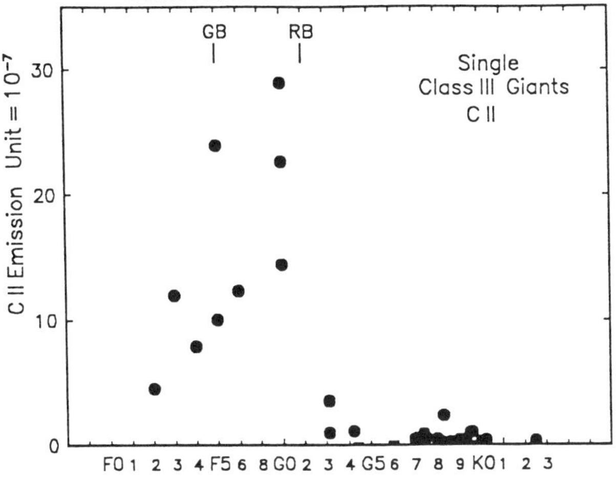

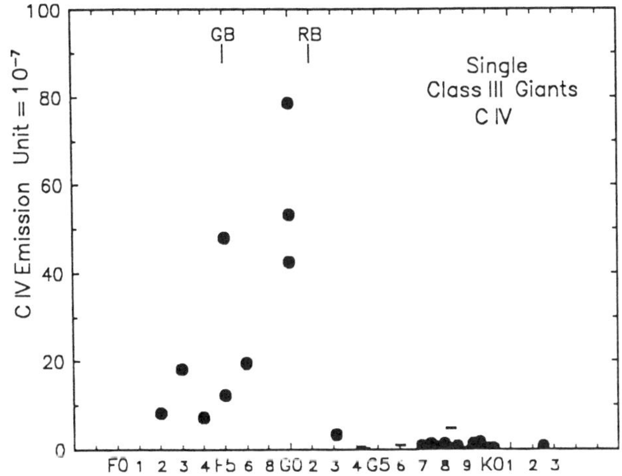

Fig. 5: Non-thermal emission increases rapidly as stars giants evolve across the H-R diagram between the granulation boundary (GB) and the rotation boundary (RB). On the cool side of the rotation boundary, the emission is greatly reduced. Data are from Simon and Drake (1989).

DOES CONVECTION CHANGE DURING A STELLAR CYCLE?

The magnetic field certainly changes during stellar cycles, and we know the magnetic field can alter convection: witness sunspots. There is also clear evidence of granulation being sensitivity to magnetic fields in the solar photosphere (Livingston 1982, 1991, Macri *et al.* 1984, Cavallini *et al.* 1985, Immerschitt & Schrter 1989, Brandt & Solanki 1990). In all cases seen on the sun, granulation seems to be inhibited by the presence of magnetic field. In a spot, one can see how the very large field simply ties up the convective flow trying to cross it. But why is granulation inhibited by much weaker fields? And is it universally true that magnetic fields can never enhance convection?

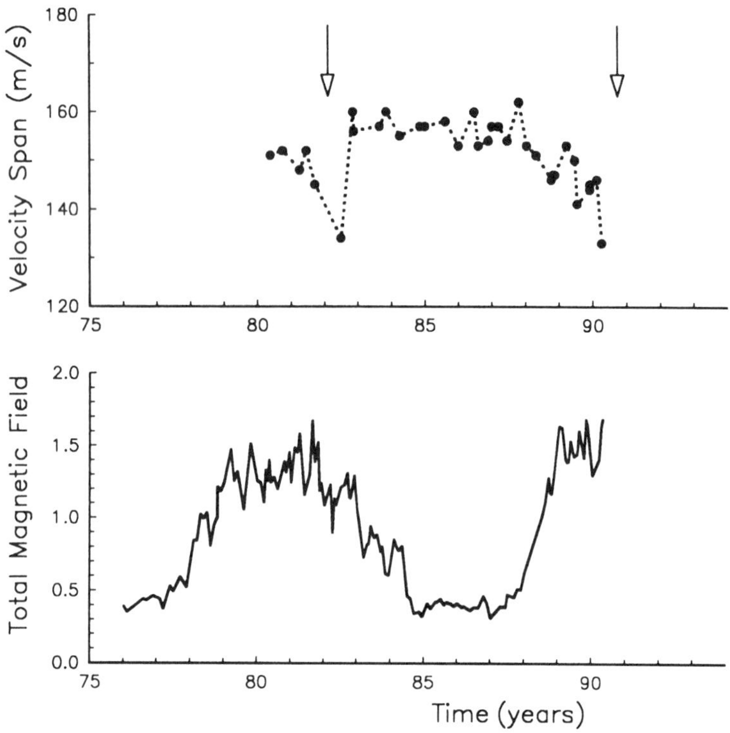

Fig. 6: On top, the solar disk-integrated spectrum show reduced granulation during part of the solar cycle strong strong magnetic activity as indicated by the relative magnetic flux in the lower graph. (Based on data in Livingston 1989.)

Convection carries most of the energy through the envelope. Does the magnetic cycle modulate this flow? Or to state it another way, does the power output vary during a magnetic cycle? And if we were to see variations in the power output, could we muster evidence that the magnetic fields, by altering convection, are the cause of the variation rather than a parametric variable? In Figure 6, you see Livingston's observations of the variation in the strength of solar granulation. Here the velocity span, measured from the bottom of the bisector to the maximum blueward displacement, is plotted as a function of time. It is a strange behavior, not varying in step with the magnetic variation, but showing a more abrupt change over a restricted phase during the time of strong magnetic flux, but not centered on it. The solar power output is slightly reduced during times of stronger magnetic field, presumably because of increased numbers of sunspots (Willson & Hudson 1988, Wolff & Hickey 1987). The sun is a rather weak magnetic star, and so we naturally ask what other stars show us. I want to concentrate on two parameters for stars. The first is a temperature index. I take the ratio of line depths for two lines showing different sensitivity to temperature,

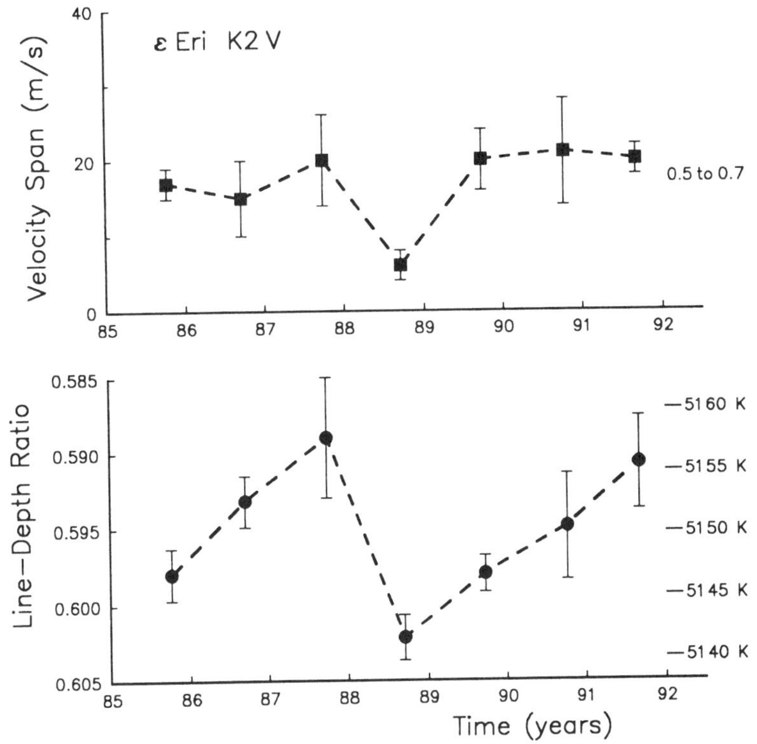

Fig. 7: Granulation and temperature both vary during the 4.15-year cycle of this dwarf (ϵ Eri = HR 10).

in this case, V I λ6261.83 to Fe I λ6252.57. The suitability and sensitivity of this line-depth ratio as a thermometer is discussed elsewhere (Gray & Johanson 1991). Suffice it to say here that a single high-quality exposure allows a star's temperature to be fixed to about $\pm$ 10 K, and combining several exposures together allows the errors to be pushed down to $\approx$ 1-2 K. The second parameter is the granulation strength indicated by the velocity span of the mean line bisector for the star. These particular velocity spans are measured between 0.5 and 0.7 of the continuum, and their yearly means typically have random errors of $\approx \pm$5-10 m/s. First, consider the active K2 dwarf ϵ Eri. The parameters are shown as a function of year in Figure 7. One always wishes the time span were longer and the number of observations were larger. The size of the error bars is largely determined by the number of exposures entering the mean (some 36 exposures over the time span shown). The errors are close to what is expected from the signal-to-noise ratios of the original data. The apparent temperature (temperature scale is on the right) shows a range of $\pm$10 K with a steep drop in the 1988 to 1989 interval and relatively more shallow rise on either side. I say apparent temperature because at the level of a few degrees I can't be sure that

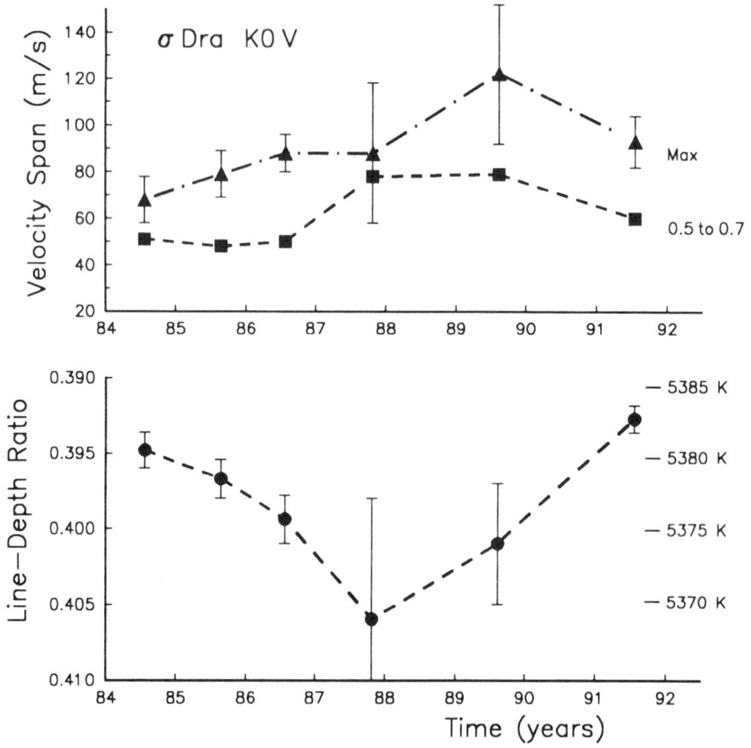

Fig. 8: Both temperature and granulation changes are indicated in σ Dra = HR 7462 = HD 185144 during a cycle of 15 years.

changes in atmospheric structure aren't affecting the line-depth ratio. On the other hand, the flux near 5560Å is approximately proportional to the effective temperature for stars like these, and so we would be talking about a change of about 2% or 0.02 magnitudes. This is not far from the kinds of variations seen in cool stars (Lockwood and Skiff 1988, Radick *et al.* 1990). Phasing of the two linear sections in the figure indicates a cycle period of 4.15 years. The bisector velocity span however is nearly constant except during the small interval near 1989 where a sudden reduction occurs. This is reminiscent of the solar behavior seen in the previous figure. But what is the phase of the magnetic field over this interval? One measure of the chromospheric activity is the one-Å-band equivalent-width measurements of the Ca II λ 8662 published by Campbell *et al.* (1988). These data cover the interval from 1981 to 1987, and they show the same general pattern of a slow build up in chromospheric activity interspersed with relatively short intervals of more rapid decline. There is an increase in chromospheric activity from 1984 right through mid-1987 where the published data end. Looking some 4.15 years earlier in their data, one sees a gap of a full year, and consequently in is not possible to be certain of the relative phasing.

Still, one is tempted to associate the general increase in chromospheric activity shown in their data from 1984 to 1987 with the general increase in temperature seen in this time interval in Figure 7. That would place the maximum magnetic field near 1988.0 at the time of maximum temperature. Immediately thereafter, the temperature dropped and the granulation showed reduced vigor. Can we presume the chromospheric activity and magnetic field also declined at this time? If so, that places the reduced granulation at the phase of magnetic minimum, just opposite the solar behavior. Next, consider the results shown in Figure 8 for σ Dra (K0 V). This stars is generally classes as an "old" star from its H&K line emission, but it is near the boundary between the "old" and "young" categories. A cycle-type variation is seen in both parameters, indicating a period of something like 15 years. The velocity span ranges over tens of meters per second, much larger than for ϵ Eri, but the apparent temperature variation is probably only half as much. Here the granulation is strongest when the temperature is least, just opposite the result for ϵ Eri. There may be a hint of a phase difference between the maximum in granulation strength and the minimum in temperature, but the data are not good enough to be sure. The Campbell *et al.* data show a reduced chromospheric index in 1984 when the temperature is larger. A gradual increase in activity index from late 1984 to 1987 goes with the decline in temperature and the rise in granulation over this same interval. So e Dra may be showing stronger magnetic activity going hand-in-hand with stronger granulation. Puzzling. It was some consolation to my confusion to find that the photometry of stellar cycles (Radick *et al.* 1990) also shows a wide variety of behavior, and in particular, the correlation between photometric brightness and H&K emission reverses sign between "young" and "old" dwarfs. How this ties in with the temperature and granulation changes has yet to be determined.

CONCLUSION AND SPECULATION

We can observe the effects of convection in several different ways. The direct observations of stellar granulation is particularly powerful and is revealing some very basic information not available from solar studies. Understanding what is happening between the granulation and rotation boundaries may open our minds to a different type of magnetic activity, possibly one depending exclusively on convection. Changes in convection apparently occur during the course of stellar cycles. Some interesting physics will have to be worked our to explain it. Does the magnetic field modulate the convective flow of power through the convective envelope, with starspots acting as thermal resistance? Or do magnetic flux tubes physically displace the convective flows inside the envelope enough to lessen the power carried to the surface? Separating cause from effect will be a real challenge. Does reduced convection alter the magnetic field or does the field force reduced convection? Or could it be both? For example, looking at Figures 6, one might wonder if the magnetic field continues to build to a level at which it finally disrupts the convection. The "reduced" convection may then supply too little helicity for regeneration of the field, and that cycle of the dynamo is consequently terminated and winds down. Convection winds up and twists magnetic field high above the photosphere leading to re-connection and flares. With an active dynamo, the internal magnetic field would be substantially stronger than those outside the surface. When convection garbles the internal field, there should be internal "flares." Could these be a significant source of heat at certain restricted phases of a magnetic cycle?

REFERENCES

Brandt, P.N. and S.K. Solanki 1990. *Astr. Ap.*, **231**, 221.

Campbell, B., G.A.H. Walker, and S. Yang 1988. *Ap. J.*, **331**, 902.

Cavallini, F. G. Ceppatelli, and A. Righini 1985. *Astr. Ap.*, **143**, 116.

de Jager, C. 1990. *Solar Phys.* **126**, 201.

de Loore, C. 1970. *Astrophys. Space Sci.* **6**, 60.

Dravins, D., L. Lindegren, and Nordlund 1981. *Astr. Ap.*, **96**, 345.

Fox, P. and R. Van der Borght 1985. *Proc. Astron. Soc. Australia* **6**, 60.

Glebocki, R. and A. Stawikowski 1980. Stellar Turbulence, IAU Coll. 51, (Springer: Heildelberg), D.F. Gray & J.L. Linsky, eds., p. 55.

Gray, D.F. 1980. *Ap. J.*, **235**, 508.

Gray, D.F. 1981a. *Ap. J.*, **251**, 155.

Gray, D.F. 1981b. *Ap. J.*, **251**, 583.

Gray, D.F. 1982a. *Ap. J.*, **255**, 200.

Gray, D.F. 1982b. *Ap. J.*, **262**, 682.

Gray, D.F. 1986. *Publ. Astron. Soc. Pacific* **98**, 319.

Gray, D.F. 1988. Lectures on Spectral-Line Analysis: F, G, and K Stars (The Publisher: Arva, Ontario), Chapter 4.

Gray, D.F. 1989. *Ap. J.*, **347**, 1021.

Gray, D.F. 1991. Angular Momentum Evolution of Young Stars (Kluwer: Dordrecht), S. Catalano and J.R. Stauffer, eds., p. 183.

Gray, D.F. and H.L. Johanson 1991. *Publ. Astron. Soc. Pacific* **103**, 439.

Gray, D.F. and T. Nagel 1989. *Ap. J.*, **341**, 421.

Gray, D.F. and C.G. Toner 1985. *Publ. Astron. Soc. Pacific* **97**, 543.

Gray, D.F. and C.G. Toner 1986. *Publ. Astron. Soc. Pacific* **98**, 499.

Immerschitt, S and E.-H. Schrter 1989. *Astr. Ap.*, **208**, 307.

Kaisig, M. and E.H. Schrter 1983. *Astr. Ap.*, **117**, 305.

Livingston, W. 1991. The Sun and Cool Stars: Activity, Magnetism, Dynamos, IAU Coll. 130, (Springer-Verlag: Berlin), I. Tuominen and G. Rudiger, eds., p. 246.

Lockwood, G.W. and B.A. Skiff 1988. Luminosity Variations of Stars Similar to the Sun, (Air Force Geophysics Laboratory: Hanscom AFB), AFGL-TR-0221.

Macri, C.J., R. Mller, J. Rsch, and T. Roudier 1984. Small Scale Dynamical Processes in Quiet Stellar Atmospheres (Sacramento Peak: Sacramento), S.L. Keil, ed., p. 371.

Nordlund, 1985. *Solar Physics* **100**, 209.

Nordlund, and D. Dravins 1990. *Astr. Ap.*, **228**, 155.

Radick, R.R., G.W. Lockwood, and S.L. Baliunas 1990. *Science* **247**, 39.

Renzini, A., C. Cacciari, P. Ulmschneider, and F. Schmitz 1977. *Astr. Ap.*, **61**, 39.

Simon, T. and S.A. Drake 1989. *Ap. J.*, **346**, 303.

Spruit, H.C., , Nordlund, and A.M. Title 1990. *Ann. Rev. Astr. Ap.*, **28**, 263.

Steffen, M. H.-G. Ludwig, and A. Frw 1989. *Astr. Ap.*, **213**, 371.

Stein, R.F., Nordlund, and J.R. Kuhn 1989. Solar and Stellar Granulation (Kluwer: Dordrecht), R.J. Rutten and G. Severino, eds., p. 381.

Willson, R.C and H.S. Hudson 1988. *Nature* **332**, 810.

Wolff, C.L. and J.R. Hickey 1987. *Science* **235**, 1631.

Seventh Cambridge Workshop on Cool Stars, Stellar Systems, and the Sun
ASP Conference Series, Vol. 26, 1992
Mark S. Giampapa and Jay A. Bookbinder (eds.)

NUMERICAL MODELS OF STELLAR CONVECTION

FAUSTO CATTANEO
Department of Astronomy and Astrophysics, University of Chicago, 5460 S. Ellis Ave., Chicago, IL 60615

ANDREA MALAGOLI
Department of Astronomy and Astrophysics, University of Chicago, 5460 S. Ellis Ave., Chicago, IL 60615

ABSTRACT Three–dimensional numerical models are used to study turbulent, compressible convection. The results show that the flow naturally organizes itself into a granular network near the upper boundary. The horizontal velocities in this upper layers can exceed the speed of sound. Nonstationary shock systems form where the supersonically moving fluid is decelerated to subsonic velocities. In the interior, the motions consist of strong, vertically coherent downflows embedded in a weaker, disorganized flow. Although these strong downflows only involve a small fraction of the total convecting fluid, they give the largest contribution to the transport of the various energetic components.

Keywords: Turbulent convection; shocks; turbulent transport

INTRODUCTION

The study of stellar convection has undergone a radical change in the last decade. To a large extent, the reason for this change is the dramatic increase in speed and availability of supercomputers. Before the advent of the fast supercomputer the mathematical description of compressible convection was limited to models that allowed analytical treatment (Spiegel 1965), to simplified one-dimensional models (Vitense 1953) or to low–order modal expansions (Latour, Toomre and Zahn 1981). One of the dominant motivations for studying turbulent convection was the need to parameterize convective energy transport so that it could be incorporated into stellar models. With supercomputers it is now possible to study more realistic two and three-dimensional models and to address more detailed questions concerning the nature of convection in stars. A wealth of numerical studies has increased our understanding of the effects of stratification (Hurlburt, Toomre and Massaguer 1984; Yamaguchi 1984; Chan and Sofia 1986; 1987; Stein and Nordlund 1989; Cattaneo *et al.* 1991), penetration (Hurlburt, Toomre and Massaguer 1986), sphericity and rotation (Gilman and Glatzmaier 1981; Gilman and Miller 1986; Glatzmaier and Gilman 1981) radiative transfer (Nordlund 1982, 1983, 1984 1985; Edwards 1990), the coupling to the acoustic modes (Stein, Nordlund and Kuhn 1989; Hossain and Mullan 1990), and so on. What made these studies particularly valuable was that they could investigate some specific aspect of the convective flow structure and how it was influenced by the various effects.

In the present paper we illustrate this approach by considering the effects of compressibility and stratification. We recall that compressibility becomes important when the flow velocity approaches or exceeds the sound speed. Recent numerical simulations suggest that such flows regimes might be realized in the uppermost layers of stellar convective zones where the background stratification changes most rapidly with depth and non-adiabatic effects are important (Cattaneo, Hurlburt and Toomre 1989; 1990; Stein *et al.* 1989; Malagoli, Cattaneo and Brummell 1990). Likewise the effects of stratification become important when the vertical extent of the motions is comparable to the local scale height. Again, recent numerical simulations show that the volume changes associated with stratification can lead to flows with complicated transport properties and geometry (Stein and Nordlund 1989; Cattaneo *et al.* 1991).

FORMULATION AND MODEL

We consider a plane parallel layer of fluid stratified under the action of a constant gravitational acceleration. We assume that the fluid is a perfect gas with constant thermal conductivity, dynamic viscosity and principal specific heats. The layer is confined above and below by impenetrable, stress-free boundaries. A fixed temperature is prescribed on the upper boundary; a fixed heat flux is prescribed on the lower one. These boundary conditions imply that the mechanical energy flux (kinetic plus acoustic) vanishes on the upper and lower boundaries, and that the energy enters and leaves the layer through conduction alone. Further, we specify that all quantities be periodic in the two horizontal directions.

The above specifications are consistent with an initial polytropic (static) equilibrium which we choose as a reference state. In this initial polytropic state the temperature T increases linearly with depth while the density and pressure increase as T^m and T^{m+1} respectively, where m is the polytropic index. The temporal evolution of the system is followed by integrating the equations of conservation of mass, momentum and energy for a fully compressible fluid (Cattaneo *et al.* 1991). The numerical integration is carried out using a code based on the Piecewise Parabolic Method (Colella and Woodward 1984). The main reasons for using this method are its particular suitability for low dissipation, compressible hydrodynamics as well as excellent numerical stability even in the presence of discontinuities (Woodward and Colella 1984).

For the present calculation the geometry was that of a rectangular "box" of unit depth and length and width equal to 10. The polytropic index and initial temperature gradient were chosen so that, in the polytropic state, the ratio of the density at the upper boundary to that at the lower one was 11, and the layer spanned approximately 5 pressure scale heights in the vertical. We found that adequate numerical resolution was provided by a grid with 56 points in the vertical by 192 points in each horizontal direction.

SURFACE FLOWS AND SHOCKS

After an initial transient lasting approximately one thermal relaxation time (several tens of turnover times) the system approaches a statistically steady state where the mean (*i.e.* horizontally averaged) temperature gradient in the interior is close to that of an adiabat. Two superadiabatic thermal boundary layers form at the two horizontal boundaries with the upper one considerably thicker than the lower one. In this asymptotic state the convection is vigorously time dependent evolving on a time scale similar to the turnover time.

The surface manifestation of the convection is the familiar granulation pattern. The flow (near the upper boundary) is organized into interlocking cells with downflow lanes surrounding regions of upflow in the granular centers. This property of compressible convection, namely the formation of a surface flow pattern consisting of *connected* downflows and *isolated* upflows (and never the other way around) seems to be one of the most robust features of numerical experiments having been reproduced by most numerical models (Stein and Nordlund 1989; Stein *et al.* 1989; Malagoli *et al.* 1990; Cattaneo *et al.* 1991).

The flow structure at a representative time is illustrated in Fig. 1. In the upper panel we show the distribution of vertical velocity as a function of horizontal position on a plane near the upper boundary. Cold, downflowing fluid is rendered in dark tones while hot, upflowing material is rendered in lighter ones. Between the upflows and downflow lanes the fluid moves predominantly horizontally. One intriguing features of compressible convection is the spontaneous formation of regions where the horizontal velocity exceeds the (local) sound speed (Cattaneo *et al* 1989; 1990; Stein *et al.* 1989; Malagoli *et al.* 1990). In Fig. 1 the supersonic patches are shown in light gray. Though surprising, the appearance of supersonic flows is quite natural and follows from the interaction between horizontal pressure gradients and non–adiabatic cooling. In its trajectory from granular center to downflow lane a fluid element is accelerated by pressure gradients while non–adiabatic processes reduce its temperature and hence its local sound speed. If the cooling is sufficiently efficient the sound speed can fall below the fluid velocity. Because the overall convection is subsonic, a fluid element moving supersonically must be decelerated to subsonic speeds before it reaches a downflow lane. Although the transition from subsonic to supersonic velocity is spatially smooth, the opposite transition, namely from supersonic to subsonic, is abrupt, *i.e.* it corresponds to a shock. This point is illustrated in the lower panel in Fig. 1 showing the profile of the Mach number along the line $a\ b$; the shock discontinuities are evident at the points x and x'.

The geometry of the supersonic regions can be summarized as follows. The regions form between the granular centers and the downflow lanes, relatively close to the lanes themselves (see Fig. 1). their vertical extent is limited to the thermal boundary layer since non–adiabatic effects are crucial for their formation. On the upflow side, a supersonic region is bounded by a smooth, sonic surface where fluid is accelerated to supersonic speed. On the downflow side, it is bounded by a shock front where the fluid is decelerated.

The non–stationarity of the shock fronts gives rise to an interesting lifecycle for supersonic regions (Cattaneo *et al.* 1989; 1990). The shocks typically propagate upflow—away from the downflow lanes—decreasing in strength as they approach regions of smaller inflow Mach number, eventually to disappear in the vicinity of the smooth sonic surface. At that point, the supersonic region ceases to exist, possibly to be replaced by a new supersonic event. Because the propagation velocity of a shock front is near sonic, a specific patch of supersonic

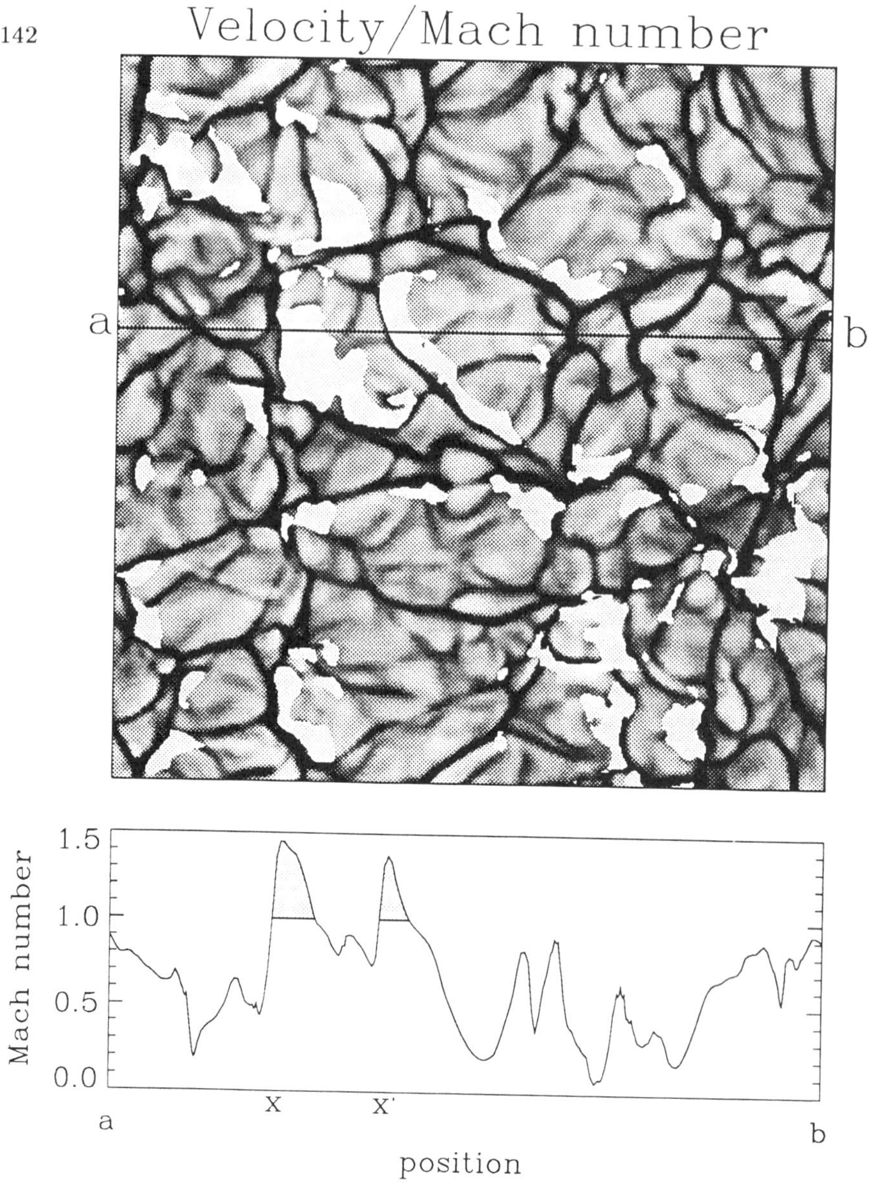

Fig. 1: Vertical velocity distribution on a horizontal plane near the upper boundary. The darker tones represent strong, downflowing fluid elements while lighter tones represent fast, upward moving ones. The white patches correspond to regions in which the fluid is moving supersonically (mostly in the horizontal direction). The shock fronts are located at the boundaries of the supersonic patches nearest to the strong downflow lanes. The plot (lower panel) shows the profile of the (total) Mach number along the line $a\ b$. The steep transitions between supersonic and subsonic motions at the shock fronts are evident at x and x'.

flow only lasts for a time comparable to the sound crossing time from cell centers to cell boundaries.

The appearance of supersonic flows and shocks has some interesting observational consequences. It is well known that propagating shocks can effectively amplify local fluctuations. We can expect that as a shock propagates away from a downflow site it will generate a gradient of post–shock turbulence with the strongest fluctuations nearest to the downflow. Although these fluctuations might always be too small for direct observation, they might contribute to the non thermal broadening of spectral lines formed in the vicinity of the downflows. Indeed, high resolution spectral observations of the solar granulation exist which are consistent with the above conclusion (Holweger and Kneer 1989; Nesis *et al.* 1991).

INTERIOR FLOWS AND TRANSPORT

One question that immediately arises from studying the granulation pattern is how representative it is of the flows in the bulk of the convective region. There seems to be a general agreement that the granular pattern is shallow phenomenon which only characterizes the uppermost layers. We can gather some knowledge of the flow geometry in the interior by examining Fig. 2. The three pairs of panels show the horizontal distribution of vertical velocity, vertical vorticity and enthalpy flux ($\sim \rho w \delta T$, where ρw is the vertical momentum, and δT is the fluctuating temperature) at a representative time. The panels in the left column correspond to a horizontal plane near the upper boundary, those on the right to a plane at a depth of 0.6. The enthalpy flux has been rendered so that upward directed fluxes are in darker tones.

The downflow lanes that characterize the flow near the surface break up into isolated sheets and columns at greater depths. The downflow sheets themselves have a corrugated and crinkled appearance. Indeed, time animation of the flow show that the downward advection of vertical vorticity probably causes the corrugations that ultimately lead to the break up of the downflow lanes. Vertical vorticity is generated in the upper layers by the non radially symmetric convergence of horizontal flows in the granular corners (see middle panels in Fig. 2). Notice how the vertical vorticity in the interior is no longer correlated (or anticorrelated) with the downflows but, rather, as a "salt and pepper" appearance. The flow in the interior seems to be characterized by regions of small scale disordered flow surrounding narrow sheets, or columns of strong vertical downflow with a long range vertical coherence. In fact, the existence of strong long range downflows surrounded by broader regions of gentler motions seems to be a common feature of all simulations with substantial background stratification (Stein and Nordlund 1989; Cattaneo *et al.* 1991). The asymmetry between the downflows and the upflows is most likely related to the asymmetry between the contraction of downward moving fluid elements and the expansion of upward moving ones. One of the consequences of this type of flow asymmetry is a net downward directed kinetic energy flux which distinguishes flows where the upflows and downflows, though giving no net mass flux across horizontal planes, have different filling factors (Hurlburt *et al.* 1984; 1986).

Given that the flow in the interior consists of strong coherent downflows and weaker disordered motions it is natural to inquire how each component in

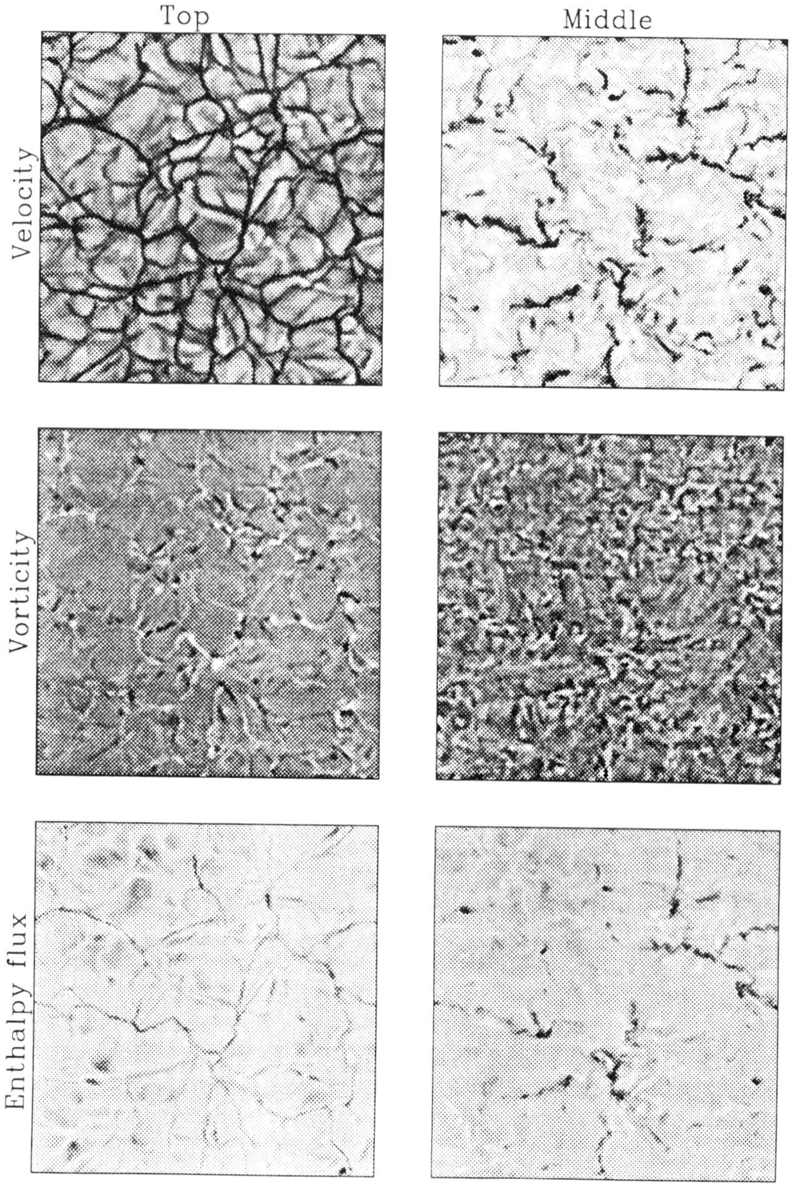

Fig. 2: Distribution of vertical velocity, vertical vorticity, and enthalpy flux at a given instant on two horizontal planes. One plane (Top) is near the upper boundary and the other (Middle) is near the middle. The darker (lighter) tones represent downward (upward) velocity, upward (downward) directed vorticity, and upward (downward) directed enthalpy flux respectively.

turn contributes to the turbulent transport of energy. The panel in the lower right hand corner of Fig. 2 shows that areas of strongest upward enthalpy flux (darkest tones) are associated with strong downflows, suggesting that these flows have the highest degree of thermal correlation. In order to pursue this matter further, we examine the distribution of thermal and kinetic correlation with vertical velocity distribution on horizontal planes. To this end we define the (vertical) enthalpy flux F_e and the kinetic flux F_k by

$$F_e = \frac{\gamma}{(\gamma - 1)}\ \rho w\ \delta T, \qquad F_k = \frac{1}{2}\ \rho w\ |u|^2,$$

where u is the velocity field. The horizontally averaged value of F_e gives the net enthalpy flux through a horizontal plane and also a measure of the correlation between the vertical velocity and the temperature fluctuations. It is easy to see that the enthalpy flux is positive (upward directed) both when hot fluid moves upwards and cold fluid moves downwards. Similarly, F_k, when horizontally averaged, gives the net flux of kinetic energy across a horizontal plane and a measure of the correlation between vertical flows and regions of high kinetic energy density. Because the kinetic energy is positive definite, the kinetic energy flux has the same direction as w.

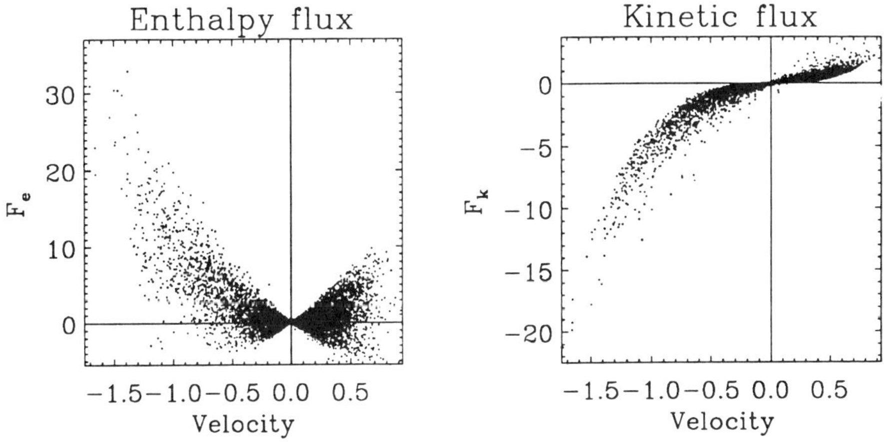

Fig. 3: Phase diagrams of the enthalpy and kinetic fluxes as function of vertical velocity at a depth of 0.6. Negative (positive) fluxes and velocities are directed downwards (upwards). Each point correspond to a gridpoint on the plane.

The distribution of F_e and F_k on a horizontal plane at a depth of 0.6 is shown in Fig. 3. Each dot in the scatter diagrams represents a gridpoint. A horizontal average corresponds to summing over the dots, therefore in order to have a net flux some asymmetry about the (zero) horizontal line must exist. For instance, in the enthalpy flux the contributions from the upflows (on the right of the vertical line) are considerably more symmetrical than those from the strong downflows. Another more physically intuitive way to describe this, is to note that whereas strong downflows always carry cold fluid, upflows are almost as likely to carry hot fluid than cold. The up–down asymmetry in the velocity is most apparent in the right hand panel in Fig. 3 showing the kinetic energy flux. It is clear that the net downward transport of kinetic energy is associated once again with the strong downflows. In fact the magnitude of the enthalpy and kinetic energy fluxes for the strong downflows are similar indicating that a strong downflow will carry enthalpy upwards at roughly the same rate as kinetic energy downwards. That being the case, the role of the strong downflows is to redistribute the energetic components throughout the convective domain but to contribute little to the total energy transport (Cattaneo *et al.* 1991).

CONCLUSIONS

The results of the numerical studies show a rather complex picture of compressible convection. Near the upper surface the convection displays the familiar granular pattern with connected downflow lanes surrounding isolated upflows. The horizontal flows can occasionally become supersonic because of the interaction between pressure gradients and non–adiabatic effects. The resulting shock structures might give rise to detectable micro–turbulent gradients with the strongest fluctuations near the downflows. The most striking and robust features characterizing the flow in the interior are the up–down asymmetries in the motions and the coexistence of ordered and disordered components. The disordered component consists of small scale, random motions with weak thermal correlation. The ordered component is given by strong, isolated downflows that originate in upper thermal boundary layer and penetrate into the bulk of the convection. These strong downflows seem to be a well established property of compressible convection that manifests itself under a variety of initial conditions and models. While it is still not clear what determines their spatial and temporal coherence, it seems that they are very efficient carriers of the energetic ingredients of the convection.

ACKNOWLEDGMENTS

The authors would like to thank Roberts Rosner and Stein for stimulating conversations. This work was partially supported by the NASA Space Physics Theory Program at the University of Chicago.

REFERENCES

Cattaneo, F., Brummell, N.H., Toomre, J., Malagoli, A. and Hurlburt, N.E. 1991, *Ap. J.*, **370**, 282.

Cattaneo, F., Hurlburt, N.E. and Toomre, J. 1989, in Solar and Stellar Granulation, eds. R.J. Rutten and G. Severino (Dordrecht: Kluwer), p. 415.

Cattaneo, F., Hurburt, N.E. and Toomre, J. 1990, *Ap. J.*, **349**, L63.

Chan, K.L. and Sofia, S. 1986, *Ap. J.*, **307**, 222.

Chan, K.L. and Sofia, S. 1987, *Science*, **235**, 465.

Colella, P. and Woodward, P.R. 1984, *J. Comp. Phys.*, **54**, 174.

Edwards, J.M. 1990, *M.N.R.A.S.*, *242*, 224.

Gilman, P.A. and Glatzmaier, G.A. 1981, *Ap. J. Suppl.*, **45**, 351.
Gilman, P.A. and Miller, J. 1986, *Ap. J. Suppl.*, **61**, 585.
Glatzmaier, G.A. and Gilman, P.A. 1981, *Ap. J. Suppl.*, **45**, 335.
Holweger, H. and Kneer, F. 1n Solar and Stellar Granulation, eds. R.J. Rutten and G. Severino (Dordrecht: Kluwer), p. 425.
Hossain, M. and Mullan, DJ. 1990, *Ap. J.*, **354**, L33.
Hurlburt, N.E., Toomre, J. and Massaguer, J. 1984, *Ap. J.*, **282**, 557.
Hurlburt, N.E., Toomre, J. and Massaguer, J. 1986, *Ap. J.*, **311**, 563.
Latour, J., Toomre, J. and Zahn, J-P. 1981, *Ap. J.*, **248**, 1081.
Malagoli, A., Cattaneo, F. and Brummell, N.H, 1990, *Ap. J.*, **361**, L33.
Nesis, A., Hanslmeier, A. Hammer, R., Komm, R., Mattig, W. and Staiger, J. 1991, *Astr. Ap.*, in press.
Nordlund, Å. 1982, *Astr. Ap.*, **107**, 1.
Nordlund, Å. 1983, in Solar and Stellar Magnetic Fields: origins and Coronal Effects, ed. J.O. Stenflo, (Dordrecht: Reidel), p. 79.
Nordlund, Å. 1984 in The Hydromagnetics of the Sun, ed. G.A. Guyenne, (ESA SP–220), p. 37.
Nordlund Å. 1985 in Theoretical Problems in High Resolution Solar Physics, ed. H.U. Schmitt, (Münich: Max–Plank Institute für Astrophysik), p. 1
Sofia, S. and Chan, K.L. 1984, *Ap. J.*, **282**, 550.
Spiegel, E.A. 1965, *Ap. J.*, **141**, 1068.
Stein, R. and Nordlund, Å. 1989, *Ap. J.*, **342**, L95.
Stein, R., Nordlund, Å. and Kuhn, J.R. 1989, in Solar and Stellar Granulation, eds. R.J. Rutten and G. Severino (Dordrecht: Kluwer), p. 381.
Vitense, E. 1953, *Z. Ap.*, **32**, 135.
Woodward, P.R. and Colella, P. 1984, *J. Comp. Phys.*, **54**, 115.
Yamaguchi, S. 1984, *PASP*, **36**, 613.

Seventh Cambridge Workshop on Cool Stars, Stellar Systems, and the Sun
ASP Conference Series, Vol. 26, 1992
Mark S. Giampapa and Jay A. Bookbinder (eds.)

MAGNETO-CONVECTION

ROBERT F. STEIN Michigan State University, East Lansing, MI 48824, USA

AXEL BRANDENBURG NORDITA, Blegdamsvej 17, DK-2100 Copenhagen Ø, Denmark; and Observatory and Astrophys. Lab., University of Helsinki, Tähtitorninmäki, SF-00130 Helsinki, Finland

ÅKE NORDLUND Copenhagen University Observatory, Østervoldgade 40, DK-1350, Copenhagen, Denmark

ABSTRACT Results of two simulations of magneto-convection are presented: an idealized calculation of a convective layer between two stable layers and a more realistic calculation of the upper solar convection zone applicable to a strong plage. The velocity field consists primarily of vertical vortex tubes corresponding to strong downdrafts, embedded in slow broad upflows. The magnetic field consists primarily of vertical flux tubes in the unstable layer, but is mainly horizontal in the stable layer below the convection zone. Comparison of solar magnetic and non-magnetic simulations shows that the magnetic field inhibits the magnitude of the velocity fluctuations and reduces the size of granules, in accordance with observations.

Keywords: Convection; Magneto-hydrodynamics; magnetic fields; the sun

INTRODUCTION

Convection is the mechanism of energy transport through the outer layers of all cool stars. Many such stars exhibit activity which is believed to be intimately connected with magnetic fields. We have therefore begun to explore the nature of magneto-convection using numerical simulations. Here we discuss the results of two such simulations, one idealized and the other a realistic simulation of a solar plage. The idealized simulation is of a perfect gas, in the presence of rotation, with a unstable layer sandwiched between two stable layers above and below it. The solar simulation is of the outer megameter (Mm) of the convection zone and includes radiative energy losses at the surface and an equation of state with the effects of ionization and excitation.

IDEALIZED SIMULATION

We solve the equations for conservation of mass, momentum and energy and the induction equation in a cartesian box of dimensions $63 \times 126 \times 37$ grid points. The x-axis points north, the y-axis points east and the z-axis points down. The horizontal boundaries are periodic, while the top and bottom boundaries are impenetrable stress-free perfect conductors. A fixed incident flux is imposed at the bottom boundary and the top few layers are cooled. The equations are cast in dimensionless units with

$$d = g = \bar{\rho} = c_p = 1,$$

where d is the depth of the unstable region, time is measured in units of $(d/g)^{1/2}$ ($\approx$ the free fall time through the unstable layer), and the magnetic field in units of $(\mu_0 \bar{\rho} g d)^{1/2}$. In these units the computational domain extends 6×3 horizontally and from -0.15 to 1.5 vertically. The radiative conductivity is small in the unstable layer ($0 \leq z \leq 1$) and large in the stable layers ($1 \leq z \leq 1.5$). The rotation period is 43 time units, corresponding to about 30 sound crossing times. Figure 1 shows the (horizontally and temporally averaged) mean atmosphere, which extends four pressure scale heights.

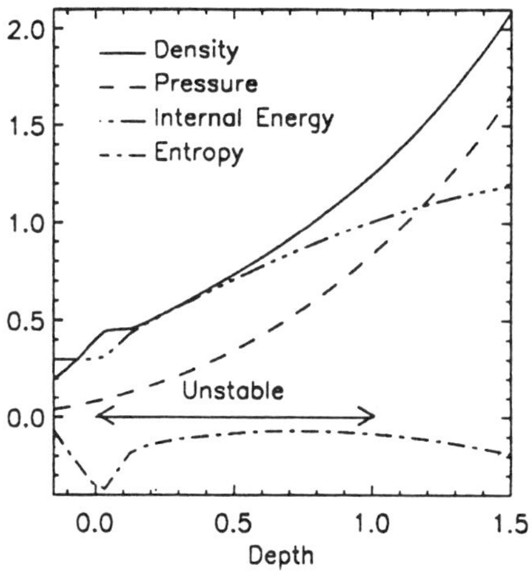

Fig. 1: Mean atmosphere of the idealized simulation.

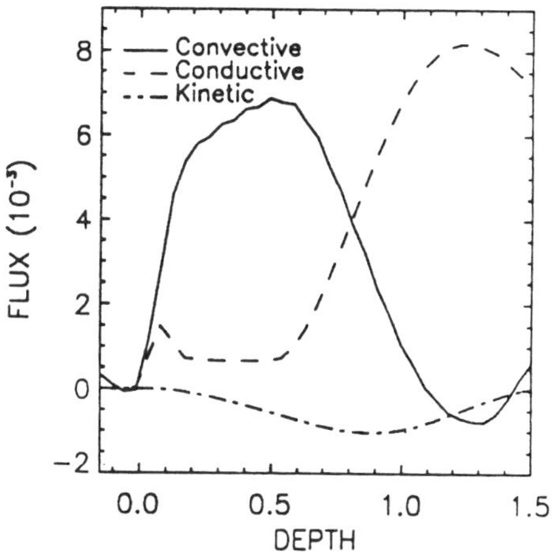

Fig. 2: Average enthalpy, kinetic energy and conductive fluxes in the atmosphere.

ENERGETICS

In the convectively unstable layer, where the thermal conductivity is small, convection carries 90% of the total flux and conduction about 10% (figure 2). There is a gradual switch over at the bottom of the convection zone and the entropy gradient becomes sub-adiabatic below the depth (0.7) where the conductive flux predominates (figure 1). Buoyancy work, which drives the motions, is positive in both the upflows and downdrafts, with most of the work being done in the downdrafts. Downdrafts as the primary site of the buoyancy work is a robust result of the various convective simulations. In the stable layer below the convection zone, the thermal conductivity is large and conduction carries most of the energy flux. In fact, the convective flux is downward. The upflows carry no net enthalpy plus kinetic energy flux, while the downflows carry both enthalpy and kinetic energy downward. Hence, below the convection zone, the conductive flux exceeds the total flux through the atmosphere in order to compensate for the downward transport of energy by the overshooting motions. In this overshoot layer, the downflows are being braked by buoyancy (negative buoyancy work), while the upflows produce no net buoyancy work (Hurlburt *et al.* 1984). At the top of the convection zone the convective flux again goes to zero and the conductive flux starts to increase, but this increase is terminated by the removal of energy by volume cooling which is imposed in the upper 10% of the computational box. (The energy flux removed by this cooling is not shown.)

MAGNETIC FIELD

This simulation was started with a horizontal seed magnetic field which reversed its sign, so there is no net flux. Dynamo action occurs (Brandenburg *et al.* 1991). After saturation, the magnetic field in the unstable layer consists primarily of vertical flux tubes as is seen in a view of the magnetic field vectors for those locations where the field is large (figure 3). The horizontally averaged rms magnitude of the field components (figure 4) shows that the field is mainly vertical in the unstable layer and horizontal in the stable region. The magnetic energy is a maximum in the lower part of the convection zone. However, the mean magnetic field is approximately zero in the convection zone and accumulates in the stable layers above and below (figure 5). Note that the toroidal (y-) and poloidal (x-) components are approximately equal in magnitude.

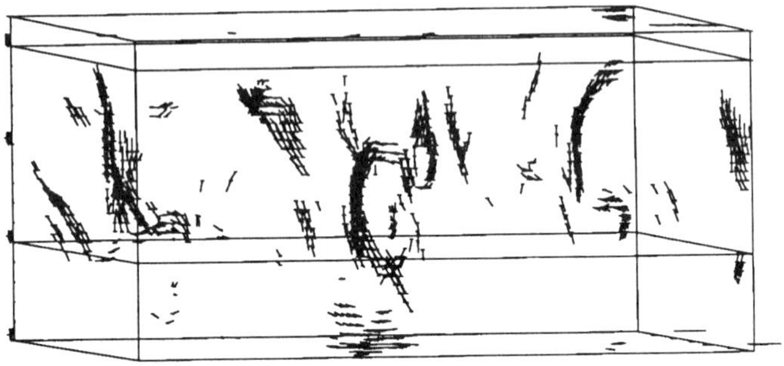

Fig. 3: Magnetic field vectors with magnitude > 43% of the maximum field strength, as seen in a 3D perspective view through the long side. The levels $z = 0$ and $z = 1$ are shown by lines around the circumference of the box.

The peakedness of the magnetic field distribution is measured by its kurtosis, $< B_i^4 > / < B_i^2 >^2$, the inverse of which is a measure of the filling factor. A gaussian distribution would have a kurtosis of 3. Here we find the magnetic field has a kurtosis of 10, so it is much more peaked than a gaussian distribution and its filling factor is about 10%.

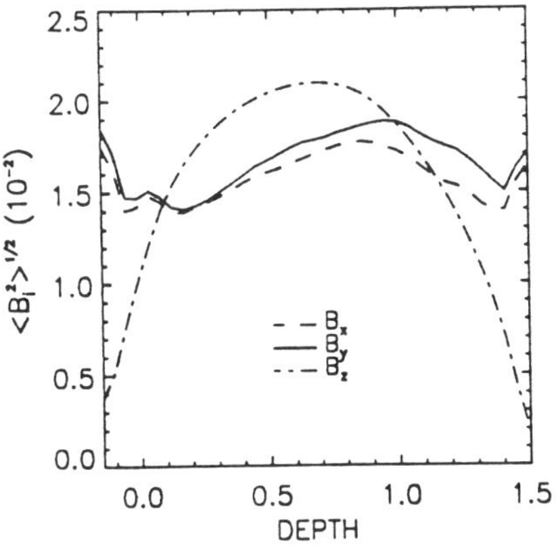

Fig. 4: Average rms magnitudes of the magnetic field components.

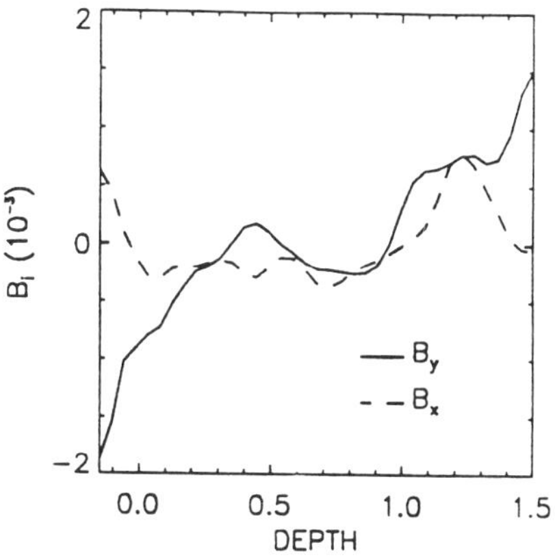

Fig. 5: Horizontal and temporal average of the magnetic field components.

VELOCITY FIELD

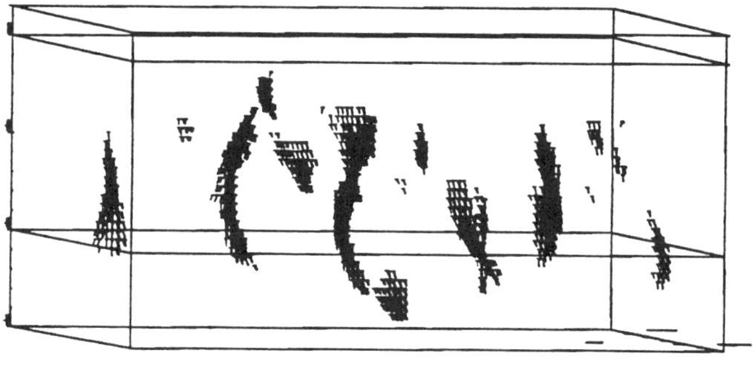

Fig. 6: Kinetic energy flux vectors for the same snapshot as fig 3. Those vectors with magnitude > 0.25% of the maximum flux are shown.

One of the most robust results of all convection simulations is that the velocity field is composed of broad regions of upflow and in the interior isolated strong downdrafts. Vectors of the kinetic energy flux (figure 6) reveal the downdrafts. Although both the velocity and magnetic field exhibit vertical structures in the convective zone, the velocity field has a much larger filling factor (its kurtosis ≈ 3, so its filling factor is about 30%).

This model has a mean rotation and it is interesting to look at its differential rotation (v_y). Figure 7 shows that the differential rotation decreases inward in the upper part of the convection zone, then increases through the bottom three quarters of the convection zone, and finally in the lower stable layer decreases again. Note also that the meridional (north-south, v_x) flow is roughly comparable to the differential rotation.

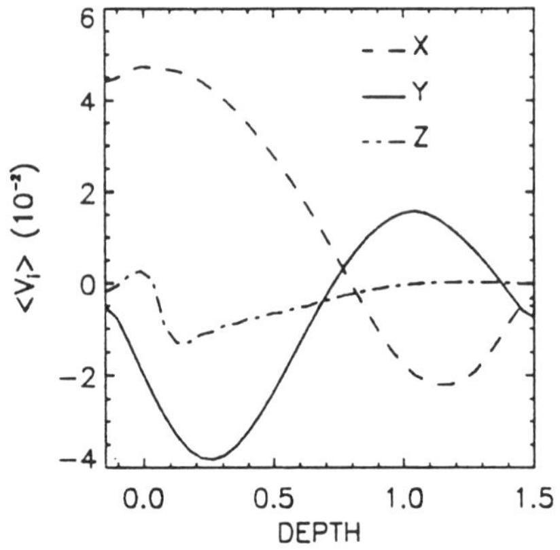

Fig. 7: Horizontally and temporally averaged velocity components. The x-component is the meridional flow, the y-component is the differential rotation and the z-component is the vertical motion.

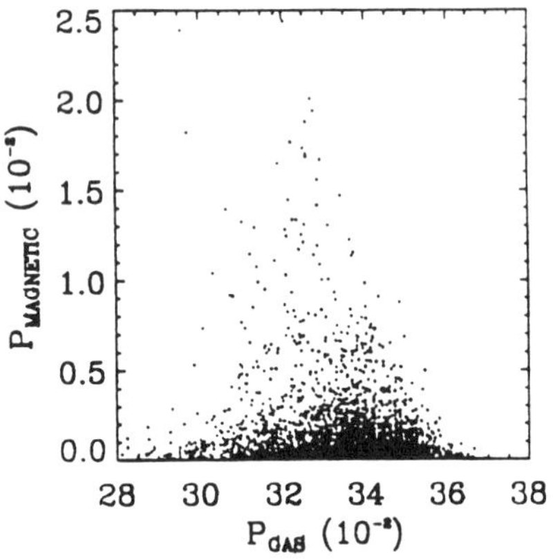

Fig. 8: Correlation of magnetic and gas pressure at $z = 0.5$.

MAGNETIC FIELD CORRELATIONS

Next we present the correlations of the magnetic field with various properties of the flow. One expects there to be a correlation between the magnetic field strength and the gas pressure and in this toy model there is, but it is weak. Where the field is strong the gas pressure is below average and where the gas pressure is above average the field is weak (figure 8). However, there is very little correlation with the density which suggests only minimal effects of magnetic buoyancy. Vorticity is organized into nearly vertical vortex tubes (figure 9), and the magnetic field tends to be wrapped around them but avoids their cores – the field is largest where the vertical vorticity is small. From movies of the vorticity and magnetic field it is easy to see that vortices wrap the magnetic field around and push it down toward the stable layer at the bottom.

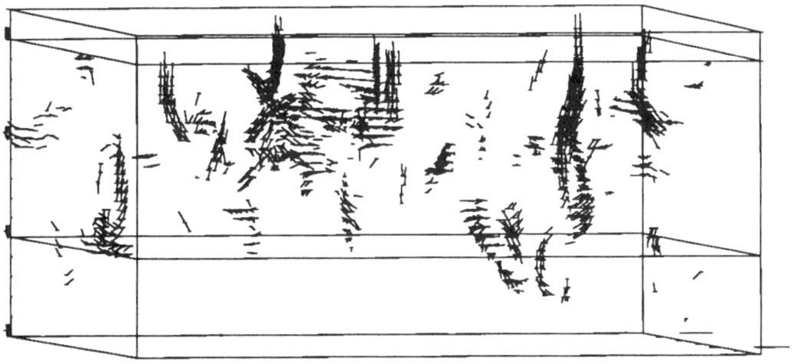

Fig. 9: Vorticity vectors scaled by $\overline{\rho}^{1/3}$ for locations where their magnitude is > 50% of the maximum value. The snapshot and viewing angle are the same as for the magnetic field vectors (fig 3) and the kinetic energy flux vectors (fig 6). (The vorticity has been scaled because the velocity decreases with depth.)

SOLAR MAGNETO-CONVECTION

The solar simulation is interesting both because it can be compared with observations of the sun and because we have run nearly the same simulation both with and without magnetic fields, so the effect of the field on the atmosphere and the convection can be seen by a differential analysis.

The magneto-convection simulation was performed on a grid 63 × 63 × 32 covering 3 × 3 × 1.5 Mm from the temperature minimum to 1 Mm below the surface. The mean vertical magnetic field is 500 gauss, which corresponds to a very strong plage region. There was no rotation in this case. The hydrodynamic simulation has the same resolution but is twice as big, covering a region 6 × 6 × 3 Mm from the temperature minimum to 2.5 Mm below the surface using a grid of 125 × 125 × 82. These calculations were not started from the identical conditions, but from similar, and statistically equivalent, convective states. The magneto-convection simulation was begun by imposing a uniform vertical field of 500 gauss on a snapshot of non-magnetic convection. Convective motions quickly swept the field to the granulation boundaries. Thereafter it slowly evolved as the granulation pattern changed (Nordlund and Stein 1990). Further discussion of the hydrodynamic convection case can be found in Stein and Nordlund (1989, 1991). We now compare these two cases.

First, consider the mean atmosphere. The temperature and entropy structure is shown in figure 10. One can see that the adiabat is about the same. In both calculations we fed in the same entropy fluid from below. The temperature is slightly higher and the region of steep temperature gradient lies slightly higher up in the atmosphere in the magneto-convection case.

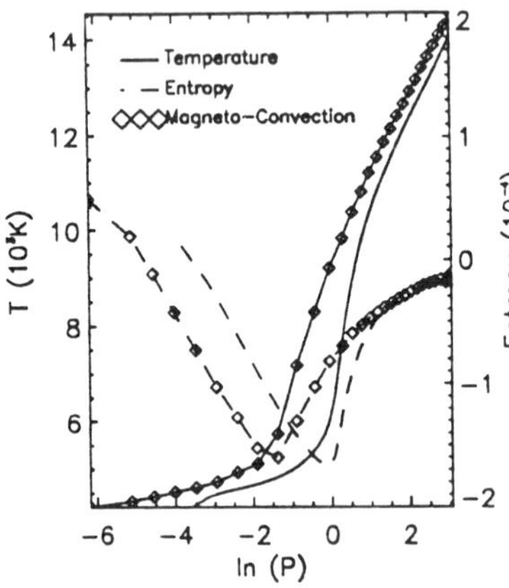

Fig. 10: Average temperature and entropy structure of the solar magneto- and hydrodynamic- convection simulations.

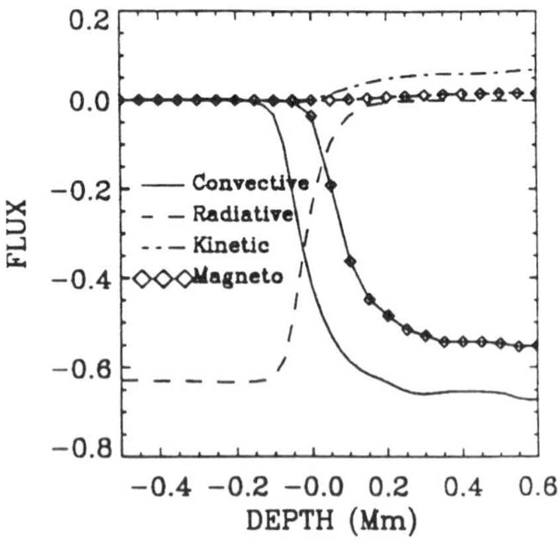

Fig. 11: Average enthalpy, radiative and kinetic energy fluxes for the solar magneto- and hydrodynamic- convection simulations.

ENERGETICS

Next consider the changes in the energetics. The magnetic field reduces the convective flux a little (figure 11). Both the enthalpy and the kinetic energy fluxes are noticeably smaller, and their sum (the net flux) is slightly reduced. This is due to a reduction in the velocity fluctuations in the magnetic case by about a factor of two in both the upward and downward flows (figure 12).

Most of the flux in this case is carried by the downflows. The upflows (negative velocities) carry the same flux in both the magnetic and non-magnetic cases. The flux carried by the downflows is however slightly smaller in the magnetic case. Which portion of the fluid, upflows or downdrafts, carries the net flux is **not** a robust result of convective simulations. It depends upon the amount of diffusion in the numerical scheme, the equation of state used and the vigor of the convection. These realistic calculations have the downdrafts carrying 70% of the flux. We have also used our code to simulate convection of an ideal gas in a rigid box with strong thermal conduction as described in Cattaneo *et al.* (1991). In that case, we too found cancellation of the enthalpy and kinetic energy fluxes in the downdrafts, with most (70%) of the net flux carried in the upflows (just the reverse of our realistic simulations). The extent of cancellation is not as complete in our case as in Cattaneo *et al.* because our code was more diffusive. The inescapable conclusion is that where the flux in a simulation is transported depends on the physics included in the calculation.

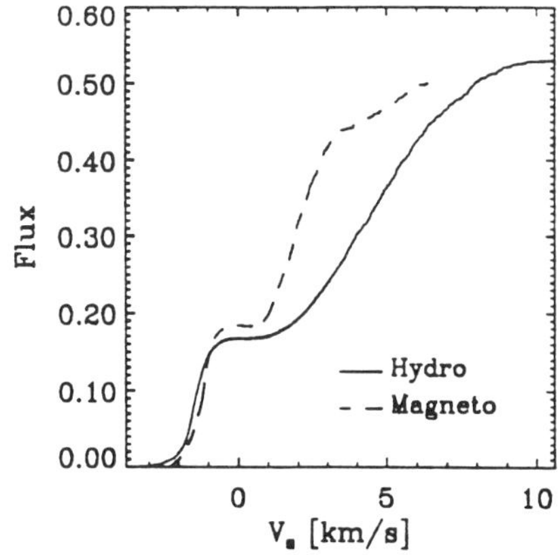

Fig. 12: Sum of the net flux $F_{net} = F_{enthalpy} + F_{kinetic\ energy}$ contributions as a function of velocity fluctuation at $z = 0.5$ Mm. Downflows are positive velocity and upflows negative.

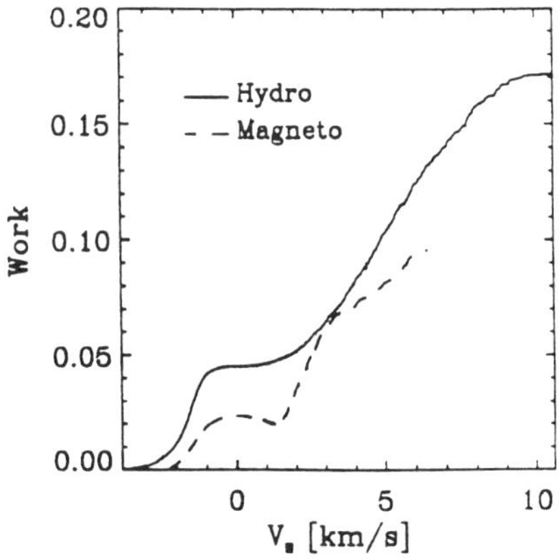

Fig. 13: Sum of the buoyancy work contributions as a function of velocity fluctuation at $z = 0.5$ Mm.

The major difference between the magnetic and non-magnetic cases is in the buoyancy work. The magnetic field significantly reduces the buoyancy work in both the upflows and downflows (figure 13).

MAGNETIC FIELD

The magnetic field is concentrated into the downflows by the horizontal velocity field. The horizontal motions of the granules sweep the field out of the granules into the intergranule lanes. Below the surface there is a significant difference in the flow topology. In the non-magnetic case the connected downflows break up into isolated downdrafts within a short distance below the surface. In the magnetic case, with its strong (500 gauss average) field, the magnetic field tends to act as rigid walls to the granule flow and make the connected downflows extend much deeper than they do in the field free case.

Again, we can look at the correlations of the magnetic field and gas properties. There is a very tight anti-correlation between the magnetic and gas pressure. Where the magnetic field is strong the gas pressure is low and vice versa (figure 14). This is much tighter than in the idealized simulation, because the magnetic pressure is closer in magnitude to the gas pressure here.

An anti-correlation with the density also occurs, because that is the primary means for reducing the pressure. The magnetic field is swept by the diverging upflows toward the downflows where it accumulates and impedes the horizontal flow (across field lines). This reduced inflow quenches the downflow as well. As a consequence, magnetic fields tend to correlate with low temperatures, low densities and small downward velocities.

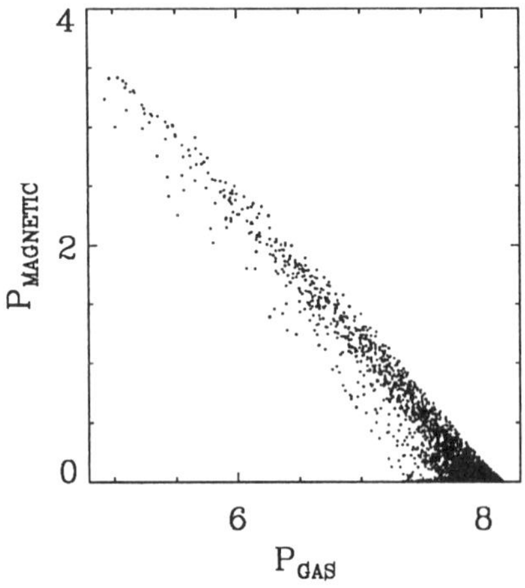

Fig. 14: Magnetic and gas pressure correlation for the solar simulations at a depth of 0.5 Mm.

OBSERVABLE EFFECTS

A magnetic field has a readily observable effect on the size of the granulation. Figure 15 shows a comparison of the emergent intensity in the magnetic and non-magnetic cases. The granules are smaller in the presence of magnetic field as described by Title (1992). This together with the inhibition of velocity fluctuations by the magnetic field are two easily observable effects of magnetic fields on convection as determined by a comparison of our two simulations.

ACKNOWLEDGEMENTS

This work was supported in part by grants from the National Aeronautics and Space Administration (NAGW 1695), the Danish Natural Science Research Council and the Danish Space Board. The simulations were performed on the Alliant FX-8 at the University of Colorado, the Cray-2 at the National Center for Supercomputer Applications, the Convex 240 at MSU, and the Cray X-MP/432 of the Centre for Scientific Computing in Finland. The authors wish to express their appreciation for the support from these agencies. We also thank Juha Ruokolainen for help in preparing videos of the simulations.

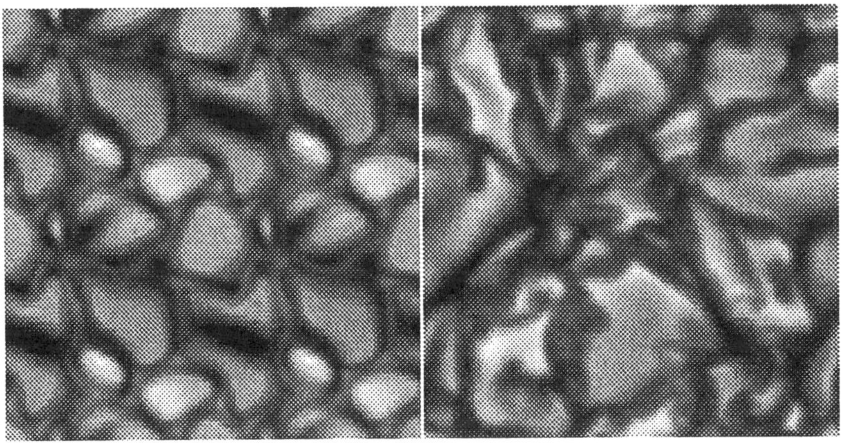

Fig. 15: Comparison of emergent intensity patterns (granulation) for the solar magneto- and hydrodynamic- convection simulations.

REFERENCES

Brandenburg, A., Jennings, R. L., Nordlund, Å., Stein, R. F. and Tuominen, I. 1991. "The role of overshoot in solar activity: A direct simulation of the dynamo," in The Sun and Cool Stars: Activity, Magnetism, Dynamos, IAU Coll. 130, eds. I. Tuominen, D. Moss & G. Rüdiger, Lecture Notes in Physics, Springer-Verlag, p. 86.

Cattaneo, F., Brummell, N.H., Toomre, J., Malagoli, A. and Hurlburt, N.E. 1991. "Turbulent Compressible Convection", *Ap. J.*, **370**, 282.

Hurlburt, N.E., Toomre, J. and Massaguer, J.M. 1984. "Two-Dimensional Compressible Convection Extending Over Multiple Scale Heights", *Ap. J.*, **282**, 557.

Nordlund, Å. and Stein, R. F. 1990. "Solar Magnetoconvection", in Solar Photosphere: Structure, Convection, Magnetic Fields, IAU Symp. 138, ed. J. O. Stenflo, p. 191.

Stein, R. F. and Nordlund, Å. 1989. "Topology of Convection Beneath the Solar Surface", *Ap. J. Letters*, **342**, L95.

Stein, R. F. and Nordlund, Å. 1991. "Convection and its Influence on Oscillations", in Challenges to Theories of the Structure of Moderate Mass Stars, eds. J. Toomre and D. Gough, Lecture Notes in Physics **388**, (Springer-Verlag: Berlin), p. 195.

Title, A. 1992 this conference.

Seventh Cambridge Workshop on Cool Stars, Stellar Systems, and the Sun
ASP Conference Series, Vol. 26, 1992
Mark S. Giampapa and Jay A. Bookbinder (eds.)

CONVECTIVE PENETRATION IN THE SUN

GABRIELLE BERTHOMIEU, PIERRE MOREL, JANINE PROVOST
Observatoire de la Côte d'Azur, BP 139, 06003 Nice Cedex, France

JEAN-PAUL ZAHN
Observatoire Midi-Pyrénées, 31400 Toulouse, France
Astronomy Department, Columbia University, New York, NY 10027
Joint Institute for Laboratory Astrophysics, Boulder, CO 80309, U.S.A.

ABSTRACT Convective motions penetrate beyond the level where the Schwarzschild criterion predicts the base of a convective envelope. Such nearly adiabatic penetration is expected to be of the order of the scale height of the radiative conductivity (Zahn 1991). When comparing models, with and without penetration, that are calibrated with the present Sun, we find that they have differences in their sound speed, and therefore in their acoustic frequencies, which are characteristic of penetration, and which can be distinguished from those due to other causes. We use this property to estimate the extent of penetration by matching the eigenfrequencies of models including convective penetration with those derived by Libbrecht *et al.* (1990) from their solar observations.

Keywords: Solar convection zone; convective penetration; helioseismology.

INTRODUCTION

Stellar models rarely include the effect of convective penetration, although there is overwhelming evidence for it to occur: in the atmosphere, the oceans, in the laboratory, and now also in numerical simulations. The problem is thus not whether there is penetration in stellar interiors, but how important it is.

A theoretical prediction for the extent of penetration has been made recently by one of us (Zahn 1991): it should be of the order of a scale height of the radiative conductivity χ ($= 16\sigma T^3/3\rho\kappa$ with the usual notations). We thus write $L_p = \zeta/(\partial \log \chi\ /\partial z)_{ad}$, where ζ is a parameter which depends on some properties of the convective motions. It cannot be determined *ab initio*, but it is presumably of order unity. Our goal is to calibrate it with the Sun, so that it may be implemented in other stellar models possessing a convective envelope.

THE EFFECT OF PENETRATION

A series of solar models have been computed with the code CESAM built by one of us (see Morel *et al.* 1990), which is particularly well suited to deal with discontinuities. These models include an increasing amount penetration at the base of the convection zone, the parameter ζ varying from 0 to 1. They were evolved from a ZAMS homogeneous model to 4.75 Gyrs, and both the mixing length parameter and the helium abundance were iterated until the radius and the luminosity fitted those of the Sun within better than 10^{-4}.

One consequence of that calibration is that these models have also nearly the same sound velocity in the deep convection zone, which is a well known property of polytropic stars. But they differ of course at the base of the convective envelope, as illustrated in fig. 1, where we compare models for $\zeta = 0.5$ and 1.0 with one that has no penetration. Note that the differences in sound speed $\delta c/c$ are localized in a rather small region, and that they taper off quickly with depth, to vanish at the center of the Sun. This property too is related to a well known tendency of an atmosphere in radiative equilibrium to settle in a polytropic stratification. Consequently, we find virtually the same helium abundance in all models; therefore, had we included penetration without allowing a change in chemical composition, the adjustments in radius and luminosity would have been of order 10^{-4}, which is the accuracy of our calibration.

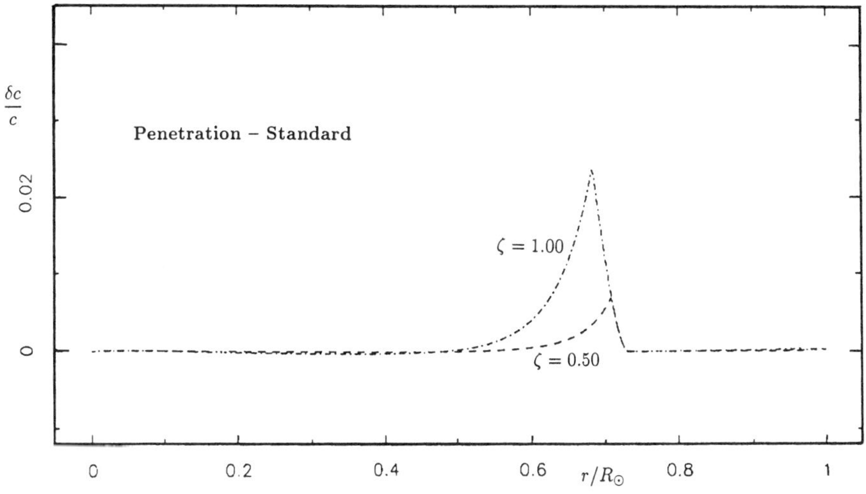

Fig. 1: The differences in sound velocity between models with penetration ($\zeta = 0.5$ and 1) and a standard model without penetration. These differences occur only at the base of the convection zone.

The modest variations in the solar model that we have seen when implementing convective penetration are in sharp contrast with the impact of an opacity change on such a model, which has been analyzed by Christensen-Dalsgaard (1988). An opacity increase below the convection zone induces there a steeper temperature gradient, and hence it would lead to a temperature rise in the central region. That rise would produce a luminosity increase, which is prohibited by the calibration, and can only be prevented by an adjustment in the chemical composition. But such change in the heliumq content is felt in the whole star, and therefore the resulting model is quite different everywhere. In particular, the sound velocity of the more opaque model is higher below the convection zone, as one expects, but it becomes smaller in the central region, due to the increase in molecular weight.

In order to reproduce the differences displayed in fig. 1, one would have to increase the opacity in the convection zone (including the penetration region) by a suitable amount, while keeping it unchanged in the radiation zone just below. Such a modification is unlikely to occur. Thus we conclude that convective penetration has a distinct signature, and that one should be able to separate it from an opacity increase when comparing the sound speed profiles of two models.

ESTIMATING THE EXTENT OF PENETRATION

The acoustic eigenfrequencies of a stellar model reflect the variation of the sound speed with depth. Therefore the comparison of the eigenfrequencies of two models reveals the differences in their sound speed. This powerful diagnostic tool has been developed by Christensen-Dalsgaard *et al.* (1988).

Likewise, one may compare the eigenfrequencies of a given model with those observed on the Sun, and again interpret the differences in terms of their respective sound velocity. We have done this with our models including convective penetration, and using the acoustic frequencies measured by Libbrecht *et al.* (1990). The details of the method will be explained elsewhere. Let us just state here that when we compare the solar frequencies with those of a model that has no convective penetration, we find the unmistakable signature of penetration, implying that such penetration indeed occurs in the Sun. But not all differences can be interpreted by the lack of penetration in that model; a small fraction seems still to be due to a deficiency in the opacity, although we have taken the latest data of Iglesias and Rogers (1991). We have also examined the effect of an imperfect equation of state.

At this writing, our best estimate for the penetration parameter is $\zeta \approx 0.70$, and we are working to refine it. That number would mean that the extent of penetration amounts to about 20,000 km. The depth of the convection zone (including penetration) would vary accordingly.

ACKNOWLEDGEMENTS

This work was supported by the Centre National de la Recherche Scientifique (GDR 131), by the U.S. Air Force through grant AFORS 89.0012 to Columbia University and by NASA through grant NSG-7511 to the University of Colorado.

REFERENCES

Christensen–Dalsgaard, J. 1988, in Seismology of the Sun & Sun-like Stars (ESA publ. SP-286), 431

Christensen-Dalsgaard, J., Gough, D.O. and Pérez Hernández, F. 1988, *M.N.R.A.S.*, **235**, 875

Iglesias, C.A. and Rogers, F.J. 1991, *Ap. J.*, **371**, 408

Libbrecht, K.G., Woodard, M.F. and Kaufman, J.M. 1990, *Ap. J. Suppl.*, **74**, 1129

Morel, P., Provost, J. and Berthomieu, G. 1990, *Solar Phys.* **128**, 7

Zahn, J.-P. 1991, *Astr. Ap.*, (in print)

Seventh Cambridge Workshop on Cool Stars, Stellar Systems, and the Sun
ASP Conference Series, Vol. 26, 1992
Mark S. Giampapa and Jay A. Bookbinder (eds.)

DYNAMICS OF THE QUIET SOLAR ATMOSPHERE: K_{2V} CELL GRAINS VERSUS MAGNETIC ELEMENTS

PETER N. BRANDT Kiepenheuer–Institut für Sonnenphysik, Schöneckstr. 6, D–7800 Freiburg, Germany

ROBERT J. RUTTEN Sterrekundig Instituut, Postbus 80000, NL 3508–TA Utrecht, The Netherlands

RICHARD A. SHINE Lockheed Palo Alto Research Laboratories, Code 913O, Bldg 252, 3251 Hanover Street, Palo Alto CA 94304, USA

JAVIER TRUJILLO BUENO Instituto de Astrofisica de Canarias, E–38200, La Laguna, Tenerife, Spain

ABSTRACT We employ high-resolution Ca II K_{2V} imaging to solve the controversy regarding Sivaraman and Livingston's identification of K_{2V} cell grains with magnetic elements.

Keywords: chromospheres; magnetism; hydrodynamics; stellar atmospheres

INTRODUCTION

The lower solar chromosphere is split into magnetic and nonmagnetic components. Both regimes are heated. The classical theory of acoustical heating now enjoys a revival: propagating sound waves and shocks may heat the nonmagnetic quiet chromosphere and so explain the "basal flux" observed in other cool stars (Schrijver *et al.* 1989). A pivot phenomenon in this context is the presence in quiet areas of the solar surface of tiny, short-lived emission features in a narrow wavelength band just to the blue of the Ca II K core, the "K_{2V} cell grains". They seem to be a direct manifestation of acoustical wave heating in the low chromosphere, but their nature remains elusive (review by Rutten & Uitenbroek 1991b).

An issue of controversy is whether these grains are of magnetic origin. Direct correspondence between K_{2V} cell grains and intranetwork magnetic field concentrations was observed by Sivaraman & Livingston (1982, see also Sivaraman 1991), but others doubt this result (Deubner 1991, Rutten & Uitenbroek 1991a, 1991b). We solve this dispute here with new observations; it turns out that everybody has been right. There are *two* distinct classes of Ca II cell-interior grains: *(i)*—oscillatory K_{2V} flashes, with 3–min periodicity, brief spatial memory, 5–10 per cell; *(ii)*—persistent K–line grains, retaining long-term identity while traveling through the cell and flashing with 3–5 min periodicity, about 1 per cell.

LA PALMA OBSERVATIONS

We use the Swedish Solar Vacuum Telescope on La Palma to study K_{2V} cell grains, combining the prototype OSL camera and the Lockheed narrow-band Ca II K filter with the Swedish video image grabbing system to obtain simultaneous and co-spatial chromospheric K_{2V} and photospheric granulation image sequences. Figure 1a is an image from a four-hour K_{2V} movie. It shows network as extended bright structures with smaller, grainy enhancements, and quiet cell interiors as darker areas at upper left and lower right. The latter contain K_{2V} cell grains, a bright one being marked by the crosshair intersection. Only bright K_{2V} grains are visible here; they would stand out more strongly for a narrower passband than our $\Delta\lambda = 0.03$ nm as is clear from the spectra published by Cram & Damé (1983).

Figure 1b is a "timeslice" along the horizontal crosshair, showing the evolution along the corresponding linear surface segment during a 56 min segment of the K_{2V} movie. The K_{2V} grain marked in the frame above is again at the crosshair intersection; the grain to the left of it appears at the same moment. Note the marked distinction between network and quiet areas in K_{2V} time behavior. The piece of network across the middle of the crosshair shows emission throughout this sequence, resulting in a columnar brightness streak with superimposed intensity oscillations of 5–min and longer periodicities. The quiet cell interiors show wispy brightness patterns on a dark background which mark the 3–min oscillation, generally present throughout quiet areas but reaching large brightness only in the more localized K_{2V} grains.

THE PERSISTENT FLASHER

Figures 2a and b are another K_{2V} image and timeslice combination. There is a bright grain below the center of the quiet cell in the upper-left quadrant of the image. On such single frames, it appears as just another K_{2V} oscillation flash, but our movie and timeslices indicate that this is a wholly different type of feature. On the movie it is seen to be a bright moment of a persistent feature which maintains its identity throughout the movie while flashing on and off in brightness. At the start, it is located higher up in this cell; it then travels towards the network at the lower part of the frame, which it reaches after about three hours. The timeslice was constructed along the track of this persistent flasher, during the hour ending when it reaches the network. The flasher demonstrates its longevity by producing a streak of brightness in the timeslice that is much more alike to the network brightness columns then to the wispy K_{2V} oscillation patterns.

Is this persistent flasher magnetic? There is no direct magnetic field diagnostic in our data; therefore, we turn the question around and ask whether similar features are seen in magnetogram movies. That is indeed the case. For example, the magnetogram movies taken by the Lockheed group at La Palma with the *SOUP* tunable filtergraph (Tarbell *et al.* 1990, Title *et al.* 1991) contain much plage but also cells without strong fields. The latter contain a few isolated magnetic features, typically one per cell, which also travel from cell center to nearby plage within a few hours.

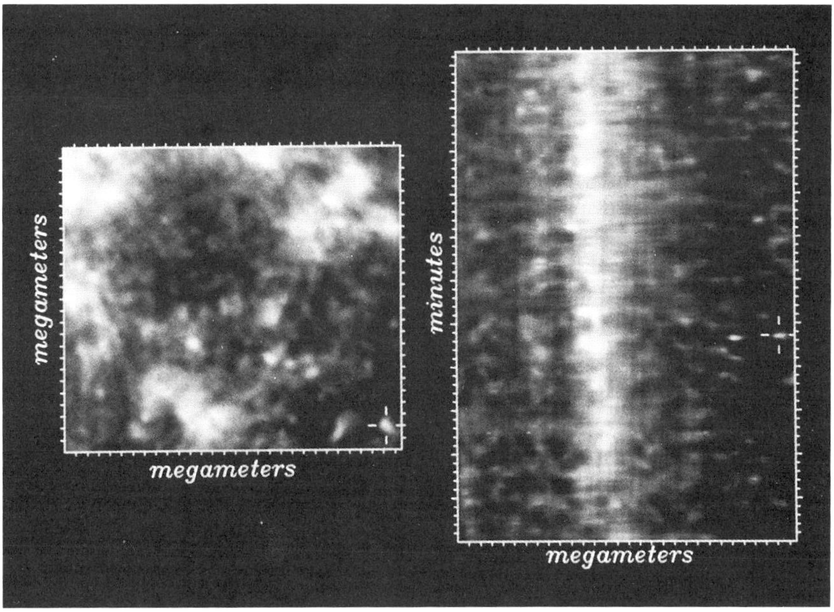

Figure 1. (a) K_{2V} image with a bright K_{2V} grain marked in the lower right hand corner. Each thick mark is 1000 km. (b) Space/time slice through the mark in (a) along the horizontal direction. The vertical axis is time with 1 minute tickmarks.

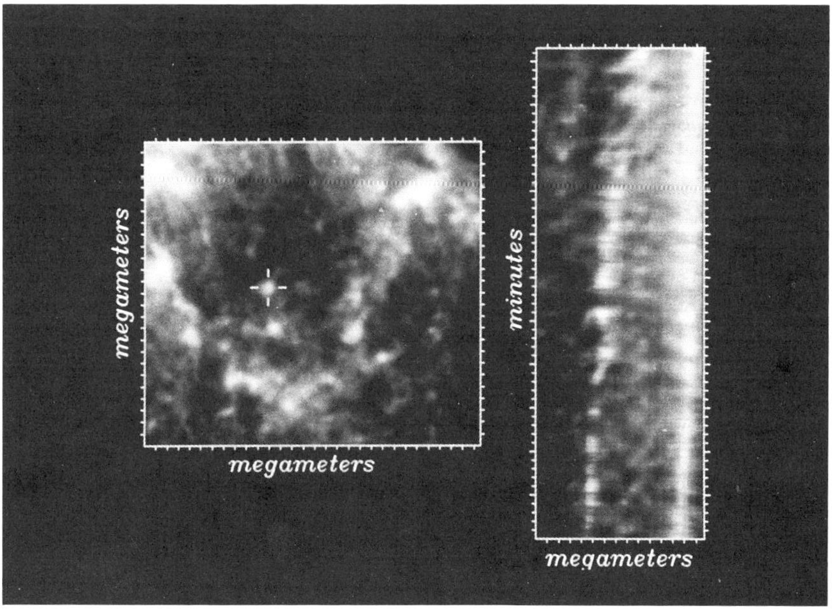

Figure 2. (a) K_{2V} image with a persistent flasher marked. (b) Space/time slice along the track of the flasher. Scales as in figure 1.

DISCUSSION

Sivaraman & Livingston used a 0.1 nm wide passband, equaling the integrated H-index of Cram & Damé (1983) in which only the very brightest K_{2V} features are seen. The best existing K_{2V} spectroheliograms have $\Delta\lambda$ = 0.01 nm and show 10–20 K_{2V} grains per cell. Our $\Delta\lambda$ = 0.03 nm passband is already too wide to observe the weaker grains properly; at $\Delta\lambda$ = 0.1 nm, most are unobservable. Sivaraman & Livingston indeed observed only few grains per cell. The migration and long-livedness which these possessed indicate that they were of the persistent-flasher type, not regular K_{2V} oscillation grains. The observed spatial correspondence with intranetwork magnetic features then indicates that persistent flashers are indeed of magnetic nature.

ACKNOWLEDGEMENTS

We gratefully acknowledge travel support from NATO.

REFERENCES

Cram, L. E. and Damé, L.: 1983, *Ap. J.*, **272**, 355

Deubner, F.-L.: 1991, in P. Ulmschneider, E. Priest, and B. Rosner (Eds.), Mechanisms of Chromospheric and Coronal Heating, Heidelberg Conference, Springer Verlag, Berlin, in press

Rutten, R. J. and Uitenbroek, H.: 1991a, *Solar Phys.*, **134**, 15

Rutten, R. J. and Uitenbroek, H.: 1991b, in P. Ulmschneider, E. R. Priest, and R. Rosner (Eds.), Mechanisms of Chromospheric and Coronal Heating, Proc. Heidelberg Conference, Springer Verlag, Berlin, p. 48

Schrijver, C. J., Dobson, A. K., and Radick, R. R.: 1989, *Ap. J.*, **341**, 1035

Sivaraman, K. R.: 1991, in P. Ulmschneider, E. Priest, and B. Rosner (Eds.), Mechanisms of Chromospheric and Coronal Heating, Heidelberg Conference, Springer Verlag, Berlin, in press

Sivaraman, K. R. and Livingston, W. C.: 1982, *Solar Phys.* **80**, 227

Tarbell, T., Ferguson, S., Frank, Z., Shine, R., Title, A., and Topka, K.: 1990, in J.-O. Stenflo (Ed.), The Solar Photosphere: Structure, Convection and Magnetic Fields, Proceedings IAU Symposium 138 (Kiev), Kluwer, Dordrecht, p. 147

Title, A. M., Topka, K. P., Tarbell, T. D., Schmidt, W., Balke, C. A., and Scharmer, G.: 1991, *Ap. J.*, in press

Seventh Cambridge Workshop on Cool Stars, Stellar Systems, and the Sun
ASP Conference Series, Vol. 26, 1992
Mark S. Giampapa and Jay A. Bookbinder (eds.)

COMPRESSIBLE CONVECTION: DEEPER LAYERS

KWING L. CHAN Applied Research Corporation, Landover, Maryland

ABSTRACT Numerical experiment shows that large subadiabatic regions exist inside deeply stratified convection zones.

Keywords: stellar structure; convection

INTRODUCTION

The appearance of strong, columnar downdrafts penetrating many scale heights in simulations of compressible convection (Chan and Sofia 1986, Stein and Nordlund 1989, Cattaneo *et al.* 1991) marks an important digression from the standard picture of the mixing length theory (MLT). Due to the large amount of kinetic energy carried by these downdrafts, the enthalpy flux is no more close to the total flux through the convection zone, as assumed by the MLT. Additionally, the concentrated downward momentum associated with these columns may inflict significant erosion of the lower radiative-convective interface. To understand overshooting and mixing near the bottoms of convection zones, it is necessary to study the behavior of these downdrafts in the very deep layers. We report here some results of a numerical experiment to investigate this problem.

THE NUMERICAL MODEL

In order to minimize the effects of imposing boundaries cutting the convection zone artificially in the middle, we study a configuration which contains radiative layers both above and below the convection zone, thus providing a less intrusive means to stop the convective motions at the two ends. The radiative and convective layers are created by setting the radiative diffusivity to large and small values respectively. (The radiative flux is proportional to the negative temperature gradient). The domain of computation contains about 8.5 pressure scale heights (p.s.h.) within which 7 p.s.h. are convective. The salient features of the model are: (i) ideal gas with ratio of specific heats = 5/3; (ii) Smagorinsky sub-grid-scale viscosity and diffusivity; effective Prandtl number = 1/3; (iii) stress-free, impenetrable upper and lower boundaries; periodic side boundaries; (iv) constant temperature at the top; constant input flux at the bottom; total flux = 0.25 (dimensionless units with initial pressure, temperature, and density at the top set to 1); (v) aspect ratio (width/depth) = 1.5; (vi) solve Navier Stokes equations in strong conservation form with a 69x69x51 mesh.

RESULTS

1. The flux of kinetic energy carried by the downdrafts first increases with depth in the upper few p.s.h., then decreases in deeper layers, indicating that the mean strength of the downdrafts eventually dwindles in deep convection zones (Fig. 1).

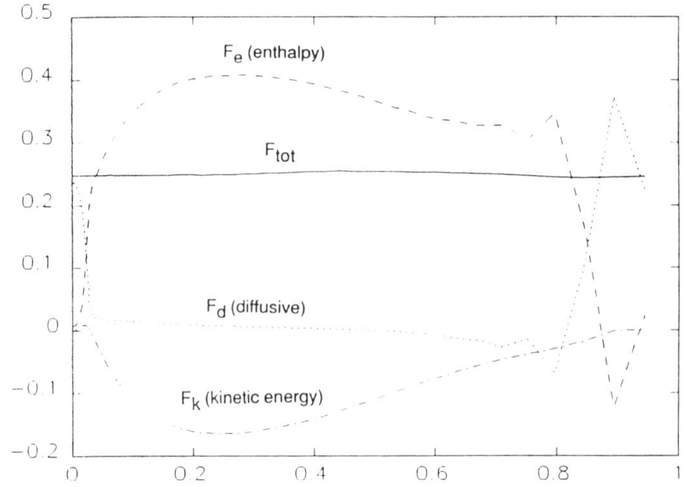

Fig. 1: Distributions of the enthalpy flux (dashed curve), flux of kinetic energy (dot-dashed), diffusive flux (dotted), and total flux (solid) vs. depth.

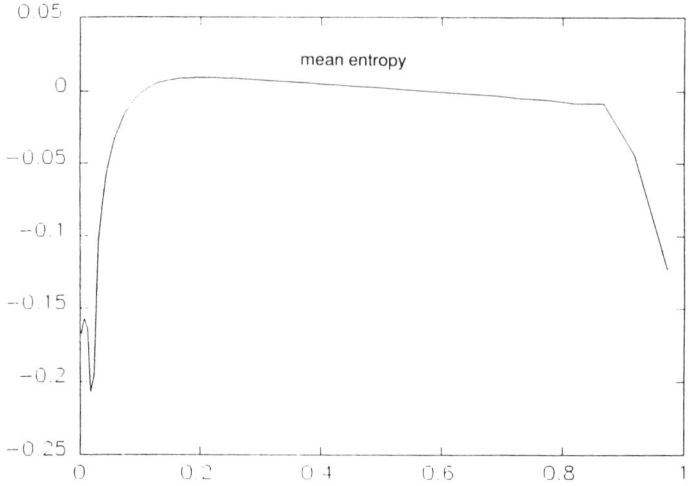

Fig. 2: Distribution of the mean specific entropy vs. depth.

2. While the convection zone is superadiabatic in the upper few p.s.h., it turns slightly subadiabatic in the lower region (Fig. 2). Therefore, the standard MLT cannot be applied in this region. Nevertheless, the lower convection zone remains quite close to being isentropic. The entropy inversion is created by the downdrafts which carry cool, low entropy fluid parcels from the very top layer to lower layers and pile them there. The subadiabaticity generates a counteraction to the downdrafts and thus the flux of kinetic energy is slowly dissipated in this region.

3. We investigated the cancellation of kinetic and enthalpy fluxes carried by the downflows discovered by Cattaneo *et al.* . Our situation is somewhat different because the Prandtl number is substantially larger and the convection is very effective (*i.e.* convective flux >> diffusive flux). The two fluxes do not cancel exactly, but the net flux is only about 30% of the total flux and the distribution is indeed rather flat (Fig. 3).

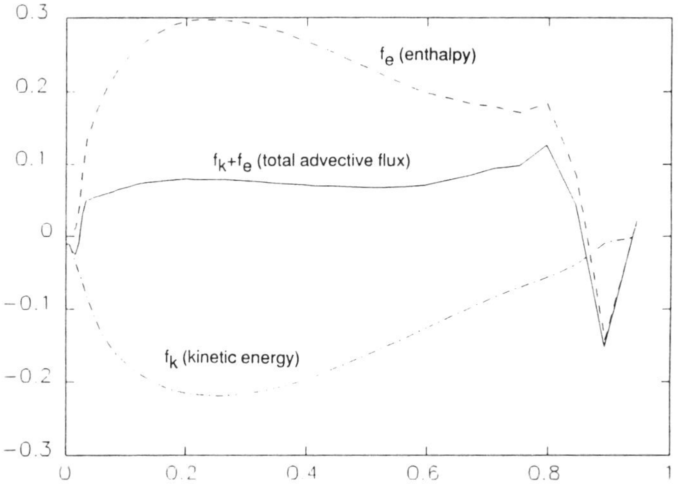

Fig. 3: Depth distributions of fluxes carried by the downflows: enthalpy (dashed curve), kinetic energy (dot-dashed), and the sum (solid).

ACKNOWLEDGEMENTS

The author would like to acknowledge support from NSF and computing support from the Pittsburgh Supercomputing Center.

REFERENCES

Cattaneo *et al.* 1991, *Ap. J.*, **370**, 282.
Chan, K. L., and Sofia, S. 1986, *Ap. J.*, **307**, 222.
Stein, R. F., and Nordlund, Å., *Ap. J. Letters*, **342**, L95.

Seventh Cambridge Workshop on Cool Stars, Stellar Systems, and the Sun
ASP Conference Series, Vol. 26, 1992
Mark S. Giampapa and Jay A. Bookbinder (eds.)

BISECTOR AND LINE PARAMETER VARIATION OVER A SINGLE SOLAR GRANULUM

ARNOLD HANSLMEIER Institut f. Astronomie, Univ. -Platz 5, A-8010 Graz, Austria

WOLFGANG MATTIG and ANASTASIOS NESIS Kiepenheuer Institut, Schoeneckstr. 6, D-7800 Freiburg, Germany

ABSTRACT We give two examples of bisector and line parameter variations over a single solar granular intergranular area. The variation of these parameters with photospheric height is investigated in detail.

Keywords: Convection; solar granulation; line profiles

INTRODUCTION

The convective overshoot motions in the solar photosphere can be observed as granulation. Because of the small spatial diameter of these elements (about 1Mm) it is very difficult to study spatially resolved line profiles originating from the hotter ascending granulum and the cooler descending intergranular lanes. These observations are needed for a direct comparison between various theoretical models (Cloutman (1979), Nelson (1978), Nordlund (1982, 1985), Steffen (1989)). The statistical variation of the line parameters over many granular intergranular areas with high spatial resolution has been given by Hanslmeier *et al.* (1990) and Guenter and Mattig (1990). This paper presents line profiles and line parameter variations from two selected large granular-intergranular areas in a non-active region near the solar disk center.

OBSERVATIONS

The observations have been performed with two different telescopes at the German solar observatory at Izaña, Teneriffe: i) Gregory Coudé telescope: 0.45-m; f/56; image scale on the spectrograph entrance slit 8.25 arc-sec/mm and ii) Vacuum Tower Telescope: 0.70-m; f/66; image scale on the spectrograph entrance slit 4.5 arc-sec/mm. In both cases a spectrograph slit width of 60μ has been used. The spectra were obtained photographically with an exposure time of 3 sec.

For a symmetric line profile, the bisectors, halving the line profile at equal intensity levels, are straight lines. Since the line center is formed at higher levels in the solar photosphere, asymmetric line profiles leading to curved bisectors, represent velocity gradients.

We present results from the analysis of the line at 6494.994 Å which had a line core formation height about 500 km. Therefore, nearly the whole photosphere up to the temperature minimum was covered. More results from observations with the Gregory Coudé telescope can be found in Hanslmeier *et al.* (1991).

RESULTS

The following examples give the line bisector variation (a), the continuum intensity variation (b) and the full width at half maximum variation (c) over a selected single granular-intergranular area. The data from Fig. 1 are from the Gregory Coudé telescope, the data from Figs. 2 from the VTT. Two neighboring bisectors in Fig. 1 have a spatial separation of 0.16 arc-sec, in Fig. 2 of 0.137 arc-sec (these values arise from the digitizing of the spectrograms).

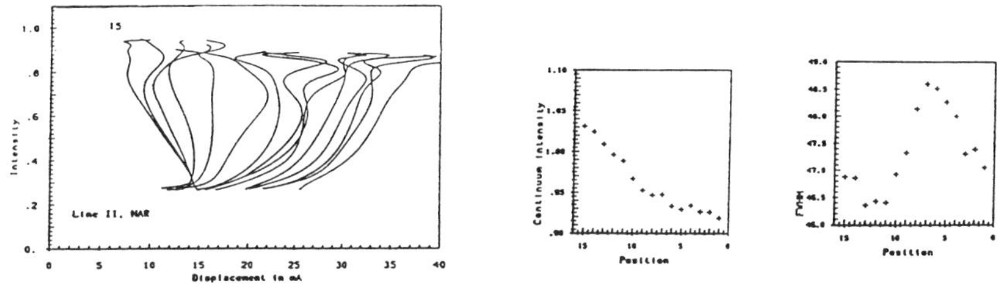

Fig. 1: Line parameter variation from an intergranular lane (position #1) to the center of a bright granulum (position #15).

The Bisectors have blue (granulum) and red asymmetries (intergranular lane). The greatest red asymmetries about 15 mÅ are found for bisector #1 in Fig. 1a and Fig. 2a. The maximal blue asymmetric bisectors are found in Fig. 2a (bisector #10) reaching nearly the same values of 15 mÅ.

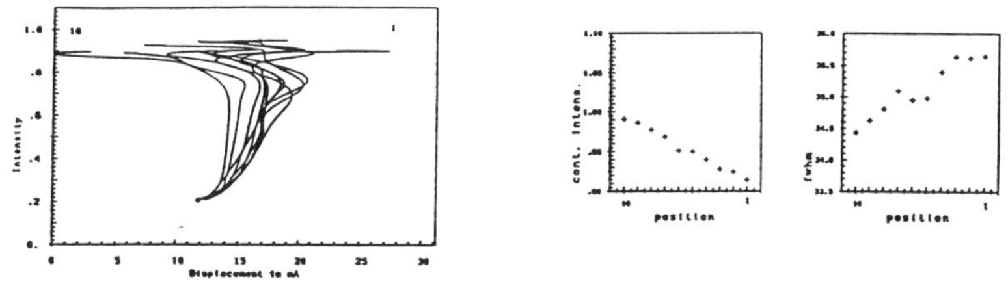

Fig. 2: Line parameter variation from an intergranular lane (position #1) to the center of a bright granulum (position #10).

The examples show a well defined transition of the continuum intensity I_0 from lower to higher values (granulum) with a difference of about $0.1 I_0$ between intergranulum and granulum. The full width at half maximum *FWHM* values are increased in the intergranulum. However, as is illustrated by Figs. 1c and 2c, this variation is rather complex. In all examples, the difference between granular and intergranular areas is about 4%. Bisector #11 in Fig. 1a and bisector #6 in Fig. 2a have the smallest asymmetry. It can be shown, that these bisectors coincide with the minima found in the variations of the *FWHM*.

DISCUSSION

The resolution of the larger granulum is better than the resolution of the smaller intergranular lanes. Therefore we get about the same red and blue asymmetries. From the argument of mass conservation the downward velocity gradients should be greater. For spatially resolved line profiles the bisectors no longer have a C shape, but are blue and red asymmetric according to their origination in the granular or intergranular areas. As can be seen from Figs. 1a and 2a, the shape of the bisectors for continuum intensity I_0 *geq* 0.8 differ from the shape for I_0 ¡ 0.8. Since the contribution to the line profile near the continuum value occurs from the deepest photospheric levels, a greater turbulence at these layers cause these variations. The bisectors of Fig. 1a are asymmetric at even higher photospheric levels with lower continuum intensity values. In this example, the convective motions may have penetrated nearly the whole photosphere. The selected area given by Fig. 2a shows no velocity fluctuations at greater photospheric heights, but again strong velocity fluctuations and asymmetry variations in the deeper layers. The analysis of the corresponding continuum intensity variations and full width at half maximum variations show, that on the one hand well known facts are confirmed (enhanced *FWHM* in the intergranulum which is darker), but that the transitions from granulum to intergranulum is not linear. The point with the smallest line asymmetry seems to be connected with a minimum found in the *FWHM* curve. The increased *FWHM* values in the intergranular areas can be explained by non-thermal line broadening in these regions, caused by pressure waves. The asymmetric transition from granular to intergranular lanes is also studied in Nesis *et al.* (1991a,b).

ACKNOWLEDGEMENTS

This project is supported by the Austrian Fonds zur Foerderung der wissenschaftlichen Forschung (Proj. Nr. 8064-PHY).

REFERENCES

Cloutman, L.D. 1979, *Ap. J.*, **227**, 614.
Guenter, E., Mattig, W. 1990, *Astr. Ap.*, **243**, 244.
Hanslmeier, A., Mattig, W. and Nesis, A. 1990, *Astr. Ap.*, **238**, 354.
Hanslmeier, A., Mattig, W. and Nesis, A. 1991, *Astr. Ap.*, in press.
Nelson, G.D. 1978, *Solar Phys.* **60**, 5.
Nesis, A., Hanslmeier, A., Hammer, R., Komm, R., Mattig, W. and Staiger, J. 1991a, *Astr. Ap.*, in press.

Nesis, A., Hanslmeier, A., Hammer, R., Komm, R., Mattig, W. and Staiger, J. 1991b, Cool Stars, Stellar Systems, and the Sun.

Nordlund, A. 1982, *Astr. Ap.*, **107**, 1.

Nordlund, A. 1985, *Solar Phys.* **100**, 209.

Steffen, M., Ludwig, H.G. and Kruess, A. 1989, *Astr. Ap.*, **213**, 371.

Seventh Cambridge Workshop on Cool Stars, Stellar Systems, and the Sun
ASP Conference Series, Vol. 26, 1992
Mark S. Giampapa and Jay A. Bookbinder (eds.)

STELLAR PHOTOSPHERIC CONVECTION

YONG -CHEOL KIM & PETER A. FOX Department of Astronomy, and CSSR, Yale University, P.O. Box 6666, New Haven, CT 06511.

ABSTRACT Using a combination of three fields of astrophysics (stellar structure and evolution, hydrodynamics, and spectral line synthesis), a "C"-shaped spectral line bisector for the Sun and a reversed "C" shaped line bisector for an A5 star can be reproduced.

Keywords: Sun; convection; granulation; spectral line asymmetry; stellar structure

INTRODUCTION

A systematic study of the asymmetry of spectral lines provides an important probe of convective motions in stellar photospheres. For the Sun, the surface convection, called granulation, has broader regions of hot rising convective elements, and narrower areas of cooler sinking elements. The net effect of the unresolved granulation is a depressed red shoulder of the solar absorption lines. As a result, the line bisector, which connects the midpoints of horizontal line segments bounded by the spectral line profile, has a "C" shape. An important discovery was that there are stars whose characteristic bisectors are a reversed-C shape as well. It is now known observationally that stars with different line bisector shapes (which, therefore, have different characteristic convective motions), are located in different regions on the H-R diagram. The boundary between these two regions is called the "granulation boundary" (Gray and Nagel 1989). The stars on the cooler side of the boundary have "C" shaped bisectors, the others have the reversed "C" shape.

In this paper we explore the possibility that the accuracy of detailed stellar convection simulations can be tested using the observed line asymmetry. An ultimate goal of this project is to reproduce the granulation boundary as well as the asymmetric spectral lines of individual stars. After these tests, the information in our detailed simulations of stellar photospheric convection can provide a deeper understanding of convective energy transportation in stars. As a consequence, certain theoretical approximations for convection, *i.e.*, the mixing length theory, can be evaluated using our hydrodynamic simulations.

OVERVIEW OF THE PROJECT

First, the detailed structure of a whole star is constructed. The stellar model is constructed using Yale Rotating Evolution Code (YREC) with the non-rotating configuration. For the equation of state, Mihalas, Hummer, and Däppen's (M-H-D equation of state) table was used for the Sun. For stars, the built-in equation of state in YREC has been used. This is basically the solution of the Saha equation taking into account only the single ionization of the elements, except He whose single and double ionization are considered. For the opacity, Los Alamos opacities are used. Then, a part of the model structure is used as an initial condition for the detailed convection simulation. This way, the convective region of the star we analyze is consistent with the interior. This ensures that the simulation is not of a convection zone in an artificial environment. For these calculations, the Anders-Grevesse mixture (1989) was used.

TABLE 1 STELLAR MODELS

	$M/M_\odot$	$\log L/L_\odot$	$\log R/R_\odot$	$\log T_{eff}$	Age(Gyr)	α	X	Z
Sun	1.00	0.00	0.00	3.76	4.5	1.230	0.7071	0.0188
A5	1.86	1.15	0.24	3.93	0.45	1.230	0.7071	0.0188

The next step is to solve the hydrodynamic equations to simulate stellar convection zones, which is one of the most challenging tasks. Since the convection is turbulent, simulation with any kind of approximation may easily fail to depict the detailed characteristics of the convection to the degree that we want. Furthermore, the effect we want to see, likely occurs at the region where the radiative energy transfer is not totally negligible. Therefore, the fully compressible, radiation-coupled hydrodynamic equations (Chan and Sofia 1986) with realistic equation of state and opacity have to be solved.

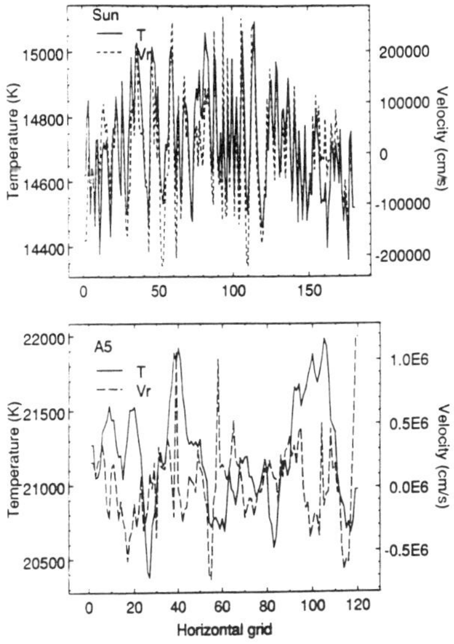

Fig. 1: The temperature and the radial velocity fluctuation.

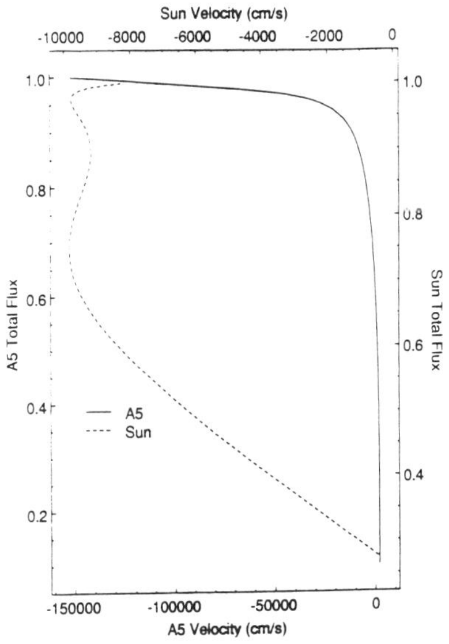

Fig. 2: The line bisector calculated using the fluctuating quantities shown in Fig. 1.

The present results are for the two dimensional case using spherical coordinates. The numerical technique is the alternate direction implicit on staggered mesh (Chan and Wolff 1982). The equation of state and opacity is consistent with that for the initial stellar models. Since in the present models, the energy transfer by radiation was treated using the radiative diffusion approximation, regions with small optical depth were avoided.

Finally, using the characteristics of the convection from the simulation (Fig. 1 shows the fluctuating quantities at a horizontal level), disk integrated synthetic spectral lines are calculated and then compared with an observed spectrum. The velocity and temperature fluctuations from the simulation cause the characteristic spectral line asymmetry. The asymmetry depends on the convection, which depends, again, on the condition of the stellar convection zone.

SUMMARY

The simulation for convection in an A5 star shows that the temperature fluctuations are larger than that of the Sun (Fig. 1). Therefore, the reversed-C shaped bisector can be understood in terms of the large intensity contrast between the hot rising elements and the cool sinking elements. When the contribution from sinking element on the integrated intensity is much smaller, the effect of the rising elements is dominant on spectral lines. Therefore, the unresolved convective motion in the disk integrated image may look like only up-flows on the whole visible sphere. This *expanding star effect* may be responsible for the reversed C shaped bisector (Fig. 2).

This simulation reveals that there are several characteristics of the convective flow on A5 stars which are different from that of the Sun. First, the mean radial velocity is larger that that of the Sun. Second, velocity of sinking elements is comparable to that of rising element. And, finally, the radial velocity fluctuation is comparable to that of the horizontal velocity fluctuation.

A more rigorous analysis of this simulation will provide interesting and important information on the stellar photospheric convection of A5 stars.

The qualitative agreement with observation in the present two simulations can be further quantified by a detailed comparison of an observed stellar spectrum at a given wavelength.

ACKNOWLEDGEMENTS

This work was supported in part by NASA grants NAGW-778 and NAG5-1486 to Yale University. Computer time was provided by NCCS at NASA/GSFC.

REFERENCES

Anders, E. and Grevesse, N. 1989, *Geochimica et Cosmochimica Acta*, **53**, 197.
Chan, K. L. and Sofia, S. 1986, *Ap. J.*, **307**, 222.
Chan, K. L. and Wolff, C. L. 1982, *Journal of Computational Physics*, **47**, 109.
Gray, D. F. and Nagel, T. 1989, *Ap. J.*, **341**, 421.

Seventh Cambridge Workshop on Cool Stars, Stellar Systems, and the Sun
ASP Conference Series, Vol. 26, 1992
Mark S. Giampapa and Jay A. Bookbinder (eds.)

VELOCITY FLUCTUATIONS AND ENERGY DISSIPATION IN THE SOLAR PHOTOSPHERE

R. KOMM National Solar Observatory, National Optical Astronomy Observatories* , Tucson, Az 85726, USA

W. MATTIG, A. NESIS Kiepenheuer Institut für Sonnenphysik, D-7800 Freiburg, Germany

ABSTRACT Cross-correlations of velocity fluctuations are used to define lengthscales characterizing the 'size' of motions which are coherent between different photospheric layers. The coherent lengthscale of lower to higher layers is rather small ($\approx$ 600 km), while the lengthscale of high to higher layers is more than two times larger ($\approx$ 1500 km). Hence the larger structures of the granular motions can not reach higher photospheric layers, and the secondary motions of the higher photosphere are of the same size than gravity waves. In addition, the dissipation rate ϵ is very large in the lower photosphere $\approx 10^8$ cm^2s^{-3} at about 0 km and decreases with height to $\approx 10^6$ cm^2s^{-3} at about 170 km, reflecting so the decay of the granular motions, and remains almost constant in the higher photospheric layers.

Keywords: Solar photosphere; cross-correlation; dissipation; granules

The dynamics of the solar photosphere has been the topic of our recent investigations. There we find that the granular motions are highly turbulent and decay with height (up to about 170 km) due to the strong density stratification of the solar photosphere. In the present paper, we further analyze this decay with height. By means of a cross-correlation analysis we derive a lengthscale L_{cc} characterizing the maximum size of velocity structures which are coherent with height. The height variation of this lengthscale must reflect the decay since the large structures provide the energy input for the turbulent energy cascade. The disintegration of granules or the decay of the granular motions with height must also result in a large energy dissipation rate in the lower photospheric layers. While the actual dissipation takes place at the very small scales, the amount of energy which is dissipated is determined by the large structures. Furthermore, the dissipation rate can provide information about the nature of the small-scale motions in the higher photospheric layers.

We used high resolution spectrograms obtained during the partial eclipse on 30 May 1984 with the echelle spectrograph of the Vacuum Tower Telescope at Sacramento Peak Observatory. The observations and the data reduction are described by Komm *et al.* (1991a). For the present paper, we calculated the cross-correlation between velocity fluctuations derived from the lines C I

---

* Operated by the Association of Universities for Research in Astronomy, Inc. under cooperative agreement with the National Science Foundation.

5380.3Å representing the lower photosphere, Fe I 5379.6Å representing mid-higher photosphere, and Fe I 5576.1Å representing the higher photosphere. Since there are several ways to define a lengthscale from correlation measurements (*cf.* Tritton, 1977), we define the lengthscale L_{cc} as the distance between the two points (for positive and for negative lag) where the cross-correlation falls to 1/e multiplied by the value of the cross-correlation at zero lag.

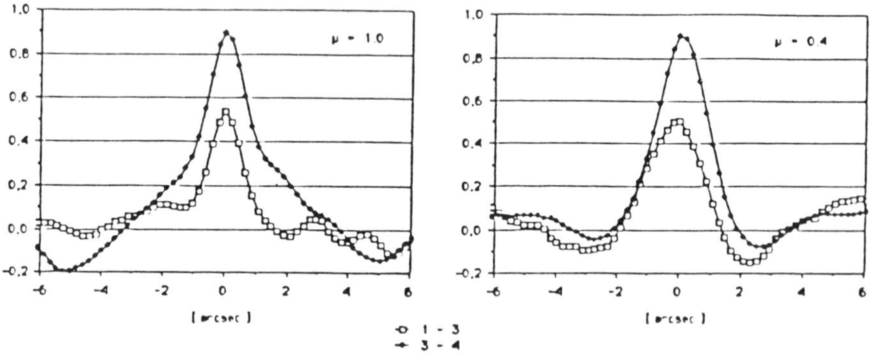

Fig. 1: The cross-correlations of velocity fluctuations from three different spectral lines. 1 - C I 5380 Å , 3 - Fe I 5379 Å , 4 - Fe I 5576 Å .

TABLE I L_{cc} in km for vertical ($\mu = 1.0$) and horizontal ($\mu = 0.4$) velocity

lines	$\mu = 1.0$	$\mu = 0.4$	layers
1 - 3	604	824	low-high
1 - 4	630	743	low-high
3 - 4	1492	1432	high-high

The coherent lengthscale L_{cc} of lower to higher photosphere is smaller than the average size of granules which is about 1000km (Bray *et al.*, 1984) and smaller than the autocorrelation length in the lower photosphere which is also $\approx 1000km$ (Komm *et al.*, 1991b). Therefore, the larger granular motions which contribute most to the turbulent energy do not reach the higher photospheric layers. The small-scale motions of the higher photosphere are of secondary non-convective origin. The coherent lengthscale L_{cc} within the higher photospheric layers is large and of the same size as the autocorrelation lengthscale which is $\approx 1400km$ (Komm *et al.*, 1991b). That simply implies that the secondary motions do not change in these photospheric layers. This lengthscale is also equal to the mean granulation cell size of $\approx$ 1400km (Bray *et al.*, 1984) and demonstrates thus that the secondary motions are generated by the granular motions ('piston' action of granules). This lengthscale is too small for oscillations but is of the right size for gravity waves (*cf.* Thomas *et al.*, 1971).

The mean dissipation of energy (erg/g·s = cm^2s^{-3}) is determined in a turbulent flow by the large eddies (*cf.* Landau and Lifshitz, 1987), and can be calculated for very large Reynolds numbers (*cf.* Rotta, 1972) as follows:

$\epsilon = 0.2(3/2)^{3/2}(\Delta u)^3/L$, Δu : velocity fluctuation, L : macro-lengthscale. The Reynolds number is indeed very large in the photosphere (Komm *et al.*, 1991b). The height dependence of the velocity field was deduced by Komm *et al.*, (1991a), and the macro-lengthscale was derived by Komm *et al.* (1991b).

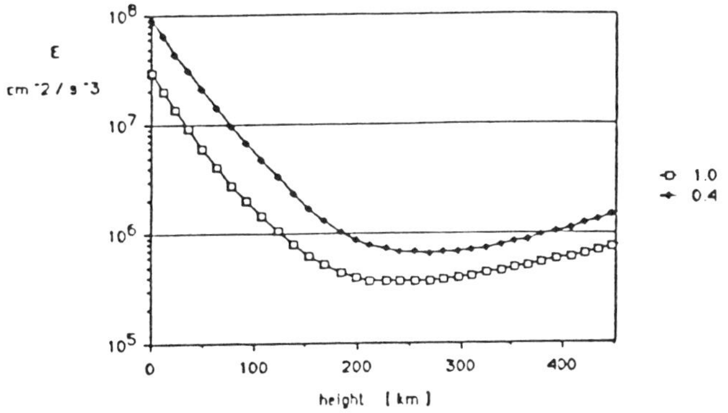

Fig. 2: The energy dissipation rate ϵ derived from the vertical and the horizontal velocity component.

In the lower photosphere (< 170 km), the energy dissipation rate ϵ is very large of $\approx 10^8 cm^2 s^{-3}$ at 0 km and decreases by two orders of magnitude to $\approx 10^6 cm^2 s^{-3}$ at 170 km. This reflects quantitatively that the granular motions decay up to this height. The horizontal velocity leads to a 2 to 3 times larger dissipation rate than the vertical component which shows its importance for the dynamics of the photosphere. However, this large amount of energy must be dissipated in the small scales. Komm *et al.* (1991b) show that granules are able to decay due to turbulent mixing which occurs at the boundary of turbulent structures. This agrees perfectly with the results of Nesis *et al.* (1991) finding enhanced nonthermal velocities in intergranular regions and interpreting this as enhanced energy dissipation. In the higher photosphere (> 170 km), the energy dissipation ϵ is comparably small and almost constant of about $10^6 cm^2 s^{-3}$. This implies that the major energy dissipation of the secondary motions occurs not in these photospheric layers. Hence the secondary motions must be some kind of wavefield, and together with the result of the correlation analysis this leads to the conclusion that the secondary motions are gravity waves.

REFERENCES

Bray, R.J., Loughhead, R.E., and Durrant, C.J. 1984, The Solar Granulation, Cambridge University Press

Komm, R., Mattig, W., and Nesis, A. 1991a, *Astr. Ap.*, **243**, 251

Komm, R., Mattig, W., and Nesis, A. 1991b, *Astr. Ap.*, **252**, 812.

Landau, L.D., Lifshitz, E.M. 1987, Theoretical Physics Vol. 6, Pergamon Press

Nesis, A., *et al.* 1991, (this volume)

Rotta, J.C. 1972 Turbulente Strömungen, B.C. Teubner, Stuttgart

Thomas, J.H., Clark, P.A., and Clark, A. 1971, Solar Phys. **16**, 51

Tritton, D.J. 1977, Physical Fluid Dynamics, Van Nostrand Reinhold (UK)

Seventh Cambridge Workshop on Cool Stars, Stellar Systems, and the Sun
ASP Conference Series, Vol. 26, 1992
Mark S. Giampapa and Jay A. Bookbinder (eds.)

CHROMOSPHERIC LYMAN ALPHA EMISSION IN A-F STARS*

ETTORE MARILLI AND SANTO CATALANO Osservatorio Astrofisico di Catania, Viale A. Doria 6, I-95125 Catania, Italy

RUBENS FREIRE FERRERO Observatoire Astronomique, 11 Rue de l'Universitè, 67000 Strasbourg, France

PIERRE GOUTTEBROZE Institute d'Astrophysique Spatiale, IAS/LPSP CNRS, BP No. 10, 91371 Vèrrières le Buisson, France

FREDERICK C. BRUHWEILER Physics Department, The Catholic University of America, Washington DC 20064, USA

ANTONIO TALAVERA ESA IUE Observatory, Apartado 54065, 28080 Madrid, Spain

ABSTRACT Within a program aimed at locating the boundary for the occurrence of chromospheric emission in the H-R diagram, we report here on the detection and behavior of Ly-α emission in late A – early F stars, using the IUE. We show that chromospheric Ly-α emission is detected in stars as early as A7 (B-V = 0.19).

Keywords: Lyman alpha; chromospheres; stellar convection

INTRODUCTION

The location of the spectral boundary for the onset of chromospheric emission in the H-R diagram is one of the major challenges in the study of stellar structure. The detection of H I Lyman alpha (Ly-α) emission in Altair (A7 V, B-V = 0.22) (Blanco *et al.* 1980, Catalano *et al.* 1991a) has risen the question whether Altair is a special case or the general domain for chromospheric activity should include stars of earlier spectral type. We have shown that also α Cep (A7 IV-V, B-V = 0.22-0.24), a star very similar to Altair, clearly displays the presence of a chromosphere (Catalano *et al.* 1991b). However, in an effort to define the problem on a more general basis we have searched for chromospheric Ly-α emission in a small sample of A-F stars, using the IUE SWP camera in the high resolution mode. Here we report on the detection and behavior of the Ly-α emission in our program stars as well as on the analysis of archive data.

IUE OBSERVATIONS AND ANALYSIS

Investigations of the Ly-α region with IUE require very long exposures, an appropriate observation strategy and an accurate decontamination of the stellar

* Based on observations by the IUE collected at VILSPA and Goddard stations.

spectra from the geocoronal background (Freire Ferrero *et al.* 1990, Catalano *et al.* 1991b).

We have obtained 9 spectra with the SWP camera of IUE in the high dispersion mode of 5 A-F stars. Including spectra taken from the IUE archive we have extracted Ly-α profile for Altair (A7 V-IV, B-V=0.22), α Cep (A7 V-IV, B-V=0.22), ι UMa (A7 IV, B-V=0.19) γ Boo (A7 III, B-V=0.19), α Hyi (F0 V, B-V=0.28), β TrA (F2 III, B-V=0.29), β Cas (F2 III-IV, B-V=0.34). Ly-α emission has been seen in all the stars, even if in some cases the detection was very marginal, or even questionable as in the case of α Hyi. The average profiles of the new extracted spectra plotted in figure 1 clearly shows that chromospheres are present in A7 stars of luminosity classes V-III, with the bluest being γ Boo and ι UMa (B-V = 0.19). Ly-α profiles for Altair, α Cep and β TrA, have been already reported (Catalano *et al.* 1991a).

We have done preliminary estimates of the energy released in the line, fitting a gaussian profile on the outer emission wings. Since the effect of the interstellar hydrogen absorption may extend to the inner emission wings, the true gaussian profile may be brighter and larger, hence the present should be considered a lower limit estimate of the emission.

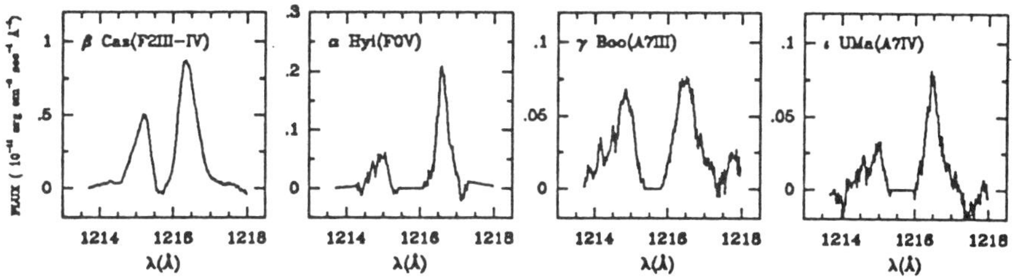

Fig. 1: IUE extracted Ly-α profile decontaminated for the geocoronal background and smoothed to 11 pixel or more.

DISCUSSION AND CONCLUSIONS

Considering stellar surface flux and luminosity of Ly-α emission derived from the present observations together with values determined from low dispersion IUE spectra by Bruno *et al.* (1991) and values taken from the literature we found that Ly-α luminosity, on the average, is larger for giants than for dwarfs of the same spectral type, thus suggesting a constant flux behavior. As a matter of fact, the surface flux plotted in figure 2, shows no clear differentiation of the emission between main sequence and giants at the same B-V. The average Ly-α surface flux has a curved trend with a maximum at B-V equal to 0.3–0.4, and a clear drop for B-V smaller than 0.3. This trend is consistent with the decline of X–ray emission found by Micela *et al.* (1988) for B-V < 0.5.

Our data refer to stars of very different rotational velocity. The fastest rotators, *e.g.* Altair and α Cep, have *vsin i* about 2/3 of the break-up velocity, which could mean that the true equatorial velocity may be very close to the limiting velocity. However, the Ly-α flux data on A-F stars do not show any clear dependence from the rotation speed, as already found for other activity diagnostics in the same spectral type range (*e.g.* Simon and Drake 1989). Although some of the stars are δ Scuti variables, no peculiar behavior is seen in their emission , in agreement with previous results on the Mg II (Fracassini and Pasinetti, 1982).

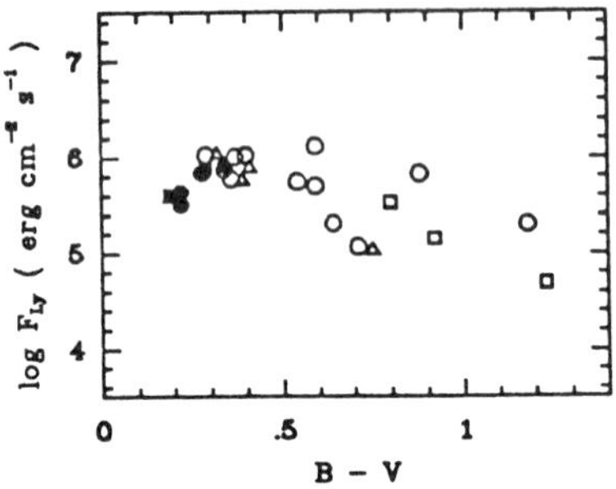

Fig. 2: Ly-α surface flux as a function of B-V. Filled symbols are present data, open symbols are IUE low dispersion (Bruno *et al.* 1991) and literature high dispersion data. Circles, triangles, squares refer to main sequence, subgiant and giant stars respectively.

From the present limited survey of Ly-α emission in relatively nearby stars we have set a new limit for the detection of chromospheric emission to B-V = 0.19. There is strong evidence for a decline in the chromospheric activity for B-V < 0.3 indicative of the drop-off in non-thermal heating, that can be consistent with a drop in magnetic energy generation due to the decrease of the convection depth, or with the trend of acoustic energy flux (Bohn 1984). We have also shown that Ly-α can be used as a very efficient chromospheric diagnostic, although the strong interstellar absorption imposes serious limitations.

ACKNOWLEDGEMENTS

This work was supported by an exchange program between the CNR(Italy) and DRCI-CNRS(France). The authors would also like to acknowledge support from Catania Astrophysical Observatory.

REFERENCES

Blanco, C., Catalano, S., and Marilli E. 1980, in Proc. Second European IUE Conference, ESA SP-157, p. 67.

Bohn, H.U. 1984, *Astr. Ap.*, **136**, 338.

Bruno, M., Catalano, S., and Marilli, E. 1991, Private communication.

Catalano, S., Marilli, E., Freire Ferrero, R. and Gouttebroze, P. 1991a, *Astr. Ap.*, in press.

Catalano, S., Gouttebroze, P., Marilli, E., Freire Ferrero, R. 1991b, in The Sun and Cool Stars: activity, magnetism and dynamos, eds. I. Tuominen *et al.* , (Springer-Verlag), p. 466.

Fracassini, M., and Pasinetti, L.E. 1982 *Astr. Ap.*, **107**, 326.

Freire Ferrero, R., Catalano, S., Gouttebroze, P. and Marilli, E. 1990, in Evolution in Astrophysics, Toulouse, France, ESA SP-310, p. 315.

Micela, G. *et al.* 1988, *Ap. J.*, **325**, 798.

Simon, T., and Drake, S.A. 1989, *Ap. J.*, **346**, 303.

Seventh Cambridge Workshop on Cool Stars, Stellar Systems, and the Sun
ASP Conference Series, Vol. 26, 1992
Mark S. Giampapa and Jay A. Bookbinder (eds.)

SOLAR GRANULATION SPECTROSCOPY: DYNAMICS OF THE INTERGRANULAR SPACE

A. NESIS Kiepenheuer-Institut für Sonnenphysik, Schöneckstr. 6, D-7800 Freiburg, F.R.G.

A. HANSLMEIER Institut für Astronomie, Universitätsplatz 5, A-8010, Graz, Austria

R. HAMMER Kiepenheuer-Institut für Sonnenphysik, Schöneckstr. 6, D-7800 Freiburg, F.R.G. and NRC, NASA - Marshall Space Flight Center, Huntsville, Alabama, U.S.A.

R. KOMM, W. MATTIG, J. STAIGER Kiepenheuer-Institut für Sonnenphysik, Schöneckstr. 6, D-7800 Freiburg, F.R.G.

ABSTRACT Using spectrograms of high spatial resolution, we investigated the functional interdependence between the convective intensity fluctuations at $\tau_{5000} = 1$ and the line broadening of magnetically insensitive weak absorption lines in nonactive regions. We found an extra broadening of the absorption lines in the intergranular regions. We attribute this increase to enhanced nonthermal velocities due to the decay process of the granulation associated with the dark regions. The variation of this broadening with height in the photosphere can be attributed to the variation of the nonthermal velocities of convective origin in the photosphere. Moreover, our findings could be relevant to the diagnostics of the recently proposed shock phenomena at the borders of solar granules.

Keywords: Sun: photosphere – Sun: granulation – line profiles.

INTRODUCTION

By means of high spatial resolution spectrograms we are now able to investigate in substantial detail the granulation, especially the transition from the granules to the intergranular space. What changes does a line profile undergo in the intergranular space?

This investigation is based on spectrograms taken with the vacuum tower telescope (VTT) in Izana (Tenerife), in July 1990. The spectrograms are taken at the center of the solar disk and include the wavelength range $\lambda\lambda$ 4910 — 4914 Å. In this study we investigate the lines λ 4911.54 and 4912.03 Åwith the equivalent widths 24 and 47 respectively. The line λ 4912.03 Å is **magnetically nonsensitive** and it is formed about 140 km above $\tau_{5000} = 1$. We digitized the spectrograms and processed the data in order to get four sets of data points: (FWHM$_i$, I$_i$), (Equivalent-Width$_i$, I$_i$), (Center-Line-Depth$_i$, I$_i$) and (Wing-Line-Depth$_i$, I$_i$), along the slit of the spectrograph for each absorption line.

Fig. 1 shows the results of fitting the two sets ($FWHM_i, I_i$) and (Equivalent $Width_i, I_i$), to a straight-line model $y = a + b * x$ (regression line) where, x represents the fluctuations of the continuum, and y the line broadening or the equivalent width of the line, respectively. In Fig. 1 we realize immediately 1) the negative gradient of the FWHM regression line, 2) the significant difference in the gradient for the two absorption lines, where the stronger gradient was obtained for the magnetically nonsensitive line λ 4912.03 Å and 3) practically the missing gradient in the regression line of the equivalent width. The stronger gradient of the magnetically nonsensitive line λ 4912.03 Å signifies an enhancement of the line broadening, FWHM, in the intergranular space of ≈20 mÅ compared to the line broadening in the bright elements. On the other hand, the weaker gradient of the weaker absorption line λ 4311.54 Å exhibits a difference between the FWHM values in the granules and those in the intergranular space which can be up to ≈12 mÅ.

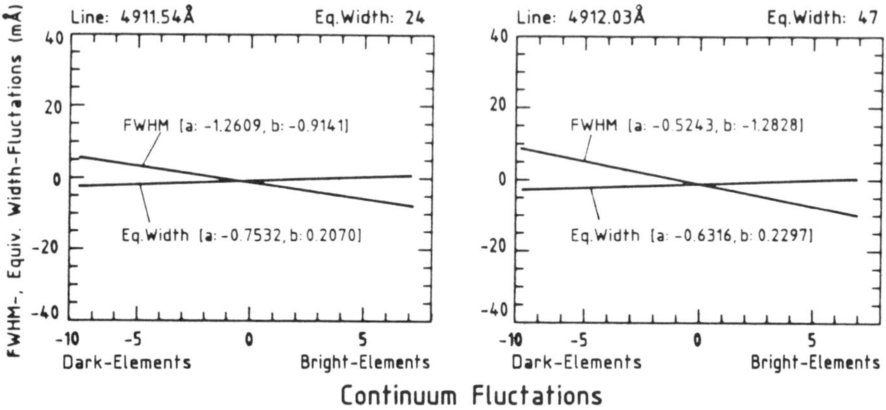

Fig. 1: Regression lines of ($FWHM_i, I_i$), (Equivalent-$Width_i, I_i$) for the lines: λ 4911.54 Å, and the **magnetically nonsensitive** line λ 4912.03 Å.

What mechanism can explain the extra broadening of the absorption lines in the intergranular space? Because the center of the magnetically nonsensitive line is formed at a height of ≈150 km its broadening reflects certainly the dynamical state of these layers which are turbulent (*cf.* Nesis and Mattig 1989; Komm *et al.* 1990).Furthermore, in spectrograms of high spatial resolution, Nesis *et al.* (1992) observed rapid changes of the line wiggles (from blueshift to redshift) at the border of granules to the intergranular space indicative of shear instability which represents another source of turbulence in the intergranular space. Accordingly, we believe that the extra broadening of the **magnetically nonsensitive** line λ 4912.03 Å in the intergranular space reveals the existence of enhanced nonthermal velocities in the dark regions of the overshoot layers (*cf.* Nesis *et al.* 1990). This statement is corroborated by our findings concerning the broadening of the line λ 4911.54 Å. This line is a weak one, and reflects, therefore, the dynamical state of the layers just above the convective layers. Here, we find still the ordered large-scale granulation pattern (*cf.* Nesis et al. 1988) *i.e.* practically the absence of turbulence. The extra broadening of the weaker line has to be, therefore, smaller than that of the stronger line, which corresponds to our findings. According to Holweger and Kneer (1990) a steep gradient of the line-of-sight velocity in the intergranular space causes the extra broadening of lines. However, in a first glance, the absorption lines used in this investigation do not demonstrate any well marked asymmetry.

Fig. 2 shows the results of fitting the two sets (Center-Line-Depth$_i$, I$_i$) and (Wing-Line-Depth$_i$, I$_i$), to a straight-line model $y = a + b * x$ (regression line) where x represents again the continuum fluctuations and y the center and wing line depth, respectively. The regression lines in Fig. 2 show 1) a positive gradient and 2) a strong difference between the gradients of both absorption lines. The stronger gradient corresponds to the magnetically nonsensitive stronger absorption line **λ 4912.03 Å**. The positive gradient signifies the decrease of the line depth in the dark regions compared to the line depth in the bright elements; the line becomes weaker in the intergranular space. Beyond that we recognize that the regression lines of the line wing depth show a gradient which is practically half the gradient of the regression lines of the center line depth. This is reasonable if we take into account that the wing line depth is measured at the same position as the FWHM.

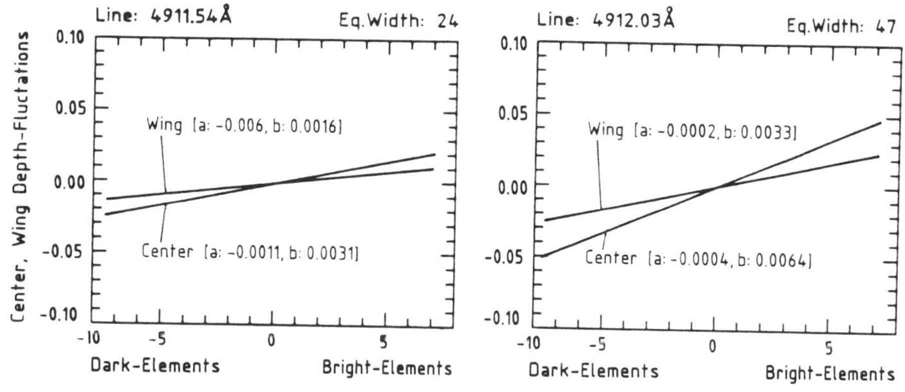

Fig. 2: Regression lines of (Center-Line-Depth$_i$, I$_i$) and (Wing-Line-Depth$_i$, I$_i$), for the lines: λ 4911.54 Å, and the **magnetically nonsensitive** line λ 4912.03 Å.

SUMMARY

Summarizing we can say that the stronger **magnetically nonsensitive** line λ 4912.03 Å and the weaker line λ 4911.54 Å demonstrate in the intergranulum the same behavior concerning the line broadening the line depth and the equivalent width: 1) an extra line broadening compared to the granulum 2) an extra reduction of the center line depth compared to the granulum followed by 3) an extra reduction of the line wing depth (half of the reduction of the line center depth), and 4) practically the same equivalent width in granulum and intergranular space. Because the equivalent width of the lines remains practically the same in the granulum and intergranular space, the depth reduction of the absorption line in the intergranular space has to be compensated by an enhancement of its broadening. The enhancement of the FWHM of the lines due to unambiguous existence of turbulence (nonthermal velocities) in the intergranular space fulfills this compensation.

REFERENCES

Holweger, H., Kneer, F. 1989, in Solar and Stellar Granulation, eds. R.J. Rutten and G. Severino, NATO-ASI Series 263, Kluwer, p. 425.

Komm, R., Mattig, W., and Nesis, A. 1991, *Astr. Ap.* **243**, 251.

Nesis, A., Durrant, C. J., and Mattig, W. 1988, *Astr. Ap.* **201**, 153.

Nesis, A., and Mattig, W. 1989, *Astr. Ap.* **221**, 136.

Nesis, A., Hanslmeier, A., Hammer, R., Komm, R., and Mattig, W. 1990, in The Dynamic Sun, ed. L. Dezső, Proc. of the 6th European Meeting on Solar Physics.

Nesis, A., Hanslmeier, A., Hammer, R., Komm, R., Mattig, W., and Staiger, J. 1992, *Astr. Ap.*, **253**, 561.

Seventh Cambridge Workshop on Cool Stars, Stellar Systems, and the Sun
ASP Conference Series, Vol. 26, 1992
Mark S. Giampapa and Jay A. Bookbinder (eds.)

SOLAR TORSIONAL OSCILLATIONS AS DUE TO MAGNETIC SUPPRESSION AND DEFORMATION OF TURBULENCE

GÜNTHER RÜDIGER Astrophysikalisches Observatorium Potsdam und Sternwarte Babelsberg, Rosa-Luxemburg-Str. 17a, D-O-1591 Potsdam, FRG

ABSTRACT Turbulence fields in rotating media are transporting angular momentum. It is widely accepted that non-uniform rotation laws in stellar convection zones are maintained by this effect. On the other hand, strong magnetic fields are known to suppress turbulence and to deform correlation structures. Strong magnetic fields exist in the solar convection zone. We ask under which conditions the observed low-mode torsional oscillations can be explained by the magnetic influence on the turbulence.

Keywords: solar differential rotation; magnetic field

1. THE REYNOLDS STRESS

The solar torsional oscillations are normally considered as directly driven by the dynamo-generated large-scale Lorentz force. In contrast, we test here the idea that the turbulence is affected by the large-scale magnetism and that the consequences for the turbulent angular momentum transport are essential for the cyclic modulation of the solar rotation law. To this end the influence must be known of a mean-field magnetic field on the "total" Reynolds stress, which is the difference between normal Reynolds stress and Maxwell stress,

$$Q_{ij}^{tot} = Q_{ij} - \frac{1}{\mu\rho} < B_i' B_j' > . \tag{1}$$

This tensor is strongly modified by mean-field quantities such as flow, rotation and magnetism:

$$Q_{ij}^{tot} = -N_{ijkl}\bar{u}_{k,l} + \Lambda_{ijk}\Omega_k + \kappa\bar{B}_i\bar{B}_j + p\delta_{ij}, \tag{2}$$

p being a magnetically modified pressure term, κ a positive scalar quantity (Rüdiger and Kichatinov 1990), N a viscosity tensor and Λ the Λ- effect, which describes the ability of rotating anisotropic turbulence fields to transport angular momentum. In detail the Λ-tensor is studied in Rüdiger (1989) and in Tuominen and Rüdiger (1989).

Now the $\bar{B}$-dependence of N and Λ contributing to the off-diagonal components $Q_{r\phi}^{tot}$ and $Q_{\theta\phi}^{tot}$ must be known. The latter quantities are usually be parameterized by

$$Q_{r\phi}^{tot} = -\nu_T r\frac{\partial\Omega}{\partial r}\sin\theta + \nu_T^0 V\Omega\sin\theta, \qquad Q_{\theta\phi}^{tot} = -\nu_T\frac{\partial\Omega}{\partial\theta}\sin\theta + \nu_T^0 H\Omega\cos\theta. \tag{3}$$

This adopted, the dependence of ν_T, V and H on Ω and $\bar{B}$ provides the profiles of the rotation laws in turbulent rotating media.

2. THE TURBULENCE MODEL

In Rüdiger and Kichatinov (1990) a series expansion is given for the dependence of the turbulence parameters on the magnetic field. The magnetic field enters the relations mainly in form of its intensity $\bar{B}^2$. In the Sun the azimuthal component, $\bar{B}_\phi = \hat{B}\Theta(\theta)$, dominates by orders of magnitude. The function $\Theta(\theta)$ provides the latitudinal dependence of the azimuthal large-scale magnetic field. The equatorially antisymmetric representation

$$\Theta = b_2 P_2^1 + b_4 P_4^1 \tag{4}$$

is used: $b_2 = 0$ gives (as realized in the Sun) 2 magnetic belts in each hemisphere and $b_4 = 0$ gives only 1 magnetic belt in each hemisphere. In our calculations the field is concentrated near the bottom of the convection zone: x = 0.7....0.85.

We considered turbulence fields with a global anisotropy in the turbulence intensities. The respective spectral tensor contains 2 parts, where one of them, $q^{(2)}$, stands for the anisotropy. Negative $q^{(2)}$ describe turbulences with dominating motions in the radial direction, which is, of course, the preferred case for our computations.

With the magnetic Strouhal number

$$S = \frac{\hat{B}\tau_{corr}}{\sqrt{\mu\rho}\, l_{corr}} \tag{5}$$

it follows

$$\nu_T = \nu_T^{(0)}(1 - \frac{8}{15}\mathcal{K}S^2\Theta^2), \qquad V = V^{(0)} - \frac{16}{35}\mathcal{J}S^2\Theta^2, \qquad H = \frac{16}{105}\mathcal{J}S^2\Theta^2 \tag{6}$$

with

$$\mathcal{J} = \frac{l_{corr}^2}{\nu_T}\int k^4 q^{(2)} dk, \qquad \mathcal{K} = \frac{l_{corr}^2}{\nu_T}\int k^4 q^{(3)} dk. \tag{7}$$

While $\mathcal{K}$ is always positive, $\mathcal{J}$ is negative for radial turbulence and positive for horizontal turbulence.

3. THE REYNOLDS EQUATION

Only the simplest form of the Reynolds equation,

$$\frac{\partial}{\partial x_i}(r\sin\theta\rho Q_{i\phi}^{tot}) = 0, \tag{8}$$

is solved with stress-free boundaries. Density stratification and the nonmagnetic Λ-mode $V^{(0)}$ are chosen as $d\log\rho/dx = -G$ and $V^{(0)} = 0$. With the latter choice we assume all deviations from the solid-body rotation as magnetically induced.

A grid-point method is used to solve the Eq. (8) with the usual stress-free boundaries. The resulting rotation laws are developed after Gegenbauer polynomials. That is important as the observations of torsional oscillations are best represented by

$$\Omega = \Omega_\odot(1 + \omega_2 P_3^1/\sin\theta + \omega_4 P_5^1/\sin\theta + ...). \quad (9)$$

Snodgrass and Howard (1985) found a variation of ω_2 and ω_4 during the solar cycle so that

$$\delta\omega_2 \cong 0.0024 \qquad \delta\omega_4 \cong 0.0010, \quad (10)$$

appeared for the difference of the ω's during activity maximum and minimum. If only magnetic parts of the Λ-effect are involved, then our ω's are identical with the $\delta\omega$'s.

The Table gives the results obtained with our numerical code.

Table: Influence of magnetic belts on the torsional oscillation

b_2	b_4	G	$\mathcal{K}$	ω_4/ω_2
0	1	0	0	0.48
1	0	0	0	-0.02
0	1	10	0	0.53
0	1	0	0.2	0.63
0	1	10	0.2	0.87

For all models we find positive ω_2. their amount is of less importance as it depends on the choice of $\mathcal{J}$.

The ω_4 is highly sensitive to the magnetic field structure. It is positive for "quadrupoles" (2 belts per hemisphere) and negative for "dipoles" (1 belt per hemisphere). This property does not change if the model is stratified (G = 10) or the magnetic quenching of viscosity is involved ($\mathcal{K}$ = 0.2).

Only for magnetic fields with 2 belts in each hemisphere (as realized in the Sun) both the rotation coefficients ω_4 and ω_2 are in phase and form the observed ratio of about 0.5.

REFERENCES

Kichatinov, L.L. 1990, *Geophys. Astrophys. Fluid Dynamics*, **54**, 145.

Rüdiger, G. 1989, Differential rotation and stellar convection: Sun and solar-type stars, Gordon and Breach, New York

Rüdiger, G. and Kichatinov, L.L. 1990, *Astr. Ap.*, **236**, 503.

Snodgrass, H.B. and Howard, R. 1985, *Sol. Phys.*, **95**, 221.

Tuominen,I. and Rüdiger,G. 1989 *Astr. Ap.*, **217**, 217.

Seventh Cambridge Workshop on Cool Stars, Stellar Systems, and the Sun
ASP Conference Series, Vol. 26, 1992
Mark S. Giampapa and Jay A. Bookbinder (eds.)

THE TURBULENT TACHYCLINE

EDWARD A. SPIEGEL
Astronomy Department, Columbia University, NY, NY 10027, U.S.A.

JEAN-PAUL ZAHN
Observatoire Midi-Pyrénées, 14 ave. E. Belin, 31400 Toulouse, France

ABSTRACT Between the outer convection zone of the sun with its pronounced differential rotation and the rigidly rotating radiative interior, there is a transition layer that observations reveal but do not yet resolve. We argue that the fluid motions in this layer are turbulent and are likely to produce dynamo action. Evolution of this structure may therefore be reflected in the change of stellar activity as cool stars evolve.

Keywords: Solar rotation; solar dynamo; tachycline.

INTRODUCTION

Helioseismological sounding of the sun has revealed the distribution of azimuthal velocity as a function of depth and latitude. The known surface rotation, with its equatorial acceleration, carries on to the bottom of the convection zone with little radius dependence. In the radiation zone below, the equatorial and polar rotation rates appear to approach a common value, in a radial increment too small to be resolved as yet (Brown *et al.* 1989; Goode *et al.* 1991).

For a theory of this transition layer in the solar rotation, we adopt the observed rotation law in the convection zone as given. We then inquire into the flow that will be driven when we require that the rotation law change smoothly on crossing the outer edge of the radiative zone. This outer edge is defined as the depth at which convective transport of heat first becomes negligible after due allowance for the effects of penetration (Zahn 1991). This approach is suggested by the manner in which oceanographers study ocean circulation subject to observed wind stress at the sea surface. In acknowledgement of that source of inspiration we call the transition layer the tachycline.

THE RADIATIVE TACHYCLINE

We use the classical equations of motion and thermal balance in approximations normally applied to situations in thin layers with rapid rotation and strong stratification. These equations predict the behavior of the tachycline. We shall discuss our solutions elsewhere; here we simply give the physical gist of the solutions.

The velocity boundary conditions applied to the top of the radiation zone induce motions that enforce a horizontal pressure gradient, leading in turn to a horizontal temperature gradient in the pole-equator direction. This strong link between the temperature and differential rotation is known in geophysical fluid dynamics as the thermal wind. In the sun, because of radiation, the horizontal temperature distribution diffuses downwards, and it is accompanied by the differential rotation, which, in the inviscid case, can be modified only by advection of angular momentum. But the meridional motions required for this are hindered by the stable stratification, so the tachycline advances inward at a much slower rate. Instead of growing in time as in the normal diffusive way like $t^{1/2}$, the thickness h of the layer increases as

$$h \approx r \left(\frac{t}{t_{ES}} \right)^{1/4} , \tag{1}$$

where the local Eddington-Sweet time is given by $t_{ES} = (\Omega/N)^{-2}\,\tau$, in terms of the mean rotation rate Ω, the buoyancy frequency N, and where $\tau = r^2/\kappa$, with r = the radial coordinate and κ = the thermal diffusivity. A similar temporal evolution is found in the stratified spin-down problem (*e.g.* Sakurai 1970; Spiegel 1972). For the sun, $h \approx 2\;10^5$ km after $4.6\;10^9$ yrs.

The actual calculation confirms this estimate. If the mean rotation rate is assumed to be constant in time, the tachycline thickness will be 180,000 km. When we allow for the variation of rotation rate with time, the thickness is even greater. Hence, we conclude that the differential rotation of the convection zone would have extended well into the sun, if radiative diffusion were alone responsible for its spread. The apparent discontinuity represented by the observed tachycline must therefore be ascribed to another physical process, capable of ensuring a quasi-steady state.

THE TURBULENT TACHYCLINE

In the process just described, the solar interior rotation tracks the swirl imposed by the convection zone with a lag time, or spindown time, that is essentially the Eddington-Sweet time. But this is a laminar process while it is generally believed among fluid dynamicists that, at sufficiently high Reynolds numbers, virtually any flow becomes turbulent. Though it is not always known which type of shear flow instability plays the major role in any given case, it is clear that such instabilities are generated by the horizontal shear of the differential rotation (Zahn 1975). However the occurrence of turbulence in the swirl below the convection zone is not by itself alone to generate the thin boundary suggested by observations; an enhanced viscosity would spread the tachycline even more quickly into the interior than before.

Hence, it is only with anisotropic turbulence that we can produce solutions that resemble the observed tachycline. Thus we require a predominantly horizontal turbulence such as one sees in the geophysical context (Rhines 1979). This produces an engaging self-consistency in which the tachycline is thin because the turbulence is mainly horizontal and, as in geophysics, the turbulence is mainly horizontal because the tachycline is thin. In this regime, the horizontal advection of angular momentum is balanced by the turbulent stresses, and a stationary flow is achieved in a layer whose thickness can be well within the bounds set by observations. In particular, if σ is the ratio between the (horizontal) eddy viscosity and the thermal (radiative) diffusivity, the tachycline thickness is

$$h = 20,000\,\sigma^{-1/4} \quad \text{km}. \tag{2}$$

The theory also predicts that the interior uniform rotation is intermediate between the polar and equatorial rotation rates of the convection zone, in agreement with the observations. We find that the vertical differential rotation changes sign at a latitude of about 45^0.

DISCUSSION

The picture of the sun that emerges from our study recalls the image of the terrestrial weather system. The solar convection zone, like the earth's atmospheres drives circulation in the fluid below, which in our metaphor corresponds to the oceans. The outer tachycline corresponds to what a planetary scientist would call the weather layer where the lively things happen while the deeper interior remains more passive and provides support and inertia. The "weather" to which we allude is of course dynamo action.

If we may be guided by simulations of the fluid dynamics in weather layers on planets, we may anticipate the formation of intense vortices in the large scale turbulence. As in the atmospheres of large planets, these may be relatively long lived and will have a dramatic influence on the magnetic structures that form.

Penetrative flow from the overlying convection zone will bring down tongues of magnetic field, as shown in the simulations reported by Stein in this conference. The differential motion in the tachycline will stretch the field into a generally toroidal configuration. According to a preliminary study by Meacham, the vortices will wind up their field locally, and we can foresee how the locally enhanced field will be buoyed up to the surface, pulling out behind it a long helical magnetic extrusion. On emergence from the surface, the tube will fall back into the sun, thus forming a firmly anchored spot with a loosely connected companion. We hope to develop these ideas in a quantitative way from the tachycline structure described here.

ACKNOWLEDGEMENTS

Support for this work was provided by the U.S. Air Force Office of Scientific Research under grant AFOSR89-0012, and by the Centre National de la Recherche Scientifique (GDR "Magnétisme dans les étoiles de type solaire").

REFERENCES

Brown, T.M., Christensen-Dalsgaard, J., Dziembowski, W.A., Goode, P., Gough, D.O. and Morrow, C.A., 1989, *Ap. J.*, **343**, 526
Goode, P.R., Dziembowski, W.A., Korzennick, S.G. and Rhodes, E.J. 1991, *Ap. J.*, **367**, 649
Rhines, P.B., 1979, *Ann. Rev. Fluid Mech.*, **11**, 401
Rüdiger, G., 1989, Differential Rotation and Stellar Convection (Berlin: Akademie-V.)
Sakurai, T., 1970, in Stellar Rotation ed. A. Slettebak (Dordrecht: Reidel), 329
Spiegel, E.A., 1972, in Physics of the Solar System ed. S.I. Rasool (Washington: NASA)
Zahn, J.-P., 1975, *Mém. Soc. Roy. Sci. Liège*, 6e série **8**, 31
Zahn, J.-P., 1991, *Astr. Ap.*, (in print)

PART IV

Magnetic Fields

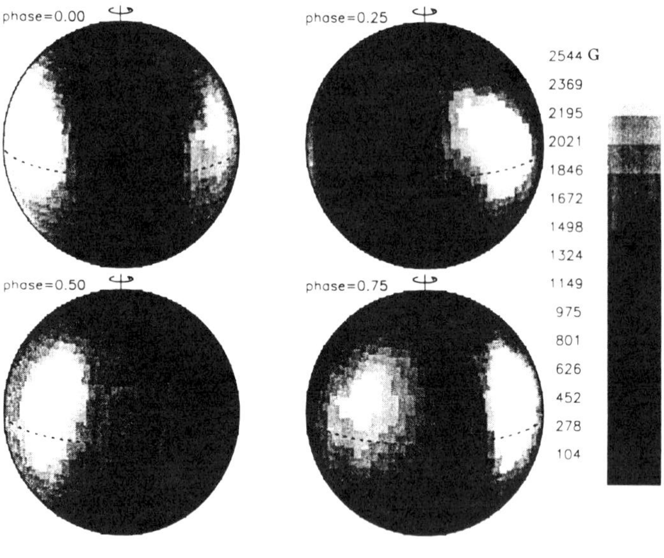

Figure 3 from "Magnetic Surface Images of HD 82558" by S. Saar, N.E. Piskunov, and I. Tuominen.

Seventh Cambridge Workshop on Cool Stars, Stellar Systems, and the Sun
ASP Conference Series, Vol. 26, 1992
Mark S. Giampapa and Jay A. Bookbinder (eds.)

STRUCTURE OF THE SOLAR X-RAY CORONA

LEON GOLUB Smithsonian Astrophysical Observatory, 60 Garden Street, Cambridge, Mass. 02138

ABSTRACT Recent high resolution observations of the Solar X-ray corona, such as those from the NIXT sounding rocket payload, show that the emission is resolved into tightly-packed bundles of structures with very high aspect ratio (L/d $\gtrsim$ 100). Such structures appear to be inherently short-lived, in contrast to the post-Skylab view of relatively stationary loops as the fundamental constituents of the inner corona. We suggest that it is the overall magnetic configuration which is stable on long timescales, but that "coronal threads" are formed and destroyed rapidly in a continual process. The processes of energy input, plasma heating, disruption of the heated region, chromospheric evaporation, and radiative and conductive cooling of the plasma all take place on similar time scales, of order 10^3-10^4 seconds. On this view, coronal structures are fundamentally transient and only in quasi-equilibrium. One observes long-lived stable loops only when the spatial and/or temporal resolutions are inadequate.

Keywords: Solar corona; Coronal structure; Solar X-rays

INTRODUCTION

In the late 1960's and early 1970's high resolution imaging instruments functioning at UV and shorter wavelengths became available. Prior to this time, the commonly accepted view of the outer Solar atmosphere was that, to a first approximation this atmosphere was homogeneous with significant but only occasional structures present (the work of Billings being a notable exception). Coronal studies of that era were thus carried out in a manner similar to much of the stellar work being done at the present time, by subdividing the work into a portion focusing on the homogeneous aspects and a portion focusing on the additional complications due to some type of activity.

This view changed completely once it became possible to view the corona on the disk from sounding rockets (Vaiana, Krieger and Timothy 1973), and one can argue that the long duration observations carried out by the Solar telescopes flown on *Skylab* in 1973-74 did more to change our perception of the corona than any other instrument or collection of instruments. The key feature which was brought home by these observations is that essentially all of the emission seen in the corona comes from structures which we have called "loops" (Vaiana and Rosner 1978). There is no evidence for either open structure emission or for any diffuse emission – on the contrary, any emission which has at one time been thought to be diffuse has invariably been resolved into loops as better observations became available (Golub *et al.* 1990).

An example of the highly structured nature of the Solar corona is shown in Figure 1, an X-ray image taken from a recent flight of the NIXT sounding rocket payload. This figure illustrates the basic point, that the hot corona is observed to consist entirely of closed structures, *i.e.*, "loops".

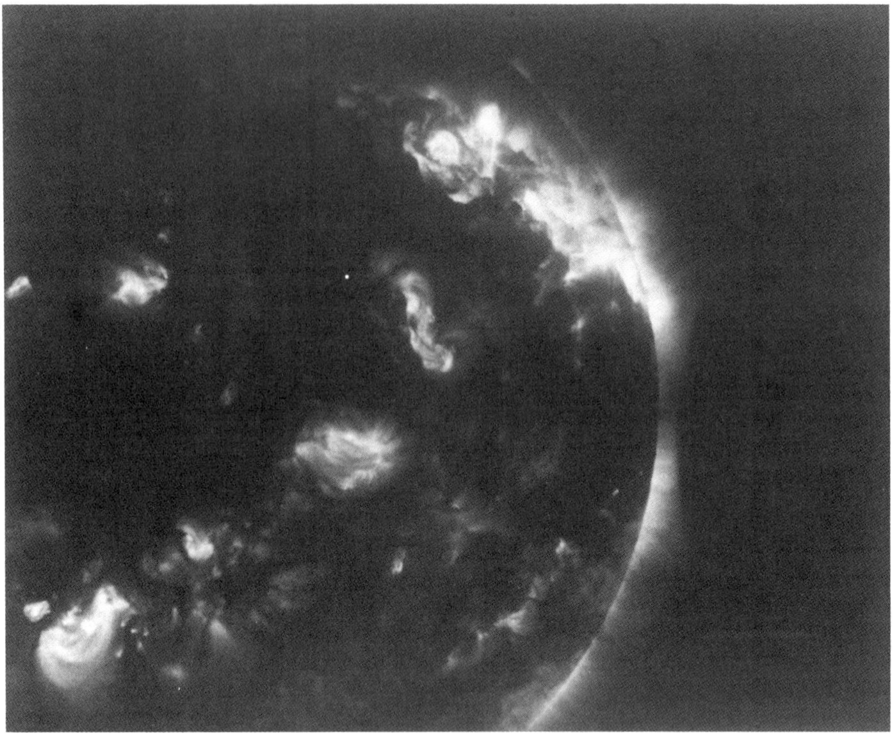

Figure 1. NIXT X-ray image of Solar corona, 11 July 1991.

LOOP ATMOSPHERES

The first clear statement of the new view appeared in Rosner, Tucker and Vaiana (1978a) and in Craig, McClymont and Underwood (1978). These authors took the position that loops are the fundamental building block of the corona, and that each loop could be viewed as an independent "mini-atmosphere". From this position they calculated some general stability criteria for loops and the well-known thermodynamic scaling relation $T \sim (pL)^{1/3}$. In the same year Rosner *et al.* (1978b) proposed a model in which the magnetic field is directly involved in transferring mechanical energy directly from the photospheric convective motions up into the corona and depositing that energy for *in situ* heating of the coronal plasma. Numerical simulations of loop atmospheres (Peres *et al.* 1982) confirmed the RTV scaling law, extended it to loops larger than the pressure scale height (Serio *et al.* 1981), and are being used to examine the dynamic behavior of loops in response to sudden changes in heating rate (Jakimiec *et al.* 1991).

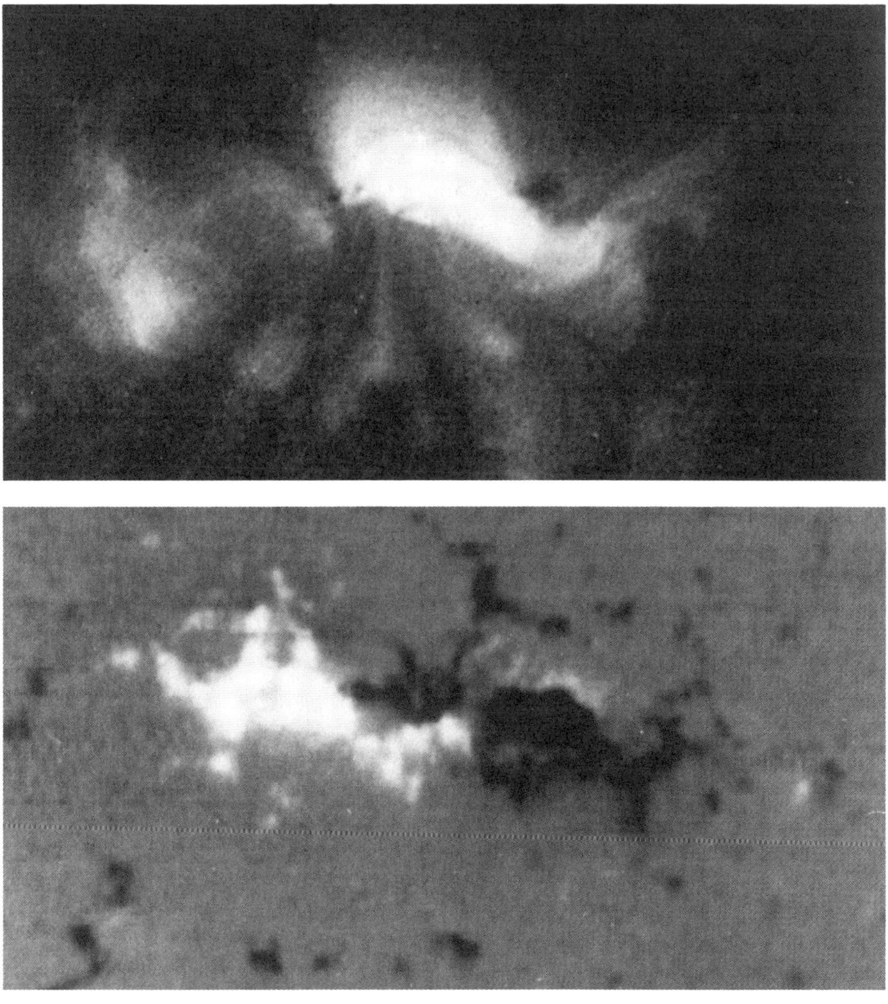

Figure 2. a) NIXT X-ray image of active region structure, compared with b) photospheric magnetic field; 22 Feb. 91.

Thus, by the late 1970's the view that structure was only an intrusion into an otherwise quiescent and homogeneous atmosphere, was completely reversed. The loop structure and its heating were seen as the fundamental element, reflecting the emergence and diffusion of the magnetic field and the subsequent transfer of energy into the outer atmosphere via the field. An example of the close connection between the surface magnetic field and the coronal emission is shown in Figure 2, in which closed loops of X-ray emission, or in some cases the lower portions of large loop structure are seen to be intimately related to the locations of B. Within the context of this view, the discussion has come to center around magnetic field structures which will provide stable loops, pressure, energy and mass balance within stable structures, and possible mechanisms for providing the energy required to balance radiative and conductive losses from coronal loops.

The critical role played by the magnetic field in the formation of coronae was made even more evident by the launch of the *Einstein* Observatory, which made possible for the first time the detection of X-ray emission from a large sample of ordinary solar-like stars (Vaiana *et al.* 1981). What was observed was completely contradictory to pre-launch expectations: rather than the Sun being near the upper end of the X-ray luminosity distribution because of its high acoustic flux level, the Sun turns out to be a rather weak X-ray emitter. Most other main sequence stars of spectral type F through M are stronger than the Sun and the determinant of X-ray level is rotation rate rather than spectral type (Pallavicini *et al.* 1981). The dramatic failure of standard acoustic heating theories and the availability of alternative magnetic field-related heating models marked a dramatic turning point in our view of coronal physics (cf. Rosner, Golub and Vaiana 1985).

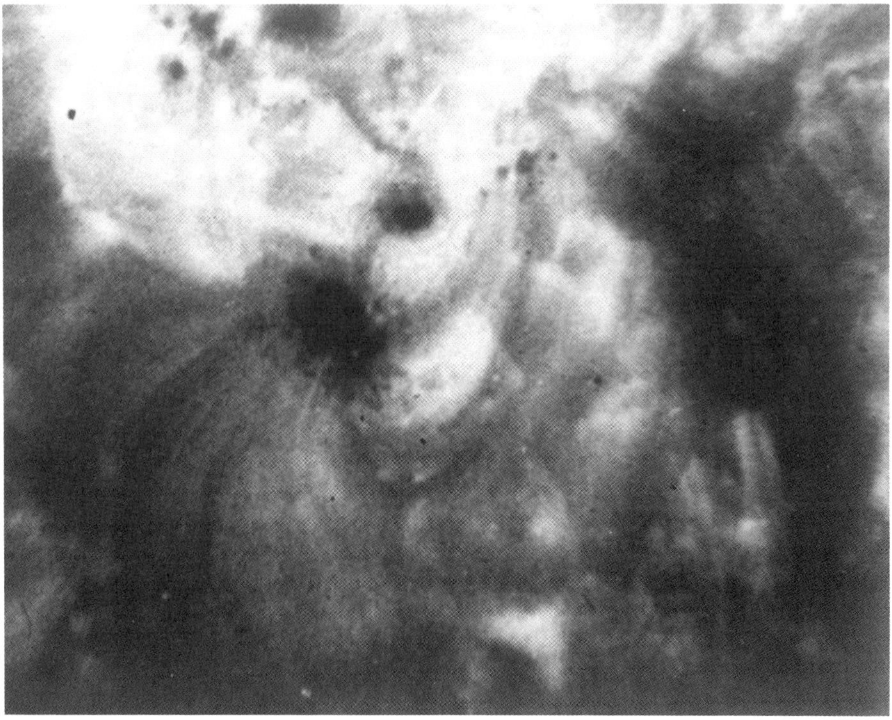

Figure 3. NIXT image of coronal structure above a sunspot, 22 Feb 1991.

RECENT OBSERVATIONAL RESULTS

A. Sunspot Fields

Before discussing the new results related to the structure of the X-ray emitting corona, we briefly discuss a related but separate observation of the emission associated with sunspots. Figure 3 shows a portion of an X-ray image taken from the NIXT payload on 22 Feb. 1991. What is unusual about this flight is that we obtained perfectly coaligned and simultaneous X-ray and white light photos (for a detailed explanation see Sams, Golub and Weiss 1992), which show the X-ray coronal structures *and* the sunspot umbra/penumbra structure. We are thus able to trace the coronal fieldlines down into the spot without ambiguity.

When we carry out the comparison of coronal to surface fields, the following general rules become evident:

1. There are *no* hot loops seen coming out of the spot umbrae; this result has been known since the *Skylab* days, but it is now seen with much greater precision and clarity.

2. There is a class of loops which emanate from the spot penumbrae, apparently in association with the steeply inclined field lines coming from the bright (as opposed to dark) penumbral fibrils (see Title *et al.* 1991).

3. The brightest loops in active regions do not emanate from sunspots, but originate in enhanced network or plage, and connect to similar areas near but not in the sunspots.

Thus the correlation between brightness of X-ray emission and magnetic field strength is not monotonic: the weakest magnetic fields far away from active regions have weak X-ray emission as one would expect. However, the *strongest* magnetic fields also have little or no X-ray emission. Presumably, as argued in Rosner *et al.* (1978), what is needed to produce a hot corona is both a magnetic field and an input of mechanical energy into the field. The field lines ending in spot penumbrae are relatively strong, but show comparatively weak X-ray emission; this indicates that they form an intermediate class which emit less than they ought to for their field strength, possibly because they are anchored in a pattern of convection weaker than that of the enhanced network areas.

B. Dynamics of Coronal Loops

In the area of coronal heating, the past decade has seen the accumulation of several important new observational results which we consider directly relevant to the problem of constructing a viable coronal heating model. Briefly, these are:

– The observations by Martens *et al.* (1985) and Schadee *et al.* (1983) of hot $> 10^7$ K plasma in nonflaring coronal loops, with small filling factor ($< 10^{-3}$);

– Variability of "normal" loops on short times scales (Sheeley and Golub 1979; also see below) which suggest that static models are not appropriate;

– Vector magnetogram data extrapolated into the corona argue for the presence of large field-aligned currents flowing along the loops (Hagyard *et al.* 1984).

– Observation of hard X–ray microflares associated with active regions (Lin *et al.* 1984) and microwave bursts with similar properties (Bastian 1991; note that these are *not* Type III bursts).

– High resolution imaging results (Golub *et al.* 1990) which show that coronal "loops" are much narrower and tightly-packed than was previously thought.

If we attempt to assemble these recent results into a coherent picture, we are led to a model in which coronal structures are heated in very localized regions, with some disruptive process occurring to spread the heat throughout the larger loop volume. The heating is likely to be abrupt or at least rapid compared to the lifetime of the loop and the timescale for attaining equilibrium in response to the heat input; a series of related processes are all occurring more or less simultaneously, as indicated schematically in Figure 4.

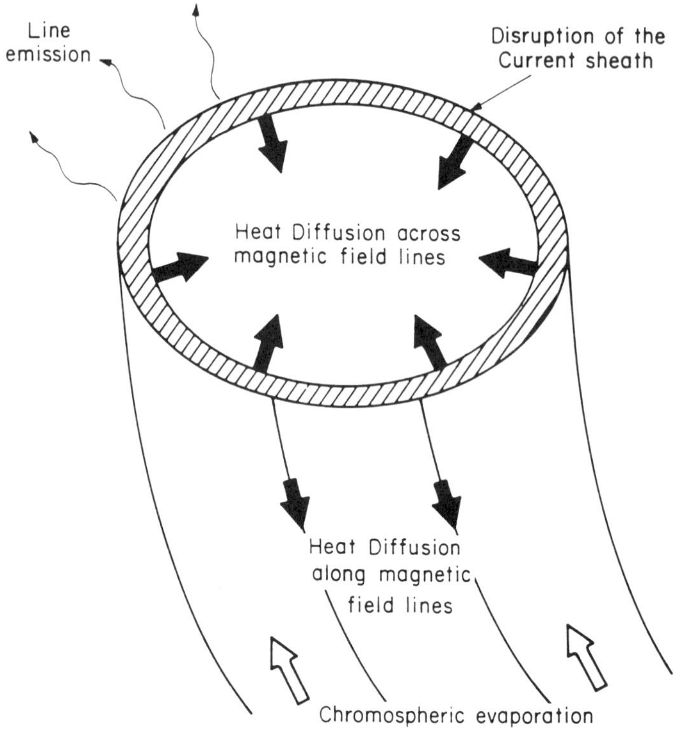

Figure 4. Proposed series of events occurring within a coronal thread.

1. The high-T component occupying a small fraction of the volume indicates that the heating mechanism is strongly localized; it may be a separate physical process (see 3. below) which later disperses the energy throughout the remaining volume.

2. The transient nature of coronal structures is of fundamental importance, but unfortunately the published literature on this subject is inadequate. Based on available data and on examination of time-resolved coronal imaging data, we take the view that coronal variability in active regions is relatively rapid (10^3 sec) and is embedded within an overall magnetic field configuration which typically evolves on much longer (10^5 sec) timescales.

3. The hard X-ray and microwave bursts are seen to be a prevalent feature of active region, and may be argued to arise out of the dynamic processes occurring in the coronal structures. We will assume that they are in some way fundamentally linked to the formation and heating of coronal structures, either as cause or effect. Either case implies that the heating process is sporadic and intermittent.

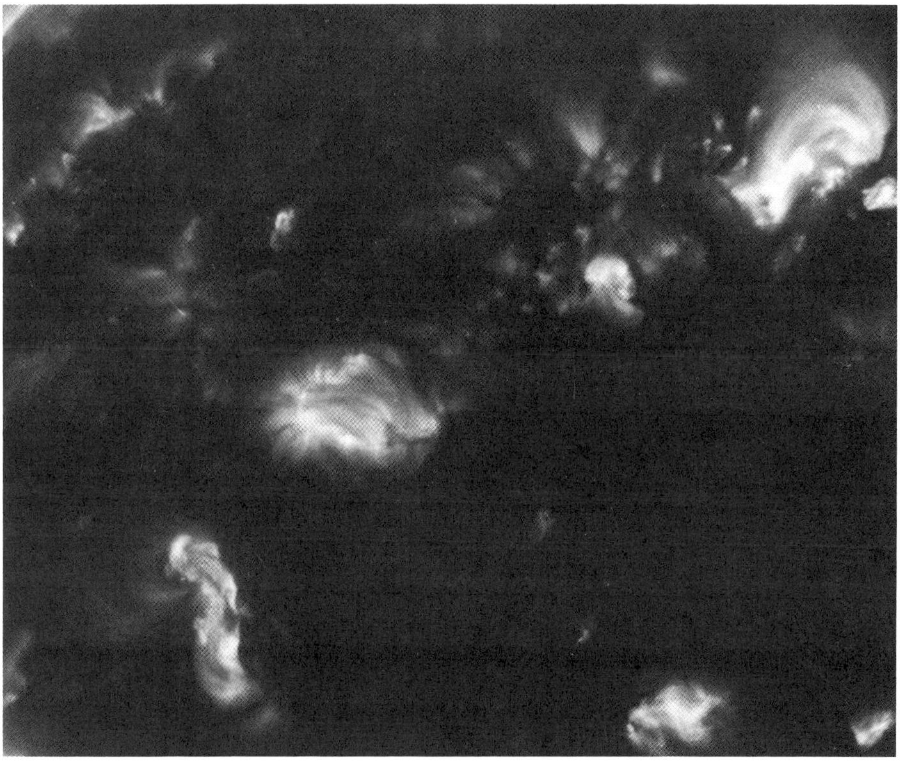

Figure 5. NIXT coronal X-ray image, 11 July 1991, showing thread-like fine structure.

4. The presumed presence of strong currents in the corona leads to an examination of their effect. As argued in Beaufume, Golub and Coppi (1992), the result is strongly localized heating in short-lived structures.

5. The extremely high aspect ratio of coronal structures seen with the NIXT (Figure 5) leads to the conclusion that any current sheaths which form will disrupt very quickly (Beaufume *et al.* 1992). Thus the heating will proceed via brief bursts having a small duty cycle.

ACKNOWLEDGEMENTS

I would like to thank Bob Rosner for many years of discussions related to this subject. In addition, Bob would certainly agree that none of this work would have happened without the late Pippo Vaiana, whose presence is felt in every paragraph of this paper.

This project was supported in part by NASA Grants NAGW-112 and NAG5-626 to the Smithsonian Institution.

REFERENCES

Bastian, T.: *Ap. J.*, **370** L49 (1991).

Beaufume, P., Coppi, B. and Golub, L.: *Ap. J.*, in press (1992).

Craig, I.J.D., McClymont, A.N. and Underwood, J.H.: *Astr. Ap.*, **70** 1 (1978).

Golub, L. *et al.* : *Nature* **344**, 842 (1990).

Hagyard, M.J., Smith, J.B., Jr., Teuber, D. and West, E.A.: *Sol. Phys.* **91**, 115 (1984).

Jakimiec, J., Sylwester, B., Sylwester, J., Serio, S., Peres, G. and Reale, F.: *Astr. Ap.*, in press (1991).

Martens, P.C.H., Van den Oord, G.H.J. and Hoyng, P.: *Sol. Phys.* **96**, 253 (1985).

Pallavicini, R., Golub, L., Rosner, R., Vaiana, G.S., Ayres, T. and Linsky, J.L.: *Ap. J.*, **248**, 279 (1981).

Peres, G., Rosner, R., Serio, S. and Vaiana, G.S.: *Ap. J.*, **252**, 791 (1982).

Rosner, R., Golub, L., Coppi, B. and Vaiana, G.S.: *Ap. J.*, **222** 317 (1978b).

Rosner, R., Tucker, W.H. and Vaiana, G.S.: *Ap. J.*, **220** 643 (1978a).

Rosner, R., Golub, L. and Vaiana, G.S.: *Ann. Rev. Astron. Astroph.* **23**, 413 (1985).

Sams, B., Golub, L. and Weiss, N.O.: *Ap. J.*, submitted (1991).

Schadee, A., de Jager, C. and Svestka, Z.: *Sol. Phys.* **89**, 287 (1983).

Serio, S., Peres, G., Vaiana, G.S., Golub, L. and Rosner, R.: *Ap. J.*, **243**, 288 (1981).

Sheeley, N.R. and Golub, L.: *Sol. Phys.* **63**, 119 (1979).

Title, A.M., Frank, Z.A., Shine, R.A., Tarbell, T.D., Topka, K.P., Scharmer, G. and Schmidt, W.: preprint (1991).

Vaiana, G.S., Krieger, A., and Timothy, A.F.: *Sol. Phys.* **32**, 81 (1973).

Vaiana, G.S. and Rosner, R.: *Ann. Rev. Astron. Astroph.* **16**, 393 (1978).

Vaiana *et al.* : *Ap. J.*, **245** 163 (1981).

Seventh Cambridge Workshop on Cool Stars, Stellar Systems, and the Sun
ASP Conference Series, Vol. 26, 1992
Mark S. Giampapa and Jay A. Bookbinder (eds.)

FINE-SCALE MAGNETIC FIELDS IN THE SOLAR PHOTOSPHERE

DOUGLAS RABIN
National Solar Observatory, National Optical Astronomy Observatories,†
P. O. Box 26732, Tucson, AZ 85726

ABSTRACT Intense magnetic flux tubes are now recognized as a basic constituent of the solar photosphere. Recent observations use the high Zeeman sensitivity of some infrared spectral lines to directly measure the true magnetic field strength within spatially unresolved flux tubes. With infrared and visible array detectors, variations of the magnetic and thermal properties of flux tubes can be studied in three dimensions. Small-scale magnetic elements significantly influence the observed properties of the solar magnetic field on intermediate and large spatial scales.

Keywords: solar photosphere; magnetic field measurements; magnetic flux tubes

INTRODUCTION

The systematic investigation of the photospheric magnetic field outside of sunspots originated with the development of a sensitive photoelectric magnetograph by Babcock and Babcock (1955). Visible-light magnetographs continue to play a key role in studying the evolution of surface magnetic flux through the solar activity cycle. Some 25 years ago, however, evidence began to accumulate that the field strength in certain non-spot areas could exceed several hundred gauss even where a magnetograph signal would be rather weak (Sheeley 1967; Beckers and Schröter 1968). A further surprise came when the ratio between the magnetograph signals in lines of different Zeeman sensitivities could best be explained by supposing that *most* of the magnetic flux originated in elements with an intrinsic field strength of about 1000 gauss (Howard and Stenflo 1972; Frazier and Stenflo 1972). The reality of these intense, small-scale fields was confirmed by infrared measurements that explicitly showed Zeeman splitting corresponding to $B > 1000$ gauss (Harvey and Hall 1975). Fortunately, theory had prepared for reality the notion of *magnetic flux tubes* (Parker 1955; Spruit 1976): elongated structures of small transverse scale ($\lesssim$ 500 km) in the photosphere, nearly vertical when standing alone, partially evacuated to maintain horizontal pressure balance with the surrounding atmosphere, spreading out rapidly above the photosphere.

Although the evidence for the existence of kilogauss flux tubes was convincing, the puzzle of how to study these sub-arcsecond structures slowed observational progress for some time. Part of the solution proved to be a polarimeter attached to a Fourier transform spectrometer (FTS), which provided

† Operated by the Association of Universities for Research in Astronomy, Inc., under cooperative agreement with the National Science Foundation.

high spectral resolution, great spectral range, a high signal-to-noise ratio, and could also be used in the infrared (Stenflo *et al.* 1984; Stenflo, Solanki, and Harvey 1987). The disadvantages of this approach are coarse angular resolution, poor temporal resolution, and single-element detection. The inability (with present FT spectrometers) to make spatial maps is a particular hindrance because so much of our knowledge of the solar atmosphere is tied to its intricate spatial structure. This part of the puzzle can now be investigated with the aid of infrared array detectors. A final piece of the puzzle will fall into place when flux tubes are spatially resolved, either from space or from the ground using adaptive optics.

Here I emphasize recent progress in obtaining spatially extended, direct measurements of magnetic field strength in flux tubes. The equally important subject of thermal structure is briefly considered. I conclude by indicating how small-scale magnetic elements can have important consequences for observations at intermediate and large scales.

WEAK AND STRONG ZEEMAN SPLITTING

The Zeeman shift of a spectral line with frequency ν (cm^{-1}) and Landé factor g in a magnetic field B (gauss) is $\Delta\nu_B = 10^{-4.33} g B$. In wavelength units, $\Delta\lambda_B \propto \lambda^2$ whereas Doppler width scales only as λ; hence the advantage of observing at longer wavelengths. Compared to the commonly-used visible line Fe I 5250.2 Å ($g = 3$), the infrared line discussed below, Fe I 6388.6 cm^{-1} (15648.5 Å, $g = 3$), is more sensitive by a factor of 3.0; the atomic emission line Mg I 811.6 cm^{-1} (12.32 μm, $g = 1$) is more sensitive by a factor of 7.8 (Brault and Noyes 1983). However, the advantage of observing flux tube magnetic fields in the near infrared instead of the visible is greater than this comparison suggests. Figure 1(a)–(b) shows the effect on intensity (Stokes I) and circular polarization (Stokes V) of splitting a simple Zeeman triplet by various fractions of the zero-field Doppler width. Near one Doppler width there is a rather sharp transition between a weak splitting regime, in which the separation of the Stokes V peaks changes very little with the underlying splitting, and a strong splitting regime, in which the separation of the V peaks is essentially equal to the underlying splitting and therefore directly measures B, the modulus of the vector magnetic field (irrespective of direction). The amplitude of the V peaks behaves in the opposite sense: directly proportional to splitting for weak fields, independent of splitting for strong fields. Unfortunately, the V amplitude cannot be used in a simple way to infer weak-field splitting because at least two other factors affect the amplitude. First, in practice V is usually normalized by I_c, the local unpolarized continuum intensity; so any relatively nonmagnetic areas within the angular resolution element will reduce V. Second, if the temperature inside a flux tube differs from the temperature outside, the V amplitude of a temperature-sensitive line will increase or decrease accordingly. The zero-order approach to these effects is a two-component model in which a fraction f of the resolution element is occupied by flux tubes and a fraction $1-f$ is field-free. Although field strength is decoupled from V amplitude in the strong splitting regime, at least two components are still necessary to interpret the amplitude.

Figure 1(c) places Fe I 5250.2 Å and Fe I 6388.6 cm^{-1} on the true-versus-apparent splitting diagram for B = 1000 gauss, assuming fully resolved

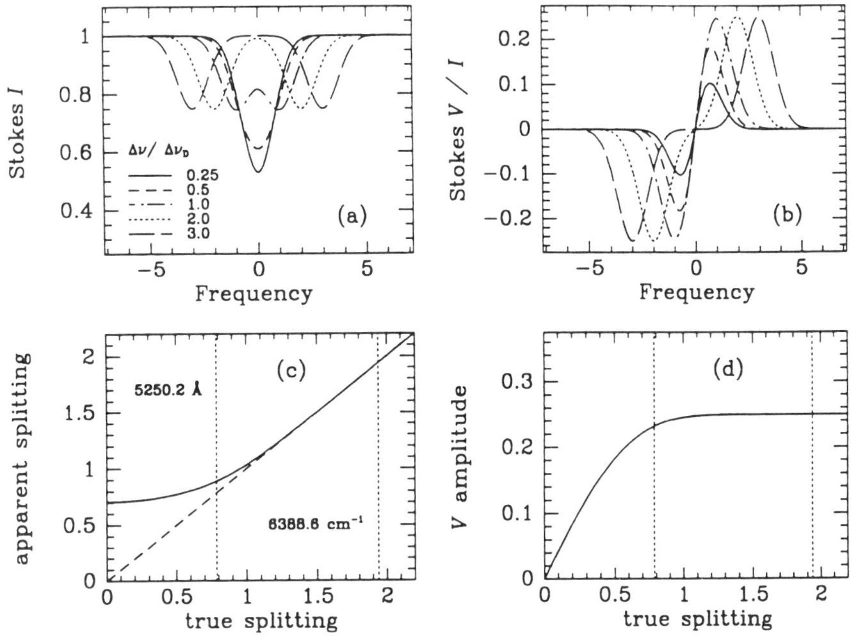

Fig. 1: Idealized splitting of an optically thin Zeeman triplet. Frequency and splitting are in units of the Doppler width of the unsplit line.

line profiles. The infrared line is securely in the strong-splitting regime, but the visible line is approaching weak splitting — even more so if the line is instrumentally broadened. Thus it happens that the characteristic field strength for which there is strong splitting in the infrared but transitional or weak splitting in the visible is also a typical field strength in solar magnetic flux tubes. Kilogauss splitting in visible lines manifests itself through relatively slight broadening, desaturation, and magneto-optic effects, for which a careful treatment of line excitation and polarized radiative transfer is necessary. The same splitting in a sensitive infrared line can be straightforwardly measured with a precision of about ±5%. That does not mean that infrared magnetic measurements replace visible diagnostics or make careful analysis unnecessary: there is much more to learn about a magnetic flux tube than its magnetic field strength at one level in the atmosphere.

SPATIALLY EXTENDED MEASUREMENTS OF FIELD STRENGTH

Figure 2 shows examples of Stokes V spectra of active-region plages near the center of the solar disk. The data are drawn from a sequence of long-slit

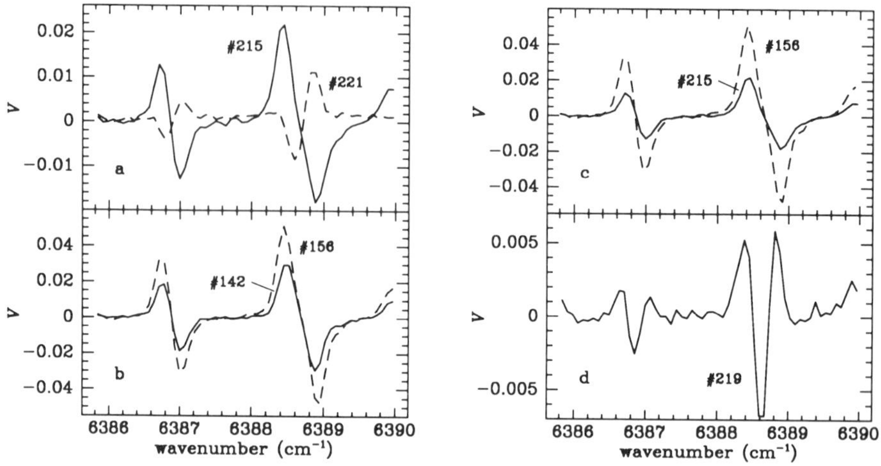

Fig. 2: Circularly polarized infrared spectra in solar plages. The spectral lines are Fe I 6388.6 cm^{-1} ($g = 3$) and Fe I 6386.9 cm^{-1} ($g_{\rm eff} = 1.53$). (a) 1700 gauss and 800 gauss; (b) 1700 gauss and 1450 gauss; (c) similar field strengths but different V amplitudes; (d) complex profile requiring two magnetic components.

infrared array spectra with 2–3″ resolution (Rabin 1991). In panels (a) and (b), differences in field strength are apparent to the eye; formal fitting detects differences down to 80 gauss (2σ). Panel (c) illustrates that locations with nearly the same field strength can have markedly different V amplitudes. The spectrum in panel (d) shows both polarities simultaneously and must be fitted with at least two magnetic components (Rüedi, Solanki, and Rabin 1991).

Figure 3 is obtained by fitting the full sequence of spectra. The left panel shows the variation of B along a 4′ east-west slice that extends from the following sunspot (visible at the left edge of the plot) toward (but not including) the leading spot of the active region. The field strength lies mostly in the range 1200–1700 gauss; fields exceeding 1000 gauss account for more than 90% of the flux. These values are consistent with the earlier observations summarized above. The new result is that the true field strength is coherently organized on spatial scales from 1′ to 10″ and probably smaller. If this comes as no surprise — because almost everything in the solar atmosphere exhibits patterns down to the limit of resolution — it nevertheless represents a first point of contact between measurements of flux tube magnetic fields, hitherto only scattered samples, and the imaging observations that constitute so much of solar physics.

The right panel of Figure 3 shows that field strength and V amplitude are positively correlated, but the scatter is greater than can be explained by the 2σ error bars; the spectra in Figure 2(c) show that intrinsic scatter is clearly detected. In more recent data with two-dimensional spatial resolution and many more samples, the lower triangle of the relationship in Figure 3 (right) is almost completely filled in: *i.e.*, the strongest fields have V amplitudes ranging from

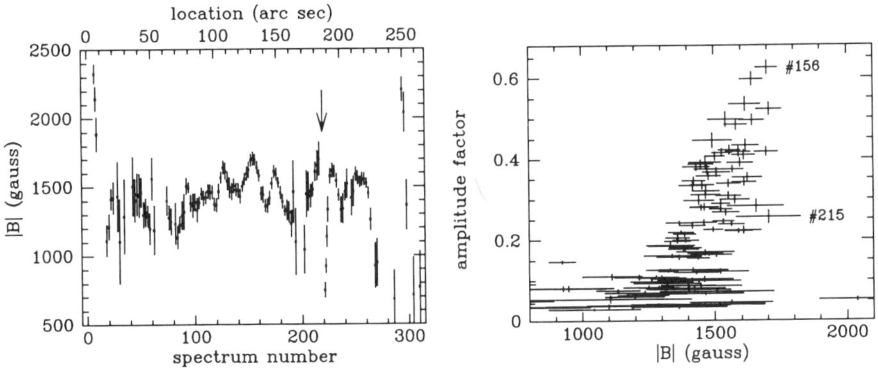

Fig. 3: Left: variation of magnetic field strength along one spatial dimension in active-region plages. The error bars are $\pm 2\sigma$ uncertainties in the fits to B. Three points at the left edge are in a sunspot. An arrow marks the position of the main polarity inversion line in the active region. Right: Stokes V amplitude plotted against field strength. The amplitude is normalized to the largest value it can attain if the spectral line does not change in strength inside the flux tube.

weak to strong, but the weakest measured fields always have weak V amplitudes. These recent data, still from active-region plages, show a distribution of field strength that peaks about 150 gauss lower than the data shown here, with lower V amplitudes. It is clear that we are just beginning to learn about the systematic spatial variations of flux tube magnetic fields.

THERMAL PROPERTIES OF FLUX TUBES

The thermal and magnetic properties of flux tubes are linked both physically and in their observational diagnostics. For a thin flux tube, magnetic field and gas pressure are simply related by $P_e - P_i = B^2/8\pi$, where the subscripts e and i label external and internal conditions. A quick estimate for typical measured field strengths indicates that the gas pressure in the tube must be at least several times smaller than in the surrounding atmosphere. Since the dominant (H^-) opacity is directly proportional to electron pressure, small-scale flux tubes must have a Wilson depression. As the plasma β is less than one, the gas pressure, opacity, and hence the size of the Wilson depression can vary substantially with only small changes in the field strength at a given geometrical height ($\delta P_i/P_i = -2\beta^{-1}\delta B/B$). Pursuing the analogy with sunspots, we might also anticipate that the strong magnetic field in the tube locally reduces convective energy transport, with the result that the interior of the tube is cooler than its surroundings. On the other hand, if the tube is only a few hundred kilometers thick, lateral radiative exchange with the ambient atmosphere will tend to equalize the internal and external temperatures. Evidently a flux tube is an

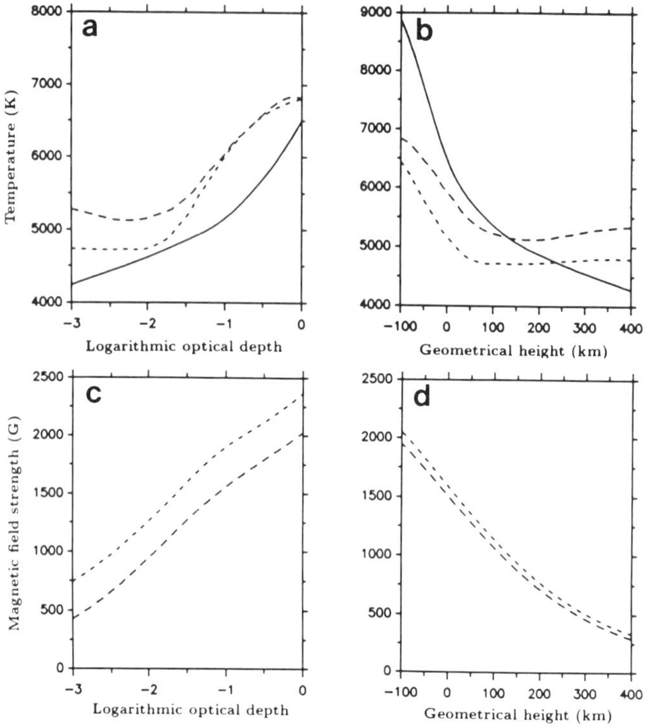

Fig. 4: Empirical models of temperature and field strength in plage (short dashes) and network (long dashes) flux tubes compared with the quiet photosphere (solid line), from Keller *et al.* (1990).

interesting place! Here I can only sketch qualitatively how the thermal structure of a tube can affect the interpretation of its measured magnetic field strengths.

Figure 4 shows the run of temperature and magnetic field strength against both continuum optical depth (τ_{5000}) and geometrical height (h) for two empirical models of flux tubes derived from an extensive set of visible diagnostics by Keller *et al.* (1990). The two models have almost the same magnetic structure, but the network model is hotter, denser (not shown), and more optically thick than the plage model at every height. A given value of τ_{5000} occurs higher in the atmosphere, and thus at lower field strengths, in the network model. High-excitation infrared lines such as Fe I 6388.6 cm^{-1} are formed at *roughly* $\tau_{5000} = -1$ in both models, which implies that the field strength sensed by such lines will be several hundred gauss lower in the hotter model, even if $B(h)$ is unchanged. For a more temperature-sensitive line such as Fe I 5250.2 Å, line weakening in the hotter model displaces the line contribution function downward, in opposition to the rise in the $\tau_{5000} = 1$ surface, with the net result that the line is formed at almost the same height and senses almost the same field strength in both models (Zayer *et al.* 1990). It is clear, therefore, that after infrared measurements have revealed systematic spatial variations of B in plage flux tubes, the road to inferring the corresponding variations in the internal structure of the tubes is still a long one. It is also clear that simultaneous and cospatial measurements using *both* infrared and visible diagnostics will be a powerful vehicle for traveling down that road.

SMALL-SCALE FIELDS IN THE LARGE

Are flux tubes a "detail" of the solar photosphere that can be safely ignored when we take a larger view of the atmosphere? This section sketches how small-scale magnetic features may produce consequences on larger scales.

First, the original stimulus for proposing the existence of photospheric flux tubes — disagreements between magnetograph flux measurements in different spectral lines — remains a matter of practical concern. Saturation effects only now coming to light lead to factor-of-two changes in some estimates of the photospheric flux and inferred magnetic field strengths in the corona and the interplanetary medium (Ulrich 1992).

A second important group of effects centers around the dispersal and disappearance of small-scale magnetic elements. In 1964, Leighton proposed a diffusion ansatz for simulating the large-scale evolution of photospheric magnetic flux (Leighton 1964). Positive- and negative-polarity elements were presumed to spread over the solar surface under the combined influence of random and systematic velocity fields. When positive and negative elements met, they simply cancelled arithmetically. Simulations based on the diffusion ansatz were in general agreement with observations then and work even better now with further refinements (see below). Nevertheless, the key assumption of flux "cancellation" had no physical or observational basis. It was something of a surprise when a process that *acted* like cancellation was actually observed in the quiet network (Martin 1984). Figure 5 (left) shows an example of this phenomenon, which is better seen in time-lapse movies. The network patches seen in the figure are too large to be individual flux tubes; but the fact that they cancel gradually over the course of a day suggests that the underlying process, probably involving the reconnection and submergence of Ω-shaped semiloops, occurs on the scale of the tubes themselves.

The effects of photospheric flux tubes can reach into the outer atmosphere. Figure 5 (right) illustrates the inhomogeneous structure and time-variability of EUV line emission from the quiet supergranular network. A careful analysis of these data shows that, in a line with a high signal-to-noise ratio, over 95% of the raster elements are intrinsically time-variable, typically by $\pm$20–30% in intensity (Rabin and Dowdy 1991). The most natural framework for such pervasive variability is a picture of the lower transition region in which the emission is modulated not by the large-scale corona but by the jostling and cancellation (reconnection/submergence) of network elements below (Dowdy, Rabin, and Moore 1986).

Cancellation is not the only way flux is observed to disappear from the photosphere. Sometimes a network element or even a sizable, dominantly unipolar piece of an active region just seems to fade away. Simple subduction of semiloops is one mechanism. Another possibility involves the thermal structure of flux tubes. As discussed above, empirical models indicate that tubes in a sparsely filled environment (such as network) are hotter at all optical depths

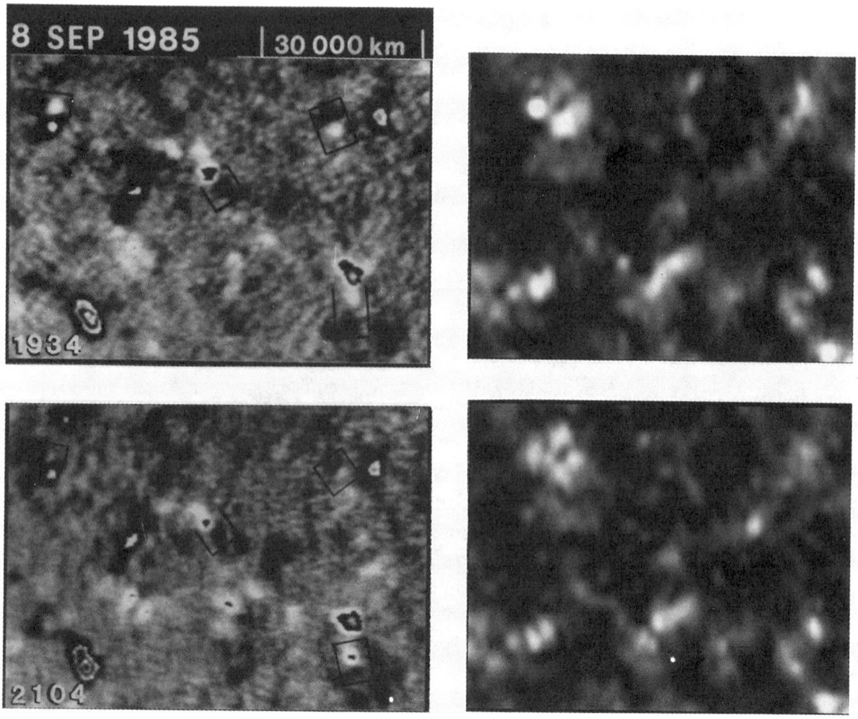

Fig. 5: Left: flux cancellation in network magnetic elements (photo courtesy of S. Martin). Right: network structure and time variability in transition-region emission from O VI 1032 Å. The EUV images are $5' x 5'$, taken 5.5 min apart.

than tubes in a highly filled area (such as strong plages or sunspots). Even if there is no difference in magnetic structure, a magnetograph employing a temperature-sensitive line such as Fe I 5250.2 Å will observe a weaker V signal in a hotter structure because the line is shallower (only partially offset by an increase in continuum intensity within the tube). Quantitatively, a Babcock-type magnetograph observing the breakup of a sunspot in a highly temperature-sensitive line may measure an apparent loss of flux by a factor of 2–10 in the absence of any true loss (Grossmann-Doerth, Pahlke, and Schüssler 1987). It is not clear that this phenomenon is sufficient to explain the fading of network elements in less temperature-sensitive lines, but there is no doubt that the internal temperature of an unresolved flux tube has macroscopic consequences for the measurement of magnetic flux in the photosphere.

Let us return finally to the evolution of photospheric flux over the whole Sun. Figure 6 shows the result of using observed bipolar magnetic regions as source terms in a flux transport simulation that includes diffusion, differential rotation, and meridional flow (DeVore, Sheeley, and Boris 1984). The qualitative agreement in the figure is bolstered by quantitative comparisons with the behavior of coronal holes, the strength of the polar field, the quasi-rigid rotation of photospheric flux patterns, and the structure of the interplanetary magnetic

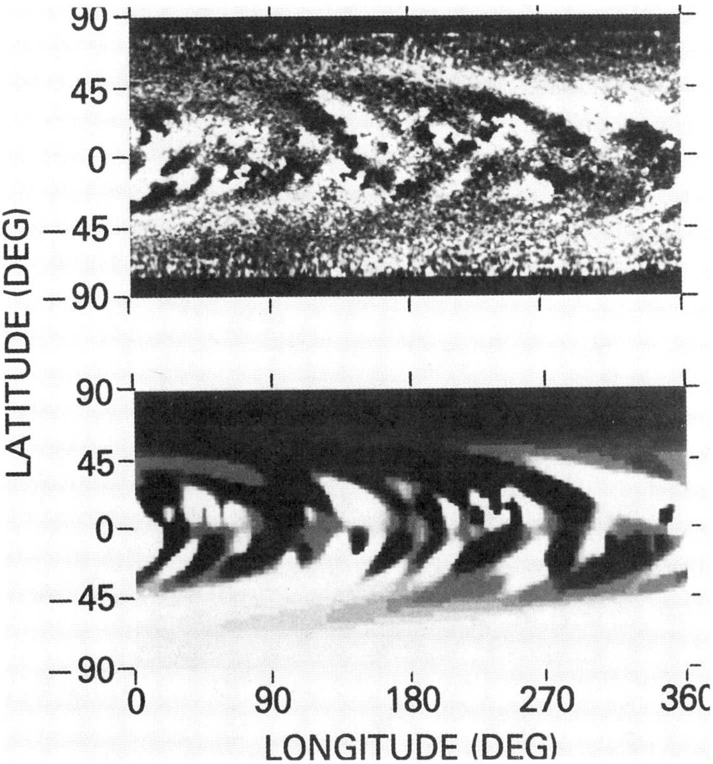

Fig. 6: Evolution of magnetic flux in the photosphere (from Wang, Nash, and Sheeley 1989). Top: observed flux during Carrington rotation 1728 constructed from daily NSO magnetograms. Bottom: simulated distribution from a flux transport model. Photo courtesy of N. Sheeley.

field (Wang, Nash, and Sheeley 1989). The agreement is *not* perfect, reflecting incomplete initial conditions, numerical approximations, and the possibility of other contributions to the large-scale field (Stenflo 1990; Wilson 1992); but the overall success of the model is impressive. What is remarkable from our present perspective is that the linchpin of the model, underlying phenomena that evolve over a solar cycle and reach into interplanetary space, is a process of mixing and disappearance that must apparently be understood on the scale of sub-arcsecond magnetic structures.

ACKNOWLEDGEMENTS

I am grateful to S. Martin and N. Sheeley for allowing me to use their grayscale images. This work was partially supported by NASA under Task 170-38-51-01-10.

REFERENCES

Babcock, H. W., and Babcock, H. D. 1955, *Ap. J.*, **121**, 349.
Beckers, J. M., and Schröter, E. H. 1968, *Solar Phys.*, **4**, 142.
Brault, J. W., and Noyes, R. W. 1983, *Ap. J. Letters*, **269**, L61.
DeVore, C. R., Sheeley, N. R., Jr., and Boris, J. P. 1984, *Solar Phys.*, **92**, 1.
Dowdy, J. F., Jr., Rabin, D., and Moore, R. 1986, *Solar Phys.*, **105**, 35.
Frazier, E. N., and Stenflo, J. O. 1972, *Solar Phys.*, **27**, 330.
Grossmann-Doerth, U., Pahlke, K.-D., and Schüssler, M. 1987, *Astr. Ap.*, **176**, 139.
Harvey, J. W., and Hall, D. 1975, *Bull. A.A.S.*, **7**, 459.
Howard, R., and Stenflo, J. O. 1972, *Solar Phys.*, **22**, 402.
Keller, C. U., Solanki, S. K., Steiner, O., and Stenflo, J. O. 1990, *Astr. Ap.*, **233**, 583.
Leighton, R. B. 1964, *Ap. J.*, **140**, 1547.
Martin, S. F. 1984, in Small-Scale Dynamical Processes in Quiet Stellar Atmospheres, ed. S. Keil (Sunspot: NSO), p. 30.
Parker, E. N. 1955, *Ap. J.*, **121**, 491.
Rabin, D. 1991, *Ap. J.*, submitted.
Rabin, D., and Dowdy, J. F., Jr. 1991, in preparation.
Rüedi, I., Solanki, S. K., and Rabin, D. 1991, in preparation.
Sheeley, N. R., Jr. 1967, *Solar Phys.*, **1**, 171.
Stenflo, J. O. 1990, in JOSO Annual Report, ed. A. von Alvensleben, p. 49.
Stenflo, J. O., Harvey, J. W., Brault, J. W., and Solanki, S. K. 1984, *Astr. Ap.*, **131**, 333.
Stenflo, J. O., Solanki, S. K., and Harvey, J. W. 1987, *Astr. Ap.*, **173**, 167.
Spruit, H. C. 1976, *Solar Phys.*, **50**, 269.
Ulrich, R. K. 1992, in Cool Stars, Stellar Systems, and the Sun, ed. M. Giampapa and J. Bookbinder (ASP Conference Series), these proceedings.
Wang, Y.-M., Nash, A. G., and Sheeley, N. R., Jr. 1989, *Science*, **245**, 712.
Wilson, P. R. 1992, in Cool Stars, Stellar Systems, and the Sun, ed. M. Giampapa and J. Bookbinder (ASP Conference Series), these proceedings.
Zayer, I., Solanki, S. K., Stenflo, J. O., and Keller, C. U. 1990, *Astr. Ap.*, **239**, 356.

Seventh Cambridge Workshop on Cool Stars, Stellar Systems, and the Sun
ASP Conference Series, Vol. 26, 1992
Mark S. Giampapa and Jay A. Bookbinder (eds.)

MAGNETIC FIELDS IN STELLAR PHOTOSPHERES

SAMI K. SOLANKI
Institute of Astronomy, ETH-Zentrum, CH-8092 Zürich, Switzerland

ABSTRACT An overview is given of the basic classes of techniques for measuring magnetic fields on cool stars. The main results of magnetic field measurements on late-type stars are then briefly summarized before the assumptions underlying the interpretation of the measurements are considered in detail. Their validity is discussed and methods of improving the interpretation are presented. Finally, it is argued that current measurements overestimate the magnetic fluxes on at least some cool stars.

Keywords: Stellar magnetism, stellar activity

1. INTRODUCTION

One of the key parameters underlying stellar magnetic activity is the magnetic field. Unfortunately, accurate observations of the magnetic field are difficult. Stars which freely exhibit an abundance of exotic behaviour — such as strong variations in luminosity or in Ca II H and K core flux, strong X–rays and microwave emission, energetic flares, etc. — suddenly become shy and secretive when it comes to revealing their magnetic fields. Special, and often quite involved, techniques have to be applied in order to trick cool stars into revealing their magnetic secrets.

In Sect. 2 I give a rough overview of the classes of techniques developed to measure solar and stellar fields, together with their relative advantages and disadvantages. No details about the individual techniques are given here. For more details see, *e.g.*, Robinson (1980), Giampapa *et al.* (1983), Marcy (1983, 1984), Gray (1988), Saar (1988, 1991b), Basri and Marcy (1988), Mathys (1989) and Solanki (1991).

Underlying every measuring technique is a model of the structure of the stellar magnetic field. The commonly used model assumes that for the purposes of magnetic measurements the stellar atmosphere is described with sufficient accuracy by 2-components: a magnetic component with a fixed field strength B covering a fraction f (the filling factor) of the stellar surface and a non-magnetic component ($B = 0$) covering a fraction $1-f$. Furthermore, it is assumed that the magnetic component is composed of many individual magnetic features distributed evenly over the stellar surface (*cf.* Sect. 5).

2. CLASSES OF TECHNIQUES FOR MEASURING MAGNETIC FIELDS

In addition to the magnetic field many stellar, atomic and instrumental parameters affect a spectral line profile (*e.g.* atomic abundance, oscillator strength, excitation potential, temperature, velocity gradients, stellar rotation, instrumental broadening). The main difficulty in measuring stellar magnetic fields is due to the fact that it is not straightforward to separate the effects of these parameters from the often subtle influence of the magnetic field. Of the techniques developed to circumvent such problems we consider only those based on the Zeeman effect.

A. The most certain signature of a magnetic field is the detection of net polarization (Stokes Q, U, V) of the distinctive spectral form produced by the Zeeman effect. Due to its sensitivity, reliability and ease of measurement Stokes V (*i.e.* net circular polarization) is the most common spectral diagnostic of solar magnetic fields. Unfortunately, Stokes Q, U and V depend sensitively on the geometry of the field. In particular, Stokes V responds only to the longitudinal component of the field and changes sign if the polarity of the field reverses. Therefore, if both polarities are present in equal amounts on the visible hemisphere of a star, then the Stokes V flux profile is completely cancelled. Since this is generally the case Stokes V profiles have been measured on only few stars (see Donati *et al.* 1992). Stokes Q and U, *i.e.* net linear polarization profiles, are only visible if the transverse component of the field has a preferred orientation (*e.g.* a bipolar configuration).

B. Somewhat less reliable, but in the case of Stokes V still an almost certain indicator of a magnetic field, is the detection of a net polarization with limited spectral information (*e.g.* broad-band polarization). Note that the problem of cancellation of polarities remains equally acute for Stokes V. Due to the increased S/N ratio and less stringent instrumental requirements many more such observations have been carried out. In the vast majority of the cases a null result was obtained (*e.g.* Borra *et al.* 1984), again due to flux cancellation, although a few detections do exist.

C. If the net polarization cannot be used then the most reliable magnetic field diagnostic is the profile shape of Stokes I. If fB is not very large then the influence of the field on Stokes I is quite subtle and much more difficult to detect than in Stokes V. To detect a field at least two lines must be observed, a magnetically sensitive and an otherwise similar insensitive one (it is also possible to use the same line on an active and an inactive star). Almost all magnetic measurements on cool stars have been made using a technique of this type, first developed by Robinson (1980). Often B and f can be obtained separately, although for incomplete Zeeman splitting it is fB^2 which is most reliably determined (Stenflo and Lindegren 1977, Gray 1984). Although the Stokes I profile shape is insensitive to the polarity of the field, it is heavily influenced by noise, blends and stellar rotation — it only works for stars with small $v\sin i$.

D. The final and least sensitive direct diagnostic of cool-star fields is based on Zeeman desaturation: In the presence of a field the individual Zeeman components move apart, so that some of them absorb in what is virgin continuum for $B = 0$. Thus the equivalent width of a saturated line is increased by a field. Zeeman desaturation has been applied to the measurement of stellar fields by Mathys and Lanz (1990) and to cool stars by Basri *et al.* (1992). Zeeman desaturation is insensitive to stellar rotation or instrumental broadening, but for slow rotators it is considerably less sensitive than the Robinson type techniques (Basri *et al.* 1992) In addition, this technique does not allow f and B to be separated, but only gives their product fB.

3. SELECTED RESULTS

In the years since the first definite detection of a magnetic field on a late-type star other than the sun (ξ Boo A, G8V) by Robinson et al. (1980) the number of stars on which a magnetic field has been reliably detected has risen to approximately 30 (see *e.g.* Saar 1990, 1991a). For a sample of similar size only upper limits have been obtained. The relatively high percentage of detections reflects the bias in the total observed sample towards stars showing high levels of activity. The stars for which definite detections have been reported range in spectral type from M4.5 to G0. All but three stars with detected fields are dwarfs (one of the exceptions being a T Tauri star). Reasons for the non-detection of further giants have been discussed by Marcy and Bruning (1984).

One of the first applications of the magnetic measurements has been the search for correlations between magnetic and other stellar parameters (*e.g.* stellar rotation frequency, inverse Rossby number, gas pressure, X-ray flux, Ca II H and K flux). Such correlations have been presented by *e.g.* Marcy (1983, 1984), Schrijver *et al.* (1989), Marcy and Basri (1989) and Saar (1990, 1991a). Two examples are shown in Figs. 1 and 2. The figures are based on the compilations of Saar (1990, 1991a).

Fig. 1 relates the surface-averaged magnetic field strength fB to the rotational frequency Ω. Dynamo theories predict relations between fB and Ω or fB and $\tau_c\Omega$ (where τ_c is the convective turn-over time). In agreement with these predictions the observed fB increases with Ω, although the scatter in the observed relation is at present too large to distinguish clearly between rivalling models.

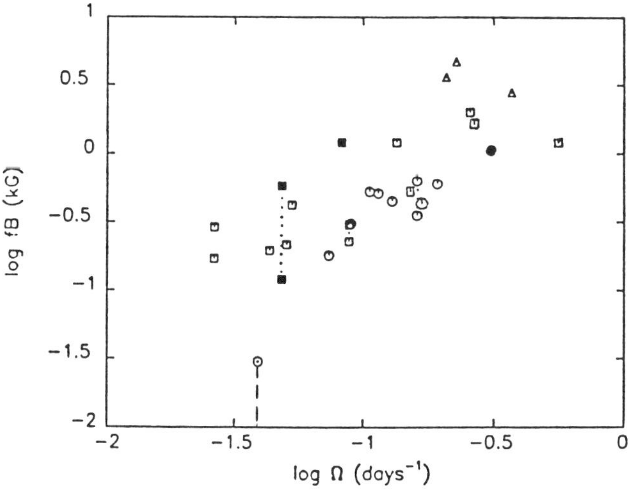

Fig. 1: Magnetic flux density $\log fB$ vs. angular rotation frequency Ω (circles: G, squares: K, triangles: M stars. Open symbols: dwarfs, filled symbols: giants).

Fig. 2 shows the measured B vs. the "equipartition" value B_{eq}, where B_{eq} is the field strength at unit continuum optical depth in the non-magnetic atmosphere, scaled such that $B_{eq}(\text{sun}) = 1500$ G. The solid diagonal line represents $B = B_{eq}$. The two dashed lines give the limits within which solar features carrying a non-negligible amount of the solar magnetic flux are observed. For most stars $B \approx B_{eq}$ appears to be satisfied, but for 5–6 stars measured B values are substantially smaller than B_{eq}. I see three possibilities to explain this discrepant behaviour: 1. Since f and B are less reliably measured then $\sqrt{f}B$ it is possible that f has been systematically over- and B underestimated for these stars. 2. For the majority of the anomalous dwarfs the B field has been measured using lines around 8000Å, which are formed higher in the atmosphere than the lines used to measure the rest of the stellar sample (Grossmann-Doerth and Solanki 1990). The field strength is expected to decrease with height, so that lines formed higher measure a smaller field strength (*cf.* Sect. 4.2). 3. The processes leading to the formation of strong filamented magnetic fields need not be equally efficient or even the same in stars of different types. Therefore, not all stars need fulfill a single scaling relationship.

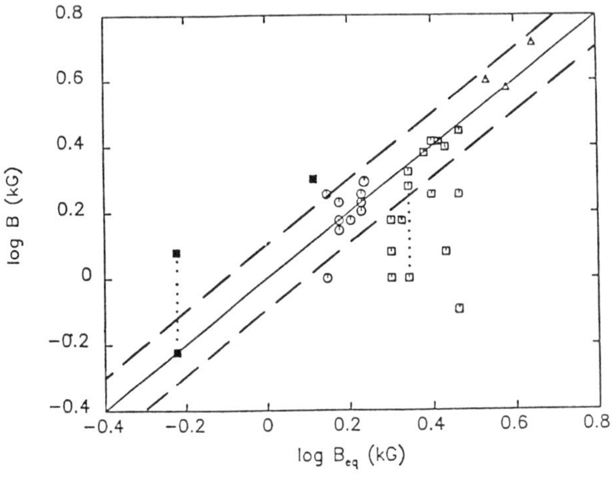

Fig. 2: Measured field strength B vs. equipartition field strength B_{eq}. Symbols as in Fig. 1. See the text for details.

4. INTERPRETATION OF STELLAR MAGNETIC MEASUREMENTS

Consider the question of how the measured B and f values relate to the true field on the star. In order to simplify the discussion we assume that the B and f values obtained from the data are uncorrupted by noise, blends, insufficient Zeeman sensitivity of the lines and insufficient realism of the radiative transfer. The influence of these parameters has been discussed by *e.g.* Gondoin *et al.* (1985), Hartmann (1987), Saar (1988) and Landolfi *et al.* (1989). In other words we assume that f and B are correct *within the constraints of the model underlying the analysis of the stellar observational data.* The main assumptions underlying the current interpretations of magnetic measurements are:

1. The structure of the field is described by a simple 2-component model (briefly outlined in Sect. 1), *i.e.* there is no horizontal distribution of field strength.
2. Thre are no vertical gradients of B (and f).
3. The magnetic features (flux tubes) are distributed homogeneously over the star.
4. The atmosphere within the magnetic features is the same as the non-magnetic atmosphere of the star.

Some techniques make additional assumptions, but we do not consider these here. Although the main assumptions appear reasonable and some of them are at present unavoidable for most stars, it is nevertheless worthwhile to consider how these assumptions influence the derived B and f values. We test their validity by confronting them with solar observations, with theory and with observations of cool stars.

4.1. Assumption 1: 2-Component Model

The 2-component description, *i.e.* no significant horizontal distribution of field strengths, is to first order a good representation of magnetic fields in solar plages. Recent observations confirm the theoretical picture that most of the flux is concentrated into small flux tubes, *i.e.* discrete bundles of field lines emerging through the stellar surface, with similar field strength and with little variation of the field strength across the diameter of a tube (Zayer *et al.* 1988, 1990, Solanki *et al.* 1992a). Theory does not rule out strong horizontal variations within individual flux tubes, but the simplest and least artificial models have a horizontally almost constant field bounded by a sharp current sheet. Furthermore, convective collapse calculations by Spruit (1979) predict that $1280 \lesssim B(z=0) \lesssim 1650$ G for different flux tubes. Therefore, at least for solar plages, theory supports the 2-component view.

Large flux tubes, such as sunspots, on the other hand, do not have a horizontally homogeneous field. The field strength drops from 2000–3000 G in the umbra to 750–1000 G at the outer penumbral boundary, as suggested by theoretical models (*e.g.* Pizzo 1986, Jahn 1989) and observations (*e.g.* Lites and Skumanich 1990, Solanki *et al.* 1992b). However, the quantity most relevant for comparison with stellar measurements is the field strength averaged over a whole sunspot. Such average values lie in the range 1000–1600 G (Solanki and Schmidt 1992) and are not too different from the field measured in plages (1200–1700 G, *e.g.* Rabin 1992, Solanki *et al.* 1992a). For stars of other spectral types the existence of multiple magnetic components cannot be ruled out (Saar 1992), although vertical field gradients may also be able to reproduce the observations.

Even if the 2-component model is not correct, this should not affect the measured magnetic parameters of most stars significantly. For example, the measured field strength would simply be a weighted average of the two or more magnetic components. There is currently no compelling reason to drop the 2-component assumption for the field. Note that although the magnetic field may be well described by two components the thermal structure of the magnetic features requires at least three components (hot plage flux tubes and cold spots), but more about temperature later.

4.2. Assumption 2: No Vertical Gradient of the Field Strength

The field strength averaged over a magnetic flux tube which is confined by pressure balance decreases exponentially with height, since in the simplest case (isothermal atmosphere, no tension) $B(z) \sim \sqrt{p(z)} \sim exp(-z/2H_p)$, where H_p is the pressure scale height. On the sun this model can account for all data (Zayer *et al.* 1988, Solanki *et al.* 1992a). The detection of broad-band circular polarization on λ And (Kemp *et al.* 1987) and on HD 129333 (Elias and Dorren 1990) indicates the presence of a vertical field-strength gradient on these stars. A longitudinal gradient is needed to produce a broad-band circular polarization of the observed magnitude, as calculations for the solar case show (*e.g.* Illing *et al.* 1975, Grossmann-Doerth *et al.* 1989, Solanki 1989).

A vertical field-strength gradient affects the interpretation of magnetic measurements in mainly three ways:

1. Lines formed at different heights do not give the same B and f, although Bf remains constant. Lines formed lower in the atmosphere should give larger B (and smaller f) than lines formed higher (Grossmann-Doerth and Solanki 1990).
2. A vertical B gradient affects line shapes in the same way as a horizontal distribution of field strengths. If the vertical gradient is not taken into account in the data interpretation it can lead to a false detection of a horizontal field-strength distribution.
3. A vertical gradient of the field, unlike a horizontal distribution, affects the equivalent width of the σ-components (it enhances their Zeeman desaturation). Consequently filling factors determined without taking a dB/dz into account are too large compared with the true values, particularly for cool stars on which dB/dz is larger.

4.3. Assumption 3: Homogeneous Surface Distribution of Magnetic Features

On the sun a homogeneous distribution of the field is an acceptable first-order approximation, although during activity maximum the presence of large active regions concentrates the field near the equator. The amount of theoretical work predicting the distribution of magnetic flux on a stellar surface is limited to studies of the solar case. And even here the theoretical models have difficulty in explaining the emergence of (large scale) flux near the equator (Choudhuri and Gilman 1987). In these calculations most of the flux emerges near the poles, being pulled there by the Coriolis force. Obviously some mechanism overrides the Coriolis force on the sun. According to Choudhuri and D'Silva (1990) one possibility of making the flux rise towards the equator is to put it into small flux tubes which can exchange angular momentum through interaction with turbulence. But in a fast rotator the Coriolis force can be over an order of magnitude stronger than on the sun, and it is proportionately more difficult to keep the field from appearing predominantly at the poles of these stars. A homogeneous surface distribution may then well be a poor representation of the field on such stars. This idea is in good agreement with the frequently observed polar spots on RS CVn or BY Dra stars. The concentration of the field at the

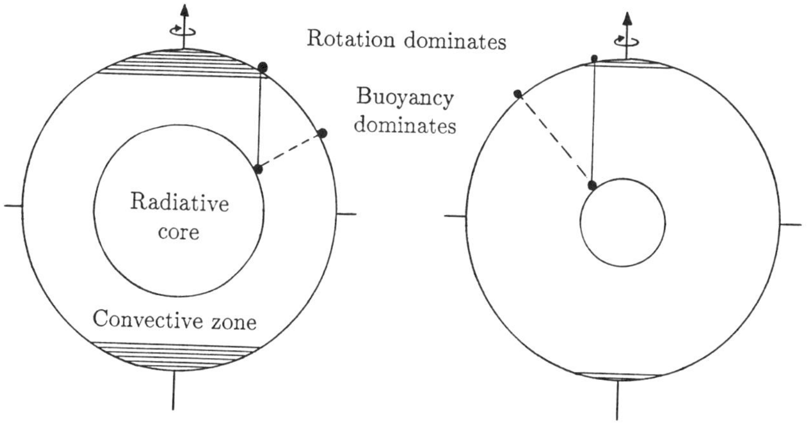

Fig. 3: Cross-section through a late G (left) and late K (right) star showing the rise of a magnetic flux tube (or toroidal ring) from the base of the convection zone if buoyancy (dashed line) and if rotation (solid line) dominates. The shaded regions near the stellar poles represent the areas of magnetic flux emergence if rotation dominates.

poles of rapid rotators should become even stronger for later spectral types as the depths of their convection zones increase (Fig. 3).

What is the directly measured distribution of magnetic fields on other late-type stars? Rotation remains the key to mapping the magnetic distribution of stellar surfaces. For slow rotators information on the distribution can be obtained, mainly by noting modulations of f and B over a stellar rotation period. The ensuing curves of f and B vs. stellar longitude can be used as an input to improve the determination of f and B in a second iteration. Further iterations may be carried out if necessary.

Rotational modulation of f and B on ξ Boo A is seen by Saar *et al.* (1987), who conclude that it has four main active areas that are roughly equally spaced in longitude, but have different sizes. On the other hand, Basri and Marcy (1988) find no changes in f and B for ϵ Eri and ξ Boo A monitored at a number of epochs.

For a few stars Stokes V can be used. It has the advantage that one can obtain information on the distribution of magnetic polarities, but, on the other hand, flux cancellation implies that only the *net* uncancelled field contributes to the measured signal. The broad-band Stokes V signal of λ And was followed over a number of rotation periods by Kemp *et al.* (1987). They saw clear temporal variations of Stokes V on the general time-scale of the rotation period. By taking measurements of solar broad-band circular polarization measurements as a guide Mürset *et al.* (1988) could roughly reconstruct the position of the 1–2 fairly localized features giving rise to the Stokes V signal.

For fast rotators more detailed information on the distribution of the field can be obtained by using Stokes V or Stokes I Zeeman Doppler imaging

(ZDI). Stokes V ZDI relies on the fact that rapid rotation can keep the Stokes V signals from regions of opposite polarity from canceling if the regions are sufficiently separated in stellar longitude. Stokes V ZDI was developed by Semel (1989), tested by Donati *et al.* (1989) and applied to the RS CVn system HR 1099 by Donati et al. (1990, 1992), who find that the uncancelled net field is rather inhomogeneously distributed. Stokes I ZDI uses the difference between the equivalent widths of magnetic and non-magnetic spectral lines to derive information on the surface distribution of the product fB. The equivalent width of the Zeeman sensitive line is larger (due to Zeeman desaturation, Sect. 2) in those parts of the rotationally broadened profile formed at a longitude with a substantial field. Since Zeeman desaturation is less sensitive to fB than other techniques, Stokes I ZDI is expected to work only for the most active stars. On the other hand, it, unlike Stokes V ZDI, is not restricted to fields whose polarities are sufficiently separated. Saar and Piskunov (1992) have applied Stokes I ZDI to the BY Dra variable HD 82558. They also find a relatively inhomogeneous distribution of the field.

4.4. Assumption 4: Horizontally Homogeneous Temperature

The temperature of most solar magnetic features is not equal to that of the quiet sun — sunspot umbrae can be up to 2000K cooler, while small flux tubes generally are 250–500K hotter. Both types of features carry substantial amounts of magnetic flux. The temperature near $\tau = 1$ within a stellar magnetic feature depends on the ratio of horizontal photon mean-free-path to diameter of the feature: The larger this ratio, the hotter the magnetic feature. According to theory only a small fraction of all magnetic features should have a temperature similar to the non-magnetic atmosphere. However, predicting the quantitatively correct temperature of magnetic features from first principles is extremely complex (*e.g.* Spruit 1976, Deinzer *et al.* 1984, Steiner and Stenflo 1990, Knölker *et al.* 1990) and requires complex model calculations. Such calculations have so far been carried out only for the sun.

There are three ways of observationally determining whether the magnetic fields on active cool stars are concentrated mainly in hot and bright or in cool and dark features.

A. **Common sense:** The continuum intensity of spot umbrae is so low, that they should be practically invisible in the flux spectrum of a star (Saar *et al.* 1986). This argument is confirmed by test calculations (Basri *et al.* 1990, Saar and Solanki 1992). However, not-so-dark features, such as penumbrae, may still give a significant contribution.

B. **High and low excitation lines:** On some stars with low $v \sin i$ the temperature in the magnetic features can be estimated by comparing fB determined from temperature sensitive lines with fB derived from temperature insensitive lines. If the two fB are equal, then the temperatures assumed for the magnetic and non-magnetic atmospheres are correct relative to each other, if not, then the assumed temperature of the magnetic relative to the non-magnetic atmosphere is wrong. It is possible to derive the correct relative temperature by changing it until both high and low excitation lines give the same fB. The additional constraint that the flux spectrum of the whole star must give the observed $B - V$ fixes the absolute temperature. A discrepancy between high and low excitation lines has been seen in ϵ Eri (Solanki and Mathys 1987, Mathys and Solanki 1989). It suggests that the temperature in the magnetic features of this star is higher than in its field-free part.

C. **Zeeman Doppler Imaging:** The ZDI techniques give a correlation between temperature and magnetic field. Note, however, that the higher or lower temperature correlated to large fB need not correspond to the temperature within the magnetic features. *e.g.*, if the true size of the stellar magnetic features is smaller than the spatial resolution of the "Doppler image", then the average temperature derived from Doppler imaging corresponds to neither the magnetic nor the non-magnetic parts of the atmosphere in the "Doppler resolution element". For such features Doppler imaging gives something akin to 1-component models of solar plages, which bear only limited resemblance to the atmosphere in the magnetic features (*e.g.* Solanki 1990).
For HR 1099 Donati *et al.* (1990) find from Stokes V ZDI that the net magnetic flux is concentrated outside the dark spots, *i.e.* probably in plage-like structures, while Donati *et al.* (1992) find the largest signal in and around dark patches at a later epoch. Saar and Piskunov (1992) find from an application of Stokes I ZDI to HD 82558 that the darker patches on the star are correlated with a magnetic field.

In summary, all the evidence seems to suggest that the temperature in stellar magnetic features is different from the temperature of the non-magnetic photosphere.

The only way to empirically obtain the *true* temperature in magnetic features from Stokes I flux is by applying method B. It is, in principle, possible to combine methods B and C and thus to determine the true temperature in the magnetic features on fast rotators as well.

Dropping the assumption of equal temperature can affect the derived f and B values considerably (*cf.* Grossmann-Doerth *et al.* 1987, Basri *et al.* 1990, Saar and Solanki 1992). Fig. 4 illustrates the expected enhancement, respectively reduction, of the spectral contribution of plage magnetic features to the measured spectral lines, for two hypothetical lines of neutral iron. An effective line strengthening greater than unity implies that measured fB values are larger than the true values, while a value below unity signifies an underestimate of fB. The Kurucz (1991) models for $\log g = 4.5$ and $3500 \le T_{eff} \le 5750$ K describe the non-magnetic atmosphere, while the magnetic atmosphere is given by Kurucz models that are 250 K hotter than the non-magnetic ones. Although the detailed dependence on T_{eff} obviously is a function of the line and of the details of the models used, the trend is probably quite general: For K and M type stars we expect plage f values to be overestimated. For G and F stars a slight underestimate of f is expected. The Ti I lines near $2.2\,\mu$, which have been used to measure magnetic fields on M stars should behave similarly. Note that spot fields are generally underestimated.

4.5. Are Very High Filling Factors Real?

Almost a third of the stars with measured filling factors have $f \gtrsim 50\%$. The main argument in favour of the reality of such large filling factors is that, even if the measured values are uncertain, there is still an unknown amount of magnetic flux hidden in the form of stellar spots. This implies that the true filling factor is even larger than the measured values. On the other hand, a number of arguments suggest that at least some measurements overestimate the true plage f values. For example:

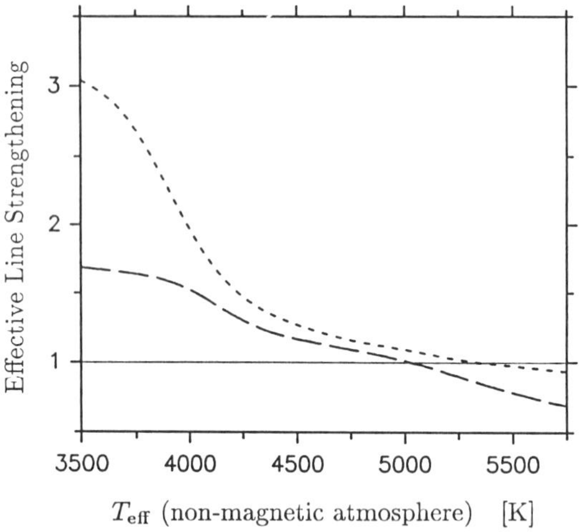

Fig. 4: Effective line strengthening in magnetic features vs. the effective temperature T_{eff} of the non-magnetic part of the star. Short dashes: Fe I line with excitation potential $\chi_e = 4$eV, long dashes: Fe I line with $\chi_e = 0$ eV.

1. The neglect of the vertical B gradient present on cool stars leads to an overestimate of f. Since B and dB/dz increase towards later spectral types this effect is expected to be larger for cooler stars.
2. If magnetic features are hotter than field-free regions then f is slightly underestimated on warm stars, but is overestimated (possibly by a large factor) on cooler stars.
3. We can measure f and B individually only on stars with small $v\sin i$. Now, a star with $f = 50\%$ can hardly be a truly slow rotator. Indeed rotation periods, measured from *e.g.* Ca II H and K variations, are small. Therefore, we probably see such stars nearly pole on. An extrapolation of the work of Choudhuri and co-workers to rapid rotators shows that conservation of angular momentum forces most of the stellar magnetic flux to appear near the poles. Therefore the true total filling factor of these stars is expected to be considerably lower than the measured values, which are strongly weighted towards the heavily magnetized poles.

In summary, the currently measured filling factor of the magnetic field in the form of plages are expected to be overestimated, particularly for K and M dwarfs. However, an unknown amount of spot contribution is also present.

ACKNOWLEDGEMENTS

Discussions with M. Giampapa, S. Saar and M. Schüssler have been very fruitful.

REFERENCES

Basri, G.S., Marcy, G.W. 1988, *Ap. J.* **330**, 274

Basri, G., Marcy, G.W., Valenti, J.A. 1990, *Ap. J.* **360**, 650

Basri, G., Marcy, G.W., Valenti, J.A. 1992, *Ap. J.* in press

Borra, E.F., Edwards, G., Mayor, M. 1984, *Ap. J.* **284**, 211

Choudhuri, A.R., Gilman, P.A. 1987, *Ap. J.* **316**, 788

Choudhuri, A.R., D'Silva, S. 1990, *Astr. Ap.* **239**, 326

Deinzer, W., Hensler, G., Schüssler, M., Weisshaar, E.: 1984, *Astr. Ap.* **139**, 426

Donati, J.-F., Semel, M., Praderie, F. 1989, *Astr. Ap.* **225**, 467

Donati, J.-F., Semel, M., Rees, D.E., Taylor, K., Robinson, R.D.: 1990, *Astr. Ap.* **232**, L1

Donati, J.-F., Brown, S.F., Semel, M., Rees, D.E. 1992, these proceedings

Elias, N.N., Dorren, J.D. 1990, *Astr. J.* **100**, 818

Giampapa, M.S., Golub, L., Worden, S.P. 1983, *Ap. J.* **268**, L121

Gondoin, P., Giampapa, M.S., Bookbinder, J.A. 1985, *Ap. J.* **297**, 710

Gray, D.F. 1984, *Ap. J.* **277**, 640

Gray, D.F. 1988, Lectures on Spectral-Line Analysis: F, G and K Stars, The Publisher, Arva

Grossmann-Doerth, U., Pahlke, K.-D., Schüssler, M. 1987, *Astr. Ap.* **176**, 139

Grossmann-Doerth, U., Schüssler, M., Solanki, S.K. 1989, *Astr. Ap.* **221**, 338

Grossmann-Doerth, U., Solanki, S.K. 1990, *Astr. Ap.* **238**, 279

Hartmann, L. 1987, in Cool Stars, Stellar Systems, and the Sun, V., J.L. Linsky, R.E. Stencel (Eds.), Lecture Notes in Physics Vol. 291, Springer-Verlag, Berlin, p. 1

Illing, R.M.E., Landman, D.A., Mickey, D.L. 1975, *Astr. Ap.* **41**, 183

Jahn, K. 1989, *Astr. Ap.* **222**, 264

Kemp, J.C., Henson, G.D., Kraus, D.J., Dunaway, M.H., Hall, D.S., Boyd, L.J., Genet, R.M., Guinan, E.F., Wacker, S.W., McCook, G.P.: 1987, *Ap. J.* **317**, L29

Knölker, M., Grossmann-Doerth, U., Schüssler, M., Weisshaar, E.: 1990, *Adv. Space Res.* **11**, 285

Kurucz, R.L. 1991, in Stellar Atmospheres: Beyond Classical Models L. Crivellari, I. Hubeny (Eds.), Kluwer, Dordrecht, in press

Landolfi, M., Landi Degl'Innocenti, M., Landi Degl'Innocenti, E.: 1989, *Astr. Ap.* **216**, 113

Lites, B.W., Skumanich, A. 1990, *Ap. J.* **348**, 747

Marcy, G.W. 1983, in Solar and Stellar Magnetic Fields: Origins and Coronal Effects, J.O. Stenflo (Ed.), *IAU Symp.* **102**, 3

Marcy, G.W. 1984, *Ap. J.* **276**, 286

Marcy, G.W., Basri,G. 1989, *Ap. J.* **345**, 480

Marcy, G.W., Bruning, D.H. 1984, *Ap. J.* **281**, 286

Mathys, G. 1989, Fundam. Cosmic Phys. **168**, 184

Mathys, G., Lanz, T. 1990, *Astr. Ap.* **230**, L21

Mathys, G., Solanki, S.K. 1989, *Astr. Ap.* **208**, 189

Mürset, U., Solanki, S.K., Stenflo, J.O. 1988, *Astr. Ap.* **204**, 279

Pizzo, V.J. 1986, *Ap. J.* **302**, 785

Rabin, D. 1992, these proceedings

Robinson, R.D. 1980, *Ap. J.* **239**, 961

Robinson, R.D., Worden, S.P., Harvey, J.W. 1980, *Ap. J.* **236**, L155

Saar, S.H. 1988, *Ap. J.* **324**, 441

Saar, S.H. 1990, in Solar Photosphere: Structure, Convection and Magnetic Fields, J.O. Stenflo (Ed.), Kluwer, Dordrecht, *IAU Symp.* **138**, 427

Saar, S.H. 1991a, in The Sun and Cool Stars: Activity, Magnetism, Dynamos I. Tuominen, D. Moss, G. Rüdiger (Eds.), Springer-Verlag, Berlin, *IAU Coll.* **130**, 389
Saar, S.H. 1991b, in Memorie della Società Astronomia Italiana, R. Pallavicini (Ed.), **61**, 559
Saar, S.H. 1992, these proceedings
Saar, S.H., Huovelin, J., Giampapa, M.S., Linsky, J.L., Jordan, C. 1987, in Activity in Cool Star Envelopes, O. Havnes *et al.* (Eds.), Reidel, Dordrecht, p. 45
Saar, S.H., Linsky, J.L., Beckers, J.M. 1986, *Ap. J.* **302**, 777
Saar, S.H., Piskunov, N.E. 1992, these proceedings
Saar, S.H., Solanki, S.K. 1992, these proceedings
Semel, M. 1989, *Astr. Ap.* **225**, 456
Schrijver, C.J., Coté, J., Zwaan, C., Saar, S.H. 1989, *Ap. J.* **337**, 964
Solanki, S.K. 1989, *Astr. Ap.* **224**, 225
Solanki, S.K. 1990, in Solar Photosphere: Structure, Convection and Magnetic Fields, J.O Stenflo (Ed.), *IAU Symp.* **138**, 103
Solanki, S.K. 1991, *Rev. Modern Astron.* **4**, in press
Solanki, S.K., Mathys, G. 1987, in Activity in Cool Star Envelopes, O. Havnes *et al.* (Eds.), Reidel, Dordrecht, p. 39
Solanki, S.K., Rüedi, I., Livingston, W., Stenflo, J.O. 1992a, these proceedings
Solanki, S.K., Rüedi, I., Livingston, W. 1992b, *Astr. Ap.* to be submitted
Solanki, S.K., Schmidt, H.U. 1992, *Astr. Ap.* to be submitted
Spruit, H.C. 1976, *Solar Phys.* **50**, 269
Spruit, H.C. 1979, *Solar Phys.* **61**, 363
Steiner, O., Stenflo, J.O. 1990, in Solar Photosphere: Structure, Convection and Magnetic Fields, J.O. Stenflo (Ed.), Kluwer, Dordrecht , *IAU Symp.* **138**, 181
Stenflo, J.O., Lindegren, L. 1977, *Astr. Ap.* **59**, 367
Zayer, I., Solanki, S.K., Stenflo, J.O. 1988, *Astr. Ap.* **211**, 463
Zayer, I., Solanki, S.K., Stenflo, J.O., Keller, C.U. 1990, *Astr. Ap.*

Seventh Cambridge Workshop on Cool Stars, Stellar Systems, and the Sun
ASP Conference Series, Vol. 26, 1992
Mark S. Giampapa and Jay A. Bookbinder (eds.)

PHYSICAL ASSESSMENT OF STELLAR DYNAMO THEORY

E. H. LEVY
Lunar and Planetary Laboratory, Department of Planetary Sciences
The University of Arizona, Tucson, Arizona 85721 USA

ABSTRACT Magnetic fields of many stars are thought to be generated by magnetohydrodynamic dynamos. Because the Sun is the only star for which detailed observations are possible, the Sun is the crucial example against which most ideas about stellar physics must be measured. Solar observations and theoretical work carried out over the past two decades have provoked an especially sharp confrontation between theoretical ideas about the generation of stellar magnetic fields and the actual physical behavior of the Sun. This confrontation centers around a few physical questions and uncertainties that lie at the very core of our ideas about dynamo magnetic field generation. This paper focuses on the central conceptual questions facing the theory of dynamo magnetic field generation in stars, especially as our understanding of the underlying physical processes is informed and challenged by our knowledge of the Sun.

Keywords: stellar magnetic fields; dynamo theory; solar cycle

INTRODUCTION

Magnetism pervades the universe, filling large volumes of interstellar—and perhaps intergalactic—space with weak, though often dynamically significant magnetic fields. In compact objects, such as stars, planets, and many varieties of nebulae, magnetic field strengths reach much higher values: for example, of the order of one to a hundred Gauss in planets, one Gauss to thousands of Gauss in main-sequence stars (Borra *et al.* 1982), as high as 10^8 or more Gauss in white dwarfs (Angel 1978), and 10^{12} or more Gauss in neutron-star pulsars (Ruderman 1972). The origin and persistence of the magnetic fields in this wide variety of objects is a puzzle that has been pieced together over approximately the past 35 years. In view of the fact that the Sun is the closest star, and the only star for which detailed observations—with very high spatial and temporal resolution—are available, the Sun plays a seminal role in all of our ideas about the behavior of stars. During the past two decades the confrontation between theoretical ideas about magnetic field generation and observations of the Sun has become especially sharply focused. In addition, significant questions have been raised about the internal consistency of certain physical aspects of the dynamo theory. As a consequence, we face several questions about the ability of 'standard' dynamo theory to account for stellar magnetic field generation. This paper reviews the basic dynamo theory of stellar magnetic fields with the specific objective of emphasizing those elemental questions that seem to attack the very heart of the theory. In this paper we do not take up a description of the stellar observations; for overviews of the pertinent main-sequence stellar magnetic-field observations themselves, the reader is referred to Borra *et al.* (1982) and Hartmann and Noyes (1987).

Why a Dynamo Theory?

At the outset it is instructive to ask why a dynamo theory for the generation (or, more accurately, *regeneration*) of stellar magnetic fields is considered to be necessary at all. In particular, it is important to understand that the mere existence of strong stellar magnetic fields is not the reason why dynamo theories are considered. Under typical cosmic conditions magnetic fields obey the magnetohydrodynamic induction equation:

$$\frac{\partial B}{\partial t} = \nabla \times (v \times B) + \eta \nabla^2 B, \tag{1}$$

where the magnetic diffusivity is $\eta = c^2(4\pi\sigma)^{-1}$, with electrical conductivity σ; B is the magnetic field and v is the fluid velocity. In Equation (1), the first term on the righthand side describes the tendency of magnetic field lines to be convected with the fluid by virtue of the fluid's high electrical conductivity. The second term describes the diffusive dissipation of the magnetic field that occurs because of the finite electrical resistivity of the fluid.

Letting ℓ be the characteristic spatial scale in equation (1), and taking v to be the characteristic amplitude of the fluid velocity, one can form the dimensionless ratio of the first term on the left-hand side of equation (1) to the second term. This ratio is commonly called the magnetic Reynolds number:

$$\mathcal{R}_m \equiv \frac{v\ell}{\eta}.$$

For low values of the magnetic Reynolds number, $\mathcal{R}_m \lesssim 1$, dissipation dominates the behavior of the magnetic field, and the effect of the fluid motion is only a minor perturbation. For high values of the magnetic Reynolds number, $\mathcal{R}_m \gg 1$, the magnetic field lines are said to be 'frozen into' the fluid and move with it. In astrophysical systems, the combination of high electrical conductivity (small η), large spatial scales, ℓ, and often high fluid velocities, typically result in high values of $\mathcal{R}_m$.

Consider a static star of radius R (or specifically the nonconvective radiative interior of such a star). The timescale for a large-scale magnetic field to diffuse out of the star, decaying as a result of resistive dissipation, is $\tau_r \sim R^2/\eta$. For ionized hydrogen at temperature T, $\sigma \approx 10^7 T^{3/2}$ esu. Thus for temperatures in excess of 10^6K, $\sigma \gtrsim 10^{16}$ s^{-1}, and $\eta \lesssim 7 \times 10^3$ cm^2 s^{-1}. So, for $R \sim 5 \times 10^{10}$ cm, $\tau_r \gtrsim 10^{10}$ years. This estimate shows that the time to dissipate the large-scale magnetic field in a star is of the order of a solar-type star's main-sequence lifetime. Therefore any initial magnetic field threading a star's interior at the time of its formation could be expected to persist for the life of the star.

This raises questions about the possibility that stars might form with significant magnetic field threading their interiors, something which could, in fact, be the case. Indeed, considerable theoretical efforts have been expended over many years to the understand the physical mechanisms by which a solar mass of prestellar material might shed the enormous quantity of magnetic flux with which it starts. The present-day Sun has a poloidal magnetic field of a few Gauss, corresponding to a total magnetic flux that approaches 10^{23} Maxwells. Consider a solar mass of material spread through space at a density corresponding to some smoothed, average interstellar value, say about one atom per cubic centimeter. The strength of the galactic magnetic field is about 3×10^{-6} Gauss; the tenuously-distributed solar mass of material would have had a spatial scale of about 10^{19} cm, entraining in excess of 10^{32} Maxwells, some nine orders of magnitude greater magnetic flux than the star that might form from it.

This idealized concept of a star forming from smoothly distributed interstellar matter is too simplified to be taken seriously. However observations of interstellar clouds and cloud cores, which are believed to be the more temporally proximate precursors of star formation, show a similar discordancy between the amount of magnetic flux entrained in the precursor material and the flux that ultimately ends up in the star. Cloud cores, the latest protostellar-cloud stage currently observable, show magnetic fluxes of around 10^{31} Maxwells per solar mass (see the discussion in Levy *et al.* 1992 and references therein). Thus, retention of a part in 10^6 or a part in 10^7 of this protostellar flux could easily account for the weak solar poloidal magnetic field. At the present time there is no way to rule out retention of this amount of magnetic flux, or even substantially larger amounts. Early evolutionary phases of a star could significantly alter its magnetic field, especially during convective, turbulent phases. But to take up this question would carry us far afield of the main subject that occupies us here.

Altogether, the *existence* of stellar magnetic fields would not have been, by itself, sufficient to drive us toward a theory of stellar dynamos. Rather it is the *behaviors* of the Sun's magnetic field—that is to say the global spatial and temporal structures of the field—that provide the greatest provocation to consider active generation mechanisms for stellar magnetism. In particular, consider the Sun. The solar magnetic cycle displays four essential characteristics, which any successful theory of stellar magnetic field generation must reproduce: 1) the overall spatial structure of the magnetic field; 2) the 22-year solar magnetic cycle; 3) the largely equatorward migration of the magnetic mode, as reflected in the usual sunspot 'butterfly' diagram; and 4) the polarity relationship between the poloidal magnetic field and the toroidal magnetic field, as reflected in the east-west ordering of sunspot-pair polarities. (For a current review of the solar cycle, see Rabin *et al.* 1991; and for a detailed discussion of solar-surface magnetic field structures see Zwaan 1987.) Moreover, it goes almost without saying that a successful theory of magnetic field generation needs to be consistent with the observed fluid motions on the Sun, as well as it needs to be internally consistent. As we discuss presently, the current dynamo theories can account for all of the observed characteristics of the Sun's magnetic field in a self-consistent way; but so far no single self-consistent model has been constructed in which all of the essential magnetic field characteristics have been reproduced in the context of picture that also agrees with other observations of the Sun.

To close this introductory discussion, we note that evidence from the observed solar magnetic cycle indicates that very little primordial magnetic flux is penetrating from the Sun's radiative interior into the convective zone (Levy and Boyer 1982; Boyer and Levy 1984; Pudovkin and Benevolenskaya 1984). The fact that the Sun seems to retain little, if any, fossil remnant of a primordial magnetic field provokes at least the speculation that few if any other stars preserve such fossil magnetic fields. In that case, we are confronted by the possible need to account for all stellar magnetism—leaving aside, for this discussion, the magnetic fields of white dwarf stars and pulsars—on the basis of contemporary generation processes, presumably by some variety of a dynamo.

BASIC DYNAMO THEORY

The basic theory of dynamo magnetic field generation was initially developed by Parker (1955), following earlier work by Elsässer (eg., 1950), and subsequently elaborated by several workers (Braginskii 1963; Steenbeck and Krause 1966; Parker 1971, among others). Magnetic field generation in cosmic objects seems to be a result of convection of highly electrically conducting fluids in rotating objects. In particular, convection in rotating a body has two properties that are essential to effective regeneration of magnetic fields: Firstly, advective angular momentum transport in the convection induces a state of differential rotation; secondly, the Coriolis force acting on the convection induces the convective motions to be cyclonic, or helical, so that $(\nabla \times v) \cdot v \neq 0$.

The discussion in this paper is couched in the language of 'kinematic' dynamo theory, wherein we assume the fluid motions to be asserted rather than solved for. This approach describes the most direct path to understanding the basis of the theory and to seeing where the most basic problems lie in applying the theory to understanding magnetic field generation in real objects. Numerical investigations of the more complete 'dynamical' theory, wherein one attempts to solve the full coupled fluid and electromagnetic equations, have been attempted by a number of workers (for example, Gilman 1981, 1983a; DeLuca and Gilman 1986, 1988; also Glatzmaier 1985 and references therein).

In a dynamo, motion of an electrically conducting fluid through a magnetic field generates an electrical current that reinforces the original magnetic field, compensating for the field decay through electrical resistance and other dissipative processes. Modern dynamo theory is usually cast in terms of a large-scale mean magnetic field; the regenerative fluid motions can produce both large-scale and small-scale magnetic-field components in which the correlated nature of the small-scale field components is crucial to the overall regeneration of the large-scale mean magnetic field.

For the purpose of this discussion, we focus first on the simplest dynamo model—the so-called $\alpha\omega$ dynamo—which seems to capture the essential conceptual basis of magnetic field generation in cosmic objects. The mean magnetic field is fully described by its 'poloidal' (B_p) and 'toroidal' (B_ϕ) components. The mean poloidal magnetic field is that part of the mean field that protrudes out of the main body of the star. The mean toroidal magnetic field winds about the axis of symmetry and is confined beneath the surface of the star in the region of high electrical conductivity.* (For simplicity we confine this discussion to

* The question of toroidal magnetic field confinement within a star is much more complicated than we have so far indicated here. In particular, the exterior of a star—unlike, for example, the exterior of a metallic planetary core—is generally a good conductor of electricity, rather than an electrical insulator. Therefore, it is possible for a toroidal magnetic field to exist in the star's atmosphere. However, for toroidal field, generated beneath the star's surface to rise into the atmosphere, it would be necessary for

axisymmetric mean fields.) In the dynamo regeneration cycle the two magnetic field components are coupled through the fluid motions discussed earlier: the differential rotation stretches the poloidal lines of force, winding them to produce a strong toroidal magnetic field; the helical convection distorts the toroidal magnetic field, generating small-scale loops of poloidal magnetic field which, owing to their correlation, reinforce the mean poloidal magnetic field. It is the Coriolis force that imposes on the fluid flow the requisite global correlation that results in efficient magnetic-field generation.

As illustrated in Figure 1, the differential rotation winds the poloidal field lines around the rotation axis to produce the toroidal magnetic field. This process is described mathematically from the magnetohydrodynamic induction equation (1) given earlier:

$$\frac{\partial B_\phi}{\partial t} - \eta \left(\nabla^2 - \frac{1}{r^2 \sin^2\theta} \right) B_\phi = \left(\frac{\partial v_\theta}{\partial r} - \frac{v_\theta}{r} \right) B_r + \cdots \qquad [\equiv \gamma B_p]. \tag{2}$$

The dots on the right side of equation (2) stand for additional terms associated with meridional gradients of azimuthal flow and the θ-component of the magnetic field. Equation (2) is written for the specific case in which the differential rotation dominates over other mechanisms for the generation of toroidal magnetic field. As we will mention later, the helical convection also so contributes to the generation of toroidal field, but with less efficacy.

The coupling of toroidal magnetic field into the poloidal magnetic field is also illustrated in Figure 1, where small poloidal magnetic loops are generated by the rising and twisting motions of the convection. The Coriolis force acting on the expansion of a highly compressible parcel of fluid, as it rises, produces the twisting distortion of the field lines. Downward moving parcels of stellar fluid, which are compressed to smaller spatial scales, twist in the opposite direction, producing similar loops of magnetic field. This process can be described mathematically by (Parker 1955, 1971, 1979; Steenbeck and Krause 1966; Krause and Rädler 1980)

$$\frac{\partial A_\phi}{\partial t} - \eta \left(\nabla^2 - \frac{1}{r^2 \sin^2\theta} \right) A_\phi = \Gamma B_\phi \qquad [\approx v_{cyc} B_\phi], \tag{3}$$

where we have written the poloidal field in terms of a toroidal vector potential, so that

$$B_p \equiv (B_r, B_\theta) = \nabla \times A_\phi. \tag{4}$$

---

the field to unburden itself of the entrained mass. It is likely that a dynamical restraint such as this is primarily responsible for confining the mean toroidal field beneath a star's surface. However, little in the current discussion is changed by this. For more detailed discussion of the possibility of toroidal field escape from stars the reader is referred, for example, to Parker (1984) and Vainshtein and Rosner (1991), and for discussion of the effect of the toroidal field outer boundary condition on the detailed character of dynamo modes the reader is referred to Chouduri (1984).

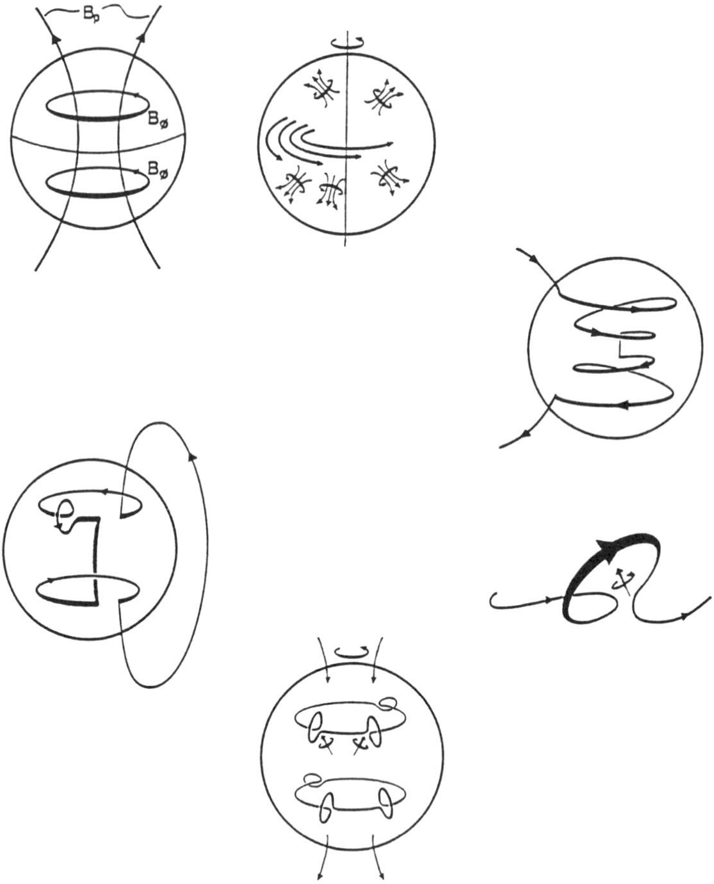

Figure 1. A cartoon representation of the basic elements of the regenerative dynamo cycle thought to produce the Sun's magnetic field. Proceeding clockwise from the upper left, a large-scale, axisymmetric, mean magnetic field, consisting of poloidal and toroidal components is regenerated by fluid motions composed of differential rotation and helical convection. The differential rotation stretches the poloidal magnetic field lines to generate an azimuthal toroidal magnetic field. The toroidal field is distorted by helical convection to produce loops of poloidal magnetic field, which 'diffuse' outward and combine to regenerate the original large-scale poloidal magnetic field. As described in the text, and although it is not obvious from the cartoon, this process gives rise to a variety of magnetic mode structures and behaviors.

In equation (3) Γ is, in general, a functional of the fluid velocity field. However, for a fully convective fluid, $\Gamma \sim v_{cyc}$, where v_{cyc} is the cyclonic part of the convective motion, i.e., that part of the motion corresponding to the screw-like motion turning around the axis of a convective eddy (see Levy 1976 for a simple physical discussion.) The specific form of the right-hand side of equation (3) is based on the approximation of high magnetic Reynolds number in a convective flow with idealized, short-lived, convective eddies (Parker 1970, 1979). Other, also idealized, derivations lead to essentially the same result (see, example, Steenbeck and Krause 1966; Levy 1978; Parker 1979; Krause and Rädler 1980). In writing equations (2) and (3), we have confined ourselves to only the dominant right-hand-side terms; the reader is encouraged to consult standard review texts on the subject for a more complete discussion (Parker 1979; Moffatt 1978; Krause and Rädler 1980).

The set of equations (2)–(4) mathematically summarize the usual $\alpha\omega$ dynamo, with equation (2) describing the so-called ω effect and equation (3) describing the so-called α effect. It is instructive to note that these equations are linear and homogeneous in the field quantities B_ϕ and A_ϕ. Consequently, neither the amplitude nor the sign of the generated magnetic field is determined by these equations; this reflects the incompleteness of the kinematical theory. A full dynamical dynamo theory would also include the equations of motion, as well as the associated constitutive relations and the relevant additional boundary conditions. This full set of governing equations would contain the Lorentz force: $(4\pi)^{-1}(\nabla \times B) \times B$. Thus, because then the fluid velocity would inherently depend on field strength, the amplitude of the generated magnetic field would be determined mathematically. On the other hand, because the equation of motion contains the magnetic field only quadratically, the fluid velocity could depend only on the amplitude of the magnetic field and not on the field's polarity. As a consequence of this, even the nonlinear dynamical dynamo equations would still have the magnetic field polarity as an exact symmetry, and either polarity of dynamo field could equally well be regenerated. This polarity symmetry is a satisfying attribute of the underlying theory inasmuch as the evidence shows that the magnetic fields of both the Sun and Earth change their sign relatively frequently and with little apparent inhibition.

The restriction of our present attention to the kinematical dynamo is not so confining as might at first seem. Indeed, although some very important aspects of dynamos are not accessible in the kinematical theory, it is important to realize that, logically, the kinematical theory is a complete theory of magnetic-field generation. Indeed, if one knew the fluid motions precisely and *a priori*, then using those fluid motions in equation (1) would provide a complete description of the behavior of the generated magnetic field. What we lack is knowledge of the physically real fluid motions.

It should be kept in mind that equations (2)–(4) are shown in their simplest form, and the reader is referred to references cited earlier for more complete discussions of the underlying theory and the full physical and mathematical elaboration of dynamo theory. The most significant simplification used in this discussion is the neglect of the contribution of the helical convection to the generation of toroidal magnetic field. Thus the form of the equations written here corresponds to the case in which the nonuniform rotation is sufficiently rapid that the generation of toroidal magnetic field by the large-scale shearing motions overwhelms the generally less efficient generation of toroidal field by the convection. This is the case for which the term $\alpha\omega$ dynamo is commonly used. However, because there is, in general, a finite contribution to the toroidal magnetic field by the helical convection, a similar regenerative dynamo cycle can be envisioned in which nonuniform rotation plays no role; this is frequently referred to as the α^2 dynamo. Finally, the case in which both contributions to the toroidal field are included is often referred to as the $\alpha^2\omega$ dynamo. The magnetic-field modes produced by these several dynamo cycles differ from one another in detail. But the differences are slight in comparison with the effects of the basic questions with which we deal in this paper; we will not refer further to these possibilities.

The Character of the Generated Dynamo Modes

In order to explore the characteristics of the generated magnetic fields, we first look at the dynamo equations (2)–(4) from a simple dimensional point of view. Consider equations (2)–(4) in the case of purely stationary magnetic field modes, so that $\partial_t \equiv 0$. Taking ℓ to be the characteristic spatial scale of the generated mean magnetic field and of the fluid flow, we find:

$$\eta \frac{B_\phi}{\ell^2} \sim \gamma B_p \qquad \left[\sim \frac{v_\phi}{\ell} B_p\right], \tag{5}$$

$$\eta \frac{A_\phi}{\ell^2} \sim \Gamma B_p \qquad \left[\sim v_{cyc} B_\phi\right], \tag{6}$$

and

$$B_p \sim \frac{A_\phi}{\ell}. \tag{7}$$

Equations (5)–(6) are a homogeneous, algebraic set, and the existence of a solution requires that the determinant of the coefficients vanishes. Thus we find:

$$\mathcal{N} \equiv \frac{\gamma \Gamma \ell^3}{\eta^2} = 1 \qquad \left[\frac{v_\phi \; v_{cyc} \ell^2}{\eta^2}\right]. \tag{8}$$

$\mathcal{N}$ is known as the 'dynamo number'. The alternative form of $\mathcal{N}$ is given in square brackets to call attention to the fact that the dynamo number is simply the product of two magnetic Reynolds numbers, one for each of the essential components of the fluid motion: the nonuniform rotation, v_ϕ, and the cyclonic component of the convection, v_{cyc}.

The requirement that the dynamo number be unity (in the crude dimensional analysis given here) simply states that, in order for a magnetic field to be regenerated, the rate of magnetic field generation induced by fluid flow must balance the decay of the field—in this case through resistive dissipation. To sharpen this point, it is instructive to rewrite the dynamo number as

$$\mathcal{N} = \left[\frac{\ell^2}{\eta}\left(\frac{\ell}{v_\phi}\,\frac{\ell}{v_{cyc}}\right)^{-1/2}\right]^2,$$

so that the dynamo number is cast explicitly in terms of the ratio of the diffusive to convective timescales. In this form, it is clear that, for a magnetic field to be generated, the rate of diffusive dissipation of the magnetic field must balance the 'typical' rate at which the field lines are distorted by the fluid flow to generate new magnetic flux. Actual solution of equations (2)–(4) for stationary dynamo modes yields a similar result, but for the precise value of the dynamo number at which this generation- dissipation balance occurs. In typical systems, the lowest value of $\mathcal{N}$ at which this balance occurs is of the order of 10–100. Moreover, in a finite body, stationary dynamo modes occur for a discrete set of dynamo-number values, with each higher value of the dynamo number corresponding to a stationary magnetic mode of smaller spatial scale. In addition to the stationary dynamo modes referred to here, dynamos also generate migrating, oscillatory magnetic fields, a subject to which will return presently.

Turning to the actual form of the dynamo number, notice the appearance of the product $\gamma\Gamma$. In particular, the magnitude (measuring the vigor of the fluid motions) and the sign (measuring the relative sense of the differential rotation and the helicity of the convection) of this product are decisive in determining the character of the generated modes—something that we will see carry over to the periodic dynamo modes also.

Turning our attention now to the time-dependent dynamo modes, it is sufficient for our present purposes to consider the characteristics of simplified dynamo equations in the a flat, Cartesian space (following Parker 1979). Thus, flattening the spherical geometry used earlier, such that $\phi \rightarrow y$, $r \rightarrow x$, and $\theta \rightarrow z$, we can write for the three dynamo equations

$$\frac{\partial B_y}{\partial t} - \eta \nabla^2 B_y = \gamma B_x, \tag{9}$$

$$\frac{\partial A_y}{\partial t} - \eta \nabla^2 A_y = \Gamma B_y, \tag{10}$$

and

$$B_x = -\frac{\partial A_y}{\partial z}. \tag{11}$$

The set of equations (9)–(11) has solutions in the form of propagating waves, which we can write in the form

$$\begin{pmatrix} A_y \\ B_y \end{pmatrix} = \begin{pmatrix} A \\ B \end{pmatrix} exp(ikz + i\omega t). \tag{12}$$

Inserting the *ansatz*, equation (12), into equations (9)–(11), and solving the resulting dispersion relation, we find solutions in the following form:

$$\begin{pmatrix} A_y \\ B_y \end{pmatrix} = \begin{pmatrix} A \\ B \end{pmatrix} exp\left[ikz \pm i\sqrt{\frac{|\gamma\Gamma k|}{2}}\, t\right] exp\left[\left(-\eta k^2 + \sqrt{\frac{|\gamma\Gamma k|}{2}}\right) t\right]. \tag{13}$$

As can be seen from equation (13), the basic dynamo modes are propagating waves with amplitudes that can either grow, decay or be stationary. Notice that there are two branches to the solution given in equation (13). By carrying the calculation through from the original dispersion relation, the reader can easily see that one branch, with dynamo-wave propagation toward $+z$, corresponds to the case with $\gamma\Gamma > 0$, while the other branch, with dynamo-wave propagation toward $-z$, corresponds to the case with $\gamma\Gamma < 0$.

For a stationary amplitude oscillating dynamo, it is necessary that the argument of the second exponential function vanish. Thus, the condition for a stationary amplitude dynamo oscillation in this simple illustration reduces to

$$\frac{\gamma \Gamma k^{-3}}{2\eta^2} = 1, \tag{14}$$

qualitatively similar, as should be expected, to the condition for stationary dynamo mode generation in the dimensional analysis leading to equation (8). More significantly, the stationary-amplitude condition allows us to rewrite equation (13) as

$$\begin{pmatrix} A_y \\ B_y \end{pmatrix} = \begin{pmatrix} A \\ B \end{pmatrix} exp(ikz \pm i\nu_d t), \tag{15}$$

where ν_d is the free-decay dissipation rate,

$$\nu_d = \eta k^2. \tag{16}$$

Equations (15) and (16) emphasize that the characteristic oscillation time of a dynamo mode with spatial wavenumber k is of the order of the free decay time of a magnetic field with the same spatial scale. This interesting result is not only intuitively satisfying, but it also works a significant simplification in estimating the oscillation period of such a dynamo mode. While calculating the details of the fluid motion with sufficient fidelity to reliably estimate the product $\gamma\Gamma$ is a formidable challenge, estimating the free decay rate—in principal, at least—seems a simpler task.

STELLAR AND SOLAR DYNAMOS

The propagating normal modes of dynamo equations given above constitute the elementary solutions. Applying these solutions to models of real objects requires that the solutions be obtained for the appropriate geometries and boundary conditions. Once this is done, the solutions that emerge show that the dynamo process described by these equations generates both oscillatory and stationary magnetic fields. Comparing these solutions with the magnetic field of the real Sun, and applying the four criteria listed earlier, we find the solutions to these dynamo equations 1) have gross spatial structures similar to the magnetic field observed on the Sun, 2) that a cycle period reasonably close to the observed 22-year period can be produced, 3) that the equatorward migration of the main magnetic mode—reflecting the sunspot butterfly diagram—can be produced, and 4) that the relationship between the polarity of the solar poloidal magnetic field and the and sunspot-pair polarities can also be reproduced. Unfortunately, while each of these principal magnetic field characteristics shows up as a feature of the dynamo solutions, the solutions that best display these characteristics seem not to be consistent with other measured characteristics of the Sun, and raise important questions about the underlying magnetohydrodynamic behavior of the solar convection zone.

The Oscillation Period

The solar magnetic cycle has an oscillation period of approximately 22 years.

Although a more accurate and complete analysis of the dynamo modes would require using a geometrically correct solution obtained with the proper boundary conditions, the major issues can be revealed, and considerable insight gained, by following the simple analysis of the previous section. Using equation (16), the oscillation period becomes

$$P \sim \lambda^2/2\pi\eta, \tag{17}$$

where, for the Sun, the wavelength $\lambda \sim 10^{11}$ cm. Taking $\sigma \sim 10^{15}$ s^{-1} for the electrical conductivity of the fluid in the convection, so that $\eta \sim 10^5$ cm^2 s^{-1}, we find $P \sim 5 \times 10^8$ years. Even in comparison with the frequently loose standards used for astrophysical estimation, this number does not compare favorably with the observed solar cycle period of 22 years, or with the similar lengths of magnetic-cycle periods inferred for other stars (Baliunas and Vaughan 1985; Gilliland and Baliunas 1987).

This estimate of the magnetic-cycle period neglects the fact that the fluid velocity influences the magnetic field in other ways than generating new toroidal and magnetic field. The convection of magnetic field lines by the fluid motion enhances the transport and mixing of the magnetic field, increasing the rate at which the field can dissipate. In the simplest case, this enhanced diffusive mixing is similar to the mixing that occurs in any stirred fluid system—cream into coffee, for example. The resulting turbulent diffusion coefficient can be estimated in analogy to diffusion in a gas. The classical gas diffusion coefficient can be worked out readily from the kinetic theory of gases, the diffusion coefficient is found to be $D = \frac{1}{3}v_{th}\lambda$, where v_{th} is the molecular thermal speed and λ is the mean free path in the gas. A passive contaminant in a turbulent fluid is carried about by the random turbulent motions in a manner that—grossly at least—is reminiscent of the random-walk motions that produce gas diffusion. By analogy then, where λ_t is the length scale typical of the largest turbulent eddies and v_t is the velocity typical of those eddies, we can write, for the turbulent magnetic diffusivity,

$$\eta_t = \kappa v_t \lambda_t, \tag{18}$$

where κ is a dimensionless number of the order of a tenth. It is important to realize that this expression is, at best, only a very crude representation of turbulent diffusion. (For discussions of turbulent diffusion, see Krause and Rädler 1978; Parker 1979)

Taking, in equation (18), values for the turbulent velocity and eddy size from the solar supergranulation, so that $v_t \sim 10^5$ cm s^{-1} and $\lambda_t \sim 10^9$ cm, we find $\eta_t \sim 10^{13}$ cm^2 s^{-1}. Using η_t in equation (17), we find $P \sim 5$ years. Although not perfect agreement, this estimate of the solar-magnetic-cycle period is startlingly—and satisfyingly—close to the observed 22-year period. Turbulent diffusion must also play another critical role in the operation of a stellar magnetohydrodynamic dynamo. Implicit in the derivation of the dynamo equations discussed earlier, is the assumption that the magnetic field lines can reconnect and change their topology on a timescale short in comparison with the timescales characteristic of the large-scale mean magnetic fields. This can be seen most directly in the requirement (Figure 1) that the small loops of poloidal magnetic field, generated by the helical convection, diffuse and reconnect on timescales short in comparison with the overall regeneration cycle. This requires that the diffusion and reconnection process proceed faster than would happen in the presence of classical resistive dissipation alone.

Neither the relatively close coincidence of this result with the actual solar cycle period, nor the lack of precise agreement, are as significant as might immediately seem. Firstly, the crude values adopted for the turbulent velocity and length scales are not obviously pertinent to the question at hand. While the values used are adopted from the observable solar photosphere, the magnetic field is generated in deeper layers of the Sun, perhaps at the bottom of the convection zone. The fluid motions in the region of the Sun where the magnetic field as actually generated are not reliably understood. Secondly, the factor-of-a-few deviation from the 22-year solar cycle could reasonably be expected to be subsumed in more accurate representations of the dynamo geometry, fluid motions, details of turbulent transport, and proper boundary conditions.

Altogether, the idea that the solar magnetic field is generated in a convection-zone dynamo, with magnetic-field transport dominated by turbulent diffusion in the convective motions is provocative and seductive, and probably captures a meaningful part of the actual physics. However, as we now proceed to take up, there remain significant and basic unresolved problems still standing between us and a final understanding of stellar dynamos.

The Physical Reality of Turbulent Mixing of Strong Magnetic Fields

The picture of turbulent mixing of magnetic fields appealed to in the previous section relies on the assumption that the field can be treated as a 'passive contaminant' in the flow, carried around passively by the moving fluid, but not exerting any forces or otherwise influencing the fluid. While this is a very good approximation to the wafting of perfume by the stirred air in a room, its applicability to strong magnetic fields in electrically conducting fluids is arguable.

Magnetic fields exert (Lorentz) forces on the fluids to which they are tied. And because these forces are inevitably correlated with the fluid motions themselves, it is not clear that the mixing of magnetic field lines, essential for the diffusion, can proceed unencumbered. In particular, if the magnetic field energy density is comparable to the energy density of the fluid motions, then it is easy to see that the mixing might be significantly inhibited. Consider a parcel of fluid—in the inertial range of the turbulence, so that the fluid is dominated by the inertial forces of the flow itself, rather than by any force imposed on the fluid—of spatial scale ℓ, so that the total kinetic energy of the parcel is $\frac{1}{2}\rho v^2 \ell^3$. If the fluid kinetic energy density is the comparable the magnetic-field energy density ($B^2/8\pi \sim \rho v^2/2$), then by the time the fluid parcel has moved a distance ℓ, comparable its own size, it will have produced a distortion in the magnetic field with associated energy $\ell^3 B^2/8\pi$, and the energy associated with the distorted magnetic field will equal the energy originally in the fluid parcel. Thus the fluid parcel will have been brought to rest and the magnetic field distortion will begin to reverse its motion. The ensuing motion is more like an Alfvénic oscillation than convective mixing. Fluid elements, instead of wandering away from one another in the manner characteristic of diffusion, simply tend to oscillate about their original locations, significantly suppressing any mixing.

In the Sun, the magnetic fields associated with the generation process seem to be of the order of 10^3 Gauss near the solar surface, as seen in the strong magnetized flux tubes associated with sunspots. The magnetic energy density associated with these fields is crudely comparable to the kinetic energy of the fluid motions, a relationship expected to persist at depth in the convection zone. Thus it is problematical whether turbulent motions can really mix the magnetic fields in the dynamo generation region. Cattaneo and Vainshtein (1992) have carried out numerical experiments to explore this question and, indeed, found that turbulent mixing of magnetic fields seemed to be suppressed when the magnetic field energy density exceeded a few percent of the fluid kinetic energy density, sharpening the question of whether turbulent motions can provide enough mixing to account for the generation of at least the Sun's magnetic field. This question deserves further exploration: The random body forces used in the Cattaneo and Vainshtein simulations may not fully reproduce the mixing behavior of a fluid driven by thermal buoyancy forces, in which true convective displacements must occur to account for the heat transport, which ultimately cannot be suppressed. Another physical avenue along which the Lorentz-force suppression of turbulent mixing may be evaded is that of two-dimensional turbulence. Magnetic suppression of turbulent mixing results from the forces that attend the distortion of magnetic field lines. Two-dimensional turbulence, in which the magnetic field lines are mixed without bending of the lines, is not suppressed in the same way as three-dimensional turbulence (Vainshtein and Levy 1991).

Altogether, the concept of turbulent mixing plays a central role in our present picture of stellar magnetic field generation. The dissipation of magnetic fields by classical resistive processes alone cannot account for the behavior of stellar magnetic-field cycles. Some mechanisms producing highly enhanced dissipation must be at work and remain to be fully elucidated.

Differential Rotation and Sunspot–Belt Migration

Referring once again to equation (15), recall that the direction of propagation of a dynamo mode is governed by the sign of the product $\gamma\Gamma$ and is in the direction perpendicular to $\nabla\Omega(r)$. Early in the development of the dynamo theory, it was assumed that the general trend of solar differential rotation was such that $d\Omega/dr < 0$, so that the angular velocity of rotation increased inward, a situation naively connected with conservation of angular momentum in the radially convecting fluid. Because of the compressibility of the solar fluid, rising convective parcels expand and rotate more slowly than their surroundings, leading to $\Gamma > 0$ in the northern hemisphere, as is indicated from the relative sense of the toroidal field and the poloidal loops in Figure 1. Altogether then, $\gamma\Gamma < 0$ in the Sun's northern hemisphere, so that the dynamo waves propagate in the direction of $-z$, i.e., toward the equator. (By symmetry, the behavior in the southern hemisphere will mirror that in the north, with the dynamo mode also propagating toward the equator.) Again, this satisfying result—which was consistent with naive physical ideas about the solar rotation, even though it was not easy to reconcile with the observed gradient of rotation at the solar surface, where the equatorial belt rotates some 10% faster than the polar regions—seemed to produce a consistent picture of the solar magnetic field. Similarly, the relative sense of the sunspot polarity ordering and the sign of the polar field implied by this simple picture of solar differential rotation was consistent with observations.

More detailed numerical modelling of convection-zone dynamics (eg., Gilman 1983b; Glatzmaier 1985) suggested that this naive picture of the

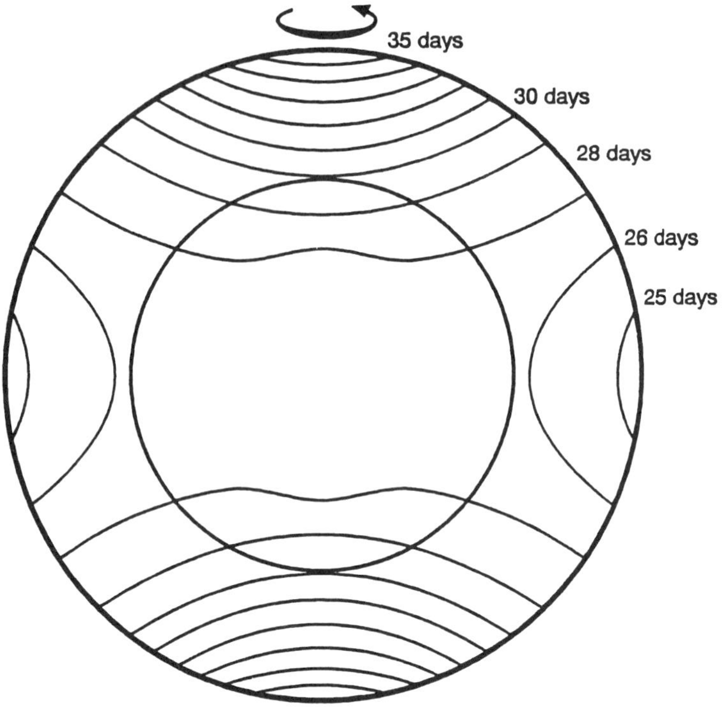

Figure 2. A crude representation of the internal rotation of the Sun as inferred from recent helioseismological measurements. The labels indicate rotation periods at several sample locations.

solar rotation might be in error; the modelled solar dynamics suggested that the observed latitudinal variation of surface rotation could simply reflect an underlying rotation on cylindrical surfaces, with $\nabla\Omega(r) > 0$ leading to poleward migration of the dynamo modes, contradicting the observed behavior of the sunspot cycle. The more recent development of helioseismology (Hill, Bos and Goode 1982; Duvall, Harvey Pomerantz 1986) has provided a way to gain direct observational insight into the Sun's interior rotation. A rotation curve, recently reduced from helioseismological measurement (Libbrecht 1989), is represented qualitatively in Figure 2. Although continuing observations and analyses are needed to refine and solidify our ideas about the Sun's interior rotation, this tentative figure suggests that neither the naive ideas nor the dynamical models captured an accurate picture of the solar rotation. Indeed, at low latitudes the indicated rotation curve has $\nabla\Omega(r) > 0$, while at mid-latitudes $\nabla\Omega(r) \sim 0$! Calculations indicate that dynamo modes generated with this form of nonuniform rotation tend not to migrate at all; instead the oscillatory magnetic field loiters in mid-latitudes (Boyer and Levy 1992).

If the solar rotation curve shown in Figure 2 withstands further scrutiny and investigation, then it will pose a definitive—though not insuperable—challenge to solar dynamo theory. The direction of migration of the sunspot-active belt over the course of a solar cycle is one of the principal behaviors of the solar magnetic field. Any tenable theory of the sunspot cycle must reproduce that behavior while, at the same time, accounting for the other major features of the solar magnetic field. Before leaving this subject, it is worth pointing out that, within the constraints of what is so far known about the fluid motions in the Sun's convective zone, there may be enough flexibility yet to account for the observed direction of sunspot-zone migration. Meridional circulations of the fluid could play a critical role, and are deserving of further exploration in that respect.

The Locale of Generation

An important question confronting our understanding of the generation of stellar magnetic fields—and especially the conceptual relationship between the solar magnetic field and stellar magnetic fields more generally—is that of the locales of the fields' primary generation. The generation of the solar magnetic field is almost certainly a consequence of processes associated with the convection zone. While in early treatments of the solar dynamo it was implicitly assumed that the regeneration was broadly distributed throughout the convection zone, more recently that assumption has been questioned because of the possible effects of magnetic buoyancy (Parker 1975). Magnetic buoyancy has the potential to drive magnetic flux out of a generation region so rapidly that the dynamo mechanism could be quenched. Although a conclusive understanding of the consequences of buoyancy is not yet in hand (eg., Parker 1984, 1987; Vainshtein and Levy 1991), there are significant dynamical reasons for suspecting that the solar magnetic field might be generated at the bottom of the convection region, near the transition between the radiative and convective zones (see, for example, Spiegel and Weiss 1980; Durney, De Young and Passot 1990, and references therein). Indeed, it may be that both the transition region and the bulk of the convection of zone play important roles in generating the solar magnetic field, with different aspects being controlled in different regions (Boyer and Levy 1992).

Should the solar magnetic field prove to be dominantly generated at the boundary between the radiative interior and the convective shell, then a provocative question will be whether magnetic fields generated in fully convective stars have physical natures significantly different natures from that of the solar magnetic field. Among the differences that could systematically be manifested are those involving temporal behavior—eg., the possible occurrence of stationary, as opposed to oscillatory, magnetic modes, and differences in the cycle periodicities. Moreover, systematic variations could occur in the strengths of the generated fields, as well as in the ratios of the poloidal to toroidal field strengths. While detailed discussion of these questions is beyond the limitations of the present paper, we note that answering these questions poses important challenges to both observational and theoretical research.

SUMMARY AND CONCLUSION

Stellar and solar dynamo theory is not yet in a satisfactory and completed state. Thus, as with any incomplete theory, dynamo theory confronts both open questions and potentially serious stumbling blocks. There are many open questions in dynamo theory, most of which I have not taken up here. Rather, I have tried to focus specifically on those question that, in my opinion, present the most important potential stumbling blocks and deep challenges to the theory. The two problems that stand out in this respect, and on which I have focused, are the efficacy of turbulent mixing of magnetic fields, and the influence of the angular velocity profile on the character of the dynamo modes. Neither of these problems is a "detail" of the theory. By working with a very basic and simple treatment of dynamo theory—staying away from the details of specific models, I have tried to emphasize that both of these questions lie at the very core of dynamo theory and are probably crucial to the success of any particular models. It is interesting that neither of these is a 'new' question; both have persisted for decades, since the very beginning of the dynamo theory nearly forty years ago.

There are fundamental and important problems left to be overcome in solidifying our understanding of the surprisingly rich and complex forms of interaction between magnetic fields and convecting stellar fluids. With these mysteries standing in our way, it is not yet obvious how the basic elements of dynamo regeneration actually operate in stars. Nonetheless, it seems to me that dynamo theory, in its present form, probably does capture the essential elements of magnetic field generation in stars, and that, in the end, our understanding will crystallize around those ideas.

ACKNOWLEDGEMENT

This work was supported in part by a grant from the National Aeronautics and Space Administration.

REFERENCES

Angel, J.R.P. 1978, *Ann. Rev. Astr. Astrophys.*, **16**, 487.

Baliunas, S.L. and Vaughan, A.H. 1985 *Ann. Rev. Astr. Astrophys.*, **23**, 379.

Borra, E.F., Landstreet, J.D. and Mestel, L. 1982. in *Ann. Rev. Astron. Astrophy.*, **20**, 191.

Boyer, D.W. and Levy, E.H. 1984, *Ap. J.*, **277**, 848.

Boyer, D.W. and Levy, E.H. 1992, *Ap. J.*, in press.

Braginskii, S.I. 1963, *Dokl. Acad. Nauk SSSR*, **145**, 1311.

Cattaneo, F. and Vainshtein, S.I. 1992, to be published.

Chouduri, A.R. 1984. *Ap. J.*, **281**, 846.

DeLuca, E.E. and Gilman, P.A. 1986, *Geophys. Astrophys. Fluid Dynamics*, **37**, 85.

DeLuca, E.E. and Gilman, P.A. 1988, *Geophys. Astrophys. Fluid Dynamics*, **43**, 119.

Durney, B.R., De Young, D.S and Passot, T.P. 1990, *Ap. J.*, **362**, 709.

Duvall, T.L., Jr., Harvey, J.W. and Pomerantz, M.A. 1986, *Nature*, **321**, 500.

Elsässer, W.M. 1950. *Rev. Mod. Phys.*, , 4.

Gilliland, R.L. and Baliunas, S.L. 1987, *Ap. J.*, **314**, 766.

Gilman, P.A. 1981, *Ap. J. Suppl.*, **46**, 211.
Gilman, P.A. 1983a, *Ap. J. Suppl.*, **53**, 243.
Gilman, P.A. 1983b, in Solar and Stellar Magnetic Fields: Origins and Coronal Effects, ed. J.O Stenflo, Reidel Publishing Company.
Glatzmaier, G.A. 1985, *J. Stat. Phys.*, **39**, 493.
Hartmann, L.W. and Noyes, R.W. 1987, *Ann. Rev. Astr. Astrophys.*, **25**, 271.
Hill, H.A., Bos, R.J. and Goode, P.R. 1982, *Phys. Rev. Lett*, **49**, 1794.
Krause, F. and Rädler, K.-H. 1980, Mean-Field Magnetohydrodynamics and Dynamo Theory, Pergamon Press.
Levy, E.H. 1978, *Ap. J.*, **220**, 329.
Levy, E.H. and Boyer, D.W. 1982, *Ap. J. Letters*, **254**, L19.
Levy, E.H., Ruzmaikin, A.A., and Ruzmaikina, T.V. 1992, in The Sun in Time, ed. C.P. Sonett, M. Giampapa and M. Matthews, University of Arizona Press.
Libbrecht, K. G. 1989, *Ap. J.*, **336**, 1092.
Moffatt, H.K. 1978. Magnetic Field Generation in Electrically Conducting Fluids, Cambridge University Press.
Parker, E.N. 1955, *Ap. J.*, **122**, 293.
Parker, E.N. 1970, *Ap. J.*, **160**, 383.
Parker, E.N. 1975, *Ap. J.*, **198**, 205.
Parker, E.N. 1979, Cosmical Magnetic Fields, Oxford University Press.
Parker, E.N. 1984, *Ap. J.*, **281**, 839.
Parker, E.N. 1987, *Ap. J.*, **312**, 868.
Pudovkin, M.I. and Benevolenskaya, E.E. 1984, *Astronom. Zh.* **61**, 783 (*Sov. Astr.*, **28** 458).
Rabin, D.M., DeVore, C.R., Hoeksma, J.T. and Sheeley, N.R., Jr. 1991, in The Solar Interior and Atmosphere, ed. A.N. Cox, W.C. Livingston and M.S. Matthews, University of Arizona Press.
Ruderman, M. 1972 *Ann. Rev. Astr. Astrophys.*, **10**, 427.
Spiegel, E.A. and Weiss, N.O. 1980, *Nature*, **287**, 616.
Steenbeck, M. and Krause, F. 1966, *Z. Naturforsch. A*, **21**, 369.
Vainshtein, S.I. and Rosner, R. 1991. to be published.
Vainshtein, S.I. and Levy, E.H. 1991. *Solar Phys.*, in press.
Zwaan, C. 1987, *Ann. Rev. Astr. Astrophys.*, **25**, 83.

Seventh Cambridge Workshop on Cool Stars, Stellar Systems, and the Sun
ASP Conference Series, Vol. 26, 1992
Mark S. Giampapa and Jay A. Bookbinder (eds.)

MAGNETIC FLUX TUBES IN A 3D CHAOTIC FLOW FIELD

GEORGE H. FISHER Institute for Astronomy, University of Hawaii, 2680 Woodlawn Drive, Honolulu, HI 96822

EDWARD E. DELUCA Institute for Astronomy, University of Hawaii, 2680 Woodlawn Drive, Honolulu, HI 96822

BRIAN M. PATTEN Institute for Astronomy, University of Hawaii, 2680 Woodlawn Drive, Honolulu, HI 96822

ABSTRACT We argue that if the magnetic fields in the solar interior are of a fibril (flux tube) form, the field strength can be amplified by fluid motions up to a limiting value B_{as}, which is determined by a balance between magnetic tension and aerodynamic drag. We find $B_{as} \simeq 3.7\ \rho^{2/3}\ l^{2/3}\ v_l^{4/3}\ \Phi^{-1/3}$, where ρ is the mass density, l is the size scale of the most vigorous motions, v_l is the velocity amplitude associated with this size scale, and Φ is the magnetic flux per fibril. We discuss implications of this limit for our understanding of the solar cycle.

Keywords: solar cycle; dynamo; magnetic flux tubes; activity

INTRODUCTION

An important component of any solar or stellar dynamo model is the amplification of magnetic fields through the stretching of magnetic field lines. In kinematic models of dynamos, the flow fields that drive field amplification are specified *a priori*; it is assumed that the magnetic fields can be carried by the flow without resistance. However, as the field strength increases, there is an increasingly important back reaction by the Lorentz force which acts to resist the fluid motions. This force is ignored in kinematic models. Under the assumption that magnetic fields are concentrated into thin flux tubes surrounded by field-free plasma, we investigate the role of magnetic tension in limiting the growth of magnetic energy.

In order to carry out this investigation, we have developed a numerical algorithm for the solution of the equation of motion of a single, closed magnetic flux tube moving in 3 dimensions. Our algorithm is based on solving the momentum equation in a novel flux tube coordinate system in which the mesh is spaced uniformly along the tube. The form of the momentum equation solved is

$$2\rho \frac{dv_\perp}{dt} = \frac{B^2}{4\pi}\frac{\partial^2 r}{\partial s^2} + C_D\ \rho\ \frac{\left|v_{f\perp} - v_\perp\right|\left(v_{f\perp} - v_\perp\right)}{\sqrt{\pi\Phi\ /B}}\ , \quad (1)$$

where $v_\perp$ is the velocity of plasma in the flux tube in the directions normal to the flux tube orientation itself, $v_{f\perp}$ is the velocity of the fluid surrounding the flux tube in the normal directions, B is the magnetic field strength, C_D is the aerodynamic drag coefficient (assumed to be unity), and Φ is the magnetic flux in the tube. Note that in this study we have neglected the buoyancy force (but see Fisher, McClymont and Chou 1991 for the effects of field amplification on flux tube equilibria near the base of the convection zone of stellar interiors). The vector $r(s)$ describes the space curve delineated by the tube, where s is distance along the tube measured from some reference point. The magnetic field strength B is related to the total length L of the closed flux tube through the equation $B/L = B_0/L_0$, where B_0 and L_0 are the initial values. Further details of the numerical code will be described in Fisher, Patten and DeLuca (1992).

The flow field v_f in which the flux tube is embedded is an "ABC" or Beltrami flow (see *e.g.* Gilbert 1991). The components of v_f are given by

$$v_{fx} = v_l \ [\ B \ sin(y/l) + C \ cos(z/l) \ ], \quad (2a)$$

$$v_{fy} = v_l \ [\ C \ cos(z/l) + A \ sin(x/l) \ ], \quad (2b)$$

$$v_{fz} = v_l \ [\ A \ cos(x/l) + B \ sin(y/l) \ ], \quad (2c)$$

where we have assumed $A = B = C = 1$.

SYNOPSIS OF SIMULATIONS

We have performed several simulations of an initially circular flux ring embedded in the flow described above. Important characteristics of the ABC flow are: (1) the fluid streamlines are generally not closed; (2) the flow has a characteristic scale size l and velocity amplitude v_l. The motion of the flux ring is governed by a competition between the magnetic tension force, whose amplitude depends both on the magnetic field strength and the radius of curvature of the field lines, and the aerodynamic drag force, whose amplitude is proportional to the square of the relative normal velocity of the tube with respect to the fluid.

Initially, the ring is stretched and distorted by fluid motions, but in so doing, the magnetic field strength of the ring increases, causing the tension force to increase more quickly than does the drag force. Eventually, an asymptotic solution is reached wherein the tension and drag forces balance. This behavior is similar to what Parker (1982a,b) found in his investigation of the motion of magnetic fibrils in steady 2-D flows and convective rolls.

The field strength achieved by the asymptotic solution can be understood by the following simple argument: The tightest bends in the asymptotic flux tube shape are presumably places where the flux tube is withstanding the full brunt of the flow field against it; the amplitude of the drag force here must be $F_d \simeq C_D \rho v_l^2/(\pi\Phi \ /B)^{1/2}$, where Φ is the magnetic flux in the tube and C_D is the drag coefficient (assumed to be unity). On the other hand, the amplitude of the magnetic tension force in the tight bends must be $F_t \simeq B^2/(4\pi l)$, where we have estimated the radius of curvature of the tube to be the characteristic length scale in the flow field. Equating the two forces and solving for asymptotic field strength B_{as} yields

$$B_{as} \simeq \frac{(4\pi C_D/\sqrt{\pi})^{2/3} \ \rho^{2/3} \ l^{2/3} \ v_l^{4/3}}{\Phi^{1/3}} \quad (3),$$

where $(4\pi C_D/\sqrt{\pi})^{2/3} \simeq 3.7$. Inserting values of the flux tube parameters from the simulation ($\Phi \ = 10^{18} \ Mx$, $\rho = 0.2 \ g \ cm^{-3}$, $l = 10^{10} \ cm$, and $v_l = 10^4 \ cm \ s^{-1}$) yields

$B_{as} \simeq 1.3 \times 10^6$ G, which compares favorably to the value of B_{as} actually reached in the simulation of 1.46×10^6 G. Other simulations with the same flux tube parameters, but a different starting position for the ring results in somewhat different equilibrium tube shapes, and values of B_{as} which differ to some degree, but the asymptotic field strength seems to be given roughly by equation (3). Values of the magnetic field strength B_{as} are in general much greater than the equipartition value, $B_{eq} \simeq (4\pi\rho v^2)^{1/2}$. Using the above values of the parameters, we find that B_{as} is roughly 50 times larger than B_{eq}.

IMPLICATIONS FOR THE SOLAR CYCLE

(1) Equation (3) shows that fluid motions on the *largest* spatial scales are most important; small scale motions are unable to move the tube because the tension forces resist most strongly at small scales and because there is less fluid energy at small scales.
(2) It is a simple matter to get field strengths far in excess of their equipartition value.
(3) Recent reconnection studies (*e.g.* DeLuca and Craig, 1992) suggest that in a collision between two flux tubes, reconnection resulting in a new flux tube configuration can happen on a time scale short compared to other dynamic time scales. This suggests that changes in magnetic topology required by a dynamo model can be accomplished easily using a flux tube paradigm.

ACKNOWLEDGEMENTS

This work is supported by Air Force grant AFOSR-90-0116, NASA grant NAGW-864, and NSF grant ATM-9106052.

REFERENCES

DeLuca E. E., and Craig I. J. D. 1992, *Ap. J.*, in press.
Fisher G. H., McClymont, A. N., and Chou, D.-Y. 1991 *Ap. J.*, 374, 766.
Fisher G. H., Patten, B. M., and DeLuca, E. E. 1992, *Ap. J.*, in prep.
Gilbert A. D., 1991, *Nature,* 350, 483.
Parker E. N. 1982a, *Ap. J.*, 256, 736.
Parker E. N. 1982b, *Ap. J.*, 256, 746.

Seventh Cambridge Workshop on Cool Stars, Stellar Systems, and the Sun
ASP Conference Series, Vol. 26, 1992
Mark S. Giampapa and Jay A. Bookbinder (eds.)

THE EAST-WEST INCLINATIONS OF MAGNETIC FIELDS IN THE SOLAR PHOTOSPHERE

ROBERT F. HOWARD
National Solar Observatory, National Optical Astronomy Observatories,*
Tucson, AZ 85726

ABSTRACT The results from three previous studies of the east-west inclinations of the magnetic fields of solar features are examined and compared. Significant differences are seen in the inclination angles of sunspots, plages, and weak magnetic features. Sunspot fields are inclined on average (to *trail* the rotation) by about 5°. The fields of leading and following spots are inclined away from each other by roughly 2°. Both spots and plages whose areas are growing and whose areas are decreasing show different inclinations. Plage fields are inclined so as to trail the rotation by only about 2° on average, but their leading and following portions are inclined toward each other by about 16°. Weak, large-scale magnetic fields are seen (from an earlier study) to have an inclination that trails the rotation by less than 1°. Leading and following polarities of these dispersed fields are inclined toward each other by several degrees. The differences in spot and plage field inclinations suggest possible differences in the subsurface configurations or dynamics of the magnetic loops which form these features at the surface. The strong inclination of young plages could result from the effects of depth-dependent rotation rate differences on the rising field lines of the young regions.

Keywords: Sunspots; Magnetic Fields; Plages

INTRODUCTION

In three papers (Howard 1974; 1991; 1992) this author has analyzed the deduced angle to the local vertical of magnetic field lines in the plane parallel to the solar equator. All three of these studies used averages determined from long synoptic data sets.

The technique for determining the east-west angle is derived in the first and second papers (Howard 1974; 1991).

RESULTS

For the determination of the inclination of sunspot magnetic fields (Howard 1992), the results from the digitization of the Mount Wilson white-light image archive (Howard, Gilman, and Gilman 1984) were used. This data set covers the

* Operated by the Association of Universities for Research in Astronomy, Inc., under Cooperative Agreement with the National Science Foundation.

interval 1917–1985. The assumption is made that the measured differences of the average areas of sunspots on the east and west sides of the CM result from the inclinations of the sunspots to the vertical in the east-west plane.

Sunspots are found to be inclined such as to trail the rotation on average by about 5°. Leading and following spots are inclined away from each other by a degree or two. Spots that are growing tend to be somewhat more inclined to trail the rotation ($\approx 8°$) than are spots that are decreasing in area ($\approx 3°$). No other characteristics of sunspots, such as residual rotation rate, polarity axis tilt angle, or distance from the average latitude of activity, showed any significant correlations with inclination angle.

It was found from 23 years of data (Howard 1991) that the average inclination of plage fields is such as to trail the rotation by $\approx 1.9°$. Leading and following fields in plages were found to be inclined toward each other by $\approx 16°$.

Regions whose magnetic flux is increasing show a very strong inclination ($\approx 24°$) in the direction to lead the rotation. Regions whose flux is decaying are inclined about 5 degree to trail the rotation. The leading and following fields derived for both these subsets of the data are inclined toward each other by about the same amount as is found for all regions.

In the earlier study (Howard 1974), which covered the interval 1967–1973, it was found that the average inclination angle of weak fields was about 0.8° trailing the rotation. On average the leading and following polarities were inclined toward each other by about 9°.

CONCLUSIONS AND DISCUSSION

We may draw the following conclusions from this review:

1. Sunspot magnetic field lines are inclined on average to trail the rotation (inclined to the east) by several degrees.
2. The magnetic fields of leading and following sunspots are on average inclined away from each other by a degree or two.
3. Growing spot fields are inclined so as to trail the rotation by nearly 8°, while decaying spot field lines trail the rotation by about 3°.
4. Plage magnetic field inclinations are about 2°, trailing the rotation.
5. The magnetic field lines of leading and following portions of plages are inclined toward each other on average by about 16°.
6. The field lines of plages whose magnetic flux is increasing are inclined on average so as to lead the rotation by about 24°.
7. The field lines of decaying plages are inclined on average so as to trail the rotation by about 5°.
8. The field lines of leading and following portions of growing and decaying plages keep their relative orientations more or less unchanged.
9. Weaker fields in the photosphere show a very slight inclination to trail the rotation ($< 1°$).
10. Leading and following polarities of weak-field features are inclined toward each other by about 9°, a quantity which is roughly half way between the sunspot and plage values.

It is reasonable to expect, because of the steep density gradient at the photosphere, that the overall east-west (and, indeed, any other direction) inclination of magnetic field lines at the photospheric level is determined entirely by forces imposed on the field lines at and below the photospheric level, not above the photosphere. The rotation gradient with depth of the sub-photospheric layers is likely to play a role in this process for the east-west component of the inclination.

Differences between sunspot and plage inclination angles are found, notably the overall difference in the inclination angles, the difference in the leading and following values, and the large difference (in the sign, even) of the inclination angles of fields in growing plages and sunspots. The latter difference suggests that for growing (most likely young) plages and spots, there is a fundamental difference in perhaps the subsurface level at which the field lines are connected, or perhaps in the physical nature or properties of that connection. It is reasonable to expect that the rising flux tubes of a newly emerging region are tipped in the east-west direction by the effects of the radial rotation gradient between the level at which the flux originates and the surface. If the level of origin of the field lines rotates faster than the surface, then the lines of force will trail the rotation by the time they emerge at the surface.

Sunspot magnetic fields are stronger than plage magnetic fields, which in turn are stronger than the fields of the weak patterns, and so it is possible that at least a part of the differences seen in the inclination angles of these features are due to the differing interactions with subsurface dynamic forces that may result from the field strength differences. It should be noted that in the sequence from strongest to weakest fields (spots, plages, weak-field features) the overall inclinations of the fields vary monotonically, with the strongest-field features having the steepest inclinations (–5°, –2°, –0.8°), but the differences in the inclination angles of the leading and following fields do not behave in this way (2°, 16°, 9°).

As features at the surface age, it may be that their magnetic connections to the subsurface source flux weaken or alter qualitatively so that they do not maintain the same east-west inclination angles. If the angle, which, for spots, starts out trailing the rotation by 8°, gradually decreases, this could be due to a shallow layer with a slower rotation rate than that at the source flux level, which tips the fields lines gradually westward. Recent results from helioseismology studies (Harvey 1988) suggest that just below the surface the rotation rate increases, so it is hard to imagine how the inclination angle of surface fields can continue to tip to the east as the fields age unless the deeper, slower rotation rate is more important in determining the inclination angle at the photosphere.

REFERENCES

Harvey, J. 1988, in Seismology of the Sun and Sun-Like Stars, ed. E. J. Rolfe, ESA Publ. SP-286, p. 55.
Howard, R. F. 1974, *Solar Phys.*, **39**, 275.
Howard, R. F. 1991, *Solar Phys.*, **134**, 233.
Howard, R. F. 1992, *Solar Phys.*, (in press).
Howard, R. F., Gilman, P. A., and Gilman, P. I. 1984, *Ap. J.*, **283**, 373.

Seventh Cambridge Workshop on Cool Stars, Stellar Systems, and the Sun
ASP Conference Series, Vol. 26, 1992
Mark S. Giampapa and Jay A. Bookbinder (eds.)

MAGNETIC FIELD STRENGTH AND CONTINUUM INTENSITY MEASUREMENTS OF SUNSPOTS AT 1.56 μm

GREG KOPP and DOUG RABIN National Solar Observatory, P.O. Box 26732, Tucson, AZ 85726-6732

ABSTRACT We present Zeeman splitting measurements of sunspots using the λ=1.5649 μm Fe I line without regard to polarization. These are compared with simultaneous intensity measurements in the adjacent continuum. We find that the relation between field strength and continuum intensity is nonlinear but has a characteristic shape.

Keywords: Sunspots; infrared; magnetic fields

INTRODUCTION

The infrared offers greater sensitivity for magnetic field measurements than the visible because 1) Zeeman splitting as a fraction of linewidth increases directly with wavelength, providing three times the sensitivity at 1.56 μm as in the visible, and 2) not only is scattering lower in the infrared than in the visible, but the effects of stray light, particularly within sunspots, are decreased due to the lower temperature sensitivity of the near infrared.

The λ=1.5649 μm (6388.6 cm^{-1}) Landé g=3 Fe I triplet is formed 110 km above the base of the quiet photosphere (Bruls *et al.* 1991). This line has proven useful for magnetic observations of flux tubes (Harvey & Hall 1975, Stenflo, Solanki & Harvey 1987, Rabin 1991) and sunspots (McPherson, Lin & Kuhn 1991). We report here on 1.56 μm observations of several sunspots and a general relation between magnetic field strength and continuum intensity.

OBSERVATIONS

We obtained Stokes I profiles of 6 sunspots using the McMath Telescope Vertical Spectrograph and a 58x62 infrared detector array on 13 May 1991. The spots varied in area from 60 to 980 millionths of a hemisphere, and all were close to disk center ($\mu \geq$0.83). Spots were not selected for their shape or unipolarity. Profiles from umbrae and penumbrae show complete splitting of the π- and σ-components at fields larger than 1400 G, allowing for the determination of magnetic field *strength*, as opposed to magnetic flux, in the spots.

We scanned a slit across each spot region, obtaining a spatial/spectral image at each slit position. From the images, we obtained both field strength B (from the split σ-components) and continuum intensity I_c (from a quiet region of the spectrum near 1.56 μm) through each sunspot. The data are seeing-limited to about 2 arcsec. No correction for stray light has been applied; we estimate it to be less than 3% of the continuum intensity.

Splittings in the nearby g=1.53 Fe triplet at λ=1.5653 μm (6386.9 cm^{-1}) (Solanki, Biémont, and Mürset 1990) are not measurable because of obscuring umbral OH lines, identified by Wallace & Livingston (1991). We use this line for spectral calibration only.

DISCUSSION

A plot of magnetic field strength vs. continuum intensity for all 6 observed spots is shown in Fig. 1. The overall shape of the plot indicates *there is a characteristic but nonlinear relation between intensity and field strength.* The range of field strengths at a given intensity (even within a single spot) implies *the relation is not single-valued.*

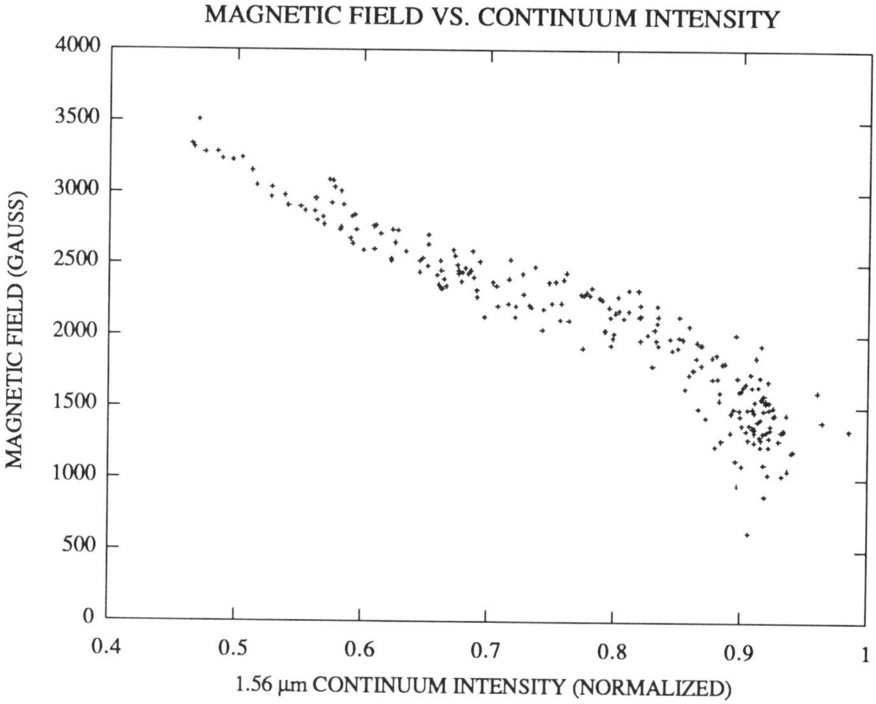

Fig. 1: Magnetic field strength vs. intensity across diameters of 6 sunspots observed on 13 May 1991. Errors are less than 100 G for fields stronger than 1500 G. Much of the scatter is intrinsic and is seen in individual spots.

The scatterplot may be divided into three roughly linear segments: 1) a strong variation of field with intensity above about 2500 G; 2) a gradual change in field with a large variation in intensity from 2000-2500 G (near the umbral/penumbral boundary of several spots); and 3) a very strong change in field below 2000 G with increased scatter.

This relation differs both from the linear relation between I_c and B reported from visible observations by Gurman & House (1981) and from the linear relation between temperature and B^2 discussed by Dicke (1970). The three segment behavior (with intrinsic scatter) seen in Fig. 1 appears in all the medium and large sunspots we have examined, as well as in similar data (not included here) taken separately by C. Lindsey, W. Livingston, and J. Kuhn.

With high magnetic sensitivity and reduced stray light effects, infrared array observations with the 1.5649 μm line are proving to be a powerful probe of magnetic conditions in sunspots.

ACKNOWLEDGEMENTS

The authors thank D. Jaksha, J. Wagner, C. Plymate, and P. Hartmann at NSO for their dedication during this observing run.

REFERENCES

Bruls, J. H. M. J., Lites, B. W., and Murphy, G. A. 1991, in L. November (ed.), Solar Polarimetry, Proc. 11th Sacramento Peak Workshop, National Solar Observatory, Sunspot, p. 444.

Dicke, R. H. 1970, *Ap. J.*, **159**, 25.

Gurman, J., and House, L. 1981, *Solar Physics*, **71**, 5.

Harvey, J. W., and Hall, D. 1975, *Bull. Amer. Astron. Soc.*, **7**, 459.

McPherson, M. R., Lin, H., and Kuhn, J. R. 1991, *Solar Physics*, submitted.

Rabin, D. 1991, *Ap. J.*, submitted.

Solanki, S. K., Biémont, E., and Mürset, U. 1990, *Astron. Astrophys. Suppl. Ser.*, **83**, 307.

Stenflo, J. O., Solanki, S. K., and Harvey, J. W. 1987, *Astron. Astrophys.*, **173**, 167.

Wallace, L., and Livingston, W. 1991, private communication.

Seventh Cambridge Workshop on Cool Stars, Stellar Systems, and the Sun
ASP Conference Series, Vol. 26, 1992
Mark S. Giampapa and Jay A. Bookbinder (eds.)

A COMPARISON OF STAR SPOT DISTRIBUTIONS FOR VARIOUS ACTIVE STARS BASED ON DOPPLER IMAGES.

MARTIN KÜRSTER MPI für extraterrestrische Physik, D–8046 Garching, FRG,

ARTIE P. HATZES McDonald Observatory, The University of Texas at Austin, TX 78712, USA,

ROBERTO PALLAVICINI, and SOFIA RANDICH Osservatorio Astrofisico di Arcetri, Largo Fermi 5, I–50125 Firenze, Italy.

ABSTRACT We present Doppler images for three active stars. Spot distributions obtained with different imaging algorithms are compared.

Keywords: star spots, Doppler imaging

INTRODUCTION

Doppler imaging is becoming an important tool for the study of stellar activity as it allows to determine the distribution of stellar magnetic surface flux and spot migration patterns ('butterfly diagram'). With Doppler imaging large polar spots have been found on three RS CVn–primaries (*e.g.* Vogt 1988, Strassmeier *et al.* 1991, Vogt & Hatzes 1991) and one FK Com–type star (Vogt 1988). Spot distributions on the FK Com–type giant YY Men=HD 32918 (Piskunov *et al.* 1990) and on the post–T Tau star AB Dor (Kürster & Schmitt 1990, Kürster 1990) seem to be dominated by an active latitude belt at low stellar latitudes.

Here we add another young star, the HD 155555 primary, to the list of imaged stars and show new images of YY Men and AB Dor. The CLEAN–like algorithm for the production of discrete 2–temperature images (Kürster & Schmitt 1990, Kürster 1991) and the maximum entropy method (MEM, Vogt *et al.* 1987) are compared. MEM reconstructions are displayed as thresholded 2–temperature images. All data were obtained with the ESO 1.4m–CAT/CES.

DOPPLER IMAGES

HD 155555=V824 Ara: Spectra from 8–18 Sep 1990 were used for the image (fig. 1) of the primary of this post–T Tau binary (*cf.* Pasquini *et al.* 1991 for evolutionary status, ephemeris, inclination). CLEAN and MEM images agree on the overall spot distribution if one considers that no smoothing is implemented in the CLEAN algorithm. CLEAN replaces the decentered polar cap seen in the MEM image by many small spots which can arise from several reasons. Limited space prohibits a discussion of these, but we note that the hypothesis of large star spots actually being groups of smaller spots agrees well with the data.

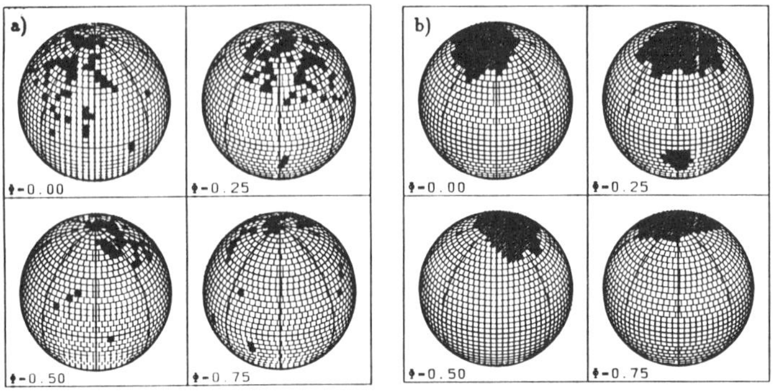

Fig. 1: HD 155555=V824 Ara. $i = 55°$. Sep 90. 12 profiles of CaIλ6439. (a) CLEAN reconstruction. (b) MEM reconstruction.

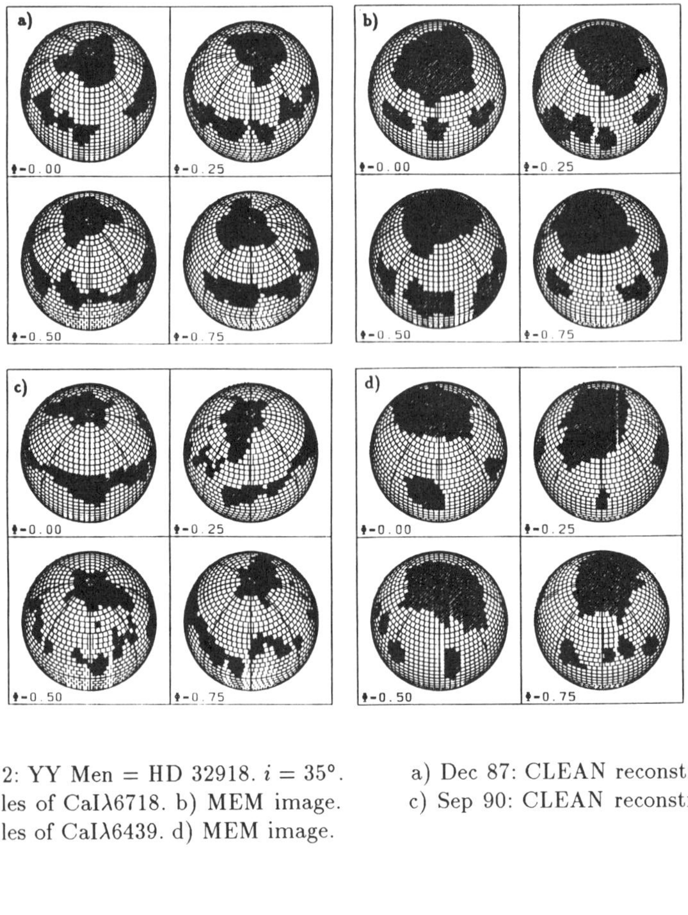

Fig. 2: YY Men = HD 32918. $i = 35°$. a) Dec 87: CLEAN reconstruction. 6 profiles of CaIλ6718. b) MEM image. c) Sep 90: CLEAN reconstruction. 5 profiles of CaIλ6439. d) MEM image.

YY Men=HD 32918: Data from 15–20 Dec 1987 and 9–18 Sep 1990 were used for two images (fig. 2). Contrary to Piskunov *et al.* (1990) who used $i = 65°$ we found best fits for $i = 35°$. CLEAN and MEM images are very similar. In both years YY Men has a polar spot and an active latitude belt. In their Jan. 1987 image Piskunov *et al.* (1990) also found a low–latitude belt, but no polar feature. However, we suspect that the polar spot is a long–lived feature much as it seems to be for the other so far imaged FK Com– or RS CVn–type stars.

AB Dor=HD 36705: Only CLEAN images (fig. 3) based on data from Dec 1987, Feb 1989 (already published by Kürster & Schmitt 1990, Kürster 1991), and Sep 1990 are shown. In 1989 and 1990 most features are close to an active latitude of $\approx +25°$. In the 1987 image high–latitude structure prevails. However, due to minor quality data we consider this image as preliminary.

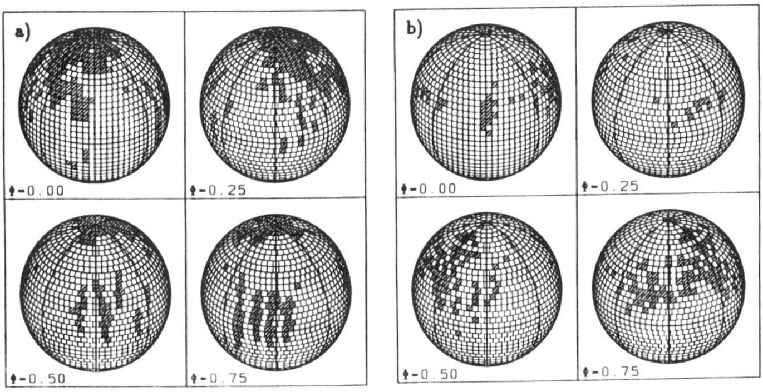

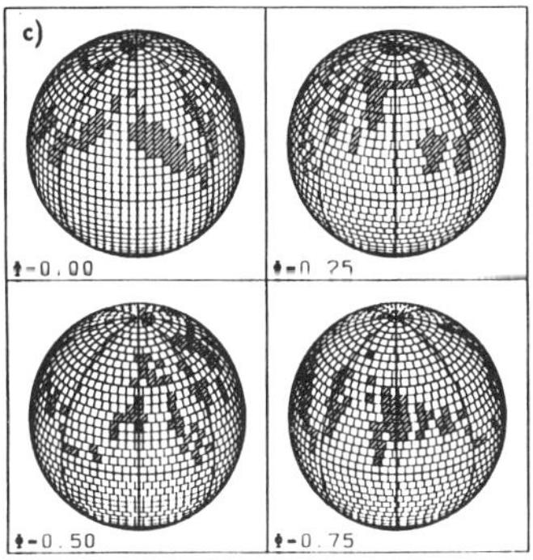

Fig. 3: AB Dor = HD 36705. $i = 60°$.
a) Dec 87: CLEAN reconstruction.
7 profiles of CaIλ6718.
b) Feb 89: CLEAN reconstruction.
26 profiles of CaIλ6439.
c) Sep 90: CLEAN reconstruction.
22 profiles of CaIλ6439.

REFERENCES

Kürster, M.: 1991, Proc. ESO workshop: Rapid Variability of OB–stars, D. Baade (ed.).

Kürster, M., and Schmitt, J.H.M.M.: 1990, Proc. Armagh Bicentennary Colloq., Surface Inhomogeneities on Late–type Stars, P.B. Byrne, J.G. Doyle (eds.).

Pasquini, L., Cutispoto, G., Gratton, R., and Mayor, M.: 1991, *Astr. Ap.*, **248,** 72.

Piskunov, N.E., Tuominen, I., and Vilhu, O.: 1990, *Astr. Ap.*, **230,** 363.

Strassmeier, *et al.* 1991, *Astr. Ap.*, **247,** 130.

Vogt, S.S.: 1988, IAU Symp. 132, G. Cayrel de Strobel, M. Spite (eds.), p. 253.

Vogt, S.S., Hatzes, A.P.: 1991, IAU Colloq. 130, Tuominen, Moss, Rüdiger (eds.).

Vogt, S.S., Penrod, G.D., and Hatzes, A.P.: 1987, *Ap. J.*, **321,** 496.

Seventh Cambridge Workshop on Cool Stars, Stellar Systems, and the Sun
ASP Conference Series, Vol. 26, 1992
Mark S. Giampapa and Jay A. Bookbinder (eds.)

EVIDENCE FOR A COMPLEX DISTRIBUTION OF MAGNETIC FIELD STRENGTHS ON THE FLARE STAR AD LEO

STEVEN H. SAAR Center for Astrophysics, 60 Garden Street, Cambridge, MA 02138, USA

ABSTRACT I present a preliminary analysis of four Ti I lines from a high S/N IR spectrum of the flare star AD Leo. The lines show consistent evidence for a complex magnetic field distribution, and may be fit with either (a) fields arising from a single atmospheric component but having a large vertical gradient, or (b) a two component field with $B_1 \approx 2 \times B_2$.

Keywords: stellar magnetic fields; stellar atmospheres; flare stars

1. INTRODUCTION AND OBSERVATIONS

The Sun exhibits two types of magnetic regions (MR): warm (plage, network) and cool (umbrae, penumbrae and pores). These regions differ in continuum brightness, field strengths (B), and in typical filling factors (f). A large range of B are present in cool MR, but hot MR show only a small dispersion in B at fixed height z (Zayer *et al.* 1990). In contrast, bright MR show significant *vertical* gradients in B, while cool MR do not. The existence of plage and spots on stars suggests these B distributions may also be present. I present here, the first *direct* evidence for such distributions – on the flare star AD Leo (M3.5Ve).

Three observations of AD Leo were made over several days with the NOAO 4m Fourier Transform Spectrometer (FTS) with an interference filter passing 4400–4600 cm^{-1} at a apodized resolution of $\lambda/\Delta\lambda$ = 45,000 (0.1 cm^{-1}). The spectra were reduced following Saar and Linsky (1985), coaligned and coadded, yielding a final spectrum (S/N $\approx$ 100) with contributions from most of the surface. The spectrum of an inactive star of similar spectral type (GL 273; M4V; Saar *et al.* 1987) was also obtained, to help in distinguishing weak blends.

2. ANALYSIS AND DISCUSSION

Comparison of AD Leo with Gl 273 and sunspot (Hall 1973) spectra near a multiplet of magnetically sensitive Ti I lines (Fig. 1) reveals a complex situation. The 4481 cm^{-1} (simple triplet) line shows a sharp central component, whose with is width consistent with $v \sin i$, embedded in a broad, shallow background. The other Ti I lines (only a factor of 2–3 more optically thick) are *much* too wide to be explained by opacity or rotational broadening. Strong magnetic broadening is the best explanation, implying the 4481 cm^{-1} "background" is due to Zeeman split σ components. To explain their anomalously large width, several scenarios are possible: (1) greatly enhanced microturbulence (ξ) in the MR; (2) multicomponent MR with different B in each component; (3) single component MR with a substantial vertical gradient; or (4) some combination of the above. As (1) required an unreasonably large $\xi \approx 10$ km s^{-1} , I focused on the other scenarios, fitting the data (F) with three models: **(A)** A three spatial component model of the form: $F = f_1 F(B_1) + f_2 F(B_2) + (1 - f_1 - f_2) F(0)$; **(B)** A two component model + gaussian B distribution with height ($\Phi(B, \sigma_B)$): $F = f F(\Phi(B, \sigma_B)) + (1 - f) F(0)$; and **(C)**, a model like **(B)**, but with a half-gaussian, truncated for $B < B_{cut}$. Model **(C)** attempts to simulate the merging of flux tubes quite low in the atmosphere when f is large (Solanki & Steiner 1990), so that part of the line formation occurs in a "pseudo–canopy" with a low, constant B value (I thank S. Solanki for suggesting this).

The models were calculated with a 15 μ angle disk–integration (including radial–tangential macroturbulence, ζ_{RT}) of local line profiles computed using the Unno–Rachkovsky radiative transfer solution, including magneto–optical effects (see Saar *et al.* 1990). I used $v \sin i = 5$ km s^{-1} and $\zeta_{RT} = 3$ km s^{-1} (Vogt *et al.* 1983). Line opacities were constrained to have their predicted (Kurucz & Peytremann 1975) ratios, and areas affected by blends (as determined from the Gl 273 spectrum) were given less weight. Results are shown in Fig. 1.

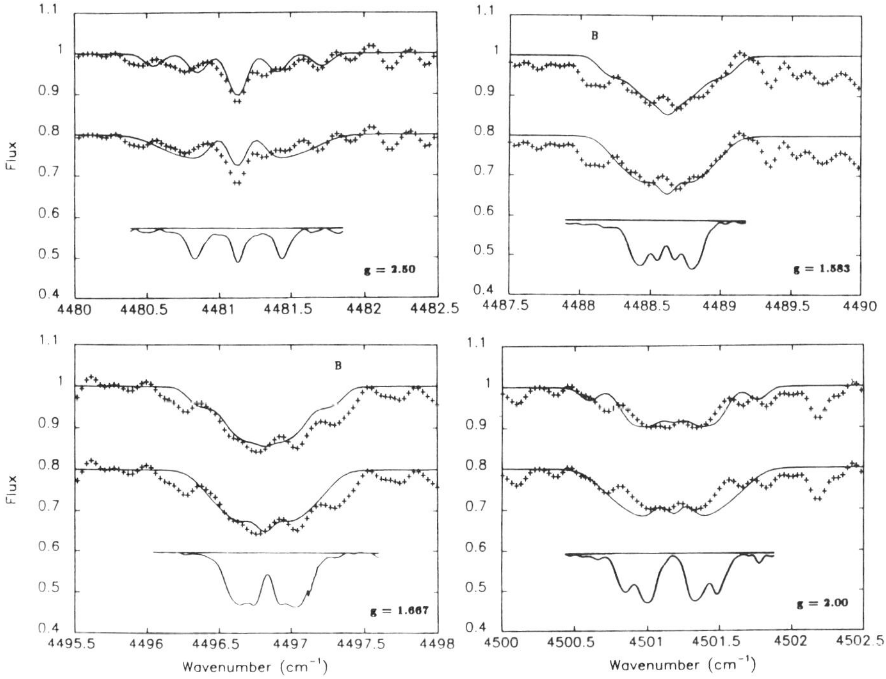

Figure 1: Ti I line profiles from AD Leo (+) and a sunspot/quiet Sun ratio (heavy solid; Hall 1973) near 2.25 μm; blends (as judged from a Gl 273 spectrum) are marked (B). Curves through the AD Leo data show the best fits with model **(A)** (top; a two spatial component model with $f_1 = 45\%$, $B_1 = 2.4$ kG, $f_2 = 30\%$, and $B_2 = 5.0$ kG) and with model **(C)** (middle; a truncated dB/dz model with $f = 45\%, \overline{B} = 3.0$ kG, $\sigma_B = 1.6$ kG).

Model (**B**) produced unacceptable fits, yielding too much absorption near line center. Model (**A**) fit the Ti I multiplet well for $f_1 = 45\%$, $B_1 = 2.4$ kG, $f_2 = 30\%$, $B_2 = 5.0$ kG. This result could be interpreted as the detection of the analog of plage/network (B_1) and spot/pore (B_2) fields on AD Leo (note that $B_2/B_1 \approx (B_2/B_1)_\odot$; Zwaan 1987). If this hypothesis is correct, it signifies the first detection of spot fields on a cool star. B_2 is smaller than the > 10 kG fields predicted by Mullan (1975). B_1 is smaller than a simple estimate a gas pressure equipartition field: $B_{eq} \propto P^{0.5}_{gas,\tau_{5000}=1} \approx 3.8$ kG (*e.g.* Saar 1991), but this may merely reflect the need to compute P_{gas} at τ_{line}. The average of B_1 and B_2 (3.7 kG) *is* close B_{eq} and is similar to previous measurements (3.8 kG, Saar & Linsky 1985; 4.0 kG, Saar *et al.* 1987). Note however, that if B_2 arises from spots like those on BY Dra, f_2 is underestimated by a factor of at least 0.7 (see Saar & Linsky 1985), implying MR cover nearly the entire star ($f_1 + f_2/0.7 = 88\%$).

Model (**C**) produced good fits for $f = 45\%$, $B_{cut} = 1.9$ kG and $\sigma_B = 1.6$ kG (Fig. 1). Here, the mean $\overline{B}$ (averaged over the distribution) is 3.0 kG – smaller than previous measurements – but the half-maximum of $\Phi(B_{cut}, \sigma_B)$ is at 3.8 kG, again similar to B_{eq}. If this model is correct, it represents the first direct detection of a field gradient on a non-solar star (see however Grossmann-Doerth & Solanki 1990). Since Φ is introduced directly into the magnetic radiative transfer, line equivalent widths are increased (as expected when $dB/dz \neq 0$) and (**C**) requires considerably lower f values than (**A**) to fit the data. (Note however, the model is only approximate: $dB/dz \neq 0$ violates the Milne-Eddington atmosphere used.) Of course, some combination of (**A**) and (**C**) is also possible. While better data and/or more sophisticated modeling will be needed to distinguish which scenario is more likely, the lower f needed if $dB/dz \neq 0$ makes model (**C**) quite attractive. Both (**A**) and (**C**) show some measure of B similar to previous results, suggesting the lower S/N of the older data may have prevented the detection of the B distributions seen here.

ACKNOWLEDGEMENTS

This work is based on data obtained at NOAO. I thank S. Solanki for enlightening discussions. This research is supported by the NASA grant NAGW-112, Interagency Transfer W-15130, and NSF grant INT-8900202.

REFERENCES

Grossmann-Doerth, U., Solanki, S.K. 1990, *Astr. Ap.*, **238**, 279
Hall, D. 1973, An Atlas of IR Spectra of the Solar Photosphere and Umbrae, KPNO
Kurucz, R., Peytremann, E. 1975, A Table of Semiempirical *gf* Values, SAO Rep. 362
Mullan, D.J. 1975, *Astr. Ap.*, **40**, 41
Saar, S.H. 1991, in IAU Colloq. 130, ed. I. Tuominen *et al.* , p. 389
Saar, S.H. *et al.* 1990, in Evolution in Astrophysics, ed. E. Rolfe, p. 431
Saar, S.H., Linsky, J.L. 1985, *Ap. J. Letters*, **299**, L47
Saar, S.H., *et al.* 1987, 27th Liege Ap. Colloq., eds. L. Delbouille, A. Monfils, p. 103
Saar, S.H., Solanki, S.K. 1991, these proceedings
Solanki, S.K., Steiner, O. 1990, *Astr. Ap.*, **234**, 519
Vogt, S.S., Soderblom, D.R., Penrod, G.D. 1983, *Ap. J.*, **269**, 250
Zayer, I. *et al.* 1990, *Astr. Ap.*, **239**, 356
Zwaan, C.J. 1987, *Ann. Rev. Astr. Ap.*, **25**, 83

Seventh Cambridge Workshop on Cool Stars, Stellar Systems, and the Sun
ASP Conference Series, Vol. 26, 1992
Mark S. Giampapa and Jay A. Bookbinder (eds.)

MAGNETIC SURFACE IMAGES OF THE BY DRA STAR HD 82558

STEVEN H. SAAR Center for Astrophysics, 60 Garden Street, Cambridge, MA 02138, USA

NIKOLAI E. PISKUNOV, ILKKA TUOMINEN Obs. and Astrophys. Lab. Univ. of Helsinki, Tähtitorninmäki, SF-00130 Helsinki, Finland

ABSTRACT We present preliminary results of a new analysis method applied to spectra of the BY Dra star HD 82558. We invert a time series of unpolarized line profiles with different Landé g_{eff} values to derive both temperature and magnetic field maps of the star. We find strong fields ($B > 2$ kG) near cool spots and weaker fields elsewhere, with $\overline{B} \approx 1$ kG.

Keywords: stellar magnetic fields; starspots; stellar atmospheres

1. INTRODUCTION AND OBSERVATIONS

HD 82558 is a single, rapidly rotating ($P_{rot} = 1.5978$ days; Strassmeier & Hall 1988), spotted K2 dwarf with high magnetic activity levels (Fekel *et al.* 1986). The star's high $v \sin i$ permits the use of surface mapping techniques, which determine the spatial temperature distribution by modeling the Doppler motions of features through absorption lines (*e.g.*, Vogt *et al.* 1987). We have developed an extension of these methods, magnetic surface imaging (MSI), which can model the full Stokes vector (I, Q, U, V). By studying lines with different Landé g_{eff}, we can simultaneously derive both T_{eff} and magnetic maps.

Here, we study Stokes I alone. Observations of HD 82558 were made with the NSO McMath stellar echelle spectrograph and a CCD detector, yielding a resolution of $\lambda/\Delta\lambda = 125{,}000$ and S/N ratios of 100–225. The CCD recorded $\approx$ 20Å centered near 6173Å. Data were obtained in two epochs: seven spectra in 2/1991 and nine in 4–5/1991. As significant line profile changes (at fixed phase ϕ) can be seen between the epochs, we focus here on the 2/1991 data only.

2. ANALYSIS AND DISCUSSION

The MSI analysis is based on the surface imaging code of Piskunov *et al.* (1990) as extended (Hackman *et al.* 1991) to permit simultaneous modeling of several lines. These inversion routines employ the Tikhonov regularization – minimizing local gradients in the solution – rather than maximum entropy (cf. Vogt *et al.* 1987). We have further extended the analysis to compute the local line profiles using the full magnetic radiative transfer equations and the DELO method (Rees *et al.* 1989). Lines are computed for a large set of model atmospheres (Kurucz 1991), with full center-to-limb effects and radial-tangential macroturbulence (ζ_{RT}) included on a grid size of 3°× 3°. We model Stokes I at each point as: $I = I\,(T_{eff}, B)$, where T_{eff} is the local temperature and B is the magnetic field (assumed radial). B can be determined at high $v \sin i$ – despite the "smearing-out" of most line *shape* clues – from the increase of local equivalent width (W_{eq}) of optically thick lines in the presence of a field (*e.g.* Leroy 1962); roughly, $\Delta W_{eq} \propto f g_{eff}{}^2 B^2$, where f is the filling factor. As we cannot uniquely determine f for each grid point, the derived B refers to the local *average* field, $\overline{B} = fB$.

We studied lines of various excitation potentials (χ_L) to enhance T_{eff} discrimination. First, two low g_{eff} lines (Fe I 6165Å, g_{eff} =1, χ_L= 4.1 eV and Ca I 6166Å, g_{eff} =0.5, χ_L= 2.5 eV) were modeled, assuming $B = 0$ and including blends, to generate a rough T_{eff} map and define $v \sin i$ (= 27 km s^{-1}) and ζ_{RT} (= 2.0 km s^{-1}). We set the microturbulence $\xi = 0.5$ km s^{-1} , and increased the damping parameter C_6 by 0.3 dex. These values produced excellent fits to an inactive, low $v \sin i$ K2V comparison star (HD 166620, B–V = 0.87); for HD 82558 they yielded Fe/H = 0.0 and an average $\overline{T}_{eff} \approx 5000$ K, consistent with the star's mean B–V (= 0.91). We next modeled five lines simultaneously (the above, plus Fe I 6173Å, g_{eff} = 2.5, χ_L= 2.2 eV; Ni I 6175Å, g_{eff} =1, χ_L =4.1 eV; Fe I 6180Å, g_{eff} = 0.6, χ_L= 2.7 eV), now allowing B also to vary. The final line models (Fig. 1) were computed assuming $i = 75°$ (Strassmeier & Hall 1988).

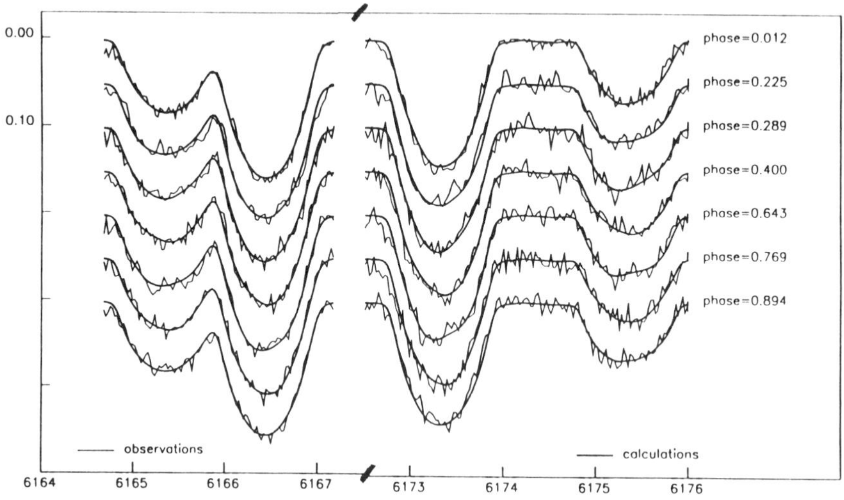

Fig. 1: Line profiles of HD 82558 in 2/91 (thin line) and inverse solutions (thick) as a function of ϕ (the Fe I 6165Å– Ca I 6166Å blend, Fe I 6173Å, Ni I 6175Å).

The resulting T_{eff} map (Fig. 2) is dominated by a large cool region at $\phi \approx$ 0.20, latitude $\theta \approx 40°$, and $\Delta T_{eff} \approx -300$ K. Other, more extended cool regions are also present. The general pattern differs from surface images of RS CVn and FK Comae stars (*e.g.* Vogt 1988; Strassmeier *et al.* 1991), which generally show larger ΔT_{eff} and more spots at higher latitudes. The spot parameters are consistent with values derived from TiO absorption (Saar & Neff 1990; $\Delta T_{eff} \leq$ 1200 K, $f_{spot} \geq 0.22$), and we note that $\Delta T_{eff} < \Delta T_{eff\odot} \approx 0.75 \times T_{eff\odot}$.

The average field on the visible surface of HD 82558 is $\overline{B} = fB = 1.02$ kG (Fig. 3). If the plage B is governed by gas pressure balance (as in slower rotators; Saar 1991), $B \propto P_{gas}^{0.5} \approx 2.2$ kG, implying $f \approx 46\%$, which is similar to the "saturation" levels of other spotted K dwarfs (*e.g.* , BD +26°730) . The greatest concentrations of magnetic flux ($fB \geq 2.5$ kG) seem to be near, but somewhat displaced from the centers of the cool regions (compare Figs. 2 & 3).

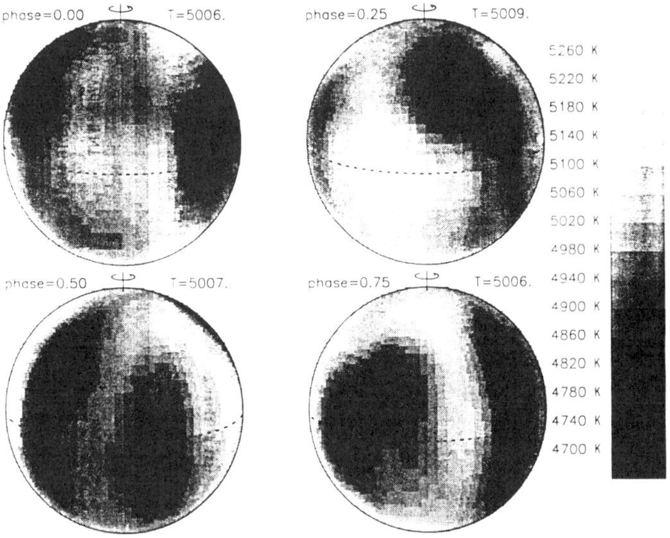

Fig. 2: Temperature maps of HD 82558 in Feb. 1991 at four phases; the T_{eff} scale runs from 4700 K (black) to 5260 K (lightest grey) in steps of 40 K.

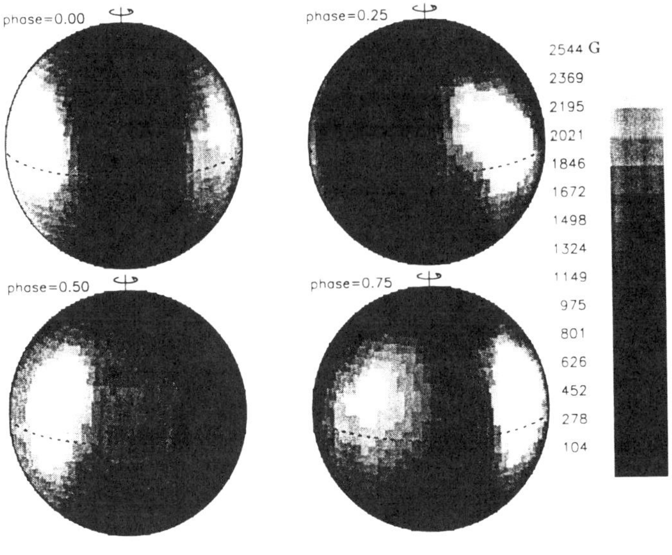

Fig. 3: Magnetic field maps of HD 82558 in Feb. 1991 at four phases; the T_{eff} scale runs from 104 G (black) to 2544 G (white) in steps of 175 G.

Donati *et al.* (1991; these proceedings) also find the main (unipolar!) flux concentrations near cool spots in their studies of circular polarization (Stokes V) from the RS CVn HR 1099 using the "Zeeman Doppler Imaging" (ZDI) technique. In contrast with the warm magnetic region they detected earlier (Donati *et al.* 1990), warm areas on HD 82558 have some of the *lowest* fB values.

Based on the above, we propose the following tentative scenario: Warm areas on HD 82558 are mostly quiet photosphere. Darker regions are areas with more magnetic flux, but with larger $R_{spot} \equiv f_{spot}/f_{plage}$ than the Sun (consistent with the much higher activity levels on HD 82558 – see Radick *et al.* 1989). Thus, plage brightness is overwhelmed by dark spots. R_{spot} is generally $\ll 1$, though, and so the *measured* fB is dominated by plage/network. Darker regions have progressively larger R_{spot} ratios, but only in the darkest areas is R_{spot} large enough that the low continuum contrast in the spots begins to depress the magnetic signal and with it, the measured fB.

We caution that our results are still rather preliminary; further study is needed to explore the degree of cross-talk between ξ, T_{eff} and fB in the solution. Since sin i is large, there is also some ambiguity near the equator. Nevertheless, the MSI technique, even using Stokes I alone, promises to be a useful tool for obtaining fB data from rapid rotators. MSI in Stokes I actually has some advantages, as it derives the *actual* fB at each point, unaffected by cancellation of unresolved polarities (cf. ZDI). Still, simultaneous imaging in Stokes I and V (or in all four Stokes parameters, as our MSI code is capable of doing) is clearly ideal, yielding information on T_{eff}, fB, *and* polarity organization. This, however, requires higher S/N: the maximum Stokes V for the fB distribution presented here is only 2.5%, for example. Simultaneous photometry will also be useful for better calibration of T_{eff} (*e.g.* , Hackman *et al.* 1991).

ACKNOWLEDGEMENTS

This work is based on data obtained at the NSO. We thank P. Avellar for obtaining several of the spectra. This research is supported by NASA grant NAGW-112, Interagency transfer W-15130, and NSF grant INT-8900202.

REFERENCES

Donati, J.-F., *et al.* : 1990, *Astr. Ap.*, **232**, L1
Fekel, F.C., *et al.* : 1986, *A. J.*, **92**, 1150
Fekel, F.C., Moffat, T.J., Henry, G.W.: 1986, *Ap. J. Suppl.*, **60**, 551
Hackman, T. *et al.* : 1991, in IAU Colloq. 130, ed. I. Tuominen *et al.* , p. 321
Kurucz, R.L.: 1991, *Bull. A.A.S.*, **23**, 967
Leroy, J.L.: 1962, *Ann. d'Ap*, **25**, 127
Piskunov, N.E., Tuominen, I., Vilhu, O.: 1990, *Astr. Ap.*, **230**, 363
Rees, D.E., Murphy, G.A., Durrant, C.J.: 1989, *Ap. J.*, **339**, 1093
Radick, R.R., Lockwood, G.W., Baliunas, S.L.: 1989, *Science* **247**, 39
Saar, S.H.: 1991, in IAU Colloq. 130, ed. I. Tuominen *et al.* , p. 389
Saar, S.H., Neff, J.E.: 1990, in Cool Stars, Stellar Systems, and the Sun, ed. G. Wallerstein, (ASP Pub. 9), p. 171
Strassmeier, K., Hall, D.S.: 1988, *Ap. J. Suppl.*, **67**, 453
Strassmeier, K., *et al.* : 1991, *Astr. Ap.*, **247**, 130
Vogt, S.S.: 1988, in IAU Symp. 132, eds. G. Cayrel, M. Spite, p. 253
Vogt, S.S., Penrod, G.D., Hatzes, A.P.: 1987, *Ap. J.*, **321**, 496

Seventh Cambridge Workshop on Cool Stars, Stellar Systems, and the Sun
ASP Conference Series, Vol. 26, 1992
Mark S. Giampapa and Jay A. Bookbinder (eds.)

HOW ACCURATE ARE STELLAR MAGNETIC FIELD MEASUREMENTS?

STEVEN H. SAAR Center for Astrophysics, 60 Garden Street, Cambridge, MA 02138, USA

SAMI K. SOLANKI Institut für Astronomie, ETH-Zentrum, CH-8092 Zürich, Switzerland

ABSTRACT We study the accuracy of two magnetic analysis methods by applying them to a large grid of lines computed using detailed 1.5–D radiative transfer (RT) calculations in a variety of stellar atmospheres embedded with fluxtubes. We find errors of $\leq \pm 20\%$ when using the more realistic analysis method if the fluxtube and external quiet atmospheres are similar; errors can increase substantially if this condition is relaxed.

Keywords: stellar magnetic fields; stellar atmospheres; stars – late-type

1. INTRODUCTION AND MODEL CALCULATIONS

Analyses of the magnetic field strengths (B) and filling factors (f) of active regions on cool stars typically fit unpolarized absorption lines (F) with a two component model: $F = fF_m(B) + (1 - f)\ F_q(0)$. Here, F_m is the profile in magnetic regions – assumed to uniformly cover a fraction f of the surface – and F_q is the quiet region profile ($B = 0$). Further simplifying assumptions used include taking the F_m and F_q atmospheres as identical, and $\nabla\ B = 0$ in F_m. Neither of these assumptions is accurate for the Sun (*e.g.*, Schüssler 1990), and much of the disagreement between fB values for some stars can be explained by a non-zero $\nabla\ B$ combined with differences in the height of the formation of the lines used (Grossmann-Doerth & Solanki, 1990). Since significant errors, especially in f, can result from these assumptions (Saar 1988; Basri *et al.* 1990), their effects on f and B determinations need to be thoroughly studied. The best work to date (Basri *et al.* 1990) has explored the effects of various atmospheres on field determinations using two Fe I lines (7748 and 8468 Å), but assuming no geometric changes with height and (usually) $\nabla\ B = 0$.

Here we expand this work, studying ten lines (5 pairs with $g_{eff} = 1.0$ and 2.5) with various f and excitation potentials (χ_L) at several wavelengths, all generated using self-consistent fluxtube models (see also Saar 1991). Moderately strong (40–250 mÅ), simple triplet Fe I lines are computed in each case. The fluxtube models (Steiner & Pizzo 1989) take into account the expansion of the field with height, as dictated by vertical and horizontal pressure balance. Several atmospheres are used to simulate F_q and F_m (see Table 1). For each atmosphere combination a grid of models with different f and B values have been calculated, summing along a number of parallel rays through the fluxtubes (1.5–D RT). Since stellar spectra are dominated by regions near disk-center, we here concentrate on models at $\mu = 1$. In spot atmospheres, we set $dB/dz = 0$ but allowed B and the angle it makes with the vertical to vary across the spot.

2. ANALYSIS AND DISCUSSION

We have applied two simple analyses to our line models (treating them as data): an Unno (1956) type code and a Milne-Eddington atmosphere (Saar 1988) and the Fourier ratio method (Robinson 1980). We adopted the following procedure for the Unno fits, since in practice, Fe/H, the oscillator strengths (log gf) and the collisional broadening must usually be adjusted. First, the low g_{eff} line was fit (minimizing χ^2) with the line opacity (η_0), source function slope, and collisional broadening (Voigt a) as free parameters. These values then became initial guesses in a fit of the high g_{eff} line with η_0, f and B free. A final fit was then made to both lines simultaneously, with η_0, a, f, and B free. Sample results (f_U and B_U) are given in Table 1, for the case of $f \approx 0.30$ at $z = 0$. The "actual" B at the level of line formation in the model has been estimated using the center-of-gravity approach (B_C; Rees & Semel 1979). For the Robinson fits, the ratio of the Fourier transforms of high and low g_{eff} lines was modeled with an equivalent width ratio, f_R and B_R as free parameters; the fits were limited to frequencies short of the first saturation zero.

A more detailed analysis is left to a future paper (which will also study center-to-limb effects and include more sophisticated analyses). Here, we briefly note: (1) The Robinson fits are almost always less accurate, especially in the K dwarfs, probably due to neglect of RT effects. (2) f is the most poorly determined parameter. B_U is typically determined to better than $\pm$ 250 G, except in the spot (H/S) models (see [6]). fB_U is usually better determined than f_U or B_U, as errors in f and B partially cancel out (cf. Gray 1984; Saar 1988; Basri *et al.* 1990). (3) Errors for Unno fits to "same"-atmosphere models (H/H and K/K) were generally $\leq$ 20%. Underestimation of fB in H/H seems to be due to an overestimation of a; overestimation for K/K arises from the inaccuracy of the simple Unno RT in the stronger K2V lines. (4) We obtain underestimates of f and fB similar to (and somewhat larger than) Basri *et al.* in the case of the solar + network model ($\approx$ 50%). Here, line weakening more than counteracts the small increase in continuum brightness in F_m. Overestimates ($\leq$ 35%) of fB in the K/H model are due the opposite effect: here, the continuum brightness of F_m dominates. These results confirm trends predicted from changes in line strengths (Solanki 1991, these proceedings). (5) For a given model, fB at 600 nm increases with χ_L, consistent with a stronger contribution to the line from F_m for lines of higher χ_L (see Grossmann-Doerth & Solanki 1990). (6) Fits were poorer for the infrared models, where the effects of ∇B on the Zeeman σ components were evident. The broad σ components led to a greatly underestimated f values (at high B), made still worse by overestimates of a due to the broad wings.

In summary, our models suggest that current Unno-based methods of stellar fB measurements are accurate only at the $\pm$ 20% level, and *then* only when differences between F_m and F_q do not change the emergent F substantially. Methods ignoring RT (*e.g.*, Robinson 1980) fare even worse. Careful estimates of a, especially in the IR, are also important for improved f_U and B_U. More realistic models applied to data on lines spanning a range of strengths, λ, and χ_L will likely be needed to disentangle the "true" f and B for cool stars.

ACKNOWLEDGEMENTS

This research is supported by an SI Fellowship, NASA grant NAGW-112, Interagency Transfer W-15130 and NSF grant INT-8900202. SHS thanks also ETH Zürich and the U. of Helsinki obs. for their hospitality and support.

REFERENCES

Basri, G.S., Marcy, G.W. 1988, *Ap. J.*, **330**, 274
Basri, G.S., Marcy, G.W., Valenti, J. 1990, *Ap. J.*, **360**, 650
Gray, D.F. 1984, *Ap. J.*, **281**, 719
Grossmann-Doerth, U., Solanki, S.K. 1990, *Astr. Ap.*, **238**, 279
Maltby, P. *et al.* 1986, *Ap. J.*, **306**, 284
Rees, D.E., Semel, M. 1979, *Astr. Ap.*, **74**, 1
Robinson, R.D. 1980, *Ap. J.*, **239**, 961
Saar, S.H. 1988, *Ap. J.*, **324**, 441
Saar, S.H. 1991, in IAU Colloq. 130, ed. I. Tuominen *et al.* , p. 389
Schüssler, M. 1990, in IAU Symp. 138, ed. J. Stenflo, p. 161
Solanki, S.K. 1986, *Astr. Ap.*, **168**, 311
Steiner, O., Pizzo, V. 1989, *Astr. Ap.*, **211**, 447

Table 1: Line Model Parameters ($f_{z=0} \approx 30\%$) and Results of the Fits

F_q/F_m atm.*	λ (nm)	$\log gf$	χ_L (eV)	B_C (G)	f_C[1] (%)	fB (G)	B_U (G)	f_U (%)	$\Delta\Phi_U$[2] (%)	B_R (G)	f_R (%)	$\Delta\Phi_R$[2] (%)
H/N	600	-5.58	0	1034	45	465	1240	13	-66	1270	12	-67
H/N	600	-3.54	2	1217	38	465	1360	15	-56	1290	15	-59
H/N	600	-1.60	4	1394	33	465	1450	18	-44	1280	18	-50
H/N	1600	0.07	6	1523	31	465	1380	17	-49	1690	16	-42
H/N	2200	-0.70	5	1333	35	465	1170	15	-58	1470	18	-42
H/H	600	-5.58	0	1248	37	464	1460	24	-26	1390	25	-26
H/H	600	-3.54	2	1409	33	464	1510	25	-20	1350	25	-28
H/H	600	-1.60	4	1551	30	464	1530	25	-18	1260	26	-30
H/H	1600	0.07	6	1671	28	464	1520	30	-1	1810	31	19
H/H	2200	-0.70	5	1518	31	464	1350	31	-11	1600	37	29
H/S	600	-5.58	0	1770[3]	30	531	...[4]	...[4]	...[4]	3120	2.8	-83
H/S	600	-3.54	2	1770[3]	30	531	2960	1.4	-92	2960	3.0	-83
H/S	600	-1.60	4	1770[3]	30	531	2730	2.0	-90	2900	2.2	-88
H/S	1600	0.07	6	1770[3]	30	531	2500	2.9	-86	2970	2.6	-86
H/S	2200	-0.70	5	1770[3]	30	531	2070	5.3	-79	2900	5.2	-72
K/H	600	-5.58	0	2159	27	591	1950	34	11	2230	37	40
K/H	600	-3.54	2	2438	24	591	2240	32	21	1560	37	-1
K/H	600	-1.60	4	2689	22	591	2480	32	35	2900	34	68
K/H	1600	0.07	6	2941	20	591	2650	29	29	3070	32	65
K/H	2200	-0.70	5	2648	22	591	2420	27	13	2740	37	72
K/K	600	-5.58	0	1915	31	585	2100	27	-3	890	26	-61
K/K	600	-3.54	2	2198	27	585	2070	33	16	1590	25	-31
K/K	600	-1.60	4	2439	24	585	2210	31	18	1540	28	-26
K/K	1600	0.07	6	2700	22	585	2500	28	20	1890	22	-28
K/K	2200	-0.70	5	2485	24	585	2240	31	17	1600	28	-23

*References: H = Harvard–Smithsonian Ref. Atm.; N = network model (Solanki 1986);
K = K2 dwarf model (Basri & Marcy 1988); S = sunspot model (Maltby *et al.* 1986)
Notes: [1] $f_C \equiv (fB)/B_C$; [2] $\Delta\Phi \equiv (fB_{\rm fit} - fB)/fB$; [3] $\overline{B}$ over spot ($dB/dz = 0$); [4] Did not converge

Seventh Cambridge Workshop on Cool Stars, Stellar Systems, and the Sun
ASP Conference Series, Vol. 26, 1992
Mark S. Giampapa and Jay A. Bookbinder (eds.)

STRONG AND WEAK SOLAR MAGNETIC FIELDS

S.K. SOLANKI[1], I. RÜEDI[1], W.C. LIVINGSTON[2], J.O. STENFLO[1]
[1] Institute of Astronomy, ETH–Zentrum, CH–8092 Zürich, Switzerland
[2] National Solar Observatory, NOAO, P.O. Box 26732, Tucson, AZ 85726, USA

ABSTRACT Stokes V (*i.e.* circularly polarized) spectra of two lines near 1.5μ are analysed using synthetic line profiles which are calculated in model flux tubes fulfilling pressure balance and having atmospheres determined empirically from Stokes V observations in the visible. We report on the derived, highly accurate field strengths in plages and present the first unequivocal measurement of weak ($B(z=0)$ <1000G) magnetic fields in the lower photosphere of solar plages.

Keywords: Solar magnetic field, Solar activity

The Stokes I and V profiles of Fe I 1.5648 μ, with Landé factor $g=3$ and of Fe I 1.5652 μ, with $g_{eff}=1.53$, have been observed in an active region near solar disk center with the McMath telescope and the main spectrograph on Kitt Peak. Due to the wavelength dependence of the Zeeman splitting the $g=3$ line at 1.5μ is completely split at field strengths as low as 400–500 G. This limit is approximately 3 times lower than corresponding values achieved by any line in the visible. Therefore, with this line it is possible to detect and measure the strength of intrinsically weak magnetic fields on the sun, besides analysing the strong flux-tube fields with unprecedented accuracy (kilogauss field strengths are accurate to within 15–30G).

Although it is possible to determine field strengths from these lines simply by measuring the wavelength difference between their peaks, considerably more information can be derived if the observed spectra are fitted by synthetic profiles. We have therefore calculated synthetic Stokes profiles in a thin flux–tube model satisfying both horizontal and vertical pressure balance. The flux tubes conserve magnetic flux and expand with height, so that the profiles are calculated along many rays passing through the tube (1.5–D radiative transfer).

The fits to 2 observed profiles are shown in Figs. 1 and 2. In approximately half the 27 analysed cases we required two distinct magnetic components to reproduce the observed lines, while the others were well reproduced by a single magnetic component. Note, however, that a vertical magnetic gradient is always required to reproduce the width of the individual σ–components of the lines. Stokes V profiles of spectral lines in the visible, calculated with exactly the same model parameters as the best–fit infrared profiles all look completely normal.

Fig. 3a shows the field strength at z=0 (corresponding to $\tau=1$ in the quiet sun). The figure suggests the presence of two different populations of magnetic features, those with strong fields between approximately 1400 and 1700 G, corresponding to traditional kG flux tubes, and those with weak fields lying in the range 400–1000 G at $z=0$. However, note that we have analysed

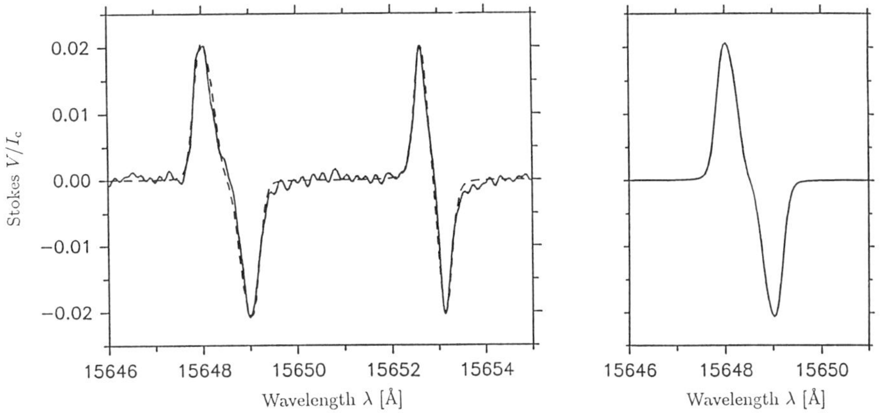

Fig. 1 Observed (solid) and best-fit (dashed) "normal" Stokes V profiles. A single thin magnetic flux tube with $B(z = 0) = 1520$ G is able to reproduce the data. The calculated profile of the $g = 3$ line is shown again in the small frame on the right.

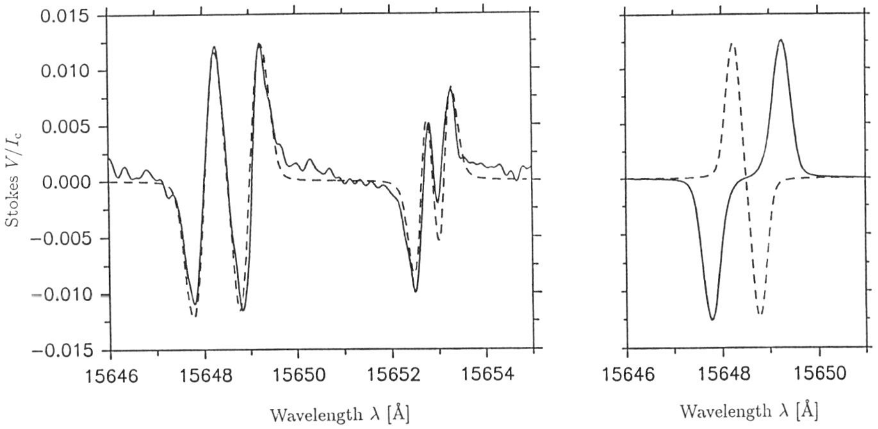

Fig. 2 Same as Fig. 1, but for a "double" profile requiring a fit based on two flux-tube components of opposite polarity. The frame on the right shows the Stokes V profiles resulting from each of the two components. $B_1(z = 0) = 1700$G, $B_2(z = 0) = -1050$G. There is no velocity shift between the two components.

only 27 regions. The lower limit to the field strength is probably not solar, since it corresponds to the smallest field strength that can be measured with the used lines. The strength of the strong fields agree very well with results obtained in the visible (*e.g.* Zayer *et al.* 1990). We estimate that approximately 10–15% of the magnetic flux is in weak field form, also in good agreement with results in the visible (*e.g.* Frazier and Stenflo 1972). Fig. 3b shows the "magnetic flux" as

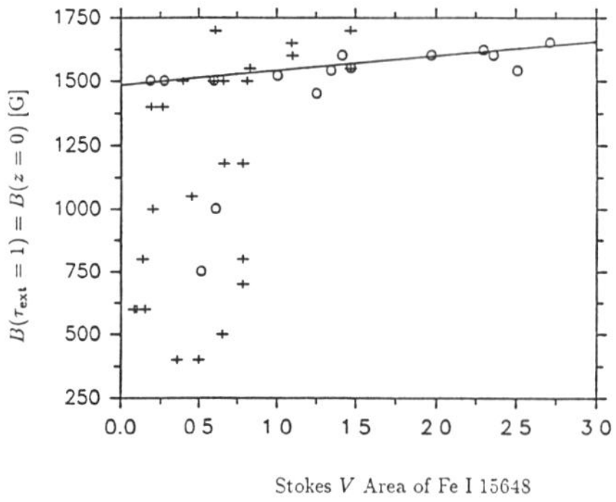

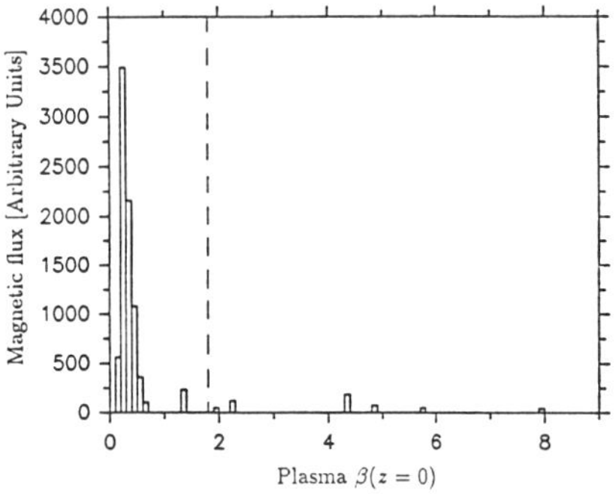

Fig. 3a $B(z=0)$ vs. Stokes V lobe area of Fe I 1.5646μ, for all the observed regions. If more than one magnetic component is needed to fit the observed profile, then B for each profile is plotted (vs. the amplitude belonging to that component alone). B resulting from such 2–component fits are denoted by a '+', while B resulting from 1-component fits are marked by a 'o'.

Fig. 3b Histogram of magnetic flux (assumed proportional to the Stokes V lobe area of the $g=3$ line) vs. plasma β. As in Fig. 3a, each magnetic component is represented separately. Fields situated to the right of the vertical dashed line are convectively unstable according to Spruit and Zweibel (1979).

a function of the plasma $\beta = 8\pi p/B^2$, which is a measure of the gas to magnetic energy ratio. The average β of the convectively stable flux tubes is 0.38.

The gas in the weak field patches does not, in general, appear to be flowing relative to the strong field. We find relative line shifts between the two components only in 3 out of 13 regions. One of the main observational signatures of a convective collapse, the mechanism by which kG flux tubes are thought to be formed (*e.g.* Parker 1978), is a downflow of the material as the flux tubes get evacuated. Consequently, most of the observed weak field features do not appear to be undergoing a convective collapse.

Models of the flux tubes containing a significant amount of weak–field return flux can be ruled out by the data, since, in contrast to the predictions by the model, the observed weak–field components do not have any preferred polarity relative to the strong–field features.

REFERENCES

1 Frazier, E.N., Stenflo, J.O.: 1972, *Solar Phys.* **13**, 118
2 Parker, E.N.: 1978, *Ap. J.* **221**, 368
3 Spruit, H.C., Zweibel, E.G.: 1979, *Solar Phys.* **62**, 15
4 Zayer, I., Solanki, S.K., Stenflo, J.O., Keller, C.U.: 1990, *Astr. Ap.* **239**, 356

Seventh Cambridge Workshop on Cool Stars, Stellar Systems, and the Sun
ASP Conference Series, Vol. 26, 1992
Mark S. Giampapa and Jay A. Bookbinder (eds.)

ANALYSIS OF MAGNETIC FLUXTUBES ON THE SOLAR SURFACE FROM OBSERVATIONS AT MT. WILSON OF λ5250 AND λ5233

ROGER K. ULRICH Department of Astronomy, University of California at Los Angeles, Los Angeles, CA 90024

ABSTRACT Simultaneous observations of λ5250 and λ5233 with spatial resolutions of 5″ × 5″, 12″ × 12″ and 20″ × 20″ have been used to determine a saturation factor which should be used to correct magnetic field measurements made with λ5250. In contrast to previous results which gave a factor of roughly 2, the present determination which takes advantage of new flexibility in the Mt. Wilson system and is based on a very widely spaced separation of the spectral acceptance bands at λ5250 yields a factor of roughly 4. This saturation factor depends on the spatial resolution and on the center-to-limb position.

Keywords: Solar Magnetograms, Interplanetary Magnetic Fields, Magnetic Fluxtubes

THE PROBLEM

It has been known since the work by Stenflo (1973) that the magnetic fields measured by systems of ordinary spatial resolution are reduced from the true values due to the concentration of the fields into magnetic elements smaller than can be easily resolved. Howard and Stenflo (1972) referred to this structure as filamentary while Stenflo *et al.* (1984) have used the term fluxtubes to describe the magnetic elements. Within each of the fluxtubes the high field strength B_{FT} shifts the Zeeman components far enough from the average line center that the signal responds non-linearly when measured with a Babcock magnetograph. A simple model of the effect is to set the measured field equal to B_{sat} for all magnetic fields above B_{sat}. The value of B_{sat} depends on the g_e value for each spectral line. As an example consider two different lines λ_1 and λ_2 which saturate at different field strengths, $B_{sat}(\lambda_1)$ and $B_{sat}(\lambda_2)$ with $B_{sat}(\lambda_1) < B_{sat}(\lambda_2)$. There are three possibilities: case 1, neither line is saturated: $B_{FT} < B_{Sat}(\lambda_1)$; case 2, the more magnetically sensitive line is saturated while the other is not: $B_{Sat}(\lambda_1) < B_{FT} < B_{Sat}(\lambda_2)$; and case 3, both lines are saturated: $B_{Sat}(\lambda_1) < B_{FT}$. In case 1, the ratio of the fields should always be unity. In case 2 if the actual fields on the solar surface have a distribution of strengths, a scatter should be observed in the ratio of the field from the saturated line to the field from the unsaturated line. Finally, in case 3 the ratio of the fields should always be the ratio of their saturation levels.

The two lines we have used in this study are λ5250 because a large database of magnetograms using this line exists and λ5233 because it is relatively magnetically insensitive and is quite broad so that it can yield a signal even for widely spaced spectral sampling. By comparing the fields on a point by point basis derived from these two lines observed simultaneously, we can study both the saturation factor for $\lambda5250$ and the distribution of actual field strengths. Previous observations with this pair of lines by Howard and Stenflo (1972) had $B_{Sat}(\lambda5250) = 220$gauss and $B_{Sat}(\lambda5233) = 500$gauss and were in case 3. their observations derived a field strength ratio $B(\lambda5233)/B(\lambda5250)$ of about 2. Frazier and Stenflo (1972) used a spectral sampling 4 times wider than Howard and Stenflo (1972) and also found a field strength ratio of about 2. their observations were case 2 so the fact that they found a the consistent value of this ratio indicates that the solar fields have a preferred value on the order of 1500 gauss. Because the work of Howard and Stenflo (1972) was case 3 instead of case 2, the Mt. Wilson line strength ratio has never been determined in a way comparable to that carried out by Frazier and Stenflo (1972). We remedy this lack with new observations using a system in which the $\lambda5233$ line saturates for a field strength of 4600 gauss. We have retained the same configuration for $\lambda5250$ as has been used for many years and which has $B_{Sat} = 200$gauss.

THE OBSERVATIONS

The observing system at the 150-foot tower on Mt. Wilson uses a dual exit stage system in which two spectral regions are monitored simultaneously. Each stage is movable under computer control. On each stage is mounted a fiber-optic image reformatter which rearranges either a pair of 0.8×25mm or 0.257×27mm rectangles into a pair of circular outputs which illuminate the cathode of a pair of photomultipliers. This system has been fully described recently by Ulrich*et al.* (1991). The polarimetric analysis to determine the longitudinal magnetic field is carried out simultaneously for the lines observed by each stage. Under normal operation we use the long-λ stage to observe $\lambda5250$ with the wider rectangle reformatter and the short-λ stage to observe $\lambda5237$ with the narrow rectangle reformatter. The newer narrow rectangle reformatter has 10 identical rectangles which abut each other while the older wide rectangle reformatter has only 2 which are separated by 0.2mm. As a special setup to study the question of the magnetic saturation from fluxtubes on June 29 and 30, 1991 we used radiation from the most widely spaced rectangles on the short-λ stage and used this stage to observe $\lambda5233$ while the long-λ stage observed $\lambda5250$ with the standard setup.

The actual distributions of field strengths in $\lambda5250$ and $\lambda5233$ are very tightly correlated even though we are certainly in case 2 for all observed fields. Figure 1 gives the slopes of the correlations as a function of $\mu = \cos\theta$ where θ is the center-to-limb angle and as a function of aperture size. The quantity δ^{-1} is a correction factor which should be applied to the measured field in $\lambda5250$ to obtain an estimate of the field strength in $\lambda5233$.

REFERENCES

Frazier, E.N. and J.O. Stenflo 1972, *Solar Phys.*, **27**, 330.

Howard, R. and J.O. Stenflo 1972, *Solar Phys.*, **22**, 402.

Stenflo, J.O. 1973, *Solar Phys.*, **32**, 41.

Stenflo, J.O., J.W. Harvey, J.W. Brault and S. Solanki 1984, *Astr. Ap.*, **131**, 333.

Ulrich, R.K., Webster, L., Boyden, J.E., Magnone, N. and Bogart, R.S. 1991, *Solar Phys.*, , in press.

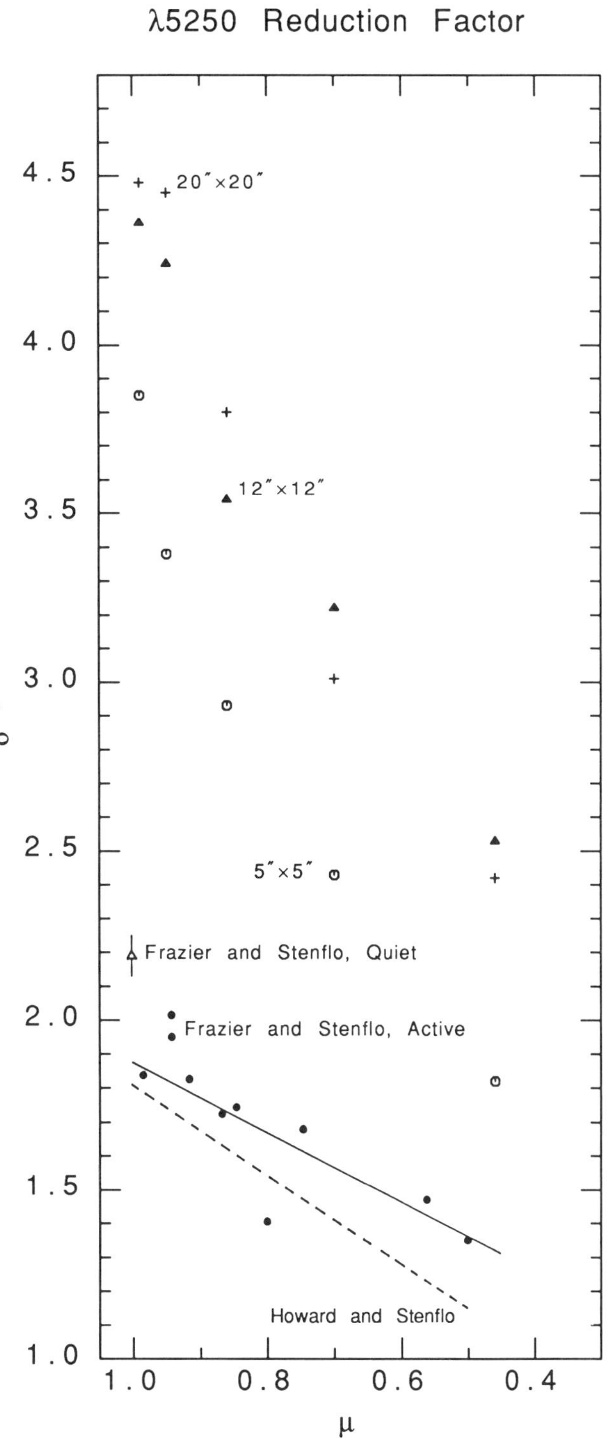

Fig. 1. This figure shows the factor by which λ5250 magnetic fields observed at Mt. Wilson should be multiplied in order to obtain unsaturated estimates which would have been measured at λ5233. Note the aperture size dependence as well as the center-to-limb angle dependence. For comparison the earlier results by Frazier and Stenflo are shown (solid line and filled circles) as well as the summary relationship found by Howard and Stenflo (dashed line).

ACKNOWLEDGEMETNS

This work has been supported by NASA grant NAGW-472, NSF grant AST90-15108 and ONR:N00014 91-J114.

Seventh Cambridge Workshop on Cool Stars, Stellar Systems, and the Sun
ASP Conference Series, Vol. 26, 1992
Mark S. Giampapa and Jay A. Bookbinder (eds.)

A COMPARISON OF MG II h AND k LINES IN THE SUN AND COOL STARS

MONICA CERRUTI-SOLA Osservatorio Astrofisico di Arcetri, Largo E. Fermi 5, 50125 Florence, Italy

CHUNG-CHIEH CHENG E.O. Hulburt Center for Space Research, Naval Research Laboratory, Washington, D.C. 85726, U.S.A.

ROBERTO PALLAVICINI Osservatorio Astrofisico di Arcetri, Largo E. Fermi 5, 50125 Florence, Italy

ABSTRACT We report the results of a detailed comparison of Mg II *h* and *k* lines in late-type stars of different activity levels and in spatially resolved solar regions of different degrees of magnetic activity. We show that the solar and stellar data are in good qualitative and quantitative agreement, thus supporting the conclusion that different levels of chromospheric activity in stars are due to different fractions of their surface covered by magnetic regions.

Keywords: Chromospheres; Mg II emission

INTRODUCTION

The Mg II *h* and *k* lines at 2802.7 Å and 2795.5 Å are important diagnostics of the chromosphere of the Sun and late-type stars. Extensive observations of these lines in cool stars have been obtained for more than a decade by the IUE satellite (Basri and Linsky 1979, Stencel *et al.* 1980, Ayres 1981, Blanco *et al.* 1982, Crivellari *et al.* 1983, Doherty 1985, Vladilo *et al.* 1987). Observations of the same lines in spatially resolved solar regions have been obtained by a number of experiments on rockets and satellites (Lemaire and Skumanich 1973, Kohl and Parkinson 1976, Doschek and Feldman 1977, Lemaire *et al.* 1981). This offers a unique opportunity for comparing Mg II lines in the Sun and stars.

In this paper we present the results of a comparison of Mg II *h* and *k* line profiles and fluxes for a sample of cool stars (observed with IUE) and for spatially resolved solar regions (observed at comparable resolution from *Skylab*). We also compare the Mg II data with other UV lines observed at shorter wavelengths (1150 - 1850 Å) from both stars and spatially resolved solar regions.

OBSERVATIONS

The Mg II stellar data were obtained with the IUE satellite and were all retrieved from the IUE Archive. We have analysed high-resolution ($\sim$ 0.2 Å) long-wavelength spectra for a sample of 55 stars of spectral type F5 to K5 and luminosity classes V and IV. Most of the stars in the sample are of solar type (*i.e.* $B - V \sim 0.6$). Six RS CVn binaries, however, were also included in the sample.

Spatially resolved Mg II solar data were obtained in 1973/74 by the NRL Slit Spectrograph on *Skylab*. These were retrieved from the original flight film. The spectral resolution of the solar data ($\sim$ 0.1 Å) is comparable to that of the high-resolution IUE spectra. We have analysed 22 different solar regions, including quiet areas, coronal holes, active regions and flares. There is a virtually continuous increase of Mg II fluxes from quiet regions to plages and flares thus making it difficult to define a characteristic flux value for different types of solar regions.

Short-wavelength (1150 - 1850 Å) low-resolution ($\sim$ 6 Å) IUE spectra of cool stars were analysed by us in a previous paper (Cappelli *et al.* 1989), and compared with spatially-resolved short-wavelength spectra from *Skylab*. Since the present sample largely overlaps that of Cappelli *et al.*, the solar and stellar Mg II fluxes can also be compared with fluxes obtained at shorter wavelengths in the lines of N V, O I, C II, Si IV, C IV, He II, C I and Si II.

RESULTS

There is a very good correspondence (in both Mg II line profiles and fluxes) between stars of different activity levels and spatially-resolved solar regions of different degrees of magnetic activity. In particular, there is a near one-to-one correspondence between quiet regions and inactive stars and, respectively, plage areas and active stars. The Mg II fluxes of RS CVn binaries are comparable to those observed in very active plages and solar flares.

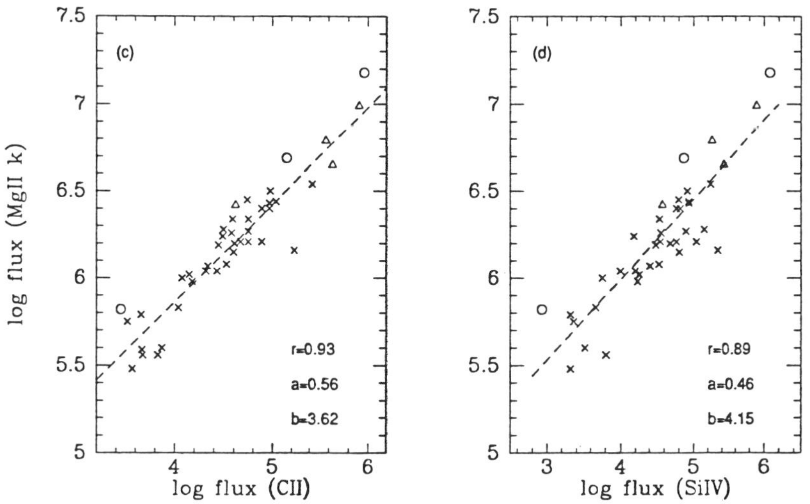

Fig. 1: Flux-flux diagrams of Mg II *k* line fluxes (this work) vs. C II and Si IV fluxes from Cappelli *et al.* (1989). Crosses indicate stars, triangles indicate RS CVn binaries and circles indicate average solar data for different types of regions. Dashed lines are best fit to the stellar data.

Flux-flux diagrams of Mg II fluxes vs. short-wavelength UV fluxes show a good correlation between different lines. Fig. 1 shows an example of such flux-flux diagrams for C II and Si IV. The three solar data points (circles) refer to *average* values for quiet areas, plages and flares, respectively. Stars are indicated by crosses, while the very active RS CVn binaries are indicated by triangles. Note that the solar fluxes obey the same flux-flux relationships as the stellar ones, with virtually the same slope. The small systematic difference that appears to be present between the absolute values of solar and stellar data is probably not significant, once we take into account the errors on the absolute calibration and the inaccuracy introduced by averaging solar data.

We have also compared some important line diagnostic parameters in the Sun and stars, including the ratio of k to h line fluxes, the separation of the k_1 minima, the separation of the k_2 peaks and the FWHM of the k line. We find no significant discrepancy between the solar and stellar data with regard to the behaviour of these line diagnostics as a function of magnetic activity.

On the basis of the above comparison, we conclude that different levels of chromospheric and transition region emission in stars are best interpreted as due to different fractions of their surface covered by magnetic regions. However, a precise determination of the magnetic coverage factor does not seem to be feasible, owing to the virtually continuous range of Mg II fluxes observed in solar regions and the highly inhomogeneous structure of solar and stellar chromospheres.

REFERENCES

Ayres, T.R.: 1981, The Universe at Ultraviolet Wavelengths, R.D. Chapman ed., NASA Conf. Publ. 2171, p. 237.

Basri, G.S., and Linsky, J.L.: 1979, *Ap. J.* **234**, 1023.

Blanco, C., Bruca, L., Catalano, S., and Marilli, E.: 1982, *Astron. Ap.* **115**, 280.

Cappelli, A., Cerruti-Sola, M., Cheng, C.-C., and Pallavicini, R.: 1989, *Astron. Ap.* **213**, 226.

Crivellari, L., Franco, M.L., Molaro, P., Vladilo, G., and Beckman, J.E.: 1983, *Astron. Ap.* **52**, 135.

Doherty, L.R.: 1985, *Montly Not. R. Astron. Soc.* **217**, 41.

Doschek, G.A., and Feldman, U.: 1977, *Ap. J. Suppl.* **35**, 471.

Kohl, J.L., and Parkinson, W.H.: 1976, *Ap. J.* **205**, 599.

Lemaire, P., Gouttebroze, P., Vial, J.C., and Artzner. G.E.: 1981, *Astron. Ap.* **103**, 160.

Lemaire, P., and Skumanich, A.: 1973, *Astron. Ap.* **22**, 61.

Vladilo, G., Molaro, P., Crivellari, L., Foing, B.H., Beckman, J.E., and Genova, R.: 1987, *Astron. Ap.* **185**, 233.

Seventh Cambridge Workshop on Cool Stars, Stellar Systems, and the Sun
ASP Conference Series, Vol. 26, 1992
Mark S. Giampapa and Jay A. Bookbinder (eds.)

SPOT MODEL OF A TEMPORARY BRIGHTENING OF SV CAM

ALEXANDER HEMPELMANN ZI Astrophysik Potsdam, D-O-1591 Potsdam-Babelsberg, Rosa-Luxemburg-Str. 17a

LASZLO PATKOS Konkoly Observatory Budapest, 1525 Budapest XII, Box 67, Hungary

ABSTRACT The unusual brightening of the RS CVn variable SV Cam observed by Patkos (1982) is analysed. We find a solution under the assumption of a temporarily existing bright spot on the primary star of the binary system. Parameters as contrast, radius and location of the spot are derived from the observations.

Keywords: SV Cam; Starspots

The system SV Cam belongs to the short period group of the RS CVn family. According to Hilditch *et al.* (1979) it consists of a G2-3, IV-V primary and a K4V secondary star.

The observed light variations result from eclipses, flare activity (Patkos 1981) and spottedness as it is typical for a RS CVn star.

From an analysis of light variations observed by Patkos (1982) and by Busso *et al.* (1985) the latter concluded that all observed deviations from the eclipsing light curve result from only one active region where size, contrast and position of the spot change cyclically with a period of about 10 years. In contrast to that model, Hempelmann and Rüdiger (1989) found that the stellar activity of SV Cam is characterized by variations in two time - scales: short - term fluctuations in the order of days up to some months are overlayed to the cycle.In agreement with Zeilik *et al.* (1988), they concluded the primary star to be the main contributor to the stellar activity of SV Cam.

Additionally to this variability quite usual for a RS CVn star, a peculiar brightening of SV Cam was observed by Patkos (1982) between JD 2442460 and JD 2442466. He proposed a temporarily existing bright spot on the primary star possibly caused by accretion of matter. Mass exchange had been indicated by a bend in the O-C diagram of the orbital period.

To test Patkos' hypothesis we asked whether the light curve of the brightening is compatible with the model of a spotted rotator or not, *i.e.* can the light curve observed at different times and in different colours be fitted by one and the same model of a single and bright spot located on the primary star?

To erect the light curve of brightening we used the observations at JD 244204/5 as a standard. Then the differences in intensities between the observations at the times of brightening and the standard curve were taken. The four light curves have the shape as it is typical for a star with one bright spot on its surface.

The four curves were analysed separately by help of a code (Hempelmann and Schöneich 1987) which was developed for light curves of Ap stars (Ap stars show bright spots). It is based on the assumption of a single and circular spot which differs from its environment only by a mean contrast, $q > 1$. The free model parameters are: the location on the primary star (longitude, ϑ_o, in relation to the meridian facing the secondary star and the angle to the rotation axis, β_o), the spot radius, α, and the mean contrast, q, between the spot and its environment and, the ground level intensity (in relation to the comparison star), I_o/I_c, of a spotless surface.

The most probable values for these parameters with estimates of errors are given in Table I. To estimate the validity of our spot model, a comparison of the solutions of the four curves analysed will be of interest. From Table I it is obvious that only two parameters have changed with time; these are the contrast, as to be expected, and the ground level intensity. The latter fact will be discussed below.

How the solutions agree with one another, is demonstrated from the fact that both the position of the spot and its radius are the same for all the four curves analysed. We conclude that the observations are consistent with our model of a temporarily existing bright spot.

Another interesting point is the darkening found for the light coming from outside the spot. This 0.02-0.03 mag decrease reflects most probably the "normal" stellar activity, in agreement with the findings of Hempelmann and Rüdiger (1989). This means, the appearance of the hot spot did not influence the usual stellar activity.

Table I Most probable values of the free model parameters

J. D. 2442400+	colour	longitude θ_0	polar distance β_0	spot radius α	contrast parameter q	ground level intensity I_0/I_c
4/5	B	spot	not	existent	1	0.5049
	V				1	0.5856
60/61	B	$162.5° \pm 2°$	$96° \pm 20°$	$70° \pm 5°$	1.104 ± 0.005	0.493 ± 0.002
	V				1.091	0.571
65/66	B	163.5°	103°	70°	1.125	0.492
	V				1.107	0.567

REFERENCES

Busso,M., Scaltriti,F., Cellino,A.: 1985, *Astr. Ap.*, **148**, 29

Hempelmann,A., Rüdiger,G.: 1989, *Astron. Nachr.* 310, 137

Hempelmann,A., Schöneich,W.: 1987, *Astron. Nachr.* 308, 201

Hilditch,R.W., Harland,D.M., Mc Lean,D.J.: 1979, *M.N.R.A.S.*, **187**, 797

Patkos,L.:1981, *Astrophys. Letters* 22, 1

Patkos,L.:1982, *Comm. Konkoly Obs. Budapest* No.80

Zeilik,M., De Blasi,C., Rhodes,M., Budding,E.:1988, *Ap. J.*, **332**, 293

Seventh Cambridge Workshop on Cool Stars, Stellar Systems, and the Sun
ASP Conference Series, Vol. 26, 1992
Mark S. Giampapa and Jay A. Bookbinder (eds.)

THE STAR-DISK CONNECTION

KEITH HORNE Space Telescope Science Institute, 3700 San Martin Drive, Baltimore, MD 21218

STEVEN H. SAAR Center for Astrophysics, 60 Garden Street, Cambridge, MA 02138.

THOMAS R. MARSH Oxford University, Nuclear and Astrophysics Laboratory, Keble Road, Oxford, OX1 3RH, UK.

ABSTRACT We find that stars and accretion disks both have emission-line surface brightnesses that scale roughly linearly with the local rotation rate. This suggests that a dynamo mechanism is responsible for powering chromospheres in accretion disks as well as in stars. If this unification of stars and disks is correct, then dynamo models can be tested over a greatly expanded range of rotation rates, and the importance of secondary dynamo parameters may become evident. The idea needs to be tested by searching for further empirical connections between magnetic activity in stars and disks.

Keywords: chromospheres; accretion disks; stars; dynamos

Cataclysmic variables are short-period interacting binary stars containing an accretion disk around a white dwarf that is fed by Roche lobe overflow from a late-type companion star. Its strong Keplerian shear makes the disk a likely site for dynamo action. Although magnetic field strengths in disks have not been directly detected by Zeeman effects, many possible indicators of magnetic activity are observed including H I, He I, Ca II, Fe II, Mg II, Si IV, C IV, N V emission lines, rapid photometric variations, winds, and x-ray emission.

The disk's emission lines exhibit striking double-peaked velocity profiles. The Keplerian velocities are highly supersonic, thus different parts of the velocity profile arise from distinct regions of the disk. The two peaks correspond to opposite rims of the disk, while the far wings of the line profile arise from regions close to the center of the disk. The detailed shape of the velocity profile can be used to reconstruct the radial distribution of the line emission.

We are using doppler tomography techniques (Marsh and Horne 1988) to map the disk's emission-line regions. Results so far published include maps of Balmer emission lines in two dwarf novae, IP Peg (Marsh and Horne 1990) and U Gem (Marsh *et al.* 1990). The doppler mapping studies confirm and strengthen the general finding from earlier line profile fitting that *disks have emission-line surface brightnesses decreasing with radius as* $R^{-3/2}$.

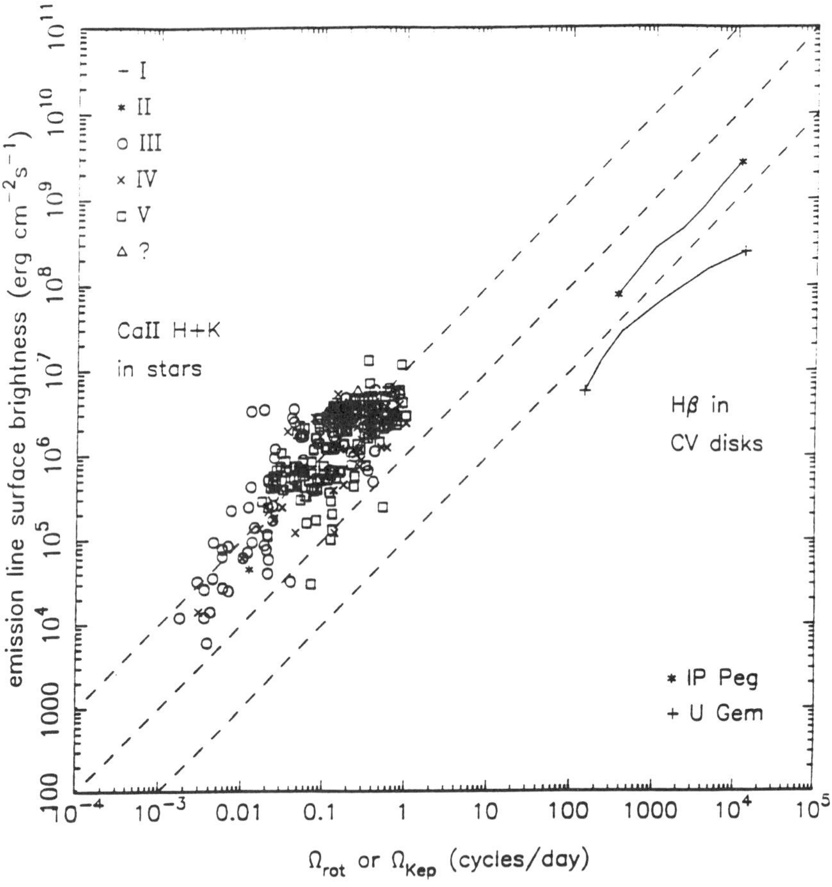

Fig. 1: A roughly linear dependence of emission-line surface brightness on rotation rate is found for Ca II H+K emission in rotating stars and for Hβ emission in two dwarf nova accretion disks.

Figure 1 compares the activity-rotation relationship observed for stars and disks. For a sample of stars taken mainly from Rütten (1987), we plot the Ca II H+K surface brightness vs the rotation rate. For two dwarf nova accretion disks, we plot lines giving the Hβ surface brightness as a function of the Keplerian rotation rate. Remarkably, the Ca II emission on stellar surfaces and the Hβ emission on disk surfaces both scale roughly linear with rotation rate.

Stars and disks exhibit the same power-law slope in Figure 1. We therefore suggest that a dynamo mechanism analogous to that which powers stellar chromospheres is also generating the emission lines in quiescent accretion disks. If this unification of stars and disks proves to be correct, then we have an excellent opportunity to test dynamo theories over a large range of rotation rates. The stellar sample spans rotation periods of 1–500 days, while in each disk the Kepler period ranges from 10 to 500 seconds.

Rotation rate appears to be the primary parameter in both stars and disks, but the geometry and other details of the dynamo must be very different in the two cases. Disks may help to establish what parameters besides rotation are important for dynamos, a program which has been difficult using stars alone because the stellar parameters tend to be correlated along the main sequence (e.g. Basri 1987).

Further studies should aim to test our proposed unification of stellar and accretion disk dynamos. In stars there appear to be simple power-law relationships among the various magnetic activity indicators. Do similar relationships hold in disks ? The photometric flickering activity seen in accretion disks may represent a superposition of flare-like events occurring at each radius at a rate proportional to the local Kepler frequency. How do the flare rates in disks compare with those in stars ? These and similar questions remain as directions for future work.

ACKNOWLEDGEMENTS

The authors would like to acknowledge support from a Smithsonian Institution Postdoctoral Fellowship (SHS), NASA grants NAGW-112 (SHS), NAGW-2264 (KH,TRM) and NAGW-2678 (KH), and the STScI Director's Research Fund (KH).

REFERENCES

Basri, G. 1987, *Ap. J.*, **316**, 377.

Horne, K. and Saar, S. H. 1991, *Ap. J. Letters*, **374**, L55.

Marsh, T. R., Horne, K., Schlegel, E. M., Honeycutt, K. and Kaitchuck, R. H. 1990, *Ap. J.*, **364**, 637.

Marsh, T. R. and Horne, K. 1990, *Ap. J.*, **349**, 593.

Marsh, T. R. and Horne, K. 1988, *M.N.R.A.S.*, **235**, 269.

Rütten, R. G. M. 1987, *Astr. Ap.*, **177**, 131.

Seventh Cambridge Workshop on Cool Stars, Stellar Systems, and the Sun
ASP Conference Series, Vol. 26, 1992
Mark S. Giampapa and Jay A. Bookbinder (eds.)

PHOTOMETRY OF 1E 1751+7046

LAURI JETSU and ILKKA TUOMINEN Observatory and Astrophysics Laboratory, University of Helsinki, Tähtitorninmäki SF–00130 Helsinki, Finland

ABSTRACT New photometry of 1E 1751+7046 is presented. The peak to peak amplitude of the rotational modulation of brightness is extremely high, nearly $0.^m3$ in V, and has a period of $13.^d9820$. The previously observed, apparently constant, radial velocity may also vary as a function of the photometric rotation period. The spectral class and the FK Comae–type candidacy of 1E 1751+7046 are discussed.

Keywords: 1E 1751+7046; photometry; variable; FK Comae–type

INTRODUCTION

Stocke *et al.* (1983) discovered the exceptionally strong X–ray flux of the K–type star 1E 1751+7046 (BD +70°959). Silva *et al.* (1985) published observations of a strong Ca II H and K emission, a rapid rotation $v\sin i \approx 30$ km s^{-1} and an apparently constant radial velocity. Furthermore, they suggested the spectral class K5IV–V, while a giant classification seemed improbable. Fleming *et al.* (1987) confirmed the constant radial velocity and also arrived at the luminosity class IV–V. They failed to find the Li I (6707 Å) feature indicative of a young object, and since the apparently constant radial velocity ruled out binarity, suggested that 1E 1751+7046 could be a FK Comae–type star. Bopp and Stencel (1981) have defined the FK Comae–type stars as apparently single, rapidly rotating, late–type giants with strong chromospheric activity.

We present some preliminary results derived from new UBVRI photometry made at the Crimean Astrophysical Observatory and the Tashkent Observatory between October 1989 and December 1990. The V magnitude observations are plotted in Fig. 1a as function of phase derived from the new ephemeris HJD_{min} = 244 7809.04 + $13.^d9820E$. A more detailed study of all available photometry of 1E 1751+7046 can be found from Jetsu *et al.* (1991a).

THE PERIOD ANALYSIS

All new photometry was normalized using the same simple procedure already described by Jetsu *et al.* (1990). The three stage period analysis by Pelt (1983) was performed [an example of the application of the method is given by Jetsu *et al.* (1991b)]. The results for the tested period interval from $0.^d5$ to $40.^d0$ were the same for the first and second order Fourier approximations of the light curve. This result was checked with the power spectrum method (Press and Teukolsky, 1988). The best period derived for the rotational modulation of brightness is $P_{phot} = 13.^d9820 \pm 0.^d0078$. Three other periods (near $0.^d93$, $1.^d07$, and $0.^d52$) would also fit to the observations. However, these "spurious periods" are predicted by the Tanner (1948) equations, *i.e.* they arise from the spacing of the observations in time. Conclusively, all "spurious periods" yield larger residuals than $13.^d9820$.

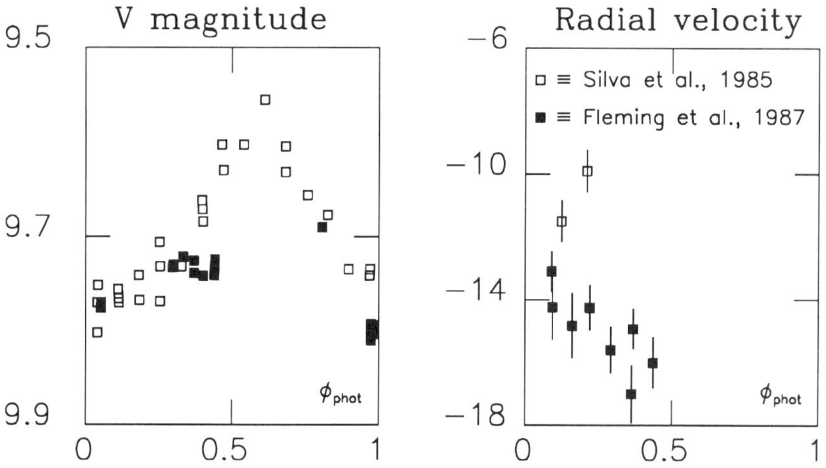

Fig. 1: **a**) The rotational modulation of the V magnitude as a function of the photometric phase derived from the ephemeris of this paper. The observations made at the Crimean Astrophysical Observatory and the Tashkent Observatory are denoted by closed and open squares, respectively. **b**) The radial velocity measurements published by Fleming *et al.* (1987) plotted as a function of the photometric phase.

THE CLASSIFICATION OF 1E 1751+7046

The radial velocities of Fleming *et al.* [1987: their Table 1 also contains the rereduced values of Silva *et al.* (1985)] plotted as a function of the photometric phase (Fig. 1b) seem to show a definite trend. A nearly synchronous rotation of the primary and orbital motion of an unseen secondary could explain this result. The shift between the measurements by Silva *et al.* (1985) and Fleming *et al.* (1987) is not totally eliminated. For the following reasons no period analysis of the radial velocity was made: the ten available measurements contain a gap of about three years, the phase coverage near $13.^d9820$ is low (less than 50 %), and a shift of the zero-point of the radial velocity measurements may have occurred as suggested by Fleming *et al.* (1987).

The photometric spectral classification was made using the intrinsic UBV colours of the luminosity classes V, IV and III by FitzGerald (1970) and the BVR colours of the luminosity classes V and III by Johnson (1966). The possible interstellar extinction was estimated from the relations by Savage and Mathis (1979). Assuming no extinction, the UBV colours would support spectral type of K0III. With an extinction of approximately $0.^m3$ in the V magnitude, the spectral class could be either K1IV or K3V. On the other hand, the BVR colours indicate K2–5V with or without interstellar extinction, while the giant class could be K0–1III. The amplitude of the variation of the colour indices is large, which complicates the possibility of a unique photometric classification. Fekel *et al.* (1986) noted that in "active" stars the $(B-V)$ index is usually in better agreement with the actual spectral type than the $(V-R)$ index. This would explain the discrepancies between *UBV* and *BVR* classification of 1E 1751+7046. Assuming $v\sin i = 30$ km s^{-1} and $P_{rot} \approx P_{phot} = 13.^{d}9820$, the simple geometrical relation $v_{rot} = 50.6\ R\ P_{rot}^{-1}$ (with $[P] = d$, $[R] = R_{\odot}$ and $[v]$ =km s^{-1}) excludes luminosity classes IV and V. Even with $i = 90°$ the stellar radius would have to exceed $8R_{\odot}$ to produce the observed value of $v\sin i$.

CONCLUSIONS

The photometric rotation period of 1E 1751+7046 has been derived. The amplitude of the rotational modulation of brightness is extremely high and the colour indices show temperature variations, typical for spot–active stars. The photometric spectral classification and the rotation period combined with the observed value of $v\sin i$ indicate a giant luminosity class. However, if one of the "spurious periods" near $0.^{d}93$ or $1.^{d}07$ were the correct photometric rotation period, then luminosity classes V and IV would be possible.

The radial velocity may vary as a function of the photometric phase. This could be interpreted as nearly synchronous rotation of the primary component with the orbital motion of a low mass secondary, *i.e.* 1E 1751+7046 may not be of FK Comae–type. Our results support the same spectral class K0III and photometric rotation period as suggested by Ambruster *et al.* (1991). their detection of the Li I line indicates that 1E 1751+7046 may be a relatively young object.

REFERENCES

Ambruster, C.W., Fekel, F.C., and Guinan, E.F. 1991, these proceedings.

Bopp, B.W., and Stencel, R.E. 1981, *Ap. J. Letters*, **247**, L131.

Fekel, F.C., Moffett, T.J., and Henry, G.W., 1986 *Ap. J. Suppl.*, **60**, 551.

FitzGerald, M.P. 1970, *Astr. Ap.*, **4**, 243.

Fleming, T.A., Gioia, I.M., Maccacaro, T., and Mereghetti, S. 1987, *A. J.*, **93**, 1502.

Jetsu, L., Huovelin, J., Tuominen, I., Vilhu, O., Bopp, B.W., and Piirola, P. 1990, *Astr. Ap.*, **236**, 423.

Jetsu, L., Anttila, R., Dmitrienko, E., Grankin, K.N., Melnikov, S.Yu., Shevchenko, V.S., and Tuominen, I. 1991a, *Astr. Ap.*, to be submitted

Jetsu, L., Kokko, M., and Tuominen, I. 1991b, *Astr. Ap.*, to be submitted

Johnson, H.L. 1966, *Ann. Rev. Astr. Ap.* **4**, 193.

Pelt, J. 1983 in Proc. Statistical Methods in Astronomy Symp., ESA SP–201, p. 37.

Press, W. H., and Teukolsky, S. A. 1988, *Computers in Physics*, Nov/Dec 1988, 77.

Savage, B.D., and Mathis, J.S. 1979, *Ann. Rev. Astr. Ap.* **17**, 73.

Silva, D.R, Liebert, J., Stocke, J.T., and Aaronson, M. 1985, *Pub. A.S.P.*, **97**, 1096.

Stocke, J.T., Liebert, J., Gioia, I.M., Griffiths, R.E., Maccacaro, T., Danziger, I.J., Kunth, D., and Lub, J. 1983, *Ap. J.*, **273**, 458.

Tanner, R. W. 1948, *Comm. of the David Dunlap Obs.*, (University of Toronto) **16**, 177.

Seventh Cambridge Workshop on Cool Stars, Stellar Systems, and the Sun
ASP Conference Series, Vol. 26, 1992
Mark S. Giampapa and Jay A. Bookbinder (eds.)

IMAGING OF THE SURFACE OF ALPHA ORI

MARGARITA KAROVSKA
Harvard-Smithsonian Center for Astrophysics, 60 Garden Street, Cambridge, MA 02138, USA.

ABSTRACT The morphology of the stellar atmosphere and the circumstellar environment of α Ori was studied using high-angular resolution speckle interferometry techniques. The results of the multi-wavelength observations include detection of bright features on the surface of α Ori and further evidence for nearby orbiting companion.

Keywords: Supergiants; stellar atmospheres; speckle imaging

EVIDENCE FOR A CLOSE COMPANION

α Ori is a cool supergiant star characterized by pronounced variability in most of its observed parameters. For more than a century it has been extensively studied by variety of techniques, including speckle interferometry. The growing interest in this star during recent years is due in large part to the detection of a bright source in the vicinity of the star (possibly of a stellar nature).

In November 1983 a bright source was detected at 0".05 from the supergiant (Karovska *et al.* 1986), using high angular resolution techniques developed at the Harvard-Smithsonian Center for Astrophysics (Nisenson, 1988). These observations are consistent with an earlier detection of a source in the vicinity of α Ori (Karovska 1984). Supporting evidence for a close companion has also been found in polarization data (Hayes 1984). Measurements of the separations and position angles of the companion, together with the modeling of the polarization data, yielded an estimation of plausible orbital parameters (Karovska *et al.* 1986): P=2.08 yrs, T=1980.4, a=4.7 a. u., Ω=66°, ω=0°, e=0.35, i=30°.

Our subsequent high angular resolution observations (August 1985 at CTIO and December 1986 at Kitt Peak) confirmed the previous detection; however, the source appeared 1 to 2 magnitudes fainter then in 1983. The measured position angles and angular distances of the companion are in good agreement with those predicted by the orbit. Observations carried out in January 1988 at Kitt Peak did not resolve the companion; however, the images of α Ori show an elongation with a position angle close to that calculated for the companion.

Similar asymmetries have been observed in the images of α Ori obtained by different groups at several epochs between 1982 and 1989 in the H α line. The observations were carried out using several telescopes with apertures ranging from 2.3-m to 6.75-m (Roddier and Roddier 1983; Goldberg *et al.* 1982; Hebden *et al.* 1986; Hebden *et al.* 1987, Buscher *et al.* 1990). The position angles of the asymmetries are close to the positions of the companion predicted by the existing orbit for each epoch of observation. The correlation between the calculated position angles of the companion and the measured position angles of the asymmetries suggests that there is a possible connection between the asymmetries and the companion. The asymmetry in the images of α Ori could be caused by the unresolved companion, by tidal distortion of the supergiant's atmosphere, or possibly by an unresolved bright spot on the stellar surface facing the companion. To determine the nature of the companion (which presently remains a puzzle), it is crucial to obtain further speckle observations using large aperture telescopes, coordinated with other ground-based observations and the observations from space.

DETECTION OF A BRIGHT FEATURE ON THE SURFACE

α Ori was observed in August 1985 under excellent seeing conditions at 656 nm (FWHM=10 nm) using the PAPA camera at the 4-m CTIO telescope. The Maximum Entropy Method (Gull and Daniell, 1978) was applied to these data in order to reconstruct an image. The reconstructed image reveals the presence of a large bright feature (or several unresolved structures) on the surface of the star (Fig. 1a). This feature is very similar to the structure which was observed using MMT in December 1983 by Hebden *et al.* (1986) (Fig. 1b) and is probably associated with the stellar convection phenomenon (Schwarzschild, 1975).

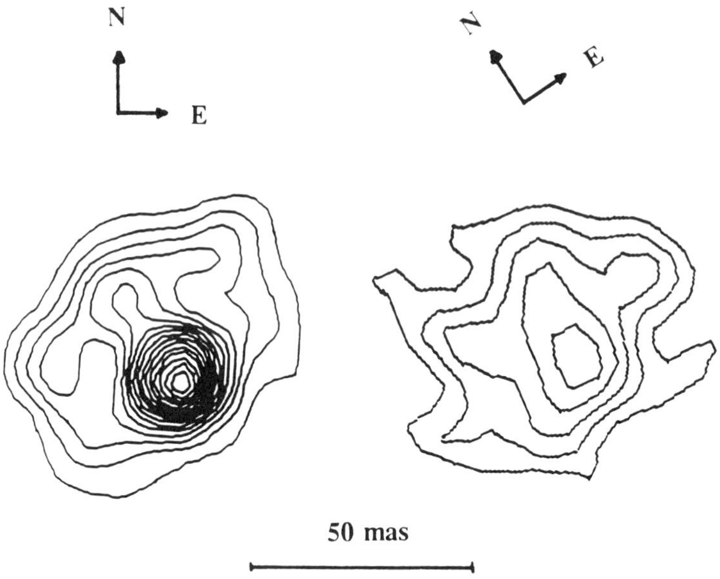

Fig. 1: (a) Image of α Ori at 656 nm reconstructed using MEM from the observations in 1985. (b) Differential speckle image of α Ori obtained in the $H\alpha$ line from the observations in 1983 (Hebden *et al.* 1986).

Changes in the position angle of the features in these images may be due to the rotation of the star. Depending on whether the rotation is in the clockwise or counter clockwise direction, several rotational periods can be calculated from the measured change in the position angle. The calculated period of the clockwise rotation - corresponding to the direction of the orbital motion of the companion - is about 2 years, which is very close to the possible orbital period of the companion.

Monitoring of similar bright structures in the atmosphere may offer a unique opportunity to determine a rotational period α Ori. Such a determination may answer the critical question of the synchronization of the rotational period of the primary with the orbital period of the close companion.

ACKNOWLEDGEMENTS

The author wish to thank P. Nisenson and S. Ebstein for their their assistance with the observations and many helpful discussions. The author would like to acknowledge support from Smithsonian Institution Scholarly Studies program.

REFERENCES

Buscher, D. F., Haniff, X. C. A., Baldwin, J. E., and Warner, P. J. 1990, *M.N.R.A.S.*, **245**, 7p.

Goldberg, L., Hege, E. K., Hubbard, E. N., Strittmater, P. A. 1986, and Cocke, W. 1982 *SAO Spec. Rep., No.*, **392**, p. 131.

Gull S. F.and Daniell, G. J., 1978, *Nature*, **272**, 686.

Hayes, D. P, 1984, *Ap. J. Suppl.*, **55**, 179.

Hebden, J., Christou, J., Cheng, A.Y.S., Hege, E. K., and Strittmater, P. A. 1986, *Ap.J.*, **309**, 745.

Hebden, J., Eckart, A., and Hege, E. K. 1987, *Ap.J.*, **314**, 690.

Karovska, M. 1984, Thesis, Nice University.

Karovska, M., Nisenson, P., and Noyes R. 1986, *Ap. J.*, **308**, 260.

Karovska, M., Nisenson, P., and Noyes R. 1987, *B.A.A.S.*, **19**, p. 755.

Nisenson, P. 1988, Proc. NATO Adv. Study Inst., (Eds. D.M. Alloin and J. -M. Mariotti, Kluwer Academic Publishers).

Roddier, F. and Roddier, C. 1983, *Ap. J. (Lett.)*, **270**, L23.

Schwarzschild, M. *Ap. J.*, **195**, 137.

Seventh Cambridge Workshop on Cool Stars, Stellar Systems, and the Sun
ASP Conference Series, Vol. 26, 1992
Mark S. Giampapa and Jay A. Bookbinder (eds.)

CHROMOSPHERIC ACTIVITY AND KINEMATICS OF VERY LOW MASS M DWARFS†

JAMES LIEBERT, REX A. SAFFER, and JANE NORSWORTHY
Steward Observatory, University of Arizona, Tucson, AZ

MARK S. GIAMPAPA,
National Solar Observatory, NOAO, Tucson, AZ

JOHN R. STAUFFER
Harvard-Smithsonian Center for Astrophysics, Cambridge, MA

ABSTRACT We discuss Multiple Mirror Telescope echelle observations primarily of Hα for M dwarf stars, including several among the lowest luminosities known with accurate trigonometric parallaxes. Equivalent widths in emission or absorption were measured, and radial velocities were used to compute the total space motions of the population.

Keywords: H-α emission; M dwarfs; Kinematics; Chromospheres

I. INTRODUCTION

The chromospheric properties of the main sequence stars of lowest mass are of special interest, since these objects are expected to have completely convective interiors so that the standard solar-like dynamo mechanism may not be applicable. Hα is the most readily accessible diagnostic for the presence of an active chromosphere. Stauffer and Hartmann (1986; hereafter SH) amassed a detailed compilation of measurements of this line and radial velocities for bright M dwarfs, while Giampapa and Liebert (1986; hereafter GL) extended this in a coarser way to the faint M dwarfs of lowest luminosity. This work extends the latter study, taking advantage of newly-discovered M dwarfs of very low luminosity and accurate new trigonometric parallaxes from the U. S. Naval Observatory (Monet et al. 1992).

II. RESULTS

In Figure 1, we plot the stars observed by SH, and by us and GL, against absolute visual magnitude (M_v), a quantity known for both samples. The newly observed stars again divide into those showing emission and absorption lines. For the faintest stars, however, an even larger scatter in Hα equivalent width (EW) is seen. One surprise is that the lowest luminosity object, LHS 2924, showed an EW in emission of 2.8 Å in an observation obtained in 1991 March, stronger than

† Observations obtained with the Multiple Mirror Telescope, operated jointly by the Smithsonian Center for Astrophysics and the University of Arizona.

had been estimated previously. The faintest star apparently lacking emission was LHS 234 (Gl 283B) at $M_v = 16.8$, which has a cool white dwarf companion and a minimum age in the billions of years.

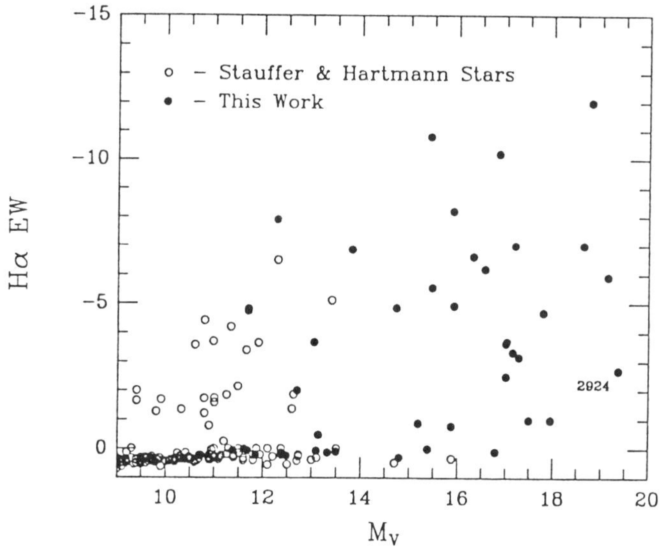

Fig. 1: EWs are plotted against M_v for stars observed by SH and by us. The new measurement for LHS 2924 is indicated.

The stars fainter than $M_v = 17$ include some of the strongest emission EWs. The star recently discovered by Schneider et al. (1991) with EW = 274 Å would extend the vertical dimension by more than an order of magnitude, but the M_v of this extremely cool stellar object is unknown.

In Figures 2a–c, we plot the U vs. V space motions for three groups binned by M_v intervals. The faintest stars ($M_v \geq 17$) show more scatter than that expected for a young disk population, but with a very different vertex deviation than the norm for the old disk. The $15 < M_v < 17$ group shows a more normal old disk distribution, while the brightest objects are divided between halo subdwarfs (deliberately chosen) and disk stars. As expected, none of the halo stars show evidence of chromospheric activity, and none are as faint as $M_v = 15$.

III. CONCLUSIONS

It now appears that all or nearly all stars fainter than $M_v = 17$ show evidence for active chromospheres, although we cannot know whether this sample includes any examples as old as the quiescent and slightly brighter LHS 234. The kinematic distribution of the faintest stars is somewhat puzzling, but suggests that they are not all young, *i.e.*, some at least are stars and not brown dwarfs.

REFERENCES

Giampapa, M.S. & Liebert, J. 1986, *Ap. J.*, **305**, 784.
Monet, D.G., Dahn, C.C., Vrba, F.J., Harris, H.C., Pier, J.R., Luginbuhl, C.B., & Ables, H.D. 1992, *A. J.*, in press.
Schneider, D.P., Greenstein, J.L., Schmidt, M., & Gunn, J.E. 1991, *A. J.*, in press.
Stauffer, J.R. & Hartmann, L.W. 1986, *Ap. J. Suppl.*, **61**, 531.

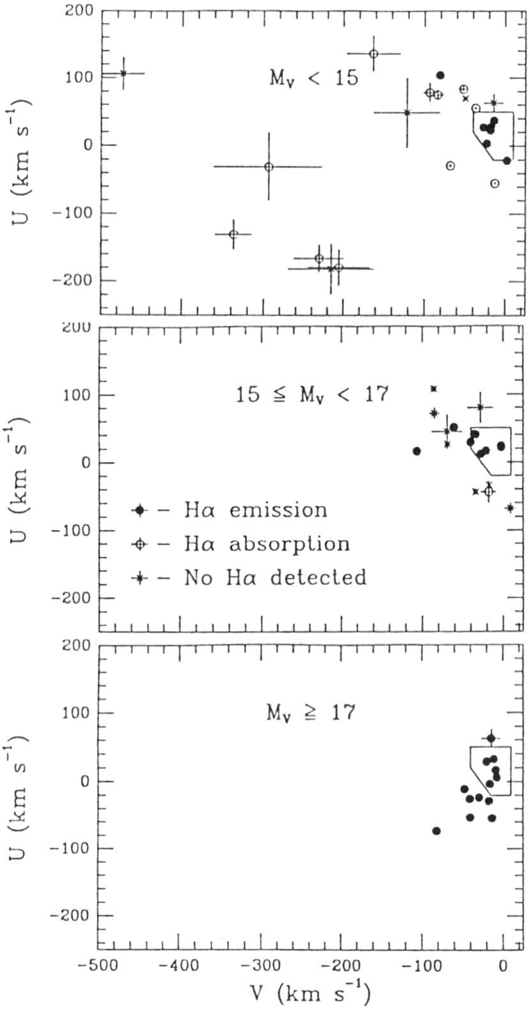

Fig. 2: U vs. V space motions in km/sec are plotted for objects showing emission (filled circles), absorption (open) and no detection (crosses), for three groups of M_v. Error bars are shown, but for the faintest stars are generally smaller than the symbol sizes due in part to the accurate trigonometric parallaxes (Monet *et al.* 1992) now available.

Seventh Cambridge Workshop on Cool Stars, Stellar Systems, and the Sun
ASP Conference Series, Vol. 26, 1992
Mark S. Giampapa and Jay A. Bookbinder (eds.)

A ZEBRA INTERPRETATION OF BY DRA STARS

BJØRN R. PETTERSEN Institute of Theoretical Astrophysics, University of Oslo, P.O. Box 1029 Blindern, N-0315 Oslo, Norway

SUZANNE L. HAWLEY IGPP, Lawrence Livermore National Laboratory, P.O. Box 808, L-413, Livermore, CA 94551

GEORGE H. FISHER Institute for Astronomy, University of Hawaii, 2680 Woodlawn Drive, Honolulu, HI 96822

ABSTRACT The currently accepted view of cool, dark starspots as the c magnetic activity on BY Dra stars is re-examined. We propose that the magnetically active regions are better described by a bright facular network, and that the dark areas which give rise to photometric rotational modulation are actually regions where the underlying quiet photosphere is seen. This interpretation is motivated by recent observations that show that bright areas covering most of the star have magnetic fields with strengths of several thousand gauss.

Keywords: stellar activity; BY Dra stars; starspots; faculae

INTRODUCTION

Photometric brightness variations on late type, magnetically active stars are usually interpreted with a starspot model. In this model dark regions, analogous to sunspots but many times larger, are postulated as the centers of magnetic activity and are presumed to be responsible for the brightness changes through rotational modulation as the spot passes in and out of view. Long term variations (*i.e.* on an activity cycle time, some few to tens of years) in the mean brightness level are explained by varying the size and location of the spots. On the Sun, however, photospheric magnetic activity manifests itself in both dark sunspots and bright facular areas. Recent satellite measurements have shown that as the Sun becomes more active during its 11 year cycle, the solar irradiance *increases* even though the spot coverage has also increased (Hudson 1988, Lean and Foukal 1988). Furthermore, although spots initially produce a large brightness deficit, the rotational modulation of an active region over several rotation periods is manifested instead as a brightening, because of the much longer lived effect of the faculae associated with it. The connection between spots and brightness variations on the Sun is therefore not as straightforward as predicted by the starspot paradigm.

Further complexity has been introduced by recent observations of solar- like stars (Radick, Lockwood and Baliunas 1990). their data suggests that relatively inactive stars show the solar effect – a brightness *increase* as activity increases, while more active stars show the opposite effect – a brightness *decrease* as activity increases (as in the starspot model). Stars at other stages of evolution also exhibit complex activity-related behavior. For example, T Tauri stars have shown *bright* spots correlated with ultraviolet emission lines (Simon, Vrba and Herbst 1990); RS CVn stars have been modeled very successfully with multi-component starspot and plage models (Rodono *et al.* 1986) but also show evidence for a strong facular contribution similar to that seen in the Sun (Dorren and Guinan 1990).

The rich variety of magnetic phenomena observed on the Sun and other stars leads us to suggest that the current model of dark spots being solely responsible for the observed photometric modulation and long term brightness variations on late type stars is too simplified. We summarize here our hypothesis that, as in the solar case, the bright facular elements play a dominant role in the observed brightness variations of BY Dra stars (see Pettersen, Hawley and Fisher 1991 for a more complete discussion).

THE BY DRA STARS

The BY Dra stars are a class of late K and M type dwarfs which exhibit photometric rotational modulation of a few hundredths to a few tenths of a magnitude (Bopp and Evans 1973). They are unique as a place to study magnetic activity because, in a few cases, the surface magnetic field strengths have now been measured directly. Values of some 2000-5000 G and large filling factors (50-80%) are typical (Saar 1990). These values refer to bright areas on the star, not to "starspots" because the "spots" are cool and hence do not contribute significantly to the total light. The nature of the bright regions is not clear; in the usual starspot model, the dark areas on the star are the sites of enhanced magnetic activity while the bright regions represent undisturbed photosphere. However, on the Sun, the filling factor of magnetic flux tubes in the quiet photosphere is very small and thus the average field from those areas is negligible. Larger filling factors are found in the bright network elements with typical field strengths of 1000-1500 G. If we assume that magnetic fields are similarly organized on the BY Dra stars, the following picture emerges:

The magnetically active star has an extended photospheric facular network similar to that seen on the Sun, which acts to raise the apparent brightness of the star. This network covers a considerable fraction of the star and replaces what was considered to be the photosphere in the historical starspot model. The remainder of the stellar surface consists of quiet photosphere, which is seen as dark in comparison to the bright facular regions. This photospheric portion is what was previously attributed to starspots. Spots and other magnetic phenomena probably exist within the facular network, but on small size scales such that their presence is not currently detectable. *In essence what we are proposing is a "Zebra Effect", where the magnetically active regions are the bright areas and the quiet photosphere constitutes the dark areas; the opposite of the current starspot model.*

Because the data and modelling techniques themselves make no *a priori* assumption about the location of the magnetic activity, the stellar surface features inferred from starspot models of BY Dra stars remain the same; it is only the interpretation that differs. With the Zebra interpretation, several problems with the starspot model are resolved, *e.g.* flux redistribution from very large spots, the existence of polar spots and spots that live for many rotations.

The Zebra model gives several observational predictions that differ from the starspot model. For example, the large filling factor of bright magnetic network elements in the Zebra model suggests that chromospheric and transition region emission lines will show only weak, if any, rotational modulation; this is also the case for coronal X–ray emission. Observations of chromospheric emission and photometric rotational modulation tend to support this view since, in most cases, no obvious correlation is found. Similarly, no preferred longitude for flaring, and hence no rotational correlation, is expected. The data are insufficient to test this prediction as yet.

Long term brightness variations are present in some BY Dra stars (*e.g.* Phillips and Hartmann 1978). In the Zebra model the star would be most active when it is brightest. The sparse data for BY Dra indicate that it has had strong Hα emission lines since 1975 (when it has been bright) and reports of Hα absorption (Vogt 1980) at earlier epochs (when it was fainter). These observations support the Zebra model, but are only suggestive. We also expect more and larger flares during the periods when the star is bright; the data are not yet sufficient to test this prediction.

Finally, the Zebra model predicts that an active main sequence star (*e.g.* a dMe star) will appear warmer and brighter than a photospherically identical dM twin without magnetic surface activity. The dMe star will thus appear to have too early a spectral type for its mass. The dMe stars are known to have brighter absolute magnitudes at the same spectral type (Joy and Abt 1974); this effect has generally been attributed to youth. Improved mass and temperature determinations for M dwarfs are needed to sort out which effects dominate, and whether there are practical applications of a possible activity-luminosity correlation.

ACKNOWLEDGEMENTS

Our thanks to H. Hudson, J. Neff, R. Radick and S. Saar for interesting and useful discussions and to O. Duhalde and W. Krzeminski for their hospitality at Las Campanas Observatory where this work was begun. This work is supported by NASA grant NAGW-2547 and NSF grant ATM-90-22385.

REFERENCES

Bopp, B.W. and Evans, D.S. 1973, *M.N.R.A.S.* **164**, 343.
Dorren, J.D. and Guinan E.F. 1990, *Ap. J.*, **348**, 703.
Hudson, H.S. 1988, *Ann. Rev. Astron. Astrophys.* **26**, 473.
Joy, A.H. and Abt, H.A. 1974, *Ap.J. Supp.* **28**, 1.
Lean, J. and Foukal, P. 1988, *Science* **240**, 906.
Pettersen, B.R., Hawley, S.L. and Fisher, G.H. 1991, preprint.
Phillips, M.J. and Hartmann, L. 1978, *Ap.J.* **224**, 182.
Radick, R.R., Lockwood, G.W., and Baliunas, S.L. 1990, *Science* **247**, 39.
Rodono, M. et al., 1986, *Astr. Ap.* **165**, 135.
Saar, S.H. 1990, in *Solar Photosphere: Structure, Convection and Magnetic Fields*, ed. J.O. Stenflo (Dordrecht: Kluwer Publ.), p. 427.
Simon, T., Vrba, F.J., and Herbst, W. 1990, *A.J.* **100**, 1957.
Vogt, S.S. 1980, *Ap.J.* **240**, 567.

Seventh Cambridge Workshop on Cool Stars, Stellar Systems, and the Sun
ASP Conference Series, Vol. 26, 1992
Mark S. Giampapa and Jay A. Bookbinder (eds.)

CHARACTERISTICS OF "MARGINAL" BY DRA STARS

STEVEN H. SAAR Center for Astrophysics, 60 Garden Street, Cambridge, MA 02138, USA

BERNARD W. BOPP Ritter Observatory and Dept. of Physics and Astronomy, Univ. of Toledo, Toledo, OH 43606 USA

ABSTRACT We analyze new UV and optical spectra of "marginal" BY Dra stars, and find support for the idea that they form a class of objects intermediate in activity (and perhaps age) between dM and dMe stars.

Keywords: stellar activity; stellar atmospheres; M stars

1. INTRODUCTION AND OBSERVATIONS

"Marginal" BY Dra stars (mBY) have been roughly defined (Bopp 1987) as K – M dwarfs having moderately strong Ca II emission, weak Hα (absorption or emission), and weak or absent photometric variability. Byrne & Doyle (1990), who call them dM(e) stars, found that the UV emission of one mBY, Gl 900, was intermediate between dM and dMe stars. Bopp (1987) suggests that these stars have intermediate rotation and magnetic activity levels, and form part of an evolutionary sequence: active, spotted dMe → moderate activity dM(e) → inactive dM. Here, we further explore Bopp's hypothesis with new data, and also search for new members of this (possible) group.

We have obtained new IUE and optical data for a number of mBY, candidate mBY, and comparison stars. The spectra were reduced using software at the CU-RDAF. Optical spectra were obtained using the NSO McMath stellar spectrograph + CCD with the echelle ($\lambda/\Delta\lambda$ = 125,000; at 6170 Å) and Milton+Roy ($\lambda/\Delta\lambda$ = 40,000; at Hα and He I D3) gratings, and using the ESO CAT+CES+CCD system ($\lambda/\Delta\lambda = 100,000$; at He I D3). These were reduced in a standard fashion using variance weighted extraction routines written in IDL.

2. ANALYSIS AND DISCUSSION

We have gathered a data base on Hα, X–ray, and Mg II and other UV emission fluxes (drawn mostly from Byrne & Doyle 1990; Doyle *et al.* 1990, Herbst & Miller 1989, Linsky *et al.* 1982, Mathioudakis & Doyle 1989, Rutten 1987, Stauffer & Hartmann 1986, Young *et al.* 1989, plus our new observations) to investigate the activity of the mBY stars relative to other K and M dwarfs. All the surface fluxes (F) have been recalibrated onto a consistent scale. When plotted against $(R\text{–}I)_K$, the $F_{Mg\ II}$ of mBY stars lie on a line almost halfway between the empirical upper and lower limits for the data set (Fig. 1). The two mBY stars so far observed with the IUE SWP camera, Gl 900 (Byrne & Doyle 1990) and Gl 425A, both exhibit intermediate levels of far UV chromospheric and TR emission (O I, C II, C IV, Si II) as well (Fig. 2). No obvious He I D3 (5876Å) feature is present in Gl 425A, Gl 900 or Gl 907.1, implying that mBY activity levels are lower than those of dMe stars (where D3 is in emission). The only mBY with a measured X-ray flux (GL 900) has F_X higher than the mean, but still below most dMe stars (Fig. 3).

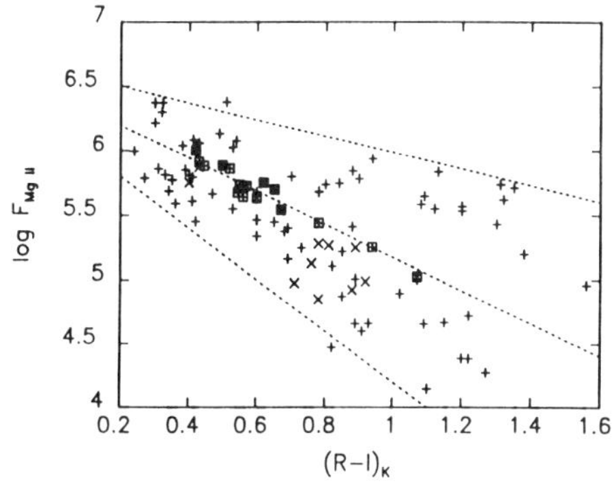

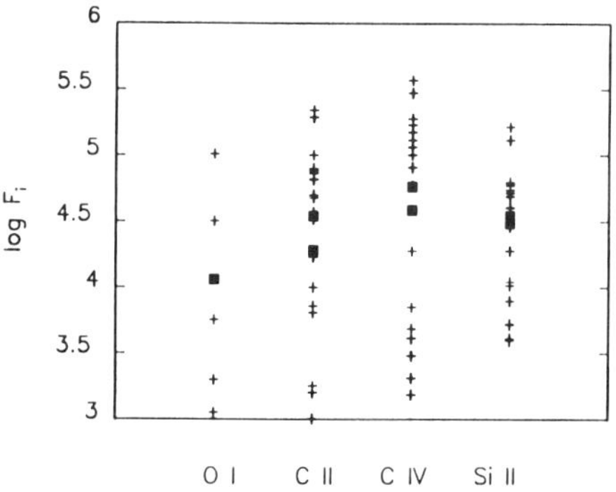

Fig. 1: Log $F_{Mg\ II}$ vs. $(R{-}I)_K$, showing mBY's (solid box), new observations (×) and data from the literature (+), with new mBY candidates surrounded by a box. Empirical upper/lower limits and the mBY locus are also drawn (dashed).

Fig. 2: Log surface fluxes in O I, C II, C IV, and Si II for a sample of K and M dwarfs. Symbols as in Fig. 1.

Theoretical models of late K and M dwarfs (Cram & Giampapa 1987) predict a nonmonotonic relation between Hα equivalent width, $W_{eq}(\mathrm{H}\alpha)$, and Ca II emission; weak Hα absorption can indicate either very low or moderately high (mBY?) activity levels. Unfortunately, most of the available Ca II data for both the mBY stars and our K – M dwarf comparison sample are on the qualitative Wilson & Woolley (1970) scale. Since Mg II scales like Ca II (Hartmann *et al.* 1984) and is formed in similar conditions, we have constructed a $W_{eq}(\mathrm{H}\alpha) - F_{Mg\ II}$ diagram, both to check the positions of the mBY stars, and to search for new mBY candidates. We find (Fig. 4) that (1) mBY's occupy the expected area of the $W_{eq}(\mathrm{H}\alpha) - F_{Mg\ II}$ plane, and (2) there appears to be a systematic shift in the locus of stars in the diagram as a function of spectral type. Thus, the mBY stars generally *do* seem to have activity intermediate between dM and dMe stars. Finally, our high resolution spectra indicate the mBY's have 3 km $\mathrm{s}^{-1} \leq v \sin i \leq 6$ km s^{-1} (detailed modeling in progress). Combining this with the hypothesis of a BY Dra "trigger velocity" of $v \sin i \approx 5$ km s^{-1} (Bopp & Fekel 1977; Vogt *et al.* 1983), it seems possible that the mBY/dM(e) stars may represent an intermediate *evolutionary* state as well – stars undergoing rotational spindown from an active dMe state to a quiescent dM old age.

Finally, we searched the K – M data for likely mBY stars, based on the stars' activity levels. The best candidates are Gl 394, 546, and 879; other good candidates include Gl 91.1A, 195, 499A, 509.1, 685, 715, 793, 889, and Wooley 9427. We recommend further study of these stars to determine their status.

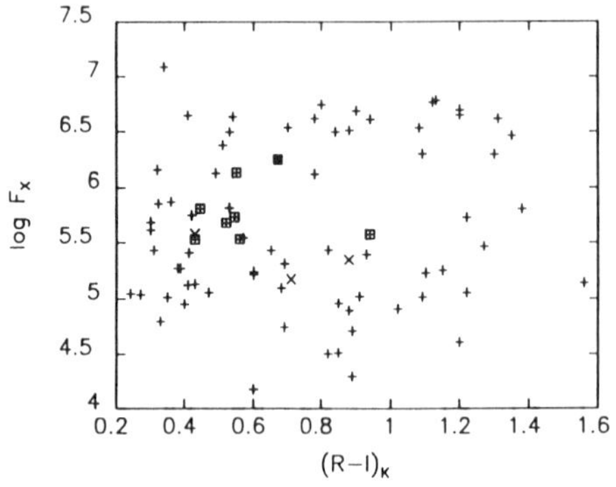

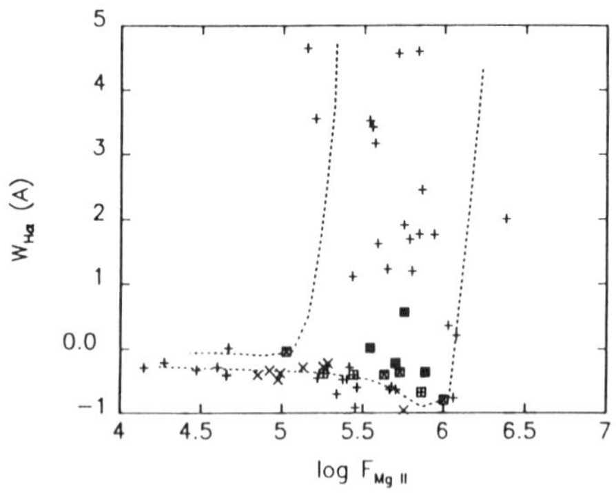

Fig. 3: Log F_X vs. $(R{-}I)_K$; symbols as in Fig. 1.

Fig. 4: $W_{eq}(H\alpha)$ vs. $F_{Mg\ II}$; symbols as in Fig. 1. Approximate loci of late K dwarfs and late M dwarfs are sketched in (dashed).

ACKNOWLEDGEMENTS

This work is based on data from the NSO, ESO, and IUE, and was supported by an SI Fellowship, and NASA grants NAGW-112, NAG 5-954, and NAG 5-87.

REFERENCES

Bopp, B. W. 1987, *Ap. J.*, **317**, 781
Bopp, B. W., Fekel, F. K. 1977, *A. J.*, **82**, 490
Byrne, P.B., Doyle, J. G. 1990, *Astr. Ap.*, **238**, 221
Cram, L. E., Giampapa, M. S. 1987, *Astr. Ap.*, **323**, 316
Doyle, J.G., Panagi, P., Byrne, P.B. 1990, *Astr. Ap.*, **228**, 443
Hartmann, L.W. *et al.* 1984, *Ap. J.*, **279**, 778
Herbst, W., Miller, J.R. 1989, *A. J.*, **97**, 891
Linsky, J.L. *et al.* 1982 *Ap. J.*, **260**, 670
Mathioudakis, M., Doyle, J. G. 1989, *Astr. Ap.*, **224**, 179
Rutten, R.G.M. 1987, PhD Thesis, U. of Utrecht
Stauffer, J. R., Hartmann, L. W. 1986, *Ap. J. Suppl.*, **61**, 531
Vogt, S.S., Soderblom, D.R., Penrod, G.D. 1983, *Ap. J. Letters*, **269**, 250
Wilson, O., Wooley, R. 1970, *M.N.R.A.S.*, **148**, 463
Young, A., *et al.* 1989, *Ap. J.*, **344**, 427

Seventh Cambridge Workshop on Cool Stars, Stellar Systems, and the Sun
ASP Conference Series, Vol. 26, 1992
Mark S. Giampapa and Jay A. Bookbinder (eds.)

MAGNETIC ACTIVITY RELATIONS

JEFFREY S. NEWMARK Computer Sciences Corp., NASA/GSFC, Code 684.9, Greenbelt, MD 20771.

ABSTRACT Echelle spectra have been obtained for 9 short period RS CVn binaries and 2 W-type W UMa contact binaries. The relationship between the activity sensitive Hα and Ca II H lines with rotation and/or tidal interaction is investigated. Some evidence is presented for a saturation in the secondary components of these systems. The relatively constant level of emission over long term periods is also shown.

Keywords: binary stars; magnetic activity; saturation

INTRODUCTION

There is evidence (Barden 1985; Vilhu and Heise 1986; Cruddace and Dupree 1984; Vilhu and Walter 1987) of a breakdown in the relationship between activity and rotation period or activity and Rossby number. Fleming *et al.* (1989) hypothesize the saturation is due either to foot points of magnetic loops covering the entire surface such that no new ones can form, or a magnetic feedback in the convection zone. Linsky (1989) suggests a slightly different interpretation of the saturation. He proposes that these stars are completely covered with strong magnetic fields and that the maximum efficiency which a stellar atmosphere converts mechanical energy into heat is reached at this point. Vilhu (1987) suggests a feedback mechanism between the subphotospheric mechanical fluxes and the total radiative losses from the CR-coronal plasma. As the radiative losses approach the generation of mechanical flux the convection would begin to be suppressed and in turn the dynamo will be suppressed.

Barden (1985) suggests the breakdown in this relationship for the W-type W UMa systems is a consequence of either their contact nature or tidal interactions suppressing the differential rotation of the secondary components. Little is understood about the magnetic field generation among W UMa systems, in particular, if their contact nature changes the convection in the envelope. Here the initial Hα investigation of Barden will be extended to more stars as well as including the Ca II H line.

OBSERVATION AND DATA REDUCTION

Fiber-Optic Echelle/CCD spectra were obtained during 1987 and 1988 using the 0.9m coude' feed and 2.1m telescopes at KPNO. Analysis of the spectra involved subtracting a synthesized stellar spectrum leaving any excess emission (or absorption). The synthesized spectra were constructed from artificially broadened, RV shifted, and intensity weighted inactive stars chosen to match the the component spectral types (Barden 1985). For each program star quadrature spectra were used to find an average excess emission level in each of the components (where possible).

Noyes *et al.* (1984) in their study of Ca II H and K emission found the Rossby number provided a better correlation with stellar surface activity than the rotational period alone. For the more moderate period RS CVn binaries it is found that the activity rises with decreasing Rossby number but there is a turnoff for the short period systems. A better understanding of the nature of the turnoff in emission can come from the relationship of line flux to the dimensionless tidal parameter $f_\star$ given by (Barden 1985)

$$f_\star = \frac{1}{4}\frac{M}{M_\star}\left(\frac{R_\star}{a}\right)^3$$

This parameter has the added benefit of being independent of orbital inclination. Barden (1985) found a trend of decreasing emission with increasing tidal amplitude which suggests that large tidal interactions may be the culprit in the suppression of the differential rotation. In particular Barden found a strong decrease in the secondary component of the contact systems.

DISCUSSION

If we look at the relationship between activity and Rossby number for these systems(not shown) we find a rather large scatter but looking at the individual components a trend emerges. Except for the secondaries on the W UMa systems and the primary for AH Vir, there is a rising trend in activity for a decreasing Rossby number. As decreasing Rossby number is a direct consequence of deeper convection zones, we see that for the short period RS CVns, the secondary components have higher emission levels while for the W UMa type systems the primary components have the higher emission. The secondaries on the RS CVns are later spectral types, therefore having deeper convection zones and as a consequence higher magnetic activity.

Understanding the emission characteristics of the W UMa type stars is more difficult. These observations suggested, at least for the primary, that dynamo driven magnetic activity is responsible for the chromospheric emission. Mullan (1975) in his starspot model found the primary component of W UMa type systems should have deeper convection zones than a single star with similar parameters while the secondary's should be shallower. Rucinski (1991) demonstrates similar behavior. This then could produce the lack of flux on the secondary component. A second factor is explored in Figure 1, showing the relation between $L_{H\alpha}/L_{bol}$ and L_{CaH}/L_{bol} and the tidal amplitude parameter $f_\star$.

Again we see a scatter among the systems but a relatively constant level of emission compared to the two orders of magnitude variation in $f_\star$. There is some implication that tidal interaction may be important in shutting down the differential rotation for stars with $f_\star > 0.02$, i.e the secondaries on the W UMa systems.

The lack of any strong trend, either increasing or decreasing activity, implies the emission originates over a large surface area. The observations are suggestive of a possible shutdown in the activity for the secondary components of the contact binaries. Most probably any shutdown in surface activity in the secondaries of W-type W UMa binaries is a consequence of both increased tidal interaction as well as the presence of a common convective envelope.

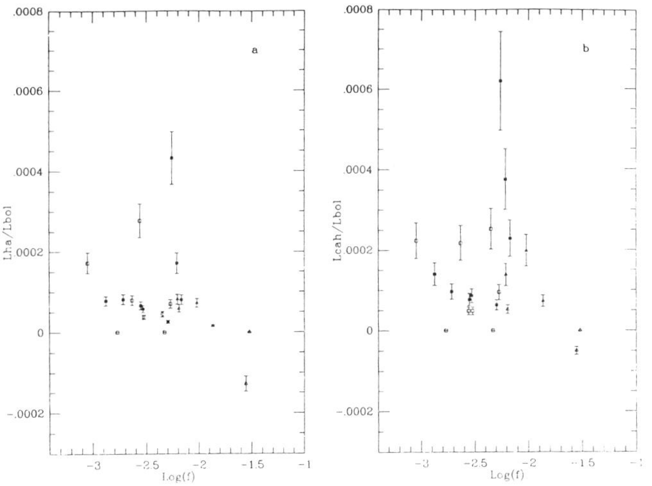

Fig. 1: L/L_{bol} versus tidal amplitude parameter ($f_\star$), (a) is for Hα and (b) for Ca II H. Filled boxes are the primary component and open boxes are the secondary component. Squares represent short period RS CVns while triangles represent W UMa systems.

ACKNOWLEDGEMENTS

This work has been partially supported by NSF grant AST 8618167 to the Pennsylvania State University and partially supported by Computer Sciences Corporation.

REFERENCES

Barden, S.C. 1985, *Ap. J.*, **295**, 162.
Bradstreet, D.H. and Guinan, E.F. 1991, private communication.
Cruddace, R.G. and Dupree, A.K. 1984, *Ap. J.*, **277**, 263.
Fleming, I.A., Gioia, I.M. and Maccacaro, T. 1989, *Ap. J.*, **340**, 1011.
Gilman, P.A. 1980, in Lecture Notes in Physics, Vol 114, ed. D.F. Gray and J.L. Linsky (Berlin: Springer Verlag), p.19.
Linsky, J.L. 1989, in Multiwavelength Astrophysics, ed. F.A. Cordova (Cambridge: Cambridge Univ. Press), p.49.
Mullan, D.J. 1975, *Ap. J.*, **198**, 563.
Noyes, R.W., Hartmann, L., Baliunas, S.L., Duncan, D.K., and Vaughan, A.H. 1984, *Ap. J.*, **279**, 763.
Rucinski, S.M. 1991, preprint.
Vilhu, O. 1987, in Cool Stars, Stellar Systems and the Sun, edited by J.L. Linsky and R.E. Stencel, (Berlin:Springer-Verlag).
Vilhu, O. and Heise, J. 1986, *Ap. J.*, **311**, 930.

Seventh Cambridge Workshop on Cool Stars, Stellar Systems, and the Sun
ASP Conference Series, Vol. 26, 1992
Mark S. Giampapa and Jay A. Bookbinder (eds.)

CA II K LINE DIAGNOSTICS OF TWO-DIMENSIONAL MODELS OF THE SOLAR CHROMOSPHERE

S.K. SOLANKI and M. BÜNTE
Institute of Astronomy, ETH Zentrum, CH-8092 Zürich, Switzerland

O. STEINER
Kiepenheuer-Inst. für Sonnenphysik, Schöneckstr. 6, 7800 Freiburg, FRG

H. UITENBROEK
Center for Astrophysics, 60 Garden Street, Cambridge, MA 02138, USA

ABSTRACT We study the influence of a thermally bifurcated atmosphere on the structure of the magnetic field and the characteristics of the Ca II K line intensity. These spectral properties are calculated using 1.5-D radiative transfer in a 2-D flux tube model and compared to a diverse set of "basic observations" of the chromosphere.

Keywords: solar magnetic fields; solar & stellar chromospheres; Ca II K line

INTRODUCTION AND 2-D MODEL CALCULATIONS

Plane-parallel models successfully reproduce *spatially averaged* properties of the solar chromosphere but they cannot account for *spatially resolved* observations. Multicomponent models (*e.g.* Vernazza *et al.* 1981; Ayres *et al.* 1986) reproduce a wider range of observations, but depend critically on the type of observations used to construct them. Furthermore, such models cannot provide any information on the geometry and location of the atmospheric components .

The classical picture of the chromospheric field-structure consists of photospheric flux tubes which continuously expand through the chromosphere and form a magnetic canopy at the base of the corona (1500 – 2500 km above $\tau_{5000} = 1$). The picture proposed in this study is based on the idea of a thermally bifurcated atmosphere with hot flux tubes (probably heated by large amplitude MHD-waves) surrounded by a CO-cooled atmosphere (Ayres *et al.* , 1986). Such a thermal structure produces low lying magnetic canopies resting upon cool non-magnetic material.

We consider a vertical flux tube in a monopolar region, whose magnetic field satisfies magnetohydrostatic equilibrium. The interior of the flux tube is described by a hot empirical atmosphere (*e.g.* FLUXT, VALP, or VALC′ of Ayres *et al.* 1986 or VALF of Vernazza *et al.* 1981). The non-magnetic atmosphere is generally described by an empirical model without a chromospheric temperature rise, COOLC of Ayres *et al.* (1986). We then calculate the profile of the Ca II K line for a 5-level plus continuum Ca^+ atom including effects of partial redistribution. The line profile is calculated along 32 vertical rays placed at various radial distances r from the flux tube axis.

RESULTS

In a thermally inhomogenous (bifurcated) chromosphere, with a cool non-magnetic gas, magnetic fields expand and merge *below* a height of $\approx$ 700 and 1200 km for *any* photospheric filling factor of the field. Therefore, a dominant part of the chromosphere is filled with magnetic fields.

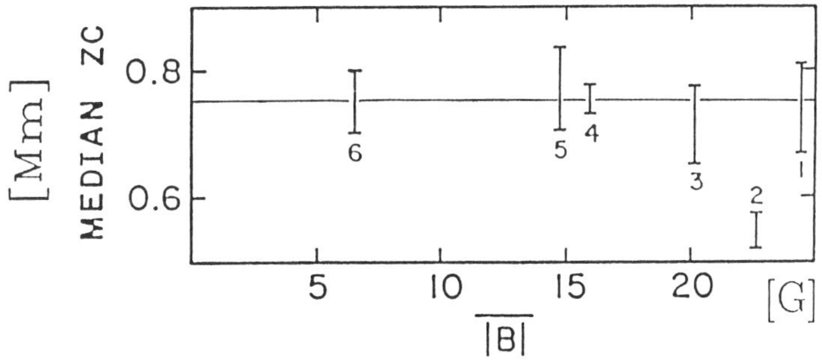

Fig. 1: Height ZC of the observed (error bars) and calculated (horizontal line) canopy base vs. mean photospheric field strength.

The rapid expansion of the field near this merging height produces a canopy with an almost horizontal base in the quiet Sun. Fig. 1 (after Jones and Giovanelli, 1983) shows the observed height of the canopy base ZC (error bars) vs. mean photospheric field strength. The horizontal line indicates ZC resulting from one of the present model–atmosphere combinations. Note the excellent agreement with the observations.

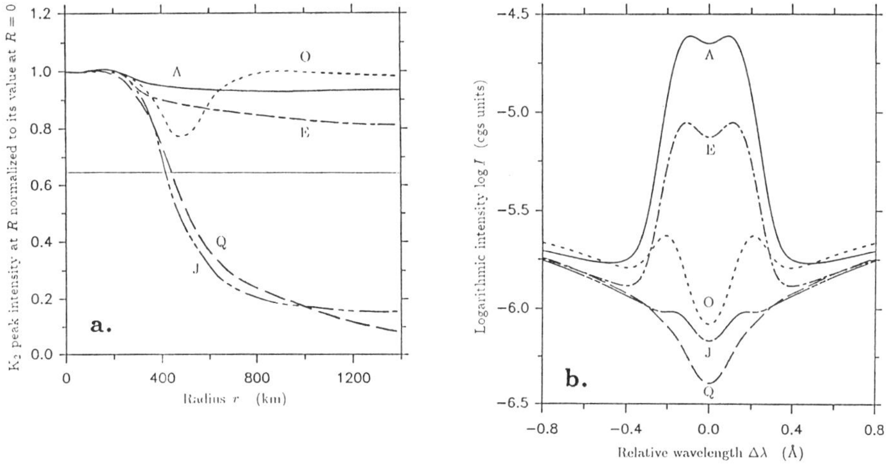

Fig. 2: **a.** K_2 peak intensity at radius r, normalized to its value at the flux tube axis ($r = 0$). **b.** Spatially averaged K line profiles for the models plotted in Fig. 2a. Profile O is closest to observations.

Although the model flux tubes have a horizontally homogenous temperature the intensity of the K_2 peak can vary strongly with r (Fig. 2a). The variation depends critically on the relative internal and external temperature: Model O has VALC′ in both components, all other models have COOLC in the non–magnetic component. The magnetic components are FLUXT for model A, VALP for model E, VALF for model J and VALC′ for model Q. The thin horizontal line near 0.64 shows the observed (normalized) K_2 intensity at large r. This diagram suggests that the flux tube atmosphere lies between VALP and VALF.

A comparison of the spatially averaged Ca II K profiles with the profile expected from the average quiet sun (marked O in Fig. 2b) confirms that the chromospheric flux tube temperature lies between VALP and VALF.

CONCLUSIONS

The 2–D models of the solar atmosphere proposed here can reproduce, at least qualitatively, the following diverse set of observations:

1. The spatially averaged Ca II K quiet sun profile.
2. The spatial scale and the amplitude of the horizontal K_2 intensity variations in the quiet sun (including the network).
3. The qualitative form of the relationship between the K line core intensity and the magnetic flux density in an active region, including the presence of a 'basal' flux, *i.e.*, a chromospheric emission in the local absence of a photospheric magnetic field.
4. The presence of a cool fraction of the solar chromosphere with temperatures below 4000 K, as suggested by observations of infrared CO lines.
5. The presence of magnetic canopies in the lower or middle chromosphere of the quiet sun and the presence of even lower canopies in active regions.
6. The highly intermittent nature of the photospheric magnetic field.

More details are to be found in Solanki *et al.* (1991). Further refinements, *e.g.* an explicit self–consistent treatment of the energetics, are being incorporated into the models in collaboration with L. Anderson (University of Toledo).

REFERENCES

Ayres, T.R., Testerman, L., Brault, J.W.: 1986, *Ap. J.*, **304**, 542
Jones, H.P., Giovanelli, R.G.: 1983, *Solar Phys.* **37**, 58
Solanki, S.K., Steiner, O., Uitenbroek, H.: 1991, A&A, in press
Vernazza, J.E., Avrett, E.H., Loeser, R.: 1981, *Ap. J. Suppl. Ser.*, **45**, 635

Seventh Cambridge Workshop on Cool Stars, Stellar Systems, and the Sun
ASP Conference Series, Vol. 26, 1992
Mark S. Giampapa and Jay A. Bookbinder (eds.)

SHORT TIME-SCALE VARIATIONS IN CA II H+K OBSERVATIONS

ANDREA K. DOBSON Whitman College, Walla Walla WA 99362

ABSTRACT Analysis of the contributions to the variance present in the long time series Ca II H+K observations of cool stars made at Mt. Wilson Observatory indicates that several active solar-type stars exhibit variation on the time scale of minutes. This variation is in excess of that expected from measurement noise and is not seen in observations of stars with activity levels comparable to that of the Sun.

Keywords: Stellar variability; Ca II H+K emission

INTRODUCTION

The study of the long-term stellar magnetic activity data from the Mt. Wilson Observatory Ca II H+K survey has clearly shown a wide range of levels of activity and variability among cool stars (*e.g.*, Baliunas and Vaughan 1985). Dobson *et al.* (1990) examined the components of the variance present in these time series in an attempt to separate the different mechanisms responsible for the observed activity by making use of the different time scales on which different components contribute to the variance. We considered the total variance to be the sum of contributions from transient phenomena, rotation, evolution of active regions, possible long-term activity cycles, and photon and instrumental noise. We broke the time series for a star into bins of a given length, calculated a variance for each bin, and calculated a pooled variance representing the variation in the data set on that particular time scale, *e.g.*, weeks, months, several years. Many stars demonstrate clear jumps in the total variance on the time scale at which a new mechanism, for instance rotation, begins to contribute. The observations are made in sets of three; it is therefore possible to consider variations on as short a time scale as 5 to 15 minutes.

SHORT TIME-SCALE VARIATION

We initially attributed the 3-point variance to noise. Further examination of a set of 15 lower main sequence stars shows, however, that the more active stars exhibit real variation in Ca II H+K emission, as indicated by fluctuations in the S index, on time scales as short as 5 to 15 minutes. Table I lists the 3-point pooled variances for these stars, along with their colors, average S values, and photometrically determined rotation periods (Baliunas 1986, Donahue 1991). Stars such as HD 39587, HD 1835, HD 131156a, HD 22049, and especially HD 129333 are clearly more variable at short time scales than less active stars of similar spectral types.

These stars are also the more active ones in the sample, as can be seen in figure 1. Figure 2 shows four representative sets of observations for

HD 129333 and HD 143761, two G stars of vastly different activity levels. (These observations were obtained on four nights, chosen because both stars were observed on all nights, during a randomly chosen five-week period in 1984.)

It is not clear whether a given star is more variable on nights when its activity is higher. Thirteen of the fifteen stars show a linear correlation coefficient between the daily standard deviations and S values which is statistically significant at the 99% confidence level. Linear least squares fits of the standard deviations as a function of S account for at most 4% of the variation in σ.

TABLE I Three-point variances

star	HD	B−V	$\sigma^2 \times 10^6$	S-index	P_{rot} (days)
τ Boo	120136	0.48	4.52	0.187	3.8
σ Per	216385	0.48	5.81	0.142	
64 Cet	13421	0.56	4.52	0.130	38?
β Com	114710	0.57	4.01	0.203	12.5
χ^1 Ori	39587	0.59	12.12	0.317	5.13
ρ CrB	143761	0.60	4.75	0.149	
	129333	0.61	100.92	0.530	3.0
	81809	0.64	6.76	0.172	40.5
	9562	0.64	6.62	0.137	37.7
9 Cet	1835	0.66	39.32	0.332	7.8
85 Peg	224930	0.67	7.67	0.179	42?
τ Cet	10700	0.72	4.16	0.171	
	103095	0.75	10.89	0.181	31.1
ξ Boo A	131156a	0.76	31.87	0.452	6.6
ϵ Eri	22049	0.88	68.10	0.507	11.1

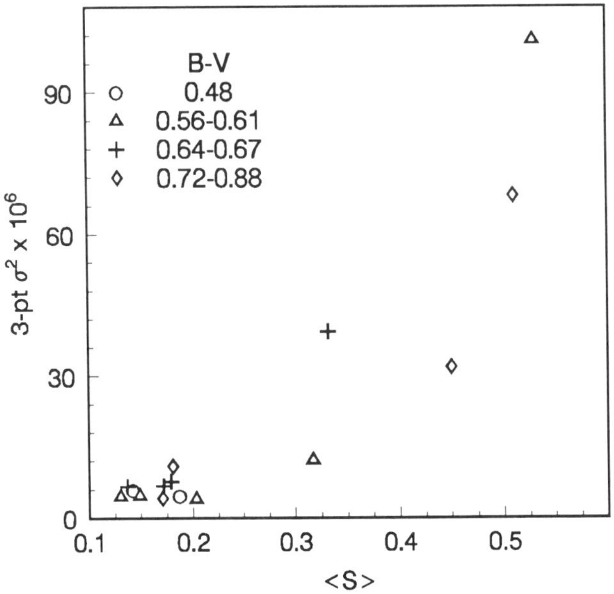

Fig. 1: Activity level *vs.* 3-point pooled variances.

As can be seen in figure 2, the active stars are not necessarily active over short times on every night. Noyes *et al.* (1984) observed HD 22049 for about 6 hours on each of four nights in 1982-3. Power spectra of two of the four nights data had definite signals at periods of about 10 minutes which the authors suggest may be due to photospheric velocity oscillations. If this is the case, the relationship between activity level and short-term variability might suggest a possible relationship between chromospheric emission levels and the amplitude of photospheric oscillations.

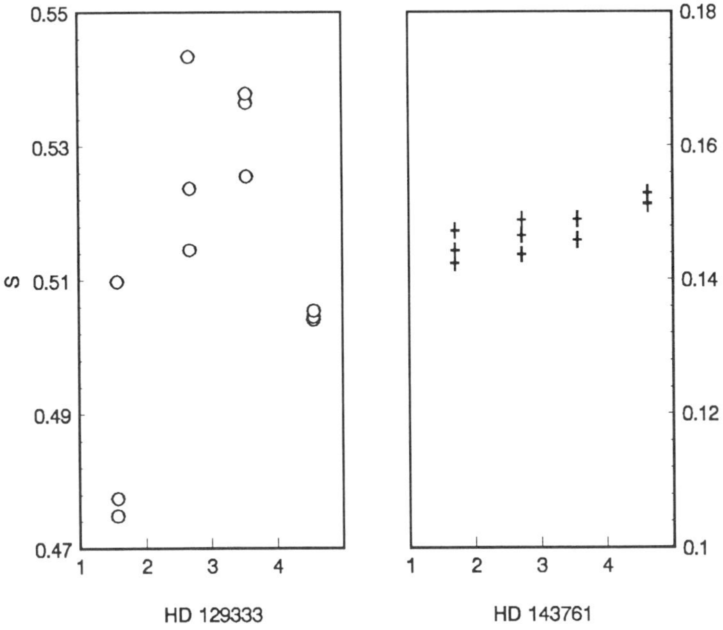

Fig. 2: Four representative nights observations for one active and one inactive star.

ACKNOWLEDGEMENTS

The author would like to thank Dr. Sallie Baliunas for making the Mt. Wilson data available and Robert Donahue for continued useful discussions regarding the activity of these stars.

REFERENCES

Baliunas, S.L. 1986, private communication

Baliunas, S.L. and Vaughan, A.H. 1985, *Ann. Rev. Astr. Ap.*, **23**, 379.

Dobson, A.K., Donahue, R.A., Radick, R.R., and Kadlec, K.L. 1990, Cool Stars, Stellar Systems, and the Sun, G. Wallerstein, ed., **6** , 132.

Donahue, R.A. 1991, private communication

Noyes, R.W., Baliunas, S.L., Belserene, E., Duncan, D.K., Horne, J., and Widrow, L. 1984, *Ap. J. Letters*, **285**, L23.

Seventh Cambridge Workshop on Cool Stars, Stellar Systems, and the Sun
ASP Conference Series, Vol. 26, 1992
Mark S. Giampapa and Jay A. Bookbinder (eds.)

A MN I EMISSION LINE AT 4032 Å DURING A FLARE ON THE RS CVN STAR II Peg

J.G. DOYLE, P.M. PANAGI, C.J. BUTLER Armagh Observatory, Armagh BT61 9DG, N.Ireland

ABSTRACT Two models are discussed in an effort to explain the appearance of an emission line at 4032Å during a large flare on the RS CVn star II Peg.

Keywords: RS CVn star; emission line; fluorescence; photoionization

INTRODUCTION

In a multi-site observing campaign in August 1989 to determine the flare activity rate of the RS CVn star II Peg, data were acquired at several different frequencies. Here, we discuss one aspect of the optical spectroscopy from the Isaac Newton telescope (INT) on La palma. During a four night observing run, observations were taken alternately in the blue (covering Ca II H&K, Hγ and Hδ) for 60 sec. and in the red (covering Hα) for 30 sec. resulting in a complete cycle every four minutes. A large flare was observed on 17 August showed a blue continuum enhancement 80% over the pre-flare level, while the enhancement of the continuum in the red was approximately 10%. The rise of the flare was very gradual, taking approximately 30 min. to reach maximum. The total duration of the flare was 4 hrs., radiating $\sim 2\,10^{36}$ ergs. A flare-only spectrum in the blue (*i.e.* with the background subtracted) is shown in the figure below.

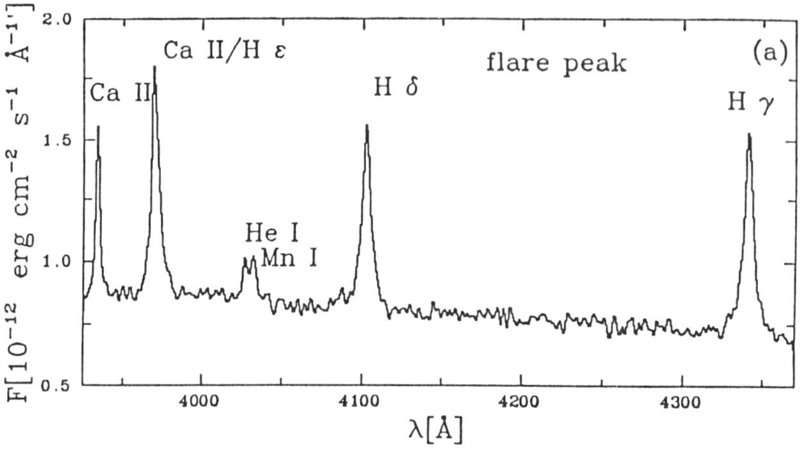

Fig. 1: The flare-only spectrum (*i.e.* a mean quiescent spectrum has been subtracted) for II Peg in the blue at around the peak of the 17 August flare

DATA REDUCTION & RESULTS

Besides the strong lines such as Ca II H & K, Hδ and Hγ that are normally observed during flares on dMe stars, two other emission lines are present. The first of these is He I 4026Å, which has been observed previously during the impulsive phase of flares on dMe stars. However the second line at 4032Å has not been observed before and the only identification possible is Mn I, a photospheric line. In fact, there are three Mn I lines which form a deep absorption feature at 4032Å in the quiescent phase, and also during flares except that during flares an emission core appears.

DISCUSSION

We are currently investigating two possible mechanisms that may give rise to the filling in of the Mn I resonance lines.

Fluorescence through photo-excitation by the strong Mg II k radiation field. Examples of such fluorescent lines are the O I 1304Å triplet, pumped by Lyβ and Cl I 1351Å pumped by C II 1335Å (see Shine 1983). Given the extreme complexity of the Mn atom, we have initially restricted our model to four bound levels and continuum, as described below. Level 4 is photo-excited by radiation from the blue wing of Mg II k line; electrons are collisional redistributed into level 3 (forbidden transition), with the emission arising from decay of level 3 to the ground state. If this mechanism proves successful, we may get a handle on the local electron and/or proton densities, depending which process dominates the collisional redistribution.

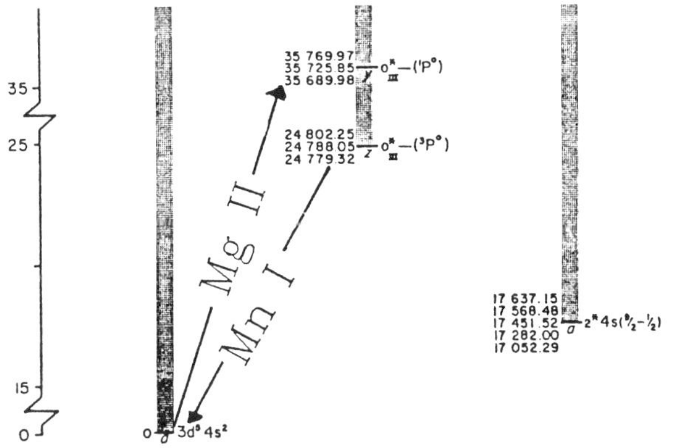

Fig. 2: Term diagrams for Mn I and Mg II

We construct an ad hoc non-LTE one-component atmosphere for an active region, by initially fitting the H_α equivalent width ($\sim$ 1.0Å) near the start of the flare. This gives a first estimate of the local electron density in the middle/upper chromosphere. The lower chromospheric structure is then adjusted to fit the Mg II flux. The thermal structure is shown in Fig. 3. We have not tried to fit the observed profiles as we have not included any velocity fields in out computations. The derived model for the thermal and electron density structures, with the

Mg II radiation field are then used to solve for the Mn I/Mn II/Mn III ionization balance, and emergent profiles of interest. Where possible, we include both electron and proton collision processes. Work is under way to include collisions with neutral Hydrogen, as these are expected to push the level populations nearer their LTE values. Given that the flare radiation field is damped very quickly, the importance of collisional processes is enhanced, given the expected high densities. Computations were performed using the efficient non-LTE code of Carlsson (1986).

The second mechanism being investigated is photoionization of inner shell Mn I, followed by radiative decay to the low lying states. This also seems a promising mechanism given the high cross-section of ionization of Mn I by X–rays ($\sim 10^{17} cm^2$) in this region. Our observations show the Mn I emission to closely follow the rise and decay of the soft X–ray flux (1-20keV). As a physical constraint, we will be using the analytical expression for soft X–ray deposition in the solar chromosphere as given by Henoux and Nakagawa (1977) to determine the flux and depth at which the soft X–rays penetrate.

Acknowledgments We would like to thank the La Palma INT staff for their help in obtaining this data. Research at Armagh Observatory is grant-aided by the Dept. of Education for N. Ireland. We also acknowledge the support provided in terms of both software and hardware by the STARLINK Project which is funded by the UK SERC.

REFERENCES

Carlsson, M., 1986, Uppsala Report, **33**.
Henoux, J-C., and Nakagawa, Y., 1977, *Astr. Ap.*, **57**, 105.
Shine, R.A., 1983, *Ap. J.*, **266**, 882.

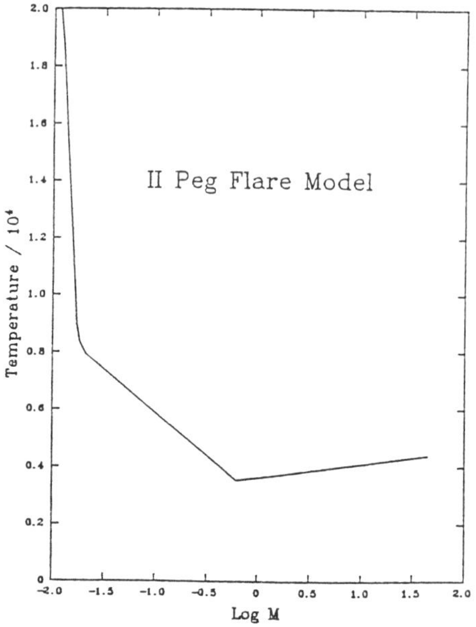

Fig. 3: The II Peg flare model

Seventh Cambridge Workshop on Cool Stars, Stellar Systems, and the Sun
ASP Conference Series, Vol. 26, 1992
Mark S. Giampapa and Jay A. Bookbinder (eds.)

OPTICAL FLARES ON II PEG

M. MATHIOUDAKIS, J.G. DOYLE
Armagh Observatory, Armagh BT61 9DG, N.Ireland

S. AVGOLOUPIS, L.N. MAVRIDIS, J.H. SEIRADAKIS
University of Thessaloniki, GR-54006, Thessaloniki, Greece

ABSTRACT During 57.4^h hours of monitoring rime in the Johnson U- and B- bands of the RS CVn binary II Peg, ten flares were detected. This implies a flare activity of 0.17 flares per hour. The absolute energies of these events are in the range $1\text{-}180\times 10^{33}$ erg. It had been generally assumed that optical flares on RS CVn stars, are an extremely rare phenomenon. The observations we report here, tend to dispel that idea.

Keywords: photometry; RS CVn stars; optical flares

INTRODUCTION

Chromospheric activity in late type stars has been established by the presence of emission in the core of Ca II H&K and the Balmer lines. The RS CVn stars are a group of active binaries in which at least one component shows evidence of spots and other phenomena associated with stellar activity. II Peg is a single line spectroscopic binary of spectral type K2-3V-IV and a period of approximately 6.72 days. Despite several attempts, its companion has yet to be observed. Based on IUE observations, II Peg appears to have a high rate of flare activity (Doyle *et al.* 1989). In the present study, we present observations of U band flares on II Peg.

OBSERVATIONS AND DATA REDUCTION

The photoelectric observations of II Peg were carried out in August 1989 and July-September 1990 at Stephanion Observatory in Peloponessos Greece, using the 30-inch Cassegrain reflector of the Department of Geodetic Astronomy of the University of Thessaloniki. The telescope, photometric system and the observing procedure has been described by Mavridis *et al.* (1982). Our observations on II Peg consist of (a) continuous monitoring in the U-band with an integration time between 2.4 and 6 sec (b) three color photoelectric observations in the international U,B,V system. Here we present the results for the continuous monitoring. During the monitoring, interruptions of 30-90 sec were performed to record sky background, while 8-13 min interruptions were required to observe nearby comparison stars. Part of our results, were taken during an international campaign and have been presented in previous works (Doyle *et al.* 1991).

RESULTS & DISCUSSION

During 57.4^h of monitoring time ten flares were observed. In fig. 1 we present the light-curve for one of the flares. Additional light-curves can be found in Doyle *et al.* (1991) and Avgoloupis *et al.* (1991). A very important parameter which indicates the level of flare activity, is the time-averaged flare luminosity (L^*) observed in a particular band. This is defined as the sum of the energies of individual flares, divided by the total monitoring time. Using the present data, we estimate a value of $1.9\,10^{30}\,erg\,s^{-1}$ for L^*_U. This value should be good to within a factor of 2.

Gershberg (1972) and Lacy *et al.* (1976) have suggested that a better estimate of the flare activity of a star can be obtained from the cumulative flare-frequency distribution diagram, *i.e.* the log ν of the frequency of flares of energy E_U or greater, against the log E_U. In fig. 2 we present such a diagram for the U band II Peg flares. Comparing the distribution of the II Peg flares with those of the classical flare stars given by Lacy *et al.* (1976), we find that for most flare stars a single power law is required to fit the part of the distribution were the S/N ratio is best. However, for II Peg a double power law distribution is required, one for the high energy events and the other for the low energy flares, with a gap of more than one order of magnitude separating the two curves. A similar two-component diagram has been presented for the very active dMe spectroscopic eclipsing binary star YY Gem, where it had been suggested that the most energetic flares could represent flares due to magnetic interaction between the two stars (Doyle and Mathioudakis 1990). Although the statistics in our study are still rather poor and obviously further monitoring is required, it is interesting to note that the two most energetic flares were observed very close to optical quadrature (phases 0.26 and 0.82, the ephemeris being that of Vogt 1981). A double power-law distribution has also been published by Mavridis and Avgoloupis (1987) for the dMe flare star EV Lac, although in this case the authors suggested the cause to be due to a 5 year periodicity in the star's flare activity rate.

REFERENCES

Avgoloupis, S., Doyle, J.G., Mavridis, M., Seiradakis, J.H., Mathioudakis, M., 1991 in Surface Inhomogeneities in Late-type stars, eds. P.B. Byrne & D.J. Mullan; Springer-Verlag, (in press).

Doyle, J.G., Byrne, P.B., van den Oord, G.H.J., 1989, *Astr. Ap.*, **224**, 153.

Doyle, J.G., Mathioudakis, M. 1990, *Astr. Ap.*, **227**, 130.

Doyle, J.G., Kellett, B.J., Byrne, P.B., Avgoloupis, S., Mavridis, M., Seiradakis, J.H., Bromage, G.E., Tsuru, T., Makishima, K., McHardy, I.M., 1991, *M.N.R.A.S.*, **248**, 503.

Gershberg, R.E., 1972, Astrophys. Space Science **19**, 75.

Lacy, C.H., Moffett, T.J., Evans, D.S., 1976 *Ap. J.*, **30**, 85.

Mavridis, L.N., Asteriadis, G., Mahmoud, F.M.: 1982, in Compendium in Astronomy, eds. E.G. Mariolopoulos, P.S. Theocaris, L.N. Mavridis, Reidel, Dordrecht, p.253.

Mavridis, L.N., Avgoloupis, S., 1987, A&A **188**,95

Vogt, S.S., 1981, *Ap. J.*, **247**, 975.

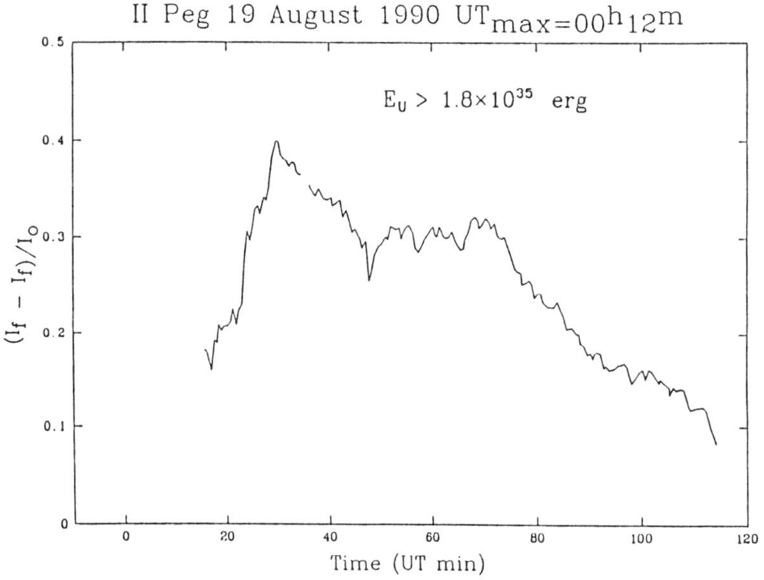

Fig. 1: Light-curve of a flare observed in the U-band

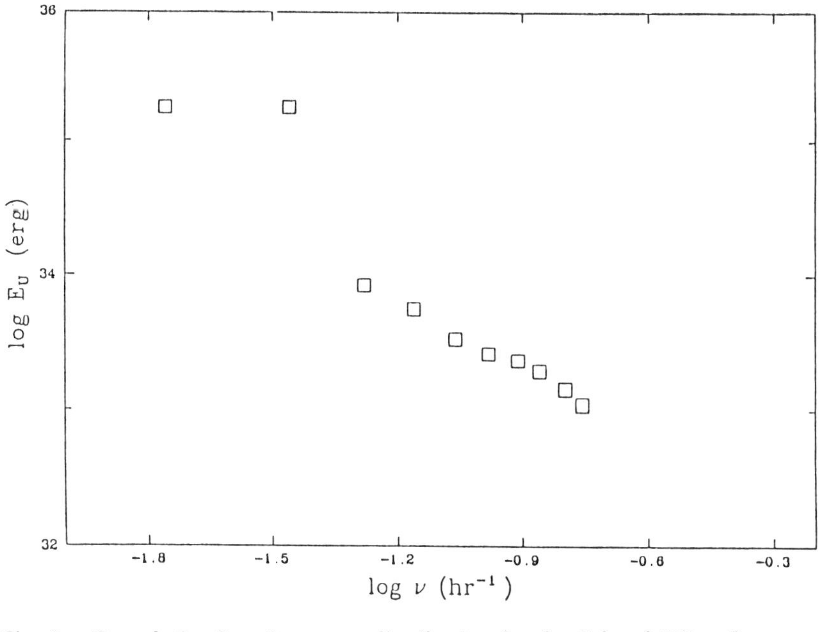

Fig. 2: Cumulative flare frequency distribution for the U-band II Peg flares

Seventh Cambridge Workshop on Cool Stars, Stellar Systems, and the Sun
ASP Conference Series, Vol. 26, 1992
Mark S. Giampapa and Jay A. Bookbinder (eds.)

TRANSIENT OPTICAL PERIODICITIES IN RED DWARFS: LOOP OSCILLATIONS IN THE CORONA

D. J. MULLAN Bartol Research Institute, University of Delaware, Newark DE 19716, USA

R. B. HERR and S. BHATTACHARYYA Dept of Physics and Astronomy, University of Delaware, Newark DE 19716, USA

ABSTRACT We have detected transient low-amplitude optical periodicities in M dwarfs with high L_X. An oscillating coronal loop model leads to N_e, T, B, and loop lengths consistent with previous X-ray modelling.

Keywords: flare stars; coronae; optical variability; magnetic loops

INTRODUCTION

Following reports by Andrews (1989) of microvariability in M dwarfs, three stars with high L_X were observed at Mount Cuba Observatory in a search for periodicity that may be related to coronal heating.

OBSERVATIONS

Unfiltered optical light is sampled by A-D conversion at 5 sec^{-1} and later binned into intervals, up to 2 sec. Applying Andrews' modified autocorrelation (MAC) and Fourier transforming (FT) the MAC, we search for periods from 10 sec to 10 min. Statistical significance of peaks in the FT is found by randomizing the data 100-1000 times and applying identical analysis: we construct a distribution function of the maximum FT peak amplitudes which is close to exponential. We estimate amplitudes of the variations by adding sine waves to a randomized data set otherwise devoid of significant frequencies. Amplitudes of 0.3% lasting throughout a 30-min observation, are detected with high confidence (P_c>0.99).

RESULTS

45% of the program star runs showed periods with $P_c > 0.95$. Only 8% of the comparison star runs had P_c>0.95.

AD LeoNIS (dM4e) was observed a total of 302 min in 8 nights. The most significant period was 4.1 min (March 6), with 1-$P_c \approx 10^{-5}$ and amplitude 0.4%. However, we cannot ignore the fact that herringbone clouds crossing the field could induce such fluctuations. A second run on AD Leo, begun 1.5 hours later, showed no significant periods, nor did a third run 1.5 hours after that. Comparison stars observed between the AD Leo sets showed nothing significant. If the first run contains stellar periodicities, their absence in the later runs suggests that the oscillatory behavior lasted <20 cycles. Other periods in AD Leo are 6.1 min (Apr. 3) ($P_c = 0.997$) and 6.0 min (May 29) ($P_c = 0.95$). A minor flare occurred during a U-filter observation on Apr 4. Careful searches in 10-min intervals before and after the flare revealed no likely periods between 10 sec and 5 min.

GLIESE 549 (dM3.5), the target with largest L_X, showed the most statistically significant variability. Monitoring time totaled 195 min, with 70 min on a comparison star. Periods with P_c>0.99 were found in four runs: 6.0 min (P_c=0.998; Apr 4), 3.4 min (P_c=0.996; May 16), 3.8 min (P_c>0.999; May 21), and 3.0 min (P_c=0.984; May 27). A. D. Andrews (priv. comm.) independently reports a period between 2.5 and 3.0 minutes from Catania on May 28. On May 27, the 3.0-min period was significant only in the second half of our 28-min data set. And the 3.0-min period was insignificant again in a 13-min run begun 8 minutes later (terminated by twilight). These data suggest the periodicity lasted no more than 5 cycles. Comparison star data showed no periods.

DT VIRGINIS (dM2e) was observed twice on May 21 (total time 56 min, plus 56 min on comparison star). A 12.6-sec period was found in the first 28-min run, but P_c was rather small (0.97). The second 28-min run had no significant periods, nor did the intervening comparison star run.

DISCUSSION

The amplitudes of our oscillations are too large to be p-modes. We propose that coronal loops may be responsible: our targets were selected on the basis of strong coronal X-ray emission. Moreover, the transient nature of our periodicities is reminiscent of the Q-factors expected of coronal resonant cavities: coronal Q-factors are predicted to be no more than about 10 (Hollweg, 1984). MHD crossing periods in coronal loops, based on Haisch's (1983) analysis of flaring loops on M dwarfs, range from <10 sec to several minutes.

We envision a loop where a disturbance of some kind (*e.g.* interaction with newly emerging flux, or disturbances from a flare elsewhere on the star) causes the loop to change its emission measure (EM), either by changing the volume of the loop, or by changing the density of material in the loop. (For an alternate "bouncing" loop model applied to radio emission, see Gibson, 1986).

Zaitsev and Stepanov (1989: hereafter ZS) have proposed a loop model to interpret transient X-ray periodicities in the Sun where the flux increases and then oscillates n_o times at period τ and amplitude $\Delta I/I$ before decaying (Fig. 1). ZS identify τ with L/v_A, and n_o with the Q-factor of the loop, where $Q \approx 2\pi M_i/(\tau \nu_{ei} m_e)$ due to $e-i$ collisions. ZS suggest $\Delta I/I \approx 4\pi NT/B^2$. Finally, we suggest that our observed optical variations originate in free-free emission, leading to $10^{-27} N_e{}^2 L(f_V/f_A)T^{-0.5} \approx 10^3$, where $f_V \approx 0.01$ and $f_A \approx 0.1$ are volume and areal filling factors. Thus $N_e{}^2 LT^{-0.5} \approx 10^{31}$ In order for us to detect optical variations against the photosphere, we need $\Delta I/I \approx 1$.

Combining the above formulas, and using observed τ ($\approx$ 200 sec), we find $N_e \approx 10^{12} cm^{-3}$. Then, $L(cm) \approx 2 \times 10^7 N_{e,12}{}^{1/3} Q^{1/3} \tau^{4/3} (\Delta I/I)^{-0.5}$. With $Q \approx 10$, L is a few times 10^{10} cm *i.e.*, comparable to the stellar radius. Further we find $T(K) \approx 3 \times 10^7 L_{10}^2 \Delta I/I$ *i.e.* a few times 10^7 K, consistent with the hot component in red dwarf coronae (Schmitt, 1991). Finally, inserting values of Q, τ, and L_{10} as quoted above, we find $B(G) \approx 30 L_{10}$. Thus, the oscillating loops have fields of order 100 G. These estimates of N_e, T, L, and B are comparable to those derived by Haisch (1983) in his analysis of X-ray flare curves.

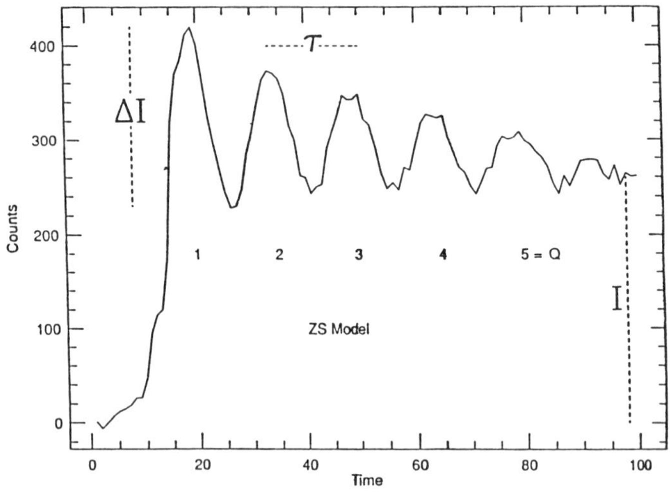

Fig. 1: Observed parameters for oscillating loop model

CONCLUSIONS

An optical observing program at Mt. Cuba has revealed significant periodicities in two red dwarfs with high X-ray luminosity. The periodicities are of low amplitude, and are transient. An oscillating coronal loop model leads to an interpretation of our data: we obtain values of loop length, temperature, density, and magnetic field strength which are comparable to previous estimates.

Significantly, those previous estimates were possible only if X-ray data were available during flare decay. Our results suggest that coronal loop parameters may be obtainable from optical data in the absence of flares.

ACKNOWLEDGEMENTS

The authors would like to acknowledge support from NASA.

REFERENCES

Andrews, A. D. 1989, *Inf. Bull. Var. Stars*, No. 3303

Gibson, D. M. 1986, in *Solar Terrestrial Physics*, eds. M. Kundu *et al.*, CSIR, New Delhi, p. 46.

Haisch, B. M. 1983, in *Activity in Red Dwarf Stars*, eds. P. Byrne and M. Rodono (Dordrecht: Reidel), 261

Hollweg, J. V. 1984, *Ap. J.*, **277**, 392

Schmitt, J. 1991, presentation at Amer. Phys. Soc. meeting, Washington DC, April 22

Zaitsev, V. V., & Stepanov, A. V. 1989, *Pis'ma Astron. Zh.*, **15**, 154

Zaitsev, V. V. 1990, presentation at US-USSR Workshop on Solar-Terrestrial Interactions, Dagomys, USSR, October 16

Seventh Cambridge Workshop on Cool Stars, Stellar Systems, and the Sun
ASP Conference Series, Vol. 26, 1992
Mark S. Giampapa and Jay A. Bookbinder (eds.)

VLBI OF SINGLE STARS: CHARACTERISTICS OF THEIR MAGNETICALLY DOMINATED RADIO EMISSION

ROBERT B. PHILLIPS MIT Haystack Observatory, Route 40, Westford, MA 01886 USA

ABSTRACT Advances in radio Very Long Baseline Interferometry (VLBI) sensitivity have permitted extension of the technique to include the faint, millijansky-level nonthermal radiation from single cool and hot stars. Since radio flares are no longer a requisite for detection, both quiescent and active states of Weak-lined T Tauri, dMe, FK Com, and magnetic A and B stars have been probed using VLBI in 1988-91. A description of the first 3 years of single star VLBI is given.

Keywords: Radio stars; pre-main sequence stars; flare stars

I. Introduction

If one revisits radio star reviews from 1984-85, the bright prospects for Very Long Baseline Array (VLBA) attacks on stellar nonthermal radio emission are always mentioned. In the intervening years, delayed funding and the stretched-out completion of the VLBA has forced radio partisans to make do with existing telescopes. Many of the advantages of the VLBA previewed in those reviews - frequency agility, wide bandwidths, and flexible scheduling among them - still hold future promise.

Still, persistent efforts have resulted in VLBI observations whose detection thresholds approach those of the Very Large Array (VLA) itself. The Haystack Mark III Correlators have provided an interim wide-bandwidth VLBI capability which puts sub-milliJansky radio stars within reach. As a result, VLBI investigations have made initial forays into the magnetic environments around single stars, where nonthermal radiation is created. As study of solitary stars removes the muddling influences of binarity, this monograph will concentrate on VLBI results from 1988 forward on stars believed to be single.

Stellar *thermal* emission, a key tool for following the winds of hot (O, B) stars or the powerful outflows from Young Stellar Objects (YSOs), will not be addressed. Stellar *nonthermal* radiation has now been observed at centimeter wavelengths from more than a dozen single, late-type stars. It will be reviewed for its potential to tell us something about the magnetic environment around cool stars. Discussion of candidate emission mechanisms will be taken up in the companion talk by R.L. Mutel to follow.

II. Lessons from early-type stars

Nonthermal radio emission is linked, both observationally and in models to explain it, to exceptional magnetic activity from the stars. Drake *et al* (1987)

identified several nonthermal-appearing radio sources through deep VLA surveys of chemically peculiar, magnetic Bp and Ap stars. They were the first to demonstrate single stars can sustain nonthermal sources around them.

The radio sources around these early stars are relatively stable, most taking days or weeks to show variations. their radio luminosities do not exceed $\sim 10^{18}$ ergs s^{-1} Hz^{-1}. The radio emission is generated within the inner portions of their apparently dipolar, kilogauss magnetic fields, believed to also play a role in their abnormal surface abundances. VLBI methods have directly measured the radio components around two Bp main-sequence stars in Orion, σ Ori E and HD 37017 (Phillips and Lestrade 1988). In the ρ Ophiuchi star forming cloud, two B- or A-stars of poorly determined spectral class show similar nonthermal components (source S1 and VSSG14; André *et al* 1991a,b). When measurements have adequate signal-to-noise, these stars show net circular polarization at levels of 6 to 10%. The radio components around the four hot B stars are of the order 6 to 12 $R_{\star}$, but VLBI measurements are not yet of sufficient coverage to differentiate any elongation or symmetry. A scaled-up "radiation belt" model, inspired by Jovian space physics, can replicate the observed emission (Havnes and Goertz, 1984; André *et al* 1991a), and the hot-star winds probably play a role in electron resupply. Still, many Bp/Ap stars with characteristics comparable to σ Ori E or HD 37017 fail to show significant radio emission at all, and a comprehensive picture of B/A star nonthermal radio activity is needed.

III. The 'A-to-F Gap'

From the coolest of the radio-loud A stars through spectral type G0, there are *no* known nonthermal radio stars. The "Gap" represents a drop in radio luminosity of at least 3 orders of magnitude. Its boundaries can be defined by effective temperature, but is probably ultimately related to the stellar interiors. Between the magnetized B stars and spectral type G0, where the convective dynamos appear to abruptly throttle up, the efficiency of converting the internal energy of stars into magnetic and particle activity drops markedly.

IV. The Rich G-to-K Feast

RS CVn binaries are well-known for their *degustation* menu of enhanced magnetic activity. Exaggerated starspots which migrate, chromospheric activity seen in the far UV, X-ray flares, and, finally, compact, nonthermal radio activity showcase the conversion of energy into various magnetically-dominated regimes. It is widely held that deep convective zones (especially in the cooler K components) combine with tidally-boosted rotation to maximize the dynamo effect, giving the RS CVn's an exceptional magnetic character. Gyrosynchrotron models, which place moderately relativistic electrons in local (1 to 3 times the size of the K subgiant, as measured by VLBI) magnetic structures (1 to 100 gauss), have been successful in reproducing most radio properties of RS CVn systems.

Still, studies of single G/K stars provide better tutorials for the native properties of late-type stars than the evolved ones within either detached or close binaries. During the 1980s, it gradually became evident that some single

G and K subgiants, both pre-main sequence and evolved, showed much of the same spotting, chromospheric activity (as seen by the *International Ultraviolet Explorer*, IUE), and X-ray flaring (as seen by *Einstein's* Imaging Proportional Counter, IPC). These active late-type stars resembled one half of a RS CVn binary, and reports of comparable radio activity, as seen using the VLA, completed the picture (*e.g.*, Feigelson and Montmerle 1985; Cohen *et al* 1982).

The current radio knowledge available on single G to K stars derives largely from studies of the Taurus-Auriga and ρ Ophiuchi star-forming regions, and their pre-main sequence (PMS) solar-type stars. Such efforts arose in 1987-88 from cross-pollination in ideas by X-ray, IUE, infrared, VLA, and VLBI investigators. Each region has properties which are assets to the studies. Taurus-Auriga has little obscuration at optical wavelengths; as a consequence, reliable spectral types, luminosities, photometric rotation periods (from spots), and, even, $v \sin i$ values are known for many of its PMS stars. In contrast, the stellar radio sources of ρ Oph, both thermal and nonthermal, are heavily obscured (A_V typically +20 to +50). For some, approximate spectral types can be estimated from black-body fits to the near-IR continua. However, the ρ Oph objects are much more concentrated towards the cloud core, allowing deep VLA surveys of unbiased, luminosity-limited samples not practical for Tau-Aur, which covers several degrees on the sky.

G/K radio structure (and a surprise)

At the distance to Taurus-Auriga or ρ Ophiuchi, 160 pc, 1 mJy corresponds to a radio luminosity of 3×10^{16} ergs s^{-1} Hz^{-1}; 1 milliarcsecond (mas) corresponds to 2.4×10^{12} cm, or about 16 $D_\odot$ (for PMS G/K stars, $D_\star \simeq 3D_\odot$). During VLA investigations, some G/K PMS stars had been seen to flare from the mJy level to several dozen mJy, or 10^{18} ergs s^{-1} Hz^{-1}(DoAr21 and V410 Tau: Feigelson and Montmerle 1985, and Bieging *et al* 1984). By comparison, the contemporary Sun achieves 10^{13} ergs s^{-1} Hz^{-1}during its greatest short-lived centimeter radio flares (Zirin 1988). Such radio luminosities are comparable to those of the most active RS CVn or Algol systems. Deep radio surveys followed, and discovered target sources for later VLBI attacks (André *et al* 1987; Stine *et al* 1988; O'Neal *et al* 1990).

VLBI studies (Feigelson *et al* 1989; Phillips *et al* 1991a; André *et al* 1991) employed large collecting area antennas and current recording bandwidths ($\sim$ 40 MHz) to make preliminary structural measurements of 8 WTT or cool PMS objects in levels of activity ranging from 0.5 to 9 mJy (Table 1). For the best cases, VLBI structural information was gathered which was comparable to the best available for RS CVn systems or Algol in much higher levels of activity.

The Cool PMS stars showed combinations of unresolved ($< 2R_\star$) and resolved (5 to 20 $R_\star$) radio-emitting regions, whose structure sometimes changed on scales as short as 1 hour. Unexpected were the larger radio sizes, not seen from RS CVn systems. Those PMS stars which showed radio components in excess of 10 $R_\star$ cause difficulties for models which generate nonthermal emission only in small magnetic loops close to the stars (0.5 to 1.0 $R_\star$). No preliminary trends or correlations were noted between rotation period, spectral type, or stellar mass and the VLBI structure. Neither did any star show evidence for multi-lobed or complex structure; VLBI closure phases indicated basically symmetric sources. Finally, no star showed evidence for the large circular polarization fractions (up to 30%) displayed by RS CVn systems. One very active WTT star (HD 283447) was also observed in followup VLBI/VLA projects in 1991.

TABLE I
A Summary of the Radio Properties of Nonthermal Single Cool Stars Detected by VLBI

Star	Sp.Type	Flux Density Range (mJy)	Log Luminosity @ 5 GHz (ergs-s^{-1} -Hz^{-1})	Measured Angular Size (mas)	*Linear Size 10^{11} cm	Approx. Brightness Temp (10^8K)
HD 283447	K3	up to 14	17.6	up to 2.5	66	10
DoAr21	K	up to 10	17.5	up to 2	56	10
HDE 283572	G5	1.5 to 3	16.9	up to 1.5	36	0.6
Hubble 4	K7	3 to 4	17.1	up to 2	40	0.7
HP Tau/G2	G1	1 to 3	17.0	<0.6	14	>6
WL 5	Cool PMS	3 to 5	17.2	2 to 5	50 to 120	0.3
ROX 39	K5	1.5	16.6	<2	50	>0.1
ROX 16	Cool PMS	3.0	16.9	<2	50	>0.2
FK Com	G5III	1 to 20	17.4	3	50	7
YZ CMi	dMe	up to 4	14.6	?	?	High
EV Lac	dMe	up to 2	14.3	<2	3	>0.2
*Previous record for late-type system: 2.6 (UX Arietis, "Halo", July 1983, Lestrade.)						

The case of HD 283447

HD 283447 (K3, $V = 10.6$, rot.period 3.4 days) appears to be in a prolonged active radio period, perhaps going back to the early 1980s (O'Neal *et al* 1990; Kutner *et al* 1986). In a survey VLBI in March 1990, HD 283447 showed distinctly different radio character on 3 separate days. First, it brightened from 5 to 7 mJy at 5 GHz (6cm) in 2 observations 65 minutes apart. The increase coincided with the appearance of an unresolved radio component which joined a larger radio "halo" present in both scans. Lestrade (1988, references therein) has described similar behavior, which he dubs "core-halo", in RS CVn systems in a rising radio state.

HD 283447 was not observed again until 3 days later (while the VLBI Network changed wavelengths). Then, the 1.7 GHz (18cm) flux density was stable at 6 mJy in two measurements 70 minutes apart, during which HD 283447 also showed stable structure. It was detected on all 10 VLBI baselines, with coverage in a broad range of position angle and resolution. Both scans on this day can be adequately fit with a single, circularly symmetric Gaussian radio component of 12 to 15 $R_\star$. The equivalent brightness temperature (T_b) is 7×10^8 K, and the two VLBI measurements set a limit (2σ) on expansion or contraction of $\sim 1600\ km\ s^{-1}$.

On the third day, HD 283447 faded from 3 to 1.5 mJy in 90 minutes. This time, both VLBI scans indicated an unresolved ($\leq 5R_\star$) source at this lower level of activity.

In June 1991, a 3-hour VLBI at the much higher frequency of 8.4 GHz (3.6cm) caught HD 283447 at the high, sustained level of 12 mJy (Deeney *et al.* 1991). While the VLBI results indicate that the radio size remained roughly constant at about 4 $R_\star$, the longer VLA integration collected in parallel with the VLBI observations permitted a sensitive search for the hitherto missing circular polarization. HD 283447 showed no circular radio polarization to a 5 σ limit of 1.1%. The limit is hard to understand, since HD 283447's slowly-varying radio structure, when seen from a RS CVn, corresponds to 10-30% polarization fraction. Most recently, VLA data taken during a September 1991 VLBI observation of HD 283447 shows 3% left-circularly polarized radiation at 5 GHz (6cm; 22 σ result) when HD 283447 was at the elevated level of 14 mJy. S.M. White has revealed that in February 1990, HD 283447 had shown a similar polarization fraction, 4%, while rising from 2.5 to 15.7 mJy (also 6cm; 4 σ result, at this lower average level: White, Pallavicini, and Kundu 1991b). their prior detection and the September 1991 value are additional unambiguous indicators that HD 283447's centimeter radiation comes from regions of strong, organized magnetic fields; moreover, the *sense* (left-circular) was seen on both occasions.

The September 1991 VLBI/VLA observation should allow direct measurement of spatial structure over its 5-hour coverage, and monitoring of radio evolution on scales as small as 1.5 $R_\star$ for comparison with models for the emission. (As of this writing, the VLBI recordings have not arrived at the processing site.)

What HD 283447 seems to tell us

Again deferring to the following speaker for physical models and their merits, I can nonetheless summarize what the young K star HD 283447 has so far told us. (1) It has yet to exhibit extremely short-lived, highly-circularly-polarized, coherent-appearing events as can Algol, but sustains moderate levels of incoherent processes for up to several hours. (2) It is up to 10 times larger in the radio than are the largest radio components seen from RS CVn systems. (3) On the other hand, its timescale for variability *and* its intrinsic luminosity are a match for any RS CVn. (4) (Two) and (three) indicate the WTT star, HD 283447, sometimes achieves such luminosities by activating a significantly *larger* emitting volume at somewhat lower T_b (under 10^9 K vs. over 10^{10} K for some RS CVn sources). (5) Low circular polarization, perhaps an order of magnitude less than that from RS CVn binaries, characterizes the Cool PMS stars. If we were innocent of the VLBI structural information, low polarization could be waved away by speculating that these unpolarized sources are the result of multiple, nondipolar coronal loop structures combining to sum to zero net polarization - the "Medusa" hypothesis. Instead, the VLBI data strongly argue that HD 283447 at times sustains nonthermal processes out at radii (10 to 20 $R_\star$) where the $\vec{B}$-field is surely beginning to take on a basically dipolar form. These "distended" radio states also take shape far above the small loop structures believed to be responsible for X-ray flares (*e.g.*, Montmerle *et al* 1983), which are expected to be $\sim 0.5\ R_\star$. (6) *Unresolved* radio components, which can apparently coexist with the distended ones, are inviting targets for future joint UV/X-ray/radio studies to investigate synergy between the chromosphere (UV), lower corona (X–ray), and intermediate corona (radio). It is this compact radio activity which may be a link with the M dwarves as well as centimeter activity seen on the Sun.

HD 283447 still shows a slight IR excess and measurable H_α equivalent width ($\sim 4\AA$). It may be an object changing from moderate- to weak-emission T Tauri. It is reasonable to speculate the magnetic/nonthermal activity is the byproduct of the stellar $\vec{B}$-field, resurgent as the convective dynamo gains potency, as it interacts with the protostellar cloud and further uncloaks the star.

It is worth contrasting HD 283447 and its consistently detected compact radio emission with the similar star V410 Tau. Also a spotted K3 WTT, and an even faster rotator at period 1.9 days, it was the first WTT star reported to vary in the centimeter radio range (Cohen *et al* 1982). Ten years ago, V410 Tau was frequently seen at 10 mJy or more (*e.g.*, Bieging *et al* 1984). More recent studies (Bieging and Cohen 1989) have found it at 2 mJy or below, and during the 1990 VLBI survey it was undetected on any of the 3 days in VLA measurements, placing it below 0.3 mJy (under 10^{16} ergs s^{-1} Hz^{-1}). Is it possible WTT stars show cycles of radio, and, by implication, magnetic activity as cool main sequence stars (and the Sun) are now known to do ?

An abrupt (temporal) gear change - HD 82558 and similar stars

An ultimate motivation in cool radio-star VLBI is a desire to sketch parallels with the Sun. Since solar-age stars are radio-quiet in the centimeter range (by 5 to 8 orders of magnitude - *e.g.*, the Sun), nearby, intermediate age cool stars are the logical next targets of VLBI/VLA nonthermal inquiries.

Puzzling cases are single Pleiades-age or younger stars which show extremely rapid rotation (less than 2 days) as deduced from photometry of spots, and matching chromospheric activity as seen by IUE on occasion (Ambruster and Fekel 1990, and Ambruster 1991). HD 82558 (K1V, age 7×10^7 y) rotates impressively every 1.7 days, and, at only 20 pc away, could be detected at absolute radio luminosities of 10^{14} ergs s^{-1} Hz^{-1}.

A 1990 VLA observation of HD 82558 showed no emission at either 20 or 6 cm to roughly that limit (Lawrence 1990). Drake and Caillault (1991) detected HD 82558 ($4\,\sigma$) at 5×10^{13} ergs s^{-1} Hz^{-1}, also using the VLA. An earlier report by Slee *et al* (1987) indicates HD 82558 achieved 13 mJy, or 2×10^{15} ergs s^{-1} Hz^{-1}, during a monitoring program. Since the detection was obtained using a single large antenna, some small chance remains that the 13 mJy was interference or confusion.

Other studies of nearby G and K main sequence stars show a dearth of radio detections. Drake *et al* (1991) found no detectable emission from 22 nearby cool stars, even though most had rotation and chromospheric properties comparable to RS CVn systems. The upper limits on their sample stars were 10^{13} to 10^{14} ergs s^{-1} Hz^{-1}. White, Pallavicini, and Kundu (1991a) find a similar absence of radio activity in a sample of young wide visual binaries, and estimate 20×10^6 y as a cutoff age for radio activity from these systems.

These early results indicate that a sharp drop in nonthermal activity takes place between WTT status ($\sim 10^{6-7}$ y) and the age of HD 82558 (7×10^7 y), even in the presence of magnetic indicators such as spots and rapid rotation. Very sensitive radio and X–ray observations will be called for to characterize the conversion of interior and chromospheric energy into nonthermal channels. Even refined upper limits should be of use in constraining the role of stellar aging in how convective energy finds its way into photospheric and chromospheric features, but not into larger scale (a few $R_\star$) zones of magnetic activity.

FK Comae Berenices - a G star in the opposite direction

FK Com is an apparently evolved star, or system, which today has taken on the identity of G0-G5 III. Different scenarios, from evolved Algol system to coalesced, extremely spun-up W UMa contact binary, have been advanced. Even the distance to FK Com is fairly uncertain, since few assumptions about it are verifiable.

Whatever its evolutionary path, FK Com mimics in many ways the active WTT PMS stars. In chromospheric activity, it resembles moderate-emission T Tauri stars. An X-ray luminosity that can reach 10^{31} ergs s^{-1} places it with the RS CVn's or WTT stars. In spotting and radio luminosity, it compares favorably with HD 283447: 2.4 day rotation period, and a (lurid) history of radio output which swings from $10^{15.5}$ to 10^{17} ergs s^{-1} Hz^{-1}(all luminosities computed for 250 pc, an adopted distance if it is a G III). VLBI (Phillips and Lestrade 1990, 1991) has confirmed a similar nonthermal, but large radio component around FK Com (12 $R_\star$, $T_b \geq 10^8$ K). Age - 10^9 vs. 10^7 y - is evidently not the fundamental indicator for nonthermal activity. Clearly, if the details of what the convective zones of FK Com and the radio WTT stars have in common were known, we would have an explicit recipe for their impressive magnetic phenomena.

V. The Challenge of M Dwarfs

Radio-loud M stars of luminosity class V, numbering among the dMe flare stars, are by far the most challenging for radio interferometrists. Their rapid demonstrations in the time domain contrast with the manageable behavior of the G and K stars. In fact, the rise and decay time for their centimeter emission can be shorter than the usual accumulation periods for the collection and fringe-fitting of VLBI systems, and great caution must be applied.

The typical pattern of a dMe radio star is to remain below cm-wavelength detection much of the time, appear for a time (usually in the tens of minutes), then fade below the detection threshold again. Just how *far* the radio dMe stars fade between detections is poorly known. These stars can also show much more rapid radio flares, with millisecond rise times, which are certainly a different mechanism than the "slow" events described above. In addition, dMe radio flares are generally the most circularly polarized of all - to the degree that some flares are only detected in one polarization. The phenomena above are described by Gary and Linsky (1981), Linsky and Gary (1983), Kundu and Shevgaonakar (1985), Lang and Willson (1986, for an extremely rapid event), Kundu *et al* (1987), and White *et al* (1989).

It is still debated whether the dMe stars, which never seem to stand still for radio instruments, have a real "quiescent" state, or merely a power spectrum of outbursts. However, the term quiescent might be defined as that emission whose characteristic time scale is comparable to the integration times for VLBI, or, several minutes. Early results of VLBI on dMe stars now follow.

In 1988, Hewitt *et al.* detected YZ CMi, EV Lac, and the dMe binary Wolf 630AB in quiescent states during an 8.4 GHz (3.6cm) VLBI astrometry program intended for ultimately measuring radio proper motion and parallax. At the typical distances to these stars, 10 pc, 1 mJy corresponds to 1×10^{14} ergs s^{-1} Hz^{-1}; the VLBI detections were made at levels from 1.8 to 5.5 mJy. Since the observing strategy often had the European sites observing different dMe stars than the two in the western U.S., all but one of the VLBI detections are relatively short baseline lengths. Hence, most of the VLBI results set only upper limits on quiescent emission size (because the radio emission appeared unresolved). For YZ CMi, Wolf 630AB, and EV Lac, upper limits of 2 to 3 mas were established for various quiescent detections. If one adopts $D_\star = 0.5\ D_\odot$, those coarse limits correspond to less than 4 to 6 $D_\star$.

In a separate detection of YZ CMi (7 hours after the first), European and U.S. antennas observed simultaneously. The shorter baselines did not appear to resolve the emission (within calibration accuracy), although securely detecting it. However, the transatlantic baseline Madrid-VLA, which had adequate sensitivity to detect the 4.2 mJy star (7 σ threshold was 2.3 mJy), failed to. Since the short-baseline detections validate that the antennas were functioning correctly, the simplest interpretation is that the 4.2 mJy quiescent component was resolved on the 7900 km baseline: or, its size fell between 0.6 $D_\star$ and 4 $D_\star$.

So, it appears that at least some of the dMe star emission occurs in structures large enough to someday be fruitfully studied via VLBI, especially since their proximity allows very high linear resolution. Future VLBI investigators must continue to exert the greatest caution to not confuse temporal variations with structural ones.

Benz *et al* (1991) are mounting a very sensitive *astronomical* dMe star project aimed specifically at the same stars, and have collected two epochs of VLBI observations at 1.7 GHz (18cm) and 5 GHz (6cm). In the case of YZ CMi, they detected it in a quiescent level, and their failure to resolve it sets a more stringent limit on T_b of 10^{10} K or more.

Because the dMe stars so often show polarized radio activity and their nearness translates to high linear resolutions from VLBI, they will be attractive candidates for VLBI polarization synthesis mapping.

VI. The next ten years

A closing section might begin with warnings to those who observe single radio stars with VLBI. For instance, is your spatial information ambiguous? Do you have enough redundancy and closure relations to tell pointing errors from structural variations? Are you absolutely sure you understand obscure effects in your VLBI correlations? Lastly, if you rely on constrained or global fringe-finding techniques for more than 15% of your baselines, get a Bigger Hammer (*i.e.*, re-design your array).

I predict that the next 5 years will see:

(1) Dynamic measurement of an expansion velocity for a plasma cloud in the corona of a cool, single star.

(2) Evolution of thinking and emission models to include particle acceleration at many $R_\star$, rather than only events in the chromosphere.

(3) Determination, through phase connection, of an offset between LCP and RCP sites of emission from WTT and dMe stars while active.

(4) VLBA observations that will jump frequencies on 1/2 hour timescales, and measure radio component spectra.

(5) Realization of simultaneous coverage all the way from X-ray to radio, in coordinated ground/space-based observations of WTT and dMe stars.

And the next 10:

(6) Successful tracking (astrometrically by VLBI) of active regions, attached at their magnetic feet to the star, through one or more rotation periods; also, similar tracking of an outward-moving disturbance in real time.

(7) Meaningful limits on the mass and existence of unseen companions around low-mass dMe stars, from VLBI astrometry.

(8) The first VLBI polarization synthesis maps of cool stars.

(9) Rapid progress in all areas of stellar VLBI when the Green Bank Telescope (*GBT*) sits in from time to time with the VLBA and phased VLA.

(10) Compilation of VLBI parallaxes and proper motions for 2 dozen PMS stars in the popular star-forming regions, refining knowledge of their distance and internal motions.

It was only 1988 when the first VLBI detections of single stars were made (Bp stars), and in rapid succession came dMe stars (1988), cool PMS stars (1989), and FK Com (1989). The challenge to VLBI workers will be to advance techniques and observing strategies beyond the demonstration stage, so that their results will test - and even add pieces to - models for replicating the sometimes ephemeral radio noise from stars.

ACKNOWLEDGEMENTS

The author would like to acknowledge support from the National Science Foundation. Research and Mark III VLBI Correlator operations at Haystack Observatory are supported by the NSF.

REFERENCES

Ambruster, C. and Fekel, F. 1990, *Bull. A.A.S.*, **22**, 857.
Ambruster, C. 1991, in preparation.
André, Ph., Montmerle, T., and Feigelson, E.D. 1987, *A.J.* **93**, 1182.
André, Ph., Lestrade, J.-F., Phillips, R.B., and Klein, K.L. 1991, *Ap. J.*, **360**, 121.

André, Ph., Deeney, B.D., Phillips, R.B., and Lestrade, J.-F. 1991, in preparation.
Benz, A.O., Guedel, M., and Alef, W. 1991, *Astr. Ap.*, , in press.
Bieging, J.H., Cohen, M., and Schwartz, P.R. 1984, *Ap. J.*, **282**, 699.
Bieging, J.H., and Cohen, M. 1989, *A.J.* **98**, 1686.
Deeney, B.D., Lonsdale, C.J., Phillips, R.B., and Feigelson, E.D. 1991, in preparation.
Drake, S.A., Abbott, D.C., Bastian, T.S., Bieging, J.H., Churchwell, E., Dulk, G., and Linsky, J.L. 1987, *Ap. J.*, **322**, 902.
Drake, S.A., Linsky, J.L., Judge, P.J., and Elitzur, M. 1991, *A.J.* **101**, 230.
Feigelson, E.D., and Montmerle, T. 1985, *Ap. J. Letters*, **289**, L19.
Feigelson, E.D., Lonsdale, C.J., and Phillips, R.B. 1989, *Bull. A.A.S.*, **21**, 1003.
Gary, D.E., and Linsky, J.L. 1981, *Ap. J.*, **250**, 284.
Havnes, O., and Goertz, C.K. 1984, *Astr. Ap.*, **138**, 421.
Hewitt, J.N., Cappallo, R.J., Corey, B.E., and Lonsdale, C.J. 1991, in preparation.
Kundu, M.R., and Shevgaonakar, R.K. 1985, *Ap. J.*, **297**, 644.
Kundu, M.R., Jackson, P.D., White, S.M., and Melozzi, M. 1987, *Ap. J.*, **312**, 822.
Kutner, M.L., Rydgren, A.E., and Vrba, F.J. 1986, *A.J.* **92**, 895.
Lang, K.R., and Willson, R.F. 1986, *Ap. J. Letters*, **302**, L17.
Lawrence, S.L. 1990, private communication.
Linsky, J.L., and Gary, D.E. 1983, *Ap. J.*, **274**, 776.
Lestrade, J.-F. 1988, in The Impact of VLBI on Astrophysics and Geophysics, eds. M.J. Reid and J.M. Moran (Dordrecht: Kluwer), p. 265.
Montmerle, T., Koch-Miramond, L., Falgarone, E., and Grindlay, J.E. 1983, *Ap. J.*, **269**, 182.
O'Neal, D., Feigelson, E.D., Mathieu, R.D., and Myers, P.C. 1990, *A.J.* **100**, 1610.
Phillips, R.B., and Lestrade, J.-F. 1988, *Nature* **334**, 329.
Phillips, R.B., and Lestrade, J.-F. 1990, *Bull. A.A.S.*, **22** 832.
Phillips, R.B., Lonsdale, C.J., and Feigelson, E.D. 1991, *Ap. J.*, in press (20 November 1991).
Phillips, R.B., and Lestrade, J.-F. 1991, in preparation.
Stine, P.C., Feigelson, E.D., André, Ph., and Montmerle, T. 1988, *A.J.* **96**, 1394.
Strom, K.M., Strom, S.E., Edwards, S., Cabrit, S., and Skrutskie, M.F. 1989, *A.J.* **97**, 1451.
White, S.M., Kundu, M.R., and Jackson, P.D. 1989, *Astr. Ap.*, **225**, 112.
White, S.M., Pallavicini, R., and Kundu, M.R. 1991a, *Ap. J.*, in press.
White, S.M., Pallavicini, R., and Kundu, M.R. 1991b, in preparation.

Seventh Cambridge Workshop on Cool Stars, Stellar Systems, and the Sun
ASP Conference Series, Vol. 26, 1992
Mark S. Giampapa and Jay A. Bookbinder (eds.)

HIGH SENSITIVITY, DYNAMIC SPECTRAL SEARCH FOR FLARE STAR RADIO BURSTS WITH THE ARECIBO SINGLE DISH

A. LECACHEUX, M. ABADA-SIMON D.A.S.O.P., Observatoire de Paris, 92195 Meudon, France

L. BELKORA, G.A. DULK Department of Astrophysical, Planetary and Atmospheric Sciences, University of Colorado, Boulder 80303

J. BOOKBINDER Harvard-Smithsonian Center for Astrophysics, 60 Garden Street, Cambridge, MA 02138

T.S. BASTIAN National Radio Astronomy Observatory, Socorro, NM 87801

P. LOUARN CNET/CRPE, Issy-les-Moulineaux, France

ABSTRACT In December 1991 and May 1991 we observed several well-known flare stars with the highly sensitive Arecibo radio telescope at 1.4 GHz and 5 GHz. We used a special observing technique to discriminate between real flares and radio frequency interference. We recorded far less activity than we had expected based on previously reported flare rates. With 7 K/Jy sensitivity, however, we were able to recognize instantaneous flux enhancements as weak as 15 mJy above the sky background variations.

Keywords: radio flares; acousto-optic spectrometry

PRELIMINARY REPORT ON OBSERVATIONS AND RESULTS

In December 1990 and May 1991 we observed several well-known flare stars (AD Leo, YZ CMi, Wolf 424, Gl 569, VW Com, YY Gem, YZ CMi, V371 Ori, EQ Peg, TZ Ari, and UX Ari) with the highly sensitive Arecibo radio telescope at 1.4 GHz and 5 GHz. We used a special observing technique in order to discriminate between real flares and radio frequency interference. We simultaneously recorded the outputs of two 1.4 GHz distinct feeds (one tracking the source, the other pointing 3 arc minutes uphill), in both circular polarizations. The four resulting IF outputs were then simultaneously analysed by means of a broadband acousto-optical spectrometer provided by the Paris Observatory. The four 50 MHz dynamic spectra were recorded with a 1 MHz frequency resolution and a 20 ms time constant.

We recorded far less activity than we had expected based on previously reported flare rates. However, with a 7 K/Jy sensitivity, we were able to recognize instantaneous flux enhancements as weak as 15 mJy above the sky background variations. Continuous tracking allowed us to accumulate more than 100 hours of stellar observations. A number of spurious events were rejected, in spite of their stellar appearance, for various reasons:

- simultaneous occurrence in both "Off" and "On" beams
- recurrence over consecutive days at the same hour angles
- evidence for narrow band interference, recognized thanks to spectral analysis capability
- comparable intensity levels in both circular polarizations.

Only a very few events (only 3 from AD Leo in more than 100 hours of observations) were unambiguously attributed to stellar flares; these events had a peak flux of 30 mJy.

We also observed at 5 GHz by using the mini-Gregorian and the wideband AOS, for about 30 hours (20 in 1989 July and 10 in 1990 December), with an effective bandwidth of 500 MHz and a 2 K/Jy sensitivity, but for these observations there was no off-source monitoring. Several events were rejected after we examined their dynamic spectrum, leaving only one possible stellar event from AD Leo.

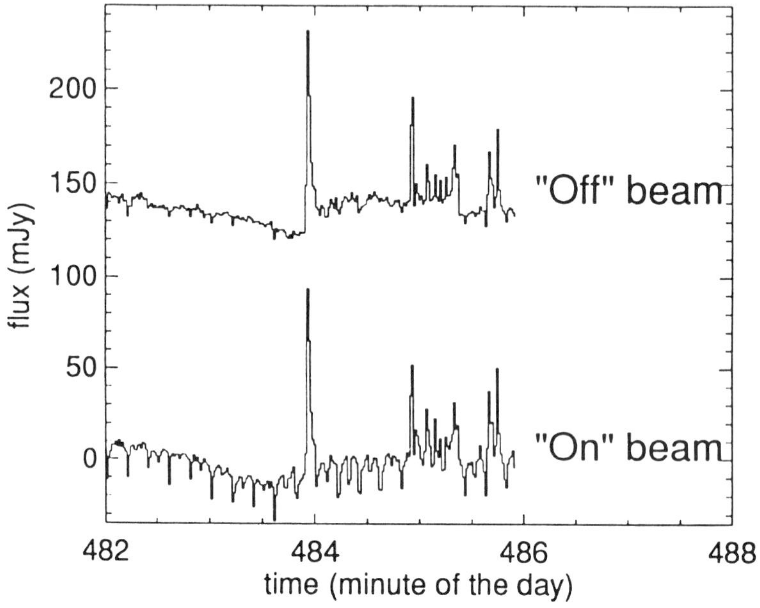

Fig. 1: Example of a "false-alarm event" recorded when tracking Wolf 424. The observation was made at 1.4 GHz with a bandwidth of 50 MHz and a time resolution of 20 msec. The observed RCP flux in the "on source" beam, averaged over 50 MHz and 1 sec. (lower curve) shows a candidate stellar event displaying fast intensity variations up to 100 mJy. The small, periodic "absorption" signal is due to off-band radar emission interacting with the telescope electronics. The positive spikes, when analyzed in the frequency-time plane, have a highly structured and polarized dynamic spectrum. The "off-source" beam intensity profile (upper curve), showing the same kind of intensity fluctuations and the same spectrum, definitely demonstrates that this event is not stellar, but is interference.

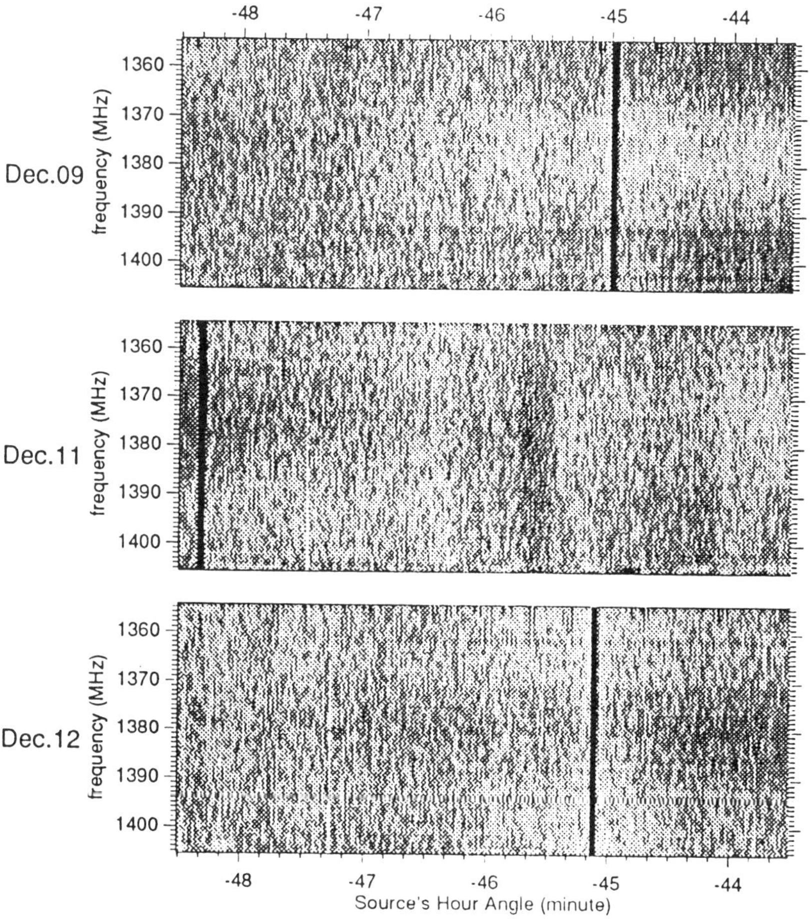

Fig. 2: Example of a probably flare event from AD Leo. The three gray-shaded panels display the dynamic spectra obtained at 1.4 GHz, during three consecutive days, when tracking AD Leo at about 46 minutes before its meridian transit time. The dark vertical streaks are due to the periodic injection of calibration noise. The faint horizontal lines are narrow-band, weak interferences. The displayed intensity dynamic range is 70 mJy (from white to black). The candidate event is the gray patch, only visible on the Dec. 11 panel at hour angle -45.6 min. It has a short duration (10 sec.) and corresponds to a smooth increase of the observed RCP flux of about 25 mJy. No fine structures are visible at the 20 msec timescale.

Seventh Cambridge Workshop on Cool Stars, Stellar Systems, and the Sun
ASP Conference Series, Vol. 26, 1992
Mark S. Giampapa and Jay A. Bookbinder (eds.)

ROTATIONAL MODULATION OF RADIO EMISSION FROM THE K1-DWARF STAR AB DORADUS (HD 36705)

JEREMY LIM Astronomy Program, University of Maryland, College Park MD 20742, USA.

ABSTRACT The first conclusive evidence for rotational modulation of stellar radio emission was reported recently by Lim, Nelson, Castro, Kilkenny, and van Wyk (1991). The observed radio intensity of the young and rapidly-rotating K1-dwarf star AB Doradus peaked during the inferred central meridian passage of large starspots across the stellar disk. This observation provides the first evidence that the radio emission of active late-type dwarf stars, like that of the Sun, originates from regions of strong coronal-magnetic-fields associated with starspots.

Keywords: Stars:late type; Stars:activity; Stars:coronae

INTRODUCTION

AB Dor is believed to be a young ($\sim$ 50 million years old) K1-dwarf star which is still evolving onto the main sequence. It displays vigorous surface-magnetic-activity in the form of large starspots, intense chromospheric Ca II *H* and *K* emission, strong X-ray emission from hot (10^7 K) coronal plasma, and strong and variable non-thermal radio emission.

Recently Lim *et al.* (1991) reported that the radio emission of AB Dor is modulated in phase with the rotation of the star. The observed stellar radio-intensity peaked near two longitudes where large starspots have been found to form preferentially. In this paper I summarize the observations and results, and discuss their implications.

OBSERVATIONS AND RESULTS

AB Dor was observed in the 6-cm wavelength band with the Australia Telescope Compact Array on July 24, Sept 27, and Oct 1, 1990. In Figure 1 the measured radio intensity of the star is plotted as a function of stellar rotation-phase; the stellar rotation-period is 0.51479 days, and the epoch of zero rotation-phase is at JD=2444296.575 (Innis *et al.* 1988). During each rotation the stellar intensity peaks at two rotation phases - approximately 0.4 and 0.9 - separated by about half a rotation period. The absolute and relative intensities of the two peaks in emission change independently with time. The width of the peaks, however, remain constant at $\sim$ 0.3 rotation phase.

During November 1990 AB Dor was monitored with optical telescopes at Siding Spring in NSW, Australia, and at the South African Astronomical Observatory. The objective of these observations was to determine the longitudes of starspots. As shown in Figure 1(d), the measured optical light-curve indicates

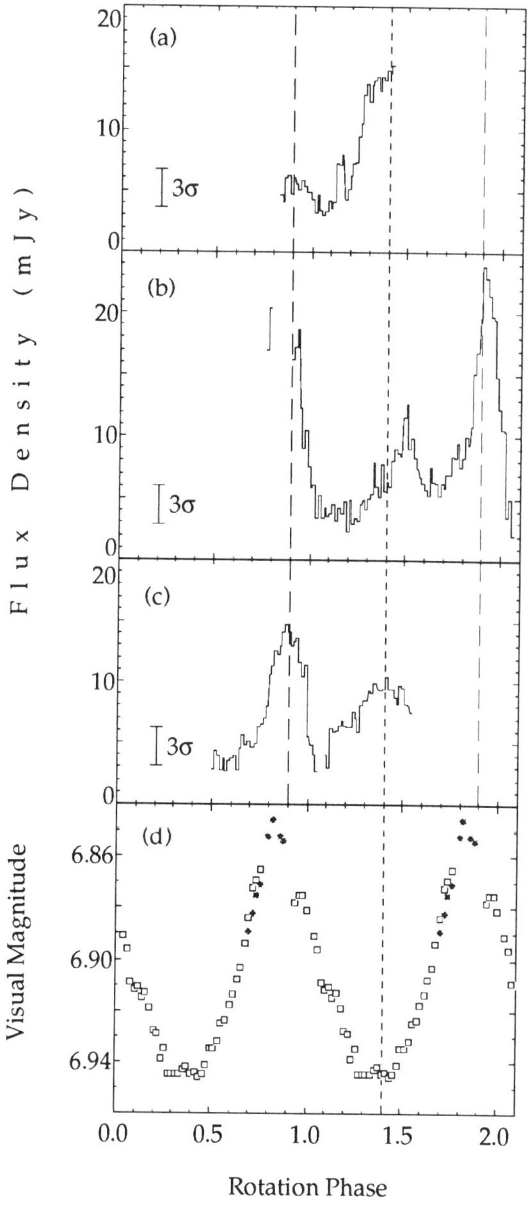

Figure 1. The radio-intensity profile of AB Dor measured on (a) Jul 24, (b) Sep 27, and (c) Oct 1, 1990. The optical light-curve measured in Nov 1990 is shown in panel (d).

the presence of a large starspot centred at phase ∼ 0.4. The peak in radio emission at phase 0.4 is coincident with the central meridian passage of this starspot. The radio peak at phase 0.9 does not appear to have an optical counterpart, but as shown in the following discussion it too is probably associated with a starspot.

Over the last decade the optical light-curve of AB Dor has been dominated by large starspots which have formed preferentially at two longitude belts (Innis *et al.* 1988). These longitude belts are centered at rotation phases ∼ 0.0 and ∼ 0.5 respectively. In the last few years starspots near phase 0.5 have dominated the optical light-curve, with the starspots near phase 0.0 only occasionally detectable. Curiously, when starspots at both longitude belts are detectable, they are invariably separated by almost exactly half a rotation period. Thus, the starspot centred at phase ∼ 0.4 in Figure 1(d) may be accompanied by a presently undetectable starspot on the other side of the stellar disk, and the latter may be associated with the peak in radio emission at phase ∼ 0.9.

DISCUSSION

The modulation in the radio emission of AB Dor is attributed to a combination of two factors: (i) occultation of the radio sources by the stellar disk, and (ii) high directivity of the radiation from these regions. The brightness temperatures of the radio sources are $\gtrsim 10^{10}$ K, and so the radiation must be produced by non-thermal electrons. The radiation mechanism is likely to be incoherent gyro-synchrotron emission, although a coherent mechanism such as plasma or electron-cyclotron maser emission cannot be ruled out. Indeed the high directivity of the radiation suggests that the emission mechanism is coherent.

The lifetime of the energetic electrons responsible for the radio emission is determined by either collisions with the ambient coronal plasma or precipitation into the chromosphere. For gyro-synchrotron emission the particle lifetime is likely to be $\lesssim 3$ hrs, whereas for a coherent emission process the particle lifetime may be an order of magnitude shorter (Lim 1991). In any case the lifetime of the radiating electrons is much less than one stellar rotation, and therefore to produce the regular modulation-profile observed electrons must be accelerated semi-continuously in the source region. I suggest that large starspots may be composed of groups of relatively small starspots, and that the frequent interaction between magnetic loops in these densely populated regions may the cause of the semi-continuous electron acceleration.

ACKNOWLEDGEMENTS

The Australia Telescope National Facility is operated in association with the Division of Radiophysics by CSIRO.

REFERENCES

Innis, J. L., *et al.* 1988, *M.N.R.A.S.*, **235**, 1411.
Lim, J. 1991, *PhD Thesis*, Macquarie University, NSW, Australia.
Lim, J., *et al.* 1991, submitted to *Ap. J. Letters*.

Seventh Cambridge Workshop on Cool Stars, Stellar Systems, and the Sun
ASP Conference Series, Vol. 26, 1992
Mark S. Giampapa and Jay A. Bookbinder (eds.)

THE CLASS OF RADIO-EMITTING MAGNETIC B STARS AND A WIND-FED MAGNETOSPHERE MODEL

JEFFREY L. LINSKY[1] Joint Institute for Laboratory Astrophysics, NIST, and Univ. of Colorado, Boulder, CO 80309-0440.

STEPHEN A. DRAKE High Energy Astrophysics Science Archive Research Center, Code 668, Goddard Space Flight Center, Greenbelt, MD 20771.

TIMOTHY S. BASTIAN National Radio Astronomy Obs., Socorro, NM 87801.

ABSTRACT We have detected a total of 16 magnetic Bp–Ap stars at 6 cm out of 61 observed, including both He-strong and He-weak/Si-strong stars with $\log L_6$ = 14.7 to 17.9, but none of the classical (SrCrEu-type) Ap stars at a detection limit of $\log L_6 < 14.9$. We believe that the radio-emitting CP (chemically peculiar) stars form a distinct class of radio stars that differs from both the hot star wind sources and the active late-type stars. For the detected CP stars we find that $L_6 \propto \dot{M}^{0.5} B_{rms}$, produced by optically thick nonthermal gyrosynchrotron emission. In our model the electrons are accelerated in current sheets located 10-20 radii from the star.

Keywords: Magnetic fields; radio sources, identifications; stars, coronae; stars, peculiar A

OBSERVATIONS

In five VLA observing runs we have extended the initial survey of radio emission from magnetic Bp–Ap stars by Drake *et al.* (1987) to include a total of 16 sources detected at 6 cm. The 11 new stars detected as radio sources have spectral types B5–A0 and are He-weak and Si-strong. We have not yet detected any of the classical (SrCrEu-type) Ap stars with detection limits of 0.2 – 0.5 mJy corresponding to $\log L_6 < 14.9$ for some stars, despite many attempts. The 16 detected sources show a wide range of radio luminosities with the early-B He-S stars on average 20 times more radio luminous than the late-B He-W stars and 1000 times more luminous than θ Aurigae, the field star with both the lowest radio luminosity and the lowest effective temperature. Multifrequency observations indicate flat spectra in all cases. Four stars have a detectable degree of circular polarization at one or more frequencies. For a full description of the observations and our model see Linsky *et al.* (1991).

We believe that the radio-emitting CP (chemically peculiar) stars form a distinct class of radio stars. For the detected CP stars we find that $L_6 \propto$

[1] Staff member, Quantum Physics Division, National Institute of Standards and Technology

$\dot{M}^{0.5}B_{rms}$, where $\dot{M}$ is the mass loss rate and B_{rms} is the rms value of the observed longitudinal field strength. The dependence on $\dot{M}$, the rapid increase in L_6 with T_{eff}, and the absence of obvious flares distinguishes the CP stars from the active cool stars. The radio emission from the CP stars is several orders of magnitude larger than the predicted thermal and nonthermal wind emission from the hot stars, and therefore not an extension of this class of radio emitters.

A WIND-FED MAGNETOSPHERE MODEL

We argue that the observed properties of radio emission from these stars may be understood as optically thick gyrosynchrotron emission from a nonthermal distribution of electrons produced in a current sheet far from the star. Our model (see Fig. 1) consists of a rotating dipolar magnetosphere in which stellar wind plasma flows out from near the magnetic poles along the field lines. Far from the star ($r \gtrsim 10R_\star$), the gas pressure exceeds the magnetic pressure and the field lines near the equator are drawn out into current sheet configurations (see Havnes and Goertz 1984), in analogy to planetary magnetotails. These current sheets are likely locations of plasma heating and of particle acceleration. In our "direct injection" scenario the electrons return to near the star by traveling along magnetic fields, where they mirror and radiate in the strong fields near the poles.

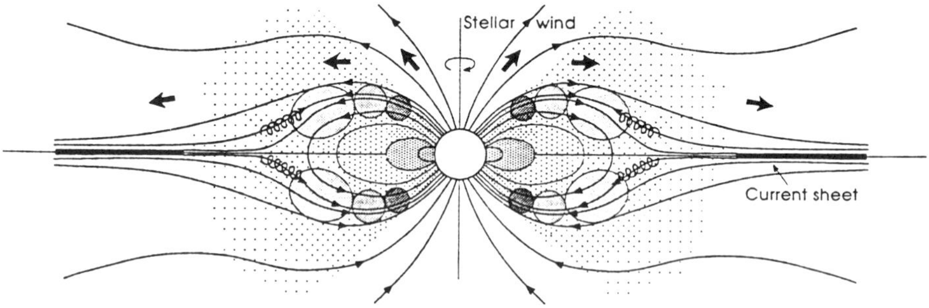

Fig. 1: A cartoon showing a cross section through a CP star and its magnetosphere. For simplicity, an aligned rotator is shown. A dense stipple indicates the presence of dense, trapped thermal plasma. A light stipple indicates the possible presence of tenuous thermal plasma. The small dark circles indicate schematically where high frequencies originate; larger, lighter circles indicate the location of the source with decreasing frequency.

Our model accounts for time variability in terms of intrinsic variability in the acceleration process and rotational modulation of the projected area of the emitting region. The flat microwave spectra result from two factors: (i) a decrease in source size with increasing frequency; and (ii) a decrease in the effective temperature of the emitting electrons with increasing frequency.

The finite degree of circular polarization observed on some sources may be the result of an opacity effect resulting from the increasing magnetic field strength with decreasing radius. Finally, the decreasing radio luminosity with decreasing stellar effective temperature along the main sequence is explained primarily as a result of the decreasing mass loss rate.

While there is some overlap between the properties of the nonthermal radio emission from CP stars and those of the RS CVn binaries (Table 1), the physical basis of the radio emission is quite different in each case. For CP stars, the radio emission is intimately related to both the magnetic field and the mass loss rate while this does not appear to be the case for the RS CVn binaries. However, the model proposed here for the nonthermal radio emission from magnetic CP stars may have some relevance to certain of the weak T Tauri stars and other PMS objects (*e.g.*, André *et al.* 1991).

ACKNOWLEDGMENTS

We thank NASA for partial support of this work through grant NGL 06-003-057 to the University of Colorado, and Interagency Transfer W-17,772 to the N.I.S.T. We also thank Dr. Henny Lamers for computing mass loss rates.

REFERENCES

André, P., Phillips, R. B., Lestrade, J.-F., and Deeney, B. 1991, this ref.

Drake, S.A., Abbott, D.C., Bastian, T.S., Bieging, J.H., Churchwell, E., Dulk, G., and Linsky, J.L. 1987 *Ap. J.*, **322**, 902.

Havnes, O., and Goertz, C.K. 1984, *Astron. J.*, **138**, 421.

Linsky, J.L., Drake, S.A., and Bastian, T.S. 1991, submitted to *Ap. J.*,

Table 1. Comparison of radio parameters of different types of stars

Parameter	Sun	dMe	RS CVn	He–S	He–W
Spectral type	G2 V	dM4e	K0 IV	B2 V	B8 V
$R_\star/R_\odot$	1.0	0.4	3	4	3
B_{photo}(kG)	1.5	3–5	≈1	1.7–3.4	0.7–17
f_{photo}	0.02	0.2–0.7	0.3?	≈1	≈1
Topology	complex	complex	complex	dipole/quad	dipole
$< fB_{photo} >$	0.03	0.6–3.5	≈0.3	1.7–3.4	0.7–17
$< fB_{photo} > R_\star^2$	0.03	0.1–0.6	≈3	27–54	6.3–153
log L_6	11.0	13.0–14.6	13.8–17.3	17.4–17.9	15.7–17.9
log L_x	27.7	27.7–29.8	29.0–31.6	30.5	—
log L_x/L_{bol}	-6	-4 to -2	-6 to -3	-7	—
flaring	yes	yes	yes	no?	no?
active regions	yes	yes	yes	—	—
$(R/R_\star)^2 T_B$	8×10^4	1×10^8	1×10^{10}	2×10^{10}	$0.04–6 \times 10^{10}$
$R/R_\star$	1	—	3	<4	—
T_B	8×10^4	—	1×10^9	$\geq 1 \times 10^9$	—

Seventh Cambridge Workshop on Cool Stars, Stellar Systems, and the Sun
ASP Conference Series, Vol. 26, 1992
Mark S. Giampapa and Jay A. Bookbinder (eds.)

MILLIMETER AND SUB-MILLIMETER EMISSION FROM FLARE STARS

D.J. MULLAN Bartol Research Institute, Univ. of Delaware, Newark DE 19716, USA

J.G. DOYLE and M. MATHIOUDAKIS Armagh Observatory, Armagh BT61 9DG, N. Ireland

R.O. REDMAN Herzberg Inst. for Astrophysics, NRC, Ottawa, ONT K1A 0R6, Canada

ABSTRACT 13 flare stars were observed with JCMT at $\lambda \approx 1$ mm, with 6 detections at 2σ or better. Free-free emission from a wind fits the data from several stars, but the implied mass loss rates are large for red dwarfs.

Keywords: flare stars; sub-millimeter; mass loss

INTRODUCTION

Mullan *et al.* (1989a: MSB) detected some 15% of 78 flare stars in IRAS at 60μ and/or 100μ with fluxes of 100 times photospheric. In a smaller sample, Mathioudakis and Doyle (1991: MD) also detected some flare stars in IRAS at 60μ. The FIR excesses may originate as synchrotron emission of fast electrons (created by continual micro-flaring) in strong B fields. Synchrotron self-absorption should set in at $\lambda \approx 1$ mm, where peak fluxes are predicted to be 10-100 Jy (for spectral index = 1-4). FIR may also be due to dust: the mean observed ratio $F(60\mu)/F(100\mu)$ is 0.25, requiring cold dust, with T = 20-30 K (White et al, 1991), and predicting 10-20 mJy at $\lambda \approx 1$ mm in Gl285. To distinguish between synchrotron and dust, flare stars were observed at $\lambda \approx 1$ mm where the models predict fluxes differing by 10^3.

OBSERVATIONS

We used the UKT14 detector at 1.1 and 0.8 mm on the James Clerk Maxwell Telescope (JCMT) to observe 11 target stars on 1990/09/30 and 1990/10/01. Sky conditions were poor with more than half of the time unusable. Primary calibrators were Uranus and Mars, with CRL 2688, W3(OH), CRL 618, OH 231.8 and G34.3 as secondaries. For each program star, one integration consisted of 10 cycles of 10 seconds each. Due to the star/sky chopping procedure, only half of this time was on the star. This sequence was repeated 10-20 times, depending on signal strength. For the stronger calibrators, 10 cycles of 4 seconds each were used. Noise was typically tens of mJy: most targets were undetectable against this background.

We found four stars which may be significant emitters.

i Gliese 207.1: F(0.8) = 150 mJy (3σ).
ii Gliese 735: F(0.8) = 50 mJy (4σ). Tsikoudi (1988) reported this star to have an IR excess.
iii Gliese 285: F(1.1) = 13 mJy (2.2σ)
iv Gliese 725: F(1.1) = 40 mJy (1.5σ). This star was previously reported as an IRAS detection at 100μm (MSB) with a flux of 260 mJy (2σ).

A few weeks before the observations reported here, JCMT data on two other dMe stars were obtained at λ = 1.1 and 2 mm (Doyle and Mathioudakis, 1991: DM). The fluxes were found to be 5-15 mJy.

DISCUSSION

Combining our sample of targets with those of DM, we conclude that among a sample of 13 dMe stars, 6 (*i.e.* almost 50%) have been detected at $\lambda \approx$ 1mm. The fluxes of the detected stars range from 5 to 150 mJy.

Synchrotron emission is clearly ruled out by the new data. Dust emission fits the millimeter data with color temperatures of 26K and 19K for emissivity spectral indices of 1 and 2 respectively. However, dust would create very large fluxes (O(1 Jy)) from Gl285 at $\lambda \approx$ 200-300μ: this should be verified.

Free-free emission from a wind is a candidate for mm emission. In Fig. 1, data from optical to radio wavelengths are shown for Gl285. A single power law can be made to fit quite well from 25μ to radio: $S_\nu \sim \nu^\alpha$, with index $\alpha = +0.7$. Millimeter data for two other red dwarfs (DM) can also be fit by a power law with $\nu^{+(0.6-0.9)}$.

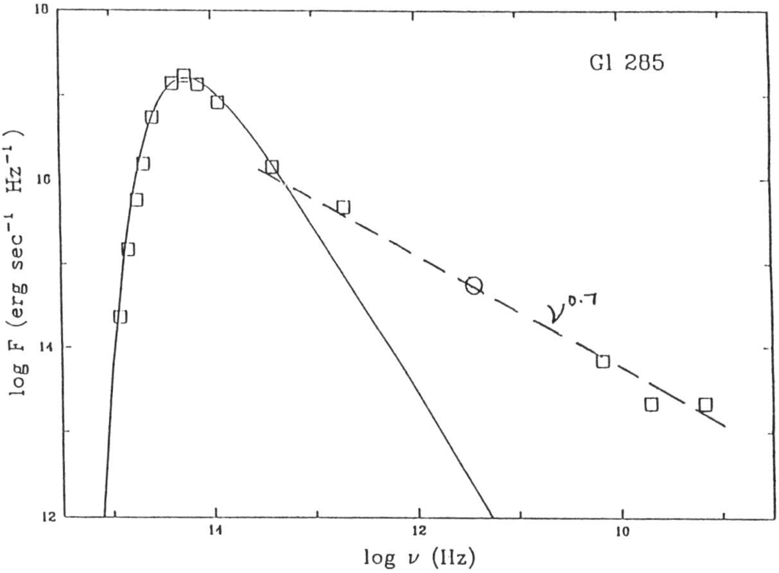

Fig. 1: Fluxes from Gliese 285 (YZ CMi) from optical to radio wavelengths. Black-body at 4000K fits data at $\lambda \leq 20\mu$. Longer wavelengths appear to be fitted quite well by a power law. The index of the power law suggests free-free emission from an ionized wind.

Now, an ionized wind with sufficient optical depth creates a power-law spectrum with $\alpha \approx +(0.6\text{-}0.7)$ or slightly larger if the wind is accelerating (Leitherer and Robert, 1991: LR). In the context of free-free emission, the flux observed at frequency ν, S_ν (in Jansky), is related to the mass loss rate $\dot{M}$ (in solar masses per year), the terminal velocity of the wind V_∞ (in km/sec), and the distance to the star D (in parsecs) by (LR):

$$S_\nu = 1.22 \times 10^8 (\dot{M}/V_\infty)^{4/3} \nu^{2/3}/D^2 \tag{1}$$

Here, we assume a hydrogen wind, fully ionized, and a free-free Gaunt factor of 3 (LR).

For YZ CMi (Gl 285), we have $S_\nu = 0.013$ Jy at $\nu = 3 \times 10^{11}$ Hz, and $D = 6$. The terminal velocity of the wind is not known. However, based on the solar wind, and on the velocity in the wind of a K2 dwarf (Mullan *et al.*, 1989b: MSBC), a plausible value is $V_\infty \approx 300$. Using these values, we find

$$\dot{M} \approx 2.7 \times 10^{-10} M_\odot/yr \tag{2}$$

For Gliese 644 and 873, the distances are $D = 7$ and 5 pc, and the fluxes observed at 1 mm are 2-3 times smaller than for YZ CMi. Thus, mass loss rates are smaller by factors of $(2-3)^{3/4}$ than for YZ CMi. Thus, $\dot{M} \approx 10^{-10} M_\odot/yr$.

CONCLUSION

The mass loss rates estimated here on the basis of mm data are 4 orders of magnitude larger than solar. They are also an order of magnitude larger than the mass loss rate reported for the K2 dwarf in V471 Tauri (MSBC). The origin of efficient mass loss in red dwarfs is not yet clear. It seems unlikely that thermal driving by a corona will be sufficient. In V471 Tauri, time-dependent wind absorptions indicate that the mass loss may occur as a series of discrete events, presumably magnetic in origin. If this is also true of red dwarfs, then the millimeter fluxes should be variable.

ACKNOWLEDGEMENTS

The authors would like to acknowledge support from NASA.

REFERENCES

Doyle, J. G. and Mathioudakis, M. 1991, *Astr. Ap.*, **241**, L41 (DM).
Leitherer, C. and Robert, C. 1991, *Ap. J.*, **377**, 629 (LR).
Mathioudakis, M. and Doyle, J. G. 1991, *Astr. Ap.*, **244**, 433 (MD).
Mullan, D. J., *et al.* 1989a, *Ap. J.*, **343**, 400 (MSB).
Mullan, D. J., *et al.* 1989b, *Ap. J. Letters*, **339**, L33 (MSBC).
Tsikoudi, V. 1988, *Astron. J.*, **95**, 1797.
White, R. *et al.* 1991, *Ap. J.*, **375**, 38.

Seventh Cambridge Workshop on Cool Stars, Stellar Systems, and the Sun
ASP Conference Series, Vol. 26, 1992
Mark S. Giampapa and Jay A. Bookbinder (eds.)

A HIGH SENSITIVITY SURVEY OF RADIO CONTINUUM EMISSION IN HERBIG Ae/Be STARS

STEPHEN L. SKINNER, A. BROWN, J.L. LINSKY
Joint Institute for Laboratory Astrophysics, Campus Box 440, Univ. of Colorado, Boulder, CO 80309

RON T. STEWART Australia Telescope Natl. Facility, PO Box 76, Epping NSW 2121, AUSTRALIA

ABSTRACT We present results of a high sensitivity VLA/Australia Telescope survey of radio continuum emission from the 57 Herbig Ae/Be stars (and candidates) cataloged by Finkenzeller and Mundt (1984). Twelve stars were detected at the primary observing wavelength of 3.6 cm, with an additional five stars possibly detected. Upper limits for undetected stars are typically 0.1 mJy (3σ). Follow-up observations in the wavelength range 2 – 20 cm give spectral index estimates for 9 stars. Preliminary analysis suggests that the radio emission is of thermal origin in most cases, although nonthermal emission is a possibility for TY CrA and MWC 137.

Keywords: stars: massive; stars: pre-main sequence; stars: radio radiation

INTRODUCTION

Herbig Ae/Be stars (Herbig 1960) are intermediate mass (2 – 20 $M_\odot$) pre-main sequence (PMS) stars undergoing gravitational contraction towards the main sequence (Strom *et al.* 1972). Although it has long been known that some objects of this type are centimeter (cm) radio sources (cf. Bertout and Thum 1982), a systematic study of their radio properties has not previously been conducted. We have thus undertaken a survey of radio continuum emission in the 57 Herbig Ae/Be stars (and candidates) cataloged by Finkenzeller and Mundt (1984, hereafter FM). Early results from this study were discussed by Skinner *et al.* (1990). Our main objectives are to determine which Herbig Ae/Be stars are cm radio sources and to identify plausible emission mechanisms.

OBSERVATIONS

The observations were conducted with the Very Large Array (VLA) and Australia Telescope (AT) between 1989 Feb 2 and 1991 Jun 4 in various configurations. To determine which stars are sources, we obtained deep 40 – 60 min VLA observations at 3.6 cm of all stars in FM except HD 97048 ($\delta = -77^o$). Also, four southern stars (HD 97048 included) were observed for 12 hours each at 6 cm with the AT. Twelve stars were detected with the VLA (Table I) above the detection limit of 70 - 100 μJy (3σ), but none were detected with the AT above a threshold of $\approx$350 μJy (3σ). An additional five stars are classified as *possible* detections, namely KK Oph, BD+40°4124, AS 442, LkHα 234, and MWC 1080. The radio and optical positions for these stars differ by more than 1″ (but by less than 1.7″), or else the emission is weak and unconfirmed.

Single-epoch multiwavelength (2, 3.6, 6, 20 cm) follow-up observations resulted in detections at more than one wavelength for 9 stars, providing estimates of the spectral index α ($S_\nu \propto \nu^\alpha$). Fifteen stars (11 detections and 4 possible detections) were observed on more than one occasion at 3.6 cm, providing a check for flux variability.

DISCUSSION

The unambiguous detection of *nonthermal* radio emission in Herbig Ae/Be stars would be significant, as it would provide *indirect* evidence for the presence of magnetic fields in these young objects. Such emission is characterized by brightness temperatures $T_b \geq 10^8$ K, flux variability on short timescales ($\leq$days), and moderate circular polarization. Our initial analysis of the data obtained here does *not* provide convincing evidence for nonthermal emission, TY CrA and MWC 137 possibly excepted.

Thermal emission is implicated in the majority of cases on the following grounds. First, circularly polarized emission is not detected in any object. Second, the emission is extended in several cases, and is partially resolved with a 0.3″ beam in Z CMa, MWC 297, V645 Cyg, and LkHα 234. This implies T_b is less than 10^8 K for these objects. Third, 9 of 12 detections and 4 of 5 possible detections show evidence for winds or outflows. Some of these objects (*e.g.* R Mon, LkHα 234) have spectral indices $\alpha \geq +0.6$, which suggests free–free wind emission. Fourth, significant flux variability is generally not detected, except for TY CrA ($\leq$days) and perhaps MWC 137 ($\leq$months). However, checks for short-term variability have not been conducted for all stars, and future observations could reveal variability in other objects.

The short-term variability and negative spectral index for TY CrA are consistent with nonthermal emission, but it is not yet known if $T_b \geq 10^8$ K. We note that the classification of MWC 137 and the eclipsing binary TY CrA as Herbig Ae/Be stars is not certain (FM, Herbig and Bell 1988).

ACKNOWLEDGEMENTS

This research is supported by NASA interagency transfer W-17772 to the National Institute of Standards and Technology.

REFERENCES

Bertout, C., and Thum, C. 1982, *Astr. Ap.*, **107**, 368.
Finkenzeller, U., and Mundt, R. 1984, *Astr. Ap. Suppl.*, **55**, 109 (FM).
Herbig, G.H. 1960, *Ap. J. Suppl.*, **4**, 337.
Herbig, G.H., and Bell, K.R. 1988, *Lick Obs. Bull.*, No. 1111.
Skinner, S.L., Brown, A., and Linsky, J.L. 1990, *Ap. J. Letters*, **357**, L39.
Strom, S.E., Strom, K.M., Yost, J., Carrasco, L., Grasdalen, G. 1972, *Ap. J.*, **173**, 353.

TABLE I Detected Herbig Ae/Be Stars

star	sp. type	distance (pc)	total 3.6cm flux[a] (mJy $\pm$ 1σ)
Elias 1	A0e - A6e	140(?)	0.41 $\pm$ 0.03
AB Aur[c]	A0e	160	0.15 $\pm$ 0.03
MWC 137[b]	cont + e	Uncertain	0.55 $\pm$ 0.03 (v)
HD 259431[b,c]	B2e - B5e	800	0.14 $\pm$ 0.02
R Mon[c]	pec , var	800	0.47 $\pm$ 0.04
Z CMa[c,d]	B?eq, pec	1150	2.07 $\pm$ 0.04 (r)
HD 150193[c]	A4e	Uncertain	0.19 $\pm$ 0.03
MWC 297[c]	Uncertain	530	14.5 $\pm$ 0.14 (r)
MWC 300[b,c]	eq	15500	0.50 $\pm$ 0.03
TY CrA[b,d]	B9e	130	0.60 $\pm$ 0.05 (v)
HD 200775[c]	B2e - B3e	600	0.59 $\pm$ 0.04
V645 Cyg[c]	O7e or A5e	3500(?)	0.73 $\pm$ 0.07 (r)

[a] A (v) indicates that radio variability is suspected, and an (r) indicates partially resolved emission.
[b] Classification as a Herbig Ae/Be star is uncertain.
[c] Evidence for mass loss (P Cygni profiles or molecular outflow).
[d] Binary.

Seventh Cambridge Workshop on Cool Stars, Stellar Systems, and the Sun
ASP Conference Series, Vol. 26, 1992
Mark S. Giampapa and Jay A. Bookbinder (eds.)

A 5 GHZ SURVEY OF WEAK-LINED T TAURI STARS

S. M. WHITE,
Dept. of Astronomy, Univ. of Maryland, College Park MD 20742

R. PALLAVICINI,
Arcetri Observatory, Largo Enrico Fermi 5, I-50125 Florence, Italy

M. R. KUNDU
Dept. of Astronomy, Univ. of Maryland, College Park MD 20742

ABSTRACT We report the results of a search for radio-continuum emission from weak-lined T Tauri stars selected on the basis of a range of criteria. A correlation is found with strong x–ray emission and with youth. All the stars in the survey older than about 20 million years were not detected as radio sources.

Keywords: stars: activity; stars: coronae; stars: pre-main sequence; radio continuum: stars

THE SURVEY

Weak-lined T Tauri stars are low-mass pre-main sequence stars which do not have the strong Hα emission line of classical T Tauri stars. They are known to be strong x–ray sources, and display solar-like activity thought to be characteristic of PMS stars. PMS stars probably spend a significant fraction of their contraction time in this state.

In this survey we look for continuum radio emission among samples of these stars chosen on the basis of a number of criteria. Many of these stars are already known to be radio sources (Bieging *et al.* 1984; Kutner *et al.* 1986; André *et al.* 1987; Stine *et al.* 1988; O'Neal *et al.* 1990). Rather than trying to completely sample the population of PMS stars in a young star-forming region, as most previous surveys have done, our survey was intended to try to identify a number of new sources suitable for further study of the nature of radio emission from these stars. We therefore chose most of the targets on the basis of known high activity levels, and particularly of strong x–ray fluxes. For comparison we included a number of stars known to be radio sources from the observations described above. However, we also included 8 targets chosen because they are late-type stars occurring in visual binaries with an early-type primary, and are therefore likely to be young stars. Such PMS stars, which are away from star forming regions, have not previously been studied in radio observations, and we felt it would be of interest to compare their radio properties with those of other PMS stars in the survey. The observations were carried out at 5 GHz with the Very Large Array in "A"-configuration, on 1990 Feb. 18 and 19. The typical 3σ detection threshold was 0.12 mJy.

Visual binaries with a young early-type primary

We observed 9 stars in this class: 8 chosen from Lindroos (1986), plus HD 28867 from Walter *et al.* (1988). Nearly all the targets from Lindroos (1986) consist of a late-type B dwarf and an F, G or K dwarf; all are in wide binaries, ensuring that the two components have evolved independently and are not Algol systems. From the large list in Lindroos (1986) we choose a small sample to include some of the closest of the Lindroos (1986) stars, with distances ranging from 23 to 157 pc. We also selected on the basis of other indicators of youth and/or activity. The primaries in all these systems should not have strong winds and are unlikely to be radio sources, and indeed none were detected.

None of the eight young-dwarf companions from Lindroos (1986) were detected. The only detection in this category is HD 28867, which was chosen on the basis of its x-ray flux (Walter *et al.* 1988). Both known components of HD 28867 are probably B stars. Normal main-sequence single B dwarf stars are not generally detected as radio sources. We believe that either HD 28867 is a Herbig Be star, or that there is an undiscovered pre-main sequence component in the system which produces the radio emission.

Post T Tauri candidates from the original list by Herbig (1978).

Three were observed, and two (V410 Tau and HD 283447, both in Taurus–Auriga) were detected. Both are short-period spotted stars (1.9 days for V410 Tau, 3.4 days for HD 283447) and had been previously detected as radio sources, and both are known to be variable. The distant K binary FK Ser was not detected, although it is otherwise known as a very active star.

Naked T Tauri candidates in Taurus–Auriga

We observed 12 sources identified as x-ray sources by Walter *et al.* (1988), ranging in spectral type from G5 to M0, with most being K stars. their x-ray luminosities range from $10^{29.6}$ to $10^{31.3}$ ergs s^{-1}. We detected 5 of the 12 targets: the rapidly-rotating G star HDE 283572 (period 1.5 days), the spectroscopic K binary V826 Tau (period 3.9 days); and three stars previously undetected as radio sources, the M2 star 040234+2143, the K5 star 045226+3013 and the K7 star 045251+3016.

Post T Tauri candidates in Ophiuchus

We observed Oph 1 – 4, identified in an x-ray survey of Ophiuchus by Walter (1986), and detected all four stars at fluxes of 0.2 – 0.6 mJy. Oph 1 – 3 are early K-type stars with strong Lithium absorption, and with a Strontium line indicating subgiant luminosity class. Given the results of the rest of the survey, it is remarkable that all four of these stars should prove to be radio sources, when they seem otherwise undistinguished.

Known radio-emitting weak-lined T Tauri's in ρ Ophiuchi

We observed the well-known radio star DoAr 21, as well as three x-ray-selected targets known to be radio sources: ROX 31, ROX 39 and a source near ROX 43

(Stine *et al.* 1988). We detected sources near the center of all four fields, but find that the ROX 43 source (which is about 40″ away from the original stellar candidate for the optical counterpart to the x-ray source; the star was not detected) is an extended jet source, and therefore confirm the speculation by Stine *et al.* (1988) that it is probably extragalactic.

DISCUSSION

Our survey is not complete in any sense, since it was biased in favour of stars already known to be active optically, in x-rays, or in radio. However, we can do some simple comparisons between the stars detected in our survey and those which were not. The mean log of the x-ray flux (ergs s^{-1}) of the detected stars was 30.5 ± 0.5 (sample of 13), compared with 30.2 ± 0.4 for the undetected stars. We have used the Student t statistic for populations of unequal variance to test the significance of the difference between the two means, and find that there is only a 5% chance that the radio-detected and radio-undetected stars actually have the same distribution of x-ray fluxes.

Ages are given for the subsample of survey stars in the Taurus-Aurigae region by both Walter *et al.* (1988) and Skrutskie *et al.* (1990). Using the Walter *et al.* ages, we find that the mean log of the age of the detected stars was 6.0 ± 0.3, while that of the undetected stars was 6.5 ± 0.6. There is an 11% probability that these two means could arise in samples from a single distribution of ages. However if we use the Skrutskie *et al.* values we find that the mean log of the age of the detected stars was 6.2 ± 0.3, while that of the undetected stars was 6.8 ± 0.4; the probability that these two means come from samples from a single distribution of ages is only 0.5%. The Lindroos (1986) stars are nearly all much older than 20 million years, and our failure to detect any of them may be further evidence of an age effect.

A flare was seen on one of the targets, and weak circular polarization was detected on two of the targets: these are discussed by White *et al.* (1990).

REFERENCES

André, P., Montmerle, T., and Feigelson, E. D.: 1987, *Astron. J.* **93**, 1182

Kutner, M. L., Rydgren, A. E., and Vrba, F.: 1986, *Astron. J.* **92**, 895

Lindroos, K. P.: 1986, *Astron. Astrophys.* **156**, 223

O'Neal, D., Feigelson, E. D., Mathieu, R. D., and Myers, P. C.: 1990, *Astron. J.* **100**, 1610

Skrutskie, M. F., Dutkevich, D., Strom, S. E., Edwards, S., Strom, K. M., and Shure, M. A.: 1990, *Astron. J.* **99**, 1187

Stine, P. C., Feigelson, E. F., André, P., and Montmerle, T.: 1988, *Astron. J.* **96**, 1394

Walter, F. M.: 1986, *Astrophys. J.* **306**, 573

Walter, F. R., Brown, A., Mathieu, R. D., Myers, P. C., and Vrba, F.: 1988, *Astron. J.* **96**, 297

White, S. M., Pallavicini, R., and Kundu, M. R.: 1991, *Astron. Astrophys.*, submitted

Seventh Cambridge Workshop on Cool Stars, Stellar Systems, and the Sun
ASP Conference Series, Vol. 26, 1992
Mark S. Giampapa and Jay A. Bookbinder (eds.)

A LOW-RESOLUTION SPECTROSCOPIC SURVEY OF POST-T TAURI CANDIDATES

ROBERTO PALLAVICINI Osservatorio Astrofisico di Arcetri, Largo E. Fermi 5, 50125 Florence, Italy

LUCA PASQUINI European Southern Observatory, Casilla 19001, Santiago 19, Chile

SOFIA RANDICH Dipartimento di Astronomia Università di Firenze, Largo E. Fermi 5, 50125 Florence, Italy

ABSTRACT We report the results of a spectroscopic survey at low-resolution (≈ 1-2 Å) of Post-T Tauri candidates. The program stars were taken from the list of Lindroos (1986) and consist of late-type secondaries of visual binaries with early-type primaries. The observations were carried out at ESO using different telescopes and instruments. From our data we conclude that only a fraction of Lindroos' stars show evidence of being pre-main sequence objects, while the others are more likely optical pairs.

Keywords: Post-T-Tauri; Lithium Abundances; Ca II Emission; PMS Stars

INTRODUCTION

Post-T Tauri (PTT) stars –as defined by Herbig (1978)– are low-mass pre-main sequence stars that are older and more evolved than Classical T-Tauri (CTT) stars. According to Herbig, the number of PTT's should be larger than that of CTT's, since the CTT stage is relatively short and represents only a small fraction of the pre-main sequence lifetime of a low-mass star. However, PTT's do not show the extreme properties of CTT's (*e.g.* strong Hα emission and infrared excess) and therefore are difficult to detect by conventional optical methods.

X–ray observations of star forming regions (*e.g.* Walter *et al.* 1988) have revealed hundreds of strong X–ray sources that are not CTT's. Given their weak Hα emission and lack of IR excess and continuum veiling, they have been termed Weak-lined T-Tauri (WTT) stars or Naked-T Tauri (NTT) stars. Many of these X–ray selected PMS stars, however, appear to be coeval with CTT's, and thus cannot be PTT's in the sense of Herbig.

Gahm *et al.* (1983) suggested an alternative method for finding PTT stars. This consists of searching for visual binaries with early-type primaries and late-type secondaries. If the binaries are physical and not just optical pairs, their late-type secondaries should still be contracting towards the ZAMS. In this way Lindroos was able to identify 78 systems with likely PTT secondaries.

In order to discriminate between physical and optical pairs we have carried out a spectroscopic survey of these PTT candidates: if the late-type companion is a genuine PMS star it should show signatures of youth, like a strong Li 6708

Å absorption line, as well as strong Ca II H and K emission lines and a filled-in Hα.

OBSERVATIONS

45 of the 78 binaries in Lindroos' list have secondaries later than F0. In total, we observed 37 of them in the red and 32 of these also in the blue. Two additional stars that were in Gahm *et al.* , but not in Lindroos were also observed.

The observations were carried out at the European Southern Observatory (ESO) at La Silla, Chile, using a variety of telescopes and instruments as shown in Table I. Additional high-resolution observations using larger telescopes are now in progress.

Table I: Summary of Observations.

Telescope	Instrument	Range	Res.	lines	Stars
2.2m MPI	B&C + CCD	5900 - 6900 Å	1.9 Å	Hα and Li	17
1.5m ESO	B&C + CCD	5900 - 6900 Å	2.2 Å	Hα and Li	22
1.5m ESO	B&C + CCD	3800 - 4300 Å	1.0 Å	Ca II H&K	32
1.4m CAT	CES + CCD	6680 - 6730 Å	0.1 Å	Li	5
1.4m CAT	CES + CCD	6538 - 6588 Å	0.1 Å	Hα	4

RESULTS

The main results emerging from our survey are the following: *i)* We have detected a very strong Li line in only 13 secondaries (plus one in Gahm *et al.* but not in Lindroos), and weak Li in 5 additional objects. For the same stars Hα is seen in emission or filled-in in only 6 cases out of 39. *ii)* We have detected Ca II H&K emission, at our resolution, in only 8 of the 32 stars observed in the blue, *i.e.* in a somewhat larger percentage than for stars showing filled-in Hα profiles. In 3 cases we were not able to establish whether Ca was in emission or not (two of them correspond to stars having a strong Li, while no Li is present in the spectrum of the third object).

We have used a few stars observed at both high and low resolution to determine the accuracy of our determinations of Li equivalent widths. For those stars for which no Li was detected, the derived upper limits are typically $\leq$ 50mÅ. We have used comparison stars with well-known Ca II K fluxes –observed at the same resolution as the stars in our sample– to estimate the chromospheric emission of our stars and/or to puts limits on the detectability of Ca II in our spectra. From this analysis we conclude that only a fraction ($\leq$ 40 %) of the late-type secondaries in Lindroos' list are likely to be PMS objects, while the other stars in the sample ($\geq$ 60%) lack both a strong Li absorption and Ca II emission.

In Fig. 1 we show an HR diagram of the observed objects. Filled symbols represent stars with Ca II H&K in emission and/or strong Li. All of them are very close to the ZAMS, thus confirming the hypothesis that they are genuine PTT's. Other stars from Lindroos which -if physical pairs- would be located well above the ZAMS (open symbols in the figure) do not show any evidence of youth. These are more likely optical pairs.

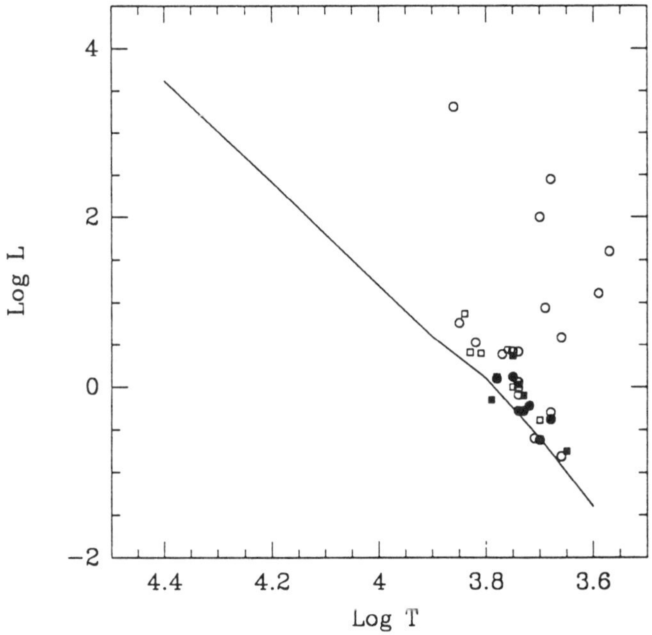

Fig. 1: HR diagram of the sample objects. Filled circles represent stars with strong Li and strong Ca II H & K; filled squares represent stars with strong Li, but no Ca II emission; empty squares represent stars with weak Li and weak or absent Ca II H & K; empty circles represent stars which lack both Li absorption and Ca II emission.

REFERENCES

Gahm, G.F., Ahlin, P., Lindroos, K.P.: 1983 *Astron.Ap.Supp.* **51**, 143.

Herbig, G.H.: in *Problems of Physics and Evolution of the Universe*, L.V. Mirzoyan ed., p. 171.

Lindroos,K.P.:1986, *Astron.Ap.* **156**, 223.

Walter, F.M., Brown, A., Mathieu, R.D., Myers, P.C., and Vrba, F.J.: 1988, *Astron.J.* **96**, 297.

Seventh Cambridge Workshop on Cool Stars, Stellar Systems, and the Sun
ASP Conference Series, Vol. 26, 1992
Mark S. Giampapa and Jay A. Bookbinder (eds.)

THE CASE FOR SPOTS ON THE NAKED T TAURI STAR LkCa-7

FREDERICK M. WALTER and SCOTT WOLK ESS Department, SUNY, Stony Brook NY 11794-2100

FREDERICK J. VRBA US Naval Observatory, Flagstaff AZ 86002

ALEXANDER BROWN JILA, University of Colorado, Boulder CO

ABSTRACT Apparent near-IR excesses around naked T Tauri stars have been interpreted as due to emission from cool circumstellar dust, or to the presence of cool photospheric starspots with large filling factors. Based on the former, it has been argued that the NTTS are not naked, but are surrounded by detectable, optically thin passive disks. We show here that that the apparent excess in LkCa-7 is best explained as a consequence of starspots with a filling factor of ≈50%.

Keywords: pre-main sequence stars; starspots

INTRODUCTION

Walter (1986) coined the term "naked T Tauri star" to describe a population of X–ray discovered, low mass pre-main sequence stars. The characteristics of this population are described by Walter *et al.* (1988). In general the naked T Tauri stars (NTTS) possess "normal" photospheres, despite ages comparable to those of the classical T Tauri stars (CTTS). The optical emission lines and the X–ray emission most likely arise in compact, saturated chromospheres and coronae. We suggested that the major difference between the NTTS and the CTTS was a lack of the warm circumstellar material which gives rise to the continuum excesses and the extreme line emission of the CTTS. In other words, the NTTS did not have circumstellar disks close in to the photosphere: they were unveiled, or naked, versions of the classical T Tauri stars.

The near-IR excesses of most CTTS arise from reprocessing of stellar radiation in the inner few AU of their optically thick circumstellar disks (Bertout, Basri, and Bouvier 1988). If the NTTS lack these circumstellar disks, clearly there should be no detectable near-IR excess from warm dust. In the J-H,H-K two-color diagrams, most of the NTTS in the Tau-Aur complex fall on the main sequence (T>2000K). We interpreted lack of evidence of warm dust as evidence of a lack of warm dust (We cannot constrain cooler, more distant material, which radiates at longer wavelengths.).

A complete lack of circumstellar material surrounding million year old stars was unexpected. A number of observers scrutinized the near-IR colors of the NTTS to search for evidence of small excesses attributable to optically thin dust disks. Strom *et al.* (1989) argued that a significant number of NTTS have V-K excesses ≥0.1 dex, and that these excesses were attributable to reprocessed radiation from optically thin circumstellar disks.

CHARACTERISTICS OF LkCa-7

LkCa-7 was discovered by Herbig, Vrba, and Rydgren (1986) in an objective prism Ca II H and K survey of the Taurus clouds. It was independently discovered as the counterpart of the X-ray source NTTS041636+2743 (Walter *et al.* 1988). This M0 star is typical of the NTTS. Walter *et al.* (1988) present spectra and mean colors of this star. There is no IRAS source at this position (Weaver and Jones 1991; Strom *et al.* 1989). Strom *et al.* 1989 claim a V-K excess of 0.14 dex, which they interpret as due to emission from optically thin circumstellar dust. Skinner, Brown, and Walter (1991) find a 3σ upper limit of $0.02M_\odot$ to the mass of any circumstellar disk.

Vrba and Weaver (unpublished) found a 5.7 day photometric modulation of the light from the star. The modulation is sinusoidal and symmetric, like the light curves of other spotted NTTS (Rydgren and Vrba 1983). The modulation appears coherent over a 4 year period (1986-1989), with a period of 5.66 days. We take this to be the stellar rotation period. The amplitude is about 0.5 mag at V, decreasing to 0.15 mag at K. No modulation is seen at U or B. Blackbody fits to the colors indicate a photospheric temperature of 4000K, with spots about 700K cooler. The spots fill about 40% of the darker hemisphere.

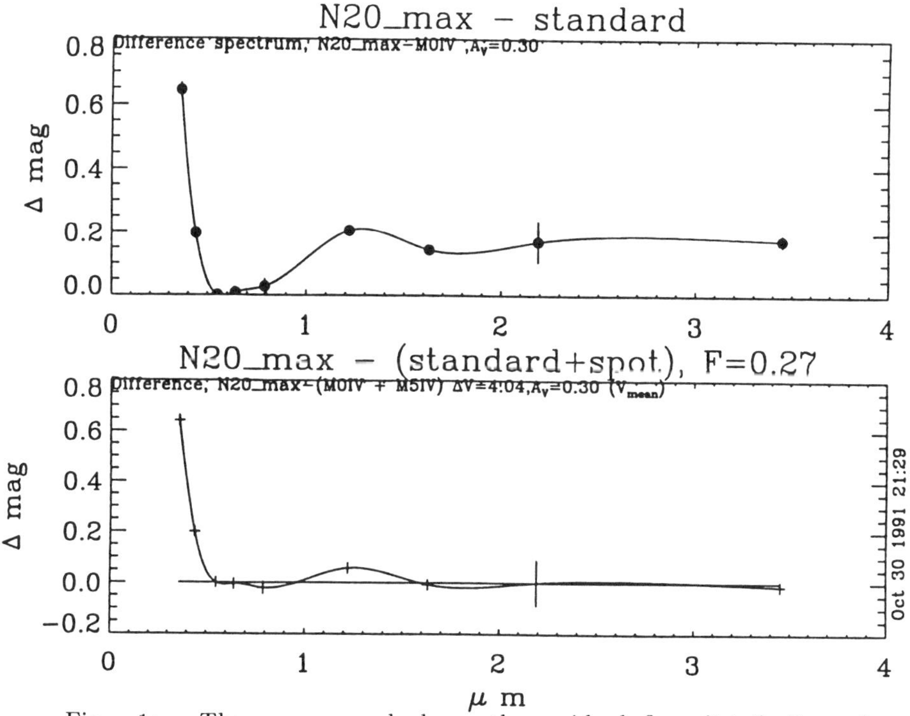

Fig. 1: The upper panel shows the residual flux distribution after subtraction of an M0IV standard, normalized at V. The lower panel shows the residual fluxes after inclusion of a cool starspot (an M5 IV star).

Figure 1 shows the residual flux from this star at maximum light, relative to M0IV standard colors, before and after adding a cool spot to the standard. The U and B excesses are comparable to those of dMe stars, and we attribute them to chromospheric line emission. The near-IR excess fluxes show no significant wavelength variation (as is also true for the excesses displayed by Strom *et al.* 1989), and must come from a region much hotter than the ≈1300K maximum for silicate dust. The best spot is 800K cooler (M5) than the photosphere, with a 27% filling factor. Note that this refers to the *bright* side of the star. The limit to V *sin i*, the stellar radius, and the rotation period force the inclination to be $< 30^o$. Unless the spots are concentrated close to the equator some will always be visible. A similar analysis of the star at minimum light, when it is redder, yields a spot filling factor of 60%.

The mean equivalent width of Hα is 3Å, comparable to that of dMe stars. The Hα emission line was observed at high dispersion simultaneous with optical photometry in October 1988. The Hα emission flux varied in antiphase to the continuum flux, as expected in the spot model. There were no significant variations in the line width or profile.

CONCLUSIONS

LkCa-7 is a prime example of a naked T Tauri star with a significant near-IR excess. But rather than indicating the presence of circumstellar material, the excess arises in cool starspots covering a significant fraction of the stellar surface. Spots have been seen in other NTTS, and indeed are expected, based on analogy with the similar-appearing stars in RS CVn and BY Dra systems. In all regards, except for its 10^6 year age and $2R_\odot$ radius, LkCa-7 is indistinguishable from a dMe star.

Similar analysis of other NTTS (Wolk *et al.* 1991, in preparation) is yielding similar results. With the exception of a handful of stars, including V836 Tau (Mundt *et al.* 1983), which have excesses indicative of 1000K dust, there is no evidence to support suggestions that the NTTS are not truly naked (at least within radii of an AU).

ACKNOWEDGEMENTS

This work has been supported by NSF grant AST89-96308, and NASA grants NAG5-1279 and NAG8-844 to the State University of New York.

REFERENCES

Bertout, C., Basri, G. S., and Bouvier, J. 1988, *Ap. J.*, **330**, 350.
Herbig, G., Vrba, F. J., and Rydgren, A. E. 1986, *A. J.*, **91**, 575.
Mundt, R. *et al.* 1983, *Ap. J.*, **269**, 229.
Rydgren, A.E., and Vrba, F.J. 1983, *Ap. J.*, **267**, 191.
Skinner, S. L., Brown, A., and Walter, F. M. 1991, *A. J.*, (in press).
Strom, K. M. *et al.* 1989, *A. J.*, **97**, 1451.
Walter, F. M. 1986, *Ap. J.*, **306**, 199.
Walter, F. M., *et al.* 1988, *A. J.*, **96**, 297.
Weaver, B., and Jones, 1991, preprint.

Seventh Cambridge Workshop on Cool Stars, Stellar Systems, and the Sun
ASP Conference Series, Vol. 26, 1992
Mark S. Giampapa and Jay A. Bookbinder (eds.)

MAGNETIC ACTIVITY AND ORBITAL PERIOD MODULATION IN CLOSE BINARIES

JAMES H. APPLEGATE Department of Astronomy, Columbia University, Pupin Hall, 538 West 120th Street, New York, N.Y., 10027

ABSTRACT Eclipsing variables often undergo orbital period modulations of amplitude $\Delta P/P \sim 10^{-5}$ over timescales of decades or longer. These modulations can be explained by the gravitational coupling of the orbit to variations in the rotational oblateness of a magnetically active star in the system. The oblateness variations are caused by variations in the star's differential rotation as it goes through its activity cycle. This mechanism typically requires that the active star be variable at the $\Delta L/L \simeq 0.1$ level and differentially rotating at the $\Omega_{dr}/\Omega \simeq 0.01$ level. The torque needed to redistribute the angular momentum can be exerted by a mean subsurface magnetic field of several kilogauss.

Keywords: eclipsing binaries; magnetic stars

INTRODUCTION

Eclipsing variables often exhibit nonmonotonic orbital period changes (modulations) of amplitude $\Delta P/P \simeq 10^{-5} - 10^{-6}$ over timescales of decades. The most striking example of this phenomenon is Algol (Soderhjelm 1980), whose eclipse record shows several cycles of a modulation with a period of roughly 30 years and a single cycle of a modulation with a 180 year period. Similar orbital period changes are seen in many other systems (Kreiner 1971; Pringle 1975; Kreiner and Ziolkowski 1978; Hall and Kreiner 1980; Glownia 1986; Warner 1988; Skillman and Patterson 1988; Hall 1989, 1990, 1991). In the systems in which more than one modulation cycle has been seen it is clear that the modulations are recurrent, but not strictly periodic. This rules out apsidal motion, third bodies, and dynamical explanations in general.

The decade timescale and lack of strict periodicity of the orbital period modulations suggests that they might be a manifestation of magnetic activity. Strong evidence in favor of this hypothesis was presented by Hall (1989) who found that of the 101 Algols he studied, all 31 systems which showed nonmonotonic orbital period changes contained at least one convective star. Since all of the stars in the Algols should be rapidly rotating, the presence or absence of a convective envelope determines whether or not all of the ingredients of a dynamo are present. Orbital period modulation only occurs in systems which possess all of the ingredients of a dynamo.

In the model described here the orbital period change is due to the gravitational coupling of the orbit to changes in the quadrupole moment (rotational oblateness) of a magnetically active star in the system. The rotational quadrupole moment of a star is determined by the rotation rate of its outer layers. If angular momentum is transferred to the outer layers of a star they rotate faster and the star becomes more oblate. If the outer layers lose angular momentum the oblateness decreases. Qualitatively, the shearing of the magnetic field by differential rotation is a key ingredient of the dynamo mechanism, so the differential rotation of the star should vary through the activity cycle at some level. Quantitatively, a subsurface magnetic field of several kilogauss can exert a large enough torque to transfer the angular momentum needed to make the observed period changes.

GRAVITATIONAL QUADRUPOLE COUPLING

Kepler's third law is not exact for close binaries because rotation and tides deform the stars away from spherical symmetry. The orbital period of a close binary depends on the internal structure of the stars in the binary; if the mass distribution in one of the stars changes, the orbital period of the binary will change. It is important to realize (Matese and Whitmire 1983; Applegate and Patterson 1987) that the orbital period changes at fixed orbital angular momentum. This is crucial because the modulation period is far too short for any known torque to operate. The details of the model sketched below are given in Applegate (1992).

A detailed calculation shows that the changes in quadrupole moment and orbital period are related by

$$\frac{\Delta P}{P} = -9\left(\frac{R}{a}\right)^2 \frac{\Delta Q}{MR^2}, \tag{1}$$

where M and R are the mass and radius of the active star, and a is the binary separation. The angular momentum ΔJ that must be transported within the active star in order to produce an orbital period change ΔP is

$$\Delta J = -\frac{GM^2}{R}\left(\frac{a}{R}\right)^2 \frac{\Delta P}{6\pi}. \tag{2}$$

The energy cost of transferring this angular momentum is

$$\Delta E = \Omega_{dr} \Delta J, \tag{3}$$

where Ω_{dr} is the difference in angular velocity between the parts of the star exchanging the angular momentum. This energy cost leads to an RMS luminosity variation of

$$\Delta L = \pi \frac{\Delta E}{P_{mod}}, \tag{4}$$

where P_{mod} is the period of orbital period modulation. If the angular momentum is transferred by a magnetic torque, the mean subsurface field strength is

$$B^2 \sim 10 \frac{GM^2}{R^4}\left(\frac{a}{R}\right)^2 \frac{\Delta P}{P_{mod}}. \tag{5}$$

RESULTS

System	P_{orb}	$\Delta P/P$	P_{MOD}	$\Delta L_{RMS}/L$	Ω_{dr}/Ω	B (kG)
Algol	2.87 d	1.6×10^{-5}	32 yr	0.2	0.014	5.5
Algol	2.87 d	1.4×10^{-5}	180 yr	0.04	0.014	2.3
SS Cam	4.8 d	3×10^{-5}	55 yr	0.10	0.012	3.5
SV Cam	0.53 d	3×10^{-6}	80 yr	0.14	0.004	9
RS CVn	4.8 d	2×10^{-5}	70 yr	0.3*	0.04	13
V471 Tau	0.52 d	10^{-6}	20 yr	0.12	0.003	11

*Requires Special Assumptions.

Results from applying the model to the orbital period modulations seen in a few systems are given in the table. The observations of Algol are taken from Soderhjelm (1980), data for SS Cam, SV Cam, and RS CVn are taken from Hall and Kreiner (1980), and the data on V471 Tau are from Skillman and Patterson (1988). The model fits the data well for all but RS CVn. The observed orbital period modulations can be explained by the changes in the distribution of angular momentum produced by the torque exerted by a subsurface magnetic field of several kilogauss. The energy budget of these changes demands that the active stars be variable, and this variability is seen in V471 Tau (Skillman and Patterson 1988) and in CG Cygni (Hall 1991). The inferred level of differential rotation is consistent with that measured for rapidly rotating spotted stars (Hall and Busby 1990).

REFERENCES

Applegate, J.H. 1992 to appear in *Ap. J.*, Feb. 1, 1992.
Applegate, J.H., and Patterson, J. 1987 *Ap. J. Letters*, 322, L99.
Glownia, Z. 1986 *Adv. Sp. Res.*, 6,211.
Hall, D.S. 1989 *Sp. Sci. Rev.*, 50, 219.
Hall, D.S. 1990 in Active Close Binaries, ed. C. Ibanoglu, Dordrecht: Kluwer, p95.
Hall, D.S. 1991 *Ap. J. Letters*, 380, L85.
Hall, D.S., and Busby, M.R. 1990 in Active Close Binaries, ed. C. Ibanoglu, Dordrecht: Kluwer, p377.
Hall, D.S., and Kreiner, J.M. 1980 *Acta Astr.*, 30, 387.
Kreiner, J.M. 1971 *Acta Astr.*, 21. 365.
Kreiner, J.M., and Ziolkowski, J. 1978 *Acta Astr.*, 28, 497.
Matese, J.J., and Whitmire, D.P. 1983 *Astr. Ap.*, 117, L7.
Pringle, J.E. 1975 *M.N.R.A.S.*, , 170, 633.
Skillman, D.R., and Patterson, J. 1988 *A.J.* , 96, 976.
Soderhjelm, S. 1980 *Astr. Ap.*, , 89, 100.
Warner, B.W. 1988 *Nature*, 336, 129.

Seventh Cambridge Workshop on Cool Stars, Stellar Systems, and the Sun
ASP Conference Series, Vol. 26, 1992
Mark S. Giampapa and Jay A. Bookbinder (eds.)

MAGNETIC ACTIVITY OF SHORT-PERIOD RS CVN STARS

MICHAEL ZEILIK Institute for Astrophysics, The University of New Mexico, Albuquerque, NM 87131 U.S.A.

ABSTRACT We examine the long-term, photometric record of starspot activity on selected members of the short-period groups of RS CVn stars. We apply our information limit optimization technique (ILOT) to extract starspot parameters (longitude, latitude, radius, and temperature contrast). In general, we find that one large (radius $\approx 10^{\circ}$) active region at a high latitude ($\approx 45^{\circ}$) in one active longitude belt ($\approx 90^{\circ}$ or 270°) on the primary star dominates the magnetic activity at one time for a period of about one year. Over about a month, the seat of activity switches from one active longitude belt to the other. However, we cannot yet identify any clear signature of the activity "cycle". Clearly, the stellar dynamos here are very different than that for the sun.

Keywords: RS CVn stars; chromospherically-active binaries

INTRODUCTION

We do not yet understand the magnetic activity cycles of cool close binary systems of sunlike stars. To arrive at the basic physics, though, requires that we first examine the phenomenology of magnetic activity for binary systems. In particular, we would like to discover if such activity has a clearly-defined *cycle*, such as the sun exhibits.

Among the proxy indicators of magnetic activity are the Ca II H and K lines. Strassmeier *et al.* (1988) used the strength of these lines as the primary criterion for the inclusion of systems in *The Catalog of Chromospherically Binary Stars.* Of the RS CVn stars in the catalog, 12 have orbital periods of one day or shorter; 9 are eclipsing systems. As part of a decade-long program, we have focused our observations and models on eight of the short-period group (Hall, 1976): XY UMa, UV Psc, SV Cam, RT And, CG Cyg, ER Vul, BH Vir, and WY Cnc. These close systems are tidally-locked in synchronous rotation and tidally-distorted into Roche lobe configurations.

A few of these systems have archival observations that span decades to supply the long time base required to probe for cyclical activity. We extend these time series with current observations with our CCD-camera at Capilla Peak Observatory and those data collected recently by other groups. We have completed a comprehensive starspot analysis on data for SV Cam (Zeilik *et al.* , 1988), RT And (Zeilik *et al.* , (1989), BH Vir (Zeilik *et al.* , 1990a), and WY Cnc (Zeilik *et al.* , 1990b). We add preliminary results for CG Cyg and XY UMa in this paper.

ANALYSIS

Our goal is to apply the simplest possible starspot model (Budding, 1977) to parameterize photometric light curves in a way that reliably and objectively extracts information relevant to photospheric magnetic activity. Using circular, dark spotted regions as the representation (Budding and Zeilik, 1987), we can optimize for the longitude, latitude, radius, and temperature (relative to the photosphere if color information is available) of the active area.

Previous criticisms of photometric modeling (*cf.* Rondonó, 1986) have aimed at the lack of uniqueness of the starspot solutions. That limitation is especially true when applied to *non-eclipsing* systems, as has usually been the case so far. *Eclipsing* systems present a different aspect; the orbital inclinations are known within a degree or so and the eclipses provide fiducial markers for longitude. Hence, eclipsing systems provide a much more constrained modeling problem: the drawback is that we need to separate eclipse and proximity effects to reveal the distortion waves. We have done so by modifying Budding's 19-parameter, optimizing curve fitter (Budding, 1973; Budding and Najim, 1980) to generate the model curves for the eclipsing system. These models use Roche lobes for the distorted stars, and include both the reflection and ellipticity effects. The difference between an optimized theoretical curve and the observed one defines the distortion effect. Another curve fitter, corresponding to the case of one or two dark, circular spots, generates an optimized starspot solution to these residuals.

Our programs (Budding and Zeilik, 1987) contain an optimizer routine that uses three methods to search efficiently for a minimum in χ^2 hyperspace (Banks, 1989). What sets our procedure apart is our *information* limit analysis, which provides an objective evaluation of the results from the optimized parameters, in the sense of the information content of the data. We call this approach the information limit optimization technique (ILOT). Our programs evaluate the *curvature Hessian* in the vicinity of the optimal solution in χ^2 hyperspace (Budding and Najim (1980). We also calculate the *error matrix*, which is the inverse of the curvature Hessian. For mathematically-determinate solutions, the curvature Hessian will have a positive eigenvalue for each optimized parameter. We approach the information limit by increasing the number of parameters until the Hessian becomes nonpositive definite; we then know that we have exceeded the information limit of the data.

RESULTS FOR THE SELECTED SHORT-PERIOD SYSTEMS

Our main results to date, which apply to the primary (hotter) star in these short-period systems, are: (1) The primary activity signature is two active longitude belts (ALBs) at the quadrature points; one belt is usually more active than the other at any given time. (2) The secondary activity signature is that the active regions appear most often at high latitudes (near 45°). (3) The spotted regions cover large areas of the primary, active stars (starspot radii are typically 10° to 20°). (4) These cool areas have temperatures about 1000 to 1500 K below that of the photosphere with a variation of only a few hundred Kelvins (for those light curves for which we have simultaneous *V* and *R* or *I* data).

A few comments about specific systems. SV Cam: over the time from 1928 to 1984, the ALB around 270° has been more active. The mean latitude value for determinate solutions is 58° The differential rotation rates within the ALBs are much smaller than the sun's. RT And: again, the ALB near 270° seems the most active (from 1920 to 1989). The mean value of determinate latitudes was 53°. The orbit does have a small eccentricity (e~0.03). BH Vir: both ALBs have roughly the same activity from 1953-1986. The mean value of the determinate latitudes was 31°. The temperature contrast averages only 1100 K. WY Cnc: The ALB at 90° seems more active in 1964-1989. The mean latitude of determinate's solutions was 16°; this is the only system that we have analyzed that resulted in mostly low-latitude spots. The temperature contrast was about 1500 K. XY Uma: From 1982-90, the ALB at 90° was the most active. Average spot latitude over all epochs was 51°. In 1989, the temperature contrast was ≈1500 K. CG Cyg: From 1922-91, the ALB at 90° has been somewhat more active. Latitudes have averaged 47°. We carefully search these data sets for instances when both ALBs were active; we found three in 1984, 1979, and 1991. These data represented about 10% of all epochs covered.

We note that the ALB activity appears binary with two major states. When one is "on", the other is "off", at least in terms of large spotted regions. (We cannot "see" the spots in good photometry if their radii is less that 5°.) We have so far found few instances in which we have a clear indication of both ALBs containing activity at the same time. The tendency is for both ALBs to be active about 10 to 15% of the time. (Nelson and Zeilik [1990] noted, for a sample of 31 RS CVn systems, that one or two spot groups accounts for the photometric periodicities in APT observations from 1983 to 1987.) The activity within an ALB lasts for about a year; the transition time is about a month. No cycle signature is evident. The mode-switching behavior may reflect a low-dimensional chaotic nature in the magnetic activity.

ACKNOWLEDGEMENT

This research was supported in part by NASA Grant AST-8903174.

REFERENCES

Budding, E., 1973, *Ap. Space Sci.* **22**, 87.

Budding, E., 1977, *Ap. Space Sci.* **48**, 287.

Budding, E., Najim, N.N., 1980, *Ap. Space Sci.* **72**, 369.

Budding, E., Zeilik, M., 1987, *Ap. J.*, **319**, 827.

Nelson, E., Zeilik, M., 1990, *Ap. J.*, **349**, 163.

Rodonó, M., 1986, in Cool Stars, Stellar Systems, and the Sun, ed. M. Zeilik and D. Gibson (Heidelberg: Springer Verlag), p.470.

Strassmeier, K.G., Hall, D.S., Zeilik, M., Nelson E., Eker, Z., Fekel, F.C., 1988, *Astr. Ap. Suppl.*, **72**, 291.

Vogt, S.S., Penrod, G.D., 1983, *Pub. A.S.P.*, **95**, 65.

Zeilik, M., De Blasi, C., Rhodes, M., Budding, E., 1988, *Ap. J.*, **332**, 293.

Zeilik, M., Cox, D.A., De Blasi, C., Rhodes, M., Budding, E., 1989, *Ap. J.*, **345**, 991.

Zeilik, M., Ledlow, M., Rhodes, M., Arevalo, M.J., Budding, E., 1990a, *Ap. J.*, **345**, 352.

Zeilik, M., Cox, D.A., Ledlow, M.J., Rhodes, M., Heckert, P.A., Budding, E., 1990b, *Ap. J.*, **363**, 647.

Seventh Cambridge Workshop on Cool Stars, Stellar Systems, and the Sun
ASP Conference Series, Vol. 26, 1992
Mark S. Giampapa and Jay A. Bookbinder (eds.)

OPTICAL AND ULTRAVIOLET ACTIVITY DIAGNOSTICS FOR THE RS CVN BINARY HK LAC

C. J. BUTLER Armagh Observatory, Armagh BT61 9DG, N.Ireland

E.R. HOUDEBINE Dunsink Observatory, Dublin 15, Ireland

K. OLAH Konkoly Observatory, 1525 Budapest XII, Box 67, Hungary

E. BUDDING Carter Observatory, Wellington, New Zealand

A GIMINEZ Inst. de Astrofisica de Andalucia, Granada, Spain

M. ZEILIK Inst. for Astronomy, UNM, Albuquerque, USA

ABSTRACT Optical photometry and IUE spectroscopy of the RS CVn star, HK Lac, show rotational modulation with an enhanced ultraviolet emission close to the phase of optical minimum. This is interpreted in terms of an association between UV emitting plages and dark photospheric spots. The Mg II *h* and *k* lines are considerably wider than would be expected from rotational modulation alone.

Keywords: RS CVn star; UV spectra; Spectral line profiles; Starspots

INTRODUCTION

HK Lacertae, with an orbital period of 24.428 days, is a member of the long-period group of RS CVn binaries. Its primary component is an evolved star, a K0 giant, and its secondary is listed by Strassmeir (1988) as an F1 main sequence star. The light curve of HK Lac has an amplitude, $\Delta V \sim 0.25$ mag, and a period 24.461 days; which suggests that the rotation of the active, K0, component is tidally locked to the orbital motion.

OPTICAL AND UV DATA AND their INTERPRETATION

Optical photometric and ultraviolet spectroscopic data have been gathered in the years 1979, 1980 and 1987. The optical light curve has been modelled by starspots using the technique described by Olah et. al (1985). It was found that one high and one low latitude spot, in a similar configuration, were sufficient to account for the light curves for the three years during which observations were obtained.

Fluxes of the prominent UV emission lines have been extracted from IUE spectra and are plotted against orbital phase in figure 1 together with the optical light curves. The most noticeable feature is that the ultraviolet emission lines are stronger at optical minimum and that this behaviour persists over the period

1979-1987. The canonical interpretation of this behaviour, which ultimately we are inclined to accept, is that it results from an association between dark spots in the photosphere of the active component of HK Lac with bright UV-emitting plages in its chromosphere.

CORRECTION OF MG II EMISSION LINE FLUXES FOR THE ORBITAL MOTION ABSORPTION EFFECT

For the Mg II *h, k* emission lines the situation is complicated by the strong and variable interstellar absorption where the variability of the absorption is due to the binary motion of the emitting star. In effect the emission line from the active binary component is doppler shifted, backwards and forwards, across the stationary absorption with a greater fraction of the emission absorbed by the interstellar medium when the peak of the absorption profile is at the same wavelength as the peak of the emission line. This variable absorption will produce a modulation of the emission line flux that will have a single, or double peaked curve depending on the relative velocity of the interstellar medium and the systemic velocity γ of the binary system. From knowledge of the value of γ and the velocity of the interstellar medium with respect to the Earth, plus the percentage absorption at line-centre, one can compute the expected modulation of the observed flux and recover the true Mg II line fluxes emitted by the star.

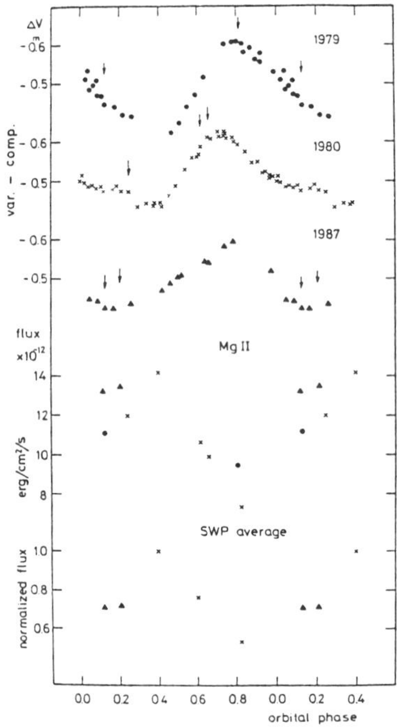

Fig. 1: Comparison of the optical light curve with UV emission line fluxes. Note that the UV emission line flux is anticorrelated with the V-band flux.

The results predict an average flux depletion of 22% for the high resolution spectra of HK Lac and are in good agreement (to within 3.4%) with those obtained by Gaussian fits to the observed profiles. In figure 2 the observed fluxes are plotted together with the computed flux modulation. The amplitude of the modulation of the interstellar correction is about 4% and the curve is double peaked. Thus, it is evident that the observed modulation of the HK Lac data cannot be explained through interstellar absorption effects. The double peaked curve occurs because of the closely similar velocities of the binary system and the interstellar medium, of −23.6 km/s and −21 km/s respectively and the relatively small Doppler motion for this star which restricts the absorption to close to the stellar line centre throughout the binary period.

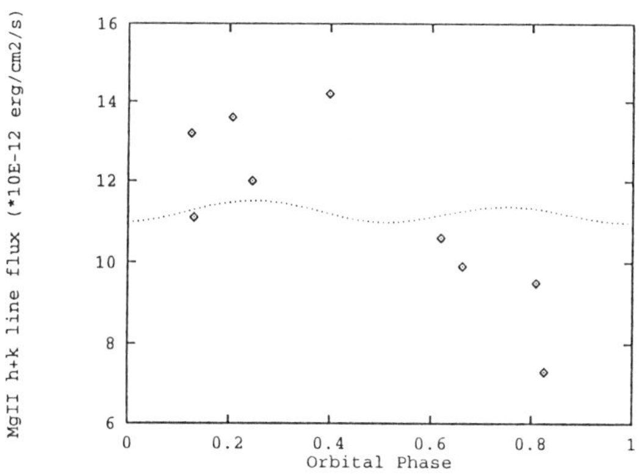

Fig. 2: The observed Mg II h+k fluxes against orbital phase. The dotted line represents the computed modulation of the stellar flux from the binary motion effect.

A CORRELATION BETWEEN THE V-BAND AND UV EMISSION LINE FLUXES

In figure 3 we show, for HK Lac, the V-band flux normalised to an assumed immaculate (unspotted) level plotted against the integrated flux in the Mg II and the SWP emission lines. Here the immaculate level we have assumed is V=6.70, the brightest value ever measured photometrically for HK Lac. In figure 3 both the sum of the SWP emission lines and the Mg II emission are well correlated with V-band flux in the expected sense, with correlation coefficients of -0.615 and -0.876 respectively. This is highly suggestive that, for HK Lac, spot and plage filling factors are correlated. Further, we may note, that the two projected lines of correlation illustrated in figure 3 intersect with the line, *emission line flux = 0*, at a single point. This is strongly indicative that, when the emission line flux is zero, there are no plages evident in either the SWP lines or in Mg II. Also, it could be assumed that the intersection on the Y-axis occurs at the immaculate V-band flux level, (V=6.51). As a corollary to this, it is seen that HK Lac must always have been observed in a spotted state.

THE STELLAR, HIGH RESOLUTION, Mg II LINE PROFILES

As mentioned earlier, the high resolution Mg II line profiles were fitted by two Gaussians, one representing the emission line from the active component, and the other, the absorption component from the IM. This enabled us to reconstruct the true, unabsorbed, stellar line profile. Our next step was to model this intrinsic line profile by convolving 'dish' and 'bell' shaped profiles, (Budding and Zeilik, 1990), to represent the broadening due to rotation and a second mechanism of unknown origin, (possibly turbulence). Since the active star rotates approximately in synchronism with the orbital motion a moderate equatorial velocity of 30-40 km/s can be expected. However the observed width of the Mg II k line of $\sim 1.4A$ FWHM is considerably greater than this and requires the broadening of unknown origin to be about twice that due to rotation. This is of some consequence for the Doppler imaging method of identifying plage regions from the shape of the Mg II lines.

This non-rotational broadening appears to be the Mg II analogue of the well known *Wilson-Bappu* correlation between the width of the Ca II K line and absolute magnitude.

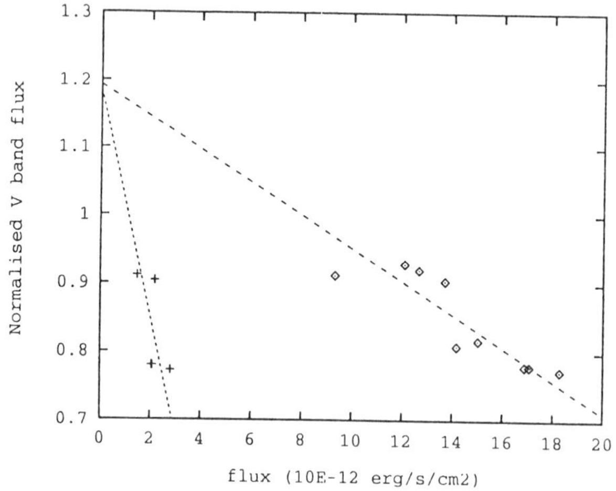

Fig. 3: Relative V-band versus: the corrected Mg II h+k flux (diamonds), and the sum of the line fluxes in the IUE SWP range (crosses).

REFERENCES

Budding, E. and Zeilik, M., 1990a, Proc of NATO Adv. Study Inst., Kusadasi, Kluwer, p. 831.

Oláh, K., Eaton, J.A., Hall, D.S., Henry, G.W., Burke, E.W., Chambliss, C.R., Fried, R.E., Landis, H.J., Louth, H., Renner, T.R., Stelzer, H.J., Wasatonic, R.P., 1985, *Astrophys. Space Sci.* **108**, 137.

Strassmeier, K.G., Hall, D.S., Zeilik, M., Nelson, E., Eker, Z., Fekel, F.C., 1988, *Astr. Ap. Suppl.*, **72**, 291.

Seventh Cambridge Workshop on Cool Stars, Stellar Systems, and the Sun
ASP Conference Series, Vol. 26, 1992
Mark S. Giampapa and Jay A. Bookbinder (eds.)

MAPPING MAGNETIC FIELDS ON RAPIDLY ROTATING STARS: APPLICATION TO THE RS CVN SYSTEM HR 1099

J.-F. DONATI Univ. of Western Ontario, London, N6A 3K7 Canada
S.F. BROWN Univ. of Sydney, Sydney NSW 2006, Australia
M. SEMEL Observatoire de Paris, F-92195 Meudon, France
D.E. REES CSIRO, Epping NSW 2121, Australia

ABSTRACT We report the first results of a campaign aimed at measuring magnetic fields on RS CVn systems using Zeeman-Doppler Imaging. On HR 1099, σ^2 CrB, II Peg and UX Ari, magnetic fields have been reliably detected. A maximum entropy image reconstruction code was applied to the Sep. 1990–Jan. 1991 data set obtained for HR 1099 to derive both temperature and magnetic maps of the cooler star photosphere at that epoch. We find that the field is mainly concentrated in dark regions and appear to be toroidal. Comparing with the results we obtained in Aug. 1989 for the same star, we infer a variation of the magnetic toplogy from a radial to a toroidal structure, probably induced by differential rotation.

Keywords: RS CVNs; Magnetic Fields; Doppler Imaging

INTRODUCTION

Due to their very strong activity and high rotation rate, RS CVn systems are probably the best targets for studying stellar magnetic activity and providing more constraints for modelling dynamo processes in the outer layers of cool stars. Zeeman-Doppler Imaging (or ZDI) was developed (Semel 1989) as a method capable of deriving some local information on the photospheric temperature and magnetic field distributions of these stars.

ZDI measures circular polarization in profiles of high Landé g lines (unambiguous signature of magnetic fields) and is best adapted to rapid rotators for which the Doppler effect acts to disclose the signatures of opposite polarities in high resolution Stokes V spectra. Ultimately, one can use ZDI as a solar like way of studying stellar magnetic activity. In this regard, ZDI constitutes a major improvement in the domain of stellar magnetic field measurements on cool stars: methods which look for Zeeman broadening of Stokes I profiles access only magnetic parameters averaged over the stellar disk and thus give little information about the topology of the stellar photospheric field.

We report herein the first results of a 2 yr. campaign aimed at measuring magnetic fields on RS CVn systems using ZDI.

OBSERVING RS CVN SYSTEMS

Seventy-six pairs of Stokes I and V spectra were recorded at both the Anglo-Australian telescope (AAT) and Canada-France Hawaii telescope (CFHT) with the instrumental device described in Semel & Donati (1991), for 9 RS CVn systems, in a wavelength domain centered at 5500 Å. For the cool component of HR 1099, II Peg, UX Ari and σ^2 CrB, Zeeman signatures are detected in about 75% of the observed spectra. On σ^2 CrB, magnetic distortions are repeatedly observed in the cooler star polarized spectrum while no Zeeman signatures are visible for the hotter companion, although both stars have almost equal masses, radii, temperatures and activity levels (Donati *et al.* 1991).

TEMPERATURE AND MAGNETIC IMAGING OF HR 1099

We have developed a maximum entropy code allowing the reconstruction of 2D temperature and magnetic maps of stellar photospheres from Stokes V profile data sets. When applied to synthetic spotted magnetic distributions like those we expect to find in cool solar like stars, this code is found to be robust and accurate at obtaining both shape and location of spots, as well as field directions and strengths in these spots (Brown *et al.* 1991).

At CFHT in Sep. 1990 and Jan. 1991, 26 pairs of I and V spectra of HR 1099 were obtained around 8 different rotational phases. New atmospheric models from Kurucz (private communication) have been used to model the 5497 Å Fe I local line profiles for various temperatures and limb angles on the stellar surface. A maximum entropy map of photospheric temperatures has then been derived for the active K1 star (Fig. 1, top). The highest contrast feature is a polar cap (800 K cooler than the photosphere) slightly off centered towards phase 0.31 with an appendage extending to lower latitudes at phase 0.66. Another spot (400 K cooler than the photosphere) sits close to the equator at phase 0.33.

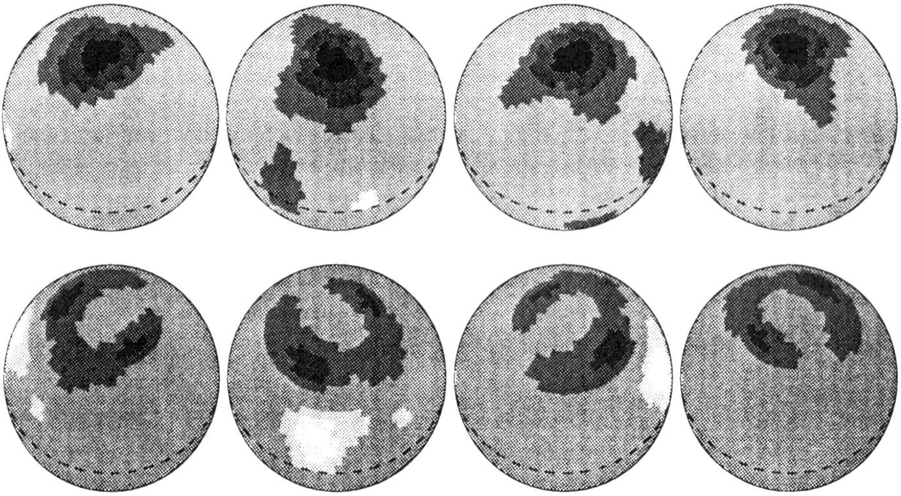

Fig. 1: Photospheric temperature (top) and magnetic (bottom) images of the K1 component of HR 1099 (as seen from the earth), displaying the star at phases 0.00 (left), 0.25, 0.50, and 0.75 (right). The rotational equator is shown by a dashed line. In the temperature image, darker grey levels indicate cooler temperatures (250 K levels), while bright and dark levels in the magnetic image indicate positive (counterclockwise) and negative (clockwise) values of the toroidal component of the field (200 G levels).

From the Stokes V data set, a maximum entropy map of photospheric magnetic field vectors was obtained. During the reconstruction process, the radial, toroidal and poloidal field components were free to vary. We find that the field concentrates in 2 main regions, within which it is toroidal (*i.e.* along the constant latitude lines). It should be noted that all attempts to reproduce the observed Stokes V profiles using only the radial and poloidal field components have failed. The first region consists of a ring of clockwise field (−300 G) encircling the rotational pole at latitude 60° (Fig. 1, bottom), similarly off centered than the cool polar spot. Moreover, this ring exactly interrupts where the cool polar spot appendage is observed, suggesting that the brightness and magnetic polar features overlap: fields may be present inside the ring as well (although not necessarily toroidal) and remain undetected in circularly polarized light due to the very low brightness of these central polar regions. Another magnetic feature of counterclockwise field (700 G) sits near the equator at phase 0.27, very close to its dark counterpart in the temperature map.

DISCUSSION

For the first time, we have found definite evidence for the presence of magnetic fields on RS CVn systems. Some stars keep showing Zeeman signatures witnessing the presence of large monopolar magnetic regions on their surface. By contrast, flat Stokes V profiles are repeatedly observed in other stars, indicating that any surface field is probably organized in small bipolar groups undetectable in circular polarization. In σ^2 CrB (F6V+G0V), both stars, although physically very similar, have different horizontal size scales for their magnetic regions: each star of the system being located on either side of the rapid deepening of the convective zone with increasing spectral types, it may indicate that different convective depths generate different dynamos modes. Alternately, it may be that each star is observed at a different phase of its own activity cycle.

The new magnetic data we obtained for HR 1099 show that the field emerges toroidally from dark photospheric regions, one polarity sitting on the rotational pole, the other one close to the equator. It firstly establishes that starspot activity on RS CVn systems has a magnetic origin. Although the derived toroidal field structure seems intriguing (since we expect photospheric field lines to emerge radially by buoyancy), it may be argued that important physical differences can exist between the huge magnetic regions we detected on HR 1099 and the classical solar flux tubes models. The preliminary results obtained for the same star in Aug. 1989 suggested that the field was emerging radially from warm equatorial regions (Donati *et al.* 1990): differential rotation (with a pole rotating faster than the equator) is likely to be responsible for the observed winding of the field lines around the star between 1989 and 1991.

REFERENCES

Brown, S.F., Donati, J.-F., Rees, D.E., Semel, M., 1991, A&A (in press)
Donati, J.-F., Semel, M., Rees, D.E., Taylor, K., Robinson, R.D., 1990, A&A 232, L1
Donati, J.-F., Semel, M., Rees, D.E., 1991, A&A (submitted)
Semel, M., 1989, A&A 225, 456
Semel, M., Donati, J.-F., 1991, A&A (submitted)

Seventh Cambridge Workshop on Cool Stars, Stellar Systems, and the Sun
ASP Conference Series, Vol. 26, 1992
Mark S. Giampapa and Jay A. Bookbinder (eds.)

THE QUIESCENT X–ray SPECTRUM OF THE RS CVN BINARY II PEG IN AUGUST 1989

J.G. DOYLE, P.B. BYRNE Armagh Observatory, Armagh BT61 9DG, N.Ireland

G.H.J. VAN DEN OORD Sterrenkundig Insituut, Postbus 8000, Utrecht, The Netherlands

B.J. KELLETT, G.E. BROMAGE Rutherford Appleton Laboratory, Chilton, Didcot, Oxon., England

ABSTRACT The quiescent X–ray spectrum of II Peg can not be fitted with a single temperature Raymond-Smith model. The best fitting results were obtained for a combination of a single temperature plasma together with a power-law to the data above 10 keV. The single temperature component implies a temperature of 12.4 10^K with an emission measure of 1.5 $10^{53} cm^{-3}$. The iron abundance in the quiescent spectrum was close to solar. We show that an interpretation in terms of free-free emission by non-thermal particles requires unrealistic large values for the number of particles involved and for the energy input required to maintain the population. The data can however be satisfactorily explained by a differential emission measure distribution of the form $\xi(T) = 2\ 10^{56}\, T^{-3/2}\ cm^{-3}\, K^{-1}$, indicating the presence of substantial amounts of plasma with temperatures in excess of $10^8 K$.

Keywords: RS CVn star; X–ray spectra; Raymond-Smith models

INTRODUCTION

II Peg (=HD 224085) is a single-line spectroscopic binary of spectral type K2 IV-V with an orbital period of approximately 6.7 days. It is photometrically variable showing a wave-like variation, of the type usually interpreted as due to cool surface spots and has been classified as an RS CVn-type star. Here we report on data obtained with the X–ray satellite GINGA during periods of non-flare activity. The spectra were accumulated every 16 sec, with II Peg being detected in quiescence throughout the observing period (approximately 60 hr.) at $6\, ct.\ s^{-1}$.

DATA REDUCTION & RESULTS

The resultant X–ray spectrum was fitted with combinations of Raymond and Smith model plasmas (Raymond and Smith 1977) and power-law photon distributions. Since II Peg is at 30 pc, the line-of-sight absorbing column density is essentially zero. The best fitting results were obtained for a combination of a single temperature plasma together with a power-law. The quiescent spectrum is incompatible with a single temperature model alone. Two-temperature models do result in a very high, but uncertain, temperature ($T_e > 10^8 K$) for the second component. The reason for this second high temperature component is the tail in the photon distribution, which extends up to 18 keV (see Fig. 1). We have made extensive efforts to check the validity of this high energy emission. We do not believe it is due to background subtraction problems: II Peg lies well away from the galactic plane ($l = 108.2°, b = -32.6°$), and it would require an excess of $\sim$ 10% over the 'average' diffuse hard X-ray background. This is equal to the galactic contribution *towards* the galactic centre (Iwan *et al.* 1982). We can also rule out a second source in the field, since neither the EINSTEIN Observatory nor EXOSAT reported any other point sources in the field of II Peg.

The 'best-fit' results were a temperature of $T_e = 12.4\ 10^6 K$ and emission measure $EM = 1.5\,10^{53} cm^{-3}$. The results for the power-law component $\{J(\epsilon) = A\epsilon^{-\gamma}\, ph\, cm^{-2}\, s^{-1}\, keV^{-1}\}$ are $A = 1.86\,10^{-3}$ and $\gamma = 2.05$. It is necessary that the power-law distribution turns over and becomes steeper somewhere above 18 keV (this being the high energy cut-off of the present observations), otherwise the total flux $\{\int_0^\infty J(\epsilon)\epsilon d\epsilon\}$ becomes infinite. The total radiative losses by the power-law component between 1 and 18 keV is $9\,10^{29}\, erg\, s^{-1}$.

DISCUSSION

In an EINSTEIN Observatory survey of RS CVn binaries, Swank *et al.* (1981) found that all their spectra (covering $\sim$ 0.5 - 4 keV) required two emission components, a plasma component of $4 - 8\,10^6 K$ and a Bremsstrahlung component of $20 - 100\,10^6 K$. The ratio of these two components varied between 0.1 - 4.0, with mean emission measures of $2.5 - 3.5\ 10^{53}\, cm^{-3}$. In a series of EXOSAT observations of II Peg [using the ME (1-6 keV) and CMA (0.05-2 keV) detectors], a similar result was found: $T_1 \sim 3.5 - 8\,10^6 K, T_2 \sim 17 - 30\,10^6 K$ (Tagliaferri *et al.* 1991). Both of these results compare very well with GINGA data *below* 6 keV. It is only GINGA's much greater collecting area and the extended observing time that allow the spectrum to be extended to $\sim$ 18 keV, and, hence to reveal the hard 'tail' of the quiescent spectrum. There are various ways to produce the observed power-law emission by both thermal and non-thermal particle distributions.

First we will assume that the power-law component is caused by collisional Bremsstrahlung of energetic electrons interacting with the thermal plasma ($T = 12.4\,10^6 K$) in the coronal volume. However, we find that $3.2\,10^{44}$ electrons have to be injected per second in order to maintain the energetic electron population against collisional losses. This represents an energy input between 1.07 and 18 keV of $\int q(E)EdE = 9.6\ 10^{35}\, erg\, s^{-1}$. Given the large energy input which goes into the fast particles, together with their low radiation efficiency, most of their energy will be transferred to the thermal plasma which, as a consequence, would heat up. This is however not observed. Therefore an interpretation of the power-law emission by non-thermal electrons interacting collisionally with the thermal plasma can not be maintained.

An alternative explanation for the power-law emission is that we are dealing with energetic particles which are only interacting amongst themselves and not with the thermal ($T_e = 12.4\,10^6 K$) plasma. Self-interaction of energetic particles will on a relatively short timescale result in the formation of a Maxwellian distribution (Trubnikov, 1965). A Maxwellian distribution will however not result in a power-law shape of the photon spectrum. Therefore we need to invoke the presence of the distribution of Maxwellians by means of a differential emission measure distribution $\xi(T)$. In this case the distinction between a thermal population and a power-law emitting population is only an artifact of our fitting procedure. A substantial amount of coronal plasma has temperatures in excess of $10^8 K$. This hot plasma could not be observed with the previously used instrumentation (*e.g.* EXOSAT, EINSTEIN) because of the low sensitivity above $\sim 6keV$. Although fits of EXOSAT data implied temperatures of tens of million Kelvin, the instrument could not be used to determine accurately temperatures above $60\,10^6 K$. Of course the present detection of emission up to $\sim 18keV$ requires an independent verification.

Acknowledgments We would like to thank the GINGA Observatory staff for their help in obtaining this data. Research at Armagh Observatory is grant-aided by the Dept. of Education for N. Ireland. We also acknowledge the support provided in terms of both software and hardware by the STARLINK Project which is funded by the UK SERC. G.H.J. van den Oord acknowledges financial support from the Royal Dutch Academy of Sciences (KNAW).

REFERENCES

Iwan, D., Marshall, F.E., Boldt, R., Mushotzky, R.F., Shafer, R.A., Stottlemyer, A., 1982, *Ap. J.*, **260**, 111.
Raymond, J.C., Smith, B.W., 1977, *Ap. J. Suppl.*, **35**, 419.
Tagliaferri, G., White, N.E., Doyle, J.G., Culhane, L.J., Hassall, B.J.M., Swank, J.H., 1991, *Astr. Ap.*, (in press).
Swank, J.H., White, N.E., Holt, S.S., Becker, R.H., 1981, *Ap. J.*, **246**, 208.
Trubnikov, B.A., 1965, Reviews of Plasma Physics, ed. M.A. Leontovich, Consultants Bureau Enterprises, New York.

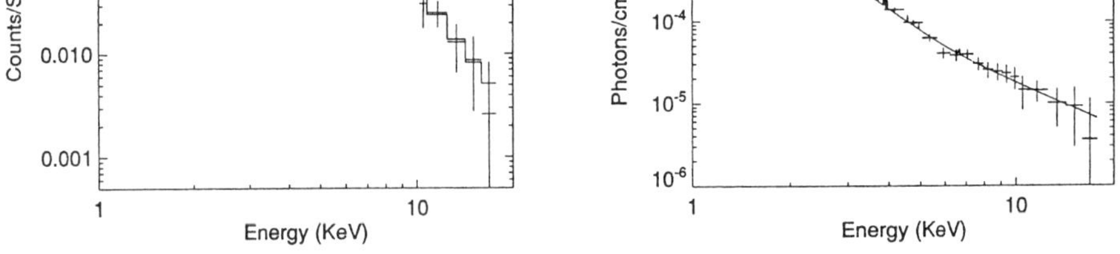

Fig. 1: The quiescent data for II Peg obtained with GINGA in August 1989, plus a model fit of a single temperature Raymond & Smith model ($T = 12.4\,10^6 K$) plus a power-law of $\gamma = 2.05$ for the high energy pts

Seventh Cambridge Workshop on Cool Stars, Stellar Systems, and the Sun
ASP Conference Series, Vol. 26, 1992
Mark S. Giampapa and Jay A. Bookbinder (eds.)

A STUDY OF EXTENDED MATTER IN RS CVn SYSTEMS

JEFFREY C. HALL and LAWRENCE W. RAMSEY
The Pennsylvania State University, Department of Astronomy and Astrophysics, University Park, PA. 16802

ABSTRACT A growing body of data suggests that extended, extrastellar matter is common in RS CVn binaries. We have surveyed eleven eclipsing RS CVn systems to detect such matter, using the eclipses to probe for extended matter associated with the eclipsing star. This matter may obscure the eclipsed star at near-eclipse phases and produce anomalous signatures in the spectrum. We report which systems have shown evidence of such material and present a first-generation model to parameterize its nature and extent.

Keywords: Chromospherically active stars; RS CVn systems; extended matter.

MOTIVATION

Evidence is accumulating both observationally and theoretically that regions of extended material may contribute significantly to the activity in RS CVn systems. Our initial work on SS Boo (Hall *et al.* 1990) revealed extended matter in that system. Neff *et al.* (1989) found possible obscuration due to extended material in AR Lac with IUE, although they interpret it differently than we. Guinan *et al.* (1986) and Young *et al.* (1991) report extended material in V471 Tau which obscures the white dwarf just before eclipse. Huenemoerder and Barden (1986) report a possibly large region of circumstellar matter above the stream impact region in RT Lac. NLTE radiative transfer models by Buzasi (1991) indicate that the ratio of excess energy emitted in Hα to that in $H\beta$ in a *subtracted spectrum* (see below) can obtain high values ($\geq$ 5) only in optically thin, tenuous plasmas akin to prominences or similar radially extended structures. Denser chromospheric structures (*e.g.*, plage) typically have much lower Hα/Hβ ratios.

METHOD

We produce a *subtracted spectrum* by artificially shifting in velocity, rotationally broadening, and weighting two quiescent stars whose spectral types match those of the active binary under scrutiny. We subtract the synthetic spectrum from that of the active binary, and the residuals reveal any excess emission or absorption in the active lines (*e.g.*, Huenemoerder *et al.*, 1989).

In our survey observing mode we take one spectrum per night, with the observations timed to maximize eclipse contact coverage. We can then establish reliable RV solutions and inspect the system for any anomalies from a number of viewpoints, including the near-eclipse spectra with their advantageous geometry. We do not obtain highly phase-resolved spectra unless there is a fortuitously timed eclipse, in which case we may take several consecutive spectra. At present we have few sequences of this "intensive mode" observing.

RESULTS OF THE SURVEY

Detections exist in 4 of the 11 systems surveyed; these are AW Her, SS Boo, AR Mon, and UV Psc. We find probable, less-thoroughly analyzed detections in an additional 4 systems, MM Her, PW Her, UX Com, and RT Lac. Two systems, LX Per, and CQ Aur, showed no apparent anomalies. The faint, possibly semidetached system AD Cap appears to have several extended components but we have no conclusive results yet.

A PARAMETRIC MODEL OF THE EXTENDED REGIONS

This model assumes a slab of extended material is standing vertically on the eclipsing star's surface. Part of the slab obscures the other star; part of it usually is seen in emission (no star behind). The slab is gridded and the effect of each grid element is added to the subtracted spectrum.

- Pseudo-LTE level populations and ionization states proved a good approximation to reality in this situation where the radiative input from the parent star's photosphere completely dominates collisional processes in the slab. A height-dependent dilution factor approximates the anisotropic irradiation of each slab element.
- We used seven parameters specifying the latitudinal extent of the vertically-oriented filament over which the slab lies, the longitude of the filament, the height h, and thickness d of the slab and the isobaric density n_e and uniform temperature T. The various projection factors render some parameters degenerate; therefore we allowed no more than four to vary on a given fit.
- The effect of each grid element was added to a disk emission profile and the parameters were adjusted iteratively to a best-fit solution. Grid elements appeared in absorption if they obscured the other star, or in emission if they did not.
- We used a Simplex algorithm to iterate to the χ^2_ν minimum and the scheme of Lampton, Margon, and Bowyer (1976) to estimate errors in the fitted parameters.

RESULTS OF THE MODEL

A sample model fit is displayed in Figure 1. The circumstances of this eclipse, at phase $\phi = 0.972$ in the system AW Her, are displayed above the model fit to the subtracted spectrum. Two extended regions are apparent, one on the blueward limb of the primary and one on the redward limb, partially obscuring the secondary. Densities in these slabs are of order $3.5 \pm 2.0 \times 10^{11}$ cm^{-3}, heights are ≈ 0.7 stellar radii, and the projected thicknesses of the slabs are several hundredths of a stellar radius. This is reminiscent of a greatly scaled-up solar quiescent prominence. The Hα/Hβ ratio for the primary in AW Her is of order 6 ± 1 at this phase, indicative of the presence of a large amount of tenuous prominence-like material.

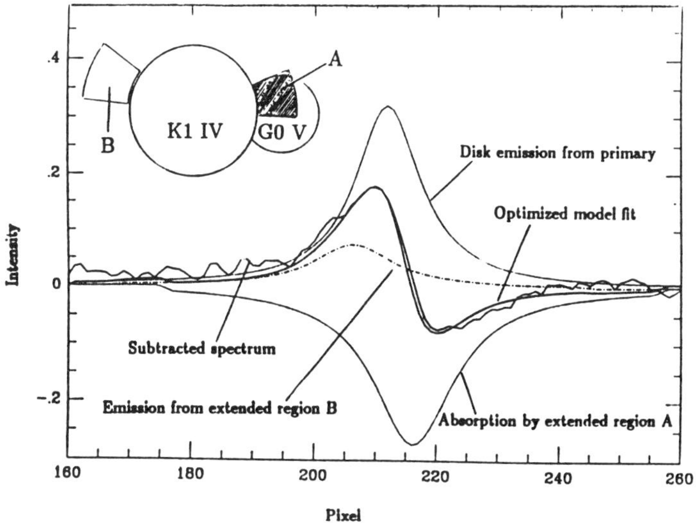

FIGURE 1. A sample model fit for the extended regions in AW Her at phase $\phi = 0.972$. The various components of the fit are labelled and a picture of the test grid generated by the model is at top.

Spectra of SS Boo and UV Psc have been satisfactorily explained with this picture as well. In all cases the Hα/Hβ ratio tracks the visibility of the extended regions, rising significantly when they rotate into view.

ACKNOWLEDGEMENTS

The authors would like to acknowledge support from NSF grant AST8919205.

REFERENCES

Buzasi, D. L. 1991 *Ap. J.*, *Submitted.*
Guinan, E. F., *et al.* 1986, *New Insights in Astrophysics*, ESA SP-263, 107.
Hall, J. C., *et al.* 1990, *Ap. J.*, **358**, 610.
Huenemoerder, D. P., and Barden, S. C. 1986, *A. J.*, **91**, 585.
Huenemoerder, D. P., *et al.* 1989, *A. J.*, **98**, 1398.
Lampton, M., Margon, B., and Bowyer, S. 1976, *Ap. J.*, **208**, 177.
Neff, J. E., *et al.* 1989, *Astr. Ap.*, **215**, 76.
Young, A. *et al.* 1991 *Ap. J. Letters*, **378**, L25.

Seventh Cambridge Workshop on Cool Stars, Stellar Systems, and the Sun
ASP Conference Series, Vol. 26, 1992
Mark S. Giampapa and Jay A. Bookbinder (eds.)

PHOTOSPHERIC AND CHROMOSPHERIC ACTIVE REGIONS ON THE CLOSE BINARY AR Lac (HD 210334)

ISABELLA PAGANO AND MARCELLO RODONÓ Institute of Astronomy of Catania University and Astrophysical Observatory, viale A.Doria 6, I-95125 Catania, Italy

JAMES E. NEFF Department of Astronomy, The Pennsylvania State University, 525 Davey Laboratory, University Park, PA 16802, USA

ABSTRACT We apply spectral imaging techniques on data obtained for the eclipsing RS CVn system AR Lac. By so doing, we can identify and study the properties of non-homogeneous, discrete regions on the stellar surface. We discuss the results that we have thus far obtained.

Keywords: stellar atmospheres, spectral imaging, RS CVn variables.

INTRODUCTION

The eclipsing RS CVn system AR Lac (HD 210334) is a valuable laboratory for studying discrete regions of stellar activity. Both stars (G2 IV - K0 III/IV) show optical, UV, X and radio solar-type variability. The out-of-eclipse optical variability is attributed to the presence of huge photospheric spots presumably on both stars. By applying the *spot model* used by Rodonó *et al.* (1986) the position and size of these active regions can be derived. An IUE program was started in 1981 (Rodonó *et al.* 1987; Walter *et al.* 1987; Neff *et al.* 1989, Pagano *et al.* 1990) to monitor strong ultraviolet emission lines for intensity and profile variations. The longitude and size of the chromospheric active regions were derived analyzing high resolution (0.24 Å) Mg II *k* line profiles, by the *Spectral Imaging Technique* (Neff *et al.* 1989).

PHOTOSPHERIC SPOTS FROM PHOTOMETRY

The almost exactly 2-day period of AR Lac limits the potential phase coverage from a single ground-based site to only two opposite phases, unless the star is observed for several months. Hence, photometric data useful to infer the spot configuration at the times of the IUE monitoring are available only for August-November 1983, July-December 1988, July-October 1989 and June-September 1990. The 1983, 1989 and 1990 UBV photometry was done with the 0.91-m and the 0.61-m reflectors at Catania Observatory. The 1988 light curve was obtained with the 25-cm APT at Mt. Hopkins. By Fourier series expansion of the observed light curve, ellipticity and other proximity effects were taken into account to extract the pure *photometric wave* for each observational period. Then, by an interactive spot model (Rodonó *et al.* 1986), we derived the spot latitudes, longitudes and radii (Table 1). The results for 1990 are not shown.

TABLE 1: Spot model results

	1983		1988		1989	
Spot	1	2	1	2	1	2
longitude (*degree*)	180	85	90	332	30	320
latitude (*degree*)	59	50	35	26	25	5
radius (*degree*)	13	9	6	15	23	18

CHROMOSPHERIC STRUCTURE FROM ULTRAVIOLET SPECTRA

The IUE observations of AR Lac we acquired in 1983 and 1985 allowed us to obtain *images* of the chromospheric active regions by using the *Spectral Imaging Technique* (Walter *et al.* 1987; Neff *et al.* 1989). More complete sets of observations, aimed at monitoring the activity evolution were obtained during the following periods: September 1987 (34 high- and 25 low-resolution spectra during two consecutive full orbital cycles); August-September 1989 (23 high- and 22 low-resolution spectra during one single orbital cycle); December 1990 (13 high- and 9 low-resolution spectra during about one orbital cycle). The *Spectral Imaging Technique* allows one to identify and study the behaviour of moving discrete features on the observed line profiles, due to the emission originating from non-homogeneous discrete regions (*plages*) brighter than the surrounding chromosphere. Each observed line profile was fitted with two symmetric stellar components, centered at the wavelengths corresponding to the stellar radial velocities, plus additional components (at most two) to account for residual emissions, as shown in Figure 1a. An absorption feature due to the interstellar medium was also included in the analysis. The primary constraints in the adopted procedure are the wavelength positions of the uniform stellar components, while secondary constraints are the rough constancy of the stellar emission line widths and a smooth point-to-point variation of the stellar flux. If the asymmetry detected on a single line profile was really due to a plage, it has to be present also in others profiles at velocities, in the star's rest frame, following a sinusoidal curve. When this occur, the detected feature is attributed to an active region passing across the visible hemisphere of the star and having a lifetime of at least one stellar rotation (Vogt and Penrod, 1983; Vogt, Penrod and Hatzes, 1987). The velocities of the discrete emission components in the 1989 spectra are shown in Figure 1b, in the K-star rest frame. These data points can be fitted by three distinct sinusoidal rotation curves, *i.e.* three distinct plages are present on the chromosphere of the K star.

From the phases and amplitudes of these sinusoidal rotation curves we can deduce the plage longitudes and the product between their height above the photosphere and the cosine of their latitude, respectively. If the amplitude is larger than the stellar $v_e sini$, only the minimum height can be inferred by assuming the plage latitude to be zero. If the amplitude is lower than $v_e sini$ the minimum latitude can be deduced by assuming the plage to lie on the photosphere. The chromospheric configurations inferred for 1987 and 1989 are shown in Figure 2 and the concerned data are listed in Table 2. Note that only for one plage we were able to infer the minimum latitude. The 1990 results are not presented because the lack of full phase coverage and the occurrence of a flare, detected also in the X–ray region by ROSAT (Linsky *et al.* 1991), require further data analysis, which is in progress.

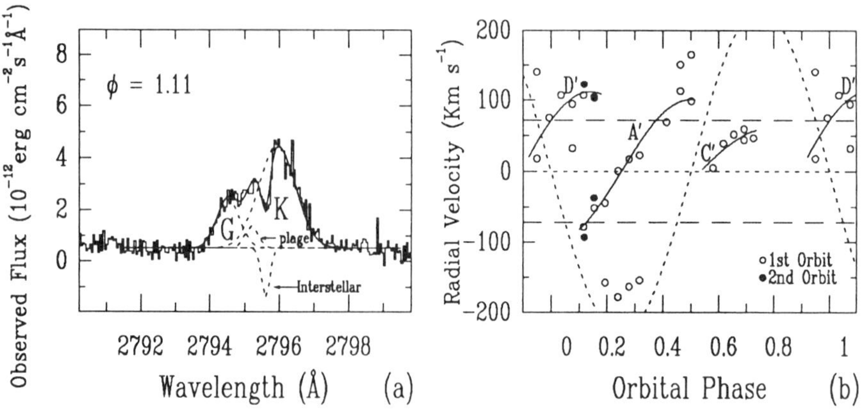

Figure 1: (a) A multi-component Gaussian fit to the high-dispersion Mg II *k* line profile. In addition to the stellar components (G and K), a good fit requires one plage on the K star and an interstellar absorption feature. (b) Radial velocity curves in the velocity rest frame of the AR Lac K-star, for the G star and three bright features (A', C', D') on the K-star, as derived from Spectral Imaging of Mg II *k* line profiles.

TABLE 2: The plage parameters

	1987				**1989**		
	A	**B**	**C**	**D**	**A'**	**C'**	**D'**
H_{min}	$1.46 \pm .10$	$1.44 \pm .15$	$1.08 \pm .08$	$1.86 \pm .10$	$1.40 \pm .10$	–	$1.56 \pm .08$
$L(deg)$	94 ± 4	148 ± 1	176 ± 7	281 ± 4	90 ± 4	191 ± 11	320 ± 7
ΔL	$\simeq 36$	$\simeq 15$	$\simeq 41$	$\simeq 23$	$\simeq 35$	≤ 15	$\simeq 32$
μ	0.10	0.02	0.12	0.04	0.09	≤ 0.02	0.08
$l_{min}(deg)$	–	–	–	–	–	$33.6 \pm .3$	–

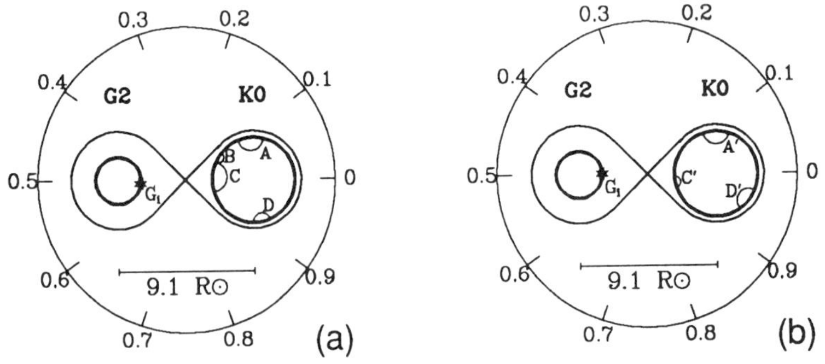

Figure 2: The chromospheric active regions on AR Lac in 1987 (a) and in 1989 (b) as derived from Spectral Imaging of Mg II *k* line profiles.

SPOT AND PLAGE EVOLUTION

The surface migration in longitude and latitude of spots and plages, are shown in Figure 3. The 1983 and 1985 data were taken from the literature (Walter *et al.* 1987; Neff *et al.* 1989). There is firm evidence of migration both in longitude and latitude: from a least square fit to the data set collected from 1983 to 1989, migration rates of about $-22 \pm 6\, deg\, y^{-1}$ and $-14 \pm 10\, deg\, y^{-1}$ for the spot and plage longitude, respectively, and $-6 \pm 2\, deg\, y^{-1}$ for the spot latitude result. Consistently with previous measurements, the differential rotation of spots turns out to be very low ($\sim 10^{-5} rad\, s^{-1}\, deg^{-1}$). Only sparse photometry is available for 1985 and 1987. No spatial correlation between photospheric and chromospheric active regions can be deduced from 1983 data, while at least a spatial coincidence is obtained from 1989 data. The 1983 spectral results are seriously limited by incomplete phase coverage, so that this negative result is not to be considered conclusive. Taking into account positive and negative results on spatial correlation between spots and plages (Rodonó *et al.* 1980; Baliunas et al., 1982; Rodonó *et al.* 1987; Butler *et al.* 1987; Andrews *et al.* 1988), we may conclude that, although well documented positive examples exist, further studies should be carried out.

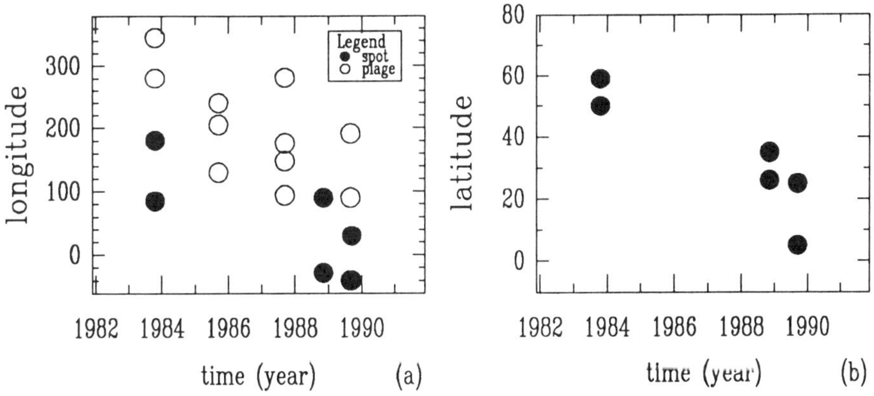

Figure 3: Spot and plage longitudes (a) and spot latitudes (b) versus time, as derived from *spot modelling* and *Spectral Imaging*, respectively.

THE MgII *k* LINE FLUX

The AR Lac stellar components are well resolved in the IUE high-resolution spectra, the difference between the two star orbital velocities at quadratures being equal to about 232 km ($\sim 2\AA$ at MgII *k* wavelength). By applying the *Spectral Imaging Technique*, in addition to the plage configuration, we have derived the fluxes emitted by the K0 III-IV and G2 IV stars alone. These residual fluxes, coming from supposedly homogeneous chromospheres, vary with the orbital phase (Figure 4a). Apart from the flux decrement during primary and secondary eclipses, there is a definite difference between the fluxes before and after each eclipse. From the 1987 data, the K-star mean flux at Earth in the phase interval 0.1–0.4 is $(6.26 \pm 0.66) \times 10^{-12} erg s^{-1} cm^{-2}$, while in the phase interval

0.6 – 0.9 is $(5.62 \pm 0.42) \times 10^{-12} erg s^{-1} cm^{-2}$. The situation is reversed for the G star, whose fluxes in the same phase intervals are $(2.07 \pm 0.23) \times 10^{-12} erg s^{-1} cm^{-2}$ and $(2.83 \pm 0.28) \times 10^{-12} erg s^{-1} cm^{-2}$, respectively. This means that the following hemispheres of the K and G stars are about 10 and 37 per cent brighter than the other hemispheres, respectively. In Figure 4b the rotational parameter $v_e sini$ from the Mg II k line profiles (1987 data) for each star is shown. The $v_e sini$ behaviour closely follows that for the flux, suggesting the presence of high-lying extended plages on the brighter hemispheres of both stars. The flux and $v_e sini$ behaviour has been derived also from 1985 (Neff *et al.* 1989) and 1989 Mg II data. The presence of long-lived active zones on one hemisphere of each star, possibly related also to mass flows between the system components, should be monitored with higher spectral resolution than allowed by IUE.

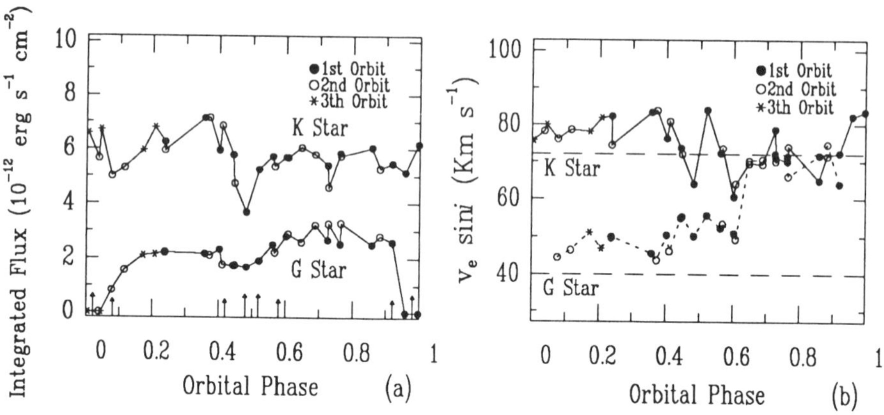

Figure 4: The integrated Mg II k line fluxes (a) and the rotational parameter $v_e sini$ (from the Mg II k line width) (b) for the K0 III-IV and G2 IV stars, as derived by the *Spectral Imaging Technique.* The arrows indicate the times of the 1st, 2nd, 3th and 4th contacts during eclipses.

REFERENCES

Andrews, A.D., *et al.* 1988, *Astr. Ap.*, **204**, 177.

Baliunas, S.L., Dupree, A.K. 1982, *Ap. J.*, **252**, 668.

Butler, C.J., *et al.* 1987, *Astr. Ap.*, **174**, 139.

Friedemann, C., Gurtler, J. 1975, *Astron. Nachr.*, **296**, 125.

Linsky, J.L., *et al.* 1991, *This conference.*

Neff, J.E., Walter, F.M., Rodonó, M., Linsky, J.L. 1989, *Astr. Ap.*, **215**, 79.

Pagano, I., Rodonó, M., Neff, J.E. 1990, in Surface Inhomogeneities in Late-Type Stars, eds. P.B. Byrne and D.J. Mullan, (Springer-Verlag), in press.

Rodonó, M., Romeo, G., Strazzulla, G. 1980, in Second European IUE Conference, ESA SP-157, p. 55.

Rodonó, M., *et al.* 1986, *Astr. Ap.*, **165**, 135.

Rodonó, M., *et al.* 1987, *Astr. Ap.*, **176**, 267.

Vogt, S.S., Penrod, G.D.: 1983, *Pub. A.S.P.*, **95**, 565.

Vogt, S.S., Penrod, G.D., Hatzes, A.P.: 1987, *Ap. J.*, **321**, 496.

Walter, F.M., *et al.* 1987, *Astr. Ap.*, **186**, 241.

Seventh Cambridge Workshop on Cool Stars, Stellar Systems, and the Sun
ASP Conference Series, Vol. 26, 1992
Mark S. Giampapa and Jay A. Bookbinder (eds.)

MAGNETIC ACTIVITY IN ALGOL-TYPE BINARIES

MERCEDES T. RICHARDS   Department of Astronomy, University of Virginia, Charlottesville, VA 22903

ABSTRACT There is growing observational evidence that the cool G – K secondaries in the short-period Algol-type binaries are magnetically active because of their rapid rotation and outer convective layers. Since the Algols are in the slow phase of mass transfer, the Hα emission from the binary is usually attributed to circumstellar gas only. However, if the secondary is magnetically active, there could also be Hα emission from the chromosphere of the secondary.

Keywords: Algol-type binaries; circumstellar material; starspot activity; magnetic activity; RS CVn binaries

INTRODUCTION

The cool G- to K-type secondaries in short-period Algol-type binaries are expected to be magnetically-active because of their rapid rotation and outer convective layers. Magnetic activity is not usually associated with these binaries because the secondaries are faint relative to the luminous B5 – A5 V primaries at visible wavelengths. However, there is growing observational evidence of magnetic activity associated with the cool secondaries in these binaries (Olson 1981; White and Marshall 1983; Hall 1989; Richards 1990).

The Algol secondaries are very similar to the doubly-active RS CVn binaries (Hall 1989) and display time-dependent magnetic properties which include weak Ca II H and K emission; brightness variations in the light curve of $\lesssim$ 0.1 mag. indicative of the motions of starspots; coronal x-ray emission with $L_X \sim 10^{30} - 10^{31}$ erg s^{-1}; weak radio flares; nonthermal gyrosynchrotron radio emission with flux densities typically less than 4 mJy at 4.9 GHz (Umana *et al.* 1989) but much stronger in the case of β Per (Lestrade *et al.* 1988); ultraviolet and infrared excesses; and alternate increases and decreases in the orbital period of the binary, $\Delta P \simeq 10^{-5} P$ which could be indicative of a magnetic cycle with a median period of $\sim$ 50 years (Hall 1989).

In the case of β Per, the x-ray and radio emission apparently arise from cospatial regions and can be understood in terms of an extended corona contained in magnetic loops around the secondary. The inferred magnetic fields at the base of the loops is a few hundred Gauss, which is about a factor of 10 weaker than the field strength of a typical RS CVn binary (Lestrade *et al.* 1988). Measurements of x-ray flares of β Per suggest that the loop heights may be comparable to the radius of the secondary, with coronal mass ejections of $\sim 10^{-13} - 10^{-12}\ M_\odot$ yr^{-1} (Stern 1991).

We have begun a project to study the evidence for magnetic activity in Algol binaries. The short-period Algol binaries ($P < 5 - 6$ days) are of special interest because they rotate more rapidly than the long-period Algols and their cool secondaries display various forms of magnetic activity. These binaries are semi-detached systems in which the gas stream from the secondary is predicted to make direct impact with the primary star (Lubow and Shu 1975). Furthermore, magnetic fields and coronal gas flows are expected to influence the mass transfer process and the formation of circumstellar gas in the binary.

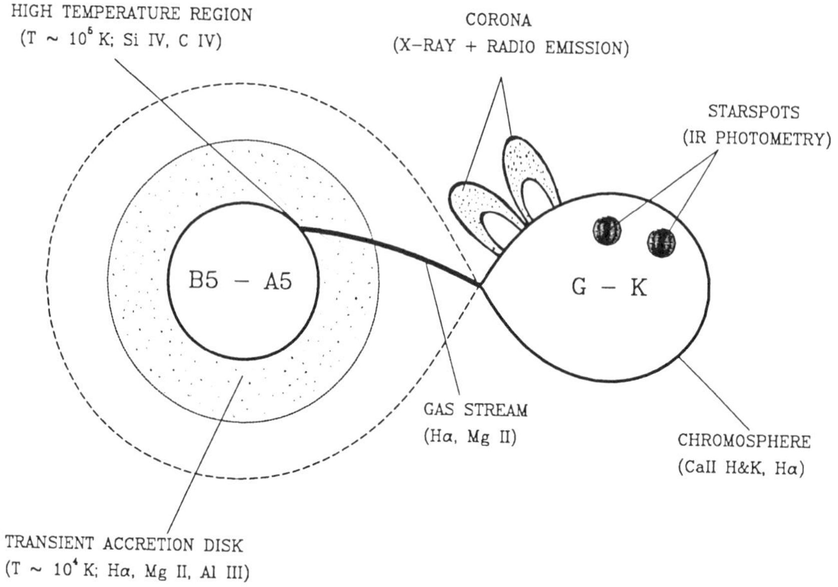

Fig. 1: A schematic diagram of a typical Algol-type binary as seen from above (or below) the orbital plane of the binary. The Roche geometry displayed is that of the prototype, β Persei.

Our primary interest was the influence of chromospheric Hα emission on the Hα difference profile produced when the combined theoretical photospheric line profile of the binary is subtracted from the observed line profile. This difference profile represents all forms of non-photospheric material which include accretion disks, gas streams, as well as chromospheric and coronal gas. In the study of β Per, Richards, Bolton, and Mochnacki (1989) and Richards (1992a,b) assumed that the chromospheric Hα emission from the K2 IV secondary would be insignificant compared to that from the circumstellar material associated with the B8 V primary. However, Richards (1992a) found that some of the Hα difference profiles seen during primary eclipse could be produced by either a transient accretion disk or a chromospheric flare. Although the lifetimes of these features suggest that the disk explanation is more likely (Richards 1992a), it is useful to examine the extent of the influence of the cool secondary star.

Emission and excess absorption from circumstellar material distort the core of the Hα line profile. Chromospheric emission will also distort the core of the Hα line profile, with negligible distortion of the line wings (LaBonte 1986; Soderblom 1989,1991). In principle, it should be possible to distinguish between the two sources of Hα emission especially since disk material would be occulted during primary eclipse. This is not a trivial exercise because both forms of emission are weak relative to the combined continuum flux of the binary, and circumstellar material has been found well above the orbital plane of the binary (Richards 1992a).

Balmer line emission has been detected in at least four Algol binaries, namely U Cep, RW Tau, β Per, RY Gem (Kaitchuck, Honeycutt, and Schlegel 1985; Richards 1992b), but this emission has traditionally been associated with the mass transfer process. Possible contributions from a chromosphere have not been considered. We will utilize the analyses of Hα emission in single and binary stars (*e.g.* Strassmeier *et al.* 1990; Barden 1985) to estimate the strength of chromospheric emission in the short-period Algols. A preliminary analysis suggests that the contribution of chromospheric Hα emission is non-trivial.

Finally, to complement the observations, there is a need for comprehensive theoretical models which would help us to understand how magnetic fields and coronal gas flows influence the mass transfer process and the formation of circumstellar gas in a close interacting binary.

ACKNOWLEDGEMENTS

This work was partially supported by NSF grants AST 8901238 and 9114214.

REFERENCES

Barden, S. C. 1985, *Ap. J.*, **295**, 162.
Hall, D. S. 1989, *Space Sci. Rev.*, **50**, 219.
Kaitchuck, R. H., Honeycutt, R. K., and Schlegel, E. M. 1985, *Pub. A.S.P.*, **97**, 1178.
La Bonte, B. J. 1986, *Ap. J. Suppl.*, **62**, 241.
Lestrade, J.-F., Mutel, R. L., Preston, R. A., and Phillips, R. B. 1988, *Ap. J.*, **328**, 232.
Lubow, S., and Shu, F. H. 1975,
Olson, E. C. 1981, *Ap. J.*, **250**, 704.
Richards, M. T. 1990, *Ap. J.*, **350**, 372.
Richards, M. T. 1992a, *Ap. J.*, **387**, in press.
Richards, M. T. 1992b, in preparation.
Richards, M. T., Bolton, C. T., and Mochnacki, S. W. 1989, *Space Sci. Rev.*, **50**, 358.
Soderblom, D. R. 1989, *Ap. J.*, **342**, 823.
Soderblom, D. R. 1991, in Cool Stars, Stellar Systems, and the Sun, eds. M. Giampapa and J. Bookbinder, in press.
Stern, R. 1991, in Frontiers of X-ray Astronomy ed. K. Koyama, in press.
Strassmeier, K. G., Fekel, F. C., Bopp, B. W., Dempsey, R. C., and Henry, G. W. 1990, *Ap. J. Suppl.*, **72**, 191.
Umana, G., Catalano, S., Rodono, M., and Gibson, D. M. 1989, *Space Sci. Rev.*, **50**, 370.
White, N. E., and Marshall, F. E. 1983, *Ap. J.*, **268**, L117.

Seventh Cambridge Workshop on Cool Stars, Stellar Systems, and the Sun
ASP Conference Series, Vol. 26, 1992
Mark S. Giampapa and Jay A. Bookbinder (eds.)

ACTIVITY IN TIDALLY INTERACTING BINARIES

C. J. SCHRIJVER and C. ZWANN
Astronomical Institute, P.O. Box 80,000, Utrecht, The Netherlands

ABSTRACT We use a multi-component analysis to study the activity of relatively close binary systems. It turns out that some property of the companion star affects stellar activity. We argue that the efficiency of stellar dynamos is enhanced in relatively close binaries possibly because large-scale flow patterns in the components are severely affected by the proximity of a companion which induces an asymmetry in the stars.

Keywords: Binaries; Tidal Interaction; Magnetic Activity

INTRODUCTION

Magnetic activity is common to stars with convective envelopes, and intimately connected with stellar rotation rate. A clear description of the rotation activity relationship is still lacking, but the data suggest that most single stars and components of wide binary systems follow the same color-dependent relations between activity and rotation period (Rutten 1987). This conclusion is based on the coincidence of relationships for dwarfs and giants at relatively long rotation periods, because rapidly rotating, single red giants do not exist owing to the increased moment of inertia as stars evolve.

Deviations from the reasonably well-defined rotation-activity relationships are seen for some components of close binaries: these may emit up to two orders of magnitude more chromospheric and coronal radiation than would be expected given their rotation period. This peculiarity has already been noted before (*e.g.*, Basri *et al.* 1985, Rutten 1987, Simon and Fekel 1987), although often it was merely noted that there was a large scatter about the mean rotation-activity relationship in the most active stars. Rutten (1987) pointed out that such overactive stars agree with relationships between radiative flux densities as defined by single stars, which suggests that the atmospheric structure shares properties with that of other cool stars. There is marginal evidence that the hot component in coronae of overactive stars is stronger than in stars of normal activity, but the coronal overactivity does not appear to be associated with only the hot coronal component, nor is the existence of a hot coronal component unique to overactive binaries (Schrijver and Zwaan 1991).

In order to clarify the rotation activity relation in relatively close binaries, we analyze a relatively small sample of single stars and binaries for which many parameters are known (for details see Schrijver and Zwaan 1991).

DESCRIPTION OF THE DATA

We selected some twenty relatively close, magnetically active binary systems, both for the availability and quality of their parameters (orbital data, fundamental stellar quantities of the components, and radiative losses from the outer atmosphere), as well as for the possibility to cover a large parameter space. For single stars we rely on earlier work by Rutten (1987), where the rotation-activity relationship is concerned, although a small sample of single stars is included in our study for the purpose of comparing flux-flux relationships and coronal structure.

Where available, optical data indicate that the bulk of the emission from the outer atmosphere is associated either with the cool component if the two stars differ substantially in size, or with both components if they are comparable. We consequently assign the radiative losses from the outer atmospheres to the cool primary or to both stars depending on their relative sizes, except in the case of Capella where the more rapidly rotating warmer component is the most active.

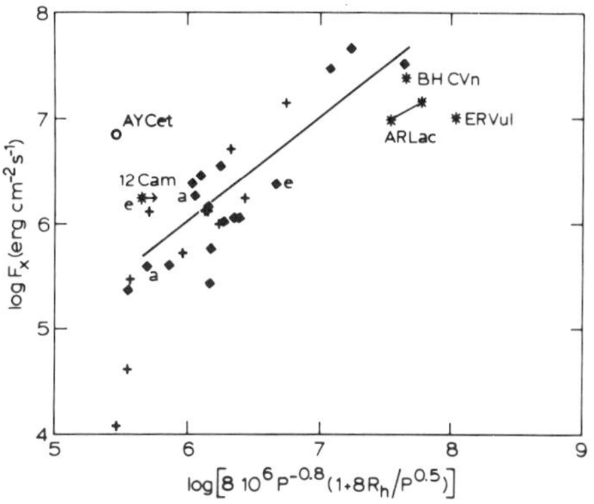

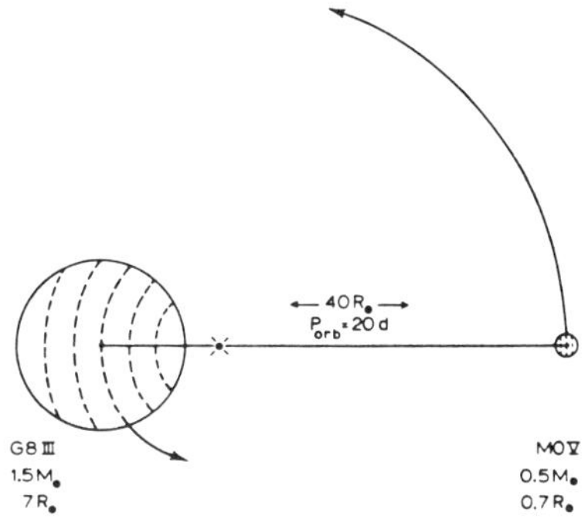

Fig. 1: (left) Observed vs. predicted (Eq.3) soft X-ray flux densities. Diamonds are close binary, * short-period system or (with arrow) unknown companion properties, + single star, o AY Cet with R_h=0.

Fig. 2: (right) An example of a relatively close binary system. Dashed lines outline surfaces of equal angular momentum (constant distance to the system's center of mass, which is shown by a cross).

ROTATION AND ACTIVITY

A principal component analysis showed that fitting a two-dimensional subspace to the data covered 87% of the variance in the sample. The tightest relationship (describing the data to within an rms scatter of about a factor of 1.7) between $\log(F_x)$ and any two other parameters contained in our tables is:

$$F_X = 8 \times 10^7 R_h^{0.94\pm0.32} / P^{1.4\pm0.2} \tag{1}$$

with F_X in erg cm^{-2} s^{-1}, R_h the companion radius in R$_\odot$, P the orbital period in days. A problem in interpreting Eq. (1) is that the stellar parameters in our sample are not independent owing to the conditions which lead to the evolutionary stage in which the active binaries are found. Thus it is possible to transform relationships into seemingly different ones. The role of the companion radius in the determination of the activity level could be a masked dependence on, say, its mass. For instance:

$$F_X = 7 \times 10^7 (D_c/R_c)^{0.47\pm0.29}/P^{1.3\pm0.2} \qquad 2$$

(with $D_c = M_c/(M_c + M_h)D$, with D the binary separation, and M_c and M_h the primary and companion masses) describes the data with an rms scatter only some 12% worse than Eq. (1). The ratio D_c/R_c may reflect the importance of the ratio of the lever arms of rotational effects in a synchronized binary component to those in a single star.

The link to single stars can be made by comparing the observed Fx for binary components with the expected value $F_{x,o}$ for a single star with the same rotation period. With $F_{x,n} \approx 8 \times 10^6/P^{0.8}$ (P< 30 days, see Rutten 1987) for single stars, one may write:

$$F_X = 8 \times 10^6 P^{-0.8}(1 + 8R_h/\sqrt{P}). \qquad 3$$

Some short-period systems lie below the expected activity, perhaps indicating saturation of activity in very close binaries.

All systems in our analysis are well separated; mass transfer by Roche lobe overflow does not occur. The exchange of radiation, accretion of a stellar wind, interaction of the companion with the magnetic field of the primary, and coupling between the fields of the components can all be excluded as mechanisms for overactivity on straightforward arguments (see Schrijver and Zwaan, 1991). Probably the dynamo efficiency in components of close binaries is enhanced. Tidal interaction may be important, because that will affect differential rotation in the components. There is another prominent distinction between single stars and binary components: in single stars rotational effects are symmetric about an axis through the stellar core, while for synchronized components of binaries the axis for rotational effects passes through the system's center of gravity. Models of single stars with convective mantles suggest that the tendency to conserve angular momentum results in a preference for the velocity patterns to be symmetric around the rotation axis (the Taylor-Proudman theorem) through the stellar center. In a synchronized binary component, however, the surfaces of equal angular momentum are cylinders centered on the revolution axis (*cf.* Fig. 2), passing through the center of gravity of the system, located well outside the star in most of the binaries in our sample. This must affect large-scale patterns in the convection and in the differential rotation, and consequently the dynamo.

REFERENCES

Basri, G., Laurent, R., Walter, F.M. 1985, *Ap. J.*, **298**, 761.

Rutten, R.G.M. 1987, *Astr. Ap.*, **177**, 131.

Schrijver, C.J., Zwaan, C. 1991, *Astr. Ap.*, , in press.

Simon, T., Fekel, F.C. 1987, *Ap. J.*, **316**, 434.

PART V

Winds and Angular Momentum Evolution

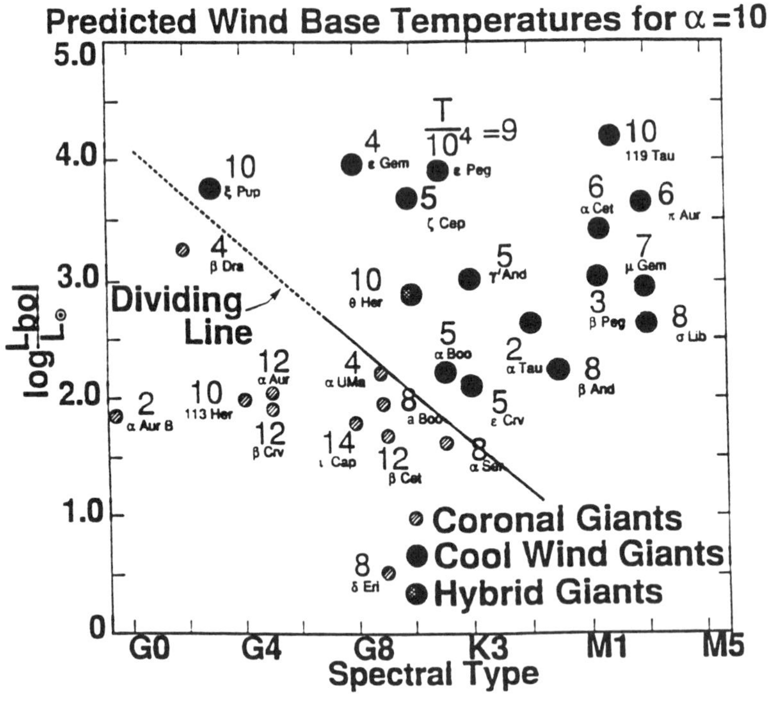

Figure 2 from "Why the Winds from Late-Type Giants and Supergiants are Cool" by R.L. Moore, Z.E. Musielak, C.-H. An, R. Rosner, and S.T. Suess.

Seventh Cambridge Workshop on Cool Stars, Stellar Systems, and the Sun
ASP Conference Series, Vol. 26, 1992
Mark S. Giampapa and Jay A. Bookbinder (eds.)

THE STRUCTURE OF THE SOLAR WIND: AS VIEWED USING RADIO PROPAGATION METHODS

Wm. A. COLES
E. C. E. Department, University of California, La Jolla, CA 92093-0407

ABSTRACT Radio propagation observations of the solar wind are very helpful in elucidating its three dimensional structure particularly in the polar regions near the sun where spacecraft have not yet penetrated. Here we review the structure of the solar wind velocity and compressive turbulence as determined by a variety of such methods.

Keywords: solar wind; plasma; turbulence; waves

INTRODUCTION

Radio waves from a distant compact source are scattered as they pass through the solar wind. Since the level of turbulence decreases rapidly with radial distance (as R^{-4}) the scattering occurs primarily where the line of sight passes closest to the sun. Measurements of the scattered wave field can be used to estimate the statistics of the electron density since the refractive index is only a function of the density. In addition one can use the density fluctuations as a *tracer* enabling one to estimate the flow velocity by observing the scattering with several spaced receivers. Here, of course, one must be careful because the density fluctuations could be propagating at some mean velocity with respect to the flow.

A variety of scattering phenomena can be used to probe the solar wind. These include: intensity scintillation, angular scattering, spectral broadening, pulse broadening, phase scintillation, ranging, and Faraday rotation. Intensity scintillation is one of the easiest measurements to make and is particularly well suited to spaced-antenna velocity measurements because the scale of the intensity scintillation pattern is reasonably small. At frequencies of interest this scale is 25 to 250 km s^{-1} , which is small enough that inter-antenna baselines comparable with the pattern scale are easily obtained and many scales can be observed in a reasonable time. This makes it feasible to observe many sources in a day, forming a map of the solar wind velocity (albeit a rather coarse map).

Angular scattering must be measured with an interferometer such as the radio heliographs which used to operate at Culgoora and Clark Lake or the VLA. It has several advantages. It is the only method by which one can directly measure the spatial anisotropy and it is the only measure for which the propagation theory involves no assumptions that are not amply satisfied. Spectral broadening is very similar in information content to angular broadening, conceptually it is just doppler broadening from the angular spectrum. It can be observed with a single antenna, but requires a monochromatic source. Only spacecraft beacons are adequately monochromatic. Phase scintillation also requires a spacecraft beacon or a long baseline interferometer, but has the great advantage that it is the only technique by which one can estimate large scale structure (greater than, say 1000 km). Pulse broadening has been used but is very limited since there are only a few strong pulsars that pass near the sun. Ranging and Faraday rotation measurements can observe the largest scales but are not made very often.

In this review we will concentrate on two types of observation. First intensity observations of the velocity, both average observations over a solar cycle and snapshot observations near the sun. Second observations of the spatial spectrum of electron density by intensity scintillation, angular scattering and phase scintillation. We will then discuss the effect of Alfvén waves on the the solar wind flow, ways in which they alter the measurements, and ways in which we may estimate their amplitude.

VELOCITY OBSERVATIONS

The multiantenna velocity measurement technique was first applied to the solar wind by Dennison and Hewish (1969). Continuous observations have been made from early 1972 to the present by groups at UC San Diego (Rickett and Coles 1991) and the Institute of Atmospherics at Toyokawa (Kojima and Kakinuma 1987). The basic idea is simply that one measures the same pattern at several places and estimates the time delays between the observations at the different locations. From these one gets a vector velocity. There are some complications due to rearrangement of the pattern as it drifts which will be discussed later. The UCSD and Toyokawa observations have been used to make latitude-longitude maps of the velocity averaged over about 6 months. These maps can be compared with similar maps of coronal density (from white light coronagraphs) and magnetic field (extrapolated from the photosphere into the solar wind). Such comparisons show clearly the relation of the high speed stream to the polar coronal holes and the slow speed wind to the magnetic neutral line. The evolution of this configuration over the solar cycle is discussed by Rickett and Coles (1991). Typical maps showing the transition from minimum to maximum activity are shown in figure 1 below.

The gross structure of the solar wind velocity at solar maximum is a spherically-symmetrical relatively-slow flow, punctuated at intervals by transients such as shock waves or coronal mass ejections. As the activity wanes a fast stream develops in the polar regions, in association with a polar coronal hole. This polar stream may not be aligned exactly with the rotation axis at first, but by the time of minimum solar activity it appears to be so aligned. The polar stream disappears rather abruptly as activity increases beyond the minimum, but this may be partially due to the fact that propagation observations are insensitive to small streams. It may be, at the time of apparent disappearance, that the stream has simply dropped below the minimum detectable width. It is interesting that the basic development and evolution of the polar high speed stream in conjunction with a polar coronal hole could not have been followed with spacecraft velocity observations because they are confined to the ecliptic plane, where there is little evidence of this evolution.

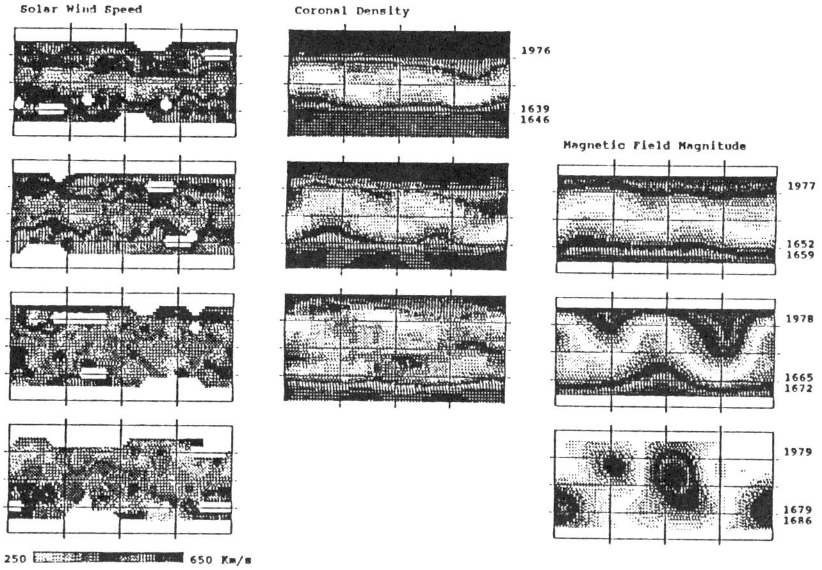

Fig. 1: Maps of the solar wind velocity, coronal density (at 1.5 R_s) and magnetic field magnitude (at 2.2 R_s) averaged over 6 rotations in 1976 through 1979. The ordinate is Carrington latitude from -90 to +90° to the top and the abscissa is Carrington longitude from 0 to 360° to the right. The grey scale is shown for the velocity, but are normalized to the map maximum and minimum for the other plots. The density scale is white for maximum density. The magnetic field scale is white for minimum field and denotes the neutral line.

The UCSD and Toyokawa observations are made mostly outside about 80 R_s where there is very little radial acceleration. Although observations have been made closer to the sun for many years, it has not been possible to use them to estimate a radial velocity profile because variations of velocity with latitude and longitude exceeded the distance variations. However recent observations by Coles *et al.* (1991) have been made with sufficient sampling density that it was possible to select measurements at the same latitude and longitude but different distances. These were compared with theoretical acceleration models in the form shown in figure 2. Two different streams are shown on this figure and several models are overplotted as continuous lines. In every case the highest velocity one could obtain with a purely thermally driven model was inadequate to match the observations. About half the time the observations could be matched with a model in which the velocity was enhanced by a flux of Alfvén waves. In the remaining cases, examples of which are shown in figure 3, the model reached sufficiently high velocity but could not match the location of the peak acceleration.

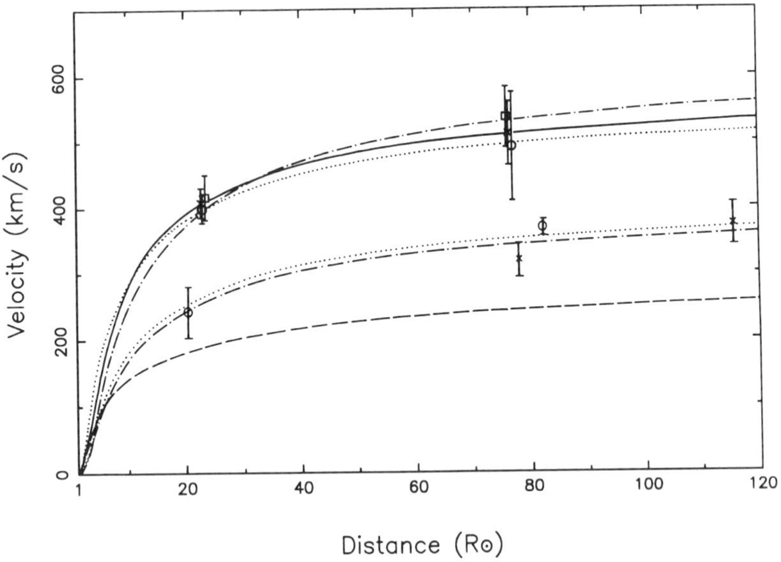

Fig. 2: A comparison of measured velocity with theoretical acceleration models. The higher velocities are from a polar stream, the lower from an equatorial stream. The dashed line is from a purely thermally driven model. The other lines are from various wave assisted models which are consistent with typical boundary conditions.

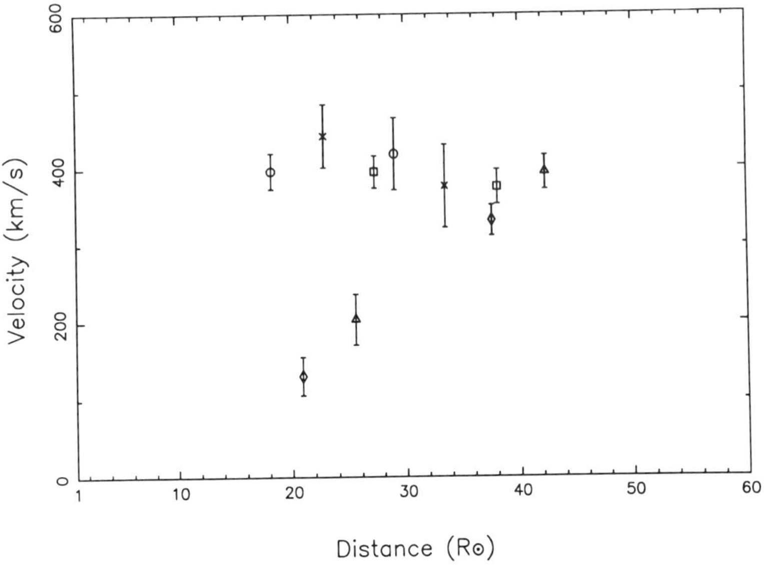

Fig. 3: Examples of measured velocity profiles which are too steep or too flat to match a wave assisted theoretical model.

SPATIAL SPECTRUM OF ELECTRON DENSITY

Scattering observations can be used to estimate the electron density spectrum at scales between 1 and 10^5 km. However this information is more conveniently displayed as a structure function. A spectrum is a distribution of power over wavenumber whereas a structure function is a distribution of power over scale. A power law spectrum of exponent $-\alpha$ corresponds to a structure function of exponent α-2. The use of structure functions in this context is discussed by Harmon and Coles (1983). Observations near the sun, discussed recently by Coles *et al.* (1991), show a distinctly flatter portion of the spatial spectrum in the scale range of 3 to 300 km as shown in figure 4. One could regard this flatter region as caused by an additional variance above a background Kolmogorov spectrum. It is interesting to note that, when a transient enhancement in the density fluctuations occurs, the Kolmogorov portion of the spectrum increases but the flatter portion does not. Thus the spectrum appears to steepen during a transient. The cause of the flatter or enhanced portion of the spectrum is not known although various mechanisms have been proposed (Coles *et al.* 1991).

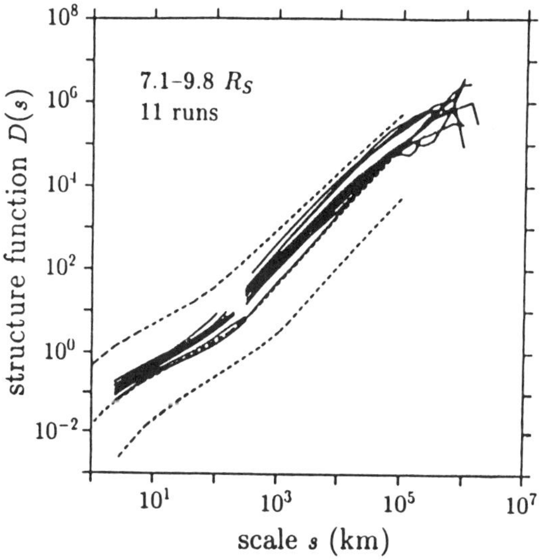

Fig. 4: Plot showing structure functions derived by Coles *et al.* (1991) from Voyager superior conjunction data between 7.1 and 9.8 R_s. The dashed lines represent model structure functions derived from a variety of observations by Coles and Harmon (1989) for distances of 5, 10 and 20 R_s (top to bottom).

Density fluctuations at scales of 1 to 30 km have recently been studied by measuring angular scattering with a radio interferometer (Armstrong *et al.* 1990). They showed that, inside of 10 R_s, the structure can be highly anisotropic. their measurements are reproduced in figure 5. In trying to understand the origin of the flatter or enhanced portion of the spectrum this anisotropy might be helpful. It could be, for example, an indicator of the filamentation instability (Kuo et al. 1988). It would also be useful to know if the flatter region arises from a different physical mechanism than does the large scale Kolmogorov portion of the spectrum. It might be possible to determine this, for example, by measuring the anisotropy of the flatter region and the larger scales simultaneously. Such a measurement is possible, using the VLA and the VLBA, but has not yet been attempted.

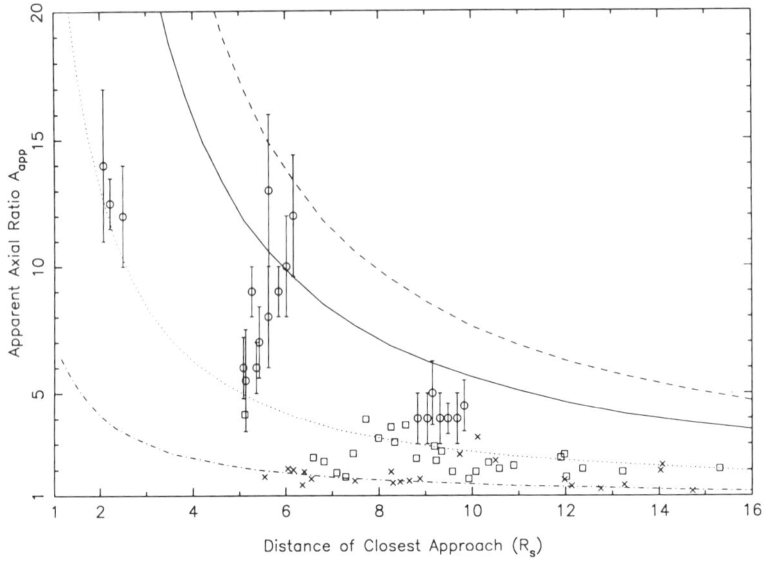

Fig. 5: Axial ratio of angular scattering observed with radio interferometers. Symbols with error bars are from the VLA, others are from Culgoora. Theoretical curves are drawn in solid lines for various models (Coles *et al.* 1991). The higher theoretical curves are from models with lower wave amplitudes and consequently the lower velocities.

The existence of Alfvén waves in the inner solar wind is an important factor. Such waves have been used to solve the *problem* that MHD models of the solar wind flow which are based entirely on thermal conduction do not give high enough velocities at 1 AU. In addition large amplitude waves can damp by various mechanisms both heating the plasma and causing compressive turbulence (Kuo *et al.* 1988, Marsch and Tu, 1990). The compressive turbulence could be related to the observed flatter region in the spatial spectrum of electron density. The presence of waves can reduce the anisotropy of angular scattering observations and they can be seen in two antenna intensity scintillation observations as they cause an apparent rearrangement of the scintillation pattern.

EFFECT OF ALFVÉN WAVES ON INTERFEROMETRY

A radio interferometer is an imaging instrument, however it actually measures the electric field correlation function $\Gamma(s) =< E(r)E * (r+s) > / < |E|^2 >$ which is the Fourier transform of the brightness distribution (the image). Although we have no direct evidence we will assume that the structure is aligned with the magnetic field (because its mean orientation is radial close to the sun). Thus if an Alfvén wave is present in the scattering volume the direction of the anisotropy will fluctuate by some angle $\delta\theta = tan^{-1}(\delta B/B)$. The scattering volume in the solar wind is much longer than the correlation length for Alfvén waves, therefore we must expect to see an incoherent superposition of many such waves. Clearly this will have the effect of *isotropizing* the apparent brightness distribution.

A calculation showing the apparent axial ratio as a function of wave amplitude, for various values of the intrinsic axial ratio, is shown in figure 6. One can see that the waves are very effective at reducing the apparent axial ratio. Measurements of the axial ratio thus set an upper bound on the wave amplitude than can be present in the scattering volume. For example if one were to observe an axial ratio of 6.5 it would imply that the wave amplitude $\delta B/B$ must be less than 0.2 because even if the intrinsic structure were completely two dimensional it would appear to have an axial ratio of 6.5.

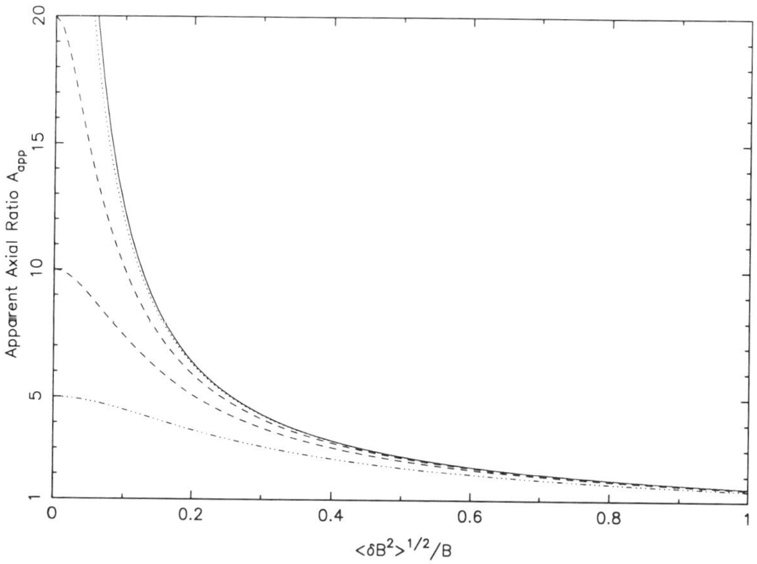

Fig. 6: Calculations of apparent axial ratio versus wave amplitude, for various intrinsic axial ratios. The intrinsic axial ratios are (top to bottom) 100, 50, 20, 10 and 5.

To compare with observations one must actually integrate the effect of waves from all locations on the line-of-sight. To do this one needs a model of the change of wave amplitude with distance. The results of a family of such calculations are overplotted on figure 5, based on the model fitting of Coles et al (1991).

The assumption in this calculation is that the intrinsic axial ratio is infinite, thus the calculated axial ratio is the largest that is allowed for that Alfvén wave amplitude. One can see that the range of theoretical models is quite consistent with the observed anisotropies. Furthermore the actual anisotropy could be much larger than observed; that is one must consider the possibility that the microstructure at scales in the vicinity of 10 km could be essentially two dimensional.

ACKNOWLEDGEMENTS

The author would like to acknowledge very helpful discussions with Ruth Esser and Barney Rickett. The work was partially supported by the NSF under grant ATM-90 08608 and partially by the Norwegian Research Council (NAVF).

REFERENCES

Armstrong, J. W., W. A. Coles, M. Kojima and B. J. Rickett, Observations of Field Aligned Density Fluctuations in the Inner Solar Wind, *Astrophys. J.* **358**, 685-692, 1990.

Coles, W. A. and J. K. Harmon, Propagation Observations of the Solar Wind Near the Sun, *Astrophys. J.* **337**, 1023, 1989.

Coles, W. A., W. Liu, J. K. Harmon and C. L. Martin, The Solar Wind Density Spectrum Near the Sun: Results from Voyager Radio Measurements, *J. Geophys. Res.* **96**, 1745-1755, 1991.

Coles, W. A., R. Esser, U-P Lovhaug and Jussi Markkanen, Comparison of Solar Wind Velocity Measurements with a Theoretical Acceleration Model, *J. Geophys. Res.* **96**, 13849-13859, 1991.

Dennison, P. A. and A. Hewish, The Solar Wind Outside the Plane of the Ecliptic, *Nature* **213**, 343, 1967.

Harmon, J. K. and W. A. Coles, Spectral Broadening of Planetary Radar Signals by the Solar Wind, *Astrophys. J* **270**, 748, 1983.

Kojima, M and T. Kakinuma, Solar Cycle Evolution of the Solar Wind Speed Structure Between 1973 and 1985 Observed with the Interplanetary Scintillation Method, *J. Geophys. Res.* **92**, 7269, 1987.

Kuo, S. P., M. H. Whang and M. C. Lee, Filamentation Instability of Large-amplitude Alfvén Waves, *J. Geophys. Res.* **93**, 9621, 1988.

Marsch, E. and C.-Y. Tu, Spectral and Spatial Evolution of Compressive Turbulence in the Inner Solar Wind, *J. Geophys. Res.* **95**, 11945, 1990.

Rickett, B. J. and W. A. Coles, Evolution of the solar wind structure over a solar cycle: interplanetary scintillation velocity measurements compared with coronal observations, *J. Geophys. Res.* **96**, 1717-1736, 1991.

Seventh Cambridge Workshop on Cool Stars, Stellar Systems, and the Sun
ASP Conference Series, Vol. 26, 1992
Mark S. Giampapa and Jay A. Bookbinder (eds.)

STOCHASTIC STELLAR WIND MODELS: EVOLVED STARS.

MANFRED CUNTZ Joint Institute for Laboratory Astrophysics [JILA], University of Colorado and NIST, Boulder, CO 80309-0440, USA.

ABSTRACT I present several remarkable observational and semiempirical results for individual cool giants and supergiants (in particular for α Orionis), which show that stochastic stellar wind flows are an extremely important feature in these stars. I also show that the status of proposed theoretical models is very poor due to the fact that appropriate models must simultaneously take into account a broad variety of physical processes which are strongly coupled and, furthermore, not well understood. Feasibility requires certain restrictive assumptions. As an example of my own research I present ab-initio models for the formation and time-dependent behavior of stochastic stellar wind flows in the chromosphere of α Orionis produced by the propagation and interaction of stochastic shocks. My wave models show time-dependent episodes of momentum and energy deposition. The direction of the flow alternates between infalling and outflowing motions depending both on the shock strengths and the hydrodynamic history of the flow. This behavior has already been suggested by observational constraints. My wave models show that the limiting shock strength behavior found in monochromatic models no longer exists, when stochastic shocks are considered. I anticipate that the stochastic variations of the energy dissipation should inevitably lead to irregular variabilities in chromospheric emission lines which are easily detectable with HST/GHRS.

Keywords: Red giants and supergiants; stochastic processes; instabilities; multiwavelength spectroscopy; stellar chromospheres; stellar winds; hydrodynamics; shock waves

OBSERVATIONAL EVIDENCE FOR STOCHASTIC PROCESSES IN THE ATMOSPHERES OF EVOLVED STARS

Based upon a broad variety of observational results it has become common knowledge that cool giants and supergiants located to the right of the coronal dividing line are characterized by complicated dynamic structures in the outer atmospheres and winds. These results are based upon multiwavelength spectroscopy which includes measurements of variabilities in optical, UV, and IR lines, and radio continuum observations. For α Orionis, for instance, Querci & Querci (1986) have summarized observations regarding the blue and UV Fe II emission core velocities given by Boesgaard & Magnan (1975), Boesgaard (1979), and Carpenter (1984). The observations have been obtained between 1970 and 1978. The results indicate complex velocity structures in the atmosphere of this star including changes from infall to outfall and from outfall to infall. The changes are significant and they are unpredictable. Due to the lack of data, however, the relevant timescales could not be determined. They might range between 10^6 and 10^8 s. An important problem which affects the interpretation of the data is that changes in overlying absorption features can easily imitate the presence of velocity fields in the line formation region. Moreover, the observations have not been performed at the same wavelengths, so they might refer to different atmospheric layers. Joras (1989) recently presented a detailed study of atmospheric features in α Orionis.

Further evidence for dynamic structures in the outer atmosphere of α Orionis is the detection of episodic chromospheric heating (Toussaint & Reimers 1989). Toussaint & Reimers reported a strong increase of the flux in the chromospheric Ca II K2 reversal from February 1988 to February 1989 which can be interpreted as an increase of the chromospheric nonradiative heating rate. The increase covers the Lorentz wings of the line out to about 5 Å from the center. This result coincides with the appearance of a bright spot on the stellar surface at wavelengths 633, 700, and 710 nm (Buscher *et al.* 1990). In this case it is extremely tempting to attribute the increase in the chromospheric activity to this photospheric bright spot because it can produce a local increase of the mechanical energy flux leading to enhanced energy dissipation in the chromosphere.

A further result which demonstrates the existence of complicated dynamic features in the atmospheres of α Orionis and other cool giants and supergiants is the presence of multiple absorption lines in Na I, K I, and CO (Goldberg *et al.* 1975, Bernat *et al.* 1979, Bernat 1981, among others). The observations indicate that separate components of shells exist in these stars. For Arcturus, evidence for atmospheric instabilities due to different chromospheric components have been described by Heasley *et al.* (1978). Tsuji (1988) has analyzed low excitation lines of the CO first overtone bands in high resolution infrared spectra of several red giants which show intensity anomalies that cannot be explained by photospheric absorption alone. Tsuji interpreted this result in favor of an extra molecular component in the outer atmospheres which shows a complex dynamics and may represent a transition between the chromospheres and the cool winds and/or chromospheric inhomogeneities in the outer atmospheres. Furthermore, a semitheoretical analysis of Jura (1984) showed that for selected objects the observed multiple absorption lines are compatible with the presence of multiple circumstellar shells of dust grains which might be formed during distinct epochs of mass ejections. The observations suggest that different atmospheric components moving at different radial velocities are present, but the direction of the flow apparently alternates between infalling and outfalling motion, possibly as a result of drastic events such as sporadic mass ejections (*e.g.* Schwarzschild 1975), or simply a consequence of stellar pulsation or waves. A detailed study for changes in UV lines and continua in α Orionis due to stellar pulsation modes was given by Dupree *et al.* (1987).

There are at least three contributed papers at this meeting which have relevance to the topic of stochastic velocity fields in cool giants and supergiant stars. First, Sada *et al.* (1992) presented CO observations for IRC +10216 obtained both at 2.3 and 4.6 μm covering a timespan from 1979 to now. Sada *et al.* identified several distinct atmospheric components having different expansion velocities and excitation temperatures. Second, Brugel *et al.* (1992) reported observational and theoretical results for L_2 Puppis, a Mira star, obtained during 1985–1987. The results provide evidence that patches of the stellar surface alternate between a quiet state (with little dust, an extended warm atmosphere, and low atmospheric TiO absorption) and a dusty wind state (with a cooler atmosphere and higher atmospheric TiO absorption). This result clearly implies the presence of complicated atmospheric velocity structures combined with hydrochemical instabilities. Third, Drake *et al.* (1992) reported results from a radio continuum observation monitoring program for α Orionis. The authors found variabilities in the form of dips most pronounced at 2 cm, evident also at 3.7 cm, and possibly present at 6 cm. The variabilities behave stochastically and the changes at different wavelengths do not appear to be correlated. The amplitude of the variations is about 20 % which shows that a large portion of the outer atmosphere must be involved. The authors also found that the variabilities can occur on a timescale of one month which is unequivocal evidence that the physical cause is not related to rotation. An attractive explanation of these observations might be thermal instabilities in the wind region as suggested by Muchmore *et al.* (1987), which can lead to molecule formation and therefore to a drastic decrease in the hydrogen ionization degree.

STATUS OF AB-INITIO MODELS

In the view of the observational results just described we have to realize that the atmospheres and winds of cool giants and supergiants possess a broad range of characteristic features: First, the atmospheres and winds in these stars contain complex velocity fields which indicate infalling and/or outflowing matter. This result provides evidence for episodic momentum and energy transfer. Second, many of these stars show variabilities in chromospheric emission lines which is evidence for episodic chromospheric heating events. Third, the atmospheres and winds contain spatial inhomogeneities with molecules and dust. These inhomogeneities often show a very complicated dynamic behavior. They are evidence for hydrochemical instabilities including the CO/SiO thermal instability.

Now I want to compile a list of physical processes, which have to be taken into account in order to obtain *realistic* ab-initio models. In the first rank, we have to consider those phenomena which give reason for the existence of the atmospheric dynamics. These phenomena include global oscillation modes, convection, and magnetic fields. They are responsible for the generation of complex wave modes, which propagate through the atmospheres and winds. The wave modes tend to form stochastic shocks, which substantially affect the outer atmospheric structures. These structures are further modified by radiative heating and cooling, which have to be computed by performing 3-D radiative transfer. The atmospheric dynamics also depend crucially on the solution for hydrogen because it controls the electron number densities, which enter the computation of the radiation damping, and it also determines the mean molecular weight. Kneer (1980) and Carlsson & Stein (1991), among others, presented theoretical arguments that the advective term in the hydrogen rate equations is of significant importance in the low density media studied here. Furthermore, Ayres (1981), Kneer (1983), Muchmore (1985), Muchmore *et al.* (1987), and Cuntz & Muchmore (1989) have pointed out that the outer atmospheres of cool stars are a suitable environment for thermal bifurcation and instabilities to occur. The latter authors presented theoretical results for the complicated interplay of thermal bifurcation through CO and SiO molecules and time-dependent radiation hydrodynamics. A further important effect not included in this work is the treatment of nonequilibrium chemistry including the formation and destruction of molecules and dust (Gail & Sedlmayr 1986, Mamon *et al.* 1987, Muchmore 1990, among many others).

This list clearly shows that *realistic* ab-initio models, which explain at least some of the observed features in cool giants and supergiants, are *not* feasible. This is in particular due to the following complication of the problem: in a certain number of cases the relevant physical processes must be solved in a fully consistent manner because they are strongly coupled and the timescales of the dominant physical processes are of similar magnitude. In many cases, the hierarchy of timescales can only be defined when ab-initio models are already available. Furthermore, the hierarchy of timescales can be different in different parts of the atmosphere and might also change with time.

A PROJECT WHICH IS FEASIBLE: AB-INITIO MODELS FOR STOCHASTIC CHROMOSPHERIC HEATING AND VELOCITY FIELDS IN ALPHA ORIONIS

Feasible projects include the computation of simplified ab-initio models for stochastic chromospheric heating and velocity fields in cool giants and supergiant stars such as α Orionis. To accomplish this, I apply the model of Cuntz (1987), which assumes deposition of mechanical energy and momentum in outer atmospheric layers due to the propagation of stochastic shocks. This is motivated by results of previous studies which show that the outer atmospheric heating in M-type supergiants is expected to be entirely dominated by nonmagnetic heating, because the relatively weak chromospheric emission of these stars depends only on their effective temperatures (Johnson 1987, Judge 1989, Judge & Stencel 1991, Dupree 1991). A similar result for slow-rotating single K giant stars was presented by Schrijver (1987 a,b). Hartmann & Avrett (1984) computed an ab-initio model for the outer atmosphere of α Orionis based on Alfvén wave dissipation, which could explain some of the observed properties. Further results based on Alfvén waves were given by An *et al.* (1990). However, the relevance of these models remain uncertain due to the lack of magnetic field measurements.

Method

I use an Eulerian one-dimensional, spherically symmetric, time-dependent radiation hydrodynamic code based upon the method of characteristics (Cuntz & Ulmschneider 1988). The code is suitable to the study of propagating shocks, which are treated as discontinuities. Boundary conditions for incoming and outgoing shocks are solved. Ionization of hydrogen is explicitly taken into account assuming a 3-level atom. The Lyman-α transition is computed using the escape probability method given by Hartmann & Avrett (1984). However I have entirely omitted the advective term in the rate equations. Radiation damping is considered in the effectively thin plasma approximation using a Cox and Tucker type law given by Judge & Neff (1990). Judge (1990) presented theoretical arguments that the effectively thin plasma approximation should work well for chromospheric layers of the inactive, low-gravity star studied here. For the α Orionis wave model I have taken the following stellar parameters: T_{eff} = 3900 K, $R_* = 860\ R_\odot$, and $\log g_* = -0.4$ (Tsuji 1989). The initial atmosphere of the model extends from 1.1 up to 1.4 R_*. The atmospheric extent has been chosen to encompass the scale length where the major part of the mechanical energy dissipation of the stochastic shock waves occur. Wave models computed to deduce the magnitude of time-averaged mass loss rates must certainly be based upon much more extended atmospheric shells.

Results and Discussion

For my study I start with a monochromatic wave model computed for a fixed wave period of 2 10^6 s and a fixed initial wave amplitude of 0.30 Mach. Then I start to introduce shock waves with stochastically changing wave periods in the short-period range. The shape of the wave period distribution is assumed to be Gaussian centered at 2 10^6 s with a standard deviation of 5 10^6 s. Negative periods have been omitted. The peak value of the wave period distribution has been chosen according to the results of traditional acoustic energy generation models (Bohn 1981, 1984). Figure 1 shows a panel of 8 snapshots of a time-dependent stochastic wave computation. It demonstrates the generation of an episodic outflow and inflow event. The numbers in the upper left corners denote the time after allowing the wave period to change stochastically. The timescale for typical changes in the wave propagation is 1 10^6 s and the time between the episodic outflow and inflow event is about 5 10^6 s. The atmospheric models are characterized by a complicated hydrodynamic structure containing a nonuniform distribution of shocks. The shock strengths and the shock speeds differ substantially and change non-monotonically with height.

This result is a clue to the basic physics which is going on: after allowing the wave period to change stochastically, shocks with different strengths are introduced into the atmosphere. Different shock strengths cause different shock speeds, which lead to interacting, overtaking and merging of shocks (*"shock-cannibalism"*). Since the strength of an overtaking shock combines with the shocks it engulfs, its speed increases, so it overtakes more and more shocks in front of it and attains an even greater strength. Consequently, the amount of momentum and energy deposition that occurs in the atmospheric layer varies drastically. The direction of the flow alternates between infalling and outflowing motions depending on the strengths of the shocks and the hydrodynamic history of the flow. In the first snapshot, shock number 5 (counted from outside to inside) is the major cannibalizing shock. This shock soon overtakes the four shocks in front of it and generates the episodic outflow event.

A further interesting result is obtained by the behavior of radiation damping. My time-dependent stochastic wave models show that the radiation damping function is strongly peaked behind the shocks and is not directly related to the structure of the time-averaged atmosphere. The radiation losses decrease with increasing atmospheric height showing that the waves behave more and more adiabatically when propagating outward. The mean thermal structure of the atmosphere is somewhat similar to that of the semiempirical chromosphere model of Basri *et al.* (1981), who computed a semiempirical chromosphere model for α Orionis based upon the Ca II *K* and Mg II *h, k* emission lines from IUE

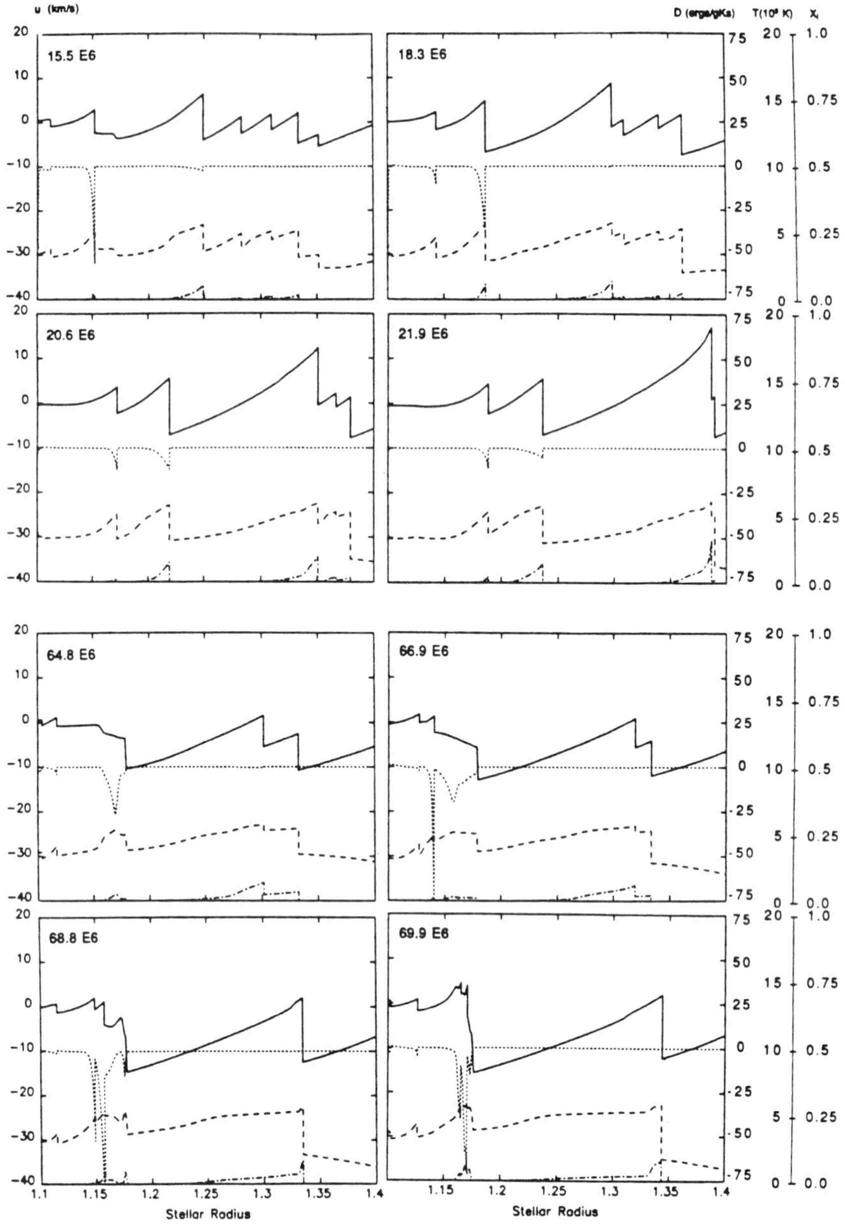

Fig. 1: Snapshots of stochastic wave computation. The flow speed u (solid line), the radiation damping function D (dotted line), the temperature T (dashed line), and the hydrogen ionization degree X_i (dashed–dotted line) are shown as functions of height.

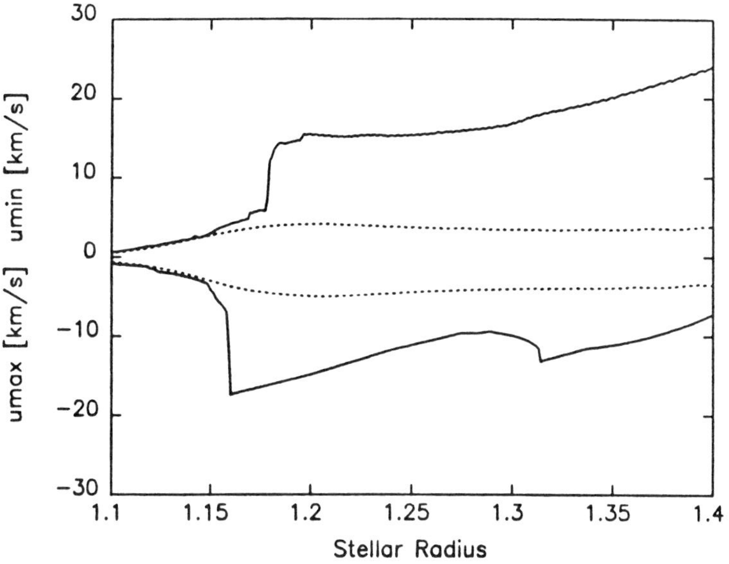

Fig. 2: Minimal and maximal flow speed in a time-dependent monochromatic (dotted lines) and stochastic (solid lines) wave computation as functions of height.

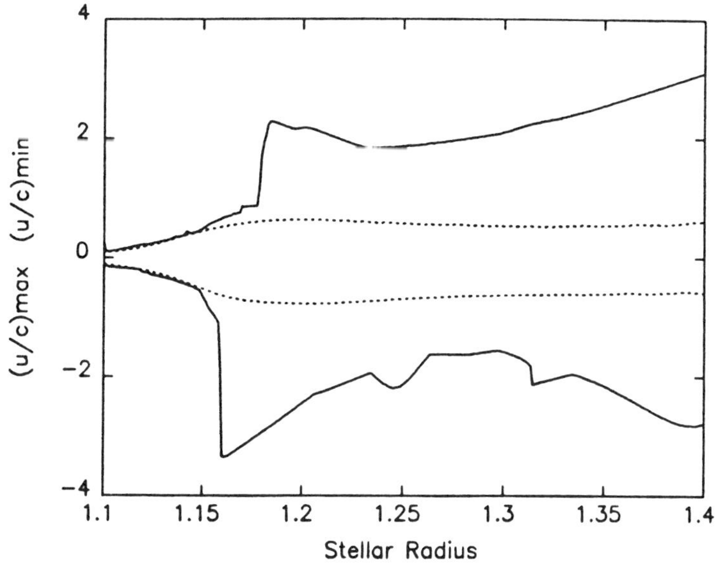

Fig. 3: Minimal and maximal ratio of flow speed to sound speed in a time-dependent monochromatic (dotted lines) and stochastic (solid lines) wave computation as functions of height.

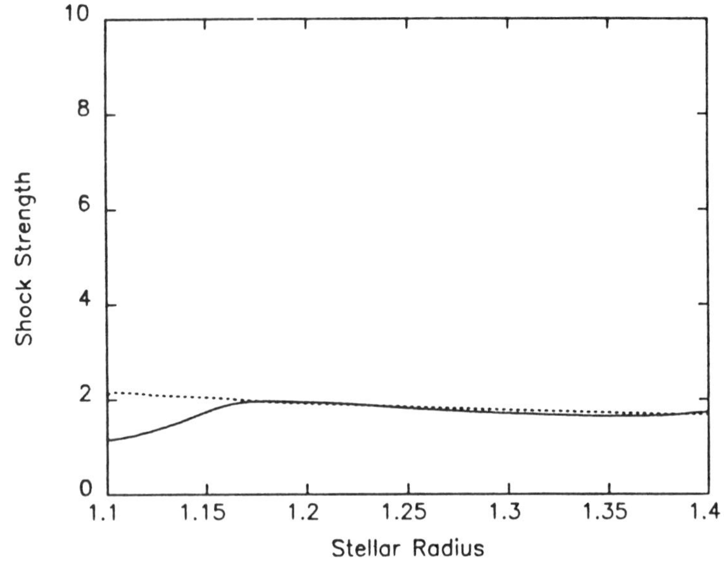

Fig. 4: Shock strength in a time-dependent monochromatic wave computation as function of height (solid line). The run of the limiting shock strength (dotted line) also is shown.

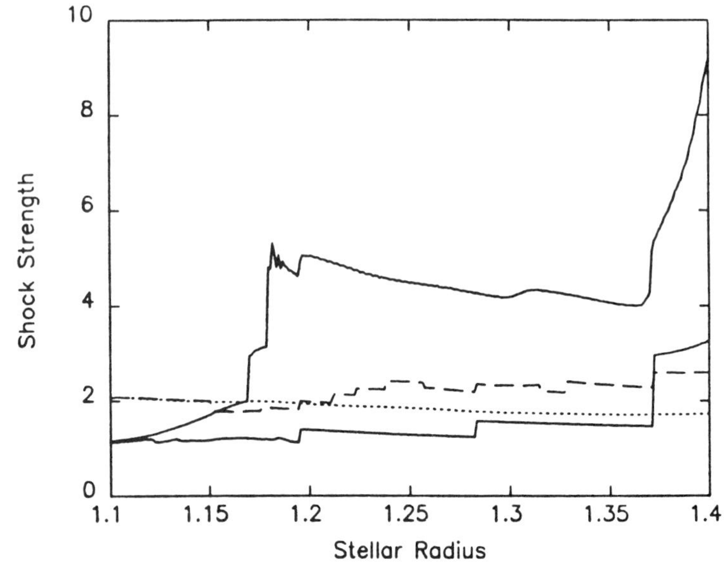

Fig. 5: Minimal and maximal shock strength (solid lines) in a time-dependent stochastic wave computation as functions of height. The run of the limiting shock strength (dotted line) is shown. A modified version of the limiting shock strength which considers the decrease of the number of shocks with height is also shown (dashed line).

spectra. They found a smooth increase of the atmospheric temperature starting from 2820 K in the temperature minimum layer up to 7000 K at a column mass density of 1.0 10^{-6} g cm^{-2}.

Figure 2 shows the minimal and maximal flow speed in a time–dependent monochromatic and time–dependent stochastic wave model as functions of the atmospheric height. Figure 3 shows the respective minimal and maximal ratio of flow speed to sound speed. The monochromatic model is based upon a wave period of 5 10^6 s. For the computation of the stochastic model I have introduced 16 stochastic wave periods into the atmospheric layer covering a timespan of 6.9 10^7 s. The minimum wave period I introduced was 1.2 10^6 s, and the maximum was 1.6 10^7 s. The results are quite impressive: in the stochastic wave model the maximal outflow velocity is 24.1 km/s and the maximal inflow velocity is – 17.4 km/s. The corresponding Mach numbers are 3.10 and 3.33, respectively, demonstrating that in stochastic wave models supersonic flow patterns can easily be generated without ad-hoc assumptions. In the monochromatic wave model the flow speeds are substantially smaller and do never exceed the sound speed.

In order to get further insight into the physics of time–dependent monochromatic and stochastic wave models, I have plotted the shock strengths for both models as functions of the atmospheric height. Figure 4 shows the results for the monochromatic model. It shows that the shock strengths remain relatively small and can accurately be predicted by the limiting shock strength formula, deduced for monochromatic waves having a fixed initial amplitude (*e.g.* Cuntz & Ulmschneider 1988). Figure 5 shows the results for the stochastic wave model. For this model I have plotted both the minimal and maximal shock strengths. Due to overtaking and merging of shocks, different shock strengths are found at the same atmospheric height. For the outer boundary point of the atmosphere the shock strengths differ from 3.26 to 9.19 demonstrating a very effective mechanism for forming strong shocks. In a model given by Cuntz (1991) an alternative sequence of wave periods has been considered. In this case the shock strengths differ from 1.22 to 3.53 at the outer atmospheric boundary point. The latter result is extremely impressive when I compute the respective energy dissipation rate for the shocks. Since the energy dissipation rate behaves as $(M_{sh}-1)^2$, a difference by a factor of 130 is produced by the stochastic nature of the flow. It is obvious that the limiting shock strength behavior found in monochromatic wave models no longer exists when stochastic shocks are considered.

CONCLUSIONS

I have presented observational, semiempirical, and theoretical results which show that the atmospheric dynamics in cool giants and supergiants are expected to be dominated by stochastic processes. The observational results are based upon multiwavelength spectroscopy. I note that the status of the theoretical models which are proposed is still very uncertain. Physical processes potentially important include: global oscillation modes, convection, magnetic fields, the generation, propagation, and interaction of stochastic waves, 3-D radiative transfer effects, time–dependent hydrogen ionization, thermal bifurcation and instabilities, and nonequilibrium chemistry, particularly the formation and destruction of molecules and dust. A further complication of the problem is that very often the relevant physical processes must be solved in a fully consistent manner because they are strongly coupled and the timescales of the dominant physical processes are of similar magnitude.

As an important first step in this process, I have presented simplified ab-initio models for the formation and time-dependent behavior of stochastic stellar wind flows in the chromosphere of α Orionis. The results I obtained are extremely interesting and reflect some of the observed features in this star. My wave models show time–dependent episodes of momentum and energy deposition produced by strong shocks which are generated by merging of shocks in the stochastic wave field. The chromospheric temperatures are found to be similar to those temperatures given by the semiempirical chromosphere model of Basri *et al.* (1981). I found that the direction of the flow alternates between infalling and outflowing motions depending both on the shock strengths and the hydrodynamic history of the flow. Furthermore, stochastic wave models are capable of generating supersonic flow patterns. My wave models also show that the limiting shock strength behavior found in monochromatic models no longer exists, when stochastic shocks are considered, because the atmospheric dynamics lead to the formation of shocks with different strengths at the same atmospheric height. The time average of the atmospheric flow speed can easily be related to the microturbulent velocity, a velocity which is required in classic stellar atmosphere models. The atmospheric dynamics of the models can best be described as being not in dynamic equilibrium, but in *time-dependent stochastic disequilibrium.*

Based on the results I have obtained I feel that the following projects should be topics of future research: First, the computation of improved chromospheric ab-initio models, which include a detailed treatment of radiative transfer and a more sophisticated hydrogen model, should be targeted. In particular, the advective term in the rate equations for hydrogen should be included, because it will significantly influence the temperatures and cooling rates behind the shocks. Second, longer wave periods should be considered to investigate the generation of mass loss. The third project I want to mention is considering the thermal bifurcation concept in ab-initio hydro models because it might lead to a theoretical explanation for the onset of molecule and dust formation without ad-hoc assumptions.

ACKNOWLEDGEMENTS

We are pleased to acknowledge support for this work through NASA grant NAG 5-1374 to the University of Colorado. The calculations were in part performed on a Cray Y-MP computer at the National Center for Atmospheric Research.

REFERENCES

An, C.H., Musielak, Z.E., Rosner, R., Moore, R.L., Suess, S.T.: 1990, in *Cool Stars, Stellar Systems, and the Sun*, Proc. Sixth Cambridge Workshop, ed. G. Wallerstein, *ASP Conf. Series* 9, p. 70

Ayres, T.R.: 1981, *Ap. J.*, 244, 1064

Basri, G.S., Linsky, J.L., & Eriksson, K.: 1981, *Ap. J.*, 251, 162

Bernat, A.P.: 1981, *Ap. J.*, 246, 184

Bernat, A.P., Hall, D.N.B., Hinkle, K.H., & Ridgway, S.T.: 1979, *Ap. J. Letters*, 233, L135

Boesgaard, A.M.: 1979, *Ap. J.*, 232, 485
Boesgaard, A.M., & Magnan, C.: 1975, *Ap. J.*, 198, 369
Bohn, H.U.: 1981, Ph.D. thesis, University of Würzburg
Bohn, H.U.: 1984, *Astr. Ap.*, 136, 338
Brugel, E.W., Willson, L.A., Bowen, G.H., & Magalhaes, M.: 1992, this volume
Buscher, D.F., Haniff, C.A., Baldwin, J.E., & Warner, P.J.: 1990, *M.N.R.A.S.*, 245, 7p
Carlsson, M., & Stein, R.E.: 1991, in *Mechanisms of Chromospheric and Coronal Heating*, ed. P. Ulmschneider *et al.* (Berlin: Springer), p. 366
Carpenter, K.G.: 1984, *Ap. J.*, 285, 181
Cuntz, M.: 1987, *Astr. Ap.*, 188, L5
Cuntz, M.: 1991, in *Instabilities in Evolved Super- and Hypergiants*, ed. C. de Jager and H. Nieuwenhuijzen (Amsterdam: North Holland Publishers), p. 134
Cuntz, M., & Muchmore, D.O.: 1989, *Astr. Ap.*, 209, 305
Cuntz, M., & Ulmschneider, P.: 1988, *Astr. Ap.*, 193, 119
Drake, S.A., Bookbinder, J.A., Florkowski, D.R., Linsky, J.L., Simon, T., & Stencel, R.E.: 1992, this volume
Dupree, A.K.: 1991, in *Mechanisms of Chromospheric and Coronal Heating*, ed. P. Ulmschneider *et al.* (Berlin: Springer), p. 185
Dupree, A.K., Baliunas, S.L., Guinan, E.F., Hartmann, L., Nassiopoulos, G.E., & Sonneborn, G.: 1987, *Ap. J. Letters*, 317, L85
Gail, H.-P., & Sedlmayr, E.: 1986, *Astr. Ap.*, 166, 225
Goldberg, L., Ramsey, R.W., Testerman, L., & Carbon, D.: 1975, *Ap. J.*, 199, 427
Hartmann, L., & Avrett, E.H.: 1984, *Ap. J.*, 284, 238
Heasley, J.N., Ridgway, S.T., Carbon, D.F., Milkey, R.W., & Hall, D.N.B.: 1978, *Ap. J.*, 219, 970
Johnson, H.R.: 1987, in *Cool Stars, Stellar Systems, and the Sun*, Proc. Fifth Cambridge Workshop, ed. J.L. Linsky and R.E. Stencel (Berlin: Springer), p. 399
Joras, P.: 1989, Ph.D. thesis, University of Oslo
Judge, P.G.: 1989, in *The Evolution of Peculiar Red Giant Stars*, ed. H.R. Johnson and B. Zuckerman (Cambridge: University Press), p. 303
Judge, P.G.: 1990, *Ap. J.*, 348, 279
Judge, P.G., & Neff, D.H.: 1990, in *Cool Stars, Stellar Systems, and the Sun*, Proc. Sixth Cambridge Workshop, ed. G. Wallerstein, *ASP Conf. Series* 9, p. 57
Judge, P.G., & Stencel, R.E.: 1991, *Ap. J.*, 371, 357
Jura, M.: 1984, *Ap. J.*, 282, 200
Kneer, F.: 1980, *Astr. Ap.*, 87, 229
Kneer, F.: 1983, *Astr. Ap.*, 128, 311
Mamon, G.A., Glassgold, A.E., & Omont, A.: 1987, *Ap. J.*, 323, 306
Muchmore, D.O.: 1986, *Astr. Ap.*, 155, 172
Muchmore, D.O.: 1990, in *Cool Stars, Stellar Systems, and the Sun*, Proc. Sixth Cambridge Workshop, ed. G. Wallerstein, *ASP Conf. Series* 9, p. 39
Muchmore, D.O., Nuth III., J.A., & Stencel, R.E.: 1987, *Ap. J. Letters*, 315, L141
Querci, M., & Querci, F.: 1986, in *Cool Stars, Stellar Systems, and the Sun*, Proc. Fourth Cambridge Workshop, ed. M. Zeilik and D.M. Gibson (Berlin: Springer), p. 492
Sada, P.V., Hinkle, K.H., & Keady, J.J.: 1992, this volume
Schrijver, C.: 1987 a, *Astr. Ap.*, 172, 111
Schrijver, C.: 1987 b, in *Cool Stars, Stellar Systems, and the Sun*, Proc. Fifth Cambridge Workshop, ed. J.L. Linsky and R.E. Stencel (Berlin: Springer), p. 135
Schwarzschild, M.: 1975, *Ap. J.*, 195, 137
Toussaint, F., & Reimers, D.: 1989, *Astr. Ap.*, 226, L17
Tsuji, T.: 1988, *Astr. Ap.*, 197, 185
Tsuji, T.: 1989, in *The Evolution of Peculiar Red Giant Stars*, ed. H.R. Johnson and B. Zuckerman (Cambridge: University Press), p. 81

Seventh Cambridge Workshop on Cool Stars, Stellar Systems, and the Sun
ASP Conference Series, Vol. 26, 1992
Mark S. Giampapa and Jay A. Bookbinder (eds.)

STOCHASTIC STELLAR WIND MODELS: T TAURI STARS

V.P. GRININ
Crimean Astrophysical Observatory, Crimea, 334413, Nauchny, USSR

A.S. MITSKEVICH
Osservatorio Astrofisico di Arcetri, Largo E.Fermi 5, 50125 Firenze, ITALY

ABSTRACT A short review of the development of T Tauri stellar wind models is given. A classical approach to this problem, based on the continuum outflow approximation, permits one to explain the observed luminosities of the most important emission lines in the spectra of these stars (hydrogen, calcium and magnesium), but not the line profiles.
Recent model calculations show that this problem can be solved in the framework of so-called stochastic wind models. This new step in the modeling of emission spectra of TTSs is based on the general idea that the outflow of matter from TTSs is due to their magnetic activity. In this case, the gas acceleration depends strongly on the local conditions on the stellar surface and the structure of the wind must reflect this inheterogeneity.
In our simplified models we assume that the stellar wind consists of separate gas structures (blobs); all blobs have the same mass and the same trajectories of motion. The random simulation of blobs and the radial symmetry of outflow are used. Our main conclusions obtained from the analysis of emission line profiles are the following:
a) Only about one-tenth of the stellar surface acts as a source of the stellar wind at any given time.
b) Two different kinematical zones exist in the stellar wind: The blobs are accelerated near the stellar surface and they are decelerated due to the gravitational field of the star in the outer region of the wind.

Keywords: winds: T Tauri stars; winds: stochastic; winds: blobs

1. INTRODUCTION

In spite of abundant data testifying to intensive mass loss from T Tauri stars (TTSs) – the emission line profiles, jets, the Herbig-Haro objects – there is no general point of view on the mechanism responsible for this process. The mass outflow may be the consequence both of the magnetic surface activity of young stars and of disc accretion of the gas on a star (Appenzeller and Mundt 1989). In the latter, the outflow is a secondary process and it occurs from a so-called boundary layer (Bertout 1989), whose properties are little known at present.

The concept of "surface magnetic activity" includes the different types of physical properties connected with the existence of magnetic spots on the surface of young stars. There are flares, MHD waves, coronal transients etc. The flare activity of TTSs was still implied as a source of stellar winds in the pioneering paper by Kuhi (1964), where the first rather simplified models of expanding envelopes of TTSs were calculated.

An important step in understanding the process of TTSs mass loss was made in the papers by De Campli (1981), Hartmann *et al.* (1982), and Lago (1984) where the basis of the MHD wind models was created. In these models, the acceleration and heating of the gas is due to MHD waves generated in magnetic spots. According to Hartmann *et al.* (1982), the mass loss rate $\dot{M} = 3 \times 10^{-8}\ M_{\odot}$/year could be provided by a magnetic field, whose strength is about 500 G under the condition that it covers the whole surface of a star.

One of the most complicated problems in these models is the problem of transforming the MHD waves into mechanical and thermal energy of the gas. It is caused by complex, nonlinear processes in MHD wave propagation in plasmas. Therefore, the set of models was calculated in the approximation of an isothermal wind. There are papers by Natta *et al.* (1988 a,b) and Grinin and Mitskevich (1990, Paper I). In the first two papers, the hydrogen lines, including the infrared lines (such as Br_{α}), were considered. The calculations have shown that it is difficult to deduce the stellar wind parameters using hydrogen lines only: the same luminosities in the lines may be obtained at different values of temperature and mass loss rate.

The better diagnostic is given by spectral lines of different elements, certainly if there is assurance that these lines are formed in the same part of the wind. The first steps in this direction were made recently in papers by Natta and Giovanardi (1990), Grinin and Mitskevich (1990) and Hartmann *et al.* (1990). In the first of these papers the formation of hydrogen and the sodium resonance doublet was considered. It was shown that a common origin of these lines is possible at a gas temperature $T_g = 6000 - 6500$ K and mass loss rate $\dot{M} = 3 \times 10^{-8} - 3 \times 10^{-7}\ M_{\odot}$/year.

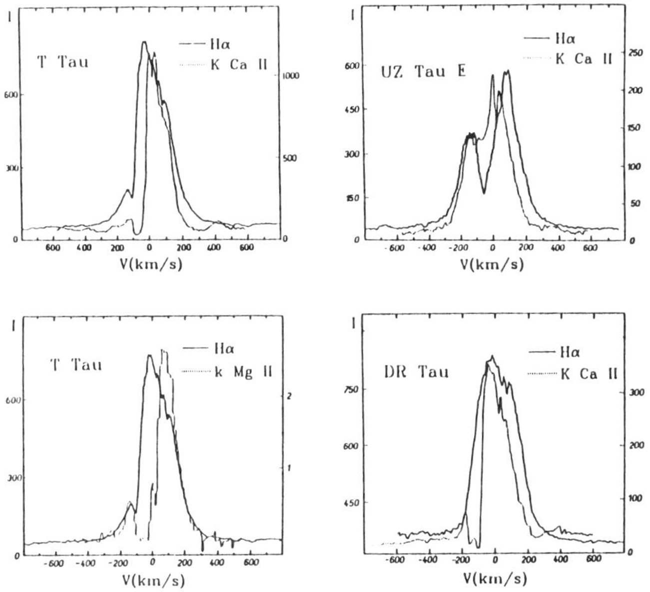

Fig. 1: The line profiles of some T Tauri stars from the papers by Mundt (1984; H_{α} and Ca II K lines) and Brown *et al.* (1981; Mg II k line).

In Paper I and in Hartmann *et al.* (1990) one and the same combination of emission lines was used: H_α, Ca II K and Mg II k. The choice of these lines is important for two reasons. First, they are a significant fraction of the luminosities of stellar winds. Defining physical conditions by these lines, we investigate the most important (i.e., the most energetic) part of the stellar wind. Second, a comparison of the line profiles of these lines shows that in many classical TTSs they are formed in about the same kinematical zones of the stellar wind (see Fig. 1).

According to Paper I, the observed luminosities in these lines correspond to the gas temperature $T_g = 7000 - 8000$ K and the mass loss rates $\dot{M} = 3 \times 10^{-8} - 3 \times 10^{-7}\ M_\odot$/year. About the same values of these parameters follow from model calculations by Hartmann *et al.* (1990).

Thus, we may consider as reliably established that T Tauri stellar winds are rather cool in their significant part and are characterized by a high mass loss rate $\dot{M}$.

2. THE EMISSION LINE PROFILES

Unlike the integral luminosities of TTSs in emission lines, which are explained well in the models given above of spherical symmetric continuous outflow of matter, formation of observed line profiles remains enigmatic (see Appenzeller and Mundt 1989).

As is well-known, more than 50 percent of all stars of this type exhibit double-peaked structure in the H_α line with an absorption component shifted towards shorter wavelengths (Herbig 1977). Similar profiles are also seen in resonance transitions of Mg II (Brown *et al.* 1984, Giampapa and Imhoff 1985). In a number of cases, the absorption component in H_α is weak or absent. In such cases, the profile is observed to be single and asymmetric. Only a few TTSs have a classical P Cygni H_α profile which can be understood in the context of current models (Hartmann *et al.* 1990).

The basic problem is the following: how can one explain the origin of double-peaked emission lines? Are dealing with an emission profile which is distorted by self-absorption in moving gas or is the observed profile the result of absence of emission at certain velocities (Appenzeller and Mundt 1990)? Among the current wind models, there is none which provides an answer to this question. The best of the MHD wind models of (Hartmann *et al.* 1990) produce double-peaked H_α profiles that are not even remotely similar to the observed profiles.

3. LINE FORMATION IN STOCHASTIC STELLAR WIND MODELS

The possible reason why no progress has been made in this old problem was pointed out in our recent paper (Grinin and Mitskevich 1991, Paper II). In particular, one of the fundamental assumptions which forms the basis of all stellar wind models mentioned above is the approximation of continuum outflow. This approximation leads to substantial simplifications in the physical modeling of the moving gas. However, in the case of T Tauri stars it is very artificial.

Indeed, if the outflow of matter from TTSs is due to surface magnetic activity, we must expect that the stellar wind will be extremely non-uniform;

mass loss will occur mainly from those parts of the stellar surface where magnetic spots are strongest, or where the magnetic configuration is most favorable. So, in order to describe the T Tauri stellar wind, we need to adopt, at the very beginning, the concept that it has an extremely non-uniform density distribution. Such an outflow should be stochastic in nature.

3.1. Basic assumptions.

For a description of the most simple model of stochastic stellar wind (S-wind), we adopt the following approach:

1. Using the Sobolev escape probability method, we calculate at the beginning the so-called basic non-LTE model (for hydrogen, calcium, magnesium atoms and their ions) for spherically-symmetric, continuum outflow. The parameters of this outflow are the gas temperature T_g and the mass loss rate $\dot{M}$. The velocity field (see Fig. 2) is determined by the acceleration force with the damping length L : $f(r) = f(r) \times exp[-(r/R_* - 1)/L]$, where R_* is the stellar radius, and by the deceleration in the gravitational field of a star. The latter should be effective since the maximum velocities of the gas motion in TTS winds (according to emission line profiles) are at about escape velocity. So, the adopted kinematical model of S-wind is nonmonotonic.

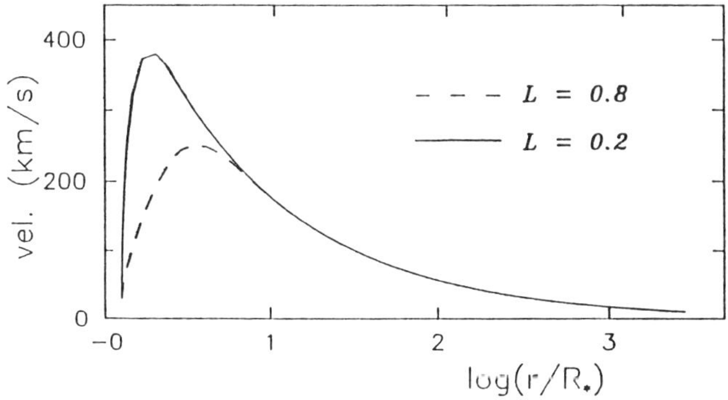

Fig. 2: The adopted outflow velocity as a function of distance: "rapid" ($L = 0.2$) and "slow" ($L = 0.8$) acceleration.

2. Using a random number generator, we separate the blobs in the continuum wind and generate the S-wind. The masses of the blobs are the same and do not change with their motion. The space between the blobs is assumed to be empty. The S-wind is radial-symmetric and quasi-steady as before.
3. The S-wind is described by a volume-filling factor q which is assumed to be constant. Hence, the effective mass loss rate now equals: $\dot{M}_{eff} = q \times \dot{M}$.

3.2. The emission line profiles in S-winds.

In calculating spectral line profiles, we used the exact expression for the intensity of the emergent radiation. The details of calculations are shown in Paper II. Here we note only one important property of S-wind models which is the key for understanding the formation of line profiles:

We have mentioned above that the filling factor q is constant through the S-wind. However, in velocity phase space, the number of blobs N per unit velocity interval v is not constant. Because of the decrease in radial velocity at the periphery of the wind, the resonant frequencies of the individual blobs become more crowded, and they begin to blend together. It means that although the optical thickness of each blob on the periphery of the wind may be small, the combined contributions may produce significant absorption at low velocities .

On the other hand, for the wings of a line, the optical thickness of the wind is dominated by its inner part, where expansion is rapid. As a result, the number of blobs intersecting the line of sight at such frequencies is greatly diminished. Hence, although each individual blob may have a significant optical depth, there is a greater probability that none of them will lie along the line of sight.

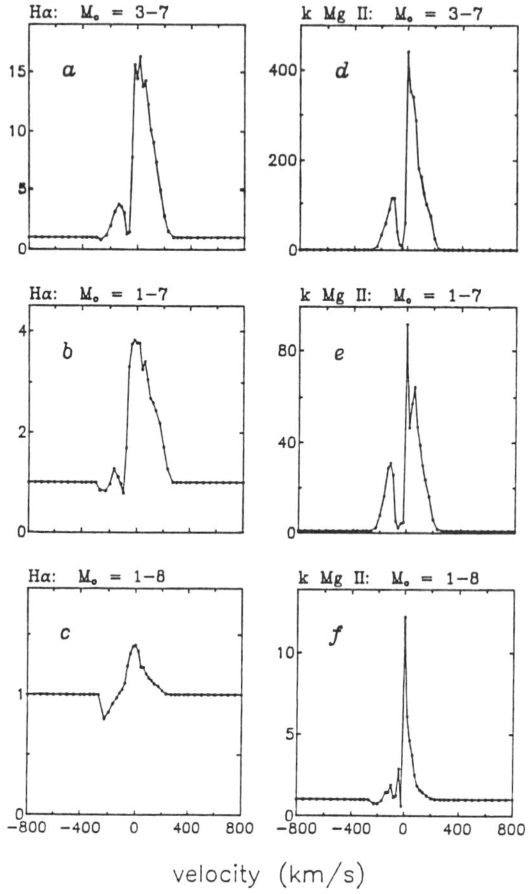

Fig. 3: The examples of theoretical profiles of the H_α line and the k Mg II line in stochastic wind models. The results are obtained at $L = 0.8$, $q = 1/8$ for mass loss rates in the base model: $\dot{M} = 3 \times 10^{-7}$, 10^{-7} and 10^{-8} $M_\odot$/year and $T_g =$ 7500 K.

So, from the viewpoint of the optical properties, the wind exhibits simultaneously the properties of a discrete medium and the properties of a continuous medium.

In order to study the effects which the parameters of the S-wind have on the line profiles, we have modeled the three strongest lines observed in TTSs: H_α, the Ca II K, and Mg II k, as well as the IR triplet of Ca II. The calculations show that the line profiles are not very sensitive to the geometry of the blobs and depend mainly on the volume filling factor q and on the mass loss rate $\dot{M}$ in the base model.

Some examples of theoretical profiles are shown in Fig. 3 (more detailed analysis is given in Paper II). One can see that all main types of observed profiles in TTSs are present among them. The two-component profiles which are typical of TTSs occur in models with $\dot{M} = (1-3) \times 10^{-7}\ M_\odot$/year with the volume filling factor $q = 0.1$ that corresponds to an effective mass loss rate $\dot{M}_{eff} = (1-3) \times 10^{-8}\ M_\odot$/year. Of the three strongest emission lines: H_α, Ca II K and Mg II k – this type of profile is seen in its purest form in the resonance line of Mg II. With progressively smaller values of $\dot{M}$ or q, the two-component profile changes over to an asymmetric single profile, which is also frequently observed in TTSs spectra.

3.3. *Diagnostics of the inner part of the stellar wind.*

According to our model calculations, the Ca II lines are formed closer to the stellar surface than the H_α line or the resonance doublet of Mg II. Consequently, if the wind contains an extended acceleration zone, the Ca II lines should be systematically narrower than the H_α line or the k line of Mg II. This suggestion is confirmed directly by the comparison of theoretical profiles presented in Fig. 4. One sees that in both cases the H_α line is definitely broader than the ionized calcium lines.

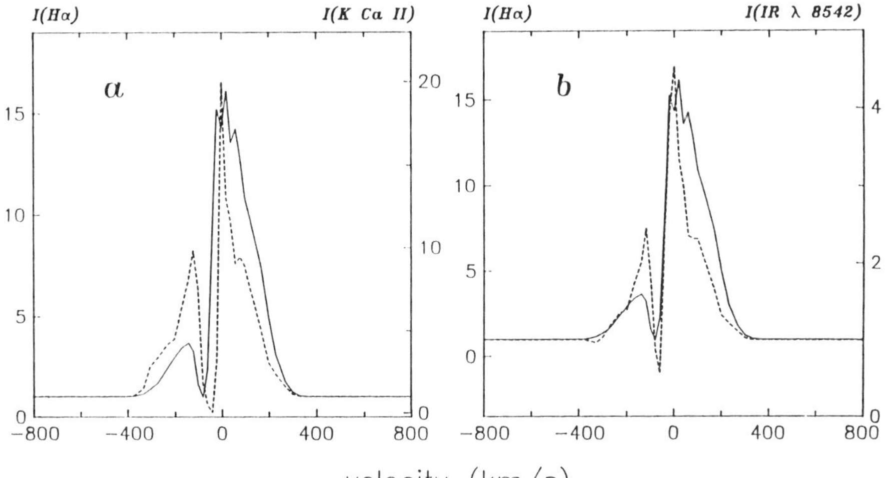

Fig. 4: The line profiles in stochastic wind models: a) the H_α line (solid line) and Ca II K line (dashed line); b) the H_α line (solid line) and Ca II IR triplet line at 8542 Å. The damping length is $L = 0.2$, $\dot{M} = 3 \times 10^{-7} M_\odot$/year, $q = 1/8$ and $T_g = 7500$ K.

In the case of rapid acceleration ($L = 0.2$), the redward emission component in the Ca II lines is deformed as a result of screening by the stellar disc of that part of the envelope moving away from the observer. The deformation is not as great in the case of slow acceleration ($L = 0.8$). The same deformations are seen clearly in the comparison of observed profiles of H_α and Ca II lines (see Fig. 1).

Hence, despite a number of simplifying assumptions, our model reproduces rather well some important details of the line profiles. These details clearly demonstrate the presence of an acceleration zone in the inner part of the stellar wind.

3.4. Conditions in the base of the stellar wind.

Can we in any way estimate the physical conditions in the base of the stellar wind, near the stellar surface? This question is extremely important for understanding the acceleration mechanism of TTS-wind.

To do this, we clearly need to use spectral lines which are formed in in the immediate vicinity of the star, where the structure of the wind still "remembers" the conditions of its acceleration. The lines of the infrared triplet of Ca belong to this class. These lines are thermalized (Shanin *et al.* 1975; Herbig and Soderblom 1980) and this means that they are formed in a dense gas which is opaque in the line frequencies. Our calculations (Paper I) indicate that the most favorable location for forming the IR triplet is at the base of the wind.

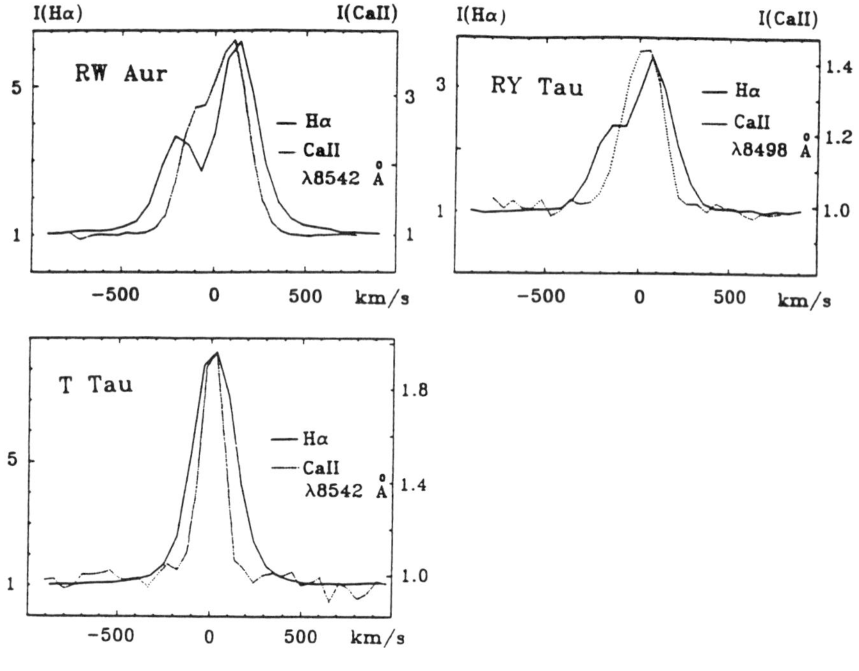

Fig. 5: The H_α and Ca II IR triplet profiles of three classical T Tauri stars (from Paper II).

If the outflow occurs uniformly over the entire surface of the star, the profiles of the triplet lines should be similar to the classical P Cygni type. However, it is obvious from Fig. 5 that the observed profiles have a single asymmetric component, without any sign of absorption on the blue side. This indicates that the region of the wind where these lines are formed does not cover the disc of the star completely.

Numerical simulations of these profiles (see Paper II) in the framework of S-wind models lead us to the following conclusion: only a small part (about 10%) of the stellar surface is responsible for the formation of stellar wind. This is in agreement with the estimation of the volume filling factor $q = 0.1$.

This means that the appreciable deposit in the stellar wind may produce the magnetic activity of one or more of the largest spots on the stellar surface. Rotating with a star, these spots may produce the rotational modulation of some parameters of emission line profiles. The evidence of such modulation was found in the H_α profile in RW Aur (Grinin *et al.* 1984) and, in the clearest form, in SU Aur (Giampapa and Basri, this Meeting).

4. THE MAIN CONCLUSION

Thus, the principal results of our numerical simulations of emission line profiles can be reduced to the following statement: the structural characteristics of T Tauri stellar winds are very different from those which are represented in all previous models based on the assumption of continuous outflow. The real stellar wind is stochastic in nature, which reflects the real inhomogeneity in magnetic activity on the stellar surface.

This conclusion has something in common with the recent idea by Mullan (1990) on the important role of reconnection processes as the source of momentum of the solar wind. If that is so, we have the reason for the following suggestion: the solar wind does not differ in its nature from winds in T Tauri stars. They differ only by the energy of their magnetic and flare activity [1].

5. SOME REMARKS

In the course of modeling a stochastic wind, we have made several simplifying assumptions. In view of these, our results should be regarded as semi-quantitative. Let us note briefly that some physical factors complicated the model description.

The model described above reflects the complex kinematic structure of T Tauri winds only in part. It is obvious that there should be a dispersion in the blob velocities, and some blobs should appear with velocities which are less than the escape velocity. Consequently, a fraction of the ejected gas should be accreted back onto the star. The observational evidence for such motions is well known, namely, there are inverse P Cygni profiles in the spectra of some TTSs (Mundt 1984). The existence of direct and opposite motions of the gas near a

[1] The estimates show that the maximum kinetic energy of a single blob in our models may be of the order of 10^{35} - 10^{36} ergs which is comparable to the energy of flares observed in TTSs (Gahm 1990).

star is one of the most complicated factors of the hydrodynamics of S-winds. The shocked multi-temperature gas structures should exist in the vicinity of TTSs.

Deviations from radial symmetry of the outflow is the other complicating factor. If the outflow of matter is due to magnetic activity on the stellar surface, then since magnetic spots are formed by the action of differential rotation, we would expect the stellar wind to be axially symmetric in general.

The existence of an extended co-rotation zone in the stellar wind (up to a few stellar radii–see Hartmann *et al.* 1986) may produce tangential velocities of about 50 - 100 kms^{-1} . As a result, the distance between emission components of two-component line profiles should be larger than in the case of purely radial motions.

Finally, the blobs may dissipate in the outer parts of the wind and the outflow of matter may be continuous in this region.

All factors mentioned above tremendously complicate the modeling of T Tauri stellar winds. Therefore, advances in our understanding of the T Tauri stellar wind will be connected first of all with new observations. Observations made with the HST will be important for a better understanding of the kinematical properties of the hot gas component of stellar wind. The well organized ground base spectral observations, such as those demonstrated by Mark Giampapa and Gibor Basri, can help us become more confident that T Tauri stellar winds originate at the surface of TTSs.

REFERENCES

Appenzeller, I., and Mundt, R.: 1989, *Astron. Astrophys. Review*, **2**, 261.

Bertout. C.: 1989, *Ann. Rev. Astron. Astrophys.*, **27**, 351.

Brown, A., Ferras, M.S.de M., and Jordan, C.: 1984, *M.N.R.A.S.*, **207**, 831.

De Campli, W.M.: 1981, *Ap.J.*, **244**, 124.

Giampapa, M.S., and Imhoff, C.L.: 1985, in Protostars and Planets II, ed. D.C. Black and M.S. Matthews (Univ. of Arizona Press, Tucson), p.386.

Gahm, G.: 1990, in Flare Stars in Star Clusters, Associations and the Solar Vicinity, IAU Symp. *No.* 137, eds. L.V.Mirzoyan, B.P.Petersen and M.K.Tsvetkov, Kluwer Acad. Publ. p. 193.

Grinin, V.P., Petrov, P.P., and Shakhovskaya, N.I.: 1984, in Activity of Red-Dwarf Stars, IAU Colloquium *N*71, eds. P.B.Byrne and M.Rodono, D.Reidel Publ. Co. p. 513.

Grinin,V.P., and Mitskevich,A.S.: 1990, *Astrofisika*, **32**, 383, Paper I.

Grinin, V.P. and Mitskevich, A.S.: 1991, *Astrophys. and Space. Sci.*, in press, Paper II.

Hartmann, L., Hewell, S.E., Stahler, S.E., and Mathieu, R.D.: 1986, *Ap. J.* , **309**, 275.

Hartmann, L., Calvet, N., Avrett, E.H., and Loeser, R.: 1990, *Ap.J.*, **349**, 168.

Hartmann, L., Edwards, S., and Avrett, E.H.: 1982, *Ap.J.*, **261**, 279.

Herbig, G.H.: 1977, *Ap.J.*, **214**, 747.

Herbig, G.H., and Soderblom, D.R.: 1980, *Ap.J.*, **242**, 628.

Kuhi, L.V.: 1964, *Ap.J.*, **140**, 1409.

Lago, M.T.V.T.: 1984, *M.N.R.A.S.*, **210**, 323.

Mullan, D.J.: 1990, *Astron.Astrophys.*, **232**, 520.

Mundt, R.: 1984, *Ap.J.*, **280**, 749.

Natta, A., Giovanardi, C. Palla, F. and Evans, N.I.: 1988a, *Ap.J.*, **327**, 817.

Natta, A., Giovanardi, C., and Palla, F.: 1988b, *Ap.J.*, **332**, 921.

Natta, A., and Giovanardi, C.: 1990, *Ap.J.*, **356**, 646.

Shanin, G.I., Shevchenko, V.S., and Shcherbakov A.G.: 1975, in IAU Symposium *N*67, Variable Stars and Stellar Evolution, ed. V.E.Sherwood and L.Plaut (Dortrecht: Reidel), p.117.

Seventh Cambridge Workshop on Cool Stars, Stellar Systems, and the Sun
ASP Conference Series, Vol. 26, 1992
Mark S. Giampapa and Jay A. Bookbinder (eds.)

ENERGETICS OF STELLAR WINDS

PHILIP G. JUDGE High Altitude Observatory, National Center for Atmospheric Research[1], Box 3000, Boulder, CO 80307, USA.

ABSTRACT I review observational constraints on the winds of cool, low-gravity stars in the context of a theoretical framework described by Holzer and MacGregor. I emphasize (i) the importance of studying a wide variety of stellar objects, (ii) the minor role of asymptotic momentum fluxes in constraining mass loss mechanisms, (iii) the energy requirements, and (iv) constraints on mechanisms available from binary systems with spatially resolved wind structure. The relative invariance of energy flux densities, low wind velocities and properties of eclipsing binary systems suggest that the essential physics of mass loss from cool stars is contained in non-linear hydrodynamical effects.

Keywords: Late Type Stars; Winds.

INTRODUCTION

It is a sobering thought that, despite several decades of research which have seen major observational and theoretical advances, we have not yet grasped a fundamental understanding of the origin of mass loss from cool, evolved stars. This is in contrast to winds of hot stars and of the Sun. This statement, based upon data discussed below, might seem (to some) extremely pessimistic or even heretical! However, by examining winds of *representative types* of cool stars in terms of *energy* as well as momentum requirements, I suggest that one must unfortunately re-think some popular ideas concerning mass loss mechanisms.

THE MUDDIED WATERS OF MASS LOSS MECHANISMS

Lafon & Berruyer (1991) present a rather complete review of cool star winds. They focus upon stars with mass loss rates $-\dot{\mathcal{M}}$ in excess of 10^{-7} $\mathcal{M}_\odot$ yr^{-1}. This (common) bias probably reflects the explosion of new data from the *IRAS* satellite and from sub-mm telescopes. Lafon & Berruyer (1991) discuss possible driving mechanisms for such stars, including linear waves (Alfvén waves, sound waves [Hartmann & MacGregor 1980]), non-linear waves (global shocks driven by stellar pulsation [Bowen 1988] or by merging of stochastically varying shocks [Cuntz 1987]), and the force of photospheric radiation upon molecules or dust (*e.g.* Kwok 1975).

The authors lamentably note that the "... circumstellar envelopes of evolved stars are more complicated than previously expected", and conclude

[1] The National Center for Atmospheric Research is sponsored by the National Science Foundation

that these mechanisms "... together with related physical or thermodynamical phenomena (such as dust condensation, grain growth and evolution, opacity variations, thermal exchanges between gas, dust and radiation, instabilities, light variation etc., ...) are important on various scales...".

Earlier work (Hartmann & MacGregor 1980; Hartmann 1981) studied driving mechanisms for stellar winds with $-\dot{\mathcal{M}} < 10^{-7}\ \mathcal{M}_{\odot}\ \mathrm{yr}^{-1}$, appealing to linear Alfvén wave models since dust is not observed around early M and K giant stars, and because short-period sound waves damp over small scale lengths. Subsequently, Holzer, Fla & Leer (1983) pointed to a fundamental problem with these models: only a very restrictive range of wave damping parameters can satisfy the observed mass loss rates and low wind velocities. Kuin & Ahmad (1989) analyzed in a semi-empirical manner the winds of ζ Aur binaries studied observationally by Schröder (1985) and Reimers & Che-Bohnenstengel (1986). They concluded that linear Alfvén wave models can account for the observed wind structures, although they were unable to provide a physical understanding of their empirical damping lengths. This again leaves open the question of the applicability of these models.

I conclude that the literature paints an incomplete and complex picture of the mechanisms driving the winds of cool, evolved stars, suggesting that we have yet to understand the essential underlying physical processes. This situation prompted an empirical study of a wide variety of giant stars which addressed the wind energy as well as momentum requirements (Judge & Stencel 1991). This work revealed interesting similarities in energy flux densities of a *wide variety of stars*, posing some basic questions: Why are wind energy flux densities so similar in stars with very different physical properties? Is there a universal "self-regulation" mechanism which can produce high mass fluxes with low wind velocities (Holzer *et al.* 1983)?

There is currently much theoretical work on specific driving mechanisms (see Lafon & Berruyer 1991). Here I adopt a different approach by reviewing the general properties of the winds of evolved cool stars in the context of a theoretical framework (Holzer & MacGregor 1985, henceforth HM85), and discussing the impact of the eclipsing binary data on mass loss mechanisms.

THEORETICAL FRAMEWORK

Following HM85, and assuming a spherical, steady, inviscid outflow from a star, the conservation equations of mass, momentum and energy yield:

$$-\dot{M} = 4\pi r^2 \rho(r) u(r) \tag{1}$$

$$u\frac{du}{dr} = -\frac{1}{\rho}\frac{dp}{dr} - \frac{GM}{r^2} + D(r) \tag{2}$$

$$F_r = \frac{-\dot{\mathcal{M}}}{4\pi}\left(\frac{1}{2}u^2 + \frac{\gamma}{\gamma-1}\frac{p}{\rho} - \frac{GM_*}{r}\right) + qr^2$$

$$= F_{r_0} + \int_{r_0}^{r} dr'\left(\frac{-\dot{\mathcal{M}}}{4\pi}D(r') + Q(r')(r')^2\right) \tag{3}$$

where symbols have their usual meanings (HM85). ρD and Q are rates (per unit volume) at which momentum and heat are added to the flow. F_r is the total energy flux associated with the advection of flow energy, enthalpy and gravitational potential energy and conduction of heat (q) through a shell of radius r. Radiative energy losses are included implicitly in Q. The terms D and Q are essentially the unknowns of the mass-loss problem. They determine the dynamics directly through these terms and indirectly through the thermal pressure gradient term.

For single stars, only spatially integrated quantities can be measured. An important quantity is the energy flux density $F_W = (F_\infty - F_{R_*})/R_*^2$ needed to drive the wind from the "base" (*i.e.* stationary photosphere at radius R_*) to infinity, given by

$$F_W \simeq 3.3 \times 10^3 (\frac{-\dot{\mathcal{M}}}{10^{-7}})(\frac{\mathcal{M}_*}{\mathcal{M}_\odot})(\frac{400 R_\odot}{R_*})^3 (1 + (\frac{v_\infty}{v_{esc}})^2) \quad erg\ cm^{-2}\ s^{-1} \tag{4}$$

where v_∞ is the observed terminal velocity (assumed purely radial), and $v_{esc}(R_*)$ is the surface escape velocity: The dominant term (not counting radiative losses) involves work done against gravity, since, observationally, $v_\infty << v_{esc}(R_*)$ in cool, low-gravity stars (HM85, although see below). The total energy flux density of the outer atmosphere must equal the wind flux density, plus F_R, the total energy flux density radiated above R_*.

It is not possible to derive a general expression for the momentum flux requirements because the momentum coupling is model-dependent (HM85). However, in the limit that a wind is driven by the force of photospheric radiation on dust (or gas) *above a radius* $r >> R_*$, one can write

$$-\dot{\mathcal{M}} v_\infty \simeq \tau L/c \tag{5}$$

where L/c is the (scalar) stellar radiation momentum flux and τ measures the (integrated) fraction of L/c which is transferred to the wind. In cases where the dust shell is optically thin (single scattering), τ is simply the flux-mean optical depth. In optically thick cases it can exceed unity (HM85). Unfortunately, some studies have focused exclusively upon this asymptotic limit of the momentum equation, ignoring both the restriction $r >> R_*$ and the energy flux requirements. For instance, Jura (1989) presents measurements of $\dot{\mathcal{M}}$, v_∞ τ and L in AGB stars, finding a good agreement with equation (5). Hence he argues that dust is dynamically important in driving the winds of AGB stars. These conclusions must therefore be treated with caution. Moreover, one could reverse the argument and conclude that the wind dynamics are determined by some other processes which then controls the formation and dynamics of the dust.

In cases where spatial information is available (*e.g.* in eclipsing binary systems), the terms $D(r)$ and (if radiative losses are known) $Q(r)$ can be studied as a function of r. Hence real comparisons of observations with specific acceleration mechanisms can be made.

SOME EMPIRICAL RESULTS

I examine two complementary datasets: (i) the stellar sample discussed by Judge & Stencel (1991), augmented by some additional stars, and (ii) stars studied through the binary eclipse technique (Table 1).

Table 1: BINARIES AND ADDITIONAL STARS

Star	Spectrum	$\mathcal{M}$ $\mathcal{M}_\odot$	R $R_\odot$	T_{EFF} K	L $L_\odot$	$-\dot{\mathcal{M}}$ $\mathcal{M}_\odot$ yr^{-1}	u_∞ km/s	F_W erg/cm^2/s
Binaries...								
22 Vul[a]	G3 Ib-II	4.3	5.5(1)[b]	5200	2.0(3)	6.0(-9)	160	6.2(5)
31 Cyg[a]	K4 Ib	6.2	2.0(2)	3800	7.7(3)	4.0(-8)	80[c]	9.8(4)
ζ Aur[a]	K4 Ib	8.3	1.4(2)	3950	4.3(3)	6.0(-9)	72[c]	4.7(4)
32 Cyg[a]	K5 Iab	8.0	1.9(2)	3800	6.6(3)	2.8(-8)	90[c]	1.1(5)
δ Sge[a]	M2 II	8.0	1.4(2)	3600	3.0(3)	2.0(-8)	28	1.3(5)
EG And[d]	M2.4 III	2.0	6.5(1)	3600	6.4(2)	1.5(-8)	30	2.5(5)
α Sco[a]	M1.5 Iab	18	6.2(2)	3540	5.5(4)	1.0(-6)	17	1.6(5)
Miras...								
R Leo[e]	M8 IIIe	1.0	1.1(3)	2230	2.8(4)	1.8(-7)	6.5	3.1(2)
RX Boo[e]	M7.5	1.0	5.2(2)	2300	6.7(3)	9.2(-7)	7.8	1.5(4)
R LMi[e]	M7e	1.0	5.2(2)	2365	7.8(3)	3.2(-7)	6.0	4.9(3)
Mira[a]	M7 IIIe	1.2	2.3(2)	2365	1.5(3)		5.7	
R Hya[e]	M7 IIIe	1.0	3.4(2)	2365	3.3(3)	7.4(-8)	7.2	4.1(3)
Others...								
ι Aur[f]	K3 II	5.5	5.5(1)	4038	7.3(2)		72[g]	
VX Sgr[h]	M4 Iae	10	3.4(3)	2500	4.2(5)	1.3(-5)	30	1.2(4)
Y CVn[i]	C 5,4	1.0	2.7(2)	2980	5.1(3)	7.1(-8)	6.3	8.1(3)

Sources: *a* Reimers (1987). *b* Kuin & Ahmad (1989). *c* Ahmad & Stencel (1988). *d* Vogel (1991). Mass guessed (the mass is unlikely to be bigger- the system has a high galactic latitude). *e* Distance, luminosity from period-luminosity relation of Glass *et al.* (1987), $\dot{\mathcal{M}}$ and v_∞ from Knapp (1989, private communication), T_{EFF} from Table of de Jager *et al.* (1988), radius from distance and Barnes, Evans & Moffett (1978). *f* Harper (1991). *g* Hartmann *et al.* 1985. *h* Chapman & Cohen (1986), wind data from Knapp (1989, private communication). *i* de Jager *et al.* (1988).

The former group contains the nearest giants (luminosity class III) of spectral classes K, M, and C, including stars of essentially all giant phases of evolution, all infrared classes and representative variability classes. Fundamental data for individual stars are of poor accuracy, owing mostly to poorly known distances. The mass of most of the stellar sample lies between 1 and 2 $\mathcal{M}_\odot$.

The latter group contains just a few stars of the ζ Aur and VV Cep eclipsing binary classes (as reviewed by Reimers 1987). ζ Aur systems consist of a hot dwarf star and a G- or K-supergiant with no detectable dust or photospheric pulsations. Fundamental data are well determined for these systems and, as importantly, the hot star companions probe the outer atmosphere of the cool star as a function of height above the photosphere. These stars are typically $\sim 7\mathcal{M}_\odot$, and are supergiants or bright giants.

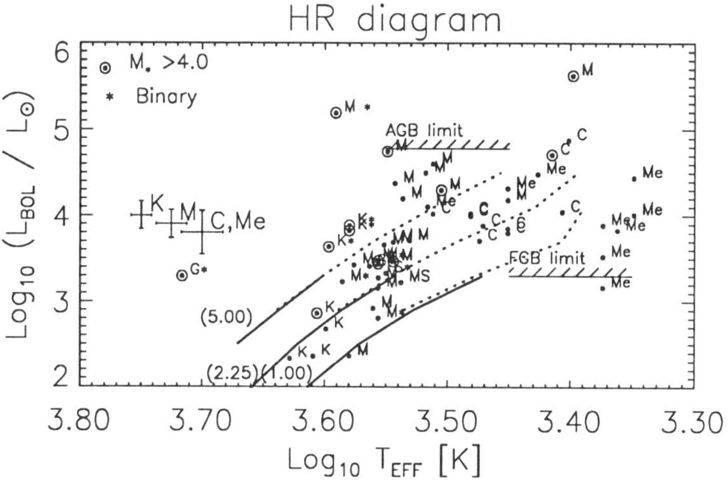

Fig. 1: HR diagram of the stellar sample. FGB and AGB evolutionary tracks shown for stars of 1, 2.25 and 5 solar masses (Bessell *et al.* , 1989). Stars more massive than $4\mathcal{M}_\odot$ are plotted using the symbol $\odot$ here and in later figures. Miras and other stars showing optical emission are designated 'Me'.

WIND VELOCITIES

Are all winds of cool, low-gravity stars, really "low-velocity" (i.e. $v_\infty << v_{esc}(R_*)$)?

Asymptotic wind velocities v_∞ are determined by absorption profiles of strong optical, infrared and/or ultraviolet lines, thermal emission profiles of sub-mm CO emission, separations of maser lines. The method applied to an individual star depends upon the nature of the circumstellar envelope (as reviewed by Judge & Stencel 1991).

In luminous stars with well developed circumstellar shells (later than spectral type M5 III), well-exposed emission profiles of CO and/or maser lines formed at many stellar radii can yield reliable terminal velocities. However, for less luminous stars, these emission features are not present and one must use absorption features. The absence of high velocity absorption features does not necessarily rule out higher wind velocities. Sensitive wind lines often lie in the UV where background continua are weak, and so v_∞ can increase with the signal-to-noise ratio of the spectrum! In several cases v_∞ for a given star has increased with the level of scrutiny afforded it: examples include the hybrid star α TrA studied by Hartmann *et al.* (1985), and the ζ Aur systems (Ahmad & Stencel 1988).

With this *caveat*, consider Fig. 2, showing v_∞ *versus* $v_{esc}(R_*)$. The figure shows (i) a (real?) lower limit of $v_\infty \sim 5$ km s^{-1} below which no stars lie; (ii) The wind velocities are scattered between 5 and 30 km s^{-1} for low gravity stars ($v_{esc}(R_*) < 200$ km s^{-1}). (iii) The majority of these stars have $v_\infty << v_{esc}(R_*)$; (iv) v_∞ approaches $v_{esc}(R_*)$ for the ζ Aur systems, the M supergiant VX Sgr and the infrared carbon stars; (v) Miras all clump near wind velocities between 5 and 10 km s^{-1} , amongst other M giants; (vi) there is no clear separation in v_∞ between stellar masses.

Point (iv) shows that the long-standing problem of producing high mass loss rates with $v_\infty << v_{esc}(R_*)$ (HM85) does not exist for the ζ Aur systems, VX Sgr and for the infrared carbon stars. Substantial momentum addition must occur in the supersonic regions of the outflows. Therefore, the winds of these systems appear to be fundamentally different from the lower mass single stars, unless wind velocities of K-M giants have been substantially underestimated. Unfortunately, the detailed information available from ζ Aur winds may not be relevant for studies of lower mass stars.

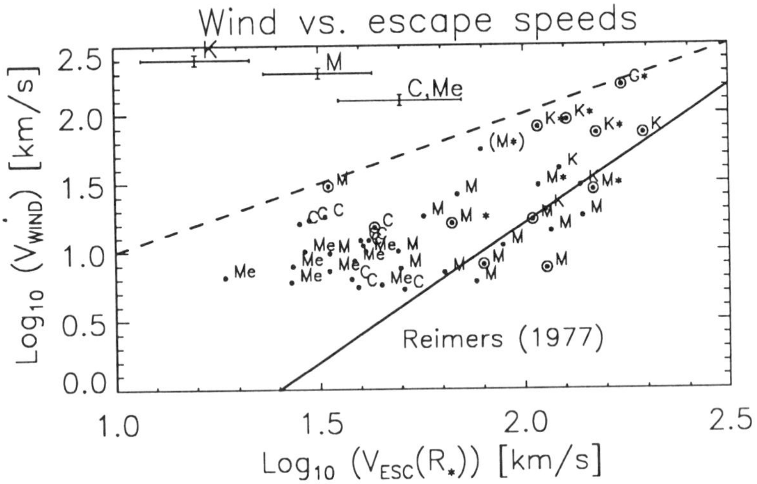

Fig 2: v_∞ *versus* $v_{esc}(R_*)$. Also shown are the lines $v_\infty = v_{esc}(R_*)$ and $v_\infty \simeq 1.6 \times 10^{-3} v_{esc}(R_*)^2$ (Reimers 1977). Uncertainties are plotted for the single stars. The binaries have much smaller uncertainties.

ENERGY FLUX DENSITIES

How do wind energy budgets depend on mass, evolutionary stage, binarity?

One might expect wind energy fluxes to depend critically on the driving mechanism (at least in linear models, *e.g.* Holzer *et al.* 1983, eqn. 39). However, Reimers (1975) showed that, if a constant fraction of the stellar luminosity is used to drive stellar winds (in our notation F_W/F_{BOL} = constant), and $v_\infty^2 << V_{esc}^2(R_*)$, then $\dot{M} \propto LR/g$. This formula was found to fit measurements of mass loss rates in ζ Aur and M supergiant systems (Reimers 1975, 1987). Judge & Stencel (1991) derived energy flux densities of their sample of lower mass stars, finding similar conclusions but with a tentative suggestion that the energy fluxes of infrared carbon stars are somewhat higher. The present dataset is shown in Figure 3.

The figure shows that most low mass stars have values of F_W/F_{BOL} spread around the value 10^{-6} *irrespective of their evolutionary status, dust shell properties, etc.* The 1σ spread appears to be close to the 1σ error bars for the relative (not absolute) uncertainties between individual stars. This suggests that most low mass stars have mass loss rates which do not critically depend on evolutionary status, variability etc. If confirmed, these data present a real and new challenge to theoretical models– why do winds of large amplitude variables require the same energy flux as the winds of field K and M stars?

More massive stars ($> 4\mathcal{M}_\odot$), including the ζ Aur systems, systematically have higher fractional energy fluxes than the low mass stars. The same is true for the infrared carbon stars. Coupled with the wind velocity data (Fig. 2), this suggests that driving mechanisms are substantially different between these groups. The symbiotic EG And and δ Sge also appear to have enhanced wind energies in comparison with the single giants.

The large absolute uncertainties on the single stars preclude the possibility of deriving firm conclusions: one acceptable possibility is that essentially all stars have F_W/F_{BOL} nearer 10^{-5}, with no difference in energy fluxes of the entire sample.

Firmer conclusions require more accurate estimates of the stellar distances, since the largest sources of uncertainty in F_W lie in distance-dependent quantities (Drake 1986; Judge & Stencel 1991).

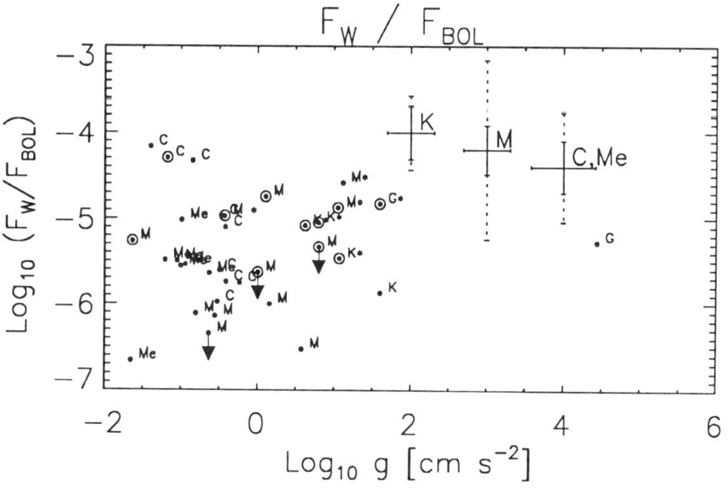

Fig. 3: Fractional energy flux (F_W/F_{BOL}) associated with the stellar winds as a function of log g. The meanings of symbols and error bars are described in earlier figure captions.

CORRELATIONS

What about observed correlations? Surely they tell us something about mass loss processes?

It has become popular to search for correlations between observed wind parameters and fundamental stellar parameters in attempts to infer causal relationships between, *e.g.* pulsation, dust formation and mass-loss. Unfortunately, in the absence of additional information, such inferences can be misleading (they would not survive scrutiny in other disciplines such as social studies)! An important property of cool evolved stars is that they exist on *branches* in parameter space, on which fundamental parameters (T_{EFF}, log g, pulsational properties...) are themselves correlated. Thus, although some authors find correlations between, *e.g.* pulsational period, amplitude and mass loss rates, an equally good correlation exists between, *e.g.* log g or T_{EFF} and mass loss rates.

I believe such correlations are actively misleading: one must examine physically relevant properties of the stars and their winds (*e.g.* energy and momentum fluxes). Furthermore, one must ensure that the sample of stars is not biased by observational techniques (*cf.* Introduction).

WIND RADIATIVE LOSSES

How can we estimate wind radiation losses?

Theoretical studies (Hartmann & MacGregor 1980; Hartmann & Avrett 1984; HM85; Bowen 1988) indicate that the accelerated gas will also be heated. In the inner wind regions of models of α Ori, for example, the energy equation is determined mostly by the balance between "chromospheric" radiative losses and wave heating (Hartmann & Avrett 1984). It is clearly important to identify these losses empirically when studying driving mechanisms. Some workers assume that observed "chromospheric" or "transition region" lines are formed in the heated regions of winds.

The observational basis for this assumption is unclear. In the Sun, the coronal wind is associated with open field regions but the coronal X-rays originate mostly in closed field regions (*e.g.* Golub, this volume). In single stars, the situation is largely ambiguous, since an observed blue-shifted absorption component overlying a "chromospheric" emission profile does not necessarily imply that the line is formed in one contiguous structure (*e.g.* Judge 1990). Indeed, Hartmann & Avrett (1984) were unable to find a model for α Ori which could simultaneously account for the "chromospheric" emission and the observed wind structure.

Eclipsing binaries provide a more direct observational test, but data from ζ Aur systems are inconclusive. In two out of three cases (31 and 32 Cyg), Schröder (1985) found that the "chromospheric" density structure observed near the stellar limb led to agreement with the wind density structure at much larger radii, assuming that the mass continuity equation is valid. For the third system (ζ Aur), structures close to the limb were of much lower density than those estimated by extrapolating inwards the continuity equation.

Recent observations of α Tau (K5 III) suggest that, in analogy with the Sun, most of the observed radiative losses are not physically associated with the stellar wind. Carpenter *et al.* (1991) obtained profiles of optically thin lines of C II], Si II] and lines of iron group elements obtained with the GHRS on the

Hubble Space Telescope . They find evidence for systematic *downflows* of ~ 4km s^{-1} from C II] λ2325Å. The significance of these observations is profound since the C II] lines are optically thin in the emitting structures (Judge 1986) and the emissivity of these lines has essentially the same dependence on the plasma properties (N_e , T_e) as the dominant radiative coolants (Mg II, Ca II).

These observations suggest interesting possibilities, which exclude spherically symmetric time-independent wind models: (i) the emitting structures are unrelated to the wind acceleration zone. This may not be in conflict with earlier statements that "wind acceleration is already well under way in the chromosphere" (Reimers 1981), which were based upon profiles of optically thick lines; (ii) the emitting structures are related to the wind acceleration zone, but in a time-dependent manner: sometimes the gas is inflowing but mostly it is outflowing (*cf.* Cuntz, this volume); (iii) The emitting structures are related to the wind acceleration zone, but the C II] profiles do not reflect the actual velocity field because timescales for atomic processes are comparable to convective and dynamical timescales (~ 10^{3-4}s for the case of α Tau). Clearly the *Hubble Space Telescope* will provide clues to this problem in future.

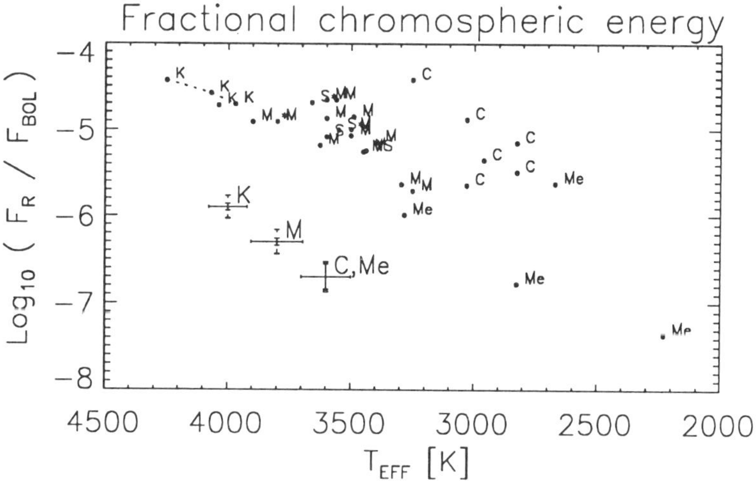

Fig. 4: Fractional "chromospheric" energy flux (F_R/F_{BOL}) associated with the stellar winds as a function of T_{EFF}. Note the rather smooth relation for stars with T_{eff} > 3500K (mostly FGB), and the increase in scatter for cooler stars (mostly AGB).

Figure 4 shows radiative losses in "chromospheric" lines only (scaled from Mg II flux densities– Judge & Stencel 1991). These are similar in magnitude to wind flux densities.

SPATIALLY RESOLVED WINDS

What can we learn from stars with spatially resolved winds?

Table 1 lists stars for which radial information the outer atmosphere is available. Flow velocities, gas densities, gas temperatures, and spectroscopic microturbulence have been derived in three ways:

(i) In ζ Aur systems, empirical curves of growth determined from the eclipse of the hot companion by the outer atmosphere yield column densities and microturbulence as a function of height (Schröder 1985; Reimers & Che-Bohnenstengel 1986). Assuming mass conservation and spherical symmetry, these yield densities and flow velocities $u(r)$. These authors find adopt of the form $u(r) = u_\infty(1 - \frac{R_*}{r})^a$, where $a \sim 3$. Data analyzed to date are restricted to radii $\leq 4R_*$. A crucial property of the winds is the relative constancy of the derived turbulent velocities $v_T(r)$.

(ii) Eclipsing systems showing Rayleigh scattering from H I (H_2) during secondary eclipse can be used to determine neutral (molecular) column densities as a function of height (Schröder 1986, Vogel 1991). Again, with the assumptions of spherical symmetry and mass conservation these yield density and velocity distributions. The symbiotic star EG And was analyzed by Vogel (1991), who derived an acceleration somewhat slower than those found for the ζ Aur systems.

(iii) Some supergiants (*e.g.* VX Sgr) have circumstellar maser lines which have been observed at various angular distances from the star (Chapman & Cohen 1986). With an assumed distance, the expansion velocity was determined at $\sim 3, 7, 20$ and 90 R_*. Unfortunately, data are unavailable between 1 and $3R_*$ where most of the energy is deposited in driving the wind. These data were discussed in the context of acceleration of dust grains by photospheric radiation by Chapman & Cohen (1986) and Netzer (1989).

Here I focus upon the ζ Aur data which have been exploited by Kuin & Ahmad (1989), using a semi-empirical model driven by linear WKB Alfvén waves, since they argue that energy fluxes from linear or non-linear sound waves (shocks) are too small to drive the observed outflows. They collect the measured $\rho(r)$, $u(r)$ and $v_T(r)$ data and study terms in the momentum and energy equations of the winds constrained by these data. They conclude that one can construct wave models satisfying the empirical hydrodynamical requirements. However, the derived empirical damping lengths behave with radius in a manner opposite to those expected from ion-frictional damping (Holzer *et al.* 1983)! In addition, the semi-empirical and theoretical models have a variety of assumptions and potential problems:

(i) Velocity functions of the form $u(r) = u_\infty(1 - \frac{R_*}{r})^a$ with $a > 1$ imply an acceleration which is very small at the base of the wind. This is a fundamentally different from *ab-initio* models (*e.g.* Hartmann & MacGregor 1980, Fig. 5), which yield a much faster wind acceleration near the photosphere. Either these models are completely inappropriate, or perhaps the derived column densities within $\sim 2R_*$ have contributions from structures such as spicules or closed magnetic loops.

(ii) The models assume linear wave modes. There are (at least) two reasons to doubt this assumption: (a) the ratio $\delta B/B$ must first decrease but then increase with radius given the empirical densities and microturbulent velocities, assuming a divergent magnetic field ($B \propto 1/r^{\alpha}$, $\alpha \geq 2$); (b) wave propagation through the steep density gradients (scale heights are $<< R_*$), will tend to steepen the waves: non-linear Alfvén waves are already observed in the solar wind, where the density scale height is $\geq R_*$.

(iii) The WKB approximation is used. This may not be valid for low gravity stars. Longer convective timescales will lead to larger wavelengths compared with scale heights. (See also Moore *et al.* [this volume]).

(iv) The wave force is evaluated using a linearized wave pressure gradient term of the form $D = \frac{1}{2\rho} d[\rho < v_T^2 >]/dr$ where $< v_T^2 >$ is a time average of the velocity^2 due to the passage of the waves.

(v) $< v_T^2 >$ is estimated from line profiles or a curve of growth through a spectroscopic microturbulence ξ. Unfortunately, depending on the assumed wave propagation mode, geometry and the "randomness" of the emitting (or absorbing) structures, the value of k assumed in $< v_T^2 > = k\xi^2$ varies by a factor of up to 8 (Jordan 1985)!

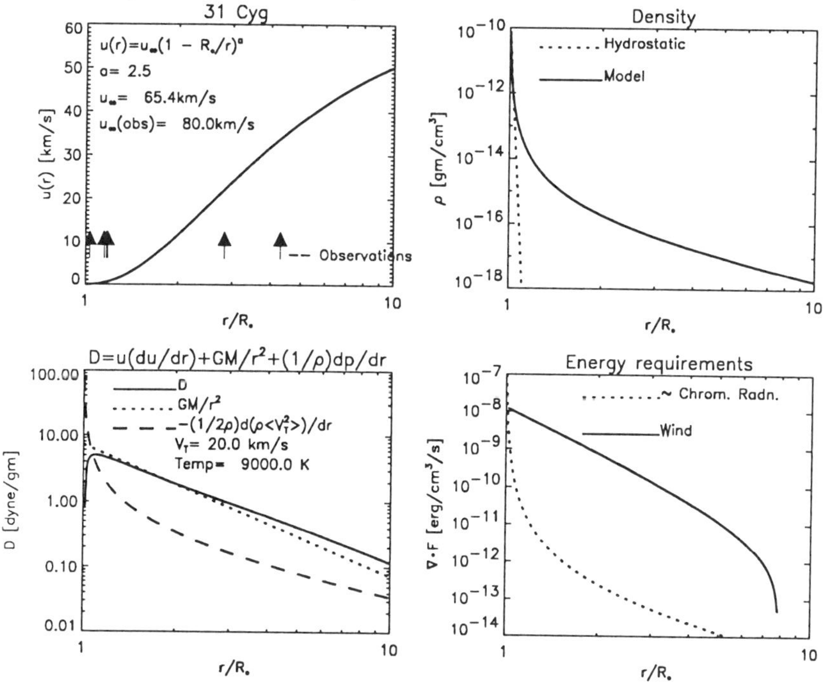

Fig. 5: Velocity and density structure of the wind of 31 Cyg derived by Schröder (1985), and terms in the momentum equation (2) plotted as a function of radius. The required driving force per unit gram, D is shown together with the force from the wave pressure gradient derived from the observed microturbulence (see text). The lower right panel shows that the scale height for "chromospheric" radiative energy losses (here assumed $\propto \rho$) is *very* much smaller than for the wind losses. Hence, here is another reason to question whether these are physically related.

Are these models valid? One empirical test is to compare the required driving force per unit mass with the force available in the form of linearized WKB waves (Fig. 5). This was first done by Kuin & Ahmad (1989) for the case of 32 Cyg, assuming that the constant k is identically unity. They concluded that linear WKB waves indeed can account for the required momentum addition. However, in Fig. 5. we see that this is not the case for 31 Cyg, and that a value of $k \sim 6$ is needed to account for the driving force.

This value is disconcertingly large (but apparently not outside the range suggested by Hartmann *et al.* 1982), suggesting that these linearized models are inapplicable. Clearly *it is of vital importance to develop a better understanding of the spectroscopic microturbulence and its dynamical implications.* This is of equal importance to studies of single stars, where observed radiative and wind energy flux densities have been compared with the energy flux density of a propagating wave (*e.g.* Jordan 1985), again derived from the spectroscopic turbulence.

There remain many important questions unanswered in these studies. Why do measured microturbulent velocities remain roughly constant in each star? This vital clue to wind dynamics appears to be outside the domain of linearized theories (*e.g.* Holzer *et al.* 1983, eqn. [27]), without some rather general form of damping. What is the nature of the observed turbulent line broadening, and what is the correct procedure to determine the dynamical effects of these motions? Are linear WKB theories applicable? What effects are expected from relaxing these assumptions? Some indications are available from the study of Holzer *et al.* (1983), and these deserve further attention.

CONCLUSIONS

Several pieces of evidence suggest that the essential physics of mass loss in cool, low gravity stars, is contained in non-linear hydrodynamical processes, including (i) the similarity in wind energy flux densities in stars of very different phases of evolution with very different physical properties; and (ii) the detailed information available from studying eclipsing binary systems. I suggest that theoretical research should focus upon finding general mechanisms which will provide the necessary "self-regulation" required by observations, instead of refining the details of existing models.

More specific new results include (i) the massive ζ Aur systems, VX Sgr and the infrared carbon stars have $v_\infty \sim v_{esc}(R_*)$. Hence problems with "fine tuning" damping lengths are less severe in these stars. (ii) These systems may also differ from "normal" giants and Miras since the winds have higher values of F_W/F_{BOL}. This probably points to different driving mechanisms. (iii) These differences between low (1-2$\mathcal{M}_\odot$) and intermediate ($\sim 7\mathcal{M}_\odot$) mass stars might indicate a different evolution of magnetic activity off the main sequence, as indicated recently by (Pasquini *et al.* 1990) on the basis of chromospheric activity.

Finally, accurate new distances from the *HIPPARCOS* mission will greatly reduce uncertainties which plague studies of nearby red giants.

ACKNOWLEDGEMENTS

I thank Keith MacGregor for helpful discussions and for a critical reading of the manuscript.

REFERENCES

Ahmad, I.A., and Stencel, R.E. 1988 *Ap. J.*, **329**, 797.
Barnes, T.G., Evans, S., and Moffett, T.J. 1978 *M.N.R.A.S.*, **183**, 285.
Bessell, M.S., Brett, J.M., Scholz, M., and Wood, P.R. 1989 *Astr. Ap.*, **77**, 1.
Bowen, G.H. 1988 *Ap. J.*, **329**, 299.
Carpenter K.G., *et al.* 1991 *Ap. J. Letters*, **377**, L45.
Chapman, J.M., and Cohen, R.J. 1986 *M.N.R.A.S.*, **220**, 513.
Cuntz, M. 1987 *Astr. Ap.*, **188**, L5.
Drake, S.A. 1986. In Cool Stars, Stellar Systems and the Sun, Proceedings, Santa Fe, New Mexico, Eds. M. Zeilik and D.M. Gibson, Springer-Verlag, p.369
Glass, I.S., Catchpole, R.M., Feast, M.W., Whitelock, P.A., and Reid, I.N. 1987. In Late Stages of Stellar Evolution, p. 51, eds. S. Kwok and S.R. Pottasch, Reidel, Dordrecht, Holland.
Harper, G.M. 1991 *M.N.R.A.S.*, in press.
Hartmann, L. 1981 in Solar Phenomena in Stars and Stellar Systems, eds. R.M. Bonnet and A.K. Dupree, Reidel, Dordrecht, p. 331.
Hartmann, L., and Avrett, H. 1984 *Ap. J.*, **284**, 238.
Hartmann, L., Edwards, S., and Avrett, H. 1982 *Ap. J.*, **261**, 279.
Hartmann, L., Jordan, C., Brown, A., & Dupree, A.K. 1985 *Ap. J.*, **296**, 576.
Hartmann, L., and MacGregor, K.B. 1980 *Ap. J.*, **242**, 260.
Holzer T.E., Fla, T., and Leer, E. 1983 *Ap. J.*, **275**, 808.
Holzer, T.E., and MacGregor, K.B. 1985. In Mass-Loss from Red Giants, p. 229, Eds. M. Morris and B. Zuckerman, D. Reidel Publ. Co., Dordrecht, Holland. (HM85).
de Jager, C., Nieuwenhuijzen, H., and van der Hucht, K.A. 1988 *Ap. J. Suppl.*, **72**, 259.
Jordan, C. 1985. *Ir. Astron. J.*, **17**, 227
Judge, P.G. 1986 *M.N.R.A.S.*, **221**, 119
Judge, P.G. 1990 *Ap. J.*, **348**, 279.
Judge, P.G., and Stencel, R.E. 1991 *Ap. J.*, **371**, 357.
Jura, M., 1986 *Ir. Astron. J.*, **17**, 322
Jura, M., 1989. In The Evolution of Peculiar Red Giant Stars, proceedings of IAU Colloquium 106, eds. H. R. Johnson and B. Zuckerman, Cambridge University Press, England, p. 339
Kuin, N.P.M., and Ahmad, I. A. 1989 *Ap. J.*, **344**, 856.
Kwok, S., 1975 *Ap. J.*, **198**, 583.
Lafon, J.-P.J., & Berruyer, N. 1991 *A. & AR*, **2**, 249
Netzer, N. 1989 *Ap. J.*, **342**, 1068.
Pasquini, L., *et al.* , 1990 *Astr. Ap.*, **234**, 277.
Reimers, D. 1975 *Mem. Soc. Roy. Liège 6e Ser.*, **8**, 369.
Reimers, D. 1977 *Astr. Ap.*, **57**, 395.
Reimers, D., 1981. In Physical Processes in Red Giants, Eds. I. Iben and A. Renzini, Reidel, Dordrecht, p. 269.
Reimers, D., 1987. In Solar and Stellar Physics, Eds. E.H. Schröter, M. Schüssler, Lecture Notes in Physics, Springer, p. 139.
Reimers, D., & Che-Bohnenstengel, A. 1986 *Astr. Ap.*, **166**, 252.
Schröder, K.P. 1985 *Astr. Ap.*, **147**, 103.
Schröder, K.P. 1986 *Astr. Ap.*, **170**, 70.
Vogel, M. 1991 *Astr. Ap.*, , in press.

Seventh Cambridge Workshop on Cool Stars, Stellar Systems, and the Sun
ASP Conference Series, Vol. 26, 1992
Mark S. Giampapa and Jay A. Bookbinder (eds.)

MODELING STÉLLAR ANGULAR MOMENTUM EVOLUTION

PAUL CHARBONNEAU High Altitude Observatory, National Center for Atmospheric Research [1], P.O. Box 3000, Boulder, CO 80307.

ABSTRACT The various facets of modeling the rotational evolution of low-mass stars are critically examined. It is shown that many parametric formulae used to model angular momentum loss compare poorly to self-consistent MHD wind models. Recent models of rotational evolution are reviewed, and some of their specific weaknesses discussed. A sample of recent numerical simulations bearing on the efficiency of internal magnetic fields in redistributing angular momentum are also presented.

Keywords: Angular Momentum Loss; Rotation; Magnetic Fields.

SCOPE OF THIS REVIEW

This review focuses on the rotational evolution of single, late-type stars, between their arrival on the zero-age main sequence (ZAMS) and the beginning of their evolution up the giant branch. For a wider view of the subject, the interested reader may wish to browse through the numerous conference and workshop proceedings devoted to specific aspects of stellar angular momentum. Note in particular the two recent NATO workshops *Angular Momentum and Mass Loss for Hot Stars*, edited by Willson & Stalio (Dordrecht: Kluwer, 1990), and *Angular Momentum Evolution of Young Stars*, edited by Catalano & Stauffer (Dordrecht: Kluwer, 1991).

1. OBSERVATIONS

1.1 The Rotation – Age relation

As soon as the spectroscopic determination of projected rotational velocities ($v \cdot \sin i$) of stars began to be carried out systematically (Struve, 1930), it was realized that early-type stars tend to show much higher $v \cdot \sin i$'s than late-type stars, a trend particularly obvious on the main sequence. However, it was Kraft (1967) who perhaps most convincingly demonstrated the discontinuous nature of the decline in rotation rates near spectral type F5 (see Fig. 1 of Kraft, 1967). He also showed (*e.g.* Kraft, 1970) that the mean angular momentum (J) along the main sequence, from the earliest spectral types down to the F5 rotational break, could be related to mass via a simple power law

$$J(M) \propto M^{1.57}. \qquad (1)$$

[1] The National Center for Atmospheric Research is sponsored by the National Science Foundation.

Stars later than the rotational break, however, seemed to have much less angular momentum than an extrapolation of eq. (1) down to low masses would suggest. Hypotheses advanced to explain this state of affair were the following: (H1) Stars later than F5 form planetary systems, earlier-type stars do not; (H2) stars later than F5 have rapidly rotating cores, earlier-type stars rotate as solid bodies; (H3) Stars earlier than F5 suffer rotational braking over time, earlier-type stars do not (for a review, see Kraft, 1970).

Although (H3) was already gaining wider acceptance in 1970, it was Skumanich (1972) who observationally derived the relation

$$\Omega(t) \propto t^{-1/2}, \tag{2}$$

while at about the same time Durney (1972) showed that such a relation could be derived from simple angular momentum loss and linear dynamo models (more on this below). Equation (2) has stood the test of time, at least in an average sense (see *e.g.* Soderblom, 1983). For a recent review see Soderblom (1991).

1.2 Young Main-Sequence Stars

The relatively simple picture drawn above broke apart when a few rapidly rotating ($v \cdot \sin i > 100$ km s^{-1}) K-stars were discovered, quite unexpectedly, in the 75 Myr old Pleiades cluster (Van Leeuwen & Alphenar, 1982). It soon became apparent that these were by no means freaks, as more rapid rotators were subsequently found among Pleiades K and M dwarfs (Stauffer *et al.* , 1984; Butler *et al.* , 1987; Stauffer & Hartmann, 1987), as well as in G, K and M dwarfs of the slightly younger α Persei cluster (Stauffer *et al.* , 1989; Prosser, 1991). The 600 Myr old Hyades cluster, on the other hand, shows a very tight rotation vs spectral type relation in the G—K range (Radick *et al.* , 1987; Stauffer *et al.* , 1987), with rotation declining with decreasing mass from about 10 km s^{-1} at spectral type G0, down to $\simeq 4$ km s^{-1} at K5. The Hyades observations are actually compatible, within error bars, with rotation being a single-valued function of mass in this spectral type range. However, the M dwarfs show a broader distribution of rotation rates, reaching values as high as 25 km s^{-1} (*e.g.* Stauffer *et al.* , 1987). Rapid rotators have recently been discovered among G-dwarfs of the Pleiades (Stauffer, private communication).

This increasing quantity of high-quality observations has also permitted the determination of the *distribution* of rotational velocities at a given spectral type. Even though observed $v \cdot \sin i$ of G- and K-dwarfs in α Persei and the Pleiades can exceed 100 km s^{-1} , in both these clusters the distributions are strongly peaked at low $v \cdot \sin i$ (in fact, often close to or at the detection limit), with a more or less flat tail extending to the highest observed $v \cdot \sin i$. The situation is not as clear for the M-dwarfs, for which $v \cdot \sin i$ distributions show no well defined peak, a tendency persisting up to the age of the Hyades. Note however that the the relative paucity of M-dwarfs for which rotation periods and/or $v \cdot \sin i$'s have been measured leads to somewhat ill-defined velocity distributions. Nevertheless, the distributions for G-, K- and M-dwarfs are in sharp contrast with earlier-type stars for which velocity distributions appear more or less maxwellian (*e.g.* Kraft, 1970; see however Wolff *et al.* , 1982). Clearly, the rotational evolution close to the ZAMS is far more intricate, and maybe less deterministic, than a naive extrapolation of the $t^{-1/2}$ relation would suggest. For a recent review of rotation in young clusters, see Stauffer (1991).

1.3 Pre-Main Sequence Stars

Measurements of rotation rates in low-mass pre-main sequence (PMS) stars is a fairly recent achievement (*e.g.* Vogel & Kuhi, 1981; Hartmann *et al.* , 1986). High mass (*i.e.*, $M/M_{\odot} \gtrsim 1.5$) PMS stars have observed $v \cdot \sin i$ often exceeding 100 km s^{-1} , while lower mass PMS rarely have $v \cdot \sin i$ in excess of 30 km s^{-1} . This may be in part an evolutionary effect, as high mass PMS stars are, on the average, closer to the zero age main-sequence (ZAMS) than their lower mass counterparts. But it is also interesting to note that the average angular momentum content of low-mass PMS stars is in rough agreement with the extrapolation at lower mass of eq. (1). Note however that a spread by almost a full order of magnitude exists at all masses for both the $v \cdot \sin i$ and rotational period distributions. An important caveat is in order here: such inferences depend sensitively on the internal structure (and radii) assumed for these stars, as well as the precise morphology of PMS evolutionary tracks used to infer masses. For recent reviews see Catala (1990) and Bouvier (1991).

1.3 Evolved Stars

As stars leave the main-sequence and start ascending the giant branch, the increase in moment of inertia due to the expansion of the outer layers is expected to lead to a significant decrease in rotation rates, a state of affairs amply confirmed by early $v \cdot \sin i$ measurements in subgiants and giants of various spectral types (For an early review see Kraft, 1970, §III). Recent work has however shown that the transition from rapid to low rotation as early- and mid-type stars cross the HR diagram on their way up the giant branch is markedly discontinuous, occurring at spectral type G0 in both class IV and III giants (see *e.g.* Gray & Nagar, 1985; Gray, 1989, and references therein). This is essentially identical to the spectral type at which the rotational break is observed in main-sequence stars, and suggests that similar mechanisms may well be responsible for the braking of young MS stars as well as evolved stars. Additional evidence for the operation of some braking mechanism in giants comes from the widely different $v \cdot \sin i$ distributions on each side of the G0 rotational break. For a recent review see Gray (1991; but also de Medeiros & Mayor, 1991).

Observed $v \cdot \sin i$ in field and cluster horizontal branch stars show again a broad distribution extending from a few km s^{-1} up to about 30 km s^{-1} , with a mean around 12 km s^{-1} (Peterson, 1983; Peterson, 1985*a,b*). If solid-body or near-solid body rotation is assumed to hold on the horizontal branch of clusters, then the corresponding angular momentum content is markedly higher than that of stars at the turnoff. This represents an interesting puzzle because it is rather difficult to imagine mechanisms by which a large fraction of stars could *increase* their angular momentum content between the main sequence and horizontal branch.

Little is known on rotational velocity distributions of white dwarfs. White dwarfs are slow rotators, in the sense that they rotate more slowly than expected from the angular momentum content of their progenitors, indicative of a substantial amount of angular momentum loss somewhere between the planetary nebula and white dwarf stages of evolution. For a brief review see Sion (1990).

1.4 The Solar Internal Rotation

The last decade has also witnessed the development of the exciting field of helioseismology, *i.e.*, the probing of the solar internal structure and velocity field via the observed spectrum of acoustic oscillations. Observational and inversion techniques have now reached a level such that a consistent picture of the internal solar angular velocity, $\Omega_\odot(r,\theta,t)$, is beginning to emerge. In the outer radiative zone, no differential rotation is observed; $\Omega_\odot$ shows, at best, a very weak dependence with latitude down to at least 0.5 $R_\odot$, and possibly a small *decrease* of $\Omega_\odot$ with depth down to 0.4 $R_\odot$ along the equatorial plane. Evidence for a time dependence of the internal rotation law below 0.5 $R_\odot$, associated with the solar cycle, has also been presented recently by Goode & Dziembowski (1991). Although the data inversion becomes rather unreliable at greater depths, there appears to be no evidence for a rapidly rotating core, although observations are not incompatible with a core rotating at 2—3 times the surface rate. For recent reviews see Gough (1991), and Goode *et al.* (1991).

2. ROTATIONAL BRAKING

Consider first a rigidly rotating, constant density, thin spherical shell of mass M and radius R; its angular momentum (J) is

$$J = \Omega I = \Omega \left[\frac{2}{3} M R^2\right] ; \tag{3}$$

consider now the (spherically symmetric) expansion of this shell from a radius R_0 to radius R_1 ($R_1 > R_0$). Conservation of mass and angular momentum immediately yields $\Omega_1 = \Omega_0(R_0/R_1)^2$, *i.e.*, the shell spins down as it expands, carrying along the amount J_0 of angular momentum it possessed initially.

Consider now an alternate situation where the shell still expands, but is forced, by external means, to maintain a constant angular velocity, *i.e.*, $\Omega_1 = \Omega_0$. The angular momentum content of the shell then increases as it expands. With mass conservation still holding, the relative excess $\Delta J/J_0$ of angular momentum carried away by the shell is

$$\frac{\Delta J}{J_0} = \frac{J_1 - J_0}{J_0} = \left(\frac{R_1}{R_0}\right)^2 - 1. \tag{4}$$

In this hypothetical situation, the excess angular momentum must ultimately be supplied by whatever is enforcing the state of corotation.

It was Schatzman (1962) who first realized that such a "toy model" was actually relevant to stellar angular momentum loss in solar-type stars. He first noted that late-type stars, having extended convective envelopes, are capable of sustaining a dynamo and a corona, while early-type stars most likely cannot do so. He then argued that given the physical conditions in the solar corona, a large-scale magnetic field anchored on the solar surface could enforce effective co-rotation out to large distances, greatly enhancing the net angular momentum loss. In this simple picture, the wind corotates with the sun out to a critical distance, and then decouples from the field, carrying away whatever angular momentum it has acquired prior to decoupling. This decoupling occurs when the kinetic energy density of the wind becomes larger than the magnetic energy density of the field. Equation (4) then applies, with R_0 identified with the stellar surface and R_1 with the maximal corotation radius. The excess angular momentum is extracted from the star, via the torque applied to the stellar surface by the magnetic field. If the field is strong enough for the corotation radius to be located many stellar radii away from the surface, a large amount of angular momentum can be extracted from the star.

Determining the position of the co-rotation point is therefore of paramount importance in such a model. The magnetic field strength is expected to be largest at the stellar surface, and to globally decrease outward. This implies that the Alfvén velocity $U_A = ||B||/(4\pi\rho)^{1/2}$ will also decrease outward. The wind velocity U_W, on the other hand, increases monotonically outward for thermally driven winds (see Parker, 1958). The corotation radius (hereafter Alfvén radius) is defined as the position r_A at which $U_A = U_W$. Since U_A is directly proportional to $||B||$, clearly r_A (and thus the angular momentum loss rate) will increases as the field strength increases, everything else being equal[2].

2.1 The Skumanich $t^{-1/2}$ relation

The flux of angular momentum carried by a stellar wind can be written as the wind's mass flux times the angular momentum per unit mass. For a spherically symmetric wind corotating up to $r = r_A$,

$$\frac{dJ}{dt} = \left(\frac{dM}{dt}\right)\left(\frac{2}{3}\Omega\, r_A^2\right). \tag{5}$$

Assuming spherically symmetry allows one to write the mass loss rate as $\partial M/\partial t = -4\pi\rho\, r^2 U_r$, and since $U_r = U_A$ at the Alfvén radius, eq. (5) becomes

$$\frac{dJ}{dt} = -\frac{(r_A B_A)^2}{U_A}\frac{2}{3}\Omega\, r_A^2. \tag{6}$$

where B_A is the magnetic field strength at the Alfvén radius. Relating B_A to the surface field B_0 requires some assumptions regarding the field geometry, subjected to the constraint $\nabla \cdot B = 0$. For an axisymmetric field without meridional component, $B(r) = B_0(R_0/r)^2$, which yields

$$\frac{dJ}{dt} = -\frac{2}{3}B_0^2 R_0^4 \Omega\, U_A^{-1}. \tag{7}$$

Assume now that the wind is (mostly) thermally driven; then $U_A \simeq c_s$, where c_s is the sound speed, which in the thermal limit is essentially independent of B_0

---

[2] More realistic models show that corotation is actually not enforced up to r_A, but that sum of angular momentum carried by wind and magnetic torques yields at the stellar surface an angular momentum loss rate equivalent to that obtained from strict corotation in a torque-free field out to r_A.

and Ω. Assume also that $B_0 \propto \Omega$, as suggested by the so-called linear dynamo "theories". By writing $J = I \times \Omega$, where I is the moment of inertia of whatever is being braked (the whole star, if rotating as a solid body, or alternately only the convective envelope), one finally obtains

$$\frac{d\Omega}{dt} \propto -\Omega^3, \tag{8}$$

which integrates to $\Omega(t) \propto t^{-1/2}$, *i.e.*, Skumanich's relation.

Although this is certainly encouraging, one must not lose track of the fact that three very fundamental assumptions entered this derivation : (A1) Spherically symmetric magnetic field and wind, (A2) (mostly) thermally-driven wind, and (A3) $B_0 \propto \Omega$, from dynamo "theory". What is annoying here is that (A1) is most probably never valid, and (A2) holds only for relatively low Ω ($P_{rot} \gtrsim$ 2–3 day) and weak B_0. As for (A3), well, in the absence of any complete stellar dynamo theory, we simply don't know.

2.2 The Weber-Davis MHD wind model

In principle, the angular momentum loss rate can be obtained from solutions of the magnetohydrodynamic (MHD) equations, *i.e.*, the three components of the Navier-Stokes equation with Lorentz force and magnetic pressure terms, the three components of the induction equation, an equation of state and an energy equation, subjected to the constraints of mass conservation and vanishing divergence of the total magnetic field. To this date, most attempts at doing so have been variations on the theme set by the very successful Weber & Davis model (1967, hereafter WD; see also Belcher & MacGregor, 1976). The standard WD model is a self-consistent solution of the MHD equations, under the following working assumptions: (A1) single-fluid model (fully ionized pure Hydrogen); (A2) ideal gas, polytropic energy equation; (A3) zero viscosity, zero thermal conduction; (A4) zero magnetic diffusivity; (A5) axisymmetric 2-D steady-state model [solves for $u_r(r)$, $r_\phi(r)$, $B_r(r)$, and $B_\phi(r)$ in equatorial plane]; (A6) radial poloidal field; (A7) meridional terms neglected in equations of motion.

Now, many of these assumptions may appear overly restrictive, which has motivated many attempts at improving the model. Most of these (Brandt *et al.* , 1969; Urch, 1969; Weber, 1969; Wolff *et al.* , 1971; Acuna and Whang 1976) have focused on (A1), (A2) and (A3). Although some disagreement and controversies persisted for some time, the consensus having emerged from these works is that the WD wind model can reproduce fairly well, without arbitrary parameter, the main characteristics of the quiet solar wind at 1 A.U. (Pizzo *et al.* , 1983, and references therein). This is especially remarkable, since (A6) is a poor approximation close to the sun and (A7) is now known to lead to formal inconsistencies in the mathematical formulation (Suess & Nerney, 1973). This has motivated various attempts at relaxing (A5), (A6) and (A7), but so far results have proven difficult to interpret and generalize (Nerney & Suess, 1975; Sakurai, 1985; on [A7] see also Yang & Schunk, 1991).

Two important points must be stressed at this juncture. *First*, the WD model yields a solution in the equatorial plane; to compute the net angular momentum loss rate over the whole stellar surface requires a latitudinal integration. The original WD calculation assumed a purely radial magnetic field all the way from the equatorial plane to the pole. Since more realistic field geometries, *i.e.,* including closed field regions, would most likely lead to lower mass loss —and thus angular momentum loss— (see *e.g.* Mestel & Spruit, 1987), the WD solution provides an *upper* limit to any full 3-D solution of the MHD equations under more realistic field geometries. *Second*, in contrast with the solar case, for which both the surface magnetic field strength and rotation rate are known, applying the WD model to the problem of rotational evolution requires an additional assumption, namely how the surface field strength is related to the rotation rate. This involves invoking again some kind of dynamo relation, and thus reintroduces a serious uncertainty in an otherwise "clean" model.

2.3 Parametric formulae for angular momentum loss

An alternative to the WD approach is to develop mathematical approximations to the MHD equations, and try to incorporate more realistic assumptions regarding field geometry and dynamics (for recent reviews see Mestel, 1984; Collier Cameron *et al.* , 1991). Recently, Kawaler (1988, following Mestel 1968, and Mestel & Spruit 1987) has proposed:

$$\frac{\partial J}{\partial t} = -K_W \Omega^{1+(4aN/3)} \left(\frac{R}{R_\odot}\right)^{2-N} \left(\frac{\dot{\mathcal{M}}}{\dot{\mathcal{M}}_\odot}\right)^{1-(2N/3)} \left(\frac{M}{M_\odot}\right)^{-N/3}, \tag{9}$$

where N, a and K_W are adjustable parameters. Mayor & Mermilliod (1991) have modeled the time-evolution of the $v \cdot \sin i$ distribution stars in young clusters using an expression equivalent to

$$I\frac{\partial \Omega}{\partial t} = -A(\Omega)\,\Omega^2. \tag{10}$$

A similar form has been obtained, in a different context, by Stepień (1988). Schatzman (1990), on the basis of a rudimentary turbulent dynamo model, proposed a formula of the form

$$I\frac{\partial \Omega}{\partial t} = -f(M)\,\Omega^{7/3}. \tag{11}$$

Note that *all* of these expressions assume a fixed value for the exponent in the power law relating $d\Omega/dt$ to Ω, *i.e.*, independent of the rotation rate, and thus of time.

2.4 Parametric formulae vs the Weber-Davis model

Figure 1 shows the rotational dependence of the braking rate ($d\Omega/dt$) as predicted by the standard WD model, with $B_0 \propto \Omega$ (solid line). It is immediately apparent that a power law with a fixed exponent (whether 2 or 3) does not provide an accurate representation of the braking rate's dependence on surface rotation rate. An exponent of 3 provides a good fit to the WD model in the range $1 \leq \Omega/\Omega_\odot \lesssim 5$, but greatly overestimates the braking rate when extrapolated to higher rotation rates ($\Omega/\Omega_\odot \gtrsim 10$, say). Likewise, an exponent of 2 seems to provide a better fit at high rotation rates, but fails to reproduce the proper dependency for

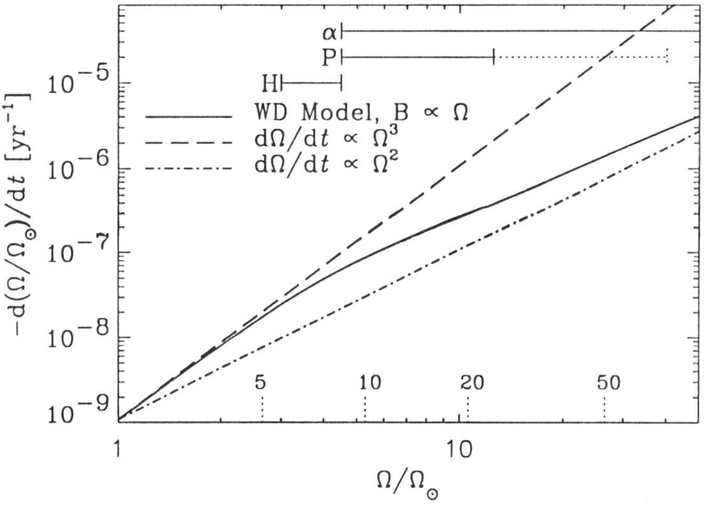

Figure 1. Dependence of the braking rate on rotation rate, as predicted by the standard Weber-Davis MHD wind model for a solar-type star (solid line). The dashed line indicates a slope of 3, corresponding to Skumanich's relation, and the dash-dotted line a slope of 2. All angular velocities are normalized to the sun. The dotted tick mark along the abscissa indicate the corresponding equatorial rotational velocities, in km s^{-1} and for $R_* = R_\odot$. The observed $v \cdot \sin i$ ranges for the α Persei (α), Pleiades (P) and Hyades (H) clusters are shown in the upper part of the figure.

solar-like rates. It then follows that *if* a fixed-exponent parameterization is used to model rotational braking, a value of 2 for the exponent is likely to provide a better description of the braking of rapidly rotating stars —as found by Mayor & Mermilliod (1991), cf. eq. (10) above—, but that a proper fit to the rotational evolution of older stars will be best reproduced with an exponent of 3, in agreement with the $t^{-1/2}$ relation (§2.2 herein).

This change in slope in the braking rate can be traced to the increasing importance of centrifugal and magnetic terms (Mestel, 1968; Belcher & MacGregor, 1976; for a succinct summary, see §5 of Mestel, 1984). Furthermore, it has also been suggested that the $B_0 \propto \Omega$ relation breaks down at a rate of about 10 times solar, due to saturation of the "dynamo"-generated surface field (Vilhu, 1984; Skumanich & MacGregor, 1986; Linsky & Saar, 1987). This would imply that the slope of the $\partial\Omega/\partial t$ vs Ω curve may well decrease even more at high rotation rates. But the important point here is that a fixed-exponent parameterization does not provide a good fit to the braking rate over rotation rates ranging from fifty times solar —characteristic of some G stars in αPersei— to solar. All parameterizations described in §2.3 become immediately suspicious.

2.5 Observational Constraints on Rotational Braking Models

Can observations provide tight enough constraints to help define an accurate angular momentum loss model ? Lets first make the maximally deterministic hypothesis that the initial angular momentum distribution $J_0(M)$ is well defined, sharply peaked at each mass, and identical from cluster to cluster. By "initial" we refer here to the angular momentum distribution at ages typical of low-mass T Tauri stars, *i.e.*, 1—10 Myr. Figure 2 illustrates the expected evolution of the surface equatorial rotation velocity for stars of 1.0, 0.8 and 0.6 solar masses. The solid lines correspond to strict solid-body rotation throughout the whole star, while the dashed lines correspond to the spinning down of the outer convective envelope only; as some level of dynamical coupling is likely to exist between the convective envelope and the radiative core, these sets of curves can be thought of as two possible extremes, "reality" sitting somewhere in between (more on this in §3 below). The adopted initial rotational velocity are typical of PMS stars, and the braking rate is computed via the standard WD model, with $B \propto \Omega$. The initial spinup is caused by the decreasing moment of inertia, while the subsequent spindown is due to rotational braking. Examination of Figure 2 in conjunction with the young star data (§1.2 above) immediately raises the following "problems":

(P1) How to "make" an order-of-magnitude spread (or more) in rotational velocities at all spectral types in α Persei and the Pleiades ?
(P2) How to "make" rapidly rotating G, K and M dwarfs in α Persei and the Pleiades ?
(P3) How to "make" a large quantity of slow rotators among α Persei G and K-dwarfs ?
(P4) How to get G-stars, slowly rotating by Pleiades age, not to spin down completely by Hyades age ?

Although "solutions" to these problems can be sought in terms of age spread, spread in J_0, incorrect cluster ages, etc., it rapidly becomes apparent that while a combination of some of these solutions can improve the fit to the data, important discrepancies remain. The large differences, in the interval $10^7 \lesssim t \lesssim 5 \times 10^9$ yrs, between the "solid-body" and "convective envelope only" spin down provide a hint at the answer, namely that the level of coupling between the outer envelope and inner radiative core can be of paramount importance in determining the surface rotational evolution[3]. From the modeling standpoint, this, unfortunately, introduces additional —and major— uncertainties in an already not-so-deterministic picture.

3. ROTATIONAL EVOLUTION

3.1 Internal Redistribution of Angular Momentum

Braking by a magnetically-confined wind is only the beginning of the story, as far as the rotational evolution of single stars goes. Structural evolution will alter the moment of inertia, leading to changes in the angular velocity. This is of minor importance during main-sequence evolution, but may dominate the

[3] This implies that rotational dating of moderately old to old late-type stars (*e.g.* Kawaler, 1989) is likely to be a sensitive function of angular momentum transfer between the core and envelope, and thus subjected to significant uncertainty.

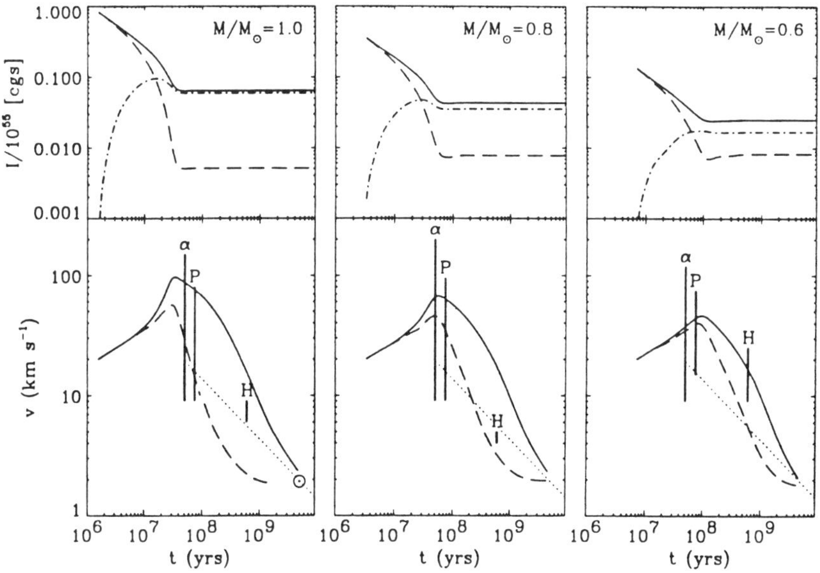

Figure 2. Surface rotational evolution, as predicted by the standard WD model with $B \propto \Omega$. The solid line is the surface rate evolution if the whole star is braked while rotating as a solid body, and the dashed line the evolution expected if only the outer convective envelope is braked. Upper panel show the time-evolution of the moments of inertia of the convective envelope (dashed line), radiative core (dash-dotted line), and the star as a whole (solid line). Thick vertical segment indicate the observed $v \cdot \sin i$ range in clusters (but recall that the distributions are strongly peaked at low rotation rates), and the dotted line is the $t^{-1/2}$ relation fitted to the solar rotation rate at the solar age.

evolution of surface rotation rates in other evolutionary stages, *e.g.* , along the giant branch or during the contraction to the main-sequence. Equally important, angular momentum may be internally redistributed by a variety of mechanisms, affecting in turn the surface rotation rate. A partial list may include (1) Viscous friction, (2) Rotationally-induced meridional circulation, (3) Convective overshoot into radiative regions, (4) Shear instabilities triggered by differential rotation, (5) Baroclinic Instabilities, (6) Ekman pumping in shear layers, and (7) Torsional Alfvén waves. The only one of these whose effect on angular momentum transport can be computed with true certainty is viscous friction... but it is completely inefficient in radiative zones.

3.2 On the Modeling of Transport by Meridional Circulation

Before describing rotational evolution models, a few comments are in order regarding one of the potentially important mechanism listed above. Rotationally-induced meridional circulation is an axisymmetric, large-scale flow arising from the inability to simultaneously satisfy the conditions of thermal and hydrostatic equilibrium in radiative zones of rotating stars (von Zeipel, 1924; for a recent review, see Tassoul, 1990). Although the circulation velocities are rather small for rotation rates of the order of that of the present day sun, the associated mass motions are expected, over the main-sequence lifetime, to affect sometimes significantly the internal distributions of angular momentum and chemical species. Quantitatively, its effects are difficult to estimate because of the inherently bidimensional nature of the flow, which in principle demands that solutions be computed over a whole meridional quadrant, a rather demanding state of affairs from the computational standpoint. As a consequence, attempts have been made to develop various approximate 1-D treatments. A prescription often used is that of Endal & Sofia (1978), who treat meridional circulation transport as a 1-D diffusive process. In practice, this means that an equation of the form

$$\frac{\partial X}{\partial t} = -U(r,\theta)\cdot\nabla\, X \qquad [2-D], \tag{12}$$

is replaced by

$$\frac{\partial X}{\partial t} = +\nabla\cdot[D(r)\nabla\, X] \qquad [1-D], \tag{13}$$

where D is a suitably chosen "diffusion coefficient". Obviously, one would hope that (13) can approximate (12) to a reasonable degree, *i.e.*, that the time-variation of X, at least in a angular average sense, are similar. Unfortunately, this is in general not the case.

This is easily illustrated with the help of a simple model. Figure 3 shows the time-evolution, in a meridional quadrant, predicted by eq. (12) for the (A) angular velocity and (B) chemical concentration, under the sole action of meridional circulation. The latter is represented here by a time-independent, quadrupolar streaming pattern, the flow rising along the rotation axis, crossing the upper envelope toward the equator and sinking back in the equatorial regions. The initial conditions are constant angular velocity (Ω_0) for part (A), and an initially discontinuous spherically symmetric concentration profile for part (B), and the contours correspond to $[\Omega(r,\theta,t)-\Omega_0]$ and chemical concentration c, respectively. The spin-up seen along the rotation axis (part A) follows from advection along the equatorial plane of material having high specific angular momentum, while the spin-down of the upper envelope results from advection from the polar regions of material initially having low specific angular momentum. The evolution of the chemical concentration profile follows the simple advection by the flow of the initially spherically symmetric profile.

Clearly, transport of angular momentum and chemicals by meridional circulation does not even remotely resemble that expected from a purely radial diffusion equation. Note in particular that a diffusion equation such as (13), starting from an initial condition of constant Ω compatible with boundary conditions, would predict $\partial\Omega/\partial t = 0$ in case (A), and a gradual — and spherically symmetric— smoothing of the initially discontinuous chemical concentration profile in case (B). Replacing (12) with (13) is thus an extremely poor approximation in this context, even over short time intervals.

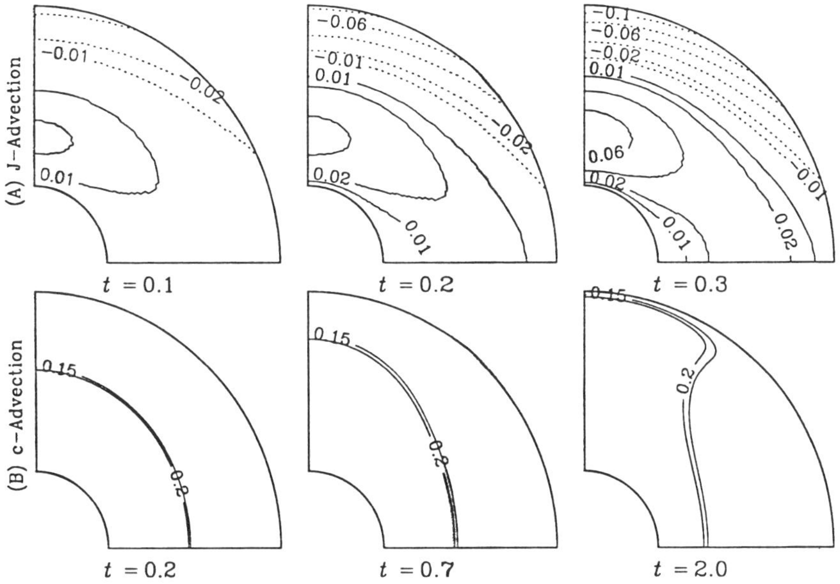

Figure 3. Evolution of angular momentum (top row) and chemical concentration (bottom row) distributions under the influence of rotationally-induced meridional circulation. The rotation axis is in the vertical direction. Time is measured in units of the Eddington Sweet time $t_{ES} = R_*/U_0$, where U_0 is a typical circulation velocity at mid-depths. For an equatorial rotational velocity of 100 km s^{-1} , U_0 is expected to be of order $10^{-4}\, cm\, s^{-1}$, yielding $t_{ES} \approx 2 \times 10^7$ yrs.

There may be a specific situation where a 1-D treatment can be justified. Chaboyer and Zahn (1991) have recently shown that if one postulates the existence of horizontal turbulence strong enough to enforce constant Ω and c along spherical surfaces, then meridional circulation transport can indeed be modeled as a 1-D process. This is confirmed by numerical simulations (Charbonneau, 1991). The transport of chemicals is then described by a pure diffusion equation similar to (16) above, but the transport of angular momentum is not; it obeys instead a 1-D advection-diffusion equation, leading to a time-evolution of $\Omega(r,t)$ differing greatly from a purely diffusive behavior (see Fig. 4 and 5 of Charbonneau, 1991). It remains to be shown how such strong horizontal turbulence can be maintained in radiative zones of stars exhibiting, by the very nature of the working postulate, very weak differential rotation. Other constraints inferred from the existence chemically peculiar stars must be also met (see *e.g.* Charbonneau & Michaud, 1991; §4 of Charbonneau, 1991). To sum up, justifying a 1-D treatment of meridional circulation transport via the existence of strong horizontal turbulence is in principle possible but still demands, at this juncture, a sizable leap of faith. But to model angular momentum transport by circulation via a pure diffusion equation is clearly incorrect.

3.3 The Yale Models

These models simultaneously follow, in a detailed fashion, the structural, chemical and rotational evolution throughout the star. They essentially follow the strategy developed by Endal & Sofia (1976, 1978, and references therein). The working assumptions are: (A1) constant angular velocity along equipotential surfaces, i.e, the model can exhibit differential rotation with depth, but not with latitude; (A2) use of Kawaler's 1988 parameterization for angular momentum loss; (A3) parametric representation of angular momentum redistribution in the interior, with concomittant material mixing, including a parametric representation of the stabilizing effects of mean molecular weight gradients; (A4) all transport processes —including meridional circulation— modeled as purely diffusive mechanisms. (A5) solid-body rotation and homogeneous abundances in convective zones. (A6) no internal magnetic field. Given the large number of physical mechanisms included, it is not surprising that quite a few adjustable parameters (between 9 and 11) need to be specified. Fortunately, a number of these can be constrained via calibration on solar models and the same parameter values then used to model other stars. The calibration for the solar model is presented in Pinsonneault *et al.* (1989), the application of the model to young low-mass stars in Pinsonneault *et al.* (1990), and to halo stars in Pinsonneault *et al.* (1991*a,b*). For a succinct review, see Demarque (1991).

These models cannot explain the large observed $v \cdot \sin i$ spread in αPersei and the Pleiades G and K dwarfs (Problem P1 in §2.5 above). Because of the adopted exponent for the wind law, they most probably cannot succeed in producing very fast rotators (*i.e.*, $v \cdot \sin i \gtrsim 100$ km s^{-1}) at ages of α Persei and the Pleiades, because of the spin down then occurring in the previous few 10^7 yrs (problem P2). Because the wind law is adjusted to reproduce the $t^{-1/2}$ relation, it is not clear if the models can satisfy P4. They do reproduce the solar beryllium abundance and the observed convergence of rotations rates by the Hyades age (Pinsonneault *et al.* , 1991*a*), as well as the observed low rotation rates of halo stars. They also reproduce the constancy of the Lithium abundance [N(Li)] in the effective temperature interval 5500 K $\lesssim T_{eff} \lesssim$ 6300 K for halo stars, but fail to yield a proper fit to the well-defined N(Li)–T_{eff} relationship among Hyades G-dwarfs (cf. Fig. 15 of Pinsonneault *et al.* , 1990). An interesting prediction of the models is the existence of a rapidly rotating core (*i.e.*, of order of ten times the surface rate), and that halo star models exhibit substantial differential rotation with depth, potentially providing an attractive explanation for the observed rotation rates of horizontal branch stars.

Overall, an unfortunate drawback of the formulation is its reliance on very poorly understood physical mechanisms for internal redistribution of angular momentum. Note in particular that the so-called secular shear instability (Zahn, 1974), for which only orders-of-magnitude estimates can be constructed for both the associated diffusion coefficient and stability criteria (cf. §II[b](ii) of Endal & Sofia, 1978), turns out to be, in these models, the dominating mass and angular momentum transport mechanism in the upper radiative envelope of solar-type stars [see §IIIe(i) of Pinsonneault *et al.* , 1989]. Two specific inadequacies of the models are the treatment of transport by meridional circulation as a purely diffusive 1-D process (cf. §3.2), and the use of a fixed-exponent braking law (cf. §2.5). The former is likely to affect predictions of CNO abundance ratios on the giant branch, while allowance for the latter may well improve the fit to the N(Li)–T_{eff} relationship observed in young clusters.

3.4 The MacGregor Models

These (MacGregor, 1991; MacGregor & Brenner, 1991) are decoupled rotation-evolution model, *i.e.*, the rotation has no effect on internal structure. Both the convective envelope and radiative core are assumed to rotate as solid bodies, and are rotationally coupled on a timescale τ_J, specified *a priori*. Rotational braking is computed using the WD model, with various hypotheses regarding the B vs Ω dynamo relationship. Although these models can make no detailed prediction of the internal angular rotational velocity profile, they can be adjusted to provide a fairly good fit to early rotational evolution on the main-sequence. Note in particular that a version of the model with $\tau_J \simeq 3 \times 10^7$ yrs, $B \propto \Omega_{core}$ with a cutoff at $40 \times \Omega_\odot$ can reproduce, starting with the J_0 spread observed on the pre-main sequence, the rotational velocity *distributions* of young clusters, without invoking an age spread (MacGregor *et al.* , in preparation).

4. INTERNAL ROTATION AND MAGNETIC FIELDS

4.1 Basic Mechanism

Consider a perfectly conducting fluid contained between two infinite plates, oriented parallel to the yx-plane and located at $z = -\ell$ and $z = +\ell$ respectively. A constant, z-directed magnetic field permeates the fluid. The fluid is then set in motion in the positive y-direction, by some external mean. What will be the time-evolution of this system ? The initial state is $B = B_z\, 1_z$, $U = U_y(z)\, 1_y$, and the evolution of the magnetic field components is described by the induction equation, which reduces here to $\partial B/\partial t = \nabla \times (U \times B)$. The only non-trivial component is in the y-direction,

$$\frac{\partial B_y}{\partial t} = B_z \frac{\partial U_y}{\partial z}, \tag{14}$$

which for a time-independent driving flow integrates to $B_y(z,t) = B_z(\partial U_y/\partial z)t$. The Lorentz force L associated with the total magnetic field is

$$4\pi L = (\nabla \times B) \times B = B_z \frac{\partial B_y}{\partial z} 1_y + B_y \frac{\partial B_y}{\partial z} 1_z. \tag{15}$$

Note that L has components in both the y and z-direction, *i.e.*, parallel *and* perpendicular to the flow. In the case of a parabolic flow,

$$U_y(z) = U_0 \left[1 - \left(\frac{z}{\ell}\right)^2\right], \tag{16}$$

the induced y-component is $B_y(z,t) = -2U_0 B_z zt/\ell^2$, with concomittant Lorentz force

$$4\pi L(z,t) = -2\,\frac{U_0\, B_z^2}{\ell^2} t\, 1_y + 4\,\frac{U_0^2\, B_z^2}{\ell^4} z\, t^2\, 1_z. \tag{17}$$

Clearly, for small t, or more precisely for $t << \ell/U_0$, the Lorentz force is in the negative y-direction, *i.e.*, it opposes the driving flow. It is also clear from eq. (14) that the only allowed steady-state is $U_y(z) = 0$, since $U_y(\pm\ell) = 0$. Note that a solution valid at all t requires the simultaneous solution of the y-induction and y-momentum equations, because of the feedback of the induced B_y component, via the associated Lorentz force, on the flow.

It is easy to verify that an equivalent toy problem in spherical polar coordinates, with $B = B_r\, 1_r$ and $U = U_\phi(r)\, 1_\phi$ at $t = 0$, behaves in a similar way, *i.e.*, the Lorentz Force is mainly in the ϕ-direction and again opposes the differential flow. What is then the expected effects of a large scale magnetic field on the internal rotation of stars ? One can easily imagine than an initially purely poloidal field will be gradually sheared by even the weakest azimuthal differential rotation, thus inducing a toroidal magnetic component with associated Lorentz force tending to oppose the differential azimuthal flow.

Although the basic physical mechanism is well understood, quantitative estimates of its efficiency are more ambiguous. Most quantitative arguments to be found in the literature are more or less sophisticated variations along the following theme; the timescale τ_A for the redistribution of angular momentum along a magnetic flux tube is of the order of the time required for an Alfvén wave to travel the length of the tube. For a large scale poloidal field of strength B_0 embedded in the radiative interior of a star of radius R_*, one can then write

$$\tau_A \approx \frac{R_*}{U_A} = \frac{R_*(4\pi\rho)^{1/2}}{B_0} \simeq \frac{10^4}{B_0}\ [yrs], \tag{18}$$

with $R_* = 10^{11}$ cm, B_0 in gauss, and for an average density of 1 $g\, cm^{-3}$. Clearly, τ_A is then much smaller than typical braking timescales on the main-sequence. Thus follows the conclusion that even milligauss magnetic fields would largely suffice to rapidly enforce solid-body rotation in an initially differentially rotating star, or to maintain a state of very near solid-body (SB) rotation despite the inexorable wind-induced surface torque (see *e.g.* Rosner & Weiss, 1985; Mestel & Weiss, 1987; Spruit, 1990; and references therein). Note that this a purely local argument, neglecting potentially important effects related to field geometry, boundary conditions, etc.

4.2 A Model Problem : Impulsive Core Spin-up

Even though allowance can be made for geometrical effects, finite resistivity, etc., in order to refine the timescale estimate obtained above, the resulting estimates become very uncertain, model-dependent, and open to controversy. Full solutions to the MHD equations would be appropriate, but this is no easy task since it involves the simultaneous solution of six strongly coupled time-dependent non-linear partial differential equations, in at least two spatial dimensions (for axisymmetric configurations).

However, Mestel *et al.* (1988; see also Tassoul & Tassoul, 1989, and Moss *et al.* , 1990) have recently proposed a simple, yet useful formulation that captures a lot of the essential physics involved, while keeping the problem computationally tractable. They consider a strictly axisymmetric configuration, and write the total velocity and magnetic field as

$$B(r,\theta,t) = B_p(r,\theta) + B_\phi(r,\theta,t)\, 1_\phi, \tag{19a}$$

$$U(r,\theta,t) = U_p(r,\theta) + U_\phi(r,\theta,t)\, 1_\phi, \tag{19b}$$

where B_p and U_p are *time-independent* poloidal magnetic field and meridional flow, both specified *a priori*. The problem then reduces to simultaneously solving, in 2-D, the ϕ-components of the induction and Navier-Stokes equations, which are now linear in U_ϕ and B_ϕ, a task well within the reach of current computational techniques and hardware.

A simple model problem, designed specifically to provide a mean of investigating the effects of magnetic field geometry and boundary conditions, has recently been studied by Charbonneau & MacGregor (1992; hereafter CM92). It consists in tracking the time-evolution of U_ϕ and B_ϕ in a meridional quadrant, from the initial state $U_\phi = 0$, $B_\phi = 0$ (*i.e.*, purely poloidal field in a non-rotating envelope), after an impulsive spin-up of the (rigidly rotating) core. Among the conclusions reached by CM92, note

(1) Whether or not SB rotation is enforced by a large-scale (axisymmetric) magnetic field depends *critically* on the behavior of the field at the boundaries (*i.e.*, anchoring on a rigidly rotating core or convection zone, etc.)
(2) There exists classes of configuration never reach a state of SB rotation. In configurations that do attain SB rotation, the timescale τ_{SB} on which they do so is in general much larger than the core-surface Alfvén transit time, and depends sensitively on field geometry/anchoring.
(3) In low viscosity/resistivity configurations, the damping of toroidal oscillations occurs mostly via *phase mixing*; writing $\tau \simeq R^2/\eta$, R^2/ν greatly overestimates damping timescales.

The dependence of τ_{SB} on field geometry and anchoring has been discussed extensively by CM92 (§3.4). We focus in the remainder of this section on the mechanism of *phase mixing*, a nice illustration of which is provided by our simulations.

4.3 Phase Mixing

It has often been argued (*e.g.*, Roxburgh 1991) that even in systems exhibiting vanishingly small viscosity and electric resistivity —which is the case for stellar interiors if microscopic values are adopted for ν and η—, the mechanism of *phase mixing* (Parker, 1991, and references therein) would lead to enhanced damping of toroidal oscillations in the present context. However, there is currently no quantitative study to be found in the literature of its effects in the context of torsional Alfvén wave damping in stellar radiative interiors.

Consider the fully core-anchored dipolar field shown on Figure 4. Following the impulsive spin-up of the core, Alfvén waves will propagate outward from the core, along each fieldline. Waves traveling along fieldlines threading the shaded portion of the envelope will reach the equatorial plane where, because of symmetry constraints, they will be effectively "reflected" back to higher latitudes. The "path length" between the core and equatorial plane is not the same along fieldlines "A" and "B", although quite similar. But this small path difference can, over time, lead to a large phase difference between waves traveling along A and B, thus leading to large *radial* gradients at a given latitude. Even though the viscosity and resistivity coefficients are quite small, they remain

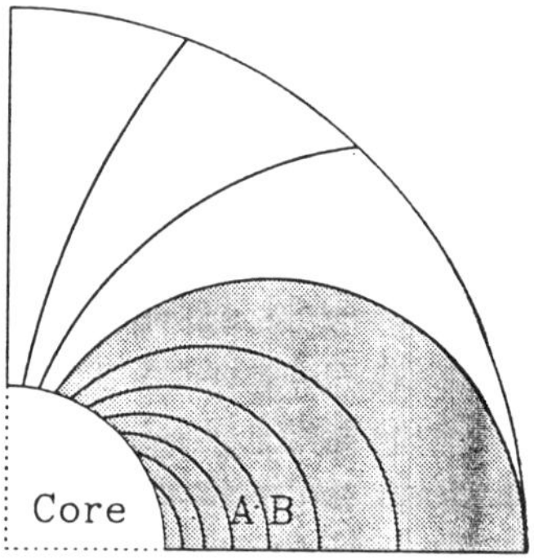

Figure 4. Fully core-anchored dipolar field. The shaded area indicates the portion of the envelope threaded by fieldlines crossing the equatorial plane, where reflections lead to enhanced dissipation by phase mixing (see text).

efficient over such short distances, leading to dissipation of the waves. This is phase mixing.

This is illustrated on Figure 5. The first two rows show, at various instants, contours of azimuthal velocity (U_ϕ) and toroidal field (B_ϕ). The bottom row shows the magnitudes of U_ϕ (dash-dotted line) and B_ϕ (solid line), along a radial "cut" 15 degrees above the equatorial plane. Note the high wavenumber radial oscillations in U_ϕ and B_ϕ close to the core at early epochs, and how rapidly they damp out. Going back to Figure 4 and to the 2-D profiles (top and middle row), it is immediately apparent that these oscillations occur *across* neighbouring fieldlines, whereas *along* fieldlines gradients are much smaller. Within a relatively short time span, the equatorial regions have, at large depths, come to a state of SB rotation (and zero toroidal field), despite the fact that global damping timescales are here of order 10^4 dimensionless time units, and that large amplitude oscillations pervaded that region at early epochs. Clearly, phase mixing will have to be reckoned with when attempting more realistic modeling.

5. FUTURE DIRECTION

This review may appear somewhat pessimistic in its assessment of our current understanding of the various facets of rotational evolution modeling. To end on a more positive note, this section identifies some specific aspects of current observations and models where significant advances appear possible —to this author at least— within the next few years.

5.1 Observations

It has become customary for theoreticians and numerical modelers to end such review talks by arguing that more observations are needed... and I unfortunately find myself in the position of having to follow the party line on this.

The eventual discovery of a rich cluster of age intermediate between the Pleiades and Hyades or younger than α Persei notwithstanding, perhaps the most interesting —from the modeling standpoint— "task" that should be

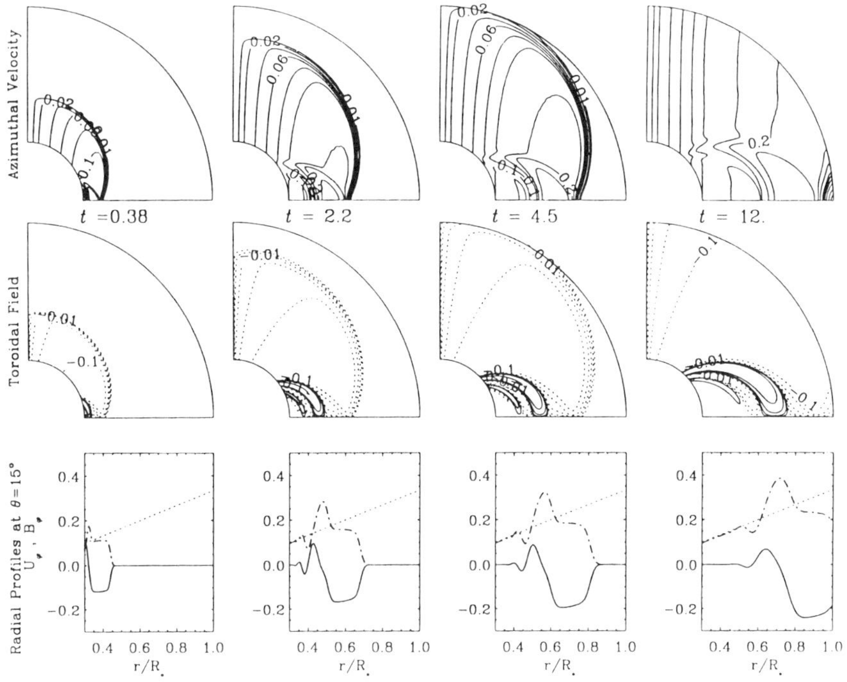

Figure 5. Illustration of the mechanism of phase mixing, for the magnetic configuration of Fig. 6. Time increases from left to right. The top row shows contours of constant azimuthal velocity (U_ϕ), the middle row contours of constant toroidal field (B_ϕ), and the bottom row radial cuts of U_ϕ (dash-dotted line) and B_ϕ (solid line) 15° above the equatorial plane. The dotted line indicates the U_ϕ profile corresponding to SB rotation.

addressed by observers is the determination of the "bottom" of the rotational distributions in the α Persei, Pleiades, and Hyades cluster. To trace the rotational evolution of the large population of slow rotators between the ages of α Persei and the Hyades is an extremely powerful test of both the braking law and the timescale(s) for internal redistribution of angular momentum. Photometric studies, yielding rotational periods as opposed to $v \cdot \sin i$, would be most appropriate in that respect.

Another important issue that can, at present, only be investigated observationally, is that of the initial angular momentum distribution. Most pre-main sequence $v \cdot \sin i$ determinations currently available are from the Taurus-Aurigae complex. Are the corresponding distributions typical of star forming regions at large ? In particular, are they typical of the denser star-forming regions that give birth to clusters?

5.2 Magnetic Braking Models

It should be clear from §2.5 that parametric braking laws with fixed exponent provide poor approximations over a large range of rotation rate. The angular momentum loss computed within the framework of the WD wind model should be preferred. The WD wind model is itself in need of improvement, particularly with regards to magnetic field geometry and consideration of meridional force balance. For axisymmetric configurations, this is within the reach of current computational techniques and hardware.

But even if this program is carried out, there remains the issue of the dependence of the surface field (in terms of both strength and geometry) on rotation rate, which enters the picture as soon as the WD model is applied to the problem of rotational braking. The B–Ω relation is in the realm of dynamo theory, a most unsatisfying state of affairs in the absence of any self-consistent such theory applicable to stars (*e.g.* Levy, these proceedings), and with none in sight for the near future. An attractive alternative is the ***observational*** determination of an empirical B—Ω relation. The work already under way in that direction (*e.g.*, Linsky & Saar, 1987; Saar, 1990; Stauffer *et al.* , 1991, and references therein) should be energetically pursued.

5.3 Rotational Evolution Models

The Yale models appear more successful in dealing with halo stars than young stars. This may well be a consequence of the braking law used, which, as discussed in §2.4 above, overestimates the angular momentum loss rate at high rotation. It would thus be interesting to implement some form of WD model in the Yale code. Allowance for internal angular momentum redistribution within the framework of the MacGregor models is also required.

In circumstances where rotationally-induced meridional circulation is likely to be important (*e.g.*, early-type stars, or rapidly rotating core in late-type stars), then a 2-D treatment of transport by circulation appears mandatory. This introduces considerable practical difficulties from the modeling standpoint, but is probably essential if the modeling of CNO isotopic ratios on the subgiant and giant branch is to be carried out with an adequate degree of realism.

Because of the inherent complexities and uncertainties involved, the effects of internal magnetic fields on the redistribution of angular momentum in stellar interiors are generally ignored, except by those who hold the opposite view that magnetic fields *always* enforce solid or very near-solid body rotation everywhere in the radiative interiors of stars. As is usually the case, "reality" probably lies somewhere in between. As a first try at this challenging problem, solar spin-down calculations, in the spirit of the Charbonneau & MacGregor (1992) spin-up calculations, should be undertaken.

ACKNOWLEDGEMENTS

It is a pleasure to thank Keith MacGregor for a number of stimulating discussion and suggestions, and for letting me play with his MHD wind code. This work was supported in part by the Natural Sciences and Engineering Research Council of Canada, in the form of a postdoctoral fellowship to the author.

REFERENCES

Acuna, M., & Whang, Y.C. 1976, *Ap. J.*, **203**, 720.
Belcher, J.W., & MacGregor, K.B. 1976, *Ap. J.*, **210**, 498.
Bouvier, J. 1991, in Angular Momentum Evolution of Young Stars, eds. S. Catalano & J.R. Stauffer (Dordrecht: Kluwer), p. 117.
Brandt, J.C., Wolff, C., & Cassinelli, J.P. 1969, *Ap. J.*, **156**, 1117.
Butler, R.P., Cohen, R.D., Duncan, D.K., & Marcy, G.W. 1987, *Ap. J.*, **319**, L19.
Catala, C. 1990, in Angular Momentum and Mass Loss for Hot Stars, eds. L.A. Willson & R. Stalio (Dordrecht: Kluwer), p. 65.
Chaboyer, B., & Zahn, J.-P. 1991, submitted to *Astr. Ap.*
Charbonneau, P. 1991, submitted to *Astr. Ap.*
Charbonneau, P., & MacGregor, K.B. 1992, *Ap. J.*, **387**, in press.
Charbonneau, P., & Michaud, G. 1991, *Ap. J.*, **370**, 693.
Collier Cameron, A., Jianke, L., & Mestel, L. 1991, in Angular Momentum Evolution of Young Stars, eds. S. Catalano & J.R. Stauffer (Dordrecht: Kluwer), p. 297.
Demarque, P. 1991, in Evolution of Stars : The Photospheric Abundance Connection, eds. G. Michaud & A. Tutukov (Dordrecht: Kluwer), p. 71.
de Medeiros, J.R., & Mayor, M. 1991, in Angular Momentum Evolution of Young Stars, eds. S. Catalano & J.R. Stauffer (Dordrecht: Kluwer), p. 201.
Durney, B.R. 1972, in Solar Wind, eds. C.P. Sonett, P.J. Coleman, & L.M. Wilcox (Washington: NASA Scientific and Technical Information Office), p. 282.
Endal, A.S., & Sofia, S. 1976, *Ap. J.*, **210**, 184.
Endal, A.S., & Sofia, S. 1978, *Ap. J.*, **220**, 279.
Goode, P.R., Dziembowski, W.A. 1991, *Nature*, **349**, 223.
Goode, P.R., Dziembowski, W.A., Korzennik, S.G., & Rhodes, E.J. Jr. 1991, *Ap. J.*, **367**, 649.
Gough, D.O. 1991, in Angular Momentum Evolution of Young Stars, eds. S. Catalano & J.R. Stauffer (Dordrecht: Kluwer), p. 271.
Gray, D.F. 1989, *Ap. J.*, **347**, 1021.
Gray, D.F. 1991, in Angular Momentum Evolution of Young Stars, eds. S. Catalano & J.R. Stauffer (Dordrecht: Kluwer), p. 183.
Gray, D.F., & Nagar, P. 1985, *Ap. J.*, **298**, 756.
Hartmann, L., Hewett, R., Stahler, S., & Mathieu, R.D. 1986, *Ap. J.*, **309**, 275.
Kawaler, S.D. 1988, *Ap. J.*, **333**, 236.
Kawaler, S.D. 1989, *Ap. J.*, **343**, L65.
Kraft, R.P. 1967, *Ap. J.*, **150**, 551.
Kraft, R.P. 1970, in Spectroscopic Astrophysics, ed. G.H. Herbig (Berkeley: University of California Press), p. 385.
Linsky, J.L., & Saar, S.H. 1987, in Cool Stars, Stellar Systems, and the Sun, eds. J.L. Linsky & R.E. Stencel (Berlin: Springer), p. 44.
MacGregor, K.B. 1991, in Angular Momentum Evolution of Young Stars, eds. S. Catalano & J.R. Stauffer (Dordrecht: Kluwer), p. 315.
MacGregor, K.B., & Brenner, M. 1991, *Ap. J.*, **376**, 204.
Mayor, M., & Mermilliod, J.-C. 1991, in Angular Momentum Evolution of Young Stars, eds. S. Catalano & J.R. Stauffer (Dordrecht: Kluwer), p. 143.
Mestel, L. 1968, *M. N. R. A. S.*, **138**, 359.
Mestel, L. 1984, in Cool Stars, Stellar Systems, and the Sun, eds. S.L. Baliunas & L. Hartmann (Berlin: Springer), p. 49.
Mestel, L., Moss, D.L., & Tayler, R.J. 1987, *M. N. R. A. S.*, **231**, 873.
Mestel, L., & Spruit, H.C. 1987, *M. N. R. A. S.*, **226**, 57.
Mestel, L., & Weiss, N.O. 1987, *M. N. R. A. S.*, **226**, 123.
Moss, D.L., Mestel, L., & Tayler, R.J. 1990, *M. N. R. A. S.*, **245**, 550.

Nerney, S.F., & Suess, S.T. 1975, *Ap. J.*, **196**, 837.
Parker, E.N. 1958, *Ap. J.*, **128**, 664.
Parker, E.N. 1991, *Ap. J.*, **376**, 355.
Peterson, R.C. 1983, *Ap. J.*, **275**, 737.
Peterson, R.C. 1985a, *Ap. J.*, **289**, 320.
Peterson, R.C. 1985b, *Ap. J.*, **294**, L35.
Pinsonneault, M.H., Deliyannis, C.P., & Demarque, P. 1991a, *Ap. J.*, **367**, 239.
Pinsonneault, M.H., Deliyannis, C.P., & Demarque, P. 1991b, *Ap. J. Suppl.*, **78**, in press.
Pinsonneault, M.H., Kawaler, S.D., & Demarque, P. 1990, *Ap. J. Suppl.*, **74**, 501.
Pinsonneault, M.H., Kawaler, S.D., Sofia, S., & Demarque, P. 1989, *Ap. J.*, **338**, 424.
Pizzo, V., Schwenn, R., Marsch, E., Rosenbauer, H., Mühlhäuser, K.-H., & Neubauer, F.M. 1983, *Ap. J.*, **271**, 335.
Prosser, 1991 *PhD Thesis*, University of California at Santa Cruz.
Radick, R.R., Thompson, D.T., Lockwood, G.W., Duncan, D.K., & Baggett, W.E. 1987, *Ap. J.*, **321**, 459.
Rosner, R., & Weiss, N.O. 1985, *Nature*, **317**, 790.
Roxburgh, I.W. 1991, in Angular Momentum Evolution of Young Stars, eds. S. Catalano & J.R. Stauffer (Dordrecht: Kluwer), p. 365.
Saar, S.H. 1990, in IAU Symp. 138, Solar Photosphere: Structure, Convection and Magnetic Fields, ed. J.O. Stenflo (Dordrecht: Kluwer), p. 427.
Sakurai, T. 1985, *M. N. R. A. S.*, **152**, 121.
Schatzman, E. 1962, *Ann. Astrophys.*, **25**, 18.
Schatzman, E. 1990, in Rotation and Mixing in Stellar Interiors, eds. M.-J. Goupil & J.-P. Zahn (Berlin: Springer), p. 3.
Sion, E.M. 1990, in Angular Momentum and Mass Loss for Hot Stars, eds. L.A. Willson & R. Stalio (Dordrecht: Kluwer), p. 307.
Skumanich, A. 1972, *Ap. J.*, **171**, 565.
Skumanich, A., & MacGregor, K.B. 1986, *Adv. Space Res.*, **6**, 151.
Soderblom, D.R. 1983, *Ap. J. Suppl.*, **53**, 1.
Soderblom, D.R. 1991, in Angular Momentum Evolution of Young Stars, eds. S. Catalano & J.R. Stauffer (Dordrecht: Kluwer), p. 151.
Spruit, H.C. 1990, in Inside the Sun, eds. G. Berthomieu & M. Cribier (Dordrecht: Kluwer), p. 415.
Stauffer, J.R. 1991, in Angular Momentum Evolution of Young Stars, eds. S. Catalano & J.R. Stauffer (Dordrecht: Kluwer), p. 117.
Stauffer, J.R., Giampapa, M.S., Herbst, W., Vincent, J.M., Hartmann, L.W., & Stern, R.A. 1991, *Ap. J.*, **374**, 142.
Stauffer, J.R., & Hartmann, L.W. 1987, *Ap. J.*, **318**, 337.
Stauffer, J.R., Hartmann, L.W., & Jones, B.F. 1989, *Ap. J.*, **346**, 160.
Stauffer, J.R., Hartmann, L.W., & Latham, D.W. 1987, *Ap. J.*, **320**, L51.
Stauffer, J.R., Hartmann, L.W., Soderblom, D.R., & Burnham, N. 1984, *Ap. J.*, **280**, 202.
Stepień, K. 1988, *Ap. J.*, **335**, 907.
Struve, O. 1930, *Ap. J.*, **72**, 1.
Suess, S.T., & Nerney, S.F. 1973, *Ap. J.*, **184**, 17.
Tassoul, J.-L. 1990, in Angular Momentum and Mass Loss for Hot Stars, eds. L.A. Willson & R. Stalio (Dordrecht: Kluwer), p. 7.
Tassoul, M., & Tassoul, J.-L. 1989, *Ap. J.*, **345**, 472.
Urch, I.H. 1969, *Solar Phys.*, **10**, 219.
Van Leeuwen, F., & Alphenaar, P. 1982, *ESO Messenger*, **28**, 15.
Vogel, S.N., & Kuhi, L.V. 1981, *Ap. J.*, **245**, 960.

Vilhu, O. 1984, *Astr. Ap.* **133**, 117.
von Zeipel, H. 1924, *M. N. R. A. S.*, **84**, 665.
Weber, E.J. 1969, *Solar Phys.*, **9**, 150.
Weber, E.J., & Davis, L. Jr. 1967, *Ap. J.*, **148**, 217.
Wolff, C.L., Brandt, J.C., & Southwick, R.G. 1971, *Ap. J.*, **165**, 181.
Wolff, S.L., Edwards, S., & Preston, G.W. 1982, *Ap. J.*, **252**, 322.
Yang, W.-H., & Schunk, R.W. 1991, *Ap. J.*, **372**, 703.
Yeh, T. 1976, *Ap. J.*, **206**, 768.
Zahn, J.-P. 1974, in Stellar Instability and Evolution, eds. P. Ledoux, A. Noels, & A.W. Rodgers (Dordrecht: Reidel), p. 185.

Seventh Cambridge Workshop on Cool Stars, Stellar Systems, and the Sun
ASP Conference Series, Vol. 26, 1992
Mark S. Giampapa and Jay A. Bookbinder (eds.)

Gl 890 - THE MAGNETIC BRAKE?

P.B. BYRNE, J.G. DOYLE and M. MATHIOUDAKIS
Armagh Observatory, Armagh BT61 9DG, N. Ireland

ABSTRACT Very high resolution Hα spectra of the 10.35 hr period BY Dra/flare star, FK Aqr (= Gl 890) provide evidence of AB Dor-like circumstellar material.

Keywords: Late-type stars; rotational modulation; circumstellar matter

1. INTRODUCTION

Cameron and Robinson (1986, 1989a,b) discovered a rapidly rotating, K0 star, AB Dor, undergoing mass-loss in the presence of closed loops. They suggest that mass-loss occurs in condensations at the apexes of coronal loops near the corotation radius. The cool material decouples from the magnetic field and removes angular momentum, providing a "magnetic brake".

Young *et al.* (1990; Y90) have discovered a rapidly rotating (P = 10.35 hr; $v_{eq} \sin i$ = 70 km/sec) M2Ve star, HK Aqr (= Gl 890). They saw shifts in the centroid of the Hα emission line over a small range of phase, with a correlated reduction in the line flux, which were attributed to non-uniform distribution of Hα-emitting plages. Byrne and McKay (1990; BMcK90) suggested that, due to HK Aqr's large rotational velocity, a partial chromospheric eclipse would cause the same effect. A coincident drop in Mg II *h* and *k* flux supported this interpretation. Doyle and Cameron (1990) proposed AB Dor-like coronal condensations as the eclipsing material.

In this paper we describe two observing campaigns on HK Aqr which throw further light on this phenomenon.

2. OBSERVATIONS

We have obtained two datasets, spaced by about $2\frac{1}{2}$ weeks. The first comprises simultaneous low-resolution (R ~ 10 000) blue (~ 3600 - 4200 Å) spectra (exposure 5 min) taken with the South African Astronomical Observatory's (SAAO) 1.9 m telescope, and UBV(RI)$_{KC}$ photometry (20 sec/cycle), taken with SAAO's 1.0 m telescope. The second comprises 16 Hα spectra (R ~ 50 000; exposure 5 min), taken using the Anglo-Australian 3.9 m telescope (AAT).

Fig. 1 is the V band lightcurve for HK Aqr excluding flares. There is a small amplitude spot modulation (ΔV ~ 0.07) with minimum at φ ~ 0.6. Fig. 2 shows the variation of Ca II K with phase on the night with best phase coverage. We note the following.

[1.] There is a weak modulation in the sense that the flux near φ ~ 0.6 (V light minimum) is higher than at φ = 0.0 – 0.4.

[2.] There is no drop in Ca K flux like that observed in Mg II h&k in 1984.
[3.] There is evidence for short-lived flaring on a time scale $\sim$ 20 min (filled point).

Fig. 3 shows the Hα spectra. The top profiles show the mean of the first 3 spectra, contrasted with the final spectrum of the series. Note the dramatic change in the blue wing and line centre while the red wing hardly changes. Fig. 3 also shows differences between the mean of the first 3 spectra and the individual spectra. We note the following.

[1.] The profile is essentially constant over the first 4 spectra.
[2.] Then the line develops a small emission feature to the blue of line centre and, possibly, a very weak absorption at line centre.
[3.] In the last 5 spectra, the absorption at line centre deepens and develops an extended blue wing, and weak emission is seen to the red of line centre.

3. DISCUSSION

From Fig. 2 we conclude that there is no equivalent in CaII K to the 40% Mg II dip in flux seen by BMcK90 and by Y90. The same conclusion is reached by considering the Ca II H/Hϵ and Hδ. Therefore it is improbable that the absorption is due to feature which are constant in time.

The variation in Hα is, however, similar to that seen by Y90. They see the Hα centroid shift, first to the red, accompanied by decreasing line intensity, and then to the blue. The red shift occurs over $\Delta\varphi \sim 0.15$, with maximum at $\sim \varphi \sim 0.65$. Our data show a blue wing *absorption* over $\Delta\varphi \geq 0.035$ at $\varphi > 0.7$, as the cause of the red shift of the Hα centroid. The maximum depression of the Hα intensity is $\sim$40% of the excess *above the local continuum* compared with $\sim$40% in Y90. Thus we have confirmed the Hα phenomenon in Y90 after $\sim$5 yr. The decrease in Hα intensity is consistent with the 1984 results as is its occurrence just after maximum spottedness ($\Delta\varphi_{84} \sim 0.05$, $\Delta\varphi_{89} \sim 0.15$) is qualitatively similar.

Our spectra, however, show much more detail than those of Y90. The flux decrease and centroid velocity shift are shown to be due to an absorption at line centre with a blue wing out to $\sim$ 120 km s^{-1}, considerably more than the rotational $v \sin i$ for HK Aqr ($\sim$ 70 km s^{-1}).

We propose a cloud of neutral hydrogen, similar perhaps to a prominence, condensing from the stellar corona close to the corotation radius, and qualitatively similar those of AB Dor, as the cause. To account for the blue wing, however, we also require that there be a substantial upflow of the material.

ACKNOWLEDGEMENTS: Research at Armagh Observatory is supported by a grant-in-aid from the Department Of Education of Northern Ireland.

REFERENCES

Byrne, P.B., McKay, D.: 1990, *Astr. Ap.*, **227**, 490.
Cameron, A.C., Robinson, R.D.: 1989a, *M.N.R.A.S.*, **236**, 57.
Cameron, A.C., Robinson, R.D.: 1989b, *M.N.R.A.S.*, **238**, 657.
Doyle, J.G., Cameron, A.C.: 1990, *M.N.R.A.S.*, **244**, 291.
Stauffer, J.R., Hartmann, L.W.: 1986, *P.A.S.P.*, **98**, 1233.
van Leeuwen, F., Alphenaar, P.: 1982, ESO Messenger, No. 28, p.15.
Young, A., Skumanich, A., MacGregor, K.B., Temple, S.: 1990, *Ap. J.*, **349**, 608.

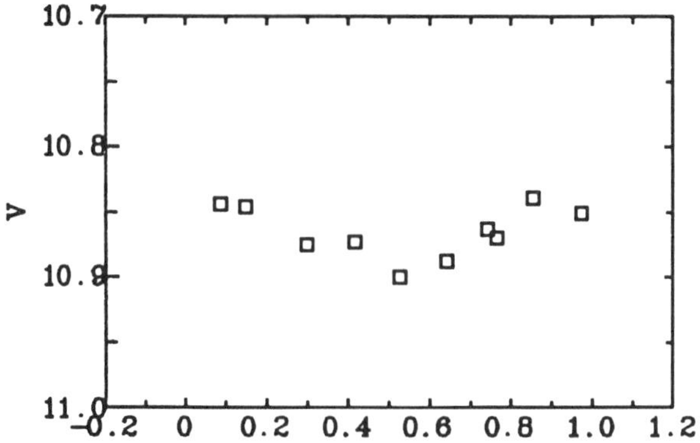

Fig. 1. The mean V light curve for HK Aqr in mid-October 1989.

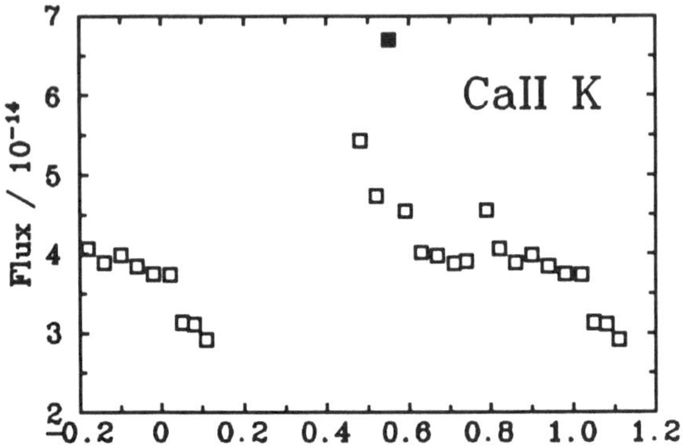

Fig. 2. 20 min averages of CaII K flux for 15 Oct 1989.

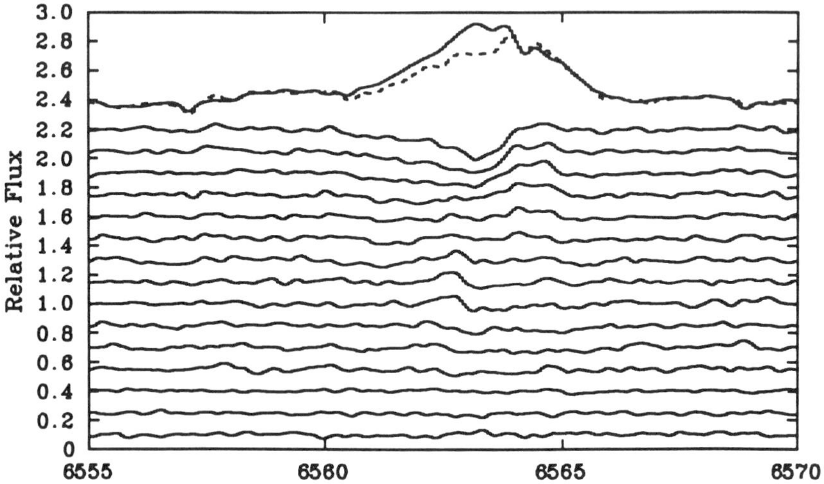

Fig. 3. The Hα line profile on 4 Oct 1989 over 80 min at $\varphi \sim 0.7$ (see text).

Seventh Cambridge Workshop on Cool Stars, Stellar Systems, and the Sun
ASP Conference Series, Vol. 26, 1992
Mark S. Giampapa and Jay A. Bookbinder (eds.)

SYNOPTIC STUDIES OF THE T TAURI STAR SU AUR

CHRISTOPHER M. JOHNS and GIBOR S. BASRI
University of California, Astronomy Department, Berkeley, CA 94709

MARK S. GIAMPAPA and ERIC DeFONSO
NOAO/NSO, 950 Cherry Av., POB 26732, Tucson, AZ 85726

ABSTRACT Eighty-three high resolution SU Aur Hα line profiles taken from Fall 1986 through 1989 are analyzed. The profiles are analysed for correlations among the profile parameters and for periodicity. Strong periodicity is found at 2.98 ± 0.4 days. A crude model is presented for the geometry of the emission region in terms of a wind emanating from a star surrounded by an accretion disk.

Keywords: T Tauri stars; Hα line profiles; stellar winds; accretion disks

INTRODUCTION

Hα in T Tauri stars (TTS) has been used as a diagnostic of circumstellar disks (for example see Basri and Bertout, 1989), and stellar winds (for example see Edwards *et al.*, 1987, and Hartmann *et al.*, 1990) These models have been very successful in accounting for the equivalent widths of the observed emission lines but the detailed line shapes do not account for all the observations.

TTS are known to be quite variable (for example see Mundt, 1984). SU Aur shows pronounced variation in its Hα profile and was selected for extensive study because of this. This should allow a determination of the constant properties of the line profile to be made and possibly interpreted in the context of one of the above models. It is also hoped that careful analysis of the variations in the line profile will help further constrain the nature of the Hα emitting region.

THE Hα PROFILES AND PERIOD ANALYSIS

The profiles were taken at Lick Observatory and Kitt Peak Observatory. Each profile was parameterized by fitting the profile to a function consisting of a flat-topped feature with from zero to five gaussian components superimposed on it. A gradient search algorithm was developed to optimize the fit. The profiles contain a clear flat top component which is slightly asymmetric with the blue side higher than the red with a smooth transition through line center on average. Eighty-one of the eighty-three profiles has a prominent absorption component, which often reaches below the stellar continuum, at about $v = -40$ km/s. Seventy-three of the profiles contain a central emission feature centered just red of line center with an average position of $v = 20$ km/s. The line profile often has an absorption feature at roughly $v = -150$ km/s and occasionally another absorption feature at roughly $v = 130$ km/s. The average observed Hα profile is shown in Figure 1. The absorption at $v = -40$ km/s and the emission at $v = 20$ km/s are clearly visible and the absorption feature at $v = -150$ km/s can barely be seen.

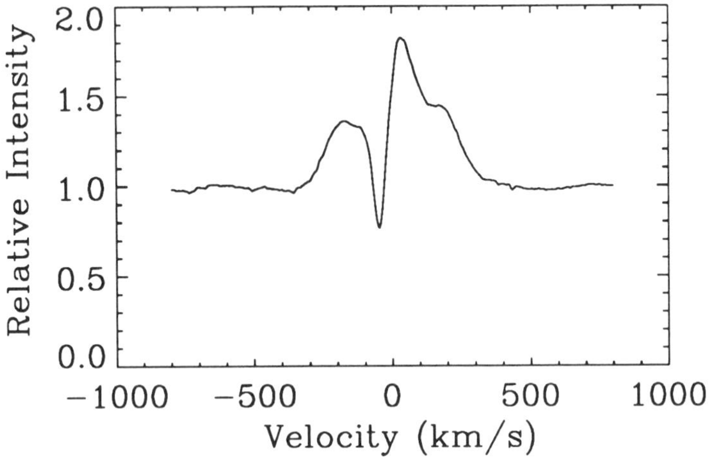

Fig. 1: The average SU Aur Hα profile.

With all the profiles parameterized as mentioned above, correlations were looked for among the profile parameters. The only correlation found is between the strength of the absorption component at $v = -40$ km/s and the strength of the emission component at $v = 20$ km/s. They correlate positively. The main absorption component does not appear to correlate in any way with the level of the flat top component.

Since the period of SU Aur is probably on the order of a few days, the period analysis was concentrated on stretches of data which have daily or better observations for several days. The best stretch of data for this purpose is shown in Figure 2 in a three dimensional surface plot. This figure clearly shows how variable the Hα profile is. Looking carefully at the blue side of the profile, it appears that this part of the profile varies periodically.

Periodogram analysis of the data confirm a period of 2.98 ± 0.4 days in the blue side of the profile caused by variations in the $v = -150$ km/s absorption feature. This is taken to represent rotational modulation of mass outflow from near the stellar surface. The rotation period of SU Aur is then 2.98 ± 0.4 days.

A GEOMETRIC MODEL OF THE Hα EMISSION REGION

From an analysis of the profiles, it is apparent that the majority of the Hα *emission* arises from the flat top feature. Optically thin, spherically symmetric stellar winds can easily produce broad flat emission features (Mihalas, 1978). Therefore, a simple model for computing line profile shapes for such winds was developed. Since SU Aur has a substantial IR and UV excess which indicates an active accretion disk is present (Bertout,Basri,and Bouvier, 1988), models were computed which have a circumstellar disk which operates as an opaque screen blocking the receding flow. The volume is assumed to be quite optically thin so that each photon produced escapes.

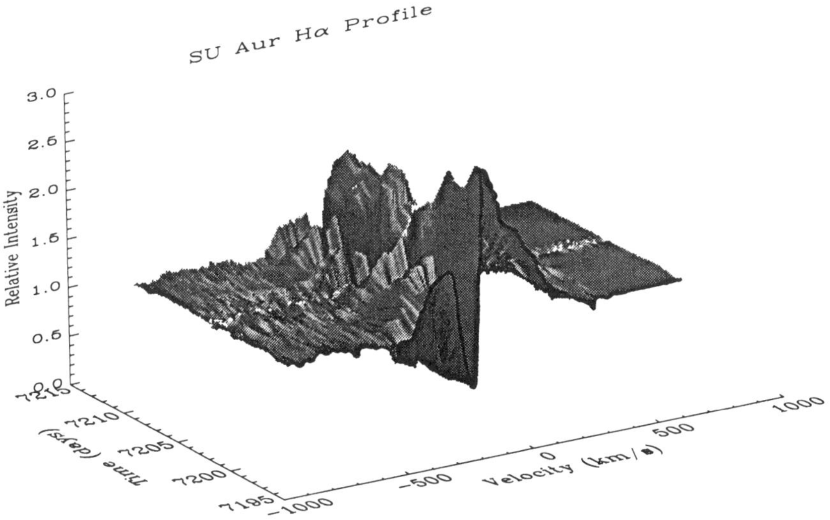

Fig. 2: The SU Aur Hα profile for the 24 observations from Julian Date 2447199.66 to 2447213.68. For clarity, the first profile in this sequence has been highlighted.

The flat top component (profile minus a single gaussian absorption feature) of the least contaminated SU Aur Hα line profile is shown in Figure 3 along with the wind model fit. The wind emission in this model begins at $r = 5R_*$ and the model calculation is stopped at $r = 10R_*$. The velocity in the wind follows a $r^{-1/2}$ law. Assuming that hydrogen is collisionally excited in such a wind, the source function was set equal to density squared. The model wind is assumed to have a constant mass loss rate with radius so the source function obeys a r^{-3} law. The asymmetry in the model is caused by the occulting disk. For an inclination of zero, the disk would block the entire redshifted half of the flow and the resulting profile would have only blue shifted emission. As the inclination increases, more and more redshifted material is on the observers side of the disk. The model shown has $i = 73°$. The results are suggestive that the Hα line emission arises in an fast moving optically thin region of the outflow which is separated from the star. Since the wind is postulated to be very optically thin, the stellar absorption line should be apparent. The line profile would just be a G2 (solar) Hα line spun up to v sin$i$ = 68.5 km/s with a flat top feature added on making the level of the absorption feature higher than normal but still apparent. The wind can not be moving at escape velocity as soon as it leaves the star so there must be an acceleration region. The continuity equation then dictates there will be a high density (possibly enough to hide the stellar line), slow moving region of the wind near the star which may be much cooler than the outer optically thin wind. The main absorption feature may form in this part of the wind and be seen through the optically thin, fast part of the wind.

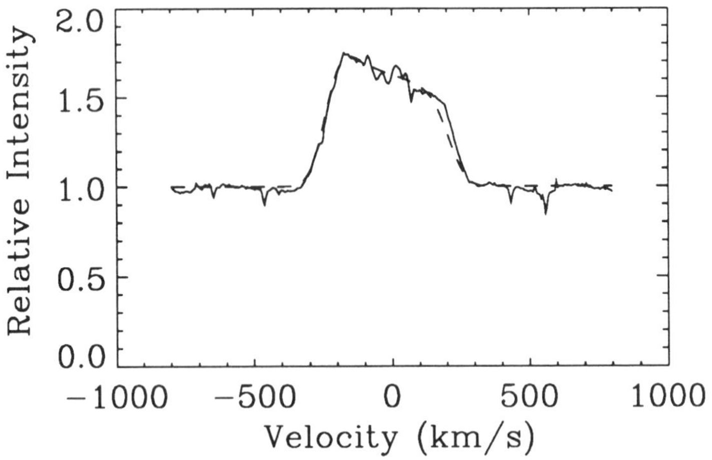

Fig. 3: SU Aur Hα line profile (solid line) and model fit (dotted line).

CONCLUSION

Analysis of the Hα profile seen in SU Aur indicates that the star rotates with a period of 2.98 ± 0.4 days. A model of the emission from SU Aur indicates that most of emission arises from a fast moving, hot, optically thin region a few stellar radii from the star. The lower velocity portion of the profile may arise from a low velocity, cool, dense, optically thick region inside the fast wind. Whatever accelerates the wind may also heat it.

ACKNOWLEDGEMENTS

This research was supported in part by NSF grant AST89-11596. C.M. Johns was supported in part by a NSF Graduate Student Fellowship. Special thanks is extended to Mr. Paul Avellar, the nighttime assistant on the McMath Telescope who aided in obtaining the Kitt Peak spectra.

REFERENCES

Basri, G., and Bertout, C. 1989, *Ap. J.*, **341**, 340.
Bertout, C., Basri, G., and Bouvier, J. 1988, *Ap. J.*, **330**, 350.
Edwards, S., Cabrit, S., Strom, S.E., Heyer, I., Strom, K., and Anderson, E. 1987, *Ap. J.*, **321**, 473.
Hartmann, L., Calvet, N., Avrett, E., and Loeser, R. 1990, *Ap. J.*, **349**, 168.
Mihalas, D. 1978, *Stellar Atmospheres*, W.H. Freeman and Co.
Mundt, R. 1984, *Ap. J.*, **280**, 749.

Seventh Cambridge Workshop on Cool Stars, Stellar Systems, and the Sun
ASP Conference Series, Vol. 26, 1992
Mark S. Giampapa and Jay A. Bookbinder (eds.)

T TAURI LINE PROFILES: SIGNATURES OF WINDS OR INFALL?

NURIA CALVET Centro de Investigaciones de Astronomia, Ap. P. 264, Mérida, Venezuela

LEE HARTMANN Harvard-Smithsonian Center for Astrophysics, 60 Garden St., Cambridge, MA 02138

ABSTRACT We present evidence that T Tauri Balmer line profiles arise from infalling material in a magnetosphere around the star.

Keywords: T Tauri stars; line profiles; magnetospheres

INTRODUCTION AND DESCRIPTION

The broad emission line profiles in the spectra of T Tauri stars (TTS) have been traditionally interpreted as arising from an expanding envelope around the central object. However, line profiles calculated with hydrodynamic wind solutions do not reproduce observations very well. Both stellar wind and disk wind models produce Hα line profiles with redshifted emission peaks (Hartmann *et al.* 1990; Calvet, Hartmann, and Hewett 1991), while most TTS stars show nearly centrally-peaked emission. Grinin and Mitskevich (1991) show that wind models with strongly decelerating flow can produce centrally-peaked emission line profiles, but constructing hydrodynamic wind models with very small asymptotic wind velocities is difficult (Holzer, Fla, and Leer 1983).

If the stellar magnetic field has a strength of the order of kG (c.f. Basri, Marcy, and Valenti, this conference), the disk around the star will be disrupted at a radius given approximately by

$$r_a \approx 1.5 R_* \left(\frac{B_0}{1kG}\right)^{2/7} \left(\frac{R_*}{2R_\odot}\right)^{5/7} \left(\frac{M_*}{1M_\odot}\right)^{-2/7} \left(\frac{\dot{M}_a}{10^{-7} M_\odot \; yr^{-1}}\right)^{-2/7}$$

where M_* and R_* are the stellar mass and radius, B_0 is the stellar magnetic field, and $\dot{M}_a$ is the mass accretion rate in the disk (Ghosh and Lamb 1977a,b; Königl 1991). In the region $r_a < r < r_a + \delta$, $\delta << r_a$, the flow of matter will be controlled by magnetic stresses; the stellar magnetic field channels disk accretion into infall through magnetic columns, producing an accretion shock at the base of the columns (Bertout, Basri, and Beauvier 1988). In this work we consider whether matter infalling in magnetic columns onto the star provide a better explanation of many T Tauri emission line profiles.

We have calculated line profiles for the Balmer lines for infalling envelopes that crudely simulate a stellar magnetosphere. We assume ballistic steady radial infall from r_{max} in restricted latitude bands, with velocity $v_p = [2GM_*/R_*(R_*/r - R_*/r_{max})]^{1/2}$. We have taken $r_{max} = 3R_*$, so that $v_p(R_*) = 290$ km s^{-1} . The envelope temperature has been taken equal to $7000K$, while the base temperature has been calculated assuming that all the kinetic energy is converted into blackbody emission. The source function has been estimated with an explicit calculation of the escape probabilities.

In Figure 1, we compare Hα line profiles calculated with spherical winds (Hartmann *et al.* 1990), non-spherical boundary layer (or inner disk) winds (Calvet, Hartmann, and Hewett 1991), and infall models, together with observed profiles for AA Tau and GG Tau. In contrast to wind models, infall in accretion columns can produce *nearly symmetric profiles*, with emission wings similar to those observed. A redshifted absorption component is present in Hα in the low density envelope. The strength of this absorption increases proceeding up the Balmer series, as observed in the YY Ori stars (Walker 1972; Appenzeller and

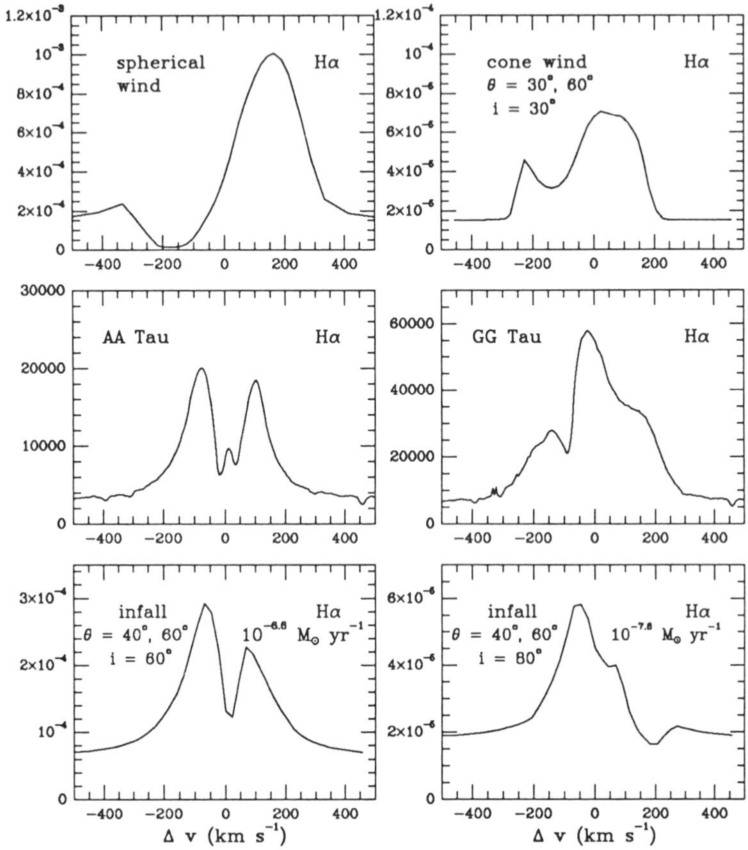

Fig.1: Line profiles for Hα for wind models (upper panel), AA Tau and GG Tau (middle panel), and infall models (lower panel). Opening angle, inclination, and mass infall rate are indicated. Infall models provide a much better fit to observations.

Wolf 1977; Bertout *et al.* 1977; Wolf *et al.* 1977; Appenzeller *et al.* 1978; Edwards 1979; Mundt 1979). We find that at high inclination and polar accretion columns, the redshifted absorption tends to disappear. Thus, if accretion is confined to thin columns (or sheets), redshifted absorption would be detected only when the line of sight intercepted the column. Magnetospheres could then be present in *all* TTS, even when redshifted absorption is not observed.

CONCLUSIONS

(1) Magnetospheric infall models produce Balmer line profiles in good agreement with observations of many T Tauri stars. The models naturally explain why redshifted absorption is rarely seen in Hα but observed in higher Balmer lines. This picture provides a simple explanation for the observed correlations between emission line strengths and disk accretion rates (Cabrit *et al.* 1990; Hartigan *et al.* 1990).
(2) The emission profiles of permitted lines in T Tauri stars may generally arise in infalling envelopes, while the high-velocity blueshifted components arise in a wind. Mass loss rates determined from emission line strengths (Natta *et al.* 1988; Hartmann *et al.* 1990) may be in error.
(3) If indeed magnetospheres are an ubiquitous phenomenon in TTS, then ultraviolet continuum emission, previously thought to be produced in the boundary layer between disk and star, would come from the accretion shock at the base of the accretion column. Disk models of TTS should be revised.
(4) If the interaction between the magnetic field and the disk in TTS can be described by the Gosh and Lamb model, then the braking of the central star can be produced by field lines threading the disk outside the magnetosphere, and the angular momentum problem could be solved (Königl 1991).

REFERENCES

Appenzeller, I., Mundt, R., and Wolf, B. 1978, *Astr. Ap.*, **63**, 289.
Appenzeller, I., and Wolf, B. 1977, *Astr. Ap.*, **54**, 713.
Bertout, C., Basri, G., and Bouvier, J. 1988, *Ap. J.*, **330**, 350.
Bertout, C., Krautter, J., Möllenhoff, C., and Wolf, B. 1977, *Astr. Ap.*, **61**, 737.
Cabrit, S., Edwards, S., Strom, S.E., and Strom, K.M. 1990, *Ap. J.*, **354**, 687.
Calvet, N., Hartmann, L., and Hewett, R. 1991, *Ap. J.*, (in press).
Edwards, S. 1979, *Pub. A.S.P.*, **91**, 329.
Ghosh, P., and Lamb, F.K. 1979a, *Ap. J.*, **232**, 259.
Ghosh, P., and Lamb, F.K. 1979b, *Ap. J.*, **234**, 296.
Grinin, V. P., and Mitskevich, A. S. 1991, preprint.
Hartmann, L., Calvet, N., Avrett, E.H., and Loeser, R. 1990, *Ap. J.*, **349**, 168.
Holzer, T.E., Fla, T., and Leer, E. 1983, *Ap. J.*, **275**, 808.
Königl, A. 1991, *Ap. J. Letters*, **370**, L39.
Mundt, R. 1979, *Astr. Ap.*, **74**, 21.
Natta, A., Giovanardi, C., and Palla, F. 1988, *Ap. J.*, **332**, 921.
Walker, M.F. 1972, *Ap. J.*, **175**, 89.
Wolf, B., Appenzeller, I., and Bertout, C. 1977, *Astr. Ap.*, **58**, 163.

Seventh Cambridge Workshop on Cool Stars, Stellar Systems, and the Sun
ASP Conference Series, Vol. 26, 1992
Mark S. Giampapa and Jay A. Bookbinder (eds.)

MASS ACCRETION AND THE INTERNAL STRUCTURE OF T-TAURI STARS

CHARLES R. PROFFITT
Space Telescope Science Institute, 3700 San Martin Drive, Baltimore MD, 21218.

ABSTRACT We have investigated the effects of substantial mass accretion on the internal structure of T-Tauri stars. At sufficiently large accretion rates the surface heating by the boundary layer and the energy released by deuterium burning can result in substantial thinning of the surface convection zone.

Keywords: stellar interiors; accretion; pre-main sequence stars

INTRODUCTION

It is believed that many T-Tauri stars are accreting material from a circumstellar disk at rates of 10^{-8} to a few$\times 10^{-7}\, M_\odot\,\mathrm{yr}^{-1}$ (*e.g.* Basri and Bertout 1989). FU Orionis stars are apparently related pre-main sequence objects accreting material at much larger rates (Hartman and Kenyon 1987).

In accretion disks, one half of the total potential energy is dissipated in bringing material to the inner edge of the accretion disk. We assume that the other half is dissipated in a shear boundary layer near the stellar surface. In some of the models presented here we will assume that a large portion of the energy dissipated in the boundary layer is absorbed by the stellar photosphere. We model this by assuming that there is a luminosity source present at the stellar surface, and adjusting the surface boundary conditions such that $L_{tot} = L_{int} + L_{surf}$, where L_{int} is the luminosity from below the photosphere, and L_{surf} is some fraction of $GM_*\dot{M}/R_*$. We will also include the energy released by deuterium burning in some of our models.

MODELS

For all of the models in this paper a starting mass of $0.88\, M_\odot$ and a final mass of $1.00\, M_\odot$ was assumed. The time dependence of the accretion rate was assumed to be $\dot{M}(t) = \dot{M}_m e^{-|t-t_o|/\tau_a}$, where $t_o = 2\times 10^6$ yrs, $\tau_a = 2\times 10^4$ yrs, and $\dot{M}_m = 3\times 10^{-6}\, M_\odot\,\mathrm{yr}^{-1}$. Models were run with and without deuterium burning, and with and without an added surface luminosity term. When it is used, the surface luminosity term assumes 20% of total accretion luminosity ($GM_*\dot{M}/R_*$) is absorbed by the photosphere.

TABLE I Models

model	deuterium	surface lum.	line in fig. 1 & 2
A	no	0	dotted
B	yes	0	long dash
C	no	$0.2 GM_* \dot{M}/R_*$	solid
D	yes	$0.2 GM_* \dot{M}/R_*$	short dash

RESULTS

The direct effects of the mass accretion are small. If deuterium burning and the surface luminosity term are not included, the model shifts smoothly from the Hayashi track for the lower mass to the track for the new mass.

When a substantial surface luminosity term is included, the increased surface temperature flattens the surface temperature gradient, reduces the super-adiabaticity of the gradient near the surface and thereby reduces the flow of convective energy (see figure 1). This reduction in the convective luminosity results in a substantially thinner convection zone (see figure 2).

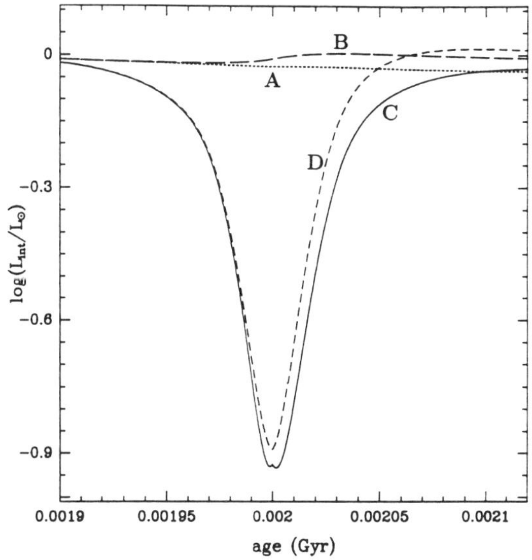

Fig. 1 The flux from the interior as a function of time during the accretion episode. Models that include deuterium burning (B and D) show a slight luminosity enhancement over those that do not, while models with substantial surface heating (C and D) show a dramatic reduction in the luminosity that escapes from the interior. In both models C and D the added surface luminosity (not shown) will completely dominate the interior luminosity.

At the peak accretion rate, the total luminosity provided by deuterium burning reaches $4.4\,L_\odot$, but this has only a modest effect on the surface luminosity. Most of the energy is absorbed in increasing the specific entropy of the surface convection zone. This also thins the convection zone.

For typical T-Tauri star accretion rates ($\approx 10^{-7}\,M_\odot\,\mathrm{yr}^{-1}$), there is little observable change in the surface luminosity, T_{eff}, or convective structure. Accretion rates must be at least an order of magnitude larger to noticeably reduce the surface convection zone or alter the interior flow of energy. At such high accretion rates the stellar photosphere will not be directly observable.

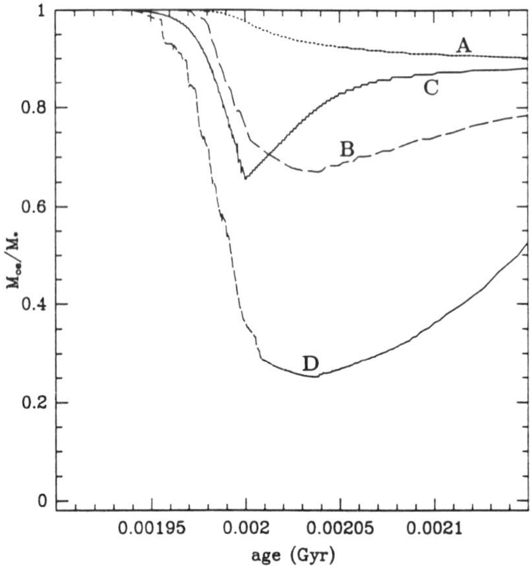

Fig. 2 The fraction of the stellar mass in the convective envelope as a function of time.

We have not yet attempted calculations at the extreme accretion rates suggested for FU Orionis stars, but we can estimate that the advection of specific entropy, the surface heating, and the energy from deuterium burning is likely to raise the specific entropy of the convective envelope, and result in a very thin surface convection zone.

ACKNOWLEDGMENTS

We wish to acknowledge support from the Ultraviolet, Visible, and Astrophysics Branch of the Astrophysics Division of NASA's Office of Space Science and Applications.

REFERENCES

Basri, G. and Bertout, C. 1989, *Ap. J.*, **341**, 340.
Hartman, L. and Kenyon, S. J. 1987, *Ap. J.*, **323**, 714.

Seventh Cambridge Workshop on Cool Stars, Stellar Systems, and the Sun
ASP Conference Series, Vol. 26, 1992
Mark S. Giampapa and Jay A. Bookbinder (eds.)

A STELLAR EVOLUTION PARADIGM BASED ON SPECIFIC MASS LOSS AND FEEDBACK MODES

MANFRED CUNTZ Joint Institute for Laboratory Astrophysics [JILA], University of Colorado and NIST, Boulder, CO 80309-0440, USA.

ROBERT E. STENCEL Center for Astrophysics and Space Astronomy [CASA], University of Colorado, Boulder, CO 80309-0391, USA.

ABSTRACT We present a new paradigm for stellar evolution which deals with a detailed treatment of mass loss and feedback modes. The paradigm is presented as a logical diagram which describes the respective dependencies of atmospheric properties relevant to mass loss generation.

Keywords: Stellar evolution; rotation; instabilities; mass loss

INTRODUCTION

"Classic" paradigms of stellar evolution describe the changes of the evolutionary status of stars by modeling the dominant nucleosynthesis processes in the stellar cores. Mass loss, which plays a crucial role for the "right" choice of the stellar evolution tracks, is therein considered as a given function of the position of the star in the H-R diagram. This scenario is somewhat incomplete because the evolution of stars in turn generates a specific set of atmospheric conditions, which allows the mass loss rates to be explained in physical terms. In general we have to consider the principal effects of interior evolution on stellar atmospheres including the generation of mass loss as well as consequences of these effects on stellar evolution. In the upcoming discussion we limit our consideration to that range of stellar masses and ages when convective envelopes are present.

PRINCIPAL EFFECTS OF INTERIOR EVOLUTION

The effects of interior evolution on stellar atmospheres are caused by changes of the internal energy generation rate. The net effect of gradual exhaustion of a given, current core fuel is a resulting increase in central temperature and corresponding eventual increase in stellar luminosity. For solar composition, the envelope and atmosphere response includes an overall increase in radius and reduced surface temperature and gravity. This change affects the generation and transmission of short and long–period wave modes and promotes atmospheric conditions for thermochemical instabilities and thermal bifurcation.

Stars undergoing *shell burning* exhibit these changes on a relatively short timescale, especially when the shell is approaching the core/envelope boundary and/or is undergoing transient *flash events*. Those stars initiating *core burning* change their interior structure in an opposite manner, leading to reduced radius, higher surface temperature and gravity boundary conditions not as conducive to thermochemical instabilities. Doom *et al.* (1986), Maeder & Meynet (1989) and Baraffe & El Eid (1991) have presented new stellar evolution models.

CONSEQUENCES OF THESE EFFECTS ON MASS LOSS

We consider four observable stellar properties (emergent stellar radiation, subsurface convective turbulence, global oscillations, and magnetic fields) which are certain consequences of stellar evolution.

Emergent stellar radiation plays a crucial role in the production of mass loss in evolved stars. Radiation forces act on molecular lines (*e.g.* Maciel 1976, 1977, Elitzur *et al.* 1989) and on dust grains (*e.g.* Kwok 1975). Reduction of the stellar effective temperature promotes conditions for the formation of molecules (*e.g.* Muchmore *et al.* 1987), which in combination with hydrodynamic enhancement of atmospheric density provides an abundance of extra opacity to capture photon momentum. These and subsequent points are diagrammed in Fig. 1.

Subsurface convective turbulence cannot directly control the mass loss behavior of a star. However, it seems to be essential for certain mass loss processes to occur, because it generates short–period acoustic waves which propagate outward and form strong stochastic shocks (*e.g.* Cuntz 1987) leading to chromospheric heating and radiative cooling by UV emission as a consequence. Cool molecular material so irradiated may form molecular ions which serve to catalyze molecular reactions, including those leading to the formation of molecular clusters and dust particles (*e.g.* Gail & Sedlmayr 1986, Mamon *et al.* 1987). On the other hand, chromospheric heating can also be provided by the dissipation of MHD waves.

Global oscillations are a direct consequence of interior structure and are most sensitive to the evolution of the internal energy generation rate. Global oscillations lead to a substantial enhancement of the outer atmospheric density and promote atmospheric dynamics due to momentum transfer. The combined influence of both effects can lead to mass loss, depending on the particulars (*e.g.* Wood 1979, Bowen 1988). In addition, global oscillations can produce hydrodynamic refrigeration, when radiative cooling and hydrodynamic interaction in time–dependent atmospheric flows reduce outer atmospheric temperatures to below radiative equilibrium values. The time evolution of the global oscillation spectrum is key in separating the relative feedback of mass loss on the interior processes (*e.g.* Guenther 1991). The onset of thermochemical instabilities can lead to the formation of molecules and dust. A paradigm for dust–driven mass loss based on observations was given by Stencel *et al.* (1986).

Magnetic fields serve largely to limit mass loss while the interior dynamo is effective enough to produce closed loop geometry, leading to coronal heating, which cannot otherwise be provided by acoustic shock wave dissipation (*e.g.* Stępień & Ulmschneider 1989). Hot coronae provide a drastic increase in the thermal pressure gradient, leading to Parker–type stellar winds with high speeds but very low mass flux. However, late–type giant and supergiant stars largely lack coronae, suggesting reduced significance of the dynamo. It is suspected that these stars may lose mass due to dissipation of Alfvén waves along mostly open magnetic field geometries (*e.g.* Hartmann & MacGregor 1980, An *et al.* 1990), although this mechanism is uncertain because the required magnetic fields are undetectable in most of these stars (Marcy & Bruning 1984).

CONSEQUENCES OF MASS LOSS ON STELLAR EVOLUTION

Fundamentally, there are two limits in which stellar mass loss could affect internal structure and possibly evolution. First, when the dynamo is effective, magnetic braking can slow the stellar rotation over a main sequence lifetime, as is strongly suggested by the observed correlations of X-ray flux with rotation rates (*e.g.* Rutten 1987). Second, when the mass loss rate grows large with respect to the total stellar mass and the nuclear timescale of internal energy generation, the entire envelope (in extreme cases) could be removed before a shift in nuclear fuels or a shift from core to shell burning (or subsequent core re-ignition) has a chance to occur. Our contention is that less extreme internal changes can cause or be affected by the onset of heavy mass loss due to stellar pulsation and dust formation epochs. The generation of extensive circumstellar shells around red supergiants (Stencel *et al.* 1989) may be an example of this behavior. The increasing pressure scale height as a star evolves up the giant branch will differentially damp and therefore change the spectrum of global oscillation modes. That change will produce additional mass loss, which in turn modifies the oscillation spectrum (Jones *et al.* 1983). A study of Alcolea & Bujarrabal (1991) for post-AGB stars provides evidence that phases of negligible and substantial mass loss alternate. Quantification of the paradigm outlined here should permit us to eventually compute improved stellar evolution models.

ACKNOWLEDGEMENTS

We are pleased to acknowledge support for this work through NASA to the University of Colorado.

REFERENCES

Alcolea, J., & Bujarrabal, V.: 1991, *Astr. Ap.*, 245, 499
An, C.H. *et al.* : 1990, in: Cool Stars VI, ed. Wallerstein, *ASP Conf. Series* 9, p. 70
Baraffe, I., & El Eid, M.F.: 1991, *Astr. Ap.*, 245, 548
Bowen, G.H.: 1988, *Ap. J.*, 329, 299
Cuntz, M.: 1987, *Astr. Ap.*, 188, L5
Doom, C., De Greve, J.P., & De Loore, C.: 1986, *Ap. J.*, 303, 136
Elitzur, M., Brown, J.A., & Johnson, H.R.: 1989, *Ap. J. Letters*, 341, L95
Gail, H.-P., & Sedlmayr, E.: 1986, *Astr. Ap.*, 166, 225
Guenther, D.B.: 1991, *Ap. J.*, 375, 352
Hartmann, L., & MacGregor, K.B.: 1980, *Ap. J.*, 242, 260
Jones, T.J., Hyland, A.R., Wood, P.R., & Gatley, I.: 1983, *Ap. J.*, 273, 669
Kwok, S.: 1975, *Ap. J.*, 198, 583
Maciel, W.J.: 1976, *Astr. Ap.*, 48, 27
Maciel, W.J.: 1977, *Astr. Ap.*, 57, 273
Maeder, A., & Meynet, G.: 1989, *Astr. Ap.*, 210, 155
Mamon, G.A., Glassgold, A.E., & Omont, A.: 1987, *Ap. J.*, 323, 306
Marcy, G.W., & Bruning, D.H.: 1984, *Ap. J.*, 281, 286
Muchmore, D.O., Nuth III., J.A., & Stencel, R.E.: 1987, *Ap. J. Letters*, 315, L141
Rutten, R.G.M.: 1987, *Astr. Ap.*, 177, 131
Stencel, R.E., Carpenter, K., & Hagen-Bauer, W.: 1986, *Ap. J.*, 308, 859
Stencel, R.E., Pesce, J.E., & Hagen-Bauer, W.: 1989, *A. J.*, 97, 1120
Stępień, K., & Ulmschneider, P.: 1989, *Astr. Ap.*, 216, 139
Wood, P.R.: 1979, *Ap. J.*, 227, 220

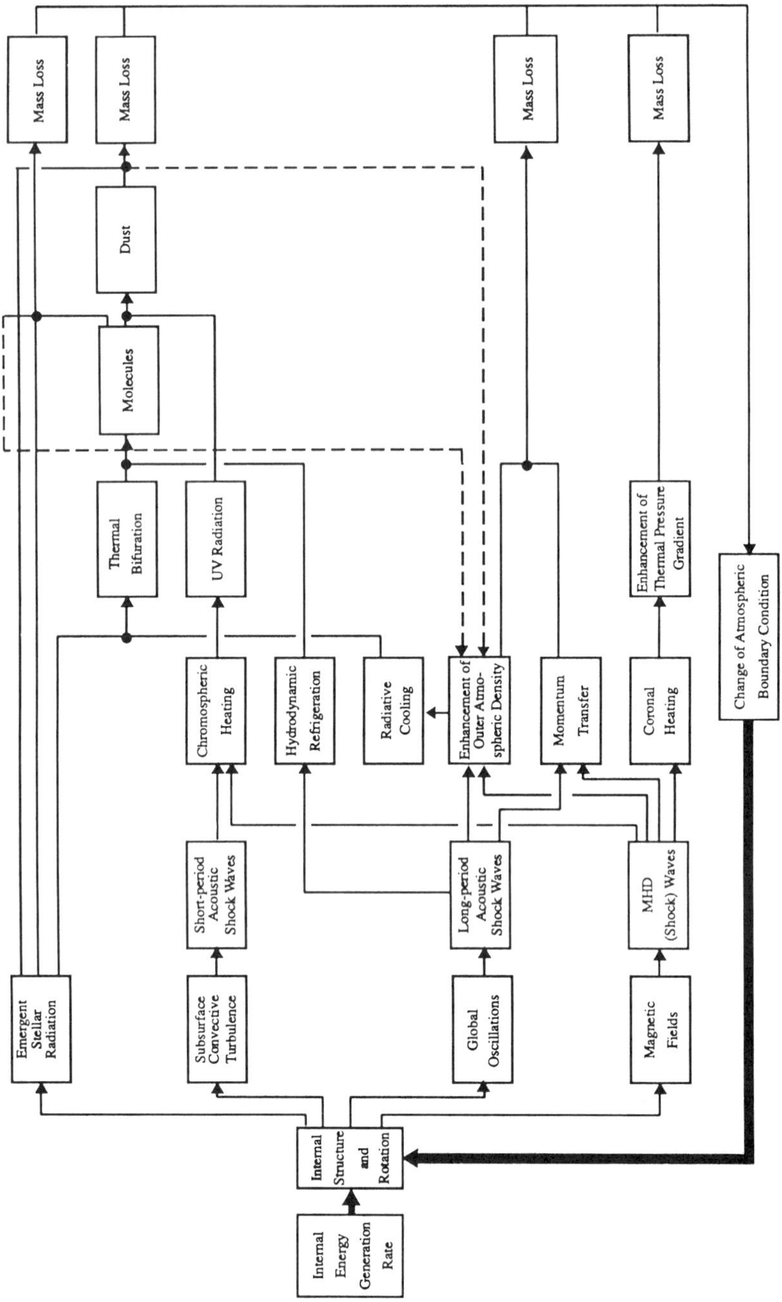

Fig. 1: The new Paradigm of Stellar Evolution

Seventh Cambridge Workshop on Cool Stars, Stellar Systems, and the Sun
ASP Conference Series, Vol. 26, 1992
Mark S. Giampapa and Jay A. Bookbinder (eds.)

FOUR YEARS OF MONITORING α ORIONIS WITH THE VLA: WHERE HAVE ALL THE FLARES GONE?

STEPHEN A. DRAKE
HEASARC, Code 668, NASA/GSFC, Greenbelt, MD 20771.

JAY A. BOOKBINDER
SAO, Center for Astrophysics, 60 Garden Street, Cambridge, MA 02138.

DAVID R. FLORKOWSKI
US Naval Observatory, 34th and Mass. Ave. NW, Washington, DC 20392.

JEFFREY L. LINSKY
JILA, Campus Box 440, University of Colorado, Boulder, CO 80309.

THEODORE SIMON
Institute for Astronomy, University of Hawaii, 2680 Woodlawn Dr., Honolulu, HI 96822.

ROBERT E. STENCEL
CASA, Campus Box 392, University of Colorado, Boulder, CO 80309.

ABSTRACT We observed this prototypical M supergiant using the VLA at centimeter wavelengths about once a month over a 4-year period from July 1986 to September 1990, resulting in a cumulative exposure time of about 120 hours. There is no evidence for radio flaring at any of the observing wavelengths. Low-level variability ($\lesssim 25\%$) does appear to be present at the shorter wavelengths (2 and 3.6 cm) and may be also occurring at 6 cm, and is usually in the form of temporary decreases of the radio flux densities from their usual levels. The timescale for these radio dips is not well-determined by the present data set, but significant variability on a a timescale of ≤ 1 month is observed on occasion. There is no correlation between the variations observed at the different wavelengths.

Keywords: radio continuum observations; stellar winds and chromospheres; red supergiants

INTRODUCTION

One of the nearest M supergiants to the Sun is α Orionis (M1-2 Ia-Iab; D = 190 pc). It has been known for a number of years to be a centimeter radio source at the 1 to 10 mJy level, with a spectral index $\alpha \sim 1.32$, where $S_\nu \sim \nu^\alpha$ (Newell and Hjellming 1982). Its radio continuum emission has been generally, but not universally, interpreted as free-free emission from the ionized component ($\sim 1\%$) of their stellar winds; however, the standard stellar wind models for radio emission (Wright and Barlow 1975) as have been extensively applied to hot stars, predict a spectral index of $\alpha = 0.6$, that is much less than this observed value. The observed spectrum can be interpreted as being due to a steeper drop-off in ionized density than the standard r^{-2} form: this, in turn, could be due to an *accelerating* stellar wind velocity *or* a *decreasing* ionization fraction with *increasing* radial distance from the star.

Newell and Hjellming (1982) gathered all the then-available radio data on α Ori - both single-dish and interferometric - to examine whether there was believable evidence for variability. They pointed out that most pre-VLA data on α Ori fitted with their VLA spectrum, but that 2 observations in 1977 and 1978 at 1.3 and 2 cm were a factor of 2 below 'normal' and that 2 measures at 2 and 2.8 cm were 12 to 50 times above the normal level, indicating that radio flaring may have occurred. They interpreted the below-normal measures as being most likely due to calibration errors in these single-dish observations.

From July 1986 to September 1990 we monitored α Ori at centimeter wavelengths using the VLA in order to confirm whether or not this variability is indeed intrinsic to α Ori, and, if so, whether such variability is periodic or irregular in nature and what the characteristic timescales T_{var} for radio variability are. The radio observations were typically made at 1-month intervals, except during the third quarter of 1986 when the supposed periastron passage of the postulated close binary companion discovered by interferometry (Karovska *et al.* 1986) occurred, when the observations were made at roughly 2-week intervals. The initial sets of observations were made at 2 and 6 cm only: commencing in early 1988 regular observations were also made at 3.6 cm. Occasional measurements were also made at 1.3 and 20 cm. All data were reduced using standard NRAO software and procedures, and the flux densities were measured using AIPS tasks IMFIT and JMFIT, being generally taken to be the average of the peak and integrated values.

RESULTS AND DISCUSSION

About 90% of the radio observations have now been reduced. The average spectrum that can be constructed from these data is $S_\nu = 0.26$ mJy $\nu^{1.26}$, compared to $S_\nu = 0.23$ mJy $\nu^{1.33}$ found by Newell and Hjellming (1982), *i.e.*, essentially the same. We note that, if we extrapolate our power law spectral fit to the mm region, we predict fluxes at 86 and 226 GHz of 72.5 and 246 mJy, respectively, compared to measured flux densities of 72.8 ± 5.5 and 353 ± 51 mJy (Altenhoff *et al.* 1986); thus, there is weak (2σ) evidence that the spectrum steepens between 86 and 226 GHz, but this should be confirmed by repeating the higher frequency measurement with a higher signal-to-noise.

The evidence for variability is strongest at 2 and 3.6 cm, where the standard deviations (s.d.) about the mean values are 6 and 7 times, respectively, the typical estimated rms errors of 0.16 mJy (2 cm) and 0.05 mJy (3.6 cm), while at 6 cm the s.d. is 0.16 mJy or 2.5 the typical rms error of a single measurement. At 2 cm, however, part of the apparent variability may be due to systematic errors: a combination of partial source resolution and bad atmospheric phases affecting the measurements in the A and B configurations, since the average 2-cm flux density in these arrays is only 89% of the average value found in the more compact C and D configurations, and the standard deviation of the A and B array measurements is 1.05 mJy compared to 0.61 mJy for the s.d. of the

C and D configurations. No such array-dependent effects are seen at 3.6 (or 6) cm, and thus the X-band (and possibly C-band) variations are almost certainly intrinsic to α Orionis.

There is no evidence for radio flaring (i.e., increases in flux density by large $\gtrsim \times 2$ amounts) at any wavelength at any epoch! At 6 cm, the maximum flux density S_ν^{max} measured is 22% above the mean, at 3.6 cm, S_ν^{max} is 15% above the mean, and at 2 cm S_ν^{max} is 21% above the mean value. This does not disprove the reality of the two radio 'flares' of α Ori observed in 1966 that involved increases in flux density of 1 or 2 orders of magnitude, but indicates that such events must be very rare, if real, and that even minor flares are very uncommon. As far as the postulated close companion is concerned, there is no evidence that the radio emission was enhanced during its proposed periastron in the summer of 1986, as might be expected by analogy with the radio properties of known interacting binaries. Thus, any increase in the volume and/or fraction of ionized material in the inner wind of the red supergiant primary due to the passage of this companion is apparently negligible.

The main type of variability seems to be small fluctuations about the mean levels, with drops to below-normal values slightly predominating. This behaviour is also apparent in the smaller and more inhomogeneous dataset assembled by Newell and Hjellming (1982). Clearly in A and B configurations poor phases and/or partial resolution of the source have sometimes led us to estimate the 2-cm flux density of α Ori by $\sim$ 10%, but such effects are not evident at the longer wavelengths, and the apparent variability at 3.6 cm and (perhaps) 6 cm thus seems to be mostly intrinsic to the star.

The variability appears to be stochastic, there is no correlation between variations at different wavelengths (not surprising since the radio emission at different wavelengths comes from physically separated regions in the star's wind), and a variety of different timescales appear to be present ranging from several months down to < 1 month (our sampling interval). A possible explanation for this behaviour given in Bookbinder *et al.* (1987) invokes the molecular catastrophe scenario of Muchmore *et al.* (1987): this hypothesis is explored in more detail in Drake *et al.* (1992).

ACKNOWLEDGEMENTS

The authors would like to acknowledge the invaluable assistance of the NRAO staff, particularly Dave Wunker and Peggy Perley, that made this project feasible.

REFERENCES

Altenhoff, W. J. *et al.* 1986, *Astr. Ap.*, 164, 227.
Bookbinder, J. A. 1987, Cool Stars, Stellar Systems, and the Sun, ed. J.L. Linsky and R.E. Stencel, p. 337.
Drake, S. A. *et al.* 1992, in preparation.
Karovska, M., Nisenson, P., and Noyes, R. 1986, *Ap. J.*, 308, 260.
Muchmore, D. O., Nuth, III, J. A., and Stencel, R. E. 1987, *Ap. J. Letters*, 315, L141.
Newell, R. T., and Hjellming, R. M. 1982, *Ap. J. Letters*, 263, L85.
Wright, A. E., and Barlow, M. J. 1975, *M.N.R.A.S.*, 170, 41.

Seventh Cambridge Workshop on Cool Stars, Stellar Systems, and the Sun
ASP Conference Series, Vol. 26, 1992
Mark S. Giampapa and Jay A. Bookbinder (eds.)

THE TRIPLE SYMBIOTIC SYSTEM CH CYGNI

KENNETH H. HINKLE National Optical Astronomy Observatories, P.O. Box 26732, Tucson, Arizona 85726-6732

FRANCIS C. FEKEL Vanderbilt University, Dept. of Physics and Astronomy, Nashville, TN 37235

DIANA S. JOHNSON National Optical Astronomy Observatories, P.O. Box 26732, Tucson, Arizona 85726-6732

WERNER W. G. SCHARLACH National Radio Astronomy Observatory, 949 N. Cherry Avenue, Tucson, Arizona 85721-0665

ABSTRACT A total of 68 infrared spectra of the bright symbiotic star CH Cygni has been obtained over the years 1979-1991. Velocities measured from these FTS spectra are of high quality and reveal a 749 day period as well as the 5257 day period previously detected. We propose that the M6 giant and the symbiotic hot companion have the 749 day orbit. We ascribe the 5257 day orbit to the motion of this pair with an unseen third star.

Keywords: CH Cyg; symbiotic stars; binaries; radial velocities; long-period variables

INTRODUCTION

Since 1979 we have been observing the 1.6-2.5μm infrared spectrum of CH Cygni at high resolution ($\lambda/\Delta\lambda \sim 10^5$) with the FTS at the KPNO 4 meter telescope. Our goal was to monitor the spectrum a few times each year to look for intrinsic variations in the M giant and to search for orbital motion. CH Cyg is the visually brightest symbiotic star (V=7) and is a bright infrared object (K=-0.7), making exposure times relatively short. Infrared spectra offer two advantages over visual spectra for this project. First, in the infrared longward of 1 μm the M giant dominates the spectrum (Ipatov, Taranova, and Yudin 1984). This is quite important because during active phases over the last 30 years, the visual spectrum has been heavily contaminated by line and continuum emission from the high excitation source. Second, spectra of bright infrared sources like CH Cygni may be observed during the daytime, making year around monitoring possible. In fact, because of the high northern declination of CH Cyg, nearly all of our observations have been taken in the daytime when access to the 4 m telescope for this kind of long term project is possible.

DISCUSSION

A total of 68 spectra were observed from February 1979 to July 1991. The velocities are shown in Figure 1. Velocities measured from 1.6 μm and 2.3 μm CO lines agree to better than ~1 km s^{-1} . Formal errors are <0.5 km s^{-1} .

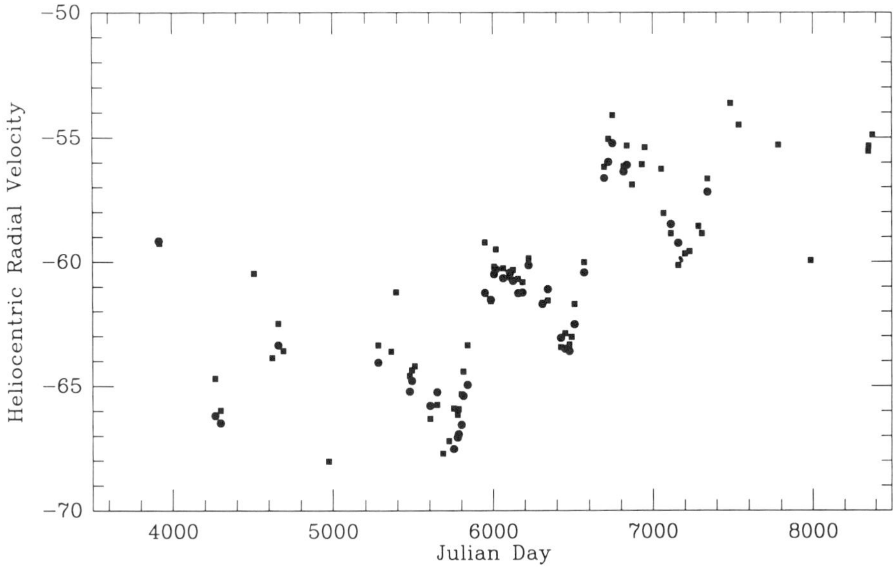

Fig. 1: Infrared radial velocities of CH Cyg as a function of Julian day number (-2440000). The squares are velocities from the 2.3 μm CO Δv=2 lines, the circles are from the 1.6 μm CO Δv=3 lines.

The velocities clearly show two periods, one of about 750 days and a second of about 15 years. The short-period velocities have an amplitude of 5.6 km s^{-1} while those of the long-period have an amplitude of 9.4 km s^{-1} . We believe that both velocity variations are due to orbital motion and are not the result of stellar pulsation because the infrared velocity curve does not look like that of any Mira or SRa variable previously observed (Hinkle, Scharlach and Hall 1984). In addition Mira and SR-type pulsation periods are typically <500 days for unobscured field stars.

The velocities are well fit by orbital motion in a triple system. The typical velocity residual from the total orbital fit is less than ~1 km s^{-1} . Such residuals are somewhat higher than the uncertainties and may result from semiregular pulsations of the M giant. The orbital elements determined from our observations are given in Table I.

Yamashita and Maehara (1979) derived an orbit with a period of 5750 days (15.8 years) for CH Cyg with radial velocities measured from visual spectra. This period is close to our period of 5257 days and their orbit, except for its moderate eccentricity, appears to be basically correct. However, we believe that this long-period orbit is for the symbiotic pair with respect to an unseen third star. Confirmed binary periods for symbiotic pairs are in the range 230-1000 days, with a mean period of 650 days (Table 3.1 of Kenyon 1986). Such data support our claim that the 749 day orbital period must be the correct one for the symbiotic binary.

TABLE I Orbital Elements

γ=-60.4±0.2 km s^{-1}	
K_S=2.8±0.1	K_L=4.7±0.2 km s^{-1}
e_S=0.30±0.05	e_L=0.06±0.05°
ω_S=230±9	ω_L=216±38°
T_S=2447294±17	T_L=2445626±533 HJD
P_S=749±4	P_L=5257±270 days

During eruptive phases, the visual spectrum has emission lines and "shell" lines that have been attributed to the hot star in the symbiotic system. For a portion of the previously proposed 5750 day orbit, the velocity of these ionized metallic absorption features appeared to be 180 degrees out of phase with velocities of the M giant. Tomov and Luud (1984) claimed that such velocities could be used to derive a double-lined spectroscopic orbit for the symbiotic system. With the identification of the 749 day orbit for the symbiotic system, the double-lined long-period orbit and results derived from it must be incorrect.

CONCLUSIONS

The CH Cyg system consists of three stars. Our radial velocities confirm the 15 year orbit proposed by Yamashita and Maehara (1979). However, we argue that it is the 749 day binary system that contains the interacting symbiotic system. The mass function for the 749 day orbit is 0.0015 $M_\odot$ and for the 15 year orbit 0.056 $M_\odot$. The requirement that the M6 giant have a mass $\lesssim$ 2 $M_\odot$ implies the inclination is $\lesssim$ 25° if the secondary has a typical white dwarf mass of $\sim$ 0.5 $M_\odot$. Since CH Cyg is a bright star, it has been extensively observed and in a future publication we will exploit the radial velocities in the literature to improve the long period orbit.

REFERENCES

Hinkle, K. H., Scharlach, W. W. G., and Hall, D. N. B. 1984 *Ap.J. Supp.* **56**, 1.
Ipatov, A. P., Taranova, O. G., and Yudin, B. F. 1984 *Astr. Ap.*, **135**, 325.
Kenyon, S. J. 1986 The Symbiotic Stars, Cambridge Univ. Press.
Tomov, T. and L. Luud 1984 *Astrofizika* **20**, 99.
Yamashita, Y. and Maehara, H. 1979 *P.A.S.J.* **31**, 307.

Seventh Cambridge Workshop on Cool Stars, Stellar Systems, and the Sun
ASP Conference Series, Vol. 26, 1992
Mark S. Giampapa and Jay A. Bookbinder (eds.)

THE ROLE OF MASS LOSS AND MIXING-LENGTH ON THE FINAL FATE OF THE SUN

UFFE GRÅE JØRGENSEN Niels Bohr Institute, Blegdamsvej 17, DK-2100 Copenhagen, Denmark

ABSTRACT Several grids of models have covered different phases of the evolution of stars with mass and composition close to that of the Sun. None, nevertheless, dealt with the entire evolution of the Sun and explored which limits the uncertainty in observations and model parameters impose on the estimated lifetime and future evolutionary phases of our Sun. We present here analytic fits to numerical models of stellar evolution, and show in a quantitative way that the present uncertainty in our knowledge of mass loss and mixing-length are the two parameters which introduce the biggest uncertainty in our estimates of the final fate of the Sun. Finally, we describe the most likely evolution of the Sun.

Keywords: Stellar evolution; the Sun; mass loss

LUMINOSITY AND EVOLUTIONARY TIME SCALES OF THE SUN

Almost every popular astronomy book presents a prediction of what and when the final state of the Sun will be, but the claims are widely different from one book to another, and at best very qualitative. In fact this is no surprise, because none of the modern grids of stellar evolution present one-solar-mass tracks that reach the solar luminosity at an age within a billion years of the age imposed by studies of the meteorites. Two recent grids (Claret and Giménez 1989;CG89 and Maeder and Meynet 1989; MM89) nevertheless compute 1-$M_\odot$ models with characteristics close enough to those of the Sun that relatively precise qualitative statements about the solar main sequence evolution can be drawn. Analytical fits to these tracks, adjusted to match the right solar age at the right luminosity and temperature, predict that the luminosity of the Sun throughout the main sequence evolution can be expressed as function of time as

$$L/L_\odot = 0.679 + 0.076\, t_9 + 5.797\, 10^{-4}\, t_9^2 - 8.233\, 10^{-4}\, t_9^3 + 1.145\, 10^{-4}\, t_9^4 \qquad (1)$$

where t_9 is the age in units of 10^9 years. Eq. (1) is valid until $t_9 \approx 11$, after which time the luminosity will decrease until the base of the red giant branch (BRGB, *i.e.*, the local luminosity minimum after which the luminosity starts increasing rapidly) is reached (for L=2.1 $L_\odot$) approximately $0.3 \cdot 10^9$ years later. Eq. (1) predicts the solar luminosity at t=0 to be 68% of the present luminosity.

Although the luminosity itself will not start increasing drastically before BRGB, it is seen that dL/dt will start increasing rapidly already at about the present time according to Eq. 1. After BRGB the Sun will evolve fast toward higher luminosities through the red giant branch (RGB) while burning hydrogen in a thin shell around a degenerate helium core. Most models [*e.g.* Sweigart and

Gross (1978; SG78), Bertelli *et al.* (1986; Be86)] predict the lifetime of the RGB to be 650±50 million years (at luminosities up to 2000 $L_\odot$). The lifetime of the following horizontal branch (HB; *i.e.*, burning helium to oxygen in the central region) is predicted to be 120±20 million years (at a luminosity of 55±15 $L_\odot$). The lifetime on the succeeding asymptotic giant branch (AGB; *i.e.* burning hydrogen and helium in two thin shells around a degenerate oxygen core) will be only of the order 10 million years according to the evolutionary predictions of Lattanzio (1986; La86) and Renzini and Voli (1981; RV81) when these tracks are treated with respect to mass loss in the manner described below.

THE ROLE OF MASS LOSS AND MIXING-LENGTH

The stellar luminosity as function of time on the RGB and AGB is qualitatively similar in most models, because they basically express the principle that the luminosity during thin shell burning phases is a function of the stellar core mass only, and that the growth in core mass per unit time is a function of the luminosity ($dM/dt = L/X\varepsilon_{H\rightarrow He}$), so that $dL/dt = f(L)$ only. The effective temperature, T_{eff}, is, nevertheless, very different in different computations, mainly due to the difference in the choice of mixing-length parameter, α. While the "correct" value of α during the main sequence evolution is the one that, for a given code, gives the right present solar luminosity and radius, there is no obvious choice for the late phases. Values between $\alpha = 1$ and $\alpha = 2$ are typical, and they give rise to very different predictions of stellar evolution. The smaller α is, the poorer is the energy transport to the surface, and the lower is therefore T_{eff}. The tracks due to SG78 and La86 are both based on α=1.0, and are therefore consistent with one another in the sense that the La86-AGB really is asymptotic to the SG78-RGB. Similarly Be86 and RV81 are mutually consistent in their choice of a high value of α ($\approx$ 1.6). The difference in T_{eff} predicted by the two set of models is $\approx$ 600 K. Other computations predict temperatures between these two values, and typical values are 5000±300 K near BRGB and 3300±500 K near the top of RGB and on AGB.

Since the luminosity during RGB and AGB is dependent only on the core mass, T_{eff} of a stellar model with a given chemical composition, mass, luminosity, and mixing-length, will depend only on the total mass. That is, the star has no memory of how it reached a certain value of (T_{eff}, L). This fact can conveniently be used to construct tracks of mass-losing stars by use of mass fixed evolutionary computations. The advantage of this method, compared to including mass loss in the original computations, is that one can easily explore the effect of a great range of mass loss values and laws. After first approximating the evolutionary tracks referred to above by analytical formulas (Jørgensen 1991), we have numerically imposed a Reimers type mass loss ($\dot{M} = \eta L^{1.5} / T_{eff}^2 M$) to the solar evolution.

The mass loss constant, η, is (just like α) model dependent, and must for each set of models be determined by observational constraints (*e.g.* prediction of the observed mass difference between RGB and HB stars in clusters). In this way we estimate η=0.3 for the tracks due to Be86-RV81, and η=0.4 for the tracks due to SG78-La86, and notice that the difference in mass loss is determined by the difference in T_{eff}, which depends mainly on the value of α. A solar model with $\alpha = 1$ (SG78-La86), calculated without mass loss (η=0), develops to a final radius of 4 AU – nearly the size of the orbit of Jupiter – whereas the choices of η argued for above will leave the final solar radius computed from the Be86-RV81 tracks at 1.3 AU and from the SG78-La86 tracks at 1.7 AU. Variations of η with ±0.1 result in a change in the final solar radius of ±0.2 AU. The total evolution time is nearly independent on α and η, due to the strong luminosity dependence of $\dot{M}$.

THE FINAL FATE OF THE SUN

The very last AGB evolution of the Sun is critically dependent on the final core mass, which depends strongly on the adopted mass loss parameter, because the evolution necessarily must terminate when $M_{core} \approx M_{total}$. If the core mass exceeds a critical value, which presumably is close to 0.6 $M_\odot$ (La86), the material produced in the helium burning shell will be mixed to the surface (following each helium shell flash), and after a few flashes the Sun will appear as a carbon star. The core mass of the Sun will, nevertheless, only exceed 0.6 $M_\odot$ if the mass loss parameter, η, is smaller than $\approx$ 0.2, which at present seems unrealistic. The most likely stage of the Sun at AGB termination is therefore an oxygen rich Mira variable star, and the period can be calculated to be around 300 days from models with $\alpha = 1.0$ and to be around 200 days if $\alpha = 1.6$. When the core mass of a Mira variable exceeds a critical value, $M_c^{crit} \approx 0.5$ - 0.7 $M_\odot$(RV81), it will blow off a planetary nebula. The core mass, M_{core}, of the Sun at the top of AGB (when $M_{core} \approx M_{total}$) is found, from the various models studied here, to be between 0.55 and 0.65 $M_\odot$, and the Sun will therefore most likely produce a planetary nebula in the end of its evolution, but the definitive answer awaits a more precise theoretical or observational determination of M_c^{crit}.

CONCLUSIONS

In conclusion, the best estimates predict the total lifetime of the Sun to be close to 12.1 billion years (*i.e.*, nearly 3 times the present age). The advanced stages of the solar evolution (*i.e.*, after BRGB) last only about 7% of the total lifetime, and the radius of the Sun will slightly exceed 1 AU at the end of AGB, where the Sun will most likely pass through the Mira variability phase (ending with a period of 250 days), cast off a planetary nebula and become a white dwarf of approximately 0.6 $M_\odot$. The main uncertainty in our knowledge of how the Sun will develop during these late phases is due to uncertainty in the value of mass loss and mixing-length during the red giant and asymptotic giant branch phase.

REFERENCES

Bertelli G., Bressan A., Chiosi C., Angerer, K.: 1986, *Astron. Astrophys. Suppl.*, **66**, 191
Claret A. and Giménez A.: 1989, *Astron. Astrophys. Suppl. Ser.*, **81**, 1.
Jørgensen U.G.: 1991, *Astron. Astrophys.*, **246**, 118.
Lattanzio J. C.: 1986, *Astrophys. J.*, **311**, 708.
Maeder A. and Meynet G.: 1989, *Astron. Astrophys.*, **210**, 155.
Renzini A. and Voli M.: 1981, *Astron. Astrophys.*, **94**, 175.
Sweigart A. V. and Gross P. G.: 1978, *Astrophys. J. Suppl. Ser.*, **36**, 405.

Seventh Cambridge Workshop on Cool Stars, Stellar Systems, and the Sun
ASP Conference Series, Vol. 26, 1992
Mark S. Giampapa and Jay A. Bookbinder (eds.)

WHY THE WINDS FROM LATE-TYPE GIANTS AND SUPERGIANTS ARE COOL

R. L. MOORE NASA Marshall Space Flight Center, Huntsville, AL 35812, U.S.A.

Z. E. MUSIELAK University of Alabama in Huntsville, Huntsville, AL 35899, U.S.A.

C. -H. AN Applied Research, Inc., Huntsville, AL 35805, U.S.A.

R. ROSNER University of Chicago, Chicago, IL 60637, U.S.A.

S. T. SUESS NASA Marshall Space Flight Center, Huntsville, AL 35812, U.S.A.

ABSTRACT We present a model for the coronae of late-type evolved stars where the coronal temperature is regulated by heating from Alfvén waves present in stellar coronal holes. Temperatures predicted from this model are in reasonable agreement with observations.

Keywords: Winds; Coronae; Evolved Stars; Alfvén Waves

INTRODUCTION

Rosner *et al.* (1991) have proposed that the massive cool ($10^4\ K < T < 10^5\ K$) winds from late-type giants and supergiants are driven by the reflection of Alfvén waves in open magnetic fields analogous to those in the solar wind that are rooted in coronal holes. They point out that the reflection can be strong because the "corona" in the open-field regions on these stars is cool ($T < 10^5\ K$), which makes the Alfvén velocity gradient steep enough for strong reflection. They also note that why these stellar coronae are so cool, *i.e.*, what determines the temperature, remains an open question. Moore *et al.* (1991) present evidence that the temperature in solar coronal holes is regulated by heating by reflected Alfvén waves. Here we point out that if in the open-field "coronal" holes (from which flow the winds) on giants and supergiants the temperature is regulated by heating by reflected Alfvén waves as in solar coronal holes, then the expected temperature of their winds is of the order of that observed.

THE MODEL AND RESULTS

In the model that we have used to compute the variation of Alfvén velocity with radius in open-field regions of stellar atmospheres, we assume: (1) that, like the solar wind, other stellar winds come from open-field regions, (2) that, like the corona in coronal holes (which are the open-field regions on the Sun), the base of

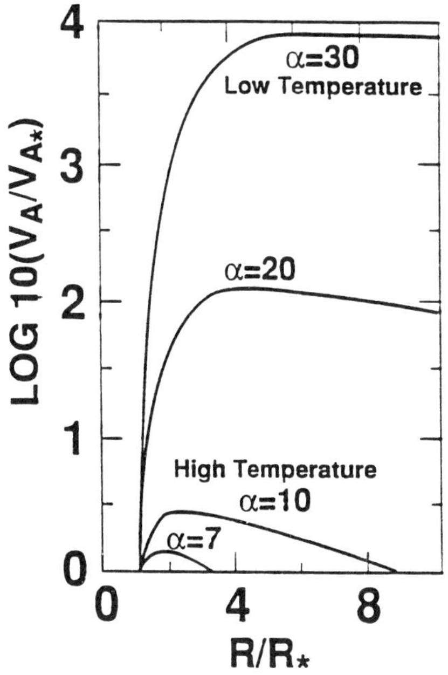

Figure 1. Variation of Alfvén velocity with radius in our model stellar wind base.

the wind within a few stellar radii of a star's surface is roughly both hydrostatic and isothermal, and (3) that, on other stars, as in the first few solar radii in a coronal hole, in most of the volume of the wind base the magnetic field is predominantly radial. We denote the stellar radius and gravity by R_* and g_*, respectively, with the subscript $*$ indicating the surface of the star. Alfvén waves of any given wavelength that are emitted up into the wind base (approximately one stellar radius above the surface) from below will be reflected back down by the wind base if the Alfvén velocity profile through the wind base is steep enough; if the profile is not steep enough, the waves will not be appreciably reflected and will pass on up through the wind base and out into the stellar wind.

Variation of Alfvén velocity with radius in our model stellar wind base is plotted in Figure 1. The parameter α is the ratio of the radius to the hydrostatic scale height at the bottom of the wind base: $\alpha = R_*/H_*$, $H_* = kT/mg_*$, $g_* = GM_*/R_*^2$. For any one star, because the mass M_* and radius R_* of the star are fixed, α is inversely proportional to the temperature of the wind base. Because the plasma density varies exponentially with temperature, a small change in temperature (or α) makes a big change in the Alfvén-velocity profile, as is seen in Figure 1. Alfvén waves with wavelengths of the order of R_* are strongly reflected by the wind base if $\alpha > 10$, but suffer little reflection and escape out into the wind if $\alpha < 10$ (Rosner *et al.* 1991). Hence, $\alpha \approx 10$ is the critical range of α in which strong reflection of these Alfvén waves sets in as α increases (as

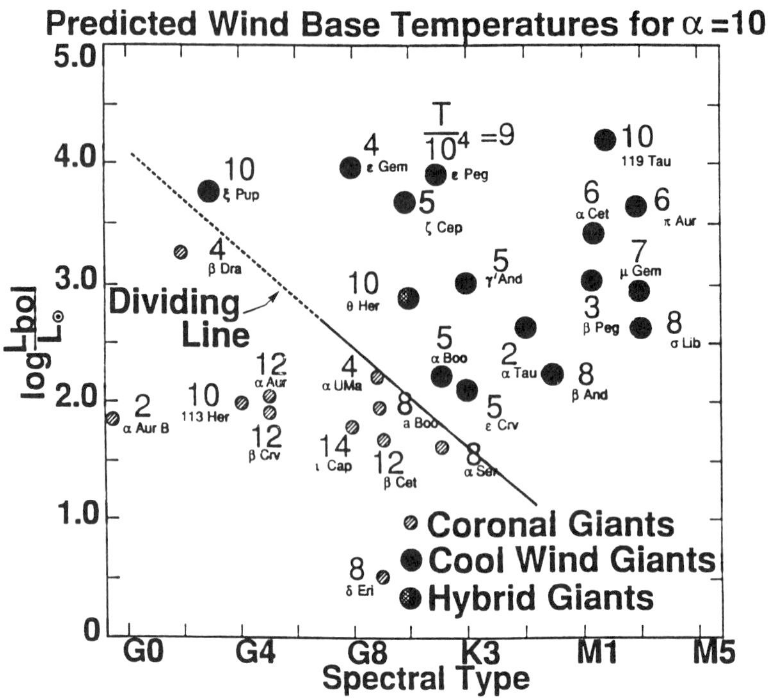

Figure 2. Predicted wind-base temperatures on giants and supergiants across the X–ray/mass-loss dividing line.

temperature decreases). Because of this effect, it appears that in solar coronal holes, the temperature is regulated by heating by the reflected Alfvén waves so that $\alpha_\odot \approx 10$: $\alpha_\odot = R_\odot / H_{coronal\ hole} \approx 10^{10.8}\ cm\ /\ 10^{9.8}\ cm = 10$ (Moore *et al.* 1991). On the basis of these results, we adopt as a working hypothesis that, as in solar coronal holes, heating by reflected Alfvén waves regulates the temperature of the base of other stellar winds, including in particular the massive cool winds from late-type giants and supergiants, so that $\alpha = 10$.

In Figure 3, we present the predicted wind-base temperatures on giants and supergiants across the X–ray/mass-loss dividing line. These temperatures were computed from the known values of mass and radius for these stars and our assumption that $\alpha = 10$. The predicted wind-base temperatures are in the range $10^4 - 10^5\ K$, of the same order as the observed temperatures of the cool winds. This agreement suggests that the winds from late-type giants and supergiants are cool because the wind-base temperature is regulated by heating by reflected Alfvén waves so that $\alpha = 10$ as in solar coronal holes. That the predicted wind-base temperatures are of the same order on both sides of the dividing line suggests that any (yet to be observed) winds from coronal giants are also cool ($T < 10^5\ K$). In Rosner *et al.* (1991), we proposed that the transition across the dividing line from coronal stars to cool-wind stars results from a transition from predominantly closed to predominantly open magnetic fields on the stars. On the coronal side, most of the star is covered by magnetically closed, hot active regions, whereas on the cool-wind side most of the star is covered by open-field cool-wind-base regions. Hence, the X–ray emission dies out and the cool winds become dominant from left to right across the dividing line. The dependence of Alfvén-wave reflection on the temperature of the wind-base region as outlined in this present paper and the resulting expected wind-base temperatures shown in Figure 2 offer an explanation for why the winds from the open-field regions on giants and supergiants are so cool. From our previous paper and the present paper together, it appears that both the acceleration of the massive cool stellar winds and the temperature of these winds result from the reflection of Alfvén waves.

REFERENCE

Moore, R. L., Musielak, Z. E., Suess, S. T. and An, C. H.: 1991, *Ap. J.* **378**, 347.

Rosner, R., An, C.H., Musielak, Z. E., Moore, R.L., and Suess, S. T.: 1991, *Ap. J. (Letters)*, **372**, L91.

Seventh Cambridge Workshop on Cool Stars, Stellar Systems, and the Sun
ASP Conference Series, Vol. 26, 1992
Mark S. Giampapa and Jay A. Bookbinder (eds.)

TEMPORAL VARIATIONS IN THE CIRCUMSTELLAR SHELL OF IRC+10216.

PEDRO V. SADA Dept. of Astronomy, New Mexico State University, Box 30001, Dept. 4500, Las Cruces, New Mexico, 88003-0001.

KENNETH H. HINKLE Kitt Peak National Observatory, 950 N. Cherry Avenue, P.O. Box 26732, Tucson, Arizona, 85726-6732.

JOHN J. KEADY Theoretical Division Group T4, MS 212, Los Alamos National Laboratory, Los Alamos, New Mexico, 87545.

ABSTRACT High resolution spectral observations of IRC+10216 at both 2.3 and 4.6 μ have been obtained occasionally since 1976 at KPNO using the 4 meter FTS. From these observations we have compiled a time series of the CO fundamental and first overtone line profiles. The CO line profiles show several components with different expansion velocities and excitation temperatures. The terminal expansion velocity relative to the center of mass velocity is near 14 km s^{-1} and a strong narrow component is seen in the CO line profile at this velocity. The line profiles also show another component expanding from the center of mass velocity at about 5 km s^{-1} . The time series shows that this last component is undergoing large changes in central depth and width. For instance, since 1979 the -5 km s^{-1} component of the (ΔV = 2-0) P(7) line increased in equivalent width by about a factor of 5 and also appears to have accelerated away from the central star at a rate of 0.23 ± 0.03 km s^{-1} per year.

Keywords: Carbon stars; circumstellar shells; IRC+10216.

INTRODUCTION

IRC+10216 is most likely a nearby luminous evolved carbon star with Mira-like variations and an exceptional mass loss rate embedded in an optically thick gas and dust envelope of its own making. For an overall summary of this type of object see Zuckermann (1980).

Carbon monoxide (CO) vibrational-rotational lines are good probes for the gas structure of carbon rich circumstellar shells because the molecule's high dissociation potential will lock up almost all the available oxygen and will also preserve the molecule from photodissociation until the gas has traveled far from the central star. The net result is that CO is fairly abundant and present throughout the shell.

OBSERVATIONS

High resolution (< 0.038 cm^{-1} spectral observations taken over the last 15 years at both 2.3 and 4.6 μ have been extracted from the NOAO 4m FTS archive in order to study and model CO vibrational-rotational absorption line profile shapes arising from the circumstellar shell.

Even though a standard assumption regarding circumstellar shells around late-type stars has been that they do not display short term (< 1000 years) variations, since 1976 noticeable changes have occurred in the line profile shapes from both the fundamental (ΔV = 0-1) and first overtone (ΔV = 0-2) bands of CO in the spectra of IRC+10216. These variations reflect changes in the structure of the circumstellar shell. See figures 1 and 2.

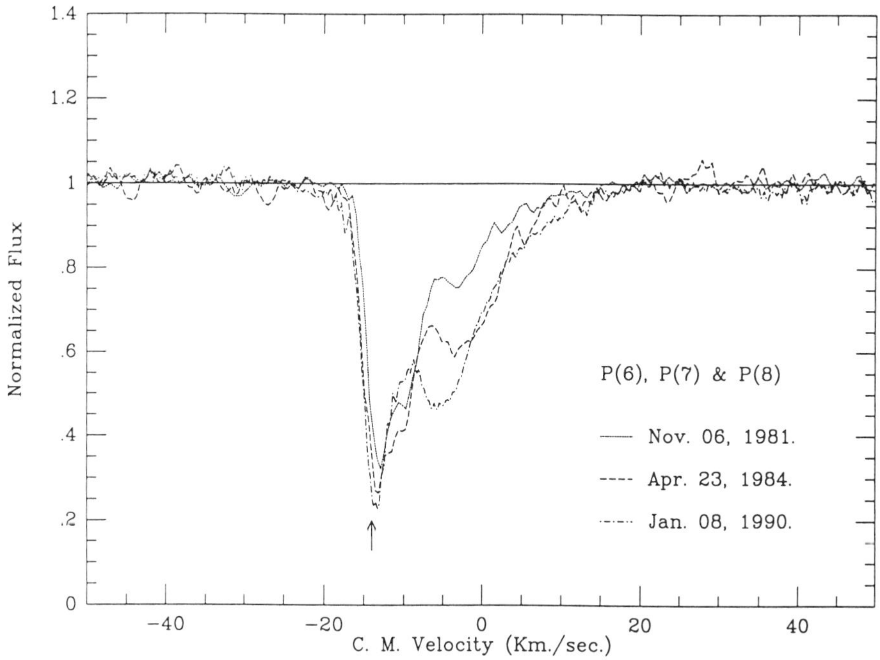

Fig. 1: Temporal variations in the first overtone of ^{12}CO (ΔV = 0-2). The profiles shown are each composites for three adjacent vibrational-rotational lines, averaged to improve the signal to noise ratio. Changes in the line profile shape during the intervening decade are clearly evident. Note that the low velocity component has increased its equivalent width by about a factor of 5, and its radial velocity has increased, indicating an acceleration away from the central star of 0.23 ± 0.03 km s^{-1} per year. Rough LTE Boltzmann relations between lines of the same velocity component produce different rotational temperatures. For example, the cold gas component (see arrow) of the 1990 line profile has a temperature of ≈ 340 K while the other one yields temperatures ranging from 650 to 1200 K.

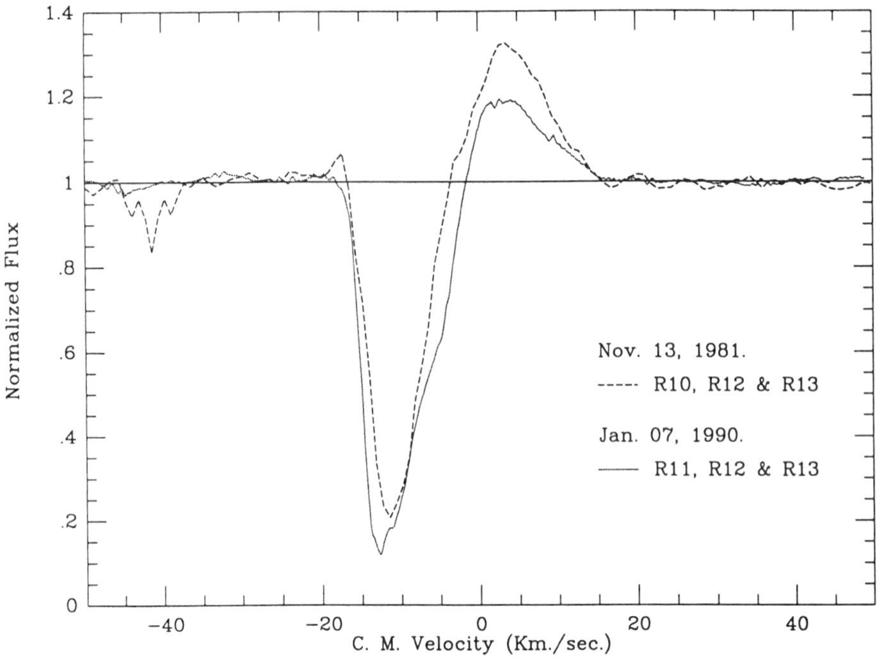

Fig 2: Temporal variations in the ^{13}CO fundamental. The figure above shows two line composites taken 10 years apart. The CO fundamental lines (ΔV = 0-1) show strong P-Cygni emission and are usually saturated. By studying the ^{13}CO isotope we can look at essentially the same region in the circumstellar shell with the advantage that the decreased abundance of ^{13}C ($^{13}C/^{12}C \approx 35$) will not saturate the line core.

RESULTS AND CONCLUSIONS

We have observed temporal changes in the infrared CO vibrational-rotational line profile shapes of IRC+10216 over a period of at least 10 years. Although the available data spans eight cycles of variability of the central star, the current sampling rate is not dense enough to distinguish between cyclic and monotonic effects in the line shapes. We have begun a parametric study of circumstellar dust and gas models that produce synthetic lines which best compare with observations to address this issue.

ACKNOWLEDGEMENTS

The authors would like to acknowledge NOAO for allowing us access to the 4m FTS archives.

REFERENCES

Zuckermann, B. 1980, *Ann. Rev. Ast. Astrophys.*, **18**, 263.

Seventh Cambridge Workshop on Cool Stars, Stellar Systems, and the Sun
ASP Conference Series, Vol. 26, 1992
Mark S. Giampapa and Jay A. Bookbinder (eds.)

ATMOSPHERIC PULSATIONS AND MASS LOSS DRIVEN BY OVERTAKING ACOUSTIC SHOCKS.

P. ULMSCHNEIDER, W. RAMMACHER, H.-P. GAIL Institut für Theoretische Astrophysik, Universität Heidelberg, Im Neuenheimer Feld 561, D–6900 Heidelberg, Germany

ABSTRACT A new pulsation mechanism was found which uses short period acoustic waves as energy source and is triggered by shock overtaking. This mechanism together with the well-known κ–mechanism could be an important mass loss driver for late-type giant stars.

Keywords: Acoustic waves; shocks; pulsation; late-type stars; mass loss

INTRODUCTION

From the analogy with the combustion engine one learns that a stellar pulsation mechanism needs an *energy source* which balances the large radiation losses of the atmosphere resulting from the pulsation and a *trigger* which releases the driving energy at the right phase in the pulsation cycle. In the *combustion engine* the spark ignites the gasoline at the moment of highest compression. The *κ–mechanism* which drives the Cepheid and Mira pulsations works by converting radiation energy into heat at the moment of highest compression, where the opacity at the highest temperature and pressure is largest. In the newly found *shock overtaking pulsation mechanism* (Rammacher and Ulmschneider 1991) short period acoustic shock waves by the process of overtaking are converted into strong shocks at the moment of largest compression. We investigate the possible significance of this mechanism for mass loss in late-type giant stars.

THE SHOCK OVERTAKING PULSATION MECHANISM

From their solar acoustic wave calculations Rammacher and Ulmschneider (1991) conclude: 1. That a single strong acoustic pulse creates pulsational oscillations in the atmosphere. These oscillations decay rapidly due to the large radiation losses (see Fig. 1). 2. Excitation by monochromatic acoustic waves with periods less than the acoustic cut-off period $P_A \approx 180$ s and greater than a period $P_O \approx P_A/5 \approx 40$ s eventually leads to atmospheric oscillations with a period forced by the wave. When these wave calculations are started, transients with periods near P_A are superposed over the shorter period waves. These transients decay rapidly similarly as in the calculation with an acoustic pulse. 3. For an excitation with monochromatic waves with periods less than $P_O \approx P_A/5 \approx 40$ s the wavelength is small enough that shock overtaking occurs in the chromosphere in regions where the temperature (and sound speed) decreases outwardly. This shock overtaking produces a strong shock which generates pulsations with periods $P_P \approx 165$ s near the cut-off period. These pulsations persist as long as the acoustic wave is present (see Fig. 1). 4. The *shock overtaking pulsation mechanism* produces first overtone pulsations (outer atmosphere expansion and inner atmosphere contraction followed by the reverse situation) P_P which are more rapid than the fundamental oscillations (expansion everywhere followed by contraction everywhere) P_A.

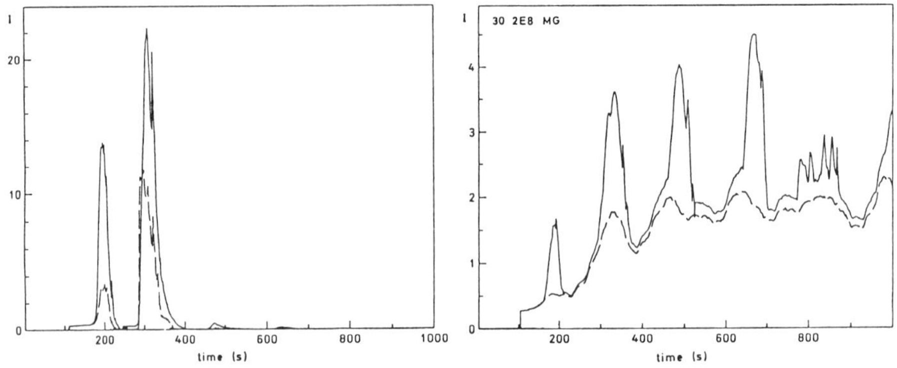

Fig. 1: The intensity of the violet (drawn) and red (dashed) emission peaks of the Mg II k-line for a single acoustic pulse (left) and a short period acoustic wave with $P = 30s$ (right) after Rammacher and Ulmschneider (1991).

MASS LOSS BY THE SHOCK OVERTAKING PULSATION MECHANISM

Fig. 2 shows empirical mass loss rates of stars in the HR-diagram as reported by Cassinelli (1979). Superposed we have indicated various mass loss mechanisms which are currently discussed in the literature. It is widely accepted that early-type stars have mass-loss by *radiation driven winds* and that late-type giants and supergiants above and to the right of a line recently given by Gail and Sedlmayr (1987) as well as Dominik *et al.* (1990) have *dust-driven winds.* Mass loss in Mira-stars is due to *pulsation driven winds.* At the main sequence, stars like the sun have weak mass loss rates due to *thermal winds.* As shown by Hammer (1981) the observed mass loss rates higher in the HR-diagram can not be explained by thermal winds. *Alfvén wave driven winds* possibly exist only below and near the Linsky-Haisch dividing line, above which, due to the very low rotation rates the dynamo action ceases and the stars are not expected to have significant magnetic fields. *Acoustic wave driven winds*, due to shock overtaking of waves of different periods in a spectrum, have been demonstrated to lead to episodic mass loss (Cuntz 1990). However, it has not been shown that this latter process can drive the large observed mass loss rates below the Dominik *et al.* line. We therefore feel that a significant mass loss process is missing to drive mass loss of red giants below the Dominik *et al.* and above the Linsky-Haisch line. Moreover near the Dominik *et al.* line for the dust driven wind mechanism to work an additional mechanism is needed to bring a lot of mass to the dust formation radius.

We propose that this missing mechanism could be the shock overtaking pulsation mechanism which produces and self-starts nonradial pulsations depending on the available short period acoustic wave power and its distribution over the stellar surface. Rammacher and Ulmschneider (1991) find that a large fraction of the short period acoustic wave energy flux is converted into pulsational wave energy. These pulsations would then be assisted and amplified by the κ-mechanism such that both the strong radiation field and the large acoustic wave flux (Bohn 1984) may be tapped for driving the wind loss. This scenario agrees strikingly well with the observational fact that these late-type giants exhibit a wide range of irregular variability. However, a final verdict over this picture can only be given when detailed computations involving the excited atmospheric and envelope oscillations have been made.

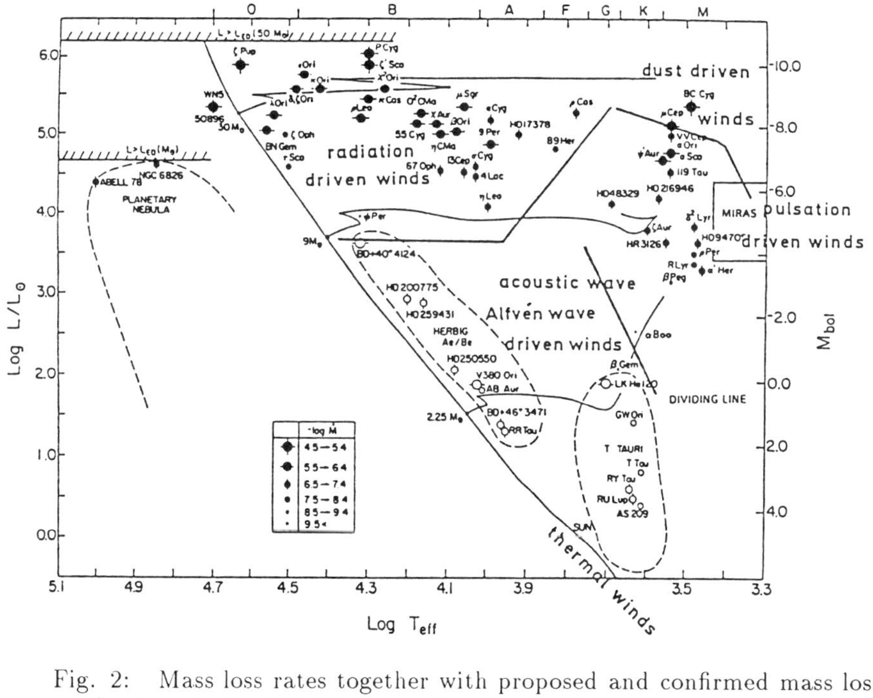

Fig. 2: Mass loss rates together with proposed and confirmed mass loss mechanisms in the HR-Diagram after Cassinelli (1979).

REFERENCES

Bohn, U.: 1984, *Astr. Ap.*, **136**, 338.

Cassinelli, P.: 1979, *Ann. Rev. Astr. Ap.*, **17**, 275.

Cuntz, M.: 1980, *Ap. J.*, **335**, 355.

Dominik, C., Gail, H.-P., Sedlmayr, E., Winters, J.M.: 1990, *Astr. Ap.*, **240**, 365.

Gail, H.-P., Sedlmayr, E.: 1987, *Astr. Ap.*, **177**, 186.

Hammer, R.: 1981, Ph. D. thesis, Univ. Würzburg, Germany.

Rammacher, W., Ulmschneider, P.: 1991, *Astr. Ap.*, in press.

Seventh Cambridge Workshop on Cool Stars, Stellar Systems, and the Sun
ASP Conference Series, Vol. 26, 1992
Mark S. Giampapa and Jay A. Bookbinder (eds.)

CIRCUMSTELLAR DUST IN MIRA VARIABLES AND THE PULSATIONAL MASS LOSS

B.G. ANANDARAO,
Physical Research Laboratory, Navrangpura, Ahmedabad-380009, India.

S. POTTASCH,
Kapteyn Astronomical Institute, P.O.Box 800, AV 9700 Groningen, The Netherlands.

P. SEEMA, AND D.B. VAIDYA
Physical Research Laboratory, Navrangpura, Ahmedabad-380009, India.

ABSTRACT IRAS data on 92 oxygen-rich Mira variables are analysed to evaluate the role of pulsations in massloss among these stars. Using a radiative model, equilibrium temperature, size, mass, and optical depth of circumstellar dust shells have been estimated. A two-component dust shell model is found necessary for fitting the photometric data. The mass of the inner shell shows a positive correlation with the pulsation period while that of the outer shell does not show any such correlation. We suggest that the outer shell is a result of the massloss during the pre-AGB stage while the inner shell represents AGB massloss caused by pulsation with the possible support of radiation pressure. Dust seems to condense at $\approx 10\ R_*$.

Keywords: Mira variables ; circumstellar dust ; massloss : mechanisms ; IRAS data

INTRODUCTION

The massloss processes among the Mira variables have been the subject of intensive study in the recent years (*e.g.* , Wood, 1990). Infrared observations especially with the advent of IRAS have been useful in constraining the theoretical predictions (De Gioia-Eastwood *et al.* 1981; Jura, 1986; Whitelock *et al.* 1987; Habing, 1990). It has been proposed that radiation pressure and the radial pulsations are the two main driving forces for the massloss during the AGB stage (Jones *et al.* 1984; Jura, 1986; Sedlmayr, 1990). Besides, it has also been proposed that the Alfvén waves as well as the acoustic waves could contribute substantially to the massloss (*e.g.*, Hearn, 1990). In order that the radiation pressure be effective it appears that the dust should form close ($\lesssim\ 3\ R_*$) to the star. Clearly the question of how close to the star does the dust condense plays a key role in determining the mechanism of massloss. Here we analyse the IRAS data on a number of oxygen-rich Mira variables using a simple radiative model to derive circumstellar dust parameters like the equilibrium temperature, the size of the shells, their optical depths and masses and correlate these with the pulsation period to estimate the relative role of the massloss mechanisms.

DATA ANALYSIS AND RADIATIVE MODEL

The IRAS photometric and LRS data on 92 oxygen-rich stars listed in Kukarkin *et al.* (1969) catalog are considered. Firstly the infrared excess has been estimated for all the 92 stars assuming that the excess is given by $E = -2.50 \log [S_{12}/S_{25}]$, where $S_{12} = \nu_{12} F_{12}$ and $S_{25} = \nu_{25} F_{25}$, with F_{12} and F_{25} the IRAS fluxes in Jy (color-corrected). A correlation of this excess with the pulsation period P of the Miras gives the best linear fit $E = 0.35 + 0.0001\,P$ with correlation coefficient of 0.5. The poor correlation could be partly due to the possible stellar contribution to 12 μm flux which is ignored here. Removing periods $P \lesssim 200$ days not only improved the fit but the line became steeper. Interestingly, though the correlation is marginal, it is clear that the excess tends to increase with P. This aspect is in line with the results of De Gioia-Eastwood *et al.* (1981) for the oxygen-rich stars and those of Jura (1986)'s for the carbon stars.

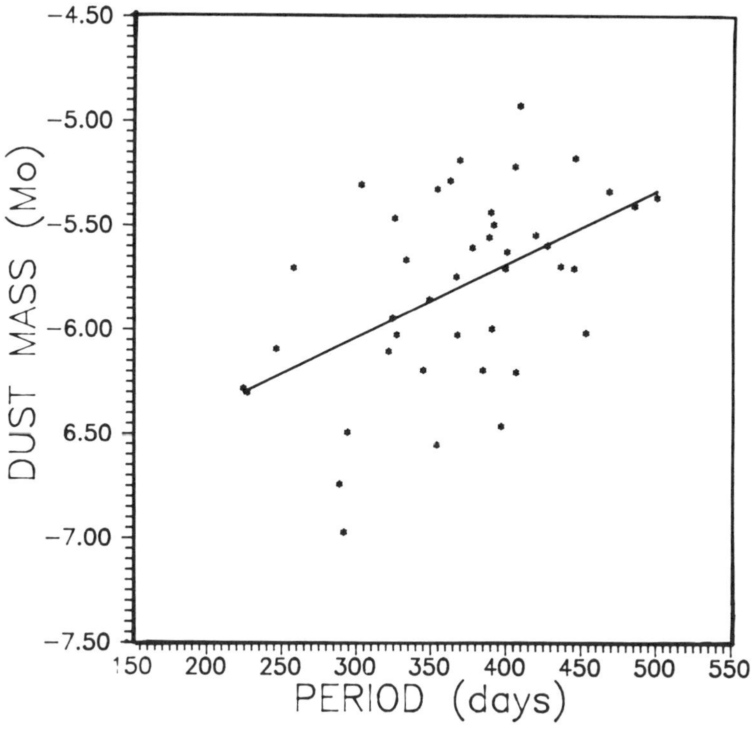

Fig. 1: Log of mass (in solar masses) of the inner dust shell vs P.

We have then made a detailed analysis of some of the sample stars (for which the IRAS flux qualities are termed good) using a radiative model for obtaining the dust parameters in which we assume that the circumstellar dust shell is a thin plane parallel isothermal layer (for details, Anandarao *et al.* 1987).

The relevant stellar parameters (inputs to the model) like the photospheric temperature and the size are computed from well-known relationships (*e.g.* , Wood, 1990). For those stars which showed a clear silicate feature at 9.7 μm in LRS we have used absorption coefficients, Q(λ) from Draine (1981) and for the others we assumed Q(λ) varies as $\lambda^{-1.5}$. A grain size (a) distribution of the form $n(a) = a^{-3.5}$ is assumed with a = 0.1 - 1.0 μm. The uncertainties in our model are about 20% in the determination of dust temperatures and about a factor of 5 in estimating the dust shell sizes and their optical depths.

RESULTS AND DISCUSSION

In general, a two-component dust shell model is required to fit the photometric data. The inner shell temperature is in the range of 250 - 450 K for different stars, while that of the outer shell is between 30 - 50 K. The sizes of the inner shell are around 0.1 arc sec or typically a few tens of stellar radii ($\lesssim$ 100 R_*) while those of the outer shell are several tens of arc secs. We estimate from our results that the dust grains would form at $\approx$ 10 R_* by assuming that the condensation temperature for silicates is $\approx$ 1000 K.

The dust shell masses are computed from the IR data using the formula given in Pottasch (1984). Fig. 1 shows a plot of log of the inner dust shell mass against P for the sample stars: a positive correlation is evident with a coefficient of 0.65 for the line $\log M = -7.64 + 0.0047 P$. This result agrees fairly with that of Jura (1986)'s on carbon stars for massloss rate. The inner shell temperatures do not particularly show any relationship with period in contrast with the result of Whitelock *et al.* (1987) which is based on a limited sample. The parameters of the outer shell do not show any correlation with the pulsation period.

We conclude from our analysis that the inner shell is a result of AGB massloss triggered by the pulsations and supported possibly by the radiation pressure. However, the condensation distance from the star perhaps suggests that there could be other mechanisms operating like sound waves (Hearn, 1990). The outer shell does not seem to represent massloss from AGB stage and hence could be the result of the fossil pre-AGB massloss.

REFERENCES

Anandarao, B.G., Taylor, A.R., and Pottasch, S.R. 1987, *Astr. Ap.*, **203**, 361.

DeGioia-Eastwood, K., Hackwell, J.A., Grasdalen, G.L., and Gehrz, R.D. 1981, *Ap. J. Letters*, **245**, L75.

Draine, B.T. 1981, *Ap. J.*, **245**, 880.

Habing, H.J. 1990, in From Miras to Planetary Nebulae, eds. M.O. Mennessier and A. Omont, (Gif-sur-Yvette:Editions Frontiers), p. 16.

Hearn, A.G. 1990, in From Miras to Planetary Nebulae, eds. M.O. Mennessier and A. Omont, (Gif-sur-Yvette:Editions Frontiers), p. 121.

Jones, T.W., Ney, E.P., and Stein, W.A. 1981, *Ap. J.*, **250**, 324.

Jura, M. 1986, *Ap. J.*, **303**, 327.

Kukarkin, B.V., *et al.* 1969,General Catalog of Variable Star, (Moscow: Academy of Sciences).

Pottasch, S.R. 1984,Planetary Nebulae, (Dordrecht: Reidel).

Sedlmayr, E. 1990, in From Miras to Planetary Nebulae, eds. M.O. Mennessier and A. Omont, (Dordrecht:Reidel), p. 179.

Whitelock, P.A., Pottasch, S.R., Feast, M.W. 1987, in Late Stages of Stellar Evolution, eds. S. Kwok and S.R. Pottasch, (Dordrecht:Reidel), p. 269.

Wood, P.R. 1990, in From Miras to Planetary Nebulae, eds. M.O. Mennessier and A. Omont, (Dordrecht:Reidel), p. 67.

Seventh Cambridge Workshop on Cool Stars, Stellar Systems, and the Sun
ASP Conference Series, Vol. 26, 1992
Mark S. Giampapa and Jay A. Bookbinder (eds.)

EXTENDED 60 μm EMISSION FROM NEARBY MIRA VARIABLES

WENDY HAGEN BAUER   Whitin Observatory, Wellesley College, Wellesley MA 02181

ROBERT E. STENCEL   CASA, Campus Box 391, University of Colorado, Boulder CO 80309

ABSTRACT IRAS has detected extended dust shells at an arc-minute scale around evolved stars. We have examined IRAS scans at 60 μm of nearby Mira variables to look for extension. Significant extension is seen for o Cet, R Leo, R Cas, U Ori, and probably R Hor. Dust shells resulting from steady mass loss appear capable of explaining the observed extension of all these sources except R Leo and R Cas.

Keywords: circumstellar dust, infrared radiation, late-type variables, mass loss, Mira stars

OBSERVED EXTENSION

IRAS has detected extended dust shells at an arc-minute scale around evolved stars (Stencel *et al.* 1989, Hawkins 1989, 1990). This extension is generally not apparent at 12 and 25 μm because the warm dust radiating at these wavelengths is relatively close to the star. At 100 μm, the prevalent cirrus generally renders the background too bumpy for the detection of faint extended emission. We have thus concentrated our work on the 60 μm bandpass.

In order to convert an observed shell extent on the sky to a physical size around the star, one must know the distance to the star. Therefore, we have selected a sample of Mira variables in the solar neighborhood for which distances have been determined by Wyatt and Cahn (1983). We further restricted the sample to those known to have silicate emission.

The individual IRAS survey scans of the stars in our sample were examined to look for extension and source confusion. Maps produced at the Infrared Processing and Analysis Center were also examined to ensure that the extension was reasonably symmetric around the Mira and not the result of alignment with an unrelated extended source.

Extension due to optically thin dust surrounding a bright central source appears as a faint pedestal at the base of the scan profile. Examination of the noise level and the smoothness of the background of the individual IRAS scans showed that peak fluxes greater than roughly 20 Jy are needed in order for the faint extension to have sufficient S/N ratio for modelling purposes.

Table I lists those Miras from the sample brighter than 20 Jy at 60 μm whose individual scans were suitable for detection of extension. A useful indicator of scan extension is the signal width at 10% of peak intensity, W(10%). As the summation of scan profiles can artificially broaden the profile, we prefer to work with the average W(10%) of the individual scans. This average W(10%) for these objects is given in Table I, along with the standard deviation.

TABLE I W(10%) for Bright Miras and Comparison Objects

Object	Flux (60 μm) (Jy)	W(10%) (arc min)	σ (arc min)
RR Aql	27.20	2.44	0.07
R Cas	102.37	2.80	0.08
T Cas	26.51	2.46	0.06
o Cet	300.79	2.76	0.07
U Her	26.92	2.51	0.07
R Hor	53.21	2.60	0.08
R Leo	114.34	2.76	0.13
U Ori	38.81	2.69	0.09
β Gru	40.65	2.45	0.09
Ceres	300	2.51	0.06
IRC+10216	5652.0	2.52	0.08

MODELLING

The dust shell model code used has been previously described by Hagen (1978). The code calculates the emission along a given line of sight. This is then convolved with the IRAS beam as the beam is stepped across the shell. The IRAS beam is divided into "pixels" of 0.1 arc min on a side, and is stepped over the emergent flux distribution in 0.1 arc min steps. One of these pixels, the "central pixel", contains the star and the inner regions of the dust shell. For a star with angular diameter $\phi = 0.05$ arc sec, this central pixel contains the shell emission out to about 135 R*. The observed extension is thus dependent on the brightness of the outer regions of the shell relative to the sum of the brightness of the star and the inner regions of the dust shell.

Input parameters for the dust shell code are: τ (optical depth at 10 μm), T* (stellar temperature), R* (stellar radius), and r_{min} (inner radius of the dust shell, in units of stellar radius). The relative flux does not depend on R*; all models have been run for R* = 500 solar radii. However, when shell extension is predicted, ϕ must be specified. Mass loss rates in terms of these parameters scale as $\tau\ r_{min}$ R*. We consider models with a steady mass loss rate, which gives rise to dust density proportional to the inverse square of the radius.

The grain absorption efficiencies (Q's) are those of Draine (1985), with Q for $\lambda > 25$ μm proportional to λ^{-2}. Although these Q's do not accurately reflect the shape of the 10 μm silicate feature, they should be useful in assessing whether models which explain the observed strength of the silicate feature can also explain the observed extension. However, it should be noted that extension does depend upon the Q's used. Furthermore, there is evidence that the Q's are not the same for all Miras (Little-Marenin and Little 1990).

The modelling shows that for a given contrast in the 10μm silicate feature, extension is greater for greater ϕ and greater r_{min}. Extension is not strongly dependent on T* for constant r_{min}, but if r_{min} increases with T* to keep a constant dust condensation temperature, extension increases as T* increases.

An acceptable model for a given star must explain both the strength of the 10 μm silicate emission feature and the broad-band point source fluxes. The combinations of τ and r_{min} for $T^* = 3250$ K and 2500 K which best reproduce the IRAS observations of our extended sources are presented in Table II. R Leo is not included, as the contrast in the silicate feature is so low that the observed extension cannot be reproduced by our models. However, R Leo does have a significant excess past 10 μm and may well be surrounded by dust which cannot be characterized by the Q's adopted in this work.

The photospheric angular diameter of Mira has been measured to be 0.028 arc sec by Bonneau *et al.* (1982). As similar measurements are not available for other Miras in the sample, we have assumed that they have the same linear diameter and have scaled the angular diameter by the relative distances of Wyatt and Cahn (1983). The extensions produced by the model dust shells are also presented in Table II.

TABLE II Model Fits to Extended Sources

Object	T^* (K)	r_{min} (R^*)	τ	W(10%) obs	W(10%) pred
R Cas	3250	20	0.025	2.80	2.73
	2500	7	0.07		2.62
o Cet	3250	10	0.035	2.76	2.81
	2500	5	0.1		2.70
R Hor	3250	20	0.025	2.60	2.68
	2500	7	0.07		2.59
U Ori	3250	10	0.05	2.69	2.72
	2500	5	0.15		2.63

For all the sources in Table II except R Cas, models for stars at typical Mira temperatures appear capable of explaining the observed extension as due to dust produced from steady mass loss from the central star.

ACKNOWLEDGEMENTS

This work was supported in part by NASA grant NAG 5-1213 to Wellesley College.

REFERENCES

Bonneau, D., Foy, R., Blazit, A., and Labeyrie, A. 1982, *Astr. Ap.*, **106**, 235.
Draine, B. T. 1985, *Ap. J. Suppl.*, **57**, 587.
Hagen, W. 1978, *Ap. J. Suppl.*, **38**, 1.
Hawkins, G. W. 1989, *Bull. A.A.S.*, **21**, 1112.
Hawkins, G. W. 1990, *Astr. Ap.*, **229**, L5.
Little-Marenin, I. R. and Little, S. J. 1990, *Astr. J.*, **99**, 1173.
Stencel, R. E., Pesce, J. E., and Bauer, W. H. 1989, *Astr. J.*, **97**, 1120.
Wyatt, S. P. and Cahn, J. H. 1983, *Ap. J.*, **275**, 225.

Seventh Cambridge Workshop on Cool Stars, Stellar Systems, and the Sun
ASP Conference Series, Vol. 26, 1992
Mark S. Giampapa and Jay A. Bookbinder (eds.)

DUST FORMATION AND CHROMOSPHERES IN R CORONAE BOREALIS STARS: HOW, WHERE, AND WHY?

GEOFFREY C. CLAYTON Center for Astrophysics and Space Astronomy, Campus Box 389, University of Colorado, Boulder, CO 80309

BARBARA A. WHITNEY Harvard-Smithsonian Center for Astrophysics, MS-15, 60 Garden St., Cambridge, MA 02138

S. ADAM STANFORD Department of Astronomy, University of California-Berkeley, Berkeley, CA 94720

ABSTRACT In the 1930's, it was deduced that the spectacular declines seen in R CrB stars were caused by the condensation of carbon-rich gas into an optically thick dust cloud lying along the line of sight. However, we still don't understand the mechanisms of mass loss or dust formation in these stars. The appearance and evolution of chromospheric emission during the declines is also not understood. In the last 5 years, large numbers of visible and ultraviolet spectroscopic, photometric and polarimetric observations have been obtained of several R CrB stars. Using these new data, models with dust formation at 2 and 20 $R_\star$ have been considered. A model with dust forming close to the photosphere of the star seems a better fit to the behavior of the light curve, the evolution of the emission spectrum and the polarization variations. However, for this model to be viable, dust must form at temperatures well above the conventional condensation temperature for carbon dust.

Keywords: R Coronae Borealis stars; chromospheres; circumstellar dust

Observed Behavior of the R CrB Stars

The general behavior of the emission spectrum in R CrB declines is well known. As the photospheric light is extinguished by the forming dust cloud, a rich narrow-line emission spectrum appears. In the visible, this spectrum consists of many lines of neutral and singly ionized metals (Alexander *et al.* 1972). This spectrum, referred to as E1, is shortlived and within a few days or weeks, most of these lines have faded. Some of the early-decline spectral lines remain strong for an extended period of time. These lines, also narrow, referred to as E2, are primarily multiplets of Sc II and Ti II. The late-decline spectrum is dominated by 5 strong broad-lines, Ca II H and K, the Na I D lines and a line at 3888 Å which may be He I. This broad-line (BL) emission spectrum seems to remain fairly constant until the star begins to return to maximum light and the photospheric continuum regains dominance.

The UV spectrum undergoes a very similar evolution (Clayton *et al.* 1992). In the early-decline spectra (E1), there is a pseudo-continuum made up of blends of many emission lines of Fe II, Ti II and other species. The Mg II doublet is present but not yet strong. As time goes on, the early-decline spectrum begins to fade. There is still much blended emission (E2) but Mg II, Mg I λ2851, and some of the Fe II lines have started to become relatively stronger. The late-decline spectrum (BL) is characterized by blended emission from multiplets of Fe II (2) λ2400, Fe II (1) λ2600, Fe II (62, 63) λ2750, as well as from Mg II and Mg I.

Spectropolarimetry of V854 Cen in a deep minimum shows very high continuum polarization which decreases to zero in the emission lines. The continuum is photospheric light scattered off the dust clouds. The emission which is the late-decline BL spectrum is not polarized so it arises in a region outside the obscuring dust cloud.

This spectroscopic and polarimetric behavior makes sense if the star is surrounded by 3 fairly distinct emission regions responsible for the E1, E2, and BL emission, with the E1 emitting region having the smallest radius, surrounded by the E2 emitting region which is in turn surrounded by the BL emitting region. Then the characteristic times for appearance and disappearance of the E1, E2 and BL emission will give information on the dust formation process. The observed behavior implies that:
i) The forming cloud is about the size of the photosphere. However, there is a lot of variation in the time and magnitude at which E1 appears so the cloud will sometimes be smaller than the photosphere or cover only part of the photosphere when it forms. Then as it moves away from the star, it will expand to cover the whole star.
ii) Since E1 is always seen, the forming cloud is never big enough to cover the E1 emitting region. Once it appears, E1 usually fades in 10-30 days. This indicates that the radius of the emitting region and the velocity of the expanding cloud are very similar from decline to decline and from star to star.
iii) The E2 emission lasts 50 -150 days. Therefore, this region is probably about 5 times as large as the E1 emitting region.
iv) The BL emitting region must be very large. If the star remains faint for a lengthy period of time, the BL emission can be seen for hundreds of days.

Models

The E1+E2 emitting region may be a true chromosphere or a region shock heated due to pulsations. The BL emission may be from a steady wind from the star. The appearance and fading of the emission lines during a decline can be described by either of two models. In the first model, gas is ejected from the star in our line-of-sight and moves out at least the escape velocity, 20 km/s. Dust forms at 10 to 20 $R_\star$ when the condensation temperature is reached and eclipses the photosphere. The cloud must be just the right size to eclipse the photosphere and not the chromosphere (E1+E2 emitting region). The emission fades as the cloud expands and covers the chromosphere, and as the dust cloud disperses the photosphere reappears. In the second model, the dust forms close to the photosphere, and then is accelerated to high velocity (100-200 km/s) by radiation pressure. The emission lines fade as the dust passes through the chromosphere, covers it, and then disperses. Both models have an eclipse of the chromosphere which obscures the lower chromosphere (E1) first and gradually expands to cover the entire chromosphere (E1+E2). The BL emitting region is never significantly obscured. Clearly the immediate objection to the second model is that dust cannot form at photospheric temperatures. However, the dust condensation temperature increases with increasing C/H ratio (Donn 1967), and R CrB stars have a C/H ratio of 10^4, compared to 10^{-4} for solar abundances (Lambert 1986). Secondly, there is evidence for shocks in the outer atmospheres of R CrB stars. These would have two effects contributing to dust formation. Shocks change the structure of the outer atmosphere, causing a shallower density law (Willson 1988). A shock propagating through could leave a high density low-temperature region in its wake.

The observational differences between the two models should lie in the timescales–for acceleration of dust, dispersal, and eclipsing of the chromosphere– and in the effect of dust formation on the chromosphere itself, in the case where dust is formed inside the chromosphere. Consider a chromosphere (E1+E2) extent of 3 $R_\star$. If dust is to form at 20 $R_\star$ covering the photospheric disk, it will take 40 weeks for it to expand enough to obscure the chromosphere, assuming a velocity of 100 km/s, and stellar radius of 85 solar radii. Dust formed between 1.1 and 1.5 $R_\star$ will take about 20 days to obscure the chromosphere, which is in good agreement with the observed timescale for chromospheric decay. Dust dispersal will happen very quickly for dust formed in the chromosphere, since density is proportional to r^{-2} for constant velocity. Thus this model must allow for the possibility of either continuous dust formation, or perhaps the formation of dust at each pulsation cycle. This is actually consistent with erratic light curves of R CrB stars. The rise to maximum light rarely is slow enough to be due to dispersal at 20 $R_\star$, but is consistent with dispersal much closer to the star. Finally the mass of dust formed, and covering factor for the star is consistent in both models. For dust forming at 20 $R_\star$, we must require that the gas from which it condensed never be larger than a size which will obscure the chromosphere. To fit the spectral evolution, it should always be just the right size that it obscures only the photosphere and not even the lower chromosphere. This would always be the case when dust is formed inside the chromosphere.

ACKNOWLEDGEMENTS

This work was supported by NASA grants NAGW-2338 and NAG5-87.

REFERENCES

Alexander, J. B., *et al.* 1972, *M.N.R.A.S.*, , 158, 305.

Clayton, G.C., Whitney, B.A., Stanford, S.A., Drilling, J.S., and Judge, P.G. 1992, *Ap. J. Letters*, , in press.

Donn, B. 1967, in Interstellar Grains, Eds. J. M. Greenberg and T. P. Roark, NASA SP-140, p. 191.

Lambert, D. L. 1986, in Hydrogen Deficient Stars and Related Objects, ed. K. Hunger (Reidel: Dordrecht), p. 127.

Whitney, B.A., Clayton, G.C., Schulte-Ladbeck, R.E., and Meade, M.R. 1992, *A. J.*, submitted.

Willson, L. A. 1988, in A Decade of Astronomy with IUE, ESA SP-281, p. 29.

Seventh Cambridge Workshop on Cool Stars, Stellar Systems, and the Sun
ASP Conference Series, Vol. 26, 1992
Mark S. Giampapa and Jay A. Bookbinder (eds.)

WIND-DRIVEN DUST: THE INTERACTION BETWEEN DUST AND GAS IN LATE-TYPE STELLAR ATMOSPHERES AND WINDS

KEITH B. MacGREGOR High Altitude Observatory, National Center for Atmospheric Research, Boulder, Colorado

ROBERT E. STENCEL APAS/CASA, University of Colorado at Boulder

ABSTRACT An assumption inherent to most models of dust-driven winds from cool, evolved stars is that the radiative and collisional drag forces acting an individual dust grain are in balance throughout the flow. We have checked the validity of this supposition of "complete momentum coupling" by comparing the grain motion obtained from such a model with that derived from solution of the full (*i.e.*, including inertia) grain equation of motion. For physical conditions typical of the circumstellar envelopes of oxygen-rich red giants, we find that silicate grains with initial radii smaller than about 5×10^{-6} cm *decouple* from the ambient gas near the base of the outflow. The implications of these results for models of dust-driven mass loss from late-type giants and supergiants are discussed.

Keywords: circumstellar envelopes; dust; mass loss; rarefied gas dynamics

INTRODUCTION

A long-standing problem in stellar physics concerns the dynamics of the extended atmospheres of cool evolved stars: namely, what processes are responsible for the initiation and maintenance of the mass loss such stars are observed to undergo? While many mechanisms have been proposed, few (if any) are capable of consistently accounting for the inferred properties of the winds from late-type giants and supergiants (Holzer and MacGregor 1985). In particular, it is difficult to devise a means for accelerating the flow such that its asymptotic speed is significantly lower than the gravitational escape speed from the stellar surface, as observations of many red giants indicate (Judge and Stencel 1991).

One mechanism which has received considerable attention and has been extensively applied relies upon the radiative acceleration of dust grains to induce overall expansion of a gaseous, circumstellar envelope. Although the physics of dust formation is presently not well-understood, once condensed, individual grains can experience a net outward acceleration by absorbing and scattering photons which originate in the stellar photosphere. The subsequent motion of a grain is impeded only by collisional encounters with the microscopic constituents of the background gas through which it drifts (viscous drag). Such collisions provide the means for transmitting a portion of the momentum acquired by the grains to the ambient atmosphere. The momentum so transferred is then diffused throughout the envelope by collisions between gas molecules. If the frequency of grain-gas collisions is sufficiently high, the collective "force" exerted on the gas by many drifting grains can be adequate to drive a continuous outflow.

Numerous wind models have been calculated which utilize the basic acceleration mechanism described above, either by itself (Kwok 1975; Tielens 1983; Dominik *et al.* 1989, etc.) or in concert with other physical processes (Jones *et al.* 1981; Bowen 1988). All of these calculations have in common the assumption that the grain motion everywhere in the flow can be computed by equating the local radiative and collisional drag forces. Physically, such a prescription of "complete momentum coupling" implies that virtually all of the momentum gained by a grain through the absorption of radiation from the stellar photosphere is transferred via collisions to the atmospheric gas (Gilman 1972). This dynamical state is in contrast to the one in which the grains are completely *uncoupled* from the background gas. If the drag force acting on a grain becomes vanishingly small, the grain retains all of the momentum imparted to it by absorbed or scattered photons, and attains its terminal speed only asymptotically (Wickramasinghe 1972). Tabak *et al.* (1976) concluded electrostatic drag to be negligible in cool, dust forming stellar atmospheres, and we have confirmed that coulomb drag is negligible for ionization fractions $< 10^{-4}$, typical of circumstellar envelopes. Berruyer and Frisch (1983) examined relaxation of the momentum coupling assumption. We will comments on their results presently.

The purpose of the this note is to summarize the results of a systematic analysis of the validity of the complete momentum coupling hypothesis in the winds of cool, evolved stars. A more complete report has been submitted to Ap.J.

DISCUSSION

We have investigated the consistency of dust-driven wind solutions by comparing the results of calculations in which the gas and dust components are assumed to be well-coupled collisionally with an independent integration of the dust grain equation of motion including inertia. For the case of a high luminosity, red giant-like model atmosphere, we find that (1) the resulting mass loss rate is a sharply increasing function of dust grain size, and (2) the assumption of complete momentum coupling appears to break down for grain sizes smaller than $\sim 8 \times 10^{-6}$ cm, suggesting that smaller grains may not be sufficiently coupled to the gas to drive the stellar wind.

These results, while interesting, are based on a number of approximations. To improve the range of validity, several improvements should be investigated, including: (1) a better model for grain formation, growth, and sputtering; (2) more accurate values for the Planck mean radiation pressure efficiency, factor, $\bar{Q}_{pr}(T_{eff}, a)$ for silicate grains; (3) inclusion of a spectrum of grain sizes; (4) an improved, physical description of the forces resulting in extension of the hydrostatic portion of the model atmosphere; (5) the effects of different grain characteristics (*e.g.*, density and shape); and (6) solving for consistent flows in cases where the grains appear to only partially couple to the gas. Ideally, these improvements will allow for the application of radiative transfer methods to then synthesize spectral features resulting from the gas and dust, in order to predict Doppler drift signatures.

The results of Berruyer and Frisch (1983) appear comparable until one notes the different assumptions used by those authors. They ignore grain growth and fix the mass loss rate. These assumptions result in suspiciously high densities at the base of the flow in their uncoupled case. Unfortunately, their solution method does not allow for an easy confirmation of their results.

Among the interesting problems raised if our initial results are born out are: (1) if small grains are selectively expelled from red giant atmospheres, where are the seed nuclei for the larger grains required for dust-driven winds; (2) will a prolonged period of small particle escape promote chemical differentiation of an atmosphere; (3) if sputtering is effective in reducing grain size, resulting in small particle escape, how can dust-driven winds be sustained, and; (4) what is the behavior of small grains that approach molecular sizes, given the prediction of Elitzur, Brown, and Johnson (1989) that radiation pressure on molecules can initiate stellar winds? These questions motivate the suggested improvements outlined above.

ACKNOWLEDGEMENTS

The National Center for Atmospheric Research is sponsored by the National Science Foundation. Partial support for this effort also has been derived from NASA grants NAG5-816 and NAS5-29337 to the University of Colorado. We acknowledge useful discussions with Sun Kwok.

REFERENCES

Berruyer, N. and Frisch, H. 1983 *Astr. Ap.*, **126**, 269.

Bowen, G. H. 1988, *Ap. J.*, **329**, 299.

Dominik, C., Gail, H.P., and Sedlmayr, E. 1989, *Astr. Ap.*, **223**, 227.

Draine, B.T. 1981, in *Physical Processes in Red Giants*, ed. I.Iben & A.Renzini (Dordrecht:Reidel), p.317.

Elitzur, M., Brown, J. A., and Johnson, H. R. 1989, *Ap. J.*, **341**, L95.

Fix, J.D., 1969, *M.N.R.A.S.*, **146**, 37.

Gail, H.P., and Sedlmayr, E. 1988, *Astr. Ap.*, **206**, 153.

Gilman, R.C. 1972, *Ap. J.*, **178**, 423.

Gilman, R.C. 1974, *Ap. J. Suppl.*, **28**, 397.

Holzer, T.E., and MacGregor, K.B. 1985, in *Mass Loss from Red Giants*, ed. M.Morris & B.Zuckerman (Dordrecht:Reidel), p.229.

Jones, T.W., Ney, E.P., and Stein, W.A., 1981 *Ap. J.*, **250**, 324.

Judge, P.G., and Stencel, R.E. 1991, *Ap. J.*, **371**, 357.

Kozasa, T., Hasegawa, H., and Seki, J. 1984, Ap.Sp.Sci. **98**, 61.

Kwok, S. 1975, *Ap. J.*, **198**, 583.

Lattimer, J.M., Schramm, D.N., and Grossman, L. 1978, *Ap. J.*, **219**, 230.

Phillips, J. 1979, *Astr. Ap.*, **71**, 115.

Tabak, R., Hirth, J., Meyrick, G. and Roark, T. 1975 *Ap. J.*, **196**, 457.

Tielens, A.G.G.M. 1983, *Ap. J.*, **271**, 702.

Wickramasinghe, N.C. 1972, *M.N.R.A.S.*, **159**, 269.

PART VI

Computing Stellar Atmospheres

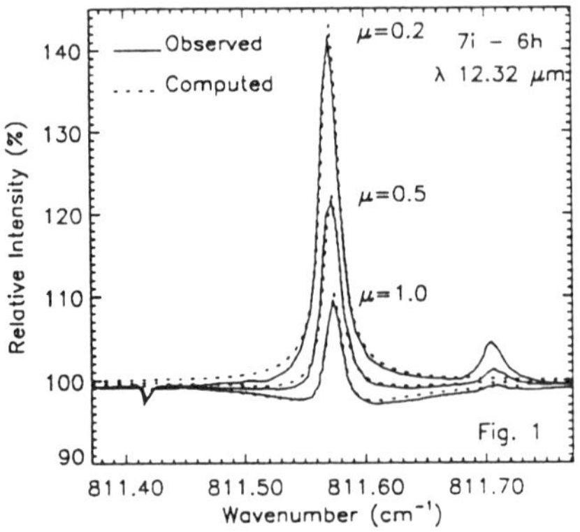

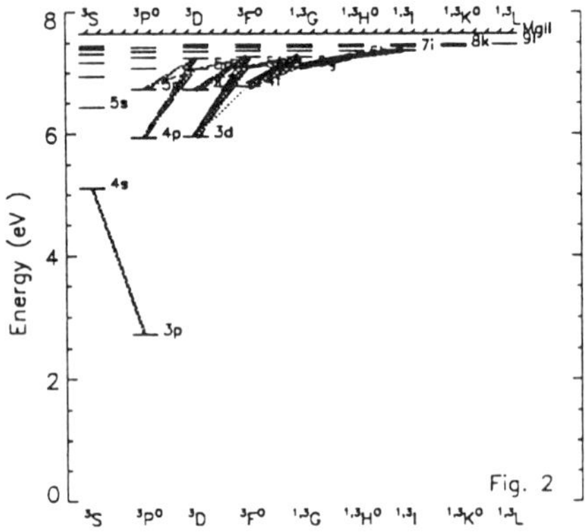

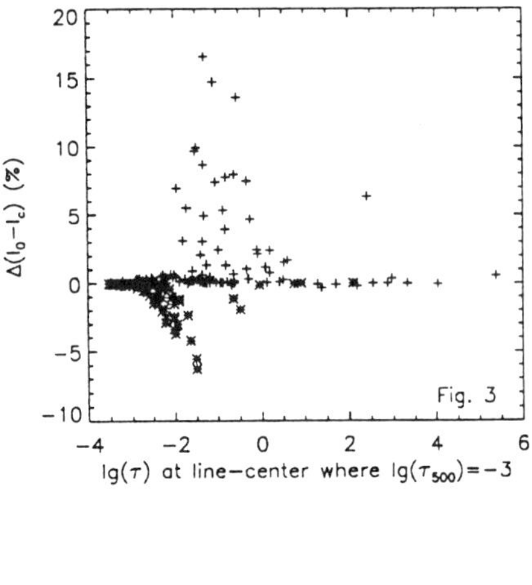

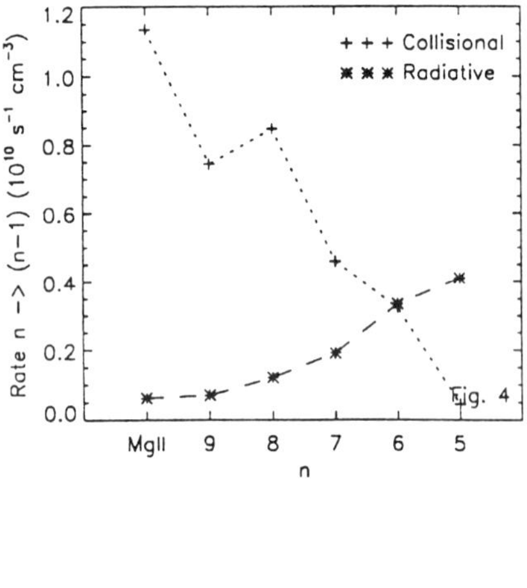

Figures 1 – 4 from "Formation of the Mg I 12 μm Lines" by M. Carlsson, R.J. Rutten, and N.G. Shchukina.

Seventh Cambridge Workshop on Cool Stars, Stellar Systems, and the Sun
ASP Conference Series, Vol. 26, 1992
Mark S. Giampapa and Jay A. Bookbinder (eds.)

THE PANDORA ATMOSPHERE PROGRAM

EUGENE H. AVRETT and RUDOLF LOESER Harvard-Smithsonian Center for Astrophysics, 60 Garden Street, Cambridge, MA 02138

ABSTRACT PANDORA is a general-purpose non-LTE computer program for calculating stellar atmospheric models and spectra. We give an overview of the development, current uses, and current capabilities of this program, compare it with other similar programs, and describe our plans for its further development.

Keywords: model atmospheres, radiative transfer

INTRODUCTION

An atmosphere program is used to calculate the detailed structure of an atmospheric region that is characterized in some general way, and then to determine its emitted spectrum. Such calculations can be used to study and identify the atmospheric conditions responsible for different spectra. The physical properties of an observed atmosphere then can be taken to be those for which the calculated and observed spectra agree in significant detail.

For the study of stellar photospheres, the ATLAS program developed by Kurucz (1970) is widely used to calculate plane-parallel model atmospheres in radiative equilibrium and LTE for given values of effective temperature, surface gravity, and chemical composition, and to calculate the resulting emitted spectra. Such model calculations depend on the availability of realistic opacities. Kurucz (1991) has compiled and computed the atomic and molecular opacities for over 58 million lines. This line list includes most of the lines in the solar spectrum and almost all of the solar line opacity as a function of wavelength. (Many of these lines have uncertain wavelengths because their energy levels have not been measured.)

Many observed stellar spectra show that the assumptions of LTE and radiative equilibrium have only limited validity. In order to treat non-LTE effects and non-radiative heating, a computer program must solve the coupled equations of radiative transfer and statistical equilibrium for an atmospheric region, subject to non-radiative energy transport effects. PANDORA was one of the first large-scale computer programs developed for this purpose (see Vernazza, Avrett, and Loeser, 1973). Another general non-LTE program from the same era is the LINEAR program of Auer, Heasley, and Milkey (1972, also see Mihalas 1978). The LINEAR program is still being used, but, as far as we know, has not been kept under active development. We have continued to improve PANDORA, to include new physical processes and capabilities. For general-purpose non-LTE calculations, PANDORA has been joined by the programs that Mats Carlsson and Lawrence Anderson describe in the next two papers in this workshop volume.

DESCRIPTION OF THE PANDORA PROGRAM

PANDORA deals with a one-dimensional atmospheric region that is either finite or semi-infinite in the plane-parallel case, or that has spherical symmetry. The atmospheric layers can be stationary or can be moving relative to each other. Illumination from an external source can be prescribed at the atmospheric boundary (front or back in the finite case).

The temperature and density stratification must be prescribed in order to carry out the calculation. Typically a temperature distribution is given and the density is determined assuming 1) hydrostatic equilibrium (balancing gas pressure and gravity), or 2) pressure equilibrium in the absence of gravity. After the radiative properties of the atmosphere are calculated, a new temperature distribution can be determined subject to energy balance constraints, using parameters that describe given non-radiative processes. For example, Avrett and Loeser (1988) in a study of quasar broad emission line regions calculated the internal structure and the emergent spectrum of a constant-pressure cloud of a given large optical thickness with given illumination incident upon one face. The temperature distribution in this case was determined from the constraint of radiative energy balance.

Given the temperature and density distributions, the non-LTE energy level populations of the various atoms and ions must be calculated. In typical problems the line and continuum optical depths can reach very large values, but collisions between atoms and electrons are too infrequent to establish LTE. The statistical equilibrium equations determine the populations at any point in the atmosphere given the properties of the radiation at that point. From the radiative transfer equation, the radiation at that point depends on the radiative sources and hence on the populations throughout a large surrounding volume. The set of coupled transfer and statistical equilibrium equations must be solved for both the populations and the radiative intensities at all points in the atmosphere. For a single line transition this can be called the two-level transfer problem. The general case of an atom or ion with many interacting line transitions can be called the multi-level transfer problem. These cases will be discussed further in the following sections. First, however, we consider some other parameters that need to be determined as part of the overall solution.

As the temperature changes throughout the atmosphere we may need to calculate the level populations of each successive ionization stage of a given element, *e.g.* Si I, Si II, Si III, Si IV, ..., and each of these stages may have a large number of discrete energy levels. For simplicity, PANDORA treats, in any one computer run, the levels of a single ionization stage together with the lowest level of the next higher stage to determine the relative populations of these levels, where the sum of the populations of these two stages is a certain fraction of the total element abundance. Thus after calculating the results for Si I-II, assuming all silicon to be in these two stages, the Si II-III calculation will exclude the Si I fraction, etc. A subsequent Si I-II calculation will exclude the fraction in Si III and higher stages. Such calculations require several iterations to get consistent results.

The electron number density needed in these calculations depends on the ionization of various constituents. Consider the simple case in which the electron density n_e is the sum of the proton density n_p and the contribution Zxn_H from all other elements, where n_H is the total hydrogen density and Z, in the simplest case, is the fraction of those elements that are once ionized, thus contributing one electron. For hydrogen we can write $n_H = n_{HI} + n_p$ (when molecular hydrogen can be ignored), and $n_{HI} = b_1 n_e n_p \psi(T)$ where $n_{HI} = n_e n_p \psi(T)$ is the LTE Saha-Boltzmann equation for the neutral hydrogen density (in the lowest level) and b_1 is the departure coefficient, or correction factor, obtained from the detailed non-LTE calculation for hydrogen. Eliminating n_{HI} and n_p gives a quadratic equation that can be solved for n_e, given Z, n_H, b_1, and T. The indirect dependence of Z and b_1 on n_e is treated iteratively.

The different ions (including neutral atoms) interact with one another not only through their contributions to n_e, but also because the different ions influence and are influenced by the common radiation field which varies with wavelength and with location in the atmosphere. In solving for the populations of a given ion we need to know in detail how other ions absorb and emit radiation at the transition frequencies of the given ion. PANDORA treats this dependence iteratively, including all important non-LTE effects. In the case of the Sun the non-LTE populations of H, H^-, He I, He II, C I, Si I, Mg I, Fe I, Al I, Ca I, and Na I are computed to account for n_e and for the observed continuum, and the influence of the very large number of lines in the spectrum is considered when calculating photoionization rates. For this purpose PANDORA uses a sampled set of Kurucz's line opacities (see Avrett, Machado, and Kurucz 1986).

THE TWO-LEVEL TRANSFER PROBLEM

Consider a line transition involving levels 1 and 2. Ignoring absorption and emission by other sources and ignoring stimulated emission for simplicity, the radiative transfer equation for frequency ν in the line is

$$\frac{dI_\nu}{dz} = \begin{cases} -\text{ absorption} + \text{emission} \\ -\frac{h\nu}{4\pi}\phi_\nu(n_1 B_{12} I_\nu - n_2 A_{21}) \end{cases} \tag{1}$$

where B_{12} and A_{21} are the Einstein coefficients for absorption and spontaneous emission, z is geometrical distance measured in the direction of the radiation intensity I_ν, n represents number density, and ϕ_ν is the normalized absorption profile that includes the effects of Doppler broadening near line center and other types of broadening in the wings. Here for simplicity we have assumed that the emission profile is also ϕ_ν so that the absorbed and emitted photons are uncorrelated. This is the assumption of complete frequency redistribution (CRD) which is a good approximation in the Doppler core of a line. When a line such as the hydrogen Lα line has substantial opacity in the wings, and when the perturbing densities are low, then the absorption of a photon in the wings is followed by the re-emission of a photon with a frequency that is likely to be close to that of the absorbed photon, rather than a frequency near line center. This is the more general case of partial frequency redistribution (PRD). PANDORA allows any line to include PRD effects (see Vernazza, Avrett, and Loeser 1981 and Avrett and Loeser 1984).

Using the simpler approximation of CRD in equation (1), we introduce the monochromatic optical depth τ_ν (again in the direction of I_ν), and write

$$\frac{dI_\nu}{d\tau_\nu} = -I_\nu + S, \tag{2}$$

where S is the line source function (in this simple case without stimulated emission)

$$S = \frac{n_2}{n_1}\frac{A_{21}}{B_{12}}. \tag{3}$$

Considering only radiative and collisional transitions between levels 1 and 2, the statistical equilibrium equation in this simple case is

$$0 = n_1(B_{12}\bar{J} + C_{12}) - n_2(A_{21} + C_{21}), \tag{4}$$

where $\bar{J}$ is the angle and frequency integral

$$\bar{J} = \frac{1}{4\pi}\int d\Omega \int d\nu\, \phi_\nu I_\nu. \tag{5}$$

One can solve equation (2) for I_ν in terms of S and substitute the result in equation (5) to obtain

$$\bar{J} = \Lambda\{S\}. \tag{6}$$

In this integral-operator expression, $\bar{J}$ at any one depth in the atmosphere depends on the values of S at all depths. This is just a formal solution of the transfer equation. The statistical equilibrium equation (4) can be written in terms of S as

$$S = \frac{\bar{J} + \epsilon B}{1 + \epsilon}, \tag{7}$$

where $\epsilon = C_{21}/A_{21}$ and $B = (2h\nu^3/c^2)\exp(-h\nu/kT)$ (the Planck function without stimulated emission). Combining equations (6) and (7) gives a set of simultaneous equations for S at each depth.

This is basically the way the two-level problem must be solved when ϵ is very small. At least $1/\sqrt{\epsilon}$ iterations would be required to obtain a solution by alternately evaluating equations (6) and (7), *i.e.*, by iterating between the radiative transfer and statistical equilibrium equations.

THE MULTI-LEVEL TRANSFER PROBLEM

When there are many radiative transitions in the model atom, equations (6) and (7) still can be applied to each transition, but ϵ and B now take a more general form based on all transition pathways between the upper and lower levels of the given transition, apart from the direct radiative transition. PANDORA is based on this generalization of the two-level solution. The problem is that the ϵ and B terms for a given radiative transition depend on the solutions of other radiative transitions, and one must iterate on these terms even though the solution for each transition, given ϵ and B, is exact. The advantage is that the individual solutions show the causes of calculated results (*e.g.* why a given source function has a particular variation with depth). The disadvantage is that this method generally requires more computer time than other methods developed in recent years. As explained above, PANDORA solves the combined transfer and statistical equilibrium equations for each radiative transition, iterating on the coupling between radiative transitions. On the other hand, the MULTI computer program of Carlsson (1986, and the following paper in this volume) uses a perturbation procedure in which the correction term can be calculated approximately and very simply. The first-order dependence of $\bar{J}$ on the depth-dependent number densities in equations (3) and (6) for each transition is substituted into the statistical equilibrium equations for the set of number densities at each depth in this correction scheme. The number of simultaneous equations then to be solved (for number-density corrections) is the number of levels N_L times the number of depths N_D. This procedure is repeated successively, iterating to account for the approximate way equation (6) is treated. PANDORA has to solve N_D simultaneous equations for each radiative transition, and then repeat these calculations iteratively because the ϵ and B terms for one radiative transition depend on the other radiative transitions. An ion with N_L levels can have as many as $N_L(N_L - 1)/2$ radiative transitions. Solving N_D simultaneous equations this many times takes longer than solving one set of $N_L x N_D$ simultaneous equations when N_L is large. (When $N_L = 2$ there is no difference.) For this reason MULTI is intrinsically faster than PANDORA when the number of radiative transitions is large.

NON-LOCAL STATISTICAL EQUILIBRIUM

We have recently incorporated velocity terms in the PANDORA statistical equilibrium equations, so that equation (4) would be written as

$$-\frac{d}{dz}(n_1 V) = n_1(B_{12}\bar{J} + C_{12}) - n_2(A_{21} + C_{21}) \tag{8}$$

to take account of the effect of a mass velocity V (increasing in the direction of the z coordinate). The derivative term causes n_1 at each depth to depend on n_1 at other depths, so that we use a finite difference procedure to solve the differential equation for $n_1(z)$ assuming given values for the other terms, including n_2. Then we evaluate

$$G = \frac{1}{n_1}\frac{d}{dz}(n_1 V), \tag{9}$$

and finally write equation (8) as

$$0 = n_1(B_{12}\bar{J} + C_{12} + G) - n_2(A_{21} + C_{21}), \tag{10}$$

which is solved as before. In this way we account for the effects of mass flows not only by the Doppler shifts of lines, but also by including such advection terms in the equations of statistical equilibrium. The term on the left side of equation (8) is more important for ionization than for excitation (*i.e.*, when n_2 represents the ion density).

We also include particle diffusion velocities when there are steep temperature and ionization gradients. The hydrogen atom and proton diffusion velocities are given by $V_H = [X/(1+X)]V_A$ and $V_p = -[1/(1+X)]V_A$ where X is the ionization fraction n_p/n_{HI} and

$$V_A = D_X d\ln X/dz + D_T d\ln T/dz \tag{11}$$

is the ambipolar diffusion velocity. The coefficients are functions of the local number densities and the temperature. The diffusion velocity enters the calculation in the same way as the mass flow velocity except that hydrogen atoms and protons diffuse in opposite directions. See Fontenla, Avrett, and Loeser (1990, 1991) for details.

RECENT SOLAR APPLICATIONS

We briefly discuss recent models of the solar chromosphere and transition region calculated with PANDORA. Figure 1 shows the calculated distribution of temperature and mass density with height for a model of the average quiet Sun, and the approximate calculated depths at which certain features of the spectrum are formed.

The results in Figure 1 can broadly be interpreted as follows. While almost all of the energy emitted by the Sun emerges from the photosphere, only about 10^{-4} of this photospheric output must be dissipated mechanically to cause the temperature to rise in the chromosphere. Hydrogen is almost completely neutral in the photosphere and low chromosphere, but starts to become ionized as the temperature increases and the density decreases. Then the temperature increases abruptly, causing hydrogen to be almost fully ionized. The temperature must rise abruptly in this partially ionized region since otherwise the strong Lα line would radiate away more energy than can be supplied by any non-localized process of mechanical energy dissipation. About 10^{-5} of the photospheric output is dissipated in the transition region and corona, and most of this energy is transported back down the steep temperature gradient, by thermal conduction and particle diffusion, to provide the energy emitted in the hydrogen and helium resonance lines. Differential mass flows also may be important in transporting energy in this region.

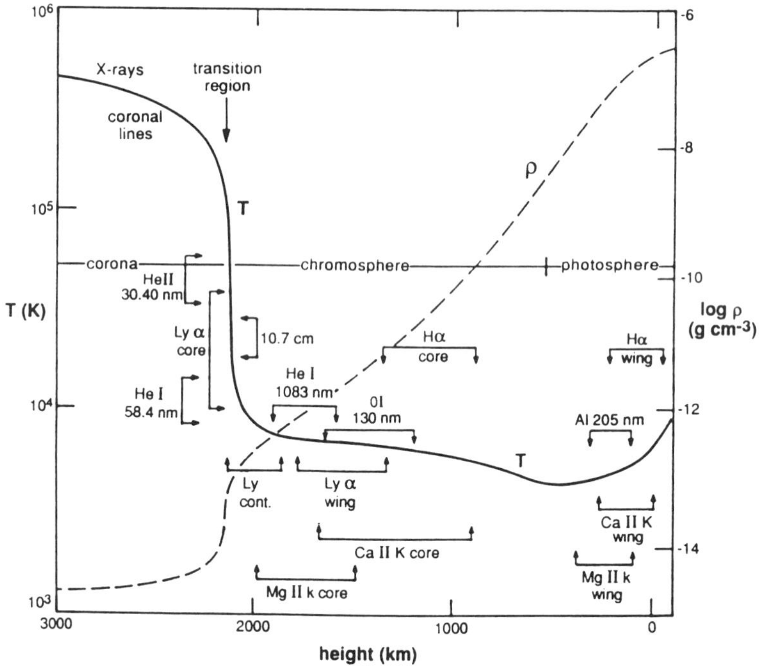

Figure 1: Temperature and density in the solar atmosphere vs. height for a model of the average quiet Sun. The calculated ranges where various emitted continua and line features originate are included.

Coronal line radiation in the He I continuum below 50 nm illuminates the upper chromospheric layers from above causing weak ionization of He I followed by recombination, thus strengthening the absorption of the chromospheric continuum by the He I lines at 587 nm and 1083 nm on the disk.

Observations show that the atmospheric structure varies from place to place on the solar disk. We can model different observed features separately because the changes of atmospheric parameters with height usually occur over vertical distances that are smaller than the horizontal scale of such features as cell and network elements of the quiet Sun. See Vernazza, Avrett, and Loeser (1981) and Fontenla, Avrett, and Loeser (1991).

The results shown in Figure 1 were obtained by computing the line and continuum spectra of H, H^-, H_2, He I-II, C I-IV, O I-II, CO, Na I, Mg I-II, Al I-II, Si I-III, S I, K I, Ca I-II, and Fe I. The density distribution is determined assuming hydrostatic equilibrium, and n_e is calculated as described in Section 2. In the photosphere and chromosphere the temperature distribution has been adjusted empirically so that the computed spectrum agrees as well as possible with observations. The temperature distribution in the transition region is calculated theoretically so that the downward flow of energy by thermal conduction and particle diffusion balances the radiative losses. The flow of energy from the corona is the value required to balance the total radiative losses in the transition region. We specify only the density, temperature, and temperature gradient at the base of the transition region, as determined from the chromospheric model. The computed intensities of transition-region lines are reasonably close to the corresponding observed values. See Fontenla, Avrett, and Loeser (1991).

PANDORA IMPLEMENTATION

Rather than going through the calculations in a fixed way, PANDORA assumes a hands-on approach by the user. There are many input switches for specifying 1) alternative methods for specific calculations, 2) amounts of printout, 3) levels of printout detail, and 4) auxiliary output files (for use by other programs needing various PANDORA-computed quantities).

The computational properties of some of the procedures PANDORA uses for various steps in the calculation have not been studied in detail for all types of applications. What has worked well for specific calculations in the past may not work well when applied to new, different regimes. We have provided many input control parameters to try to prevent unreasonable numerical behavior.

For some steps in the calculation, *e.g.* the evaluation of equation (6), PANDORA provides a menu of different methods that have been found to be well suited for specific situations. In each case the user can specify which method to use. We have begun to document what we have learned about the advantages and disadvantages of these alternatives, to help others choose.

The user is not compelled to specify any of these options, alternative methods, or numerical control parameters, because PANDORA provides defaults for all of them. There now exists an extensive collection of atomic data input files for ions of general interest; in many cases more than one version of a given ion is available, ranging from abbreviated to detailed atomic models.

PANDORA DOCUMENTATION

There are four sources of information about the program. 1) The PANDORA printout is intended to be reasonably self-explanatory. There are printing options that allow the details of almost any calculation to be printed so that any step can be verified. 2) A user's guide called "How to Run PANDORA" is available which lists all input parameters and program options along with notes explaining many of the parameters and options. We continue to add explanatory material to this guide. 3) The basic PANDORA documentation, begun in October 1966, now consists of over 2800 pages of handwritten program notes specifying every program change or addition. Finally, a readily accessible source of information consists of 4) the derivations and explanations included in the publications listed below.

PANDORA BIBLIOGRAPHY

This is an annotated list of publications that describe the computational methods used in PANDORA.

Avrett, E. H., and Loeser, R. 1969, Formation of line and continuous spectra. I. Source-function calculations, Smithsonian Astrophys. Obs. Spec. Rept. No. 303, 99pp.

[Complete derivation of the two-level and multi-level statistical equilibrium and transfer equations, with integral operators used for the formal solution of the transfer equation.]

Avrett, E. H. 1971, Solution of non-LTE transfer problems, *J. Quant. Spectrosc. Radiat. Transfer*, **11**, 511-529.

[Basic formulation and discussion of non-LTE computational methods.]

Vernazza, J. E., Avrett, E. H., and Loeser, R. 1973, Structure of the solar chromosphere. I. Basic computations and summary of results, *Ap. J.*, **184**, 605-631.

[Derivation of the equations of statistical equilibrium and radiative transfer for lines and the hydrogen Lyman-continuum; electron number density, hydrostatic equilibrium, and H^- non-LTE equations.]

Vernazza, J. E., Avrett, E. H., and Loeser, R. 1976, Structure of the solar chromosphere. II. The photosphere and the temperature-minimum region, *Ap. J. Suppl.*, **30**, 1-60.

[Effect of line opacities throughout the spectrum based on Kurucz data; continuum data calculations for wavelengths 0.13 – 500 μm.]

Vernazza, J. E., Avrett, E. H., and Loeser, R. 1981, Structure of the solar chromosphere. III. Models of the EUV brightness components of the quiet Sun, *Ap. J. Suppl.*, **45**, 635-725.

[Derivation of the energy balance equations, Lα partial redistribution, supplementary levels in the statistical equilibrium equations and in the Lyman-continuum transfer equation. Continuum data calculations for wavelengths 40 nm – 3 cm.]

Avrett, E. H., and Loeser, R. 1984, Line transfer in static and expanding spherical atmospheres, in Methods in Radiative Transfer, ed. W. Kalkofen (Cambridge: Cambridge Univ. Press), 341-379.

[Derivation of the line transfer equations in spherical geometry with radial expansion and partial frequency redistribution.]

Hartmann, L., and Avrett, E. H. 1984, On the extended chromosphere of α Orionis, *Ap. J.*, **284**, 238-249.

[Application of PANDORA calculations in spherical geometry with radial expansion.]

Avrett, E. H. 1985, Recent thermal models of the chromosphere, in Chromospheric Diagnostics and Modelling, ed. B. W. Lites (Sunspot, NM: National Solar Observatory), 67-127.

[Detailed calculations of net radiative cooling rates.]

Avrett, E. H., Machado, M. E., and Kurucz, R. L. 1986, Chromospheric flare models, in The Lower Atmosphere in Solar Flares, ed. D. F. Neidig (Sunspot, NM: National Solar Observatory), 216-281.

[Describes the use of Kurucz's sampled line opacities in photoionization rate calculations and spectrum synthesis.]

Avrett, E. H., and Loeser, R. 1987, Iterative solution of multilevel transfer problems, in Numerical Radiative Transfer, ed. W. Kalkofen (Cambridge: Cambridge Univ. Press), 135-161.

[Derivation and discussion of the equivalent two-level atom method of solving multi-level problems, with tabulated numerical solutions.]

Avrett, E. H., and Loeser, R. 1988, Radiative transfer in the broad emission-line regions of quasi-stellar objects, *Ap. J.*, **331**, 211-246.

[Energy-balance non-LTE model atmosphere calculations with incident radiation, heavy-element cooling, and X-ray absorption.]

Mauas, P. J., Avrett, E. H., and Loeser, R. 1988, Mg I as a probe of the solar chromosphere, *Ap. J.*, **330**, 1008-1021.

[Includes a discussion of how to treat multiplet transitions in a simple way.]

Luttermoser, D. G., Johnson, H. R., Avrett, E. H., and Loeser, R. 1989, Chromospheric structure of cool carbon stars, *Ap. J.*, **345**, 543-553.

[Modeling of N-type carbon stars with PRD calculations of the Mg II lines.]

Mauas, P. J., Avrett, E. H., and Loeser, R. 1989, Computed profiles of the C I multiplets at λ1561 and λ1657, *Ap. J.*, **345**, 1104-1113

[Partial frequency redistribution in multiplet lines.]

Fontenla, J. M., Avrett, E. H., and Loeser, R. 1990, Energy balance in the solar transition region. I. Hydrostatic thermal models with ambipolar diffusion, *Ap. J.*, **355**, 700-718.

[Derivation of the effects of hydrogen ambipolar diffusion on the hydrogen number density calculations and on energy transport.]

Mauas, P. J., Avrett, E. H., and Loeser, R. 1990, On carbon monoxide cooling in the solar atmosphere, *Ap. J.*, **357**, 279-287.

[CO opacity and net radiative cooling rate calculations.]

Mauas, P. J., Machado, M. E., and Avrett, E. H. 1990, The white light flare of June 15, 1982: Models, *Ap. J.*, **360**, 715-726.

[Detailed set of flare model calculations using PANDORA.]

Fontenla, J. M., Avrett, E. H., and Loeser, R. 1991, Energy balance in the solar transition region. II. Effects of pressure and energy input on hydrostatic models, *Ap. J.*, **377**, 712-725.

[Results of model calculations with hydrogen ambipolar diffusion for quiet and active regions of the solar atmosphere.]

Chang, E. S., Avrett, E. H., Mauas, P. J., Noyes, R. W., and Loeser, R. 1991, Formation of the infrared emission lines of Mg I in the solar atmosphere, *Ap. J.*, **379**, L79-L82.

[Line transfer calculation with a 41-level atomic model for Mg I, including charge exchange and collisions with hydrogen atoms.]

PLANS FOR FURTHER DEVELOPMENT

We continue to add new capabilities to PANDORA in order to include features needed for ongoing research and to improve the performance and usefulness of the program. A new version of PANDORA usually is completed every six months or so.

In collaboration with J. M. Fontenla, we are currently investigating 1) the diffusion of elements heavier than hydrogen, so that our energy balance modeling of the solar transition region can be extended to higher temperatures, 2) mass flows and related time-dependent effects, and 3) how both are affected by magnetic fields. These studies require numerous program changes and additions.

While the set of line opacities we use are derived from Kurucz's extensive line lists, the corresponding line emission is simply assumed to be a combination of thermal emission and scattering. We need to incorporate a treatment of non-LTE line blanketing similar to that developed by Anderson (1989, and subsequent paper in this volume).

As discussed in Section 4, PANDORA is slower than MULTI in multi-level calculations with many radiative transitions. We plan to adopt procedures similar to those described in the following paper by Carlsson to reduce this difference.

REFERENCES

Anderson, L. S. 1989, *Ap. J.*, **339**, 558.

Auer, L. H., Heasley, J. N., and Milkey, R. W. 1972 Kitt Peak National Observatory Contribution No. 555.

Avrett, E. H., and Loeser, R. 1984, in Methods in Radiative Transfer, ed. W. Kalkofen (Cambridge: Cambridge Univ. Press), 341.

———. 1988, *Ap. J.*, **331**, 211.

Avrett, E. H., Machado, M. E., and Kurucz, R. L. 1986 in The Lower Atmosphere in Solar Flares, ed. D. F. Neidig (Sunspot, NM: National Solar Observatory), 216.

Carlsson, M. 1986, A computer program for solving multilevel non-LTE radiative transfer problems in moving or static atmospheres, Uppsala Astronomical Observatory Report No. 33.

Fontenla, J. M., Avrett, E. H., and Loeser, R. 1990, *Ap. J.*, **355**, 700.

———. 1991, *Ap. J.*, **377**, 712.

Kurucz, R. L. 1970, Atlas: A computer program for calculating model stellar atmospheres, Smithson. Astrophys. Obs. Spec. Rept. No. 309, 291 pp.

———. 1991, in Stellar Atmospheres: Beyond Classical Models, ed. L. Crivellari, I. Hubeny, and D. G. Hummer (Dordrecht: Kluwer), 440.

Mihalas, D. 1978, Stellar Atmospheres, (San Francisco: Freeman).

Vernazza, J. E., Avrett, E. H., and Loeser, R. 1973, *Ap. J.*, **184**, 605.

———. 1981, *Ap. J. Suppl.*, **45**, 635.

Seventh Cambridge Workshop on Cool Stars, Stellar Systems, and the Sun
ASP Conference Series, Vol. 26, 1992
Mark S. Giampapa and Jay A. Bookbinder (eds.)

THE MULTI NON-LTE PROGRAM

MATS CARLSSON Institute of Theoretical Astrophysics, P.O. Box 1029, Blindern, N-0315 Oslo, Norway

ABSTRACT MULTI is a general and portable program for solving multi-level non-LTE radiative transfer problems. The computer code is described and recent developments are discussed. Special emphasis is on the basic assumptions and CPU-time requirements. With these methods it is possible to solve a 66-level problem with 380 transitions treated self-consistently in five minutes on a modern high-end workstation. A personal overview is given of application areas where more work is needed.

Keywords: model atmospheres; radiative transfer; spectral line formation

INTRODUCTION

Our primary source of information about the conditions in stellar atmospheres is the detailed wavelength variation of the radiation, the stellar spectrum. The spectrum depends on the physical conditions in the regions of the star from where photons may escape into space. The complexity of the interactions between photons and the stellar gas often makes it impossible to directly infer the physical conditions from the spectrum. Instead, a modeling technique is used.

From a small set of free parameters (*e.g.* the effective temperature, the acceleration of gravity and the chemical composition) a model atmosphere is calculated by solving the appropriate equations describing the physics of a stellar atmosphere. The emergent radiation from the model is calculated and compared with observations. The model parameters are then adjusted until the calculated spectrum resembles the observed one.

Several simplifying assumptions are made — partly to reduce the number of free parameters and expose the physical conditions that are of decisive importance and partly to make the problems numerically tractable. In stellar photospheres, the energy is mainly transported through convection and radiation. Furthermore, the distribution over energy states of an atomic species can be found from the local temperature and density (LTE, Local Thermodynamic Equilibrium). Even if millions of spectral lines have to be taken into account for the radiative energy balance, the problem is numerically tractable. Extensive sets of photospheric model atmospheres have been constructed that compare favourably with observations (*e.g.* Gustafsson *et al.* 1975). In stellar chromospheres, the temperature increases with height which means there is an energy input in addition to radiation and convection. As long as this heating is poorly known it is not possible to formulate a proper energy equation. To make matters even worse, the assumption of LTE breaks down which makes the non-linear problem also highly non-local. For a better understanding of chromospheric physics it is necessary to solve a large number of atmospheric problems with given temperature structures (to semi-empirically find the "best" atmospheric model). Other applications of non-LTE modeling are the development of spectral diagnostics, investigations of effects of dynamics, three-dimensional radiative transfer, the determination of non-LTE effects on abundance determinations etc. There is thus a need for general, efficient methods for solving non-LTE problems.

In the next section one particular such computer code, MULTI, is described in some detail. The CPU-time requirements for given problems are discussed, both for MULTI and for the class of methods known as operator perturbation techniques. Finally a personal overview is given of application areas where more work is needed to further the understanding of cool star atmospheres.

DESCRIPTION OF THE MULTI PROGRAM

MULTI is a code for solving multi-level non-LTE radiative transfer problems in moving or static atmospheres. The code in its original form implements the Scharmer (1981) operator perturbation technique. For a detailed description of the operator, the generalization to multi-level problems and the code implementation, the reader is referred to Scharmer (1981), Scharmer & Carlsson (1985) and Carlsson (1986).

The code has been written to be highly portable and modular. The philosophy behind the implementation is to provide a well tested basic module to treat the standard model atmosphere problem. The detailed physics that must be incorporated to treat specific problems is the responsibility of the user and the standard version of MULTI therefore does not provide a very large set of subroutines and options as PANDORA does (see Avrett & Loeser, this volume). The code is, however, very efficient which permits experimentation with uncertain parameters to a much larger extent.

The basic assumptions/restrictions are:

* The temperature structure of the atmosphere is given. There is thus no energy equation and any given energy balance can only be achieved through semi-empirical adjustments of the temperature structure.
* The atmosphere is assumed to be one-dimensional with lateral homogeneity. In the standard version plane parallel geometry is assumed but there are versions that can treat spherical geometry.
* All lines are assumed to be separate in frequency with no blends. There are versions without this restriction.
* Complete redistribution (CRD) is assumed for all lines. There are versions that can treat partial redistribution (PRD).
* The mass density as a function of height is either given or computed from the condition of hydrostatic equilibrium.

The main properties of the code are:

* The statistical equilibrium and the radiative transfer equations are solved for all lines and all continua *simultaneously*. This ensures consistency.
* The code can treat any given macroscopic velocity field. With high amplitude velocity fields it is necessary to have a large number of frequency points to cover the doppler shifted Doppler core of the lines but since the CPU-time scaling is at most linear with the number of frequency points this

is a small cost even for velocity fields with amplitudes of tens of the thermal Doppler width.

* MULTI is written in strict compliance with the ANSI standard FORTRAN-77 and is therefore highly portable. It has been run on some thirty different brands of computers and almost the same number of operating systems. Sixty copies of the code on tape have been distributed.
* Large numerical cancellations are avoided through analytical preconditioning of the equations. 32-bit arithmetics normally gives high enough numerical precision for the calculations. On some computers and some problems (typically involving many ionization stages of one element) it is necessary to use double precision to get an exponent range large enough.
* The time consuming parts vectorize well and are also well adapted to parallelization. Almost linear speedups have been achieved on parallel architectures with a small number of CPUs.

New features in version 2.0 compared with the version 1.0 described in Carlsson (1986) are:

* The photoionization can be calculated from a given radiation field. In that way it is possible to read in the radiation field from a calculation including line-blanketing in millions of lines.
* Photoexcitation from other lines can be included (*e.g.* pumping by Ly-α)
* The opacity package is more flexible with opacity contributions given in tables. New opacities can easily be added and data can be updated.
* Background opacity sources can be treated out of LTE.
* The collisional input is more general allowing many different formats (CI, CE, Ω, charge transfer, collisions with hydrogen).
* Collisional-Radiative switching (Hummer & Voels 1988) helps the convergence in tricky cases (typically hydrogen).
* There is a choice between the original Scharmer global operator and a local operator (Olsen *et al.* 1986). The local operator is to be preferred for very large problems since the slower convergence is then more than outweighed by the shorter CPU-time per iteration. With the local operator it is possible to solve problems with several hundred atomic energy levels.
* There is no longer a need to recompile when changing the number of atomic levels or atmospheric depth-points.
* It is simple to switch between single precision and double precision.

CPU-TIME REQUIREMENTS

Operator perturbation techniques consist of three main parts:

1) The calculation of the error using the exact equations and the present approximation to the solution (called the formal solution step)
2) The calculation of the approximate operator, usually the filling of matrix elements.
3) The inversion of the approximate operator, usually the solution of a linear set of equations.

The relative CPU-time usage of these parts as a function of the number of atomic energy levels is shown in Fig. 1. It is evident that for small problems, the calculation of the error term is dominant. The number of iterations needed for convergence is here the critical factor. If the convergence can not be speeded up we need faster formal solvers or minimize the number of grid points by a more optimal use of the quadrature points in depth and frequency. To speed up the matrix filling or the equation system solver (which is what is done by local operators) will not be cost-effective for smaller problems. For larger problems the gain in these parts can outweigh the larger number of iterations needed for convergence to result in a total gain in computing speed. This is shown in Fig. 2. It is also seen that a problem with 100 atomic energy levels can be solved in less than 30 minutes on a modern workstation.

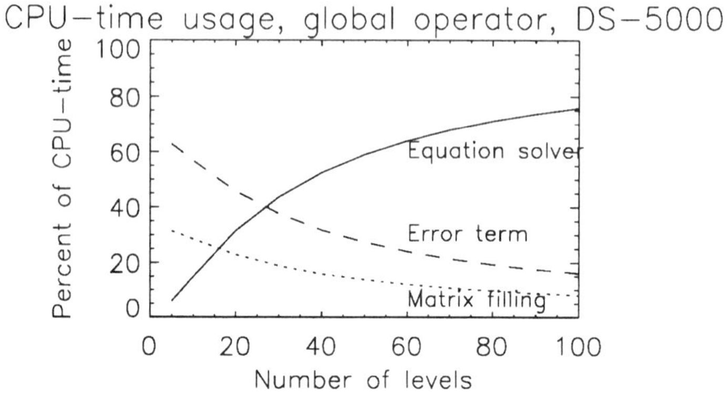

Fig. 1 The relative CPU-time usage using a global operator as a function of the number of atomic energy levels. The computer used was a DECstation-5000/200 workstation.

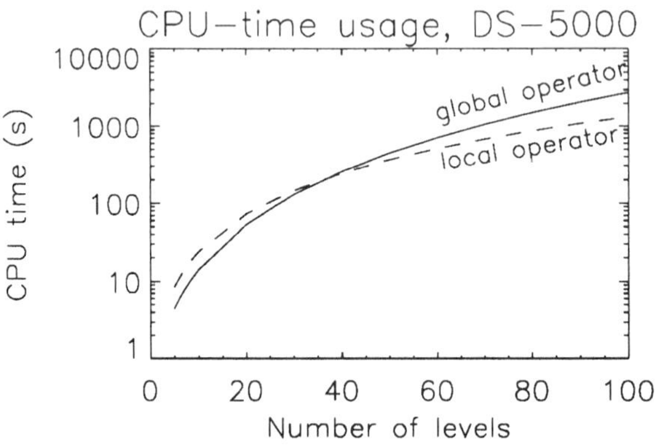

Fig. 2 The CPU-time usage using a local and a global operator as a function of the number of atomic energy levels. The computer used was a DECstation-5000/200 workstation.

APPLICATIONS

Now that we have very efficient methods to solve non-LTE radiative transfer problems, what are the areas where we need more work? The following is a personal list and is by no means complete.

* *Semiempirical modeling.* This may seem like a very outdated field – what is the point in pretending that stars are flat and without lateral structure and never changing? This is a valid point – one has to remember that a static, one-dimensional model at best is a starting point. The semiempirical approach is far from satisfactory from the point of view of uniqueness because of the large number of free parameters. A difference between the calculated and the observed spectrum is usually interpreted as an error in the temperature scale to which an adjustment is made to decrease the difference. The cause might just as well be the inadequacy of some of the initial assumptions. Inappropriate assumptions may thus lead to erroneous conclusions about the atmospheric structure and because of the large number of free parameters, the errors will not show up as a discrepancy between the calculated and the observed spectrum. With these restrictions kept in mind, these models are still useful for testing spectral diagnostics and non-LTE effects and effects of poorly known atomic data. The models may also have some meaning in representing a rough mean as a starting point for more physically realistic modeling.
* *Specific line formation problems.* Many features in stellar spectra are still poorly understood. Potential diagnostic uses of spectral lines remain unused due to the need for non-LTE modeling to explore their use. We need to explain the flux-flux relations. Comparing collisionally excited lines and photoexcited lines may lead to new diagnostic tools.
* *Partial redistribution.* Proper partial redistribution calculations have been made only for a limited set of the lines where such effects can be assumed to be important. In hydrodynamic calculations it is important to find efficient ways of computing the PRD effects in hydrogen lines. Furthermore, most PRD calculations have been made in static atmospheres. What are the effects of velocity fields on lines formed in PRD? Will macroscopic velocity fields have the same effects on the frequency redistribution as on the line broadening (so that the same free parameter, microturbulence, can be used for both)?
* *Many level problems.* Abundance analysis often relies on lines with substantial excitation potential. To calculate the non-LTE effects on these lines it is necessary to include very many energy levels. Other application areas where many-level atomic models are necessary are the formation of lines in the infrared, Rydberg line formation and the formation of lines in complicated atoms like iron. The problem is here partly a problem in atomic physics; many of the necessary cross-sections are unknown or poorly known. Astronomical spectra combined with non-LTE modeling may help in narrowing down some cross-sections. Extended experimenting with a fast code can show to which rates the solution is most sensitive (*e.g.* Carlsson *et al.* , this volume). To understand the line formation processes in complicated model atoms, new techniques are needed.

* *Dynamics.* Can micro/macroturbulence be completely explained by macroscopic velocity fields in the form of convection and waves? What are appropriate spectroscopic diagnostics of different wave modes? Will radiative cooling rates calculated from a static model bear any resemblance to cooling rates from a dynamic model? Different timescales in the dynamics and in the rates between energy levels may lead to situations far removed from the static case (Carlsson & Stein, this volume). Non-LTE radiation dynamics in chromospheric conditions is now computationally possible and we may expect to see a rapid development of the field in the near future.
* *Magnetic fields.* The full treatment of radiative transfer in the presence of magnetic fields is complex but tractable. Large non-LTE computations are necessary to interpret magnetically sensitive lines in the far infrared (like the Mg I 12μm lines).
* *Molecular lines.* Which molecules can be adequately treated in LTE? What is the role of molecules in the cooling of chromospheres? Timescales for the formation of molecules are often very long. A static treatment of the molecular dissociation equilibria and a fully dynamic computation could possibly give drastically different answers to the questions above.
* *Multi component models.* Most models have been one-dimensional with all variables being only a function of height. It is necessary to investigate the effects of inhomogeneous atmospheres by the construction of multi component models. It is important to impose criteria (like pressure balance) to minimize the number of free parameters.
* *2D – 3D radiative transfer.* Ideally, the radiative transfer in multi component models should be computed taking the full geometry (2D or 3D) into account. The radiative transfer techniques of today are fast enough to allow a full 2D or 3D treatment for small model atoms (*e.g.* Castor *et al.* 1991, Nordlund 1991) but more work is needed both in developing methods and in applications.
* *Energy balance – non-LTE line blanketing.* The classical semi-empirical model atmosphere approach assumes a given temperature structure. In the photosphere, where LTE often is a good approximation, the full energy equation can be solved. The corresponding non-LTE problem is much more complex and more work on non-LTE line blanketing is needed (see Andersen, this volume).

Finally I would like to stress the need to *understand* the calculations. It is not enough to put in some atomic and atmospheric models, run through with a non-LTE code and compare with observations. It is important to try to understand the processes involved in the particular problem to ascertain the validity of the assumptions and the atomic model used. The number of free parameters must be kept to a minimum. If an agreement between the observed spectrum and the model has been achieved by using a large number of fudge-parameters, the model has lost its value. One might as well use the star itself as a model since we there have a perfect agreement with the observations! A discrepancy between observations and computations can always be removed by the introduction of fudge-parameters but it is in insisting on physically sound parameterizations we may learn new physics.

REFERENCES

Carlsson, M., 1986, *Uppsala Astronomical Observatory Report*, **33**.

Castor, J.I., Dykema, P.G., Klein, R.I., 1991, in Stellar Atmospheres: Beyond Classical Models, eds. L.Crivellari, I.Hubeny, D.G.Hummer, Kluwer Academic Publishers, p. 49.

Gustafsson, B., Bell, R.A., Eriksson, K., Nordlund, Å., 1975, *Astr. Ap.*, **42**, 407.

Hummer, D.G., Voels, S.A., 1988, *Astr. Ap.*, **192**, 279

Nordlund, Å., 1991, in Stellar Atmospheres: Beyond Classical Models, eds. L.Crivellari, I.Hubeny, D.G.Hummer, Kluwer Academic Publishers, p. 61.

Olsen, G.L., Auer, L.H., Buchler, J.R., 1986, *J. Quant. Spectrosc. Radiat. Transfer*, **35**, 431

Scharmer, G.B., 1981, *Ap. J.*, **249**, 720.

Scharmer, G.B., Carlsson, M., 1985, *J. Comput. Phys.*, **59**, 56.

Seventh Cambridge Workshop on Cool Stars, Stellar Systems, and the Sun
ASP Conference Series, Vol. 26, 1992
Mark S. Giampapa and Jay A. Bookbinder (eds.)

COMMENTS ON THE USE OF PANDORA

DONALD G. LUTTERMOSER
Physics & Astronomy Dept., Iowa State University, Ames, Iowa 50011

ABSTRACT One of the most versatile NLTE radiative transfer codes available to the public is the PANDORA code created by E.H. Avrett and R. Loeser of the Harvard–Smithsonian Center for Astrophysics. A description of the advantages and disadvantages of using the code is presented here.

Keywords: Radiative transfer; NLTE; computer codes; PANDORA

INTRODUCTION

I describe the advantages and disadvantages of using the non–local thermodynamic equilibrium (NLTE) radiative transfer code PANDORA created by E.H. Avrett and R. Loeser (see their paper in this conference series). Under development since 1966, it is a very large and sophisticated computer program which contains over 3500 subroutines. PANDORA solves the coupled equations of radiative transfer and statistical equilibrium in a self consistent manner using the *equivalent two–level atom* approach. This is one of the major two techniques for solving these equations — the other being the *complete linearization* method. See sections 12–1 and 12–3 in Mihalas (1978) for further comments on these two competing techniques.

I first began to use PANDORA in 1986 while working on my dissertation research of cool carbon star chromospheres (Luttermoser *et al.* 1989). I have successfully operated PANDORA on VAX mainframes, a microVAX II, and on VAXstations. The bulk of my current calculations are now done on a VAXstation 3100. I will first present what I consider to be the major disadvantages of PANDORA, followed by advantages. Although users of other large NLTE codes may claim to have encountered no disadvantages in using these codes, I remind the reader that no computer code is perfect. One must balance disadvantages and advantages with respect to the goals of the research in question.

THE DISADVANTAGES

PANDORA checks the machine on which it is operating for various attributes as a safeguard against overloading the machine. As a result of these safeguards, PANDORA is operating system dependent. There is a VMS version and a Unix version. Since I only have experience with the VMS version, the comments here pertain to VAXes. PANDORA requires an extremely large virtual memory allocation to operate — 50,000 pages (1000 pages equals 0.512 megabytes (Mb) of virtual memory) are required for the page file quota on a PANDORA user account. It is often difficult to get a system manager to give this large of a quota on a multi-user machine. This of course can be circumvented by having your own workstation. Once PANDORA fills internal memory, it dumps data to disk during a run (over and above the virtual memory page swapping) to a temporary file. For example, on a VAXstation 3100 with 12 Mb of RAM, a run utilizing the background Kurucz bound-bound opacities requires 106,000 blocks of free disk space (note that 1000 blocks equals 0.512 Mb of disk space). A 5-level hydrogen model atom in spherical geometry, including a macroscopic velocity field and Lyman-α in partial redistribution requires approximately 200,000 free blocks!

The equivalent two-level approach to the solution of the transfer equation can be slow, particularly in a cool, low density medium (*i.e.,* M-type and N-type giant star atmospheric models). It is fairly quick in warmer (K type and earlier) giant models and in main sequence models. The main difficulty in the cooler, low density models results from the net radiative rate ($A_{ji}\rho_{ji}$) of a transition being negative and much greater than the collisional and other rates (Z_{ji}) associated with the transition. This can lead to negative departure coefficients and source functions, which can lead to inconsistencies in the solution (see Avrett and Loeser (1987) for a detailed description of this problem).

To operate PANDORA in its most basic form requires about a two week learning period. PANDORA can generate a tremendous amount of output regarding information of the run. Fortunately this output is well commented, so only a few extra months are usually required to fully understand it. Although PANDORA has a user's manual describing the various input parameters and switches, there is no manual that describes the physics. One must consult the various published papers by Dr. Avrett to understand the physics and mathematical techniques used in the code (see Avrett and Loeser in this workshop proceedings for a list).

THE ADVANTAGES

Perhaps the main advantage of PANDORA is the tremendous amount of physics it contains. One can perform calculations in either plane-parallel or spherical geometries; include macroscopic velocity fields in the rate, source function, and line profile calculations; assume a semi-infinite atmosphere or finite slab model; and irradiate the atmosphere with an external radiation field. Many types of opacities are included in the code, from bound-free transitions of 11 different species to various free-free opacities and scattering events, to name just a few. Atomic models are read in as external data files, so one can construct any type of arbitrary atomic model or modify older atomic models as updated cross-sections are published.

PANDORA contains many self-correcting mechanisms to overcome difficulties in convergence. Self-consistency checks are made for each iteration to confirm whether a solution is converged. The ability exists to print out as much information on a run as one may require for detailed analyses. This leads to tremendous insight on the physical mechanisms involved in the radiative transfer. Apart from the long output file, one can generate additional output files which contain a variety of information from net radiative cooling rates to the emergent spectrum. Such output files are handy when using PANDORA calculations in supplementary research. For instance, I am currently calculating NLTE net radiative cooling rates from PANDORA in a *snapshot* approximation of the Bowen (1988) hydrodynamic models of Mira–type variables. These cooling rates are then mapped back into the Bowen code to refine the models. At the same, the synthetic spectra generated by PANDORA can be compared to observations (see Luttermoser and Bowen in this conference series). In essence, the main disadvantages of the code results directly from its enormous size, which also gives rise to the primary advantage of the code.

CONCLUSION

I have outlined the advantages and disadvantages of using the NLTE radiative transfer code PANDORA. The majority of both the advantages and disadvantages arise from the large size of the code. Although the equivalent two–level approach is often criticized, it is very useful in gaining physical insight on the mechanisms controlling the excitation, ionization, and line broadening in a stellar atmosphere.

As an example, Luttermoser and Johnson (1992) have used PANDORA to investigate the ionization and excitation in late–type giant star atmospheres. Some of the interesting discoveries of this paper include: (1) The assumption of detailed balance, especially in the Lyman lines, can lead to severe over/under–excitations and ionizations in the chromospheric/photospheric regions of these models. (2) Carbon abundance plays a distinct role in the ionization and excitation of hydrogen through the carbon bound–free opacity under the Lyman lines. (3) *Backflowing* chromospheric (or shock) radiation has a major influence on the excitation and ionization of hydrogen and the neutral metals at photospheric depths. Without the detailed printout and the equivalent two–level atom approach of PANDORA, it would have been virtually impossible to extract all the complications in the transfer of radiation in these models.

ACKNOWLEDGEMENTS

Most of the work I have accomplished during my graduate and post–doctoral careers would not have been possible without PANDORA. Because of this, I owe E.H. Avrett and R. Loeser many thanks and am deeply indebted to them. I also wish to thank my dissertation advisor H.R. Johnson and my employers of the post–doc positions I have had, J.L. Linsky and L.A. Willson — their willingness to let me work on my own research projects is greatly appreciated. I also owe NASA a variety of thanks for funding these research projects on various ADP and IUE grants. Finally, I thank P.M. Marcum for useful comments on the manuscript.

REFERENCES

Avrett, E.H. and Loeser, R. 1987, in Numerical Radiative Transfer, ed. W. Kalkofen, Cambridge: Cambridge University Press, p135.

Bowen, G.H. 1988, *Ap. J.*, **329**, 299.

Luttermoser, D.G., Johnson, H.R., Avrett, E.H., and Loeser, R. 1989, *Ap. J.*, **345**, 543.

Luttermoser, D.G. and Johnson, H.R. 1992, *Ap. J.*, (to appear 20 March 1992).

Mihalas, D. 1978, Stellar Atmospheres, 2nd ed., San Francisco: Freeman.

Seventh Cambridge Workshop on Cool Stars, Stellar Systems, and the Sun
ASP Conference Series, Vol. 26, 1992
Mark S. Giampapa and Jay A. Bookbinder (eds.)

PAM: A CODE FOR NON-LTE LINE BLANKETED ATMOSPHERES

LAWRENCE S. ANDERSON Ritter Observatory, The University of Toledo, Toledo, OH 43606

ABSTRACT *PAM* is a computer program designed for calculating stellar atmospheres and their emergent radiation, including the thermodynamic and spectral effects of millions of lines, under the global constraints of non-LTE statistical equilibrium, hydrostatic equilibrium, and energy conservation.

Keywords: stellar atmospheres, radiative transfer, line blanketing

STATISTICS WITH MILLIONS OF TRANSITIONS

The goal of any non-LTE radiative transfer solution is to solve for atomic level populations and radiation fields consistent with the equations of statistical equilibrium and radiative transport. In a complex ion with many levels and transitions, the full treatment of the statistical equilibrium is prohibitive and not particularly instructive. More insight is gained when one simplifies the ion to its essentials. In structure calculations, one is interested in the total effect of an ion (related to the total oscillator strength out of any one level or assembly of levels) and the ability of electrons to recombine and cascade toward the ground state. Thus statistical approaches which account for these effects are sufficient. *PAM* incorporates a number of such approaches, itemized below.

A. Model Ions. A complex real ion may be mapped onto a simple model ion in a straightforward manner, by assigning an assembly of real levels to each model level. Model transitions become sums of real transitions between these assemblies, in the same way a multiplet is a sum of lines. The mathematics of this separation of real and model statistics appears in Anderson (1989).

B. Line Coalescence. Once the transitions have been grouped into model 'multiplet' bands, one may employ various methods to simplify their composite cross section profile. A familiar method is to coalesce the transitions into one line with the total oscillator strength and average broadening properties. This method works if there is no concern for the real emergent spectrum profile, or for the effects of real line blends (the real emergent spectrum can be recovered later via a formal solution of the radiative transfer equations with detailed profiles, given the model populations). Line coalescence is described in Anderson (1991). It works best for light ions with relatively few lines.

C. Frequency Sampling. For heavy ions with many thousand transitions contributing to each model 'multiplet,' one may *sample* the composite cross sections. The model level populations depend only on the net transition rates between model levels, and these rates may be sampled in a restricted set of frequencies as long as the total oscillator strengths and cross section distributions are correct. Frequency sampling is described in Anderson (1991). Formulae for composite cross sections and collision strengths are in Anderson (1989).

D. The multi-frequency / multi-gray algorithm. The above mapping and spectral simplification allows one to reduce the number of independent spectral frequencies in an atmosphere problem from about 3×10^6 to about 3×10^3. The multi-frequency / multi-gray algorithm developed by Anderson (1987) reduces this number by about one more factor of 10, by grouping together into radiation blocks those photons which 1) escape from the same predefined ranges of physical depth, and 2) are dominated by the same lower atomic level. Detailed formulae and prescriptions are given in Anderson (1987).

CODE STRUCTURE AND PHYSICS

The independent variables are physical depth measured column density, photon direction cosine, and photon frequency. The dependent variables are the block radiation densities and the electron temperature and pressure at each depth.

PAM makes no assumptions about model atoms or contributions to the transport coefficients; these particulars must be specified in input files. The code simply accumulates rates and coefficients for the transitions specified, and inverts the appropriate matrices for the populations and their derivatives. There are no absolute restrictions on the number of levels or transitions included. In practice, I have usually included up to three ions of each element, with up to 20 levels in each ion except the highest. The highest ion may contain several levels to account for different parents and to allow for resonant transitions in that ion.

All bound-bound, bound-free, and free-free transitions are treated alike, and must be specified in the transition list. Dielectronic recombination is best treated by insuring that the model ion contains some multiply-excited states above the ground continuum, and including auto-ionization/recombination transitions in the list. All bound-bound transitions employ depth-dependent profiles broadened by thermal Doppler, natural, and either linear Stark, quadratic Stark, or van der Waals processes. One may also select from depth-dependent microturbulence formulae. Partial redistribution for any pseudo-resonant line (including the effects of cross-redistribution through a common upper level) may be activated in the input files. At present, no direct inter-species processes (*e.g.* charge exchange) are allowed, except those cases wherein the population of one of the participants is not significantly altered.

The radiative boundary conditions at the upper surface include contributions from an analytical integration of the source function extrapolated above the surface, and optional irradiation from external sources. The former contribution means that the upper surface need not be optically thin at all frequencies; photons in an optically thick line core do not escape.

The global equilibrium is determined through iterative corrections of the linearized global equations. Energy conservation takes the form

$$\frac{4\pi}{c}\int_0^\infty \kappa_\nu\,(S_\nu - J_\nu)\,d\nu = Q(m),$$

where $Q(m)$ is a user-specified, depth-dependent source of heating. $Q(m)$ may be adjusted until the emergent spectrum matches an observed spectrum, or simply to see how the atmosphere responds to heating and/or cooling.

INPUT AND RESOURCE REQUIREMENTS

The input files are the model atoms file, the transition data file, the frequency quadrature file, files containing the previous iteration or startup model, and a control file which contains diagnostic switches and includes the effective temperature and gravity of the model. Models may be started from old models with slightly different parameters, or from a file containing all state populations. It is relatively straightforward but tedious to proceed from one model to another with different model atoms, transitions, and frequency quadrature.

A model with about 70 depths, 5000 frequencies, and 3 angles, including 10 elements with a total of 30 ions, 250 levels, 300 bound-free transitions, and 750 bound-bound transitions (60 of which are sampled composite transitions) requires approximately 4 million words of memory and uses about 800 CRAY YMP seconds for an iteration containing two statistical and formal solutions and one linearization correction. Convergence requires roughly 6-10 such iterations, depending on the error of the startup model. Required resources scale approximately linearly with the number of frequency points. Time scales linearly with the number of depths, and weakly with the number of bound-free and sampled transitions.

APPLICATIONS AND THE FUTURE

PAM, (an acronym for plane-parallel atmosphere modeling), was designed for static plane-parallel applications. There are no other restrictions, although to include magnetic fields in the radiative transfer would entail some considerable modification (particularly if the field strength was part of the solution).

Atmospheres with organized velocity fields having differential velocities larger than the thermal Doppler width cannot be treated with the multi-frequency / multi-gray algorithm. This restriction comes about because the radiation block assignments of individual line frequencies become blurred in the transformation between the atom and laboratory frames. If one does not make use of the algorithm (*e.g.* by assigning each frequency to its own unique block) one may treat problems with more general *a-priori* velocity fields.

A spherical version of the code has been written (Rosenzweig 1988). Multi-dimensional problems are prohibitive if only because of limited resources. There is nothing inherently limiting the dimensionality of the geometry (except that usually multi-dimensional geometries are accompanied by velocity fields and other difficult physics). I am developing a two-dimensional version of the code for use with relatively fixed geometries such as those encountered in thin magnetic fluxtubes (in collaboration with S. Solanki at ETH Zürich).

REFERENCES

Anderson, L.S. 1987, in Numerical Radiative Transfer, ed. W. Kalkofen (Cambridge: Cambridge Univ. Press), p. 163.

Anderson, L.S. 1989, *Ap. J.*, **339**, 558.

Anderson, L.S. 1991, in Stellar Atmospheres: Beyond Classical Models, ed. L. Crivellari, I. Hubeny, and D.G. Hummer (Dordrecht: Kluwer), p. 29.

Rosenzweig, P. 1988, *Pub. A.S.P.*, **100**, 1025.

Seventh Cambridge Workshop on Cool Stars, Stellar Systems, and the Sun
ASP Conference Series, Vol. 26, 1992
Mark S. Giampapa and Jay A. Bookbinder (eds.)

THE CHROMOSPHERIC TEMPERATURE RISE IN FLUXTUBES

J.H.M.J. BRULS Sterrekundig Instituut Utrecht, P.O. Box 80000, 3508 TA Utrecht, The Netherlands

S.K. SOLANKI Institute of Astronomy, ETH-Zentrum, CH-8092 Zürich, Switzerland

ABSTRACT We try to derive the location of the chromospheric temperature rise in fluxtubes, based on non-LTE Stokes *V* profile fits of medium-strong iron lines. Our preliminary conclusion is that the fluxtube chromosphere starts at about the same optical depth in the tube as in the quiet sun.

Keywords: fluxtubes; chromospheres; non-LTE line formation

INTRODUCTION

Several attempts have recently been made to infer semi-empirical models of the temperature and magnetic flux (or field strength and filling factor) in the magnetic part of the solar photosphere and lower chromosphere, either by fitting the Stokes *V* component of a number of line profiles assuming local thermodynamic equilibrium (LTE) to observed line profiles of network and plage regions (Stenflo 1975, Solanki 1986, Keller *et al.* 1990), or by computing non-LTE line profiles without using polarization information (Solanki *et al.* 1991).

LTE methods cannot reveal the presence of a chromospheric temperature rise because they determine the excitation temperature rather than the electron temperature. No empirical models based on non-LTE computations of Stokes *V* profiles exist so far. We undertake such an analysis using Fe I and Fe II Stokes *V* profile observations.

MODELS AND METHOD

The concept of fluxtubes makes this analysis tractable, in the sense that it allows us to use separate components for the magnetic and non-magnetic parts of the solar atmosphere in 1.5-D approximation with a limited number of vertical lines of sight sampling the composite 2-D atmosphere at different distances from the fluxtube center (*cf.* Steiner & Pizzo 1989). The external non-magnetic atmosphere around the fluxtube is represented by the Maltby *et al.* (1986) quiet sun atmosphere. We use several fluxtube models (Fig. 1), all derived from the chromosphereless NETWORK model of Solanki (1986). We employ 24 magnetically sensitive Fe I and Fe II lines that are nearly free of blends and have sufficient spread in excitation energy and equivalent width to sample different formation heights and different temperature sensitivities. For these lines we compute non-LTE Stokes profiles resulting from the 6 different fluxtube models, with 10 vertical lines of sight through each model. The radiative transfer calculations were done in a two-step procedure, employing the field-free approximation. First a non-magnetic non-LTE source function was computed with the latest version of the Carlsson (1986) code MULTI, employing a 100-level iron model atom. That source function was then integrated using the Diagonal Element Lambda Operator (DELO) method of Rees *et al.* (1989), assuming a magnetically split line profile. Finally, the computed line profiles for all lines of sight were averaged with appropriate weights to represent a cylindrical fluxtube geometry embedded in a non-magnetic environment.

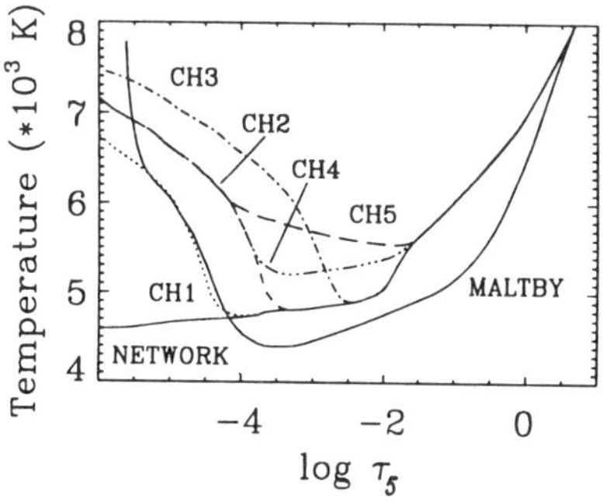

Fig. 1: Temperature stratification of fluxtube models against the optical depth at 5000 Å for each model. The chromosphereless model (solid) is the NETWORK model from Solanki (1986). The other solid curve represents the Maltby *et al.* (1986) quiet sun model. The other fluxtube models were derived by adding the chromosphere of the quiet sun model to the NETWORK model, at different heights.

DISCUSSION

Figure 2 displays observed (solid) and computed (dashed) Stokes V components for 2 stronger lines; the computed profiles are smeared with Gaussian macroturbulence. The best, but not perfect, fit to the observed V peak widths for all of these lines is reached for $v = 2$ km/s. All profiles are scaled such that the computed Stokes V amplitudes best fit the asymmetric observed peaks. This scaling factor is a free parameter corresponding to the magnetic filling factor, since the computed Stokes V amplitudes are directly proportional to this filling factor. Only a few computed profiles and none of the observed profiles display line-center inversions in the Stokes V component. However, these disappear with 2 km/s macroturbulent smearing, except for the Fe II 4923 line; this line requires 3 km/s to remove the inversion, which still seems a reasonable velocity. None of the models can be excluded on the basis of line center inversions.

Additionally, we look at Stokes V area ratios (Table I) for a number of well-chosen pairs of lines which differ in strength, excitation energy or ionization stage. This procedure removes the necessity to scale the computed Stokes V profiles, but still requires applying macroturbulence; we use 2 km/s for all lines. Excluding the line ratios involving lines 5405 and 5083 (for which we have problems fitting line profiles in the quiet sun), it is clear that model CH5 best fits the observed ratios, followed by CH4 and CH3. Increasing the ratios involving 5405 and 5083 by the percentage needed to bring their Stokes V amplitude scaling factors in agreement with the ones found for the other strong lines, all these 3 models do approximately equally well.

The current set of lines therefore cannot completely constrain the fluxtube temperature in the upper photosphere ($-3 < \log \tau_5 < -2$). Fe II 4923 needs a low temperature near $\log \tau_5 = -4$, but since CH1 and CH2 do not correctly fit the other lines, our best preliminary guess is a model like CH4 (steep chromospheric temperature rise) or CH5 (more gradual rise), but with decreasing or nearly constant T_e out to the point where it crosses CH1, and following CH1 from that point on, possibly supported by somewhat lower temperatures in the surrounding non-magnetic atmosphere as well. This agrees well with the conclusions by Solanki *et al.* (1991) based on the stronger Ca II resonance lines.

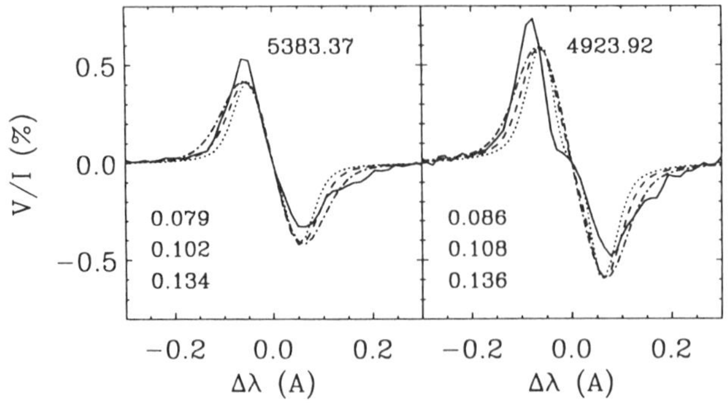

Fig. 2: Scaled Stokes V profiles. Computed profiles are smeared with 1 (dotted), 2 (dashed) and 3 km/s macroturbulence (dot-dashed); the corresponding scale factors are shown at the lower left of each panel. We plot only NETWORK results, because the scaled profiles are virtually the same for all models.

TABLE I Stokes V area ratios

Line ratio	Observed	NETWORK	CH1	CH2	CH3	CH4	CH5
5405.772 / 5083.337	0.77	0.87	0.88	0.86	0.79	0.82	0.68
5232.938 / 5217.388	1.05	1.20	1.21	1.20	1.18	1.17	1.08
4923.925 / 5197.577	2.50	2.20	2.15	1.94	1.61	1.80	1.60
5405.772 / 5217.388	0.70	0.69	0.70	0.69	0.60	0.59	0.45
5405.772 / 5383.370	0.78	0.75	0.76	0.75	0.64	0.64	0.48
5232.938 / 5383.370	1.17	1.30	1.31	1.31	1.27	1.26	1.17
5405.772 / 5232.938	0.67	0.57	0.58	0.57	0.51	0.50	0.41
5083.337 / 5383.370	1.02	0.86	0.86	0.87	0.82	0.77	0.71
5405.772 / 5197.577	1.46	1.17	1.18	1.17	1.11	1.03	0.86
5083.337 / 5197.577	1.90	1.35	1.33	1.36	1.41	1.25	1.27
5232.938 / 5197.577	2.18	2.05	2.03	2.06	2.18	2.04	2.09
5383.370 / 5197.577	1.86	1.57	1.55	1.57	1.72	1.62	1.79

REFERENCES

Carlsson, M.: 1986, *Thesis. Uppsala Report No. 33*

Keller, C. U., Solanki, S. K., Steiner, O., Stenflo, J. O.: 1990, *Astron. Astrophys.* **233**, 583

Maltby, P., Avrett, E. H., Carlsson, M., Kjeldseth-Moe, O., Kurucz, R. L., and Loeser, R: 1986, *Astrophys. J.*, **306**, 284

Rees, D. E., Murphy, G. A., Durrant, C. J.: 1989, *Astrophys. J.* **339**, 1039

Solanki, S. K.:1986, *Astron. Astrophys.* **168**, 311

Solanki, S. K., Steiner, O., Uitenbroek, H.: 1991, *Astron. Astrophys.* , in press

Steiner, O., Pizzo, V.J.: 1989, *Astron. Astrophys.* **211**, 447

Stenflo, J. O.: 1975, *Solar Phys.* **42**, 79

Seventh Cambridge Workshop on Cool Stars, Stellar Systems, and the Sun
ASP Conference Series, Vol. 26, 1992
Mark S. Giampapa and Jay A. Bookbinder (eds.)

Ca II K2V BRIGHT GRAINS FORMED BY ACOUSTIC WAVES

MATS CARLSSON Institute of Theoretical Astrophysics, P.O. Box 1029, Blindern, N–0315 Oslo, Norway

ROBERT STEIN Department of Physics & Astronomy, Michigan State University, East Lansing, MI 48824, USA

ABSTRACT We present self-consistent non-LTE radiative-hydrodynamic calculations of acoustic waves in the solar chromosphere. Calculations have been made of the vertical propagation of sinusoidal waves with periods of 30, 180 and 300 seconds.

We find that not only radiation, but the departures from LTE and non-instantaneous rates significantly alter the shocks that form in the chromosphere.

All three wave periods produce Ca II K2V bright grains which repeat with a period close to 180 seconds. The 30s and 300s driving periods are, however, not compatible with observations of the inner wing of the calcium K-line. The 180 second wave reproduce the observed profiles quite well, including the observed K2V/K2R asymmetry.

Keywords: chromospheres; radiation hydrodynamics

METHOD

The equations of hydrodynamics, radiative transfer and statistical equilibrium are set up using an implicit finite difference scheme. The equations are formulated on an adaptive grid (Dorfi & Drury 1987) — the locations of the grid-points are solved for together with all the other equations. The grid equation is set up such that the grid points gather where there are large gradients thus giving maximum resolution where it is needed. The radiative transfer is treated using Scharmer's method. This efficient non-LTE method takes into account the global nature of the radiation field. The lower boundary condition is a sinusoidal piston. Transparent boundary conditions are used for the upper boundary.

As the initial atmosphere we use the VAL3C model atmosphere (Vernazza *et al.* 1973) without a transition region.

The non-LTE radiation is treated with three bound levels plus continuum model atoms for hydrogen and calcium. The Lyman-α, Lyman-β, H-α, Ca-K and Ca-IR lines and the Lyman, Balmer and Paschen continua and all three calcium continua are all treated in detail. Voigt profiles with complete redistribution (CRD) are assumed for all lines. For the Lyman lines profiles truncated at six Doppler widths are used in order to mimic effects of partial redistribution (PRD) (Milkey & Mihalas 1973).

Other elements are included as electron donors and for the continuum opacity assuming LTE.

WAVE STRUCTURE

We have compared cases with

1. non-LTE populations and radiation
2. LTE populations and radiation
3. only collisional rates and no radiation.

The non-LTE case is similar to the LTE case in the photosphere with strong radiative damping of the wave. At the temperature minimum, the radiation starts to decouple from the local conditions and radiation does not play a dominant role in the energy balance. The hydrogen ionization is dominated by radiative rates. Due to long timescales, the ionization/recombination occurs behind the shock front and the available shock heating goes into thermal energy instead of into hydrogen ionization energy. This causes the non-LTE temperature jump over the shock to be much larger than in the LTE-case. The shock temperature profile is thus primarily sensitive to the details of the hydrogen ionization and not the radiative damping.

Ca II K2V BRIGHT GRAINS

The average quiet sun K-line profile is a strong absorption line with a centrally reversed emission feature at line center. Following Hale & Ellerman (1904), we denote the absorption at line center K3 and the emission peaks on the violet and red side of line center K2V and K2R respectively. In observations of network cell interiors only the K2V feature is present in the form of intermittent bright grains of short duration, often recurrent with a period of three minutes. It has been difficult to reproduce the absence of the K2R feature in theoretical computations.

Figure 1 shows the basic features of the observations. Figure 2 shows that these features are well reproduced by acoustic waves with a period of 180 seconds. The absence of the K2R feature is caused by the correlation between the upward velocity and the pronounced temperature spike (amplified by the slow, non-local ionization/recombination rates, see previous section) in the shock.

Thirty second period waves form shocks which coalesce in the upper atmosphere to form stronger shocks with about 180 second periodicity. The observed K2V behaviour is reproduced. The wing brightening is, however, formed before the waves coalesce and shows a non-observed 30 second periodicity.

Three hundred second period waves excite 180 second period waves in the mid chromosphere which give a K2V and K3 behaviour similar to the observations. The 3-minute signal is, however, missing in the deeper region where the wings are formed and therefore the "dark whiskers" pattern is lacking.

REFERENCES

Cram, L.E., Damé, L., 1983, *Ap. J.*, 272, 355
Dorfi, E.A., Drury, L.O., 1987, *J. Comput. Phys.*, 69, 175
Hale, G.E., Ellerman, F., 1908, *Ap. J.*, 19, 41
Milkey, R.W., Mihalas, D. 1973, *Ap. J.*, 185, 709
Rutten, R.J., Uitenbroek, H., 1991, *Solar Phys.*, 134, 15
Vernazza, J.E., Avrett, E.H., Loeser, R., 1981, *Ap. J. Suppl.*, 45, 635

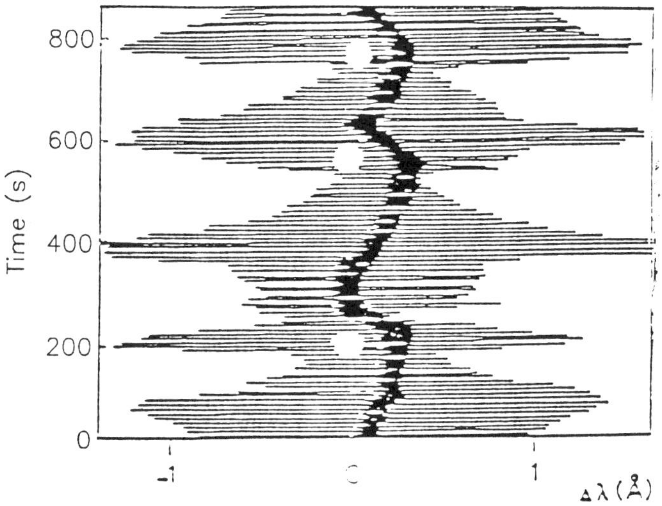

Fig. 1. Sketch of the typical K2V cell grain evolution (from Rutten & Uitenbroek 1991) based on observations by Cram & Dame (1983); the following features all repeat with about 180 seconds periodicity.

1) Symmetrical wing brightenings come in from the far wings. These wing brightenings outline the "dark whiskers" seen in the figure.

2) K2V bright flashes occur when the wing brightening reaches the position of K2V. They are much stronger than K2R.

3) The line center has its maximum redshift at the time of the K2V flash. The line center then abruptly shifts to the blue and gradually shifts back to red again.

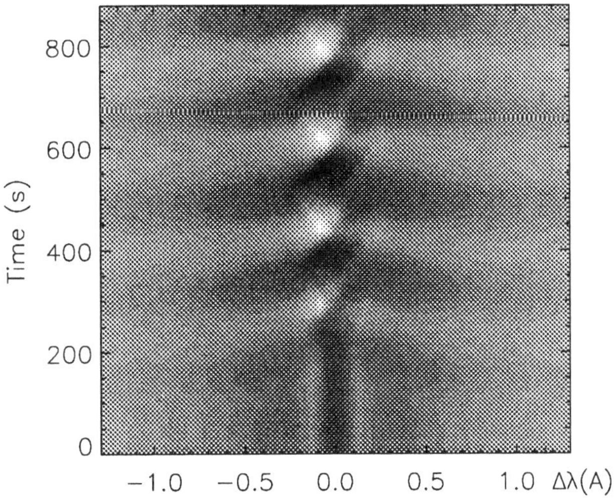

Fig. 2. The computed Ca K line intensity as a function of wavelength and time. The piston period is 180 seconds with an amplitude of 1% of the local soundspeed. The computed line behaviour is qualitatively very similar to the observed behaviour shown in Fig. 1.

Seventh Cambridge Workshop on Cool Stars, Stellar Systems, and the Sun
ASP Conference Series, Vol. 26, 1992
Mark S. Giampapa and Jay A. Bookbinder (eds.)

FORMATION OF THE Mg I 12 μm LINES

MATS CARLSSON Institute of Theoretical Astrophysics, P.O. Box 1029, Blindern, N-0315 Oslo, Norway

ROBERT J. RUTTEN Sterrekundig Instituut, Postbus 80000, NL 3508-TA Utrecht, The Netherlands

NATALYA G. SHCHUKINA Main Astronomical Observatory, Academy of Sciences of the Ukrainian SSR, 252127 Kiev, Ukrainian Republic

ABSTRACT The long-standing problem of the formation of the Mg I emission features near 12 μm is solved by recognizing that the Mg I Rydberg levels maintain a departure diffusion flow to replenish NLTE population losses lower in the Mg I term diagram. We illustrate this Rydberg flow, its driving and its collisional control.

Keywords: photospheres; chromospheres; stellar atmospheres; infrared spectrum; spectral line formation

INTRODUCTION

The riddle of the formation of the 12 μm lines of Mg I has been solved by Chang *et al.* (1991) and Carlsson *et al.* (1991) in forthcoming papers to which we refer the reader for details. Figure 1 (taken from the latter paper) demonstrates that our computed center-to-limb profiles of the strongest 12 μm line fit the observations very well; we also obtain excellent fits for weaker emission features and for Mg I Rydberg absorption lines in the infrared.

How do the emission peaks arise? As first suggested by Carlsson *et al.* (1990), they are a natural consequence of the replenishing of NLTE population depletion in a minority species from the population reservoir in the next higher ionization stage. The driving depletion is caused by photon losses in lines with large transition probability which become optically thin in the upper photosphere and by overionization from levels for which the bound-free edges are in the ultraviolet. The replenishment goes by step-wise departure diffusion along a ladder of close-lying Rydberg levels which is collisionally dominated at its top and which competes successfully with direct recombination into the depleted levels. We illustrate these properties here with further diagrams based on our detailed modeling in Carlsson *et al.* (1991).

SENSITIVITY TO RADIATIVE TRANSITIONS

The Grotrian diagram in Fig. 2 shows the levels in the triplet system included in the computations. Over 300 transitions are included in our model atom; this partial diagram shows only those to which the Rydberg departure flow is most sensitive. The plot results from a "multi-MULTI" perturbation analysis in which 2525 converged NLTE solutions were obtained with MULTI while doubling all cross-sections (bound-bound, bound-free, radiative, collisional) one by one. The resulting height of the 12.32 μm emission peak is used as sensitivity indicator; thicker lines are those of which doubling the oscillator strength results in larger emission. The 12.32 μm line itself is present because larger strength increases its height of formation, so that the outward-rising NLTE line source function produced by the departure diffusion is sampled at larger height. Other emission-increasing lines such as $6h-5g$, $6g-4f$, $6f-3d$ and $6f-4p$ drive the Rydberg flow with their photon losses. The $4s-3p$ line is present because the $3p$ ionization edge is important in closing the flow back to the continuum.

Figure 3 shows computed emission changes against the optical depth in bound-bound transitions sampled at $\lg \tau_{500} = -3$ (where the line-center emission is formed). The plusses mark transitions with either $n \leq 6$ or $n \geq 7$ for both the lower and the upper level; these are the depopulation lines. They lie in the infrared and combine large transition probability with sufficiently small opacity to become optically thin in the upper photosphere, as evident here; lines with larger or smaller opacity have less effect. The asterisks mark lines which connect an upper level $n \geq 7$ with a lower level $n \leq 6$. These transitions provide alternative paths to the Rydberg flow through $n = 7 \rightarrow 6$. If they become stronger, more of the flow skips this ladder step so that the departure divergence between the levels is reduced. The lonely plus at $\lg \tau_{500} = 2.5$ marks the $4s-3p$ line.

SENSITIVITY TO COLLISIONAL TRANSITIONS

The collisional part of our "multi-MULTI" sensitivity matrix specifies the importance of collisions for the Rydberg flow. Increasing those collisional rates which form part of the Rydberg flow (the plusses in Fig. 3) produces *larger* non-LTE departure divergence, thus larger emission in the 12.32 μm line. The increase is largest for the transitions $6g-5f$, $6h-5g$ and $8k-7i$ because enlarging their collision rates gains ladder-flow against direct recombination. The largest decrease is for $8s-7p$ because it provides a competing downward path, and for $7g-6f$, $7h-6g$ and $7i-6h$ itself because larger collision rates in these transitions maintain the same Rydberg flow for smaller departure divergence between the $n = 7$ and $n = 6$ manifolds.

NLTE THANKS TO COLLISIONAL COUPLING

Perhaps the most striking characteristic of the Rydberg departure flow is that it is dominated by collisions along the top of the Rydberg ladder: the 12 μm NLTE emission peaks arise *because* the Rydberg collision cross-sections are large. A Rydberg departure flow exists only if sufficient recombinations enter at the top of the stepwise channel which the Rydberg levels provide from the continuum reservoir down to the photon-losing $n = 6$ and $n = 5$ levels, rather than go there directly; this is the case because the level-to-level coupling is sufficiently large to provide a wide channel. Rydberg atoms are geometrically large and the coupling is therefore primarily collisional. This is illustrated in Fig. 4 which displays the dominance of collisions in the step-wise net rates between levels with $n > 5$. Thus, departures from LTE along the ladder arise because the collision rates along the ladder are high; in this case, collisional coupling maintains NLTE line formation!

REFERENCES

Carlsson, M., Rutten, R.J., Shchukina, N.G. 1990, in The Dynamic Sun, ed. L. Deszö, Publ. Debrecen Heliophys. Obs. **7**, 260

Carlsson, M., Rutten, R.J., Shchukina, N.G. 1991, *Astr. Ap.*, in press, ITA Oslo preprint

Chang, E.S., Avrett, E.H., Mauas, P.J., Noyes, R.W., Loeser, R. 1991, *Ap. J. Letters*, in press, CfA Boston preprint

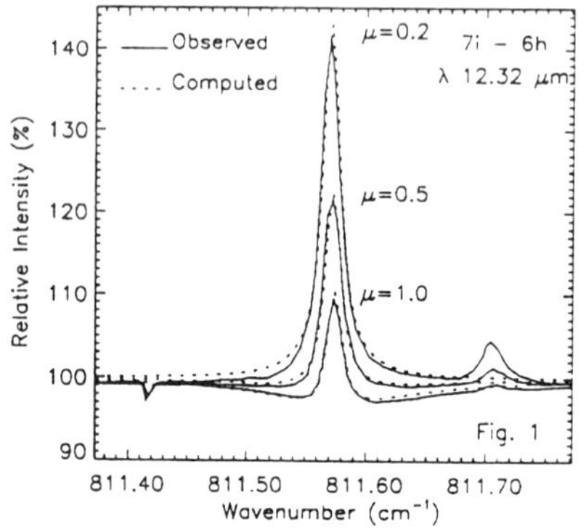

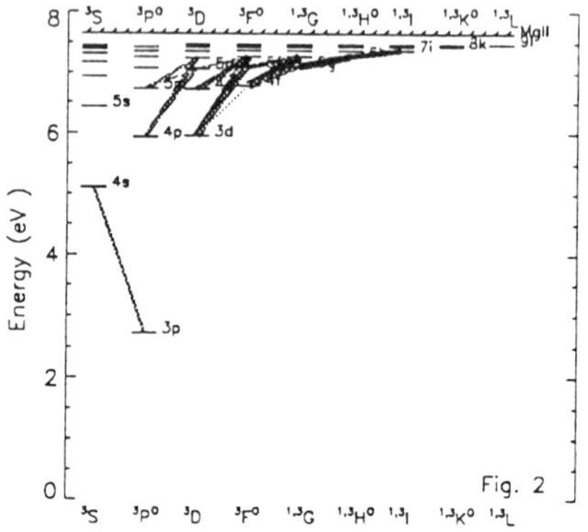

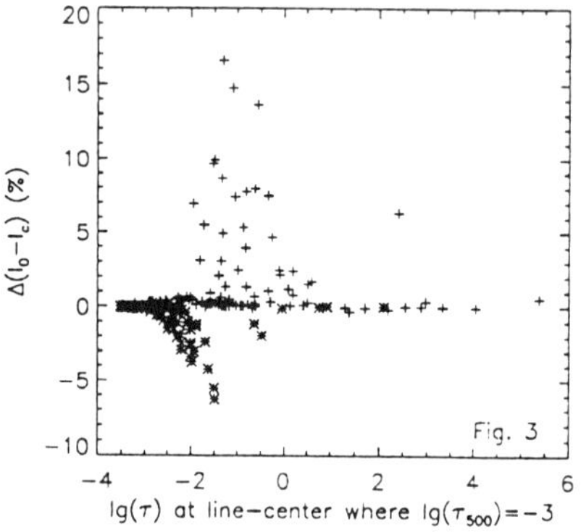

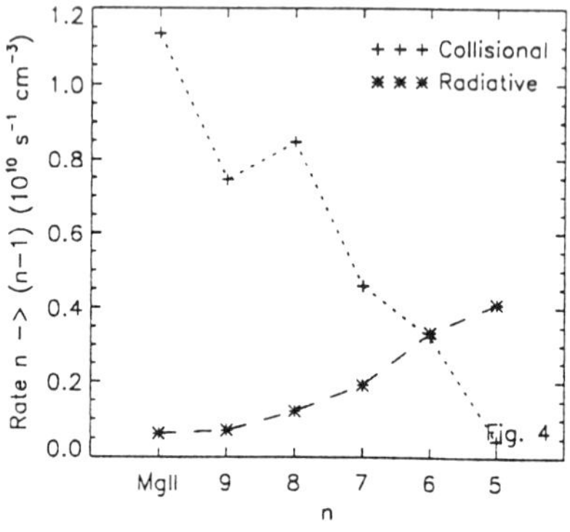

Seventh Cambridge Workshop on Cool Stars, Stellar Systems, and the Sun
ASP Conference Series, Vol. 26, 1992
Mark S. Giampapa and Jay A. Bookbinder (eds.)

NON-LTE EFFECTS ON MGI LINE PROFILES IN THE INFRARED SOLAR SPECTRUM

E.S. CHANG,[1,2] E.H. AVRETT,[1] P. J. MAUAS,[3] R.W. NOYES,[1] and R. LOESER[1]

[1] Harvard-Smithsonian Center for Astrophysics
[2] Department of Physics and Astronomy, University of Massachusetts
[3] Osservatorio Astrofisico di Arcetri, Firenze, Italy

ABSTRACT In a recent paper (Chang *et al.* 1991) we confirmed that the solar emission lines of Mg I at 7 and 12 μm are formed in the upper photosphere, and that the emission is caused by radiative depopulation of the lower transition levels together with the effects of collisional coupling of higher levels with the Mg II continuum. Here we extend this work by calculating the profiles for the first two or three members of the n=4,5,6 and 7 series and comparing the results with the ATMOS data at $\mu = 1$ and with the Brault and Noyes data at $\mu = 0.14$.

Keywords: Solar infrared spectrum; Mg I emission; solar photosphere

INTRODUCTION

In a recent paper (Chang *et al.* 1991) we solved the non-LTE equations for a 41-level Mg I atom to calculate the profiles of the solar emission lines at 12 and 7 microns. Similar results have been given by Carlsson, Rutten, and Shchukina (1991).

When compared with the observed $\mu = 1$ ATMOS profiles (Farmer and Norton 1989) and the $\mu = 0.2$ profiles of Brault and Noyes (1983), the calculated lines show the observed peak-in-trough shape at $\mu = 1$ and the observed μ^{-1} limb brightening. However, our calculated peaks are weaker and our troughs deeper than observed, suggesting that our calculated line source functions are not sufficiently enhanced relative to the Planck function in the upper photosphere where the lines are formed. Carlsson *et al.* report better agreement with observations than we obtained. Our main purpose here is not to intercompare the two independent calculations, but to show several additional calculated Mg I profiles, some of which share the same source functions with the 12 and 7 micron lines, and provide further comparisons with observed data.

CALCULATIONS

In our atomic model, all states with the same n (principal quantum number) and with $\ell > 2$ (orbital quantum number) are assumed to share the same atomic level. Thus the 7g-6f transition has the same line source function as 12 μm line transitions 7h-6g and 7i-6h.

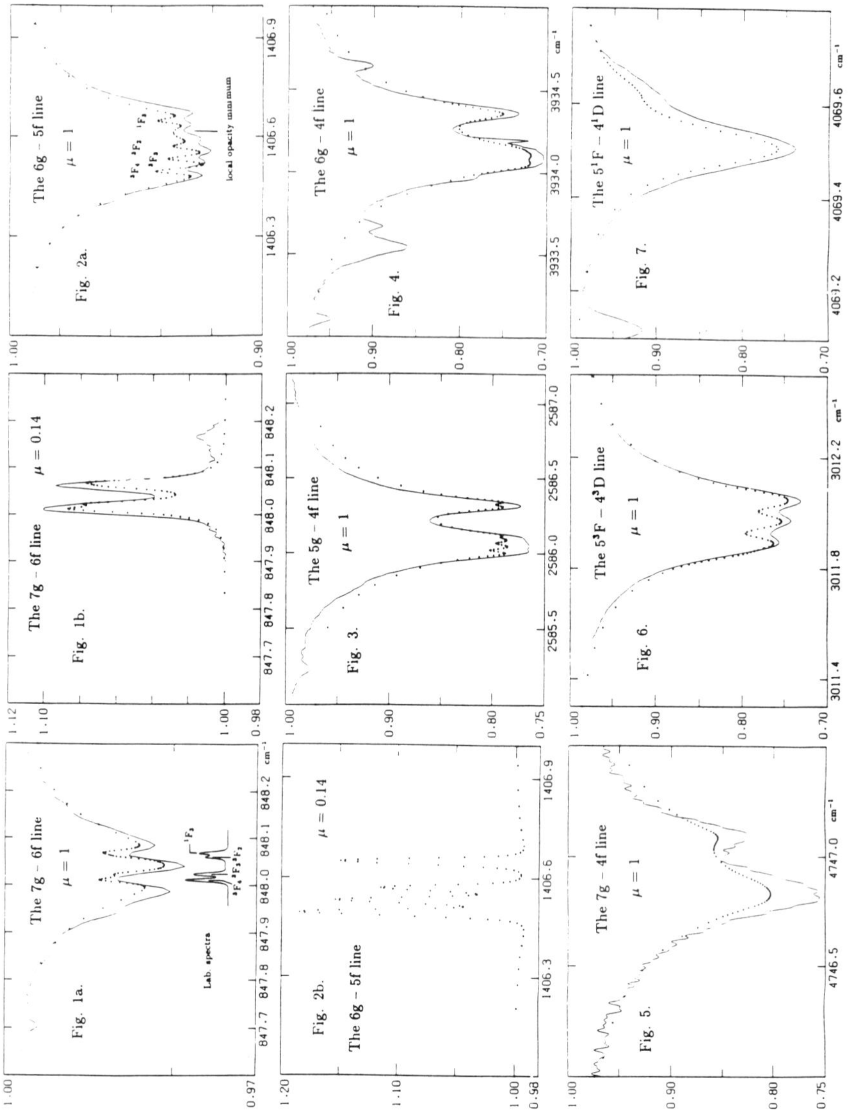
Fig. 1a.
The 7g – 6f line
μ = 1
Lab. spectra
Fig. 1b.
The 7g – 6f line
μ = 0.14
Fig. 2a.
The 6g – 5f line
μ = 1
local opacity minimum
Fig. 2b.
The 6g – 5f line
μ = 0.14
Fig. 3.
The 5g – 4f line
μ = 1
Fig. 4.
The 6g – 4f line
μ = 1
Fig. 5.
The 7g – 4f line
μ = 1
Fig. 6.
The 5³F – 4³D line
μ = 1
Fig. 7.
The 5¹F – 4¹D line
μ = 1
cm⁻¹

Figure 1a shows the 7g-6f $\mu = 1$ line profile. The solid curve represents the ATMOS observations and the dots are our calculations. Note that in contrast to the 12 μm lines, our calculated intensities here are above the observed ones. The inset is a plot of the high resolution laboratory profile showing the magnetic fine structure of the lower level (Lemoine *et al.* 1990). Note that the singlet component on the extreme right produces an emission feature in the same way as the three other triplet components. Our computed line profile at $\mu = 0.14$ is shown in Figure 1b and is compared with observations by Brault and Noyes.

The 6g-5f transition has the same source function in our calculation as the 6h-5g line at 7μm. Figure 2a shows that all 4 fine structure components are in emission. The 5th peak is just a minimum-opacity feature between the triplet and singlet troughs. Jefferies (1991), using early laboratory data that has since been corrected, identified the singlet component with the trough on the lower wavenumber side and concluded that the singlet was in absorption. Note that again our calculated intensities lie above the observed ones.

Figure 2b shows the calculated 6g-5f profile at $\mu = 0.14$. In this case there are only 4 maxima. No observed data are available for comparison.

Figure 3 shows the 5g-4f transition at $\mu = 1$. In the progression ng-(n-1)f where n = 7 for Figure 1a, n = 6 for Figure 2a, and n = 5 for Figure 3, the absorption troughs deepen while the emission peaks weaken. Simultaneously, the triplet-singlet splitting increases. Our calculation in Figure 3 shows that the singlet has a small central emission peak, but the peak to trough displacement is just under the ATMOS instrumental resolution of 0.017 cm^{-1} so that such a weak feature would not appear in the observed data. The singlet thus appears to be in absorption as noted by Jefferies. The triplet peak at 2586.021 cm^{-1} is calculated to be displaced by 0.020 cm^{-1} from the adjacent troughs and thus barely appears in the ATMOS profile. The two other triplet peaks cannot be resolved. Our results do not support Jefferies' contention that triplets are in emission and singlets in absorption.

Figures 4 and 5 show the similar profiles of the 6g-4f and 7g-4f transitions at $\mu = 1$. As line broadening increases, the triplet peaks merge into one in Figure 4 and disappear entirely in Figure 5.

In Figure 6 we show the 5^3F-4^3D line. As noted by Jefferies, the triplet fine structure components are in absorption. The corresponding singlet 5^1F-4^1D transition is also in absorption, as shown in Figure 7. The small shoulder in the singlet profile at 4069.6 cm^{-1} is caused by the intercombination line 5^3F_3-4^1D_2.

The calculations of Carlsson *et al.* match the 12 μm lines almost exactly, while we find less emission than is observed (see Chang *et al.* 1991). Two possible reasons for the difference need to be investigated: 1) We combined high ℓ levels of the same n and calculated a single departure coefficient for each combination level while Carlsson *et al.* treat these levels separately. On the other hand, we treat levels up to n=15 while they only include levels up to n=10. 2) The two calculations use different approximations in the treatment of UV line blanketing, which determine the extent of lower level departures from LTE. The emission in the infrared is sensitive to ionizing radiation in the Mg I UV continua.

Comparison with the additional line transitions considered here may help resolve the differences between their calculation and ours.

REFERENCES

Brault, J., & Noyes, R. 1983, *Ap. J.*, **269**, L61.
Carlsson, M., Rutten, R. J., & Shchukina, N. G. 1991, *Astr. Ap.*, in press.
Chang. E. S., Avrett, E. H., Mauas, P. J., Noyes, R. W., & Loeser, R. 1991, *Ap. J.*, **379**, L79.
Farmer, C. B., & Norton, R. H. 1989, A High-Resolution Atlas of the Infrared Spectrum of the Sun and the Earth Atmosphere from Space, NASA Ref. Pub. 1224, Vol. 1.
Jefferies, J. T. 1991, *Ap. J.*, **377**, 337.
Lemoine, B., Petitprez, D., Destombes, J. L., & Chang, E. S. 1990, *JPhysB*, **23**, 2217S.

Seventh Cambridge Workshop on Cool Stars, Stellar Systems, and the Sun
ASP Conference Series, Vol. 26, 1992
Mark S. Giampapa and Jay A. Bookbinder (eds.)

WIND AND ATMOSPHERE MODELS FOR THE HYBRID SUPERGIANT ALPHA AQUARII (G2 Ib)

A. K. DUPREE, B. A. WHITNEY, and E. H. AVRETT
Harvard-Smithsonian Center for Astrophysics, 60 Garden Street, Cambridge, MA 02138 USA

ABSTRACT *Ab-initio* and semi-empirical models of the atmosphere and wind of the hybrid supergiant Alpha Aqr, constructed to match chromospheric line fluxes and profiles, indicate that a rapid acceleration must occur within $\approx 1.2\ R_\star$.

Keywords: hybrid supergiants; chromospheres; stellar atmospheres; stellar winds

INTRODUCTION

The hybrid stars may represent the physically important connection between the solar-like stars with coronas and fast winds of low mass-loss rate, and the cool supergiant stars (Alpha Ori-like) with cool outer atmospheres and slow winds of high mass loss rate. Thus hybrid stars can play a pivotal role in understanding the atmospheric heating and winds of cool stars. We illustrate that two transitions (He I $\lambda 10830$ and Mg II $\lambda 2800$) show particular sensitivity to the wind structure.

THE MODEL OF THE ATMOSPHERE

An *ab-initio* model of the extended supergiant atmosphere of Alpha Aqr (in spherical geometry) was constructed by invoking a fixed magnetic field strength (10 Gauss) and Alfvén wave energy flux (3×10^6 ergs cm^{-2} s^{-1}), and calculating the response of the atmosphere following the methods of Hartmann and MacGregor (1980). A static lower chromosphere was appended (Basri *et al.* 1981). With PANDORA (Vernazza *et al.* 1981), theoretical line profiles were generated for comparison to observations of chromospheric, transition region, and coronal line fluxes and line profiles. Modifications to the atmospheric temperature, density and velocity profile are introduced iteratively back into the PANDORA code so that the model is consistent with the observations. This procedure defines the temperature, density, and velocity in a steady-state stellar atmosphere. The maximum temperature is 66000 K at $3R_\star$; a terminal velocity of 115 km s^{-1} is achieved at $\approx 5R_\star$ (*slow-wind* model). A *fast-wind* model was also constructed that rapidly reached 150 km s^{-1} at $\approx 1.2R_\star$ and then decelerated to the terminal velocity of 115 km s^{-1}. This velocity profile appears necessary to fit observed line profiles.

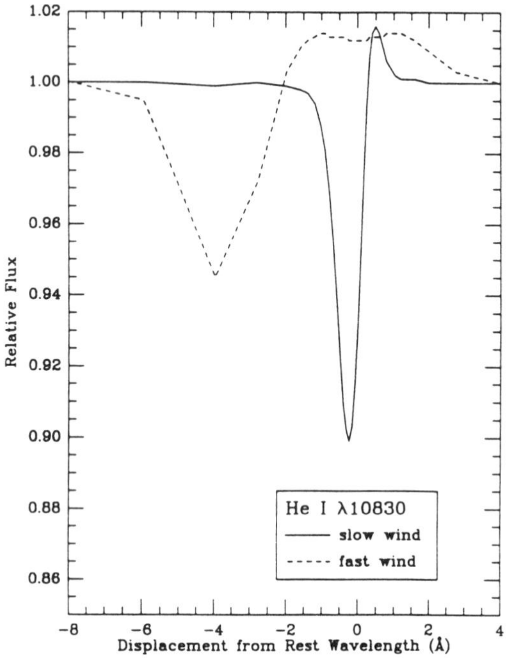

Fig. 1: The He I $\lambda10830$ profiles in Alpha Aqr calculated with a spherical atmosphere show that a rapidly accelerating chromospheric velocity profile (*fast wind* – see text), produces the observed short wavelength extension.

THE He I ($\lambda10830$) TRANSITION

This He I transition is remarkable in the spectrum of α Aqr for its P Cygni appearance and extension of the short wavelength wing of the absorption profile 5 to 7 Å shortwards of line center (O'Brien and Lambert 1986). We have demonstrated (Dupree *et al.* 1989) the sensitivity of the absorption profile to the presence of an X-ray flux in the atmosphere. The absorption increases when X-ray illumination is present, because the population of the $2s\ ^3P$ level is enhanced due to recombination following photoionization of He I. The central depth can change by 6 to 10 percent for reasonable fluxes. However, the short wavelength extension of the $\lambda10830$ line (Fig. 1) can only be produced with a rapidly accelerating velocity field in the chromosphere where line formation occurs.

THE Mg II ($\lambda2800$) TRANSITION

The Mg II doublet has strong emission, a narrow interstellar absorption feature, and a broad absorption commonly believed to mark the terminal velocity of the wind at ≈ 110 km s^{-1} where Mg II recombines. A fast wind model better reproduces the observed profile (Fig. 2), producing the broad emission, an absorption trough at the terminal velocity, and emission to shorter wavelengths.

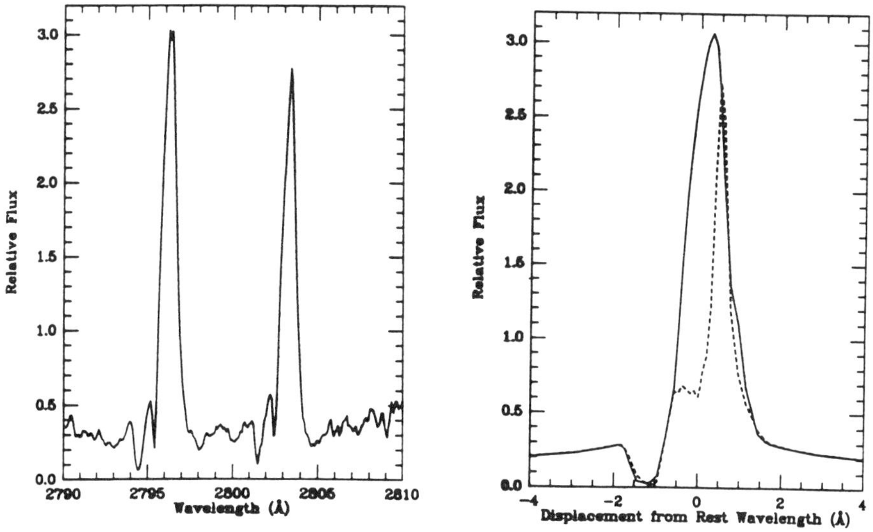

Fig. 2: Averaged profiles of Mg II *h* and *k* in α Aqr from the *IUE* (*left*); calculated profile of Mg *k*-line (*right*) for a fast (—) or slow (- - -) wind.

CONCLUSIONS

These models and simulations of line profiles in α Aqr call for a rapid acceleration deep in the chromosphere to velocities on the order of 150 km s^{-1} and suggest that shocks and transient events may be present. Such velocities could also provide a natural explanation of the chromospheric and transition line widths found in hybrid stars, supporting a previous suggestion (Hartmann *et al.* 1981). An extended source of heating may be required to accelerate the wind at distances beyond the sonic point in order to achieve velocities on the order of 100 km s^{-1} as indicated by absorption troughs in the Mg II profiles. The level of the X-ray flux in the atmosphere is not known, but the He I line might provide a sensitive discriminant of its presence.

ACKNOWLEDGEMENTS

We are grateful to D. Zucker for producing the illustrations that accompany this contribution.

REFERENCES

Basri, G.S., Linsky, J.L., and Eriksson, K. 1981, *Ap. J.*, **251**, 162.
Dupree, A. K., Whitney, B. A., and Avrett, E. H. 1989, *Bull. A.A.S.*, **21**, 115.
Hartmann, L., and MacGregor, K.B. 1980, *Ap. J.*, **242**, 260.
Hartmann, L., Dupree, A. K., and Raymond, J. C. 1981, *Ap. J.*, **246**, 193.
O'Brien, G., and Lambert, D. 1986, *Ap. J. Suppl.*, **62**, 899.
Vernazza, J., Avrett, E.H., and Loeser, R. 1981, *Ap. J. Suppl.*, **45**, 635.

Seventh Cambridge Workshop on Cool Stars, Stellar Systems, and the Sun
ASP Conference Series, Vol. 26, 1992
Mark S. Giampapa and Jay A. Bookbinder (eds.)

ON THE GLOBAL PROPERTIES OF ACTIVE REGIONS

REINER HAMMER NRC/NASA Marshall Space Flight Center, Code ES-52, Huntsville, AL 35812

ABSTRACT The area-integrated X-ray flux from solar active regions is nonlinearly correlated with the emission from lines formed in the chromosphere and transition region (hereafter TR), with a power law exponent $\beta \approx 1.5$. Recently it has been suggested that this nonlinearity can be understood if the volume V of coronal loops is related to the loop footpoint area A by $V \propto A^{3/2}$. While this argument may well be true for the chromosphere, the situation is different for the TR. In standard corona models, where the relationship between TR and corona is determined by thermal conductive coupling, homologous ($V \propto A^{3/2}$) changes of the loop geometry are shown to produce only a very mild nonlinearity, $\beta \approx 1.08$. A main point to be drawn from this analysis is that the three plausible ideas, (*i*) that the TR emission from active regions comes mainly from the feet of static coronal loops, (*ii*) that the TR in active regions is mainly powered by heat conduction from the corona, and (*iii*) that most active region coronal loops have about the same shape, cannot all be true of real active regions. In order to account for the observed magnitude of the nonlinearity by geometry effects, one would need *non*homologous volume changes in which the loop cross section at the apex varies more rapidly than at the foot. Several other effects that lead to a nonlinear relationship between coronal and TR emission are also discussed.

Keywords: Sun: activity; Sun: transition region; Sun: corona

INTRODUCTION

The total X-ray emission from solar active regions is tightly correlated with the emission in a given transition region (TR) line,

$$F_X \propto F_{TRL}^{\beta}, \tag{1}$$

with $\beta \approx 1.5$ (Schrijver *et al.* 1985, Table 7). Here F represents the integral over the area of the active region. Similar relations hold for the integrated hemispheres of cool stars. Schrijver (1988) showed that the behavior of stellar hemispheres can only be reconciled with the behavior of individual active regions if both active and quiet regions change with the level of stellar activity.

Nonlinear correlations have also been found for the relationship between the X-ray flux and the excess emission in chromospheric lines like Ca II K. The exponents β tend to be somewhat larger than 3/2. Recently Cram (1991) suggested that this might be due to a simple geometry effect: While the coronal emission is proportional to the whole volume V of magnetic loops, the chromospheric emission is controlled by their footpoint area A. If with increasing size of active regions the loops become larger, but roughly maintain their shape, then their volume scales like the 3/2th power of the footpoint area.

This suggestion may well be true for the relationship between coronal and *chromospheric* emission. The present paper discusses if geometry effects might contribute to the relation between coronal and *transition region* emission.

CORONAL AND TR EMISSION

In standard models of active regions, the X–ray emission is formed in a (more or less) isothermal plasma that occupies most of the volume of coronal loops. If such a loop is quasistatic and its maximum height is small compared to the pressure scale height, it can be assumed isobaric. Then the emission is given by

$$F_X = \int_V n_e^2 P_{Rad}(T_c)\, dV \propto V p^2 T_c^{\alpha\ -2}, \tag{2}$$

where V is the volume of the loop, T_c the coronal temperature, and $P_{Rad} \propto T_c^\alpha$ a power law fit to the emissivity. Dimensional analysis of the energy equation shows that the expression $T_c/(pL)^{1/3}$, where L is the loop length, is constant for a given type of loop. This "scaling law" can be used to eliminate T_c from (2),

$$F_X \propto V L^{(\alpha-2)/3} p^{(\alpha+4)/3}. \tag{3}$$

The emission F_{TRL} in a given TR line originates from a thin zone at the foot of the loop. The thickness Δh of this zone depends on the steepness of the temperature gradient, which in turn is controlled by the pressure p: With increasing p, the emission from the TR increases. In a conductively heated TR, this increase of the emission necessitates a steepening of the temperature gradient in order that the energy losses can still be balanced. As a result, Δh decreases with p. Detailed models yield $\Delta h \propto p^{-1}$ and

$$F_{TRL} \propto Ap. \tag{4}$$

If the TR is heated not only by conduction, but also by the enthalpy of down-flowing matter, the pressure dependence of F_{TRL} can be shown to be reduced.

GEOMETRY EFFECTS

At a first glance, the idea that the observed value $\beta \approx 1.5$ might be due to geometry effects is very attractive. However, this cannot be a complete theoretical explanation, for two reasons. First, a relation of the form $V \propto A^{1.5}$ holds only if the loops that dominate the emission have the same shape, independent of their size. There is no obvious theoretical reason for why coronal loops should be so uniform. Second, we have shown above that the emission from a given TR line is not only proportional to the loop footpoint area, but also to the height interval over which the line is formed, which depends on the pressure and thus need not be constant for all active regions.

But even if these two requirements are met, we do not get the expected result. For homologous (shape-invariant) loops with constant pressure, $F_{TRL} \propto A$

with $A \propto L^2$ and $V \propto L^3 \propto A^{3/2}$. Elimination of L from (3) and (4) then yields relationship (1) with $\beta = (7 + \alpha)/6$. For coronal temperatures not exceeding 20 MK, $\alpha \approx -0.5$ provides a good fit to the emissivity function P_{Rad}, and results in an exponent β of only 1.08. Even for extremely hot loops with temperatures above 20 MK, where $\alpha \approx +0.5$, β is only 1.25. (Such extremely hot plasma may not exist on the Sun, but likely exists on some other types of stars.)

Therefore, homologous ($V \propto A^{3/2}$) volume changes indeed lead to nonlinear flux-flux relations (1), but are, by themselves, insufficient to explain the observations quantitatively. If the observed nonlinearity is due to geometry effects alone, then the cross section of representative loops must increase with increasing activity more rapidly at the apex than at the footpoints. This is consistent with the results of Schrijver *et al.* (1985), who found that the projected coronal area of active regions increases more rapidly than the TR area.

OTHER POSSIBLE CAUSES OF NONLINEAR RELATIONS

Even without any volume change, a variation of the pressure leads already to a nonlinear flux-flux relation (Hammer and Linsky 1984). This can be shown by assuming $(L, A, V) = const$ and eliminating p from (3) and (4). This gives $F_X \propto F_{TRL}^{\beta}$ with $\beta = (4 + \alpha)/3$. For $\alpha \approx -0.5$, the exponent β is $7/6 \approx 1.17$; while for $\alpha \approx +0.5$ (*i.e.*, for $T_c > 20$ MK) one obtains $\beta \approx 1.5$. It can be shown that β increases rapidly if the enthalpy of downflowing hot matter becomes significant in the energy balance of the TR or if the loops are no longer small compared to the pressure scale height.

The time-dependent nature of active regions (Golub 1991) might also be relevant to the observed nonlinearity; but this role still needs to be explored.

Finally, several authors have suggested that a major portion of the "transition region" emission comes from plasma that is not physically connected to the corona, but rather resides in loops containing only "cool" material (Rabin and Moore 1984; Antiochos and Noci 1986). In this case, the relationship between X-ray and UV emission would be determined by the relative frequencies of cool and hot loops.

ACKNOWLEDGEMENTS

This research was conducted while the author held a National Research Council Research Associatehip. He acknowledges helpful discussions with R. L. Moore.

REFERENCES

Antiochos, S. K., and Noci, G: 1986, *Ap. J.*, **301**, 440.

Cram, L. E.: 1991, in Mechanisms of Chromospheric and Coronal Heating, eds. P. Ulmschneider, E. R. Priest, and R. Rosner. Springer, p. 282.

Golub, L.: 1991, these proceedings.

Hammer, R., and Linsky, J. L.: 1984, in Proc. Fourth European IUE Conference, ESA SP-218, p. 25.

Rabin, D., and Moore, R.: 1984, *Ap. J.*, **285**, 359.

Schrijver, C. J.: 1988, *Astr. Ap.*, **189**, 163.

Schrijver, C. J., Zwaan, C., Maxson, C. W., Noyes, R. W.: 1985, *Astr. Ap.*, **149**, 123.

Seventh Cambridge Workshop on Cool Stars, Stellar Systems, and the Sun
ASP Conference Series, Vol. 26, 1992
Mark S. Giampapa and Jay A. Bookbinder (eds.)

A NUMERICAL SCHEME FOR THE MODELING OF ELECTRIC CURRENT SHEET FORMATION IN THE SOLAR ATMOSPHERE

PAUL CHARBONNEAU and BOON CHYE LOW High Altitude Observatory, NCAR [1], P.O. Box 3000, Boulder, CO 80307.

ABSTRACT It has been demonstrated analytically that electrical current sheets can form in the solar atmosphere, in response to external perturbations on an initially smooth equilibrium state. However, the classes of configurations (and perturbations) for which analytical proofs exist are rather limited, requiring a numerical approach in more general cases. We present here preliminary results obtained with a finite element-based frictional relaxation code currently under development.

Keywords: Hydromagnetics; Numerical Methods.

INTRODUCTION

The magnetic field in electrically highly conducting plasmas has the general tendency of forming electric current sheets or magnetic tangential discontinuities, because of the topology preserving constraint associated with the frozen-in condition. Circumstances exist under which this constraint on the motion results in discontinuous or tearing plasma displacements to form electric current sheets. This fundamental theorem of magnetostatic equilibrium is due to Parker (1972, 1979, 1990, and references therein) and has been suggested to be a mechanism for heating stellar atmospheres. Many interesting analytical calculations have been presented to illustrate the formation of electric current sheets (see Low 1990, Berger 1990, Amari 1991 for recent reviews). We present here preliminary *numerical* calculations of current sheet formation. Physically, the main constraint of interest is the preservation of the magnetic field topology as described by the hydromagnetic induction equation $\partial B/\partial t = \nabla \times (v \times B)$, as the perturbed state relaxes to a new equilibrium state.

A MODEL PROBLEM: NUMERICAL SCHEME AND SAMPLE RESULT

Consider a 2-D system ($\partial/\partial x \equiv 0$) in cartesian coordinates (x, y, z), and suppose that the magnetic field has only components in the y and z-direction. Such a field can be described by a vector potential $A = (A_x[y,z], 0, 0)$, such that $B = (0, \partial A_x/\partial y, -\partial A_x/\partial z)$. In this context, the induction equation reduces to the scalar equation,

$$\frac{\partial A_x}{\partial t} = -v \cdot \nabla A_x, \tag{1}$$

[1] The National Center for Atmospheric Research is sponsored by the National Science Foundation.

A detailed tracking of the time-evolution if the perturbed state towards a new equilibrium is not in itself of great interest here, but the topology of the initial magnetic field must be strictly preserved, both in the final state and throughout the evolution towards the final state. We therefore introduce a simple model based on *frictional relaxation*, *i.e.*, the equations of motions are assumed to be always dominated by a viscous drag term which is proportional to the velocity, *i.e.*, v remains parallel to the total force at all time (see *e.g.*, Chodura & Schlüter, 1981). For a cold gas, the total force is just the Lorentz force, so we write $4\pi\gamma(y,z)\,v = -\nabla^2 A_x \nabla\, A_x$. Equation (1) becomes the nonlinear diffusion equation:

$$\frac{4\pi\gamma}{||\nabla\, A_x||^2}\frac{\partial A_x}{\partial t} = \nabla^2 A_x, \tag{2}$$

where γ controls the rate of relaxation, and is free to be prescribed. We may treat eq. (2) numerically as a *linear* diffusion equation, with $||\nabla\, A_x||^2$ evaluated from the previous step. Clearly, in the asymptotic limit our solution is identical to a formal nonlinear solution of (2), meaning that this crude linearization actually yields the true steady-state. Solutions have been obtained using the Galerkin finite element method (FEM), with a centered implicit time stepping scheme. An important advantage of the FEM is the ability to accurately model solutions exhibiting discontinuities in first (and higher) spatial derivatives, an important feature if electric current sheets occur in the final states.

Figure 1 shows results for a standard model problem, namely the stretching of a neutral point into an electric current sheet (Syrovatskii, 1981). An equilibrium initial state with a neutral point is first computed, and then disrupted by downward compression; the subsequent relaxation to equilibrium is then followed by solving eq. (2). The numerical accuracy is verified by mesh-refinement. The solution shown on Figure 1 is converged *given the grid spacing*; In all cases, once the current sheet reaches a thickness comparable to the grid spacing, the scheme no longer resolves accurately the discontinuous first derivative, and numerical dissipation follows, leading to gradual reconnection and, if the time integration is pushed further in time, to the eventual dissipation of the current sheet. The asymptotic thickness of the sheet becomes a direct function of grid spacing. Nevertheless, these simple test problems provide an unambiguous and intuitively appealing illustration of the natural tendency to relax to asymptotic states containing electrical current sheets.

Despite its simplicity, our finite-element based numerical scheme is proving to be both accurate and flexible. Even on the relatively coarse mesh used above, the scheme has no difficulty in tracking the development of electric current sheets of arbitrary shapes down to grid spacing resolution. We are currently studying a similar problem under more astrophysically realistic settings, namely the frictional relaxation of magnetic structures imbeded in an isothermal atmosphere stratified by gravity.

REFERENCES

Amari, T. 1990, in Advances in Solar System Magnetohydrodynamics, ed. E. Priest (Cambridge: Cambridge University Press), p. 3.

Berger, M.A. 1989, in Turbulence and Nonlinear Dynamics in MHD Flows, eds. M. Meneguzzi, A. Pouquet, & P.L. Sulem (Amsterdam: North Holland), p. 197.

Chodura, R., & Schlüter, A. 1981, *J. Comp. Phys.*, **41**, 68.

Low, B.C. 1990, *Ann. Rev. Astr. Ap.*, **28**, 491.
Parker, E.N. 1972, *Ap. J.*, **174**, 499.
Parker, E.N. 1979, Cosmical Magnetic Fields: Their Origin and their Activity (Oxford: Oxford University Press).
Parker, E.G. 1990, *Geophys. Astrophys. Fluid Dyn.*, **53**, 43.
Syrovatskii, S.I. 1981, *Ann. Rev. Astr. Ap.*, **19**, 163.

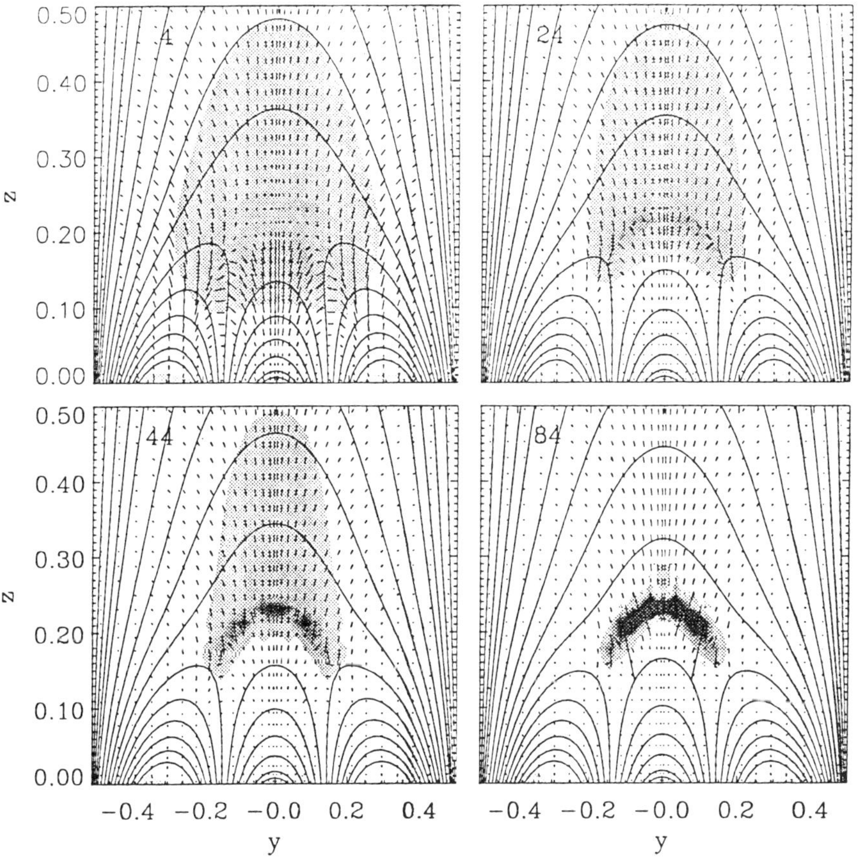

Figure 1 : Formation of an electrical current sheet by stretching of a pre-existing neutral point. Each cell of the $n_y \times n_z = 30 \times 50$ orthogonal grid is tiled with an 8-nodes quadratic element. Solid lines are magnetic fieldlines, the gray scale codes the current density, and the arrows indicate the magnitude and direction of the Lorentz force. The stretching of the neutral point initially located at $(y, z) \simeq (0, 0.15)$ into a curved current sheet is obvious. The same solution computed on a tighter mesh develops a correspondingly thinner current sheet, but the general behavior of the solution remains the same.

Seventh Cambridge Workshop on Cool Stars, Stellar Systems, and the Sun
ASP Conference Series, Vol. 26, 1992
Mark S. Giampapa and Jay A. Bookbinder (eds.)

ENERGY BALANCE MODELS OF SOLAR FLARE ATMOSPHERES

SUZANNE L. HAWLEY Institute of Geophysics and Planetary Physics, Lawrence Livermore National Laboratory, P.O. Box 808, L-413, Livermore, CA 94551

GEORGE H. FISHER Institute for Astronomy, University of Hawaii, 2680 Woodlawn Drive, Honolulu, HI 96822

ABSTRACT We present a set of solar flare atmospheric models which are in energy balance, including energy transport by nonthermal particles, thermal conduction, and important radiation processes. The models are computed within a coronal loop geometry including a self-consistent treatment of beam heating and coronal temperature. We determine the cooling in the optically thick chromosphere using a rigorous non-LTE treatment of the radiative transfer and statistical equilibrium in the important cooling transitions of H I, Ca II and Mg II. The models are in hydrostatic equilibrium and cover a range of non-thermal electron energy fluxes from 5×10^8 to 5×10^{10} erg cm^{-2} s^{-1}, corresponding to coronal apex temperatures of 5-25 million K.

Keywords: solar flares; coronal loops; model atmospheres

INTRODUCTION

Model flare atmospheres are useful for investigating how the emission observed during flares changes with atmospheric parameters. The models we present are "synthetic", in that they are computed from postulated flare heating mechanisms, in contrast to "semi-empirical" models which are constructed to match observations. By changing the amount of flare heating and the transport mechanisms by which the energy is deposited into the lower atmosphere, we compute models which then give predictions of observed quantities that can be compared with the observations. Using multi-wavelength data that sample the entire temperature range of the outer atmosphere, we hope to eventually determine the relative importance of thermal and nonthermal energy transport during solar flares.

METHOD

To calculate the models we first set up a self-consistent coronal loop, characterized by two parameters – the loop length L and the incident flux of nonthermal electrons F_c. The electrons are assumed to have a power law distribution with energy, $F(E) = F_c \times (E/E_c)^{-\delta}$ with cutoff energy $E_c = 20$ KeV and spectral index $\delta = 5$. The loop length was taken to be 10^9 cm for all models. The beam flux F_c is the principal parameter that determines the flare energy; it was allowed to vary between 5×10^8 and 5×10^{10} erg cm^{-2} sec^{-1}. The flux of electrons that is stopped in the corona is given in Fisher (1989), equations 33a and 33b, while the heating rate required to maintain a static, scaling law loop is given in Hawley and Fisher (1992) (hereafter HF92), equation 6. By equating the average beam heating rate in the corona with the static loop heating rate, we can solve for the coronal parameters that give a static, scaling law loop in energetic equilibrium. These parameters are the coronal pressure, apex temperature and total column depth; these in turn specify completely the coronal structure. From this structure we calculate the X-ray flux from 1-250Å using the Gan and Fang (1990) plane plus point source approximation for the X-ray emitting regions and the Raymond and Smith (1977) X-ray emissivities.

In the transition region, we follow HF92 in adopting an energy equation that assumes radiative losses are balanced by conductive heating. The heating is specified by the coronal model (hence is unique for given F_c, L). From the energy equation we can calculate the transition region structure down to chromospheric temperatures.

The chromosphere and photosphere are again calculated using the methods discussed in HF92. The pre-flare solar model is from Metcalf *et al.* (1990) with the addition of a 2×10^6 K corona. We used the program MULTI (Carlsson 1986) to compute the NLTE radiative transfer in H I, Ca II and Mg II. The cooling rates in the bound transitions and bound-free continua, together with the optically thin metal cooling computed according to the prescription in Ricchiazzi and Canfield (1983), were required to balance the heating rates from the flare X-ray and beam fluxes at every depth level in the lower atmosphere. The heating rates were calculated following Emslie (1978) for the beam and Henoux and Nakagawa (1977) for the X-rays.

DISCUSSION

The five values of F_c in Table 1 give five independent models. Each was computed with (1) only X-ray heating and (2) combined X-ray and beam heating to assess the effects of the different heating mechanisms. The loop length and range of beam fluxes were chosen to match those inferred from observations. The resulting model coronae have reasonable temperatures and emission measures, but the pressures are larger than observed. Flare loops probably never experience intense heating for a sufficiently long time to reach this equilibrium state. High coronal pressure forces the transition region to deeper column mass, giving high densities in the upper chromosphere. Thus, enhanced emission lines are observed though the maximum chromospheric temperature in even the M5 model is only slightly hotter than that in the pre-flare (QF1) model.

The electron beam has only a modest direct impact on the chromosphere of the least energetic model (M1) and almost none on that of the most energetic (M5) model. The corona is so dense in the M5 model that the beam energy is almost entirely deposited above the flare transition region. The flare X-rays, however, are efficient at heating the chromosphere in all of the models. In the most energetic models, backheating from the Balmer continuum heats to the temperature minimum region. However, it is not enough to heat the photosphere significantly (in contrast to the M dwarf flare models in HF92).

The models are easily distinguished by the optical emission line profiles and by the X–ray emission measure. All transition region diagnostics (*e.g.* the differential emission measure) scale directly with the coronal pressure. Therefore, the models can be used to interpret optical, UV and/or X–ray observations. We are currently in the process of computing the impulsive phase models, where the density (pressure) is frozen in pre-flare state. We anticipate large temperature enhancements and much different line profile behavior in these models.

ACKNOWLEDGEMENTS

This work is supported by NASA grant NAGW-2547 and NSF grant ATM 90-22385 (SLH) and NASA grant NAGW-864 and NSF grant ATM 88-22366 (GHF).

REFERENCES

Carlsson M. 1986, *Uppsala Obs. Report No. 33*.
Emslie A.G. 1978, *Ap. J.*, **224**, 241.
Fisher G.H. 1989, *Ap. J.*, **346**, 1019.
Gan W.Q. & Fang C. 1990, *Ap. J.*, **358**, 328.
Hawley S.L. & Fisher G.H. 1992, *Ap. J. Suppl.*, **78**, in press. (HF92)
Henoux J.-C. & Nakagawa Y. 1977, *Astr. Ap.*, **57**, 105.
Metcalf T.R., Canfield R.C., Avrett E.H. and Metcalf F.T. 1991, *Ap. J.*, **350**, 463.
Raymond J.C. & Smith B.W. 1977, *Ap. J. Suppl.*, **35**, 419.
Ricchiazzi P.J. & Canfield R.C. 1983, *Ap. J.*, **272**, 739.

Table 1

Properties of Theoretical Model Atmospheres

Model	F_c	Q_{cor}	T_A	P_{cor}	N_{tr}	F_{cond}	F_{X-rays}	F_{beam}	EM
QF1	–	0.01	2.0	3.4	6.9×10^{18}	1.7×10^{6}	–	–	45.4
M1	5×10^{8}	0.39	5.7	78	5.6×10^{19}	4.0×10^{7}	1.3×10^{8}	2.3×10^{8}	47.2
M2	1×10^{9}	0.90	7.2	158	8.9×10^{19}	8.0×10^{7}	3.2×10^{8}	2.2×10^{8}	47.6
M3	5×10^{9}	4.9	11.7	676	2.3×10^{20}	3.4×10^{8}	2.3×10^{9}	2.6×10^{8}	48.5
M4	1×10^{10}	9.9	14.2	1236	3.5×10^{20}	6.3×10^{8}	5.5×10^{9}	2.9×10^{8}	48.8
M5	5×10^{10}	49.8	22.7	4953	8.8×10^{20}	2.5×10^{9}	3.9×10^{10}	3.6×10^{8}	49.6

F_c = total electron beam flux (erg cm^{-2} sec^{-1})
Q_{cor} = coronal heating rate (erg cm^{-3} sec^{-1})
T_A = coronal apex temperature (K)
P_{cor} = coronal and transition region pressure (dyne cm^{-2})
N_{tr} = column depth at transition region ($T = 2\times10^5$K) (cm^{-2})
F_{cond} = conductive flux at 10^5K (erg cm^{-2} sec^{-1})
F_{X-rays} = total soft X–ray flux deposited in chromosphere (erg cm^{-2} sec^{-1})
F_{beam} = total beam flux deposited in chromosphere (erg cm^{-2} sec^{-1})
EM = emission measure, assuming loop radius = 1.5×10^8 cm (cm^{-6})

Seventh Cambridge Workshop on Cool Stars, Stellar Systems, and the Sun
ASP Conference Series, Vol. 26, 1992
Mark S. Giampapa and Jay A. Bookbinder (eds.)

RADIATIVE PUMPING OF CHROMOSPHERIC EMISSION LINES DURING A WHITE LIGHT FLARE

E. R. HOUDEBINE   Dunsink Observatory - Dublin Institute for Advanced Studies, Castleknock, Dublin 15, Ireland

ABSTRACT We investigated, by means of NLTE-radiation transfer computations, the effect of the flare white light on Hydrogen line formation (for dM4e type stars). We show that the Balmer continuum is extremely efficient in "pumping" the Balmer series chromospheric emission. Investigating multi-wavelength observations, we also point out that the white light and the Balmer line fluxes are correlated over orders of magnitude.

Keywords: flares ; chromospheres; white light

EVIDENCES FOR PHYSICALLY LINKED WHITE LIGHT AND BALMER EMISSIONS

With the presently available multi-wavelength datasets, it becomes possible to map out the main flare spectral signatures and propose preliminary scenarios (*e.g.*, Houdebine, 1990, 1991). We distinguish two phases; a former impulsive phase during which a main tracer is the white light, and a later gradual phase identified by slower spectral line evolutions and little continuum. Observations provide evidence that during the former phase, the Balmer line fluxes closely follow the white light flux (*e.g.*, Rodonò *et al.* , 1989, Hawley, 1989). At this stage of the flare, line profiles are composed of a narrow emission core and sometimes broad wings. This differs from the solar case, for which broad profiles arise from the kernels.

We plotted in Figure 1 the H_δ line flux versus the continuum flux at about 4200Å for the impulsive phase of four flares on stars of similar spectral types. One can depict a meaningful linear correlation over four orders of magnitude. Similar correlations were found between other lines (Balmer lines, HeI 4026Å line) and white light measurements (*e.g.* U-band). These correlations are forceful indicators that both emissions are physically related.

RADIATIVE PUMPING OF CHROMOSPHERIC EMISSION

We computed the Balmer series emission outgoing from a chromosphere irradiated by a coronal/transition region white light radiation field. We modified the Scharmer and Carlsson (1985) code so as to simulate a flare white light at about the Balmer continuum. We used the dM4.5e model chromosphere of Houdebine and Panagi (1990), and an 8 level atomic model. Hydrostatic equilibrium was solved.

We show in Figure 2 the H_α profiles when increasing the continuum intensity, up to a factor of 724 above the quiescent level. The H_δ line flux versus the Balmer continuum photoionization rate has been superimposed to the observations in Figure 1. Note that the (Xo,Yo) position of the theoretical curve is arbitrary, owing to the unknown extent of the plasma emitting the white light. One sees that the line curve of growth agrees reasonably well with the empirical correlation over almost three orders of magnitude.

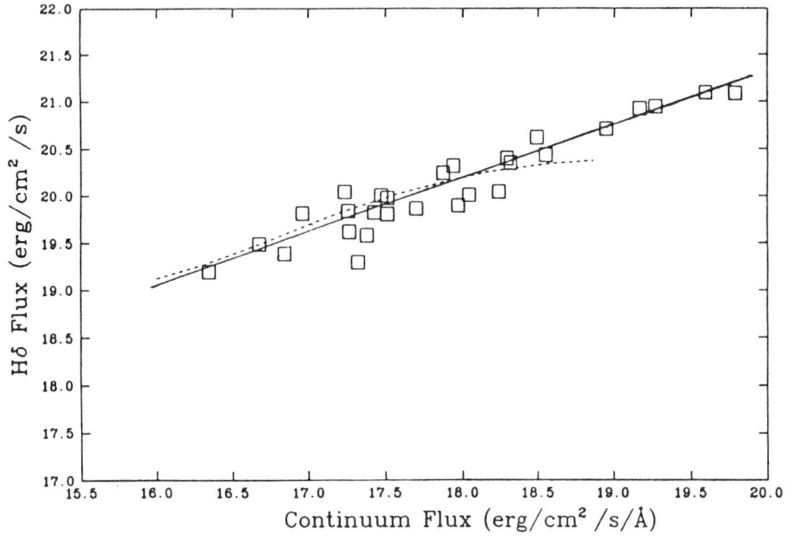

Fig. 1: H_α flux versus white light during four stellar flare on AD Leo, Prox Cen and UV Ceti (log-log scale). Note that the "quiescent component" were subtracted. This data was collected during coordinated multi-wavelength observations organized by the *Cool Star Consortium*. We superimposed to the data the results given by the model (dashed line).

The effect of an enhanced continuum radiation field at Balmer continuum frequencies is to greatly enhance the Balmer photoionization rate. This depletes the level 2 population and increases the Hydrogen ionization fraction. For temperatures less than 10^4K (where the Balmer series are formed), this raises the electron and upper level densities, the latter populations being in Partial LTE with the HII population. Consequently, collisional recombination and spontaneous de-excitation processes are magnified. The new resulting collisional-radiative equilibrium considerably depletes the Balmer series chromospheric opacities and increases the outgoing fluxes. The astonishing efficiency of this process in such an atmosphere can account for most of the observations during the impulsive phase.

Of course, the present model remains incomplete, because only the effect of the Balmer continuum was considered, and no allowance was made for possible changes in the chromospheric temperature structure. Nevertheless, the above results point to the importance, in the case of K and M type stars, of the radiative interactions between the various flare components. Furthermore, similar qualitative conclusions could be drawn for the effect of other recombination continua in the visible or the UV wavelength range (*e.g.*, HI

Lyman continuum, HeI continua, etc ...) on the formation of transition region and chromospheric lines. As far as this is concerned, the collisional-radiative equilibria of other elements may well be drastically modified by the radiation field but also changes in electron densities, thus leading for instance to incorrect emission measures.

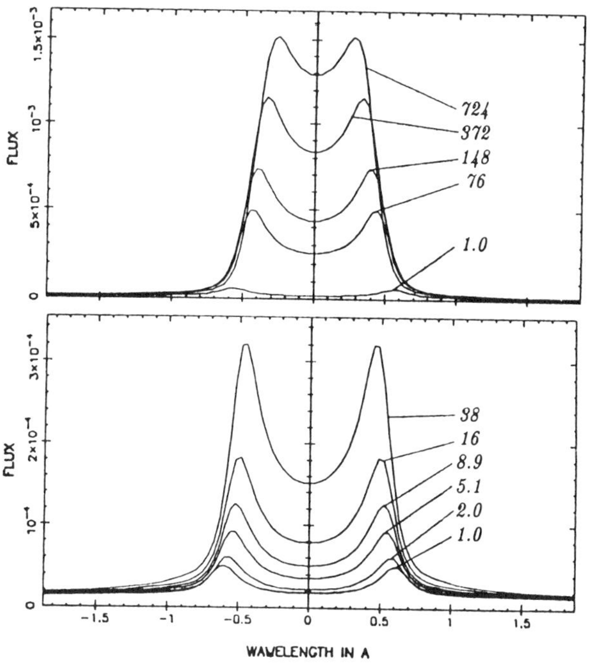

Fig. 2: H_α profiles when increasing the Balmer continuum strength. In these figures, the line flux (in $erg/cm^{-2}/s$) increases when increasing the Balmer continuum. One may notice the diminishing magnitude of H_α self-reversal due to the depleted chromospheric opacity. The magnification of the Balmer continuum intensity with respect to the quiescent level is indicated.

REFERENCES

Hawley, S.L. 1989, IAU Coll. 104 on *Solar and Stellar Flares*, Poster Papers, Catania Astrophys. Obs. Special Volume, Eds. B.M. Haisch and M. Rodonò, 49.

Houdebine, E.R. 1990, in Activité des étoiles de type dMe: Observation et modélisation des éruptions et de leur environnement chromospherique calme, Thesis of the Université d'Orsay - Paris XI, Institut d'Astrophysique Spatiale, France.

Houdebine, E.R. 1991, proceedings of the Astronomical Science Group Meeting of Ireland, September 1991, to appear in the *Irish Astron. J.*.

Houdebine, E.R. and Panagi, P.M. 1990, *Astron. Astrophys.*, **231**, 459.

Rodonò, M., Houdebine, E.R., Catalano, S., Foing, B.H., Butler, C.J., Scaltriti, F., Cutispoto, G., Gary, D.E., Gibson, D.M., Haisch, B.M. 1989, IAU coll. 104 on Solar and Stellar Flares, proceedings, Eds. B.M. Haisch and M. Rodonò.

Scharmer, G.B., Carlsson, M. 1985, *J. Comp. Phys.*, **59**, 56.

Seventh Cambridge Workshop on Cool Stars, Stellar Systems, and the Sun
ASP Conference Series, Vol. 26, 1992
Mark S. Giampapa and Jay A. Bookbinder (eds.)

SPHERICAL MODELS FOR CARBON STARS.

UFFE GRÅE JØRGENSEN Niels Bohr Institute, Blegdamsvej 17, DK-2100 Copenhagen, Denmark.

HOLLIS R. JOHNSON Astronomy Department SW 319, Indiana University, Bloomington, Indiana 47405, U.S.A.

ÅKE NORDLUND Copenhagen University Observatory, Østervoldgade 3, DK-1350 Copenhagen, Denmark.

ABSTRACT To be able study the crucial region between the outer photosphere and the chromosphere of carbon stars, we investigate effects of sphericity in about 180 photospheric models constructed for the purpose, with $2500 \leq T_{eff} \leq 3400$ K, $-1 \leq \log(g) \leq +0.5$, $1.02 \leq C/O \leq 2.0$, and a range of masses. Our spherical models are cooler (100 K) and have higher pressures (factor of 3) in the surface layers than in corresponding plane-parallel models. Among the molecular bands studied, only HCN and the stronger lines of CO show a dependence on sphericity. Through the influence of the photosphere, sphericity affects also the temperature of the temperature-minimum region and the structure of the chromosphere.

Keywords: stellar atmospheres; carbon stars; spectral lines

INTRODUCTION

As stars evolve through the red-giant stage, their atmospheres become greatly extended, temperatures in the outer layers fall, and surface gravity becomes very low. Convection generates a spectrum of acoustic waves, shocks form, chromospheres exist, stellar pulsations appear, grains may form, and mass loss becomes appreciable. To understand these phenomena requires model photospheres that realistically describe all layers, including the outermost photosphere, which generally deviates appreciably from the results of plane-parallel geometry (PPG). We report here the first atmospheric models of carbon stars that employ both spherical geometry (SG) and realistic opacities.

The models are luminosity-constant, based on LTE and hydrostatic equilibrium, and computed for various combinations of the parameters T_{eff} = 3400, 3100, 2800, 2500 K; $\log(g) = 0.5, -0.5, -1$; $C/O = 1.02, 1.35, 2.0$; $Z/Z_{\odot}=1$, 1/2, 1/10; and $M/\mathcal{M}_{\odot} = 0.7$, 1.5, 3, 10, and 100. The higher mass models, as well as corresponding PPG models, were computed to check the convergence of the SG models toward the PPG models for $M \rightarrow \infty$. The computer program used is an improved version of the MARCS program (Gustafsson *et al.* 1975) with routines for SG from Nordlund (1984), new routines written for this work, and updated opacities (described by Jørgensen 1991). Opacity from continuum sources and the molecules C_2, CN, CO, HCN, C_2H_2, and C_3 were included. The molecular opacities were treated by opacity sampling (OS) at 5365 wavelengths throughout the spectrum from 1250 Å to 25 μ.

GENERAL TRENDS OF SPHERICITY

Generally, the radiative energy density in the outermost layers is less in SG models than in PPG models, and this allows SG models to contract. On the other hand, the lower temperature in the outer layers of SG models leads to greater association and therefore increased opacity per mass unit of stellar material, which acts to extend the SG models. In M-star models, Plez (1990) found the second effect to dominate, while we find the first effect dominates in carbon stars. Spherical carbon star models will therefore be cooler, less distended, and have higher gas pressure in their surface layers than will their plane parallel counterparts.

In PPG, the structure of an atmosphere of a given chemical composition is totally determined by the effective temperature, T_{eff}, and the surface gravity, g. Particularly, the distance between two layers of given τ_{ross} is independent of the stellar mass, M, which is not the case in SG. One can therefore get an overview of the parameter space in which sphericity is important by studying the atmospheric extension, $d = d(T_{eff},g,Z,C/O)$, which we here take as $d = [r(\tau=10^{-5}) - r(\tau=1)] / R$, with the "stellar radius", R, considered as the value defined by the equation $g = GM/R^2$. Since R decreases for decreasing M and fixed g, low-mass stars are more spherical than high-mass stars. Figure 1 shows d as function of mass for a number of parameters T_{eff}, g, C/O. The biggest sphericity effect is found for models with low gravity, high T_{eff}, and high C/O.

THE RESPONSE OF DIFFERENT MOLECULES TO SPHERICITY

As the models cool in the uppermost layers, the molecular balance is shifted. General trends are: (1) An increase, at small $\log(\tau)$, of the partial pressure of the dominant molecules (CO, C_2H_2, HCN, and C_3) in the low-mass models compared to the high-mass models, and (2) a more pronounced effect for models of low gravity than for models of higher gravity. Due to cooling in the surface layers, the pressure of the polyatomic molecules are generally increased at the cost of a decrease in the partial pressure of the "donor" molecules (e.g., C_2 and CN).

The biggest changes in the partial pressures are found for CO and HCN. For CO the biggest effect of sphericity turns up outside the maximum of its partial pressure (*i.e.*, for lower values of the optical depth). Thus, with the exception of the cores of the strongest lines, we expect the overall spectrum of CO (as well as the entire spectrum of CN and C_2) to be relatively independent of mass [for a given T_{eff} and $\log(g)$]. In contrast, HCN has its maximum partial pressure much further out in the photosphere (as have also other polyatomic molecules), and big effects of sphericity are active both where the maximum partial pressure of HCN occurs and beyond.

Generally, the upper photospheric layers are more sensitive to sphericity than are the lower, and the ratio between spectral features formed in the upper layers and features formed deeper down is therefore always sphericity dependent. Since the amount of sphericity in real stars is dependent on mass as well as chemical composition (here mainly C/O), gravity, and effective temperature (see Fig. 1), the differential effect in the spectrum is dependent on all of these parameters, too.

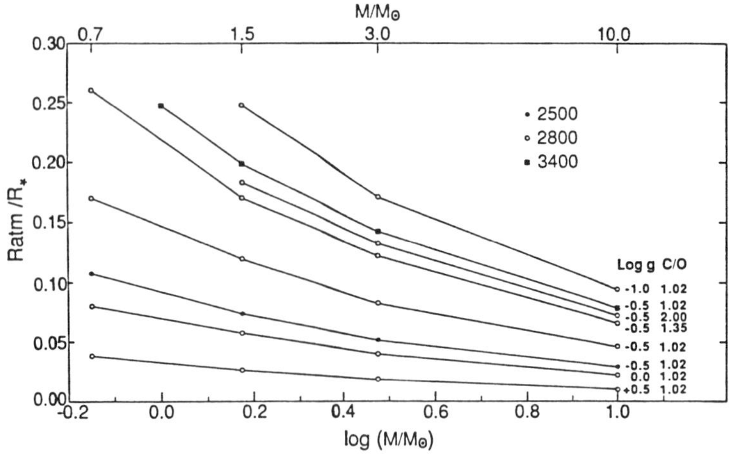

Fig. 1 Sphericity (extension) of carbon-star model atmospheres, (R_{atm}/R), as a function of stellar mass, $(M/M_{\odot})$. Different lines are drawn for different values of effective temperature (marked by symbol selection), log (g) (marked at lines), and C/O ratio (marked at lines).

CONCLUSIONS

1. Effects of sphericity are more evident in spectroscopic features formed in the outer atmosphere, and therefore show best when features formed in the outer atmosphere (HCN bands or strong CO lines) are compared to those formed more deeply (CN, C_2, or weak CO lines).

2. Bands of CN are remarkably unaffected by changes in mass and are therefore useful as standards when attempting to infer this quantity.

3. Band intensities of CO in carbon stars do show the differential effects reported for selected M stars (Plez 1990), but oppositely and much weaker. Such effects are due to the same mechanisms that affect the ratio of intensity of strong to weak lines as well as the ratio between intensities of bands from different molecules.

4. Only one previous investigation (Scholz and Tsuji 1984) has dealt with sphericity in carbon stars. Based on our much more complete opacities and detailed sampling method, we find considerably smaller sphericity effects than previously reported, but we agree on the qualitative effect that SG carbon stars compared to PPG carbon stars are cooler in the surface layers (as are M stars) and more compact (as opposed to M stars).

REFERENCES

Gustafsson, B., Bell, R.A., Eriksson, K., and Nordlund, Å. 1975, *Astr. Ap.*, **42**, 407.
Jørgensen, U.G. 1991, *Rev. Mexicana Astron. Astrof.*, in press.
Nordlund, Å. 1984, in Methods in Radiative Transfer, ed. W. Kalkofen (Cambridge: Cambridge Univ. Press), p. 211.
Plez, B. 1990, *Mem. Soc. Ast. It.*, **61**, 765.
Scholz, M. and Tsuji, T. 1984, *Astr. Ap.*, **130**, 11.

Seventh Cambridge Workshop on Cool Stars, Stellar Systems, and the Sun
ASP Conference Series, Vol. 26, 1992
Mark S. Giampapa and Jay A. Bookbinder (eds.)

THE 3 MIN OSCILLATIONS IN CHROMOSPHERIC BRIGHT POINTS

WOLFGANG KALKOFEN, PAOLA ROSSI Harvard-Smithsonian Center for Astrophysics, 60 Garden Street, Cambridge, MA 02138, USA

GIANLUIGI BODO Osservatorio Astronomico di Torino, Pino Torinese, 10025 Torino, Italy

SILVANO MASSAGLIA Istituto di Fisica Generale, Università di Torino, Via P. Giuria 1, 10125 Torino, Italy

ABSTRACT Oscillations of the quiet solar chromosphere with periods near the acoustic cutoff period of approximately 3 min can be excited in a gravitationally stratified medium by: a velocity impulse representing exploding granules, or a piston that moves with periods longer than the acoustic cutoff period representing the photospheric 5 min oscillations, or a piston that moves with periods shorter than the acoustic cutoff period representing the turbulent motions in the convection zone. — We solve the wave equation in order to infer plausible mechanisms for the generation of 3 min waves. We find that the velocity for both impulse and long-period excitation decays in a manner similar to that observed in the K_{2v} and H_{2v} emission features in bright points, but that the velocity for short-period excitation does not match the temporal variation of the line profiles.

Keywords: bright points; 3 min oscillations; wave excitation; wave equation

INTRODUCTION

In the quiet solar chromospheric we recognize three different regions. Ordered by the strength of the emission in the Ca II H and K lines they are: magnetic elements on the boundary of supergranulation cells, bright points in the cell interior, and the truly quiet chromosphere in the cell interior. This paper focusses on the generation of 3 min waves in cell interior bright points. We describe the observations, consider explanations of the 3 min waves as standing waves and as oscillations at the acoustic cutoff period, and discuss solutions of the wave equation in a stratified medium that lead to waves with 3 min period.

OBSERVATIONS AND MODELING OF 3 MIN OSCILLATIONS

The neutral chromosphere extends from the temperature minimum to the end of the 7000 K plateau. The cores of the K and H lines of Ca II are formed in these layers, and the emission features K_2 and H_2 delineate the chromospheric bright points. In the model of the average chromosphere (model C of Vernazza, Avrett & Loeser 1981), most of the energy in these layers is emitted in the lines of ionized calcium, the strongest of which is the K line. Observations of the

K and H lines (*e.g.* Liu 1974, Cram & Damé 1983) show waves traveling to the chromosphere with repetition times near 3 min (*cf.* Liu, Fig 7) and causing enhancements of the blue emission peaks K_{2v} and H_{2v}; the red peaks are much less, and much less often, enhanced. In these events, the intensity of the K_{2v} emission peak increases by typically a factor of 3 (*cf.* Liu 1974) and the central absorption feature K_3 is usually redshifted (*cf.* Grossmann-Doerth *et al.* 1974), implying downflow of the K_3-forming layers and suggesting propagating waves.

Observations of waves with periods near 3 min have been interpreted in terms of waves trapped in a cavity formed by the steep temperature rises into the corona and in the photosphere (Leibacher, Gouttebroze & Stein 1982) in which acoustic perturbations would form standing waves. But the disturbances traveling toward the line center demand an explanation in terms of propagating waves. Furthermore, the strong increase of the emergent K_{2v} intensity suggests shocks (*cf.* Cram 1974), which suffer exponential damping (*cf.* Ulmschneider 1970). Hence, only a small fraction of the initial energy flux traveling outward would be returned from the upper boundary of the cavity even if the waves were plane and the reflection were perfect. Thus, the observations implying strong damping in the K_2-forming layers also demand propagating waves.

The explanation of the wave periods in terms of standing waves in a cavity depends on the temperature structure, and, in particular, on the distance between the temperature rises that act as reflecting boundaries. An alternative explanation is in terms of oscillations at the acoustic cutoff period in the wake of an initial disturbance. This phenomenon is a property of the density stratification and exists even in an isothermal atmosphere.

In the absence of a magnetic field, the linearized equations for the vertical propagation of a disturbance in a one-dimensional, isothermal atmosphere can be cast in the form of the Klein-Gordon equation (Lamb 1908),

$$\frac{\partial^2}{\partial t^2} u = \frac{\partial^2}{\partial x^2} u - u \ , \tag{1}$$

where the dimensionless variables u, t and x for velocity, time and height, resp., are defined in terms of the corresponding physical variables v, t' and x', by

$$v(x',t') = au(x',t')\mathrm{e}^{x'/2H} \ , \quad t = \omega_{ac} t' \ , \quad x = x'/2H \ , \tag{2}$$

with a the sound speed, H the scale height, $\omega_{ac} = a/2H$ the acoustic cutoff frequency and $P_{ac} = 2\pi/\omega_{ac}$ the acoustic cutoff period.

Lamb's solution for a velocity impulse imparted to an infinite atmosphere at height $x = 0$ and time $t = 0$ can be written as

$$u(x,t) = \frac{1}{2}\delta(t - |x|) \ - \ \frac{t}{2} \frac{J_1(\sqrt{t^2 - x^2})}{\sqrt{t^2 - x^2}} H(t - |x|) \ , \tag{3}$$

where δ is the Dirac δ-function, H is the Heaviside function, and J_1 is a Bessel function. The asymptotic solution for $t \gg x$ is a sine wave that decays as $t^{-1/2}$ and whose period is equal to the acoustic cutoff period. Thus the impulse solution for the physical velocity v is an oscillation at the acoustic cutoff period that decays with time and grows exponentially with height.

The nonlinear impulse solutions of Bodo *et al.* (1991) showed velocity variations that were in qualitative agreement with the observed H and K line profiles as regards the redshift of the absorption core K_3 at the time of K_{2v} intensity enhancement and decayed in a manner similar to the linear solution (3), a feature that is frequently observed (*cf.* Cram & Damé 1983, Plate 11).

Oscillations at the acoustic cutoff period can be excited in a linear process also via a piston at the boundary $x = 0$ of a semi-infinite, isothermal, gravitationally stratified atmosphere. For piston motion beginning at $t = 0$ and described by $\exp(i\sigma t)$, the solution of the wave equation is

$$u(x,t) = \mathrm{e}^{i\sigma(t-x)} - x \int_0^{t-x} \mathrm{e}^{i\sigma\tau} \frac{J_1\left(\sqrt{(t-\tau)^2 - x^2}\,\right)}{\sqrt{(t-\tau)^2 - x^2}} d\tau \tag{4}$$

(*cf.* Doetsch 1956). The asymptotic solution ($t \gg x$) is the superposition of an oscillation at the acoustic cutoff period that decays with time as $t^{-3/2}$, and an oscillation at the piston period that is constant in time and propagates with phase speed $1/\sqrt{\sigma^2 - 1}$. The height dependence of the two waves is a function of the period: Whereas cutoff and short-period oscillations ($\sigma^2 > 1$) are depth-independent (u is constant, v grows exponentially with x), long-period oscillations ($\sigma^2 \leq 1$) are exponentially damped (as follows from the expression for the phase speed). Therefore, cutoff oscillations gain over oscillations at the piston period only for long-period excitation so that at sufficiently great height the cutoff oscillations will be more prominent (*cf.* Fleck & Schmitz 1991). But for short-period excitation, the cutoff oscillation can play a role only for a short time and equally at all heights. In either case, since the oscillation at the acoustic cutoff period is due to the *beginning* of the piston motion, only the oscillation at the piston period will remain after a sufficiently long time.

Thus, prominent oscillations at the acoustic cutoff period can be excited by an impulse or, at sufficiently great height, by a piston moving with a period longer than the acoustic cutoff period. In the linear limit studied in this paper (for the nonlinear results, see Rossi *et al.* in these proceedings), short-period excitation does not result in oscillations at the acoustic cutoff period, except for excitation very close to the cutoff period and for a very short time.

ACKNOWLEDGEMENTS

We are grateful for support by NASA through grant NAGW 1568.

REFERENCES

Bodo, G., Kalkofen, W., Massaglia, S. & Rossi, P. 1991. In: Mechanisms of Chromospheric and Coronal Heating, P. Ulmschneider, E. Priest, and R. Rosner eds., Springer Verlag, Heidelberg, 353.

Cram, L. 1974. *Solar Phys.* **37**, 75.

Cram, L. E. & Damé, L. 1983. *Astroph. J.* **272**, 355.

Doetsch, G. 1956. *Handbuch der Laplace-Transformation*, Birkhäuser Verlag, Basel.

Fleck, B. & Schmitz, F. 1991. In: Mechanisms of Chromospheric and Coronal Heating, P. Ulmschneider, E. R. Priest & R. Rosner, eds. Springer Verlag, Heidelberg, 22.

Grossmann-Doerth, U., Kneer, F. & v. Uexküll, M. 1974. *Sol. Phys.* **37**, 85.

Lamb, H. 1908. *Proc. London Math. Soc. (2)* **7**, 122.

Leibacher, J., Gouttebroze, P. & Stein, R. F. 1982. *Ap. J.*, 258393.

Liu, S.-Y. 1974. *Astroph. J.* **189**, 359.

Ulmschneider, P. 1970. *Solar Phys.* **12**, 403.

Vernazza, J. E., Avrett, E. H. & Loeser, R. 1981. *Astroph. J. Suppl.* **45**, 635.

Seventh Cambridge Workshop on Cool Stars, Stellar Systems, and the Sun
ASP Conference Series, Vol. 26, 1992
Mark S. Giampapa and Jay A. Bookbinder (eds.)

OSCILLATIONS IN A STRATIFIED ATMOSPHERE

PAOLA ROSSI, WOLFGANG KALKOFEN Harvard–Smithsonian Center for Astrophysics, 60 Garden St., Cambridge, MA 02138, USA

GIANLUIGI BODO Osservatorio Astronomico di Torino, Strada dell'Osservatorio 20, 10025 Pino Torinese, Italy

SILVANO MASSAGLIA Istituto di Fisica Generale dell'Università di Torino, Via P. Giuria 1, 10125 Torino, Italy

ABSTRACT In the context of the interpretation of 3 min oscillations in chromospheric bright points, we have studied the propagation of disturbances in a gravitationally stratified atmosphere by numerical integration of the full set of the one–dimensional hydrodynamic equations. We have examined two different ways of exciting the oscillations: by a velocity impulse and by a piston oscillating with various frequencies. In this paper we present the nonlinear solutions.

Keywords: chromospheric bright points; 3 min oscillations; wave excitation; shock waves

INTRODUCTION

Quasi–periodic oscillations in the flux of the Ca II H and K lines on time scales of the order of 3 min, *i.e.*, about the acoustic cutoff period, have been detected in several observations of chromospheric bright points, (see *e.g.* Liu 1973, 1974, Cram 1974, Cram & Damé 1983, Damé 1984). These periodic variations can be interpreted as being caused by acoustic wave trains, whose frequency is of the order of the atmospheric cutoff frequency and which are propagating upwards in the chromosphere. Therefore, we intend to investigate how it is possible to produce waves with periods close to the cutoff period. To this purpose, we excite the atmosphere using different mechanisms, namely, a velocity impulse or a piston oscillating at different frequencies.

MODEL AND RESULTS

We study the nonlinear behavior of waves propagating in a stratified, isothermal atmosphere by numerical integration of the full set of 1–D hydrodynamic equations, using a code based on the Piecewise Parabolic Method (Woodward and Colella, 1984). The domain of integration spans 40 scale heights and is divided into 1000 intervals.
The energy equation contains an artificial heating–cooling term of the form

$$Q \propto (T - T_0)^{\alpha} \quad ,$$

where T_0 is the temperature of the background atmosphere and $\alpha = 3$. This has been done in order to keep a constant average temperature in the atmosphere so that the medium, in which the waves propagate, does not change its properties with time.

In the first excitation mechanism, the atmosphere evolves from an initial condition in which the velocity is perturbed only in a very narrow region and all the other variables have their unperturbed values.
The initial velocity distribution has the form

$$v(x, t = 0) = v_0 \, exp[-(x - x_0)^2/\sigma^2] \quad ,$$

where v_0 is the maximum amplitude, x_0 is the position of the velocity perturbation in the atmosphere and σ is its spatial width. The initial perturbations of the other quantities are kept zero and, at both boundaries, we impose the conditions $\partial v/\partial z = 0$.
The velocity impulse is followed by a wake of oscillations with decreasing amplitude, the impulse forms a shock at about 5 scale heights, and the subsequent oscillations form shocks at progressively greater height because of their smaller initial amplitudes. The first oscillation after the impulse, propagating in its cool wake, travels at low speed until it is eventually overtaken by the following oscillation and the two merge.
The Fourier spectrum of the temporal velocity behavior varies with height in the atmosphere. The spectrum at low height (almost linear behavior) shows a distinct peak at the cutoff frequency, and this peak becomes more enhanced at greater heights.
At a low height the phase shift between velocity and pressure is $\sim 90^o$ at the cutoff frequency, and it goes to zero at higher frequencies, as predicted by the linear theory. Conversely, higher up in the atmosphere, at heights where shocks are fully formed, the phases become all scrambled by nonlinear effects, but, at the cutoff frequency we have a phase shift $\sim 0^o$, indicating energy transport by the shocks.

The second type of excitation mechanism is by means of a piston forming the bottom boundary whose velocity as a function of time has the form

$$v(x = 0, t) = v_0 sin(\omega t) \quad ,$$

where for excitation frequency we distinguish the two cases of ω larger or lower than the cutoff value. At the upper boundary, we use again $\partial v/\partial z = 0$.
In the case of short-period oscillations ($\omega > \omega_{co}$), in the lower region of the atmosphere, where the behavior is still linear, we have essentially an oscillation at the imposed piston frequency, modulated and distorted by frequencies near the cutoff, which are present because of the starting of the piston. Propagating upward in the atmosphere, the oscillations increase their amplitude and turn into shocks. The different shocks have different strengths and propagate in a medium with varying average temperature, therefore they move at different speeds, faster shocks will then overtake slower ones and merge with them. The effect of this process becomes clearer looking at the Fourier transform of the signal at different heights: at lower heights, we have only the piston frequency and its harmonics, at 15 scale heights begins a transfer of power from the piston frequency to lower frequencies, closer to the cutoff value. At 20 scale heights,

power at the piston frequency has disappeared and the spectrum is dominated by the cutoff frequency. The mechanism of shock overtaking and merging keeps working higher in the atmosphere and shifts the maximum of the spectrum to still lower frequencies.

Excitation by a piston oscillating exactly at the cutoff period leads to the formation of shocks separated by this period. In the Fourier spectra we can see distinct peaks at the cutoff frequency and at its harmonics along the whole domain of integration. Around 30 scale heights, some power has been transferred to a frequency which is half the cutoff frequency and its harmonics, and these tend to become more prominent higher up in the atmosphere.

In the case of long-period oscillations ($\omega < \omega_{co}$), the linear analysis predicts an exponential attenuation of the wave amplitude. In the linear regime, it is therefore expected that going up in the atmosphere a point can be reached where the cutoff frequency dominates the spectrum. Looking at the Fourier spectra, we can see that at 5 scale heights the cutoff frequency begins to be important, but the second and third harmonics of the exciting frequency have an even larger amplitude. At 10 scale heights the driving frequency is strongly attenuated and the spectrum is dominated by its harmonics. Going still higher, nonlinear effects transfer power to lower frequencies and the amplitude at the driving frequency increases again (15 H), forming a broad maximum between this and the cutoff value (20 H) and eventually leading again to the dominance of components at the acoustic cutoff frequency (30 H).

Finally, we have considered an excitation by a spectrum of frequencies centered around 30 sec. Namely, we have taken a signal with modulus of the Fourier transform given by

$$|F(\omega)| = exp[(\omega - \omega_0)^2/\sigma_\omega^2] \quad ,$$

where $\omega_0 = 6\omega_{co}$, and $\sigma_\omega = 2\omega_{co}$. The phases of the different Fourier components have been taken at random. We observe also in this case the phenomenon of shock overtaking and merging, so the maximum of the Fourier transform drifts toward lower frequencies.

ACKNOWLEDGEMENTS

The authors would like to acknowledge support from NASA through NAGW 1568. They also thank P. Ulmschneider for clarifying discussion and useful comments. P.R. wishes to thank the Smithsonian Institution for her fellowship.

REFERENCES

Cram, L.E., 1974, *Solar Phys.*, **37**, 75

Cram, L.E., and Damé, L., 1983, *Ap. J.*, **238**, 1134

Damé, L., 1984, in: Small-Scale Dynamical Processes in Quiet Stellar Atmosphere, S.L. Keil ed., Sacramento Peak

Liu, S.-Y., 1973, *Solar Phys.*, **31**, 127

Liu, S.-Y., 1974, *Ap. J.*, **189**, 359

Seventh Cambridge Workshop on Cool Stars, Stellar Systems, and the Sun
ASP Conference Series, Vol. 26, 1992
Mark S. Giampapa and Jay A. Bookbinder (eds.)

MULTI-LEVEL NON-LTE RADIATIVE TRANSFER AND HYDRODYNAMICS IN PULSATING ATMOSPHERES

DIMITAR D. SASSELOV Harvard-Smithsonian Center for Astrophysics,Cambridge, MA 02138, USA.

ALEX RAGA Department of Astronomy, University of Manchester, Manchester, UK.

ABSTRACT We describe the code HERMES which is a Godunov-type Lagrangian hydrodynamic scheme, run iteratively and simultaneously with a version of the multi-level non-LTE radiative transfer solver MULTI. The time-dependent hydrodynamic modeling involves calculations for H, Ca II, and Mg II. During the initial stabilization of the model the realistic radiative transfer is used only for determining the cooling and heating term in the energy equation. The model is constrained by varying the amplitude of the pulsation until we achieve a fit to the observed velocity curve and some observed line parameter. We derive the radial velocities from synthesized photospheric lines by non-LTE solutions, and measure the velocities in the same fashion as did the observers.

Keywords: stellar atmospheres; Cepheids; pulsation; hydrodynamics

INTRODUCTION

The atmospheres of pulsating F to K-type stars cannot be represented adequately by a sequence of static solutions, even at low amplitudes of pulsation. Such atmospheres are no longer in hydrostatic equilibrium, they expand and attain a different (on the average) pressure structure. In addition, even at low amplitudes shocks tend to form and propagate in the upper atmosphere.

While observational data on dynamic effects (line splitting, different velocity curves, emission, etc.) can be found for decades (*e.g.* Kraft 1956; Sasselov and Lester 1990; Lawson *et al.* 1991), theoretical models are few and only qualitative (Schwarzschild 1952; Hilendahl 1968; Karp 1975; Hill and Willson 1979; Bowen 1988). The models have been successful in describing bulk dynamic properties, but have all failed to produce realistic (or any!) spectral syntheses due to their unstable and inadequate hydrodynamic schemes. This problem has long been recognized (*e.g.* Hilendahl 1968, p.105; Mihalas and Mihalas 1984, p. 287). The code HERMES described here is based on a numerical scheme which is stable and without artificial dissipation.

THE NUMERICAL METHOD

Our time-dependent treatment of hydrodynamics and non-LTE radiative transfer is applicable to stars in which the formed shock waves do not dominate the energy balance of the atmosphere. Strong dominating shocks, their energy yields and radiative precursors, demand an accurate understanding of the physical processes inside the extremely thin shock front. Until such understanding is available and applicable, we can still model accurately weaker atmospheric shocks as mere discontinuities. However, a careful selection of the hydrodynamic numerical scheme is crucial. HERMES contains a 1-D explicit conservative upwind monotonic second-order (in time and space) Godunov-type Lagrangian hydrodynamic scheme. The core of Godunov's method is to consider each point of the numerical grid as a possible location of discontinuity with waves propagating in both directions from this point (Godunov 1959). An exact solution to the Riemann problem is used, which is equivalent to calculating the Rankine-Hugoniot relations at each grid point. All these features allow physically correct treatment of discontinuities, as well as monotonic second-order accuracy in the rest of the atmosphere. The scheme can be described as characteristic based and allows natural handling of the boundary conditions (see Roe 1990; Falle 1991).

The hydrodynamic scheme is explicit and Lagrangian in order to facilitate the radiative transfer solver. Most of the scheme's features which may appear as luxuries are in fact minimal requirements for correct radiative transfer solutions. HERMES contains a version of the multi-level non-LTE radiative transfer solver MULTI (Carlsson 1986). Consecutive detailed calculations are performed for model atoms of H, Ca II, Mg II, He I, and He II. It is used in three separate ways: to construct an initial semi-empirical hydrostatic atmosphere; to run as a subroutine in the hydro scheme and provide losses and gains to the energy equation (for a Ca II 6-level atom); and to run full solutions for several species once the pulsating model has stabilized.

HERMES APPLIED TO CEPHEIDS

It has been argued that multi-level non-LTE radiative transfer calculations are not needed in order to obtain the basic parameters or the bulk dynamics of a Cepheid atmosphere. This has been based on the belief that the major losses and gains in the energy equation could be accounted for in LTE. However, even if the latter were true, we are still left with no reliable observables to constrain a purely hydrodynamic Cepheid model. There are many reasons to why such an *ab initio* model will not be reliable: current pulsation theory cannot provide amplitudes; the source of chromospheric heating is not known; mass loss mechanisms are not understood, etc. The alternative is a semi-empirical model, and unless we calculate the spectral line profiles we have no means of constraint. Only a well-constrained model of the dynamic atmosphere can allow us to study the physics involved. In Figure 1 we give an example of part of the model constraining procedure for the 10-day classical Cepheid ζ Gem.

REFERENCES

Bowen, G.H. 1988, *Ap. J.*, 329, 299.

Carlsson, M. 1986, Uppsala Astron. Obs. Report No.33.

Falle, S.A.E.G. 1991, *M.N.R.A.S.*, 250, 581.

Godunov, S.K. 1959, *Matemat. Sbornik*, 47, 271.

Hill, S.,and Willson, L.A. 1979, *Ap. J.*, 229, 1029.

Hillendahl, R. 1968, Ph.D. Thesis, UC-Berkeley.

Jacobsen, T.,and Wallerstein, G. 1982, *Pub. A.S.P.*, 94, 471.
Karp, A. 1975, *Ap. J.*, 201, 641.
Kraft, R. 1956, *Pub. A.S.P.*, 68, 137.
Lawson, W.,Cottrell, P.,and Clark, M. 1991, *M.N.R.A.S.*, 251, 687.
Mihalas, D.,and Mihalas, B. 1984, Foundations of Radiation Hydrodynamics, Oxford Univ. Press.
Roe, P. 1990, in The Numerical Modelling of Nonlinear Stellar Pulsations, ed. R. Buchler, Kluwer, p.183.
Sasselov, D.,and Lester, J. 1990, *Ap. J.*, 362, 333.
Schwarzschild, M. 1952, in Trans. of the IAU VII, ed.P.Oosterhoff, Cambridge Univ. Press, p.811.

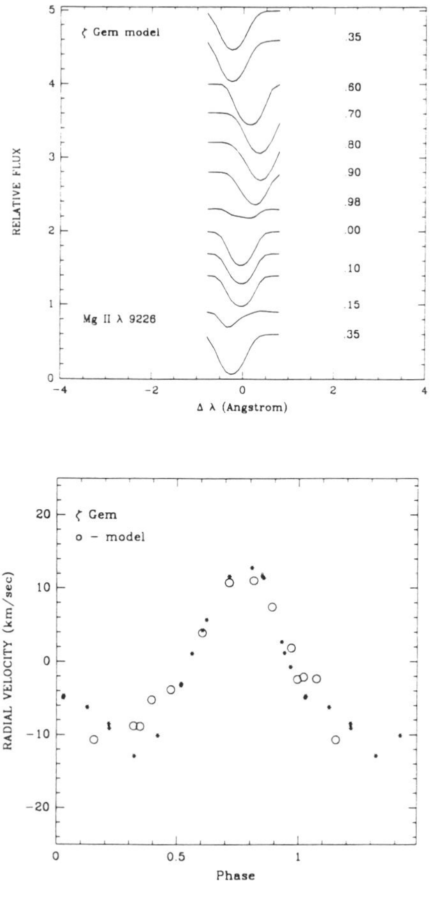

Fig. 1: (a) A sequence of synthesized line profiles by the time-dependent non-LTE solution in the dynamic model. (b) The radial velocity curve of ζ Gem (photosphere). We derive the radial velocities from the computed line profiles following the technique used by the observers in order to get a correct fit to the observed curve (*) by Jacobsen and Wallerstein (1982).

Seventh Cambridge Workshop on Cool Stars, Stellar Systems, and the Sun
ASP Conference Series, Vol. 26, 1992
Mark S. Giampapa and Jay A. Bookbinder (eds.)

USING HELIUM LINES AS A PROBE FOR ESTIMATING SURFACE FILLING FACTORS ON SUN-LIKE STARS

VINCENZO ANDRETTA Dipartimento di Scienze Fisiche dell'Università di Napoli, Mostra d'Oltremare pad. 19, 80125 Napoli

MARK S. GIAMPAPA National Solar Observatory, National Optical Astronomy Observatories, Tucson, AZ 85726-6732

ABSTRACT We estimate the maximum absorption equivalent widths for the neutral helium lines $\lambda10830$ and $\lambda5876$ (D_3) in active regions of dwarf G, F stars, thus providing a tool for estimating active region filling factor in solar type stars.

Keywords: Helium lines; chromospheres; stellar activity; radiative transfer

INTRODUCTION

Theories which attempt to explain stellar activity must confront the problem that with current technology we can only observe an average of the surface emission from dwarf stars, while, because of the presence of magnetic fields, it is likely that surface inhomogeneities play an important role in the dynamics of the stellar atmosphere. Indeed, if we interpret observations in terms of two component atmospheres ("quiet" and "active" components), we can assume that the active component is strongly related to the stellar magnetic field; thus, two otherwise identical stars with the same quiet component can exhibit different levels of activity essentially because their active atmospheres have intrinsically different radial structure, or because the fraction of the stellar surface covered by the active regions (the "filling factor") is different, or both. These considerations make clear the importance of having reliable, widely applicable methods for estimating surface filling factors.

METHOD

Spectroscopic signatures of active regions which appear markedly different in the active component, as compared to the quiet atmosphere, can be utilized to estimate filling factors. An ideal tracer could be a line which is absent in the quiet atmosphere, and has a known intrinsic strength (*i.e.*, equivalent width) in the active one. In this case the filling factor is just the ratio of the observed equivalent width to the intrinsic one (Giampapa 1985). Among those spectroscopic diagnostics that can serve as tracers of active regions on the surfaces of Sun-like stars are the He I lines, namely the D_3 line at 5876 Å and the line at 10830 Å. These features are usually very weak when observed in the quiet Sun, but appear strongly enhanced (in absorption) in active regions. The problem is that their intrinsic strength in plage-like regions on stars is in general not known.

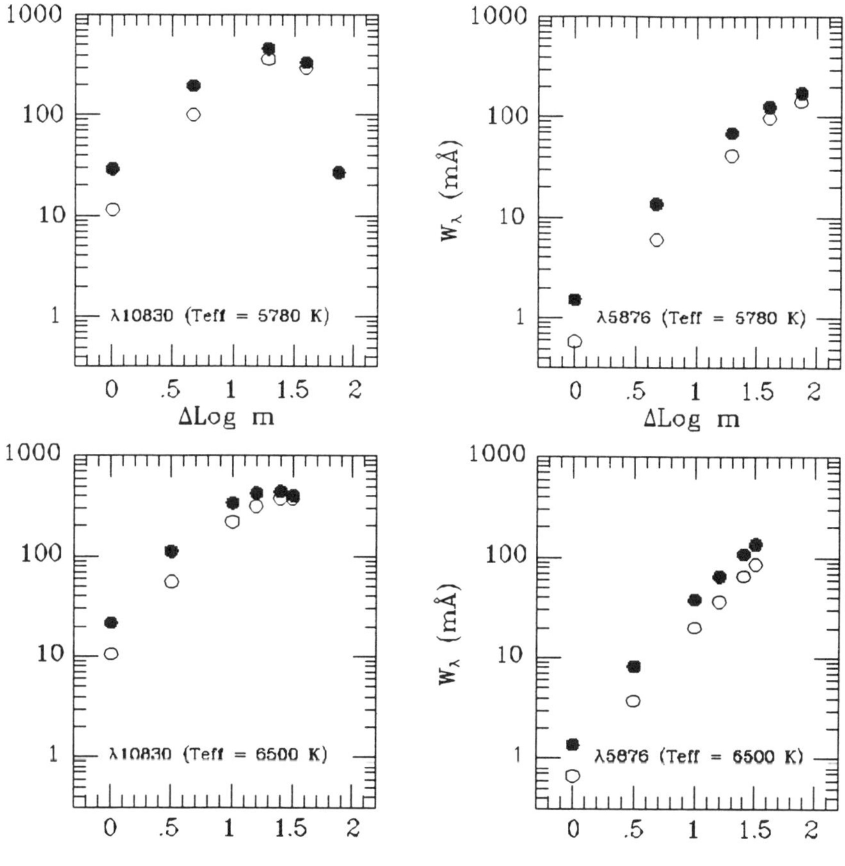

Fig. 1: Computed helium line equivalent widths (W_λ) vs. the activity parameter $\Delta\, log\, m$ in the solar case (T_{eff} = 5780 K) and for a dF star (T_{eff} = 6500 K). Results obtained ignoring coronal back-radiation (open circles) are also shown.

To overcome this problem, we adopted a schematic approach in which we explicitly computed the He I lines for a series of model atmospheres representing different degrees of activity. For each line, one expects that as the atmospheric activity increases, the He I line strength increases as well. Eventually a point is reached where the maximum in absorption equivalent width is attained and the line begins to be driven into emission. The maximum value of the absorption line equivalent width is useful because it gives a lower limit for the active region filling factor.

In our approach, each model is characterized by one activity parameter: the increment $\Delta\, log\, m$ of the mass loading at the "top" of the chromosphere. The model active atmospheres were created modifying the VAL C quiet Sun model (Vernazza *et al.* 1981) for G stars, and adding several chromospheres to a radiative equilibrium model with $T_{eff} = 6500$ K and $log\, g = 4.5$ (Kurucz 1979) for F stars. The coupled non-LTE radiative transfer and statistical equilibrium problem for hydrogen and helium was solved using a modified version of the Carlsson (1986) code.

RESULTS AND DISCUSSION

Things are not so simple for the He I lines. First, there is the problem of the coronal back-radiation which can have relevant effects on the line formation. We include this effect by assuming that the X-ray illumination is amplified as the activity increases according to the square of the coronal mass loading above the chromosphere (*i.e.*, proportionally to the coronal emission measure). In spite of that, at high levels of activity, the influence of the coronal back-radiation is not very important, thus leaving the maximum of the equivalent width substantially unaffected.

Another problem is that the line at 5876 Å (and any other relatively weak line we could consider) never reaches its maximum absorption equivalent width for reasonable active atmosphere models. While we found an upper limit for the λ10830 absorption equivalent width of $\approx$ 450 mÅ both with the G and the F model, for the D_3 equivalent width the allowable maximum must then be set by other observational constraints. In particular, we terminate our grid of model chromospheres at the point when Hα begins to show some emission features. Since the Hα line is seen in absorption in dG and dF stars, the appearance of emission features is an indication that the chromospheric model has exceeded 'normal' levels of activity outside of large flares. It is at this point that we assign the maximum value ($\approx 70 - 100$ mÅ) for the strength of λ5876 in stellar active regions, and infer active region filling factors for F and G dwarfs based on this model-dependent quantity.

ACKNOWLEDGEMENTS

The authors wish to thank E. H. Avrett for providing unpublished data and for his constructive comments and suggestions.

REFERENCES

Carlsson, M. 1986, Uppsala Astronomical Observatory Internal Report No. 33
Giampapa, M. S. 1985, *Ap. J.*, **299**, 781
Kurucz, R. L. 1979, *Ap. J. Suppl.*, **40**, 1
Vernazza, J. E., Avrett, E. H., and Loeser, R. 1981, *Ap. J. Suppl.*, **45**, 635

Seventh Cambridge Workshop on Cool Stars, Stellar Systems, and the Sun
ASP Conference Series, Vol. 26, 1992
Mark S. Giampapa and Jay A. Bookbinder (eds.)

ON THE FORMATION OF HELIUM I 5876 (D_3) AND 10830 Å LINES IN THE ATMOSPHERES OF LATE-TYPE DWARFS

ALESSANDRO C. LANZAFAME Armagh Observatory, College Hill, Armagh BT61 9DG, N. Ireland

ABSTRACT We present the results of non-LTE radiative transfer calculations for He I lines in late-type dwarfs using the model atmospheres given by Thatcher, Robinson and Rees (1991) for ϵ Eridani (dK2), Houdebine (1991) for AU Microscopii (dM2.5e) and Panagi (1990) for Gl 494 (dM2e). The coupled non-LTE equations of radiative transfer and statistical equilibrium are solved using the Carlsson (1986) code. The computed intensities and profiles of the triplet He I lines at 5876 Å (D_3) and 10830 Å for the ϵ Eridani model are in reasonable agreement with the observations of Lambert and O'Brien (1983), O'Brien and Lambert (1986) and Zirin (1982). The computed D_3 equivalent width is about 40% smaller than observed for the AU Mic model and about one order of magnitude smaller for Gl 494. The present computations do not take into account the coronal X–ray back-radiation or particle diffusion effects in the transition region.

Keywords: Helium lines; chromospheres; radiative transfer.

INTRODUCTION

The formation of He lines in solar-type stars has been discussed in the literature for many years without definite conclusion. The evidence appears to point to the need for a mechanism other than collisional excitation by local thermal electrons to account for the observations. This has led to a number of suggestions including ionization and excitation by electron collision (*e.g.* Athay, 1965), enhanced collisional excitation due to mixing (Jordan, 1975) or diffusion (Shine, Gerola, Linsky, 1975) and photoionization by coronal XUV back-radiation followed by recombination - cascade (Zirin, 1975). In either case both solar observation and theory predict much enhanced He-line emission near active regions. Since the K and M dwarfs possess coronae which are both hotter and globally more luminous in X–rays, we would expect any such effects to be more marked than in the solar case. We discuss here our preliminary results in modelling He I lines in red dwarfs.

MODEL ATMOSPHERES

The model for ϵ Eridani has been derived by Thatcher, Robinson and Rees (1991) matching flux profiles of the Ca IR triplet lines at 8498Å and 8542Å , H_α and H_β lines, the Na D lines and the Ca II K line. So far, no complete model chromosphere has been proposed for any dM or dMe star. Recently Houdebine

(1991) and Panagi (1990) have attempted to derive chromospheric models for AU Mic (dM2.5e) and Gl 494 (dM2e) respectively. The model proposed by Houdebine for AU Mic has been obtained by matching flux profiles for both H_α and H_β lines. Use has been made of the additional constraint given by the ratio of Ly_α to H_α surface fluxes of about 1.1 (Doyle, Panagi and Byrne, 1990). Unlike the others, this model is for active regions on the star. Houdebine found it very difficult to obtain a satisfactory model under the assumption of homogeneity. He considered the observed profiles as representative of the emission from plages, instead of a mean of different profiles. Under this assumption he inferred from the modelling a filling factor of active regions of about 30%. The model for Gl 494 is for a homogeneous chromosphere. This has been obtained by fitting the observed H_α profile.

RESULTS AND DISCUSSION

The calculations reported here have been carried out for an 11-level model plus continuum. All the sublevels up to n=3 have been included in the model. We refer to Lanzafame (1991) and references therein for a description of the atomic data used in these computations. In table 1 we report the calculated equivalent widths and integrated fluxes together with the equivalent widths observed by various authors. As far as we know, no observation of the He 10830Å line exists in the literature for dMe stars: even the observational data for the 10830Å line in ϵ Eridani (dK) are not very accurate. Apart from this limitation, the agreement between computed lines for ϵ Eridani chromospheric model and the observations is satisfactory; the D_3 line is also reproduced quite well. From the AU Mic model we derive EW(calculated)/EW(observed)$\approx$0.43 for the D_3 line, using the same filling factor (30%) inferred by Houdebine from his computations. This ratio is $\approx$0.11 for Gl 494. Our preliminary calculations indicate that inferring the formation of He I lines in red dwarfs from purely an electron temperature distribution with height, appears to be a reasonable approximation for the dK (Balmer absorption) model for ϵ Eridani. The models for dMe stars are still poorly constrained. We obtain better agreement between calculations and observations for the AU Mic model, but a more realistic self-consistent physical description (that should take into account the X–rays coronal back-radiation or diffusion effects) should be used for these stars. Unfortunately, very few observational data on He lines in red dwarfs exist in the literature.

ACKNOWLEDGMENTS The author would like to thank Dr. P.B. Byrne, Dr. P.M. Panagi and Dr. E.R. Houdebine for useful discussions. Research at Armagh Observatory is grant-aided by the Department of Education for N. Ireland.

REFERENCES

Athay, R.G. 1965, *Ap. J.*, **142**, 755.

Byrne, P.B., & Doyle, J.G. 1989, *Astr. Ap.*, **208**,159.

Carlsson, M. 1986, Uppsala Astronomical Observatory report No.33.

Doyle, J.G., Panagi, P.M., & Byrne, P.B. 1990, *Astr. Ap.*, **228**, 443.

Houdebine, E.R. 1990, PhD Thesis, Université Paris XI-Orsay.

Houdebine, E.R. 1991, in Mechanisms of Chromospheric and Coronal Heating, 5-9 June 1990, ed. P. Ulmschneider.

Lambert, D.M. & O'Brien, G.T. 1983, *Astr. Ap.*, **128**, 110.
Lanzafame, A,C. 1991, *Irish Astron. J.*, **20**, 84.
Jordan, C. 1975, *M.N.R.A.S.*, **170**, 429.
O'Brien, G.T.& Lambert, D.L. 1986, *Ap. J. Suppl.*, **62**, 899.
Panagi, P.M. 1990, PhD Thesis, Queen's University of Belfast.
Shine, R., Gerola, H., & Linsky, J.L.: 1974, *Ap. J. Letters*, **202**, L101.
Thatcher, J.D., Robinson, R.D., & Rees, D.E. 1991, *M.N.R.A.S.*, **250**, 14.
Zirin, H. 1975, *Ap. J. Letters*, **199**, L63.
Zirin, H. 1982, *Ap. J.*, **260**, 655.

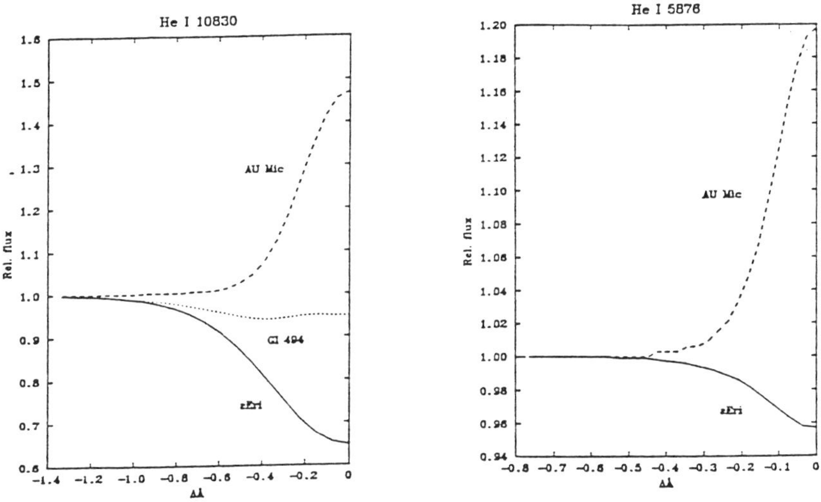

Fig. 1: The computed 10830 Å and 5876 Å line profile for the models.

Tab. 1: Observed equivalent widths, calculated equivalent widths and integrated fluxes for the ϵ Eridani, AU Microscopii and Gl 494 chromospheric models.

	$\lambda(\AA)$	Object	Obs. EW ($m\AA$)	Calc. EW($m\AA$)	Calc int. flux (erg/cm^2/s)
	10830.2	ϵ Eri	200/500[a]	320	5.5×10^4
		AU Mic	-	245*	2.4×10^5
		Gl 494	-	87	6.7×10^4
	5875.6	ϵ Eri	16/19 [b]	16	5.6×10^4
		AU Mic	129*[c]	55*	7.8×10^4
		Gl 494	90*[d]	10*	7.7×10^3

*in emission; [a]O'Brien and Lambert (1986) and Zirin (1976, 1982); [b]Lambert and O'Brien (1983); [c]Houdebine (1990); [d]Panagi (1990).

Seventh Cambridge Workshop on Cool Stars, Stellar Systems, and the Sun
ASP Conference Series, Vol. 26, 1992
Mark S. Giampapa and Jay A. Bookbinder (eds.)

NLTE SYNTHETIC SPECTRA OF THE MIRA-TYPE VARIABLE *S CAR*

DONALD G. LUTTERMOSER & GEORGE H. BOWEN
Physics & Astronomy Dept., Iowa State University, Ames, Iowa 50011

ABSTRACT The latest findings of the on-going research concerning NLTE radiative transfer in hydrodynamic models representative of Mira-type variable stars are updated. The equations of radiative transfer and statistical equilibrium are solved with PANDORA in a *snapshot* approximation at various phases for a hydrodynamic model of a pulsating AGB star. NLTE cooling and heating rates and electron densities are then used in subsequent runs with the Bowen hydrodynamics code. The resulting hydrodynamic models, modified with the NLTE cooling rates, are used to generate synthetic spectra with the PANDORA code for comparisons with observations. We report here on NLTE radiative transfer calculations for an atmospheric model representative of the Mira-type variable S Car (K7-M4 IIIe) using a 5-level hydrogen atomic model. Variations in the synthetic hydrogen-line spectra over a pulsation cycle are similar to variations in the actual stellar spectra. A fundamental discovery is made from these calculations: The hydrodynamic models produce Balmer lines in which the flux of Hα is *less* than Hβ and the flux of Hβ is *less* than Hγ. This has been observed in the spectra of Mira-type variables, and in the past, has been attributed to obscuration by overlying absorption. We show here that radiative transfer in the lines *alone* can give rise to this phenomenon.

Keywords: Radiative transfer; NLTE; hydrodynamics; Balmer lines; Mira-type variables; stellar atmospheres

INTRODUCTION

Mira variables are asymptotic giant branch stars that show evidence for pulsations in their spectra and light curves. Emission lines are seen in their spectra (*e.g.* Fe II lines and the hydrogen Balmer lines) which vary in flux over the pulsation cycle. Phase-dependent velocity measurements of these emission line features indicate the presence of shock waves propagating through the outer envelopes of these stars (Willson 1976). Hydrodynamic models for Mira variables have already been generated by Bowen (1988). Due to the intrinsic computational difficulty of handling radiative transfer in a dynamic medium, radiative cooling in these models is approximated in a simplified manner. We report the latest results of the Luttermoser/Bowen collaboration to improve these models by using the radiative transfer code PANDORA (Vernazza, Avrett, and Loeser 1981) to compute net radiative cooling rates, electron densities, and synthetic spectra of the Bowen hydrodynamic models in a *snapshot* approximation. PANDORA solves the equations of radiative transfer and statistical equilibrium in either plane parallel or spherical geometry in a self-consistent manner. Macroscopic velocity fields can be included in the calculation of the line profiles, source functions and rate equations.

HIGHLIGHTS OF THE RESULTS

The computational procedures used in this research have already been published by Luttermoser & Bowen (1990). We present highlights of the results of our 5-level hydrogen runs for various *snapshots* of a 3600 K, 0.8 $M_\odot$, 145 $R_\odot$, and 149 day period hydrodynamic model representative of the Mira-type variable S Car.

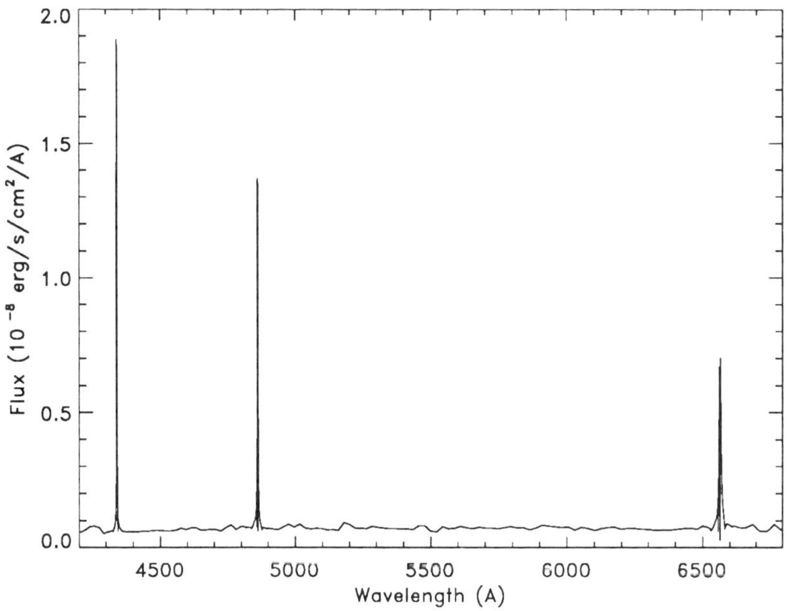

Figure 1: Balmer line series from a hydrodynamics model representative of S Car at pulsational phase 0.125. Background line calculations were included and the flux scaled to an angular diameter of 10 milli-arcsecs.

Variations in the synthetic hydrogen-line spectra over a pulsation cycle are similar to variations in the actual stellar spectra. Also, the hydrodynamic models produce Balmer lines in which the flux of Hα is *less* than Hβ and the flux of Hβ is *less* than Hγ (as seen in Figure 1). This has been observed in the spectra of Mira-type variables, and in the past, has been attributed to obscuration by overlying absorption. This Balmer line *increment* is the result of two factors in the transfer of radiation in the lines: (1) the optical depth of the line which determines the depth of formation of the line; and (2) the thermalization of the line which determines how closely the source function couples to the Planck function. The emission portion of the lines typically originates at line center optical depth of $10^2 - 10^3$. The core of the Hα line, which has the highest optical depth, forms in the 2nd innermost shock, while the emission wings originate at the front of the innermost shock at phase 0. The Hβ line center forms at the front face of the innermost shock due to its lower optical depth. As such, the emission wings are formed in the post shock region which is at a slightly higher temperature. Because the oscillator strength of the Hβ line is lower than the Hα line, the ratio of collisional rates to radiative rates is higher for Hβ, allowing Hβ to "see" more of the Planck function in comparison to the Hα line. The effect is even stronger for the Hγ line for the same reasons.

One problem with the Bowen models is that the Balmer and Paschen continuum are in emission during a good portion of the pulsation cycle. The inclusion of bound–bound opacities from other species does not *hide* this emission signature. As the innermost shock propagates outward, the continuum emission reduces until the Balmer and Paschen continuum become optically thin shortly after phase 1/2. Finally, by comparing calculations with and without the inclusion of background line opacity, we demonstrate that the notion of "continuum" in the emergent visual and ultraviolet flux of cool stars is incorrect and should not be used. **It is incorrect to call the highest flux level in the observed spectrum of a cool star "continuum".**

The Bowen hydrodynamic model of S Car reproduces many characteristics in the spectra of Mira–type variables. The main problem with the models resides with the optical thickness of the innermost shock in the Balmer and Paschen continua. Future refinements to the radiative cooling from these NLTE radiative transfer calculations and improvements to the pseudo–viscocity treatment in the hydrodynamics modelling should sharpen and cool the shocks enough to overcome the problem with the above mentioned optically thick continua.

ACKNOWLEDGEMENTS

This research was carried out under NASA grants NAG 5–1181 to Iowa State University and NAG 5–1305 to the University of Colorado. We gratefully acknowledge this support. E. H. Avrett and R. Loeser kindly gave us a copy of the PANDORA stellar atmospheres code and instructed D.G.L. in its use. We thank them as well. Finally, I wish to thank P.M. Marcum for useful comments on the manuscript.

REFERENCES

Bowen, G.H. 1988, *Ap. J.*, **339**, 299.

Luttermoser, D.G. and Bowen, G.H. 1990, in Sixth Cambridge Workshop on Cool Stars, Stellar Systems, and the Sun, ed. G. Wallerstein, A.S.P. Conference Series, Vol. 9, pg. 491.

Vernazza, J.E., Avrett, E.H., and Loeser, R. 1981, *Ap. J. Suppl.*, **45**, 635.

Willson, L.A. 1976, *Ap. J.*, **205**, 172.

Seventh Cambridge Workshop on Cool Stars, Stellar Systems, and the Sun
ASP Conference Series, Vol. 26, 1992
Mark S. Giampapa and Jay A. Bookbinder (eds.)

PHOTOIONIZATION OF CA II BY HYDROGEN LYMAN-ALPHA

ANDREW K. ROWE Department of Physics (Theoretical Physics), 1 Keble Road, Oxford, OX1 3NP, U.K.

ABSTRACT Linsky & Avrett (1970) noticed the near coincidence of the $3\,^2$D photoionization edge of Ca II at 1219 Å with the H Lyα line at 1216 Å. Here I present the results of radiative transfer calculations for Ca II taking into account the photoionization by H Lyα out of the 3d levels for quiet Sun and sunspot atmospheres. The aim is to understand the observed differences in the Ca II H & K profiles between these regions. In particular I show the effects of photoionization by H Lyα on the Ca II H-line and on the Ca II/Ca III ionization balance.

Keywords: radiative transfer; the Sun; Ca II

1 CALCULATIONS

The effects of photoionization by Lyα out of the 3d levels of Ca II was investigated for 2 model atmospheres: model C (average quiet sun) of Vernazza, Avrett and Loeser (1981), and the sunspot umbral model of Lites & Skumanich (1982).

The hydrogen atomic model used to calculate the radiation field due to the Lyα line consisted of 8 levels of hydrogen, and a single continuum level. The calcium model used consisted of five levels of Ca I, 12 levels of Ca II, and a single Ca III continuum level.

The computer code MULTI (Carlsson, 1986) was used to obtain solutions for the model atom and atmospheres. MULTI was modified, following the method of Uitenbroek (1989), to allow selected line-profiles to be calculated in PRD. The H & K lines were calculated in PRD, and all other lines were calculated in CRD.

The calculated line-profiles were compared with those observed in the active region NOAA/USAF 5698. The observations have spectral resolution of 0.0197 Å, and spatial resolution of 2″. For full details of the observations and data reduction techniques see Martínez Pillet *et al.* (1990).

2 RESULTS

2.1 Quiet Sun

Figure 1a shows the calculated H-line profiles when photoionization out of the 3d levels of Ca II by Lyα is and is not taken into account. The difference is quite small. Figure 1b compares the calculated and observed line-profiles. The calculated profile exhibits the self-reversal seen in the observed profile, but further work needs to be done to investigate what changes in the model are required to improve the agreement.

Figure 2 shows the calculated fractions of calcium in the form of Ca II. There is a large difference between the two results, the amount of Ca II being lower, and that of Ca III being higher when the photoionization is increased, as one would expect when both collisional and photoionization contribute to the total ionization.

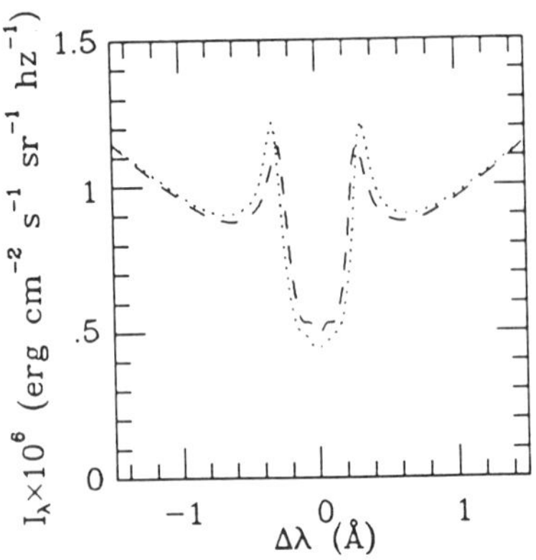

Fig. 1a: Theoretical H-line profiles with (dashed line) and without (dotted line) photoionization by Lyα taken into account.

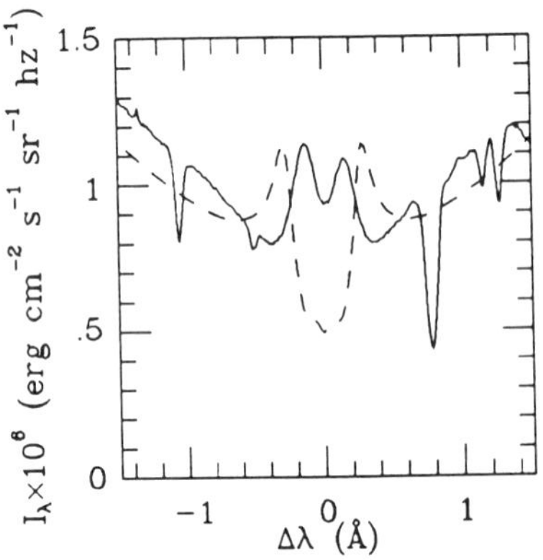

Fig. 1b: Theoretical (dashed line) and observed (solid line) H-line profiles for quiet Sun region.

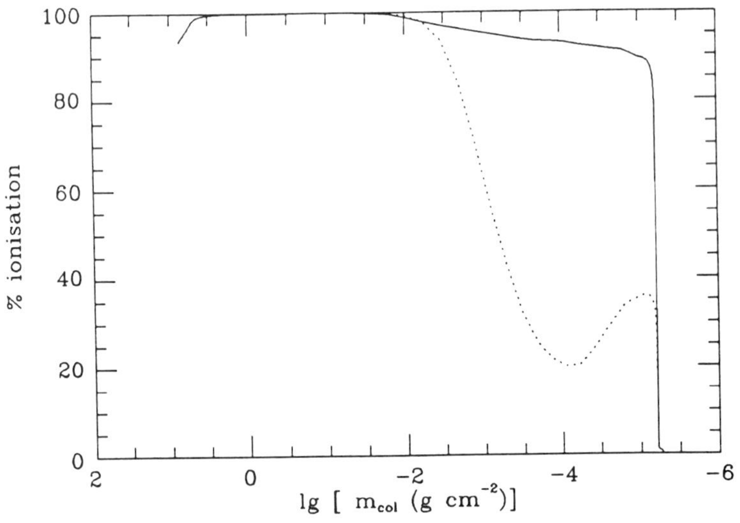

Fig. 2: Fraction of calcium as Ca II for photoionization by H Lyα included (dotted line) and not included (solid line) in the calculation.

2.2 Sunspot Umbra

Figure 3a shows the calculated H-line profiles when photoionization by Lyα is and is not included. In contrast to the quiet sun there is quite a large difference between the two profiles, and in the opposite sense; the increased photoionization increases the intensity of the core of the quiet Sun profile, but decreases the it in the umbral profile. Figure 3b compares the calculated and observed profiles. Although the calculated profile is much narrower than that for the quiet sun region, in agreement with the observations, it still exhibits a central self-reversal not seen in the observations, and further work is required to match the calculated profile to the observations. We point out that instrumental resolution of the observations exceeds the resolution of the calculations.

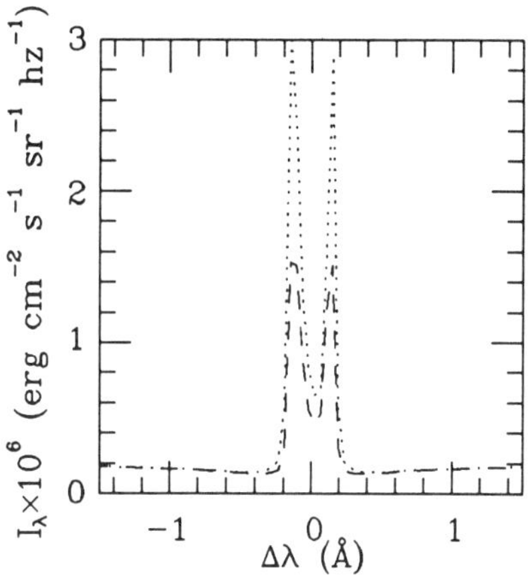

Fig. 3a: Theoretical H-line profiles with (dashed line) and without (dotted line) photoionization by Lyα taken into account.

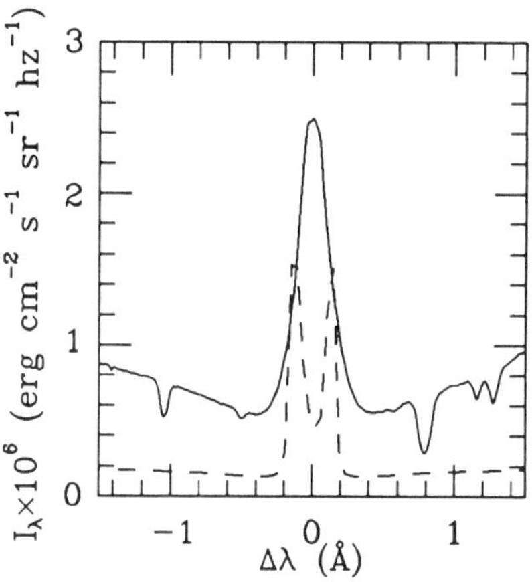

Fig. 3b: Theoretical (dashed line) and observed (solid line) H-line profiles for sunspot umbra.

ACKNOWLEDGEMENTS

I thank Drs. C. Jordan and G. Harper for their advice, and acknowledge the financial support of the SERC as a D.Phil. student.

REFERENCES

Carlsson, M., 1986, A Computer Program for Solving Multi-Level Non-LTE Radiative Transfer Problems in Moving or Static Atmospheres, Uppsala Astronomical Observatory Report No. 33.

Linsky, J.L., and Avrett, E.H., 1970, *Pub. A.S.P.*, **82**, 169.

Lites, B.W., and Skumanich, A., 1982, *Ap. J. Suppl.*, **49**, 293.

Martínez Pillet, V., García López, R.J., del Toro Iniesta, J.C., Rebolo, R., Vázquez, M., Beckman, J.E., and Char, S., 1990, *Ap. J.*, **361**, L81.

Uitenbroek, H., 1989, *Astr. Ap.*, **213**, 360.

Vernazza, J.E., Avrett, E.H., and Loeser, R., 1981, *Ap. J. Suppl.*, **45**, 635.

Seventh Cambridge Workshop on Cool Stars, Stellar Systems, and the Sun
ASP Conference Series, Vol. 26, 1992
Mark S. Giampapa and Jay A. Bookbinder (eds.)

THE Mg II *h* & *k* LINES AS DIAGNOSTICS OF THE SOLAR CHROMOSPHERE

HAN UITENBROEK Harvard–Smithsonian Center for Astrophysics, 60 Garden Street, Cambridge, MA 02138.

ABSTRACT The Mg II resonance lines are used as a probe of the thermal structure of the solar chromosphere. Observations from the UVSP/SMM data base are discussed together with their calibration onto an absolute intensity scale. Theoretical line profiles match the observed ones well except for the k_3 self reversal, which is too deep in all theoretical profiles.

Keywords: Sun; Mg II *h* & *k*; chromosphere; radiation transfer

INTRODUCTION

Avrett and co–workers have done extensive numerical calculations with the PANDORA code to construct models of different components of the solar atmosphere (*e.g.* Vernazza *et al.* 1981 and the paper by Avrett & Loeser in these proceedings). Recently, the latest of these VAL models have been modified to include the effects of differential diffusion of hydrogen and helium atoms and ions on the energy balance of the chromosphere and transition region (Fontenla, Avrett, & Loeser 1990 and 1991). The upward diffusion of neutral hydrogen in the new models makes the artificial temperature plateau at 20 000 K obsolete. Due to lack of accurate diagnostics, uncertainties in temperature and density distributions are large at the lower boundary of the transition region. To further constrain the thermal structure of the models I compare observed line profiles of the Mg II *h* & *k* lines with theoretical profiles from the present VAL models.

Due to their large opacity the Mg II resonance lines provide excellent atmospheric diagnostics over a wide range of heights if sufficient observations with high spatial and spectral resolution are available. A modest number of such observations is available in the data base of the Ultraviolet Spectrometer and Polarimeter (UVSP) that flew on board the Solar Maximum Mission (SMM) spacecraft.

UVSP/SMM OBSERVATIONS

The secondary mirror of the UVSP instrument could be tilted to scan the solar surface pixel by pixel over a range of $256'' \times 256''$, where the pixel size was determined by the field of view of the entrance slit of the spectrometer. The Mg II *h* & *k* lines at around 280 nm were observed with two different entrance slits (see Table). Slit number 20 was used to make atlas scans. Due to the large field of view $\Delta\Omega_s$ of this slit these scans contain little spatial information. They are useful here to obtain an absolute intensity calibration (see next Section).

Slit 12 was used to make both spectrograms (a wavelength scan from one specific pixel) and rastergrams (surface scans at one wavelength) in magnesium. Contiguous combinations of spectrogram and rastergram are of particular interest here.

slit no	$\Delta\Omega_S$	$\Delta\lambda$ (nm)
12	$3'' \times 3''$	0.006
20	$1'' \times 180''$	0.002

UVSP slits used for Mg II

Provided the rastergram is taken at a wavelength near the core such sequences allow us to determine whether the spectrogram was taken from a plage or quiet–Sun region, network or cell–interior. Due to the low number of counts per pixel this determination is not always conclusive. Statistical variations in the count rate allow for about half the observed intensity variations.

CALIBRATION

To compare spectrograms with calculated line profiles the observed photon counts need to be calibrated onto an absolute intensity scale. Observations in this paper were calibrated by fitting a spectrogram obtained with the large aperture slit (20) onto a well–calibrated (better than 20 %, courtesy of R. L. Kurucz) rocket spectrum. Calibrations over a large field of view have the advantage that they are largely independent of the exact location on the solar surface. When the calibration is established for slit 20 it can be converted to a calibration of the small aperture slit (12) since both slits use the same detector (number 5) and light path.

The measured intensity $\tilde{I}_{\lambda_0}$ (the average of the true intensity over viewing angle of the entrance slit S and width of the exit slit) at wavelength λ_0 is related to the number of photon counts Φ at that wavelength by:

$$\tilde{I}_{\lambda_0} = C(\lambda_0; S)\Phi(\lambda_0; S)/t; \qquad C \equiv hc/(\lambda_0 A \Delta\Omega_S \Delta\lambda_0),$$

where A is the effective telescope area, t the gate time, $\Delta\Omega_S$ the aperture of the entrance slit, and $\Delta\lambda_0$ is the width of the exit slit. We find $C(280; 20) = 375$ [erg cm^{-2} sr^{-1} nm^{-1}] (Experiment 21342). Using the different aperture and exit slit width of slit 12 we find $C(280; 12) = 2.5 \times 10^3$ [erg cm^{-2} sr^{-1} nm^{-1}].

RADIATION TRANSFER MODELING

Theoretical profiles from the given VAL model atmospheres were obtained with a Partial frequency Redistribution (PR) version (Uitenbroek 1989) of MULTI (Carlsson in these proceedings). Figure 1 shows the theoretical line profile from model C (thick line) together with the observed intensity from UVSP Experiment 22466 (thin line). From a contiguous rastergram (taken at 278.86 nm, as indicated by the vertical line) we determined that the spectrogram represents quiet–Sun cell interior with a possible network contribution. The theoretical profile (thick line) matches the observed one well apart from the central k_3 self reversal. Convolving the theoretical profile with a Gaussian of up to 6 km/s width, which is comparable to the instrumental broadening, does not improve the fit much. In all computations we have done so far the theoretical

k_3 minima are too deep compared with observed profiles, not only for model C but also in the of the hotter models VAL F and P.

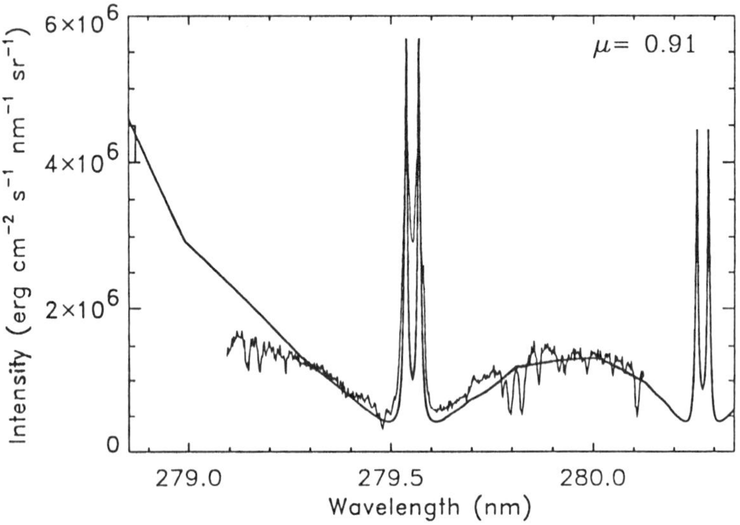

Fig. 1: Theoretical profile from atmospheric model VAL C (thick line). Thin line is observed spectrogram (Experiment 22466) taken with slit 12. Vertical line marks wavelength of contiguous rastergram; μ is the cosine of viewing angle.

CONCLUSION

The current VAL models match the quiet–Sun Mg II profiles well up to and just above the temperature minimum, judging from the fit of wings, k_1 minima and emission peaks. They do not reproduce the observed central absorption self reversal in the Mg II resonance lines (the same holds true for the Ca II resonance lines). Therefore, either the temperature structure at the top of the chromosphere has to be adjusted, or the theoretical line broadening has to be treated in a more sophisticated way. In the first case the transition temperature rise should occur at higher densities in "hot" elements, so that core line source function does not decrease as it does in very hot models. In the second case a more realistic representation of macroscopic velocity fields may be needed instead of the combination of micro– and macro turbulence that is used here.

REFERENCES

Fontenla, J. M., Avrett, E. H.,& Loeser, R. 1990, *Ap. J.*, **355**, 700
Fontenla, J. M., Avrett, E. H.,& Loeser, R. 1991, *Ap. J.*, **377**, 712
Uitenbroek, H. 1989, *Astr. Ap.*, **213**, 360
Vernazza, J. E., Avrett, E. H.,& Loeser, R. 1981, *Ap. J. Suppl.*, **45**, 635
Woodgate, B. E. *et al.* 1980, *Solar Phys.*, **65**, 73

PART VII

Further Perspectives on Cool Stars

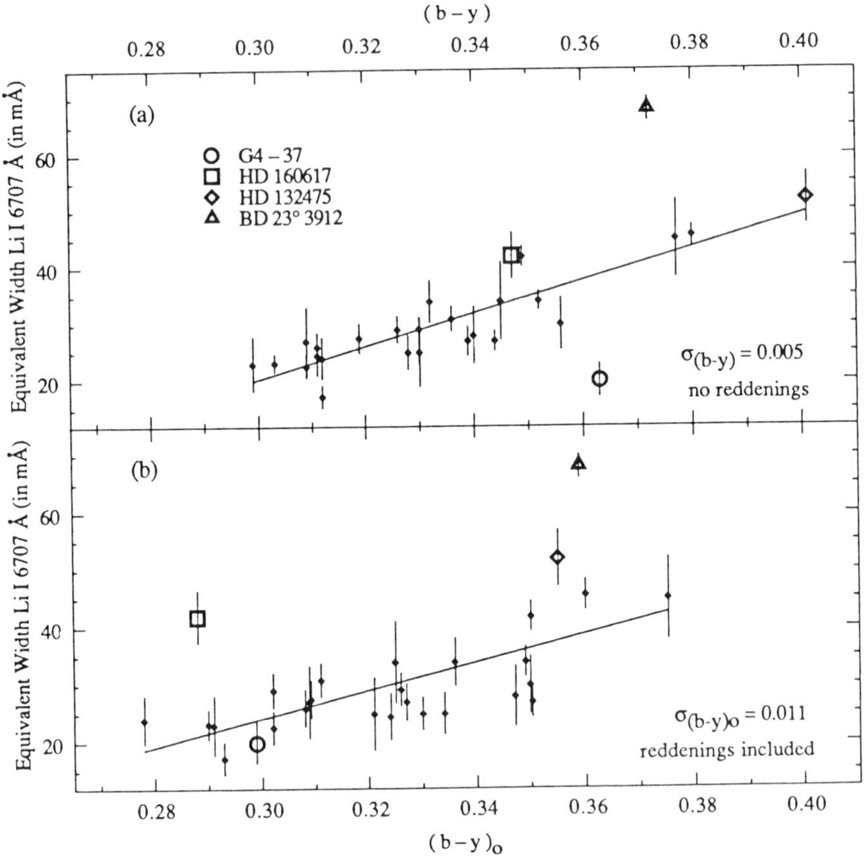

Figure 1 from "Lithium Dispersion of Halo Dwarfs" by C.P. Deliyannis and M.H. Pinsonneault.

Seventh Cambridge Workshop on Cool Stars, Stellar Systems, and the Sun
ASP Conference Series, Vol. 26, 1992
Mark S. Giampapa and Jay A. Bookbinder (eds.)

IMPLICATIONS OF LI 6707 Å DETECTION IN THE FK COM CANDIDATE 1E1751+7046

C.W. AMBRUSTER, Villanova U., Villanova, PA 19085

F.C. FEKEL, Vanderbilt U., Nashville, TN 37235

E.F. GUINAN, Villanova U., Villanova, PA 19085

ABSTRACT New observations from the KPNO coude feed show that 1E1751+7046, a spun-down FK Comae star candidate, is the most active member to date of a well-documented handful of **Li-rich (log Li = 1.6), but Old Disk**, rapidly rotating (vsini = 19 km s^{-1} ; P = 13-15 days), magnetically active, single, cool giants. As such, it cannot be a former A-F main sequence star making its first crossing of the Hertzsprung Gap. The need to explain both the relatively rapid rotation, and the Li, in single, low-mass, giants appears to point to an episode where both angular momentum and processed material are dredged up to the surface.

Keywords: Stars: FK Comae; Abundances: Lithium; Stellar Populations

INTRODUCTION

1E1751+7046, the stellar X-ray source with the highest X-ray - to - optical flux ratio in the EINSTEIN Medium Sensitivity Survey (Silva *et al.* 1985), was first proposed as a spun-down FK Comae candidate (coalesced binary) by Fleming *et al.* (1987). The observations of both the Silva and Fleming groups suggested an unusually active, rapidly rotating, single, evolved star. Table I summarizes the fundamental properties of this star, many of which have been revised as a result of our new observations. These are discussed below.

OBSERVATIONS

Lithium Detection Fig. 1 shows the Li 6707 Å line detection in the KPNO coude feed spectrum of 1E1751+7046. The Li line in 1E1751+7046 corresponds to an EW = 90 mÅ, or log Li of about 1.6 ± 0.2.

Light Curve UBVRI observations in May and June 1991 with the Four College 30-inch Automatic Photometry Telescope (APT) on Mt. Hopkins, AZ, show a large (0.2 mag) amplitude modulation in U-, B-, and V-band light curves over 45 days. The best-fit period to our limited data gives P = 15 ± 0.1 days (Fig. 2), although other periods are possible. The more extensive photometry of Jetsu and Tuominen (this vol.) indicates P = 13.98 days. The large amplitude of the light variation, and a lack of evidence for a flat maximum in Fig. 2, suggest a high inclination; we assume, conservatively, $45° \leq i \leq 90°$.

These inclination limits combined with vsini = 19 km s^{-1} and a 13.98 day period yield: $5.2R_{\odot} \leq \mathbf{R} \leq 7.4R_{\odot}$; $14.5L_{\odot} \leq \mathbf{L} \leq 29.4L_{\odot}$; $+1.9 \geq \mathbf{M}_v \geq +1.1$; 347 pc $\leq$ **d** $\leq$ 501 pc; 3.7 x 10^{31} ergs s^{-1} $\leq$ **Lx** $\leq$ 7.6 x 10^{31} ergs s^{-1} . These spreads in radius and luminosity are consistent with the K0 III spectral type. The derived L_x values are at the upper end of the range for RS CVn stars (Walter and Bowyer 1981).

Space velocities are necessary to classify 1E1751+7046 as Old Disk or Young Disk population. However, no proper motion measurement exists. Still, the distances above, combined with the star's b^{II} = 30°8 yields: 178 pc $\leq$ z $\leq$ 257 pc, similar to FK Com itself (z = 245±50 pc). Therefore, despite its Li abundance, 1E1751+7046 probably belongs to the OD population.

TABLE I: Properties of 1E1751+7046

m_v	= 9.6
sp	= K0 III (revised from K5 IV)
vsini	= 19 km s^{-1} (revised from 30-40 km s^{-1})
log Li	= 1.6 (revised from a non-detection)
P	= 13-15 days (revised from ≈ 5 d)
b^{II}	= 30°8
V_R	= -14.83 ± 0.83 km s^{-1}
log f_x/f_v	= -0.8
L_{5Ghz} ≈ 10^{17} ergs s^{-1} hz^{-1}	

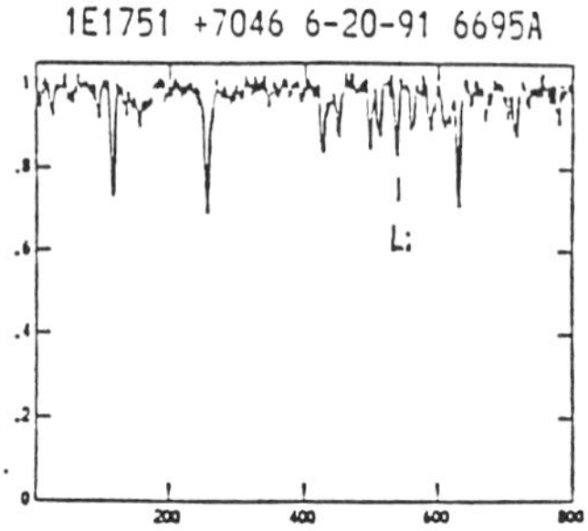

Figure 1: Spectrum of 1E1751+7046 taken at the coude feed, KPNO, showing the significant (90 mÅ) Li 6707 Å line.

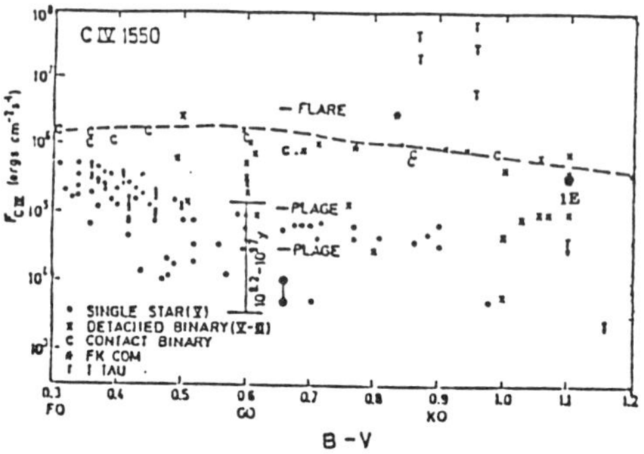

Figure 2: V-band photometric observations of 1E1751+7046 with the 30-inch APT on Mt. Hopkins, AZ, during May and June 1991, phased with a 14.95 day period. Repeat observations within a night are plotted to indicate observational scatter.

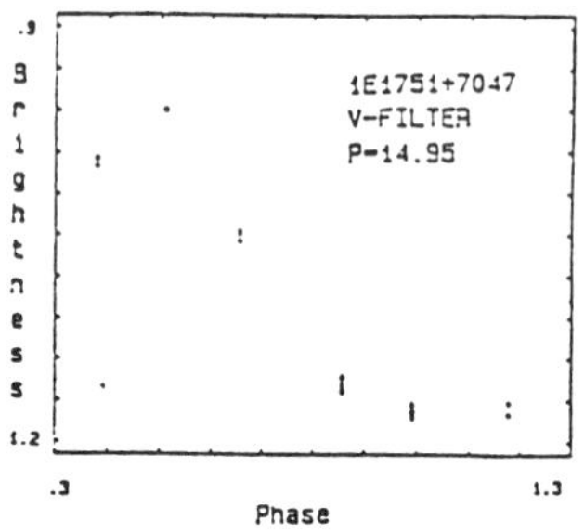

Figure 3: The quiescent C IV 1549 Å surface flux for 1E1751+7046 (2.3 x 10^5 ergs cm^{-2} s^{-1}) is plotted as a large, filled-in circle at B-V = 1.1 on Figure 4 of Vilhu (1987). The saturation level is indicated by the dashed line.

UV Flares and Surface Fluxes Our seven *IUE* observations of 1E1751+7046 and three existing spectra in the *IUE* Archives were reduced with standard *IUE* software. The UV surface fluxes for 1E1751+7046, computed from the Barnes-Evans relation (Barnes *et al.* 1978), are remarkably high, falling close to saturation levels (Fig. 3). The saturation level is defined as complete coverage of the stellar surface (*i.e.*, , filling factor = 1) by 1-2 kG magnetic fields (Vilhu 1987). A flare was also seen in the TR line fluxes of one spectrum, and Mg II (2800 Å) fluxes are conspicuously variable in observations spanning 2 years (Ambruster *et al.* 1991a,b).

DISCUSSION

Lithium is not expected in the atmospheres of cool giant stars, yet Brown *et al.* (1989) found 12 of 644 such stars (2%) to have log Li $\geq$ 1.5. In fact, it now appears that a significant fraction of chromospherically active stars, both binary and single, are lithium rich (Pallavicini *et al.* 1990; Fekel and Balachandran 1991; Randich, Giampapa, and Pallavicini, this vol.). However, only a tiny handful are OD like 1E1751+7046, whose main-sequence progenitors therefore had to be low mass stars (spectral type F or later) having convective, Li-depleting, envelopes (Fekel and Marschall 1991; Gratton and D'Antona 1989 for other examples).

The anomalous Li is the major problem in understanding 1E1751+7046 and the other OD Li-rich single giants. The evolutionary models of Pinsonneault *et al.* (1989), which include rotation, show that solar mass stars at the base of the giant branch in fact may dredge up high angular momentum material from a rapidly rotating core. This, along with the dynamo mechanism, could then explain both the rapid rotation and the high activity levels of 1E1751+7046. But whether lithium, created perhaps by the Be-transport (3He-to-7Be-to-7Li-to-4He) mechanism of Cameron and Fowler (1971), could survive dredge-up is not known.

1E1751+7046's ultraviolet fluxes typically exceed those of the other Li-rich, active giants (both OD and YD) by factors of 2-10, yet they are well below those for FK Com (Ambruster *et al.* 1991). FK Com, with its bizarre activity levels, is generally accepted as a coalesced binary; it is OD, but apparently lacks the Li 6707 Å line. Yet the only other stars in Fig. 3 with surface fluxes comparable to those of 1E1751+7046 are active binaries, NTT's, and FK Com.

Dredge-up and/or mixing, as well as rapid rotation, is clearly also involved in the coalescence of two stars, and the frequency of W UMa binaries, along with considerations of their evolution, indicates that coalescence must be taking place relatively frequently (Guinan and Robinson 1986). But whether lithium would be produced or survive during this process is unclear.

If dredge-up processes in a single star, and not coalescence, represent the recent history of 1E1751+7046, then the near-saturation surface fluxes of 1E1751+7046 provides an indication of how dramatic that episode must be.

ACKNOWLEDGEMENTS

Author C.A. gratefully acknowledges the support of NASA grant NAG 5-980.

REFERENCES

Ambruster, C., Guinan, E.F., and Siah, M.J. 1991a, *Bull. A.A.S.*, **23**, 941.

Ambruster, C., Fekel, F.C., and Guinan, E.F. 1991b, in prep.

Barnes, T.G., Evans, D.S., and Moffett, T.J. 1978, *M.N.R.A.S.*, **183**, 285.

Brown, J.A., Sneden, C., Lambert, D.L., and Dutchover, E. 1989, *Ap. J.*, Supp. **71**, 293.

Cameron, A.G.W. and Fowler, W.A. 1971, *Ap. J.*, **164**, 111.

Fekel, F.C. and Marschall, L. 1991, *A.J.,* **102**, 1439.

Fekel, F.C. and Balachandran, S. 1991, in preparation.

Fleming, T.A., Gioia, I.M., Maccacaro, T., and Mereghetti, S. 1987, *A.J.* **93**, 1502.

Gratton, R.G., and D'Antona, F. 1989, *Astr. Ap.*, **215**, 66.

Guinan, E.F., and Robinson, C.R. 1986, *A.J.* **91**, 935.

Pallavicini, R., Randich, S., Giampapa, M., and Cutispoto, G. 1990, **The Messenger**, No. 62, 51

Pinsonneault, M.H., Kawaler, S.D., Sofia, S., and Demarque, P. 1989, *Ap. J.*, **338**, 424.

Silva, D.R., Liebert, J., Stocke, J.T., and Aaronson, M. 1985, *Pub. A.S.P.*, **97**, 1096.

Vilhu, O. 1987, in Cool Stars, Stellar Systems and the Sun, ed. J. L. Linsky and R. Stencel, (Springer-Verlag), p.110.

Walter, F.M., and Bowyer, S. 1981, *Ap. J.*, **245**, 671.

Seventh Cambridge Workshop on Cool Stars, Stellar Systems, and the Sun
ASP Conference Series, Vol. 26, 1992
Mark S. Giampapa and Jay A. Bookbinder (eds.)

ON THE POSSIBILITY OF A DISPERSION IN THE LITHIUM ABUNDANCES OF EXTREME HALO DWARFS

CONSTANTINE P. DELIYANNIS Institute for Astronomy, University of Hawaii, 2680 Woodlawn Drive, Honolulu, HI 96822

M. H. PINSONNEAULT Center for Solar and Space Research, Yale University, P. O. Box 6666, New Haven, CT 06511

ABSTRACT Knowledge of the primordial Li abundance (Li_p) helps to constrain cosmological parameters (Krauss and Romanelli 1990; Yang *et al.* 1984), and to discriminate between models of primordial nucleosynthesis (Deliyannis *et al.* 1991). The best current probe of Li_p is the set of Li abundances measured in old halo stars, *provided that stellar Li depletion mechanisms are taken into account.* Standard stellar evolutionary models show little depletion (Deliyannis *et al.* 1990, DDK), consistent with standard BBN, whereas models with rotational mixing deplete about an order of magnitude (Pinsonneault *et al.* 1992, PDD), possibly consistent instead with inhomogeneous BBN, or other non-standard scenarios. The rotational models can be tested through their prediction that a small dispersion results in the Li abundances. For all extreme halo dwarf data, we have performed a detailed analysis in the equivalent width-color plane, which eliminates uncertainties associated with the transformation to the abundance-T_{eff} plane; we have also computed uncertainties in the equivalent width, based on the individual S/N and relevant instrumentation. Using a χ^2 analysis, we find evidence for a small dispersion ($\sim$ 10%) consistent with, but not uniquely supporting, the rotational depletion.

Keywords : Stellar Evolution, Halo Stars, Lithium, Cosmology

INTRODUCTION, AND ANALYSIS

In the Li-T_{eff} plane, Li observations in extreme halo dwarfs (Spite and Spite 1982) seem uniform (DDK); however, a small intrinsic dispersion could be hidden. To test this, we perform an analysis in the equivalent width-color plane, which avoids uncertainties introduced when using model atmospheres.

Li Equivalent Widths : For our sources on extreme (DDK) halo dwarf data see Deliyannis and Pinsonneault (1992). The random uncertainty σ_W in W_{Li} depends on the S/N, the (effective) spectrograph resolution, and the effective pixel size. We have calculated σ_W for each observation using (6) from Cayrel (1988) and knowledge of each instrumentation. For $\sigma_W \geq 1$ mÅ, we assume that random errors dominate; otherwise we raise σ_W to 1 mÅ ($\equiv$ 1 mÅ case). To be conservative, we also consider the case of a 2 mÅ minimum error.

Color : We use the excellent (b-y) colors of Schuster and Nissen (1988, 1989a,b; hereafter SN1, SN2, and SN3), which provide a uniform set of photometry. SN1 obtained an "internal plus mean external" uncertainty of 0.005 mag in b-y, which we conservatively adopt as a "1σ" error in our calculations.

χ^2 **Analysis :** To determine whether halo dwarfs have an intrinsic Li dispersion, we have performed a χ^2 analysis. Figure 1a shows the weighted mean W_{Li} for each star (1 mÅ case) as a function of (b-y). The chosen fitting function is a line, fitted by weighted least-squares. In computing χ^2/ν, we propagated σ_{b-y} into σ_W for each star, yielding an overall uncertainty σ_W(tot).

If we assume that a line is indeed a correct fitting function, then χ^2/ν measures the likelihood that the points are distributed randomly about the fitting line with the assumed uncertainties. If the value is much higher than 1, then either the uncertainties are underestimated, or the data exhibit an intrinsic dispersion about the fitting function. For the 1 mÅ (2 mÅ) case, $\chi^2/\nu = 4.13$ (3.34). Thus, assuming that the uncertainties estimated above are realistic, χ^2/ν suggests at the 6.8σ (5.5σ) confidence level that there is indeed some intrinsic dispersion in the halo star Li abundances ($\sim$ 12%).

Reddening : Observable halo dwarfs are nearby; thus, most studies of Li have ignored reddening, which could affect the Li abundance derivations. However, *some stars have significant reddenings, which must be considered* (see G4-37 in Fig. 1 and in Hobbs and Thorburn [1991]). We use the photometric reddenings of SN2; we use $\sigma_{E(b-y)} \sim$ 0.01 from SN3, and add this in quadrature to σ_{b-y} to obtain $\sigma_{(b-y),o}$. This may be overly conservative, since $\sigma_{E(b-y)}$ dominates over σ_{b-y} but may pertain only to those stars with non-zero reddening.

A plot of W_{Li} vs. $(b\text{-}y)_o$ shows (Fig. 1b) : 1) the sample appears shifted to the blue, 2) G4-37 has shifted significantly, marginally joining the plateau, 3) HD 132475 and HD 160617 appear to be above the plateau. However, SN did not measure the β index for these two, and their quoted E(b-y) may be misleading. Fe I line analysis suggests this is so for HD 160617, and HD 132475 may be a binary (Nissen, priv. comm.). Nevertheless, BD 23° 3912 lies above the plateau regardless of E(b-y). But its kinematics do not meet the selection criteria of DDK for extreme halo stars (conceivably it formed from material already enriched in Galactic Li), so this star *was not* included in the analysis.

χ^2/ν now implies an intrinsic dispersion at the 5.3σ (4.6σ) level (14%). To bring this below the 2σ level, $\sigma_{(b-y),o} \geq$ 0.0175 (0.0165) is required. Removing HD 132475 and HD 160617 still implies a dispersion at the 3.2σ (2.3σ) level (9%).

DISCUSSION

Given the possible implications of these results for cosmology and stellar evolution, we call for numerous new observations of Li in extreme halo dwarfs, to : 1) examine more thoroughly the possibility of a dispersion, 2) improve existing observations that have $\sigma_W \geq$ 2 mÅ, and 3) search for candidates lying clearly above the plateau (possibly rare according to PDD).

ACKNOWLEDGEMENTS

We thank : A. Boesgaard, D. Duncan, L. Hobbs, P. Nissen, R. Rebolo, S. Ryan, D. Soderblom, and F. Spite for useful discussions; NASA grants NAGW-777, NAGW-778, and the University of Hawaii Foundation for support.

REFERENCES

Cayrel, R. 1988, in IAU Symposium 132, The Impact of Very High S/N Spectroscopy on Stellar Physics, eds. G. Cayrel de Strobel and M. Spite, (Kluwer), p. 345.

Deliyannis, C.P., Demarque, P., and Kawaler, S.D. 1990, *Ap. J. Suppl.*, **73**, 21.

Deliyannis, C.P., Demarque, P., and Pinsonneault, M.H. 1991, "Primordial Lithium, Stellar Evolution, Cosmology, and Dark Matter", presented at the 177th AAS meeting in Philadelphia. Abstract in *Bull. A.A.S.*, **22**, 1214.

Deliyannis, C.P., and Pinsonneault, M.H. 1992, *Ap. J.*, , submitted.

Hobbs, L.M., and Thorburn, J.A. 1991, *Ap. J.*, **375**, 116.

Krauss, L.M., and Romanelli, P. 1990, *Ap. J.*, **358**, 47.

Pinsonneault, M.H., Deliyannis, C.P., and Demarque, P. 1992, *Ap. J. Suppl.*, **78**, 181.

Schuster, W.J., and Nissen, P.E. 1988, *Astr. Ap. Suppl.*, **73**, 225. (SN1)

Schuster, W.J., and Nissen, P.E. 1989a, *Astr. Ap.*, **221**, 65. (SN2)

Schuster, W.J., and Nissen, P.E., 1989b, *Astr. Ap.*, **222**, 69. (SN3)

Spite, F., and Spite, M. 1982, *Astr. Ap.*, **115**, 357.

Yang, J., Turner, M.S., Steigman, G., Schramm, D.N., and Olive, K.A. 1984, *Ap. J.*, **281**, 493.

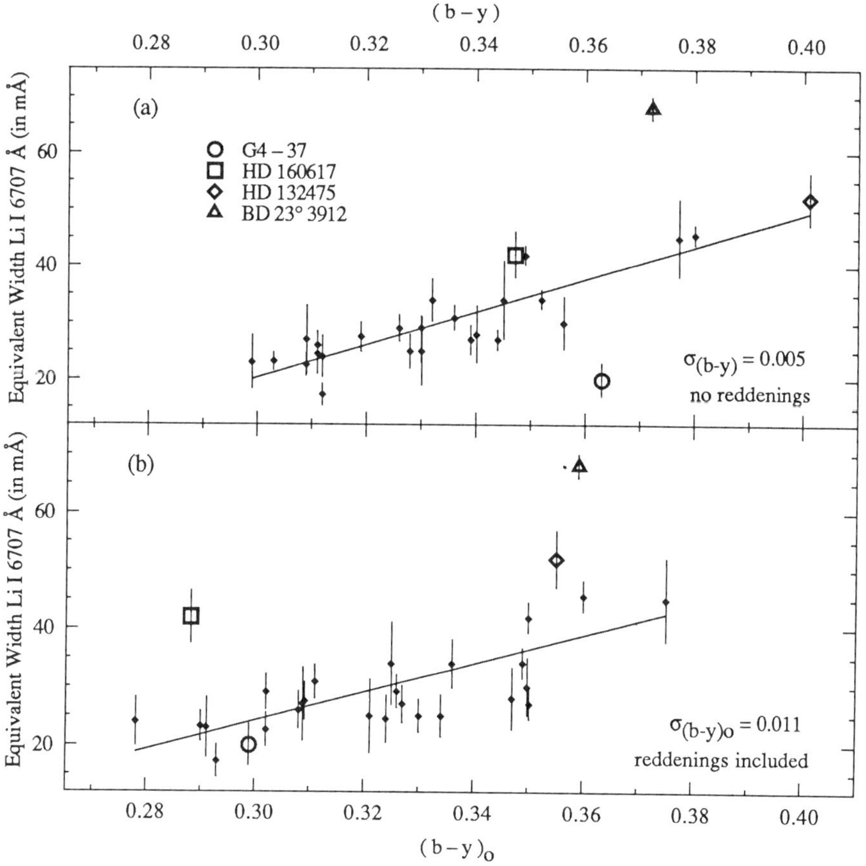

Figure 1 : Color vs. W_{Li} for extreme halo dwarfs.

Seventh Cambridge Workshop on Cool Stars, Stellar Systems, and the Sun
ASP Conference Series, Vol. 26, 1992
Mark S. Giampapa and Jay A. Bookbinder (eds.)

LITHIUM ABUNDANCES IN NORTHERN RS CVN BINARIES

SOFIA RANDICH Dipartimento di Astronomia Università di Firenze, Largo E. Fermi 5, 50125 Florence, Italy

MARK S. GIAMPAPA National Solar Observatory, P.O. Box 26732, Tucson, AZ 85726, U.S.A.

ROBERTO PALLAVICINI Osservatorio Astrofisico di Arcetri, Largo E. Fermi 5, 50125 Florence, Italy

ABSTRACT We report the results of a survey of the Lithium 6708 Å line in Northern RS CVn binaries. 37 RS CVn's taken from the list of Strassmeier *et al.* (1988) were observed at the National Solar Observatory, using the McMath Main Telescope and the Stellar Spectrograph. From a preliminary analysis of the spectra we infer the presence of a moderate, and sometimes strong, Li line in many cool stars in the sample. This confirms what previously found by Pallavicini *et al.* (1991) for the Southern hemisphere, *i.e.* that RS CVn binaries tend to have larger Li abundances than typical inactive stars of the same spectral type.

Keywords: RS CVns; Lithium Abundance; Stellar Surveys

INTRODUCTION

It is commonly believed that Lithium should not be present in large amounts in late-type evolved stars. This element, in fact, is destroyed during main sequence lifetime in cool stars, and a subsequent dilution occurs during the giant phase, leaving the star heavily depleted of Li. However, recent observations by Pallavicini *et al.* (1990, 1991) using the 1.4m CAT telescope and the Coudé Echelle Spectrometer at ESO yielded quite different and unexpected results. Many K-type giants and subgiants in their sample, including a large number of RS CVn binaries, showed an anomalously high Li abundance with respect to typical inactive stars of the same spectral type.

In order to explain such a large amount of Li, Pallavicini *et al.* considered several possibilities, involving both the evolutionary status of the stars and characteristic properties of RS CVn binaries, like their high level of chromospheric activity and large rotation. However, they could not definitively discriminate between the various alternatives because: *i)* their sample of Southern stars was rather inhomogeneous; *ii)* even those objects classified as RS CVn binaries had generally poorly determined parameters (masses, radii, absolute luminosities).

Most well-studied "classical" RS CVn systems are found at Northern declinations and they offer a better opportunity to discriminate between alternative possibilities. For this reason, we have recently carried out a Li survey of Northern RS CVn stars with the purpose of investigating Li in "classical"

RS CVn binaries and testing whether the presence of a strong Li line in several southern active stars could be due to the heterogeneous nature of the sample.

OBSERVATIONS

The observations were carried out at the National Solar Observatory (NSO) using the McMath Main Telescope with the Stellar Spectrograph and CCD detector. Using the 105mm transfer lens the nominal resolution was R=42,000. The covered spectral range was about 70 Å, centered at the Li 6708 Å resonance doublet. 37 RS CVn binaries were observed in two observing runs in February and May 1991. Short period SB2 binaries were monitored at different phases.

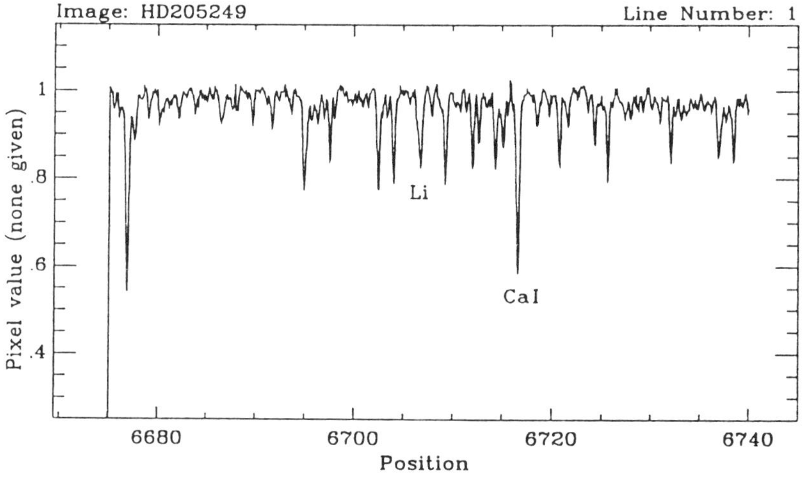

Fig. 1: Spectrum of the SB1 RS CVn binary HD 205249.

Several inactive cool giants and subgiants were also observed for comparison as well as a few stars already observed at ESO. The quality of the NSO spectra was similar to that of the ESO data, although the resolution of the latter was slightly higher. This makes us comfortable when comparing the results obtained for the Northern stars to those obtained previously at ESO. An example of the acquired data is shown in Fig. 1 where we plot the spectrum of the K1 III SB1 binary HD 205249. The Li line, though not exceptionally strong, is clearly present with an equivalent width of $\sim$ 100 mÅ of which only $\sim$ 35 mÅ can be attributed to the nearby Fe I 6707.44 Å line.

RESULTS

A preliminary analysis of the spectra confirms what previously found for the Southern hemisphere. Lithium is present in more than 50% of the late-type binaries in the sample. The strength of the line is not exceptionally high ($W_\lambda \sim 80 - 120$ mÅ), but significantly larger than typically observed for K-type giants. In a few stars, the Li line appears to be very strong (larger than the

6718 Å Ca I line). A careful spectrum synthesis analysis, however, is required for SB2 binaries, in order to take into account possible contributions to the Li line of one component from Fe lines of the other component.

Approximate Li abundances have been derived from the measured equivalent widths using curves of growth. This has been done, at the moment, for only a few stars in the sample, including RS CVn itself.

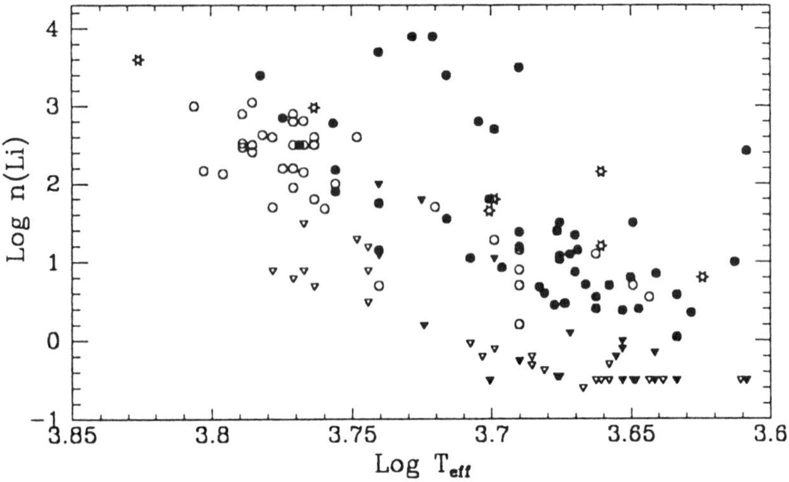

Fig. 2: Li abundances *vs.* effective temperature for inactive field stars (open symbols), the active stars from Pallavicini *et al.* (filled symbols) and the stars in the present survey (asterisks). Inverted triangles indicate upper limits.

The results are summarized in Fig. 2 where Li abundances *vs.* effective temperature are plotted for inactive field stars (open symbols), the active stars from the sample of Pallavicini *et al.* (filled symbols) and the stars in the present survey (asterisks). The positions of the Northern RS CVn's in the plot are consistent with those of Southern active stars, *i.e.* their Li abundances are typically higher than for normal inactive stars of similar spectral type. A more detailed analysis of the data is currently in progress. Besides determining Li abundances for all stars in the sample through spectrum synthesis analysis, we will also determine metallicities and rotation rates. This will allow us to investigate possible relationships between these quantities and Li abundance, in an effort to understand the presence of Li in RS CVn binaries.

REFERENCES

Pallavicini, R., Randich, S., Giampapa M.S., and Cutispoto, G.: 1990, *ESO Messenger* **62**, 51.

Pallavicini, R., Randich, S., and Giampapa, M.S.: 1991, *Astron. Ap.*, in press.

Strassmeier, K.G., Hall, D.S., Zeilik, M., Nelson, E., Eker, Z., and Fekel, F.C.: 1988, *Astron. Ap. Suppl.* **72**, 291.

Seventh Cambridge Workshop on Cool Stars, Stellar Systems, and the Sun
ASP Conference Series, Vol. 26, 1992
Mark S. Giampapa and Jay A. Bookbinder (eds.)

DISCOVERY OF A COMPANION TO THE M0S STAR HD 191589

THOMAS B. AKE, III Astronomy Programs, Computer Sciences Corporation, Code 681/CSC, Goddard Space Flight Center, Greenbelt, MD 20771

HOLLIS R. JOHNSON Astronomy Department, SW 319, Indiana University, Bloomington, IN 47405

ABSTRACT We have found with *IUE* that the M0S star HD 191589 has an early F-type companion. Comparison of the SWP spectrum with *IUE* standards shows that the secondary is an F0 star, while comparison with an M0 III + F-type composite indicates the LWP region is best matched when the secondary is an F2 star. Several arguments lead to a component magnitude difference of $\Delta V = 3.7$ and $E(B-V) = 0.0$, and hence, a value of $M_V = -1.5 \pm 0.4$ for the red giant. Because of the presence of the secondary, we cannot determine whether the system has a white dwarf companion, which under the binary scenario for Tc-deficient PRGs would have been the source of the s-process enhancement of the current primary star.

Keywords: stars: S-type – stars: individual (HD 191589) – stars: binaries

INTRODUCTION

The peculiar red giants (PRGs) include the Ba II, S, and carbons stars and are thought to involve asymptotic giant branch (AGB) stars undergoing the third dredge-up from thermal pulses during double-shell burning. There has been increasing evidence that there are a range of histories for these stars, and in particular, that Ba II and Tc-deficient MS and S stars may attribute their chemical peculiarities to mass-transfer when the secondary star, which is now a white dwarf (WD), was itself an AGB star.

HD 191589 was first noted as having S-type characteristics by Bidelman (1957), who classified it as a M0 Ib or II star with strong Sr II and Ba II lines and having ZrO. Subsequently Ake (1979) classified it as an M0S star. Smith and Lambert (1988) found that it lacked Tc lines, and Brown *et al.* (1990) have found it to be a velocity variable. These characteristics make HD 191589 a good candidate to test the binary hypothesis for PRG formation, and we observed it with *IUE* to search for the putative WD companion. Instead of a WD, we found a secondary star intrinsically much brighter.

ANALYSIS

In Figure 1, we compare the SWP spectrum of HD 191589 with late A- to early F-type *IUE* standards (Wu *et al.* 1983), and in Table I show the corresponding apparent visual magnitude of the secondary based on flux ratios with the

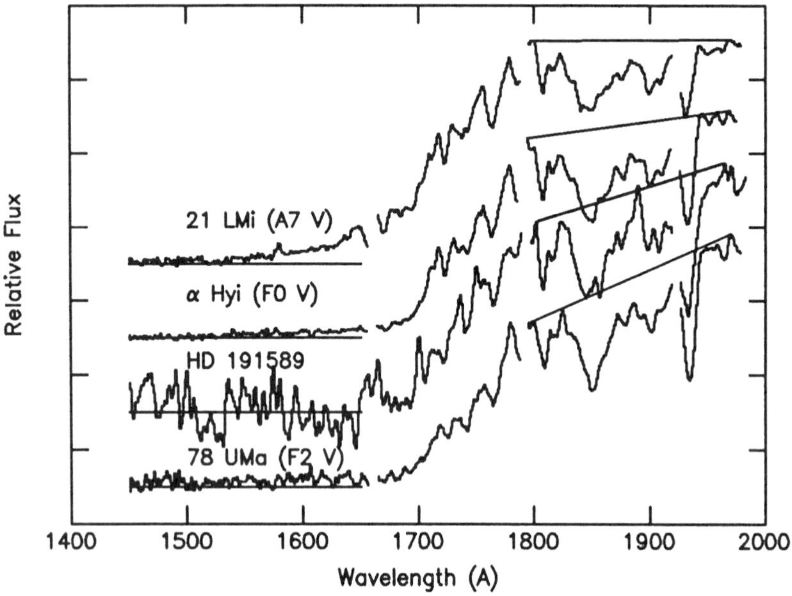

Fig. 1. *IUE* SWP spectra of HD 191589 and late A- to early F-type standard stars. The spectral slope from 1800-1950 Å and depths of the absorption features at 1850 and 1933 Å are diagnostic of temperature in this region.

standards in the 1725 – 1985 Å region. The line spectrum of HD 191589 indicates the secondary lies amongst the F0-F2 standards, or more appropriately, with $0.28 < (B - V) < 0.36$ since UV spectral types do not always match optical ones as well as with $(B-V)$. Compared to these stars, we find an apparent magnitude for the secondary of $V_{F*} = 11.0 \pm 0.2$. Using this difference, we find that light from the F star dominates the combined spectrum from 1500 to 3100 Å, and a detailed fit to the LWP spectrum using the M0 III standard μ UMa and an F2 V star works very well, including an observed filling in of the F star's Mg II absorption by emission from the PRG. Finally, using *UBV* photometry from Lee and Perry (1971) and spectrophotometry throughout the *IUE* region, we find

Table I Secondary Star V Magnitude Determinations

Comparison Star	Spectral Type	$B-V$	V_{F*}
44 Oph	A9 V	0.28	11.20
α Hyi	F0 V	0.28	11.17
η Lep	F0 IV	0.33	11.09
78 UMa	F2 V	0.36	10.79
UBV+IUE	F0 V	0.31	11.0

the energy distribution from 1500 – 5500 Å is well-matched by an M0 III + F0 V composite spectrum with no reddening and an V magnitude difference of 3.7 mag between the components, yielding component magnitudes of $V_{MS*} = 7.3$ and $V_{F*} = 11.0$ for the primary and secondary stars respectively.

Knowing the spectral type and having a secure value of the apparent magnitude of the secondary and the reddening of the system, we should be able to derive an absolute magnitude of the PRG star, which is important since only a few are known with any accuracy for these stars. Unfortunately, at *IUE* low dispersion, there are no adequate criteria to unambiguously determine the luminosity class of the secondary, although there is some evidence in the narrowness of the 1850 Å feature that it is not as bright as a giant. Assuming it is an F0-F2 IV-V star, with $1.8 < M_V < 2.6$, we derive $M_V = -1.5 \pm 0.4$ for the primary star.

DISCUSSION

Under the binary star scenario for Tc-poor PRGs, the chemical peculiarity of the primary star of HD 191589 should have been caused through mass-loss or wind accretion from a secondary star, which should now be a WD not an F-type star near the main sequence. Can HD 191589 be a triple system? Evidence for a degenerate companion to the system is difficult to assess due to the presence of the F-type spectrum. If so, is the F-star polluted with s-process elements as the primary is? The *IUE* low-dispersion spectra do not indicate any large-scale peculiarities in the absorption line spectrum, but they do not have sufficient resolution to study individual s-process features. Perhaps HD 191589 is not Tc-poor, but rather the presence of the F star weakens the Tc lines so that they cannot be seen. We calculate from the composite model fluxes that at λ4262 the F star contributes about 10% of the light from the system, which likely is not sufficient to fill in the Tc absorption lines. Further observations in the UV are needed at high resolution to answer these questions, as well as radial velocity measures to determine the orbit.

ACKNOWLEDGEMENTS

We thank the staff of the *IUE* observatory for their help in making these observations. This work was supported by NASA through contract NAS5-28749 (TBA) and grant NAS5-182 (HRJ).

REFERENCES

Ake, T. B. 1979, *Ap. J.*, **234**, 538.
Bidelman, W. P. 1957, *Pub. A.S.P.*, **69**, 326.
Brown, J. A., Smith, V. V., Lambert, D. L., Dutchover, E. Jr., Hinkle, K. H. and Johnson, H. R. 1990, *A. J.*, **99**, 1930.
Lee, P. D. and Perry, C. L. 1971, *A. J.*, **76**, 464.
Smith, V. V. and Lambert, D. L. 1988, *Ap. J.*, **333**, 219.
Wu, C.-C., Ake, T. B., Boggess, A., Bohlin, R. C., Imhoff, C. L., Holm, A. V., Levay, Z. G., Panek, R. J., Schiffer, F. H., III and Turnrose, B. E. 1983, *IUE* NASA Newsletter, No. 22.

Seventh Cambridge Workshop on Cool Stars, Stellar Systems, and the Sun
ASP Conference Series, Vol. 26, 1992
Mark S. Giampapa and Jay A. Bookbinder (eds.)

STRONG ^{13}CN LINES IN BARIUM STARS *

B. BARBUY
Universidade de São Paulo, IAG, Depto. Astronomia, CP 9638, São Paulo 01065, Brazil

A. JORISSEN
ESO, Karl-Schwarzschild Strasse, 2 D-8046 Garching bei München, Germany

S.C.F. ROSSI
Universidade de São Paulo, IAG, Depto. Astronomia, CP 9638, São Paulo 01065, Brazil

Abstract The ^{12}C:^{13}C together with ^{24}Mg:^{25}Mg:^{26}Mg isotopic ratios and the elemental abundances of Li, C, N and O are obtained for 8 barium stars, with the purpose of deriving constraints on the nucleosynthesis event responsible for the chemical peculiarities observed in the envelope of these stars.

Keywords: Isotopic abundances; Barium stars

1. Introduction

Barium stars constitute a class of peculiar red giants representing about 1% of Pop I G and K giants. They exhibit abnormally strong Ba II and Sr II lines, as well as enhanced CH and CN bands. Barium stars were found to belong to spectroscopic binary systems (McClure 1983, Jorissen & Mayor 1988, McClure & Woodsworth 1990), their companions being probably white dwarfs. Detailed abundance analyses of the barium stars revealed that heavy elements synthesized by the s-process are enhanced by a factor up to 30 with respect to normal giants, whereas carbon is overabundant by a factor of 2 in the most extreme cases. Such an overabundance might result either from a nucleosynthesis event occurring within the barium star itself, or from the accretion of heavy element-rich material from its former asymptotic giant branch companion (Boffin & Jorissen 1988). In each case, the very nature of the neutron source responsible for the s-process nucleosynthesis remains to be determined, the most likely candidates being the reactions ^{13}C(α,n)^{16}O and ^{22}Ne(α,n)^{25}Mg. That question can be answered by measuring the ^{12}C:^{13}C and ^{24}Mg:^{25}Mg:^{26}Mg isotopic ratios in barium stars, which have to bear distinct signatures from the operation of each of these neutron sources.

* Observations collected at the European Southern Observatory ESO, La Silla, Chile

2. Observations: ^{12}CN and ^{13}CN lines

High resolution spectra were obtained by means of the 1.4 m Coudé Auxiliary Telescope (CAT) at the European Southern Observatory (ESO, La Silla, Chile). The Coudé Echelle Spectrometer (CES) was used either with the long camera and a Reticon detector, or with the short camera and ESO CCD #9 (RCA 1024x512), depending on the required resolution. The observations were carried out from December 14th to 18th, 1986 and from November 29th to December 5th, 1987.

The program stars, selected from Lü et al. (1983), are given in Table 1.

The wavelength range considered is $\lambda\lambda$ 799 - 822 nm, where unblended ^{12}CN and ^{13}CN lines of the $A^2\Pi$ - $X^2\Sigma$ system are available.

The overall C abundance is obtained from C_2 lines and the N abundance is obtained by fitting ^{12}CN lines.

The resulting $^{12}C/^{13}C$ ratios for the program Barium stars and the abundances for the Sun, are shown in Table 1, together with the Li and CNO abundances, and the magnesium isotopic ratios.

The set of ^{12}CN lines at λ 8003.553 ($R_1(36)$) λ 8003.910 ($R_2(37)$) and the ^{13}CN feature at λ 8004.554, .728, .781 ($Q_2(23)$, $Q_1(21)$, $P_2(17)$) are considered the most trustful, and the basic results of Table 1 correspond to these lines. The results using ^{13}CN lines at λ 7997.8 ($P_2(16)$), λ 7998.205 ($Q_2(22)$), λ 8010.458 ($Q_1(21)$) and λ 8016.429 ($Q_1(23)$) agree with the λ 8004. feature, whereas those at λ 8015.166 ($R_1(32)$) and λ 8019.351 ($P_2(19)$) give systematically higher ^{12}C:^{13}C ratios. The errors in the ^{12}C:^{13}C indicated in Table 1 correspond to the derivation from these different lines.

3. Conclusion

An important achievement of the present work is the determination of $^{12}C/^{13}C$ ratios for more barium stars, in addition to those provided by Tomkin & Lambert (1979), Sneden, Lambert & Pilachowski (1981), Smith (1984) and Harris, Lambert & Smith (1984).

Our results, combined with previous determinations of the $^{12}C/^{13}C$ ratio of barium stars, indicate that some of these stars have abnormally low ratios, *especially when compared to their* $^{12}C/^{16}O$ *ratios.* There is however a large scatter between the various determinations of $^{12}C/^{13}C$ in a given star (e.g. HD 46407), which prevents from drawing definite conclusions. If the low ratios are indeed close to the actual values, the pollution of the barium star envelope with matter from N-type carbon stars, as is usually assumed, would not allow to match these low $^{12}C/^{13}C$ ratios, unless that pollution occurred before the first dredge-up. The interplay between the former pollution of the envelope by carbon-rich matter and the first dredge-up could possibly lead to carbon-rich envelopes with low $^{12}C/^{13}C$ ratios. This argument, which should be confirmed by detailed stellar model calculations, provides a new way to tackle the question of the formation of barium stars in the dwarf or giant phase. Alternatively, pollution of the barium star envelope by matter from carbon stars with low $^{12}C/^{13}C$ ratios (SC stars?) could also match the observed $^{12}C/^{13}C$ and $^{12}C/^{16}O$ ratios. The latter explanation has to be preferred in view of the Li content of the barium stars, which seems too large to be compatible with the deep mixing required by the low $^{12}C/^{13}C$ ratio.

Finally, the normality of the Mg isotopic ratios and the absence of large Al overabundances are not compatible with the operation of the $^{22}Ne(\alpha,n)^{25}Mg$ neutron source, and further confirms the merits of the alternate $^{13}C(\alpha,n)^{16}O$ neutron source.

Acknowledgements: The calculations were carried out at the Vax 8530 of the Astronomy Department of the University of São Paulo. This work was partially supported by Fapesp and CNPq.

References

Boffin, H., Jorissen, A. 1988, *Astr. Ap.*, **205**, 155
Harris, M.J., Lambert, D.L., Smith, V.V. 1985, *Ap. J.*, **292**, 62O
Jorissen, A., Mayor, M. 1988, *Astr. Ap.*, **198**, 187
Lü, P.K., Dawson, D.W., Upgren, A.R., Weis, E.W. 1983, *Ap. J. Suppl.*, **52**, 169
McClure, R.D. 1983, *Ap. J.*, **268**, 264
McClure, R.D., Woodsworth, A.W. 1990, *Ap. J.*, **352**, 709
Smith, V.V. 1984 *Astr. Ap.*, **132**, 326
Sneden, C., Lambert, D.L., Pilachowski, C.A. 1981, *Ap. J.*, **247**, 1052
Tomkin, J., Lambert, D.L. 1979, *Ap. J.*, **227**, 209

Table 1. Abundances of Li, C, N and O in the $log\,\epsilon(H) = 12$ scale, along with the carbon and magnesium isotopic ratios

HD	$log\,\epsilon(Li)$	$log\,\epsilon(C)$	$log\,\epsilon(N)$	$log\,\epsilon(O)$	$^{12}C:^{13}C$	$^{24:25:26}Mg$
27271	+0.7	8.25	8.2	8.48	7^{+12}_{-3}	79:10:11
44896	0.0	8.55	8.2	8.83	7 ± 3	79:10:11
46407	+0.7	8.4	8.05	8.63	7^{+12}_{-3}	79:10:11
60197	–	7.85	7.85	8.28	5 ± 3	79:10:11
65699	+0.9	8.15	8.2	8.58	20 ± 15	79:10:11
83548	+0.9	8.25	7.95	8.63	> 20	79:10:11
92626	+0.8	8.25	7.95	8.58	4^{+12}_{-3}	–
95345	0.0	8.15	7.95	8.88	7^{+12}_{-3}	79:10:11

Seventh Cambridge Workshop on Cool Stars, Stellar Systems, and the Sun
ASP Conference Series, Vol. 26, 1992
Mark S. Giampapa and Jay A. Bookbinder (eds.)

A NEW CATALOG OF COOL DWARF STARS

ERIC R. CRAINE
Western Research Company, Inc, 2127 E. Speedway, Suite 209, Tucson, AZ 85719 and Department of Physics, Colorado State University, Ft. Collins, CO 80523

ROGER B. CULVER
Department of Physics, Colorado State University, Ft. Collins, CO 80523

RONALD G. PROBST
Kitt Peak National Observatory, P.O. Box 26732, Tucson, AZ 85726

JOHN R. ENGEL
Western Research Company, Inc, 2127 E. Speedway, Suite 209, Tucson, AZ 85719

ABSTRACT We are in the process of creating a new catalog of M dwarf stars located in a large area of the sky in the direction of the North Galactic Pole. This catalog is unusual in that it represents a faint, color-selected sample, rather than a potentially biased kinematic sample. Of interest is cross correlation of this catalog with other catalogs of astronomical observations, in particular the pool of X-ray data now available. The goal here is to at least set limits on the X-ray properties of cool M stars as they may relate to various theoretical models.

Keywords: M dwarf stars; Catalogs

INTRODUCTION

We are creating a new catalog of color selected M dwarf stars in the direction of the North Galactic Pole. This catalog will provide a useful resource to address several interesting astrophysical issues:

1. It will expand the known population of nearby late M dwarfs necessary to evaluate the luminosity function of nearby dwarfs.
2. It will increase the statistical significance of X-ray properties of the M dwarf stars.
3. It will stimulate a better understanding of the physics of cool dwarf stars, in particular the nature of the dynamo process in fully convective stars.
4. It will provide an essential archive of data for use in the study of cool dwarfs.

THE CATALOG

The source survey for selection of the M dwarf candidates is the Near Infrared Photographic Sky Survey (NIPSS) by Craine (1990). The NIPSS project consists of photographs of the northern celestial hemisphere in two spectral bands, V and I, to a limit of about 19 mag and 16.5 mag respectively. Each program field has a diameter of 4.5 deg and an image scale of 110 arcsec/mm. Exposure times were matched so that images of neutral color appear with (V-I) of about 2.5 mag, making the detection of very red objects obvious.

The general approach we have taken for this project is to extract stars of V-I $\geq$ 2.5 mag and $b^{II} \geq 80$ deg from the NIPSS database. The methodology involved comparison of digitized V and I band frames from the NIPSS data archive, locating those stars which are red based on the V-I color, confirming the reality of the images, measuring equatorial coordinates and confirming the type of the star through spectroscopic or photometric observation. Spectroscopic observations (Craine *et al.* 1989) show that the probability that a star so selected is an M dwarf exceeds 95%; the probability that it is an Me dwarf exceeds 50%. At present our catalog contains about 400 stars; we expect this number to increase to about 1,200 upon completion of the project.

In addition to our spectroscopic observations we have undertaken infrared photometry of about 70 of these stars. Photometric (JHK) observations have been made with the 1.5-m telescope at Mt. Bigelow and the 2.1-m telescope at Kitt Peak National Observatory. Data frames were stored on tape and processed using IRAF at Kitt Peak National Observatory. Dark frames were combined using a median filter calculation over all dark images in each bandpass. Flat fields were similarly determined using a median filter with IRAF's *imcombine* routine. Each star image was flat fielded by subtracting the appropriate dark frame and dividing by the corresponding flat field frame. The *qphot* routine in the *apphot* package was used to compute instrumental magnitudes. Instrumental magnitudes obtained from the IRAF routines were further processed using the STARBRIGHT photometry program (Peterson and Craine 1991) to derive magnitudes and colors in the observed bands.

Of the program stars on our list all were successfully located; about 10-15% of the stars were discovered to be double (the double components appear to all be M dwarfs as well). A summary of the IRIM observational results to date appears in the J-H vs. H-K diagram in Figure 1.

CATALOG CROSS CORRELATIONS

All 400 stars presently in our catalog have been cross-correlated with the Luyten NLTT catalog of proper motion stars as well as the following X–ray catalogs: HEAO, EXOSAT, Einstein IPC and Extended Medium Sensitivity Survey (EMSS). Approximately 10% of the stars in our catalog correlate with high proper motion stars in the NLTT catalog, suggesting that our color selected sample is not a kinematically biased sample. On the other hand, none of these stars have X–ray fluxes above the detection limits of the above referenced X–ray surveys. Correlation with recently obtained ROSAT data has not yet been done.

SUMMARY

The color selection technique applied to NIPSS program stars in the vicinity of the NGP is consistently yielding M dwarf stars. We anticipate that the catalog can be expanded to 3-5 times its present size using data already collected. We are continuing photometric observations of program stars and expect to be generating photometric parallaxes for many of these stars during the coming year.

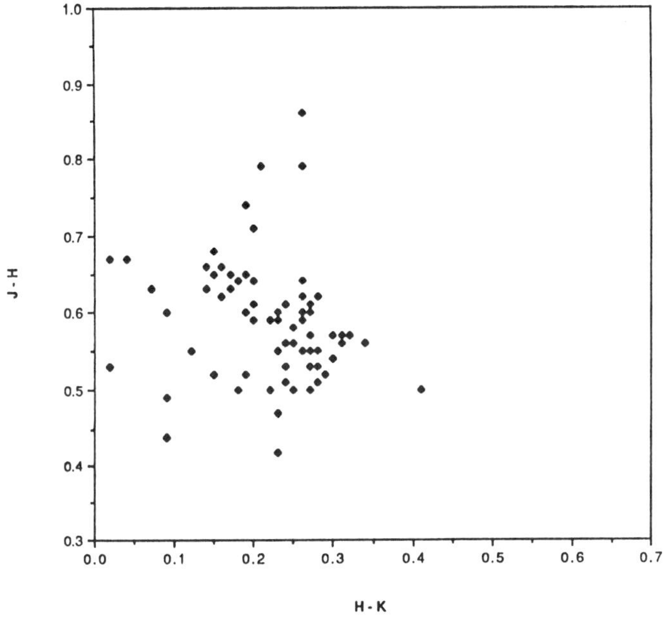

Fig. 1: Color-color diagram based on infrared photometry of stars selected from the cool star catalog.

ACKNOWLEDGEMENTS

This work has been supported in part by a NASA grant (NAS5-30923) and by observing and computer time at Kitt Peak National Observatory.

REFERENCES

Craine, E.R. 1990, Log of the Near Infrared Photographic Sky Survey, Western Research Company, Tucson, Arizona.

Craine, E.R., Tapia, S. and Culver, R.B. 1989 in Automatic Small Telescopes, ed. D.S. Hayes and R.M. Genet, Fairborn Press, Mesa, Arizona p.95

Peterson, J. and Craine, E.R. 1991, Starbright, Western Research Company, Tucson, Arizona.

Seventh Cambridge Workshop on Cool Stars, Stellar Systems, and the Sun
ASP Conference Series, Vol. 26, 1992
Mark S. Giampapa and Jay A. Bookbinder (eds.)

A COOL STAR SAMPLE AND EXTRA-SOLAR SYSTEM PLANETARY SEARCH PROGRAM

ERIC R. CRAINE
Western Research Company, Inc., 2127 E. Speedway, Suite 209, Tucson, AZ 85719 and Department of Physics, Colorado State University, Ft. Collins, CO 80523

JOHN R. ENGEL
Western Research Company, Inc., 2127 E. Speedway, Suite 209, Tucson, AZ 85719

ROY A. TUCKER
Tucker and Associates, 5500 W. Nebraska, Tucson, AZ 85746

ABSTRACT Conventional search techniques for extra-solar system planets have not yet yielded indisputable evidence of the existence of such planets. Further, because of their very small sample size most previously employed search techniques ensure that a null result provides little or no useful information on the numbers or distributions of such planets. The photometric technique, wherein monitoring is undertaken to detect light curves distinctively characteristic of a planetary transit, has been a tantalizing prospect, but because of its observational intensity has always been rejected. Recent developments make a review of this photometric technique worthwhile.

Keywords: M dwarf stars; extra-solar system planets

INTRODUCTION

The Solar System is the only planetary system in the Universe known by direct observation. Numerous observational search strategies have been devised and implemented with the goal of detecting extra-solar planetary systems. Though some have claimed positive results, none have remained unrefuted and none unequivocally represent direct observation of such planets. Because these programs have involved extensive observations of a miniscule number of program stars, failure to detect an attendant planet or planetary system tells us nothing about the statistical distributions of such planets.

One long-recognized technique, the photometric technique, offers the possibility of making a meaningful statement about the frequency of occurrence of planetary systems, but has the failing of being so observationally intensive that it has never been employed on a realistic scale. The photometric technique involves frequent photometric monitoring of a large ($\sim$ 1000) sample of stars in an effort to detect a transit event associated with an orbiting planet or planets.

The development of automated telescopes and associated instrumentation appears to be a feasible solution to the intense observational requirements.

The development of extensive catalogs of cool dwarf stars provides a useful program star list which has the advantage of producing orbital geometries and photometric properties enabling detection of planetary transits with a network of equipment of modest proportions.

TRANSIT CHARACTERISTICS

A Jupiter-size planet transiting an M0 to M8 main sequence dwarf star can be calculated to diminish the brightness of the star by roughly 0.02 to 0.08 magnitudes, neglecting effects of limb darkening. This variation is readily detectable using differential photometry. We note that an earth-size planet will, in general, produce an undetectable transit using conventional photometric techniques with telescopes of modest size. Assuming that at least 50% of the transit events will occur along chords within one half stellar radius of the center of the transited star, it can be shown that the mean transit distance across the star is greater than 0.866d, where d is the stellar diameter; or the transit time, t, is greater than 0.866t(c), where t(c) is the duration of a central transit. We shall postulate that each program star should be monitored with a frequency of one observation per time interval of 0.5 to 0.8 times t (the transit duration).

We can hypothesize several configurations of star-planet-planetary orbit configurations. For our purposes a Jupiter-size planet in a Mercury- to Earth-size orbit around a main sequence M0 to M8 star can produce a transit observable with modest equipment. For M0 to M5 stars orbital periods range from 0.4 to 2.2 years, the duration of central transits ranges from 7.2 to 13.8 hours, and the amplitude of the transit ranges from 0.03 to 0.11 magnitudes. We note that these parameters are coincident with planet formation models based on temperature constraints on planetary formation (Black 1991).

OBSERVING NETWORK

The requisite observing network for the proposed photometric planetary search is based upon the Global Network of Automatic Telescopes (GNAT) as envisioned by Crawford (1989). The planetary search GNAT would consist of three or four primary observing locations. Four sites separated by 90 deg in longitude would be able to "pass-off" simultaneous observations of program stars with a maximum HA of observation of 3^h at each site; three sites at 120 deg separation would require an HA of at least 4^h at pass-off. Each site would have two or more one meter class, fast reflecting telescopes equipped for CCD photometry. We estimate that on average each set of telescopes in the net could follow about 120 stars per year (where a set consists of one telescope at each site); with 2 to 6 telescopes per site, this would translate to 240 to 720 program stars per year, or up to 3,600 program stars in five years.

In the absence of any other selection factors for program stars (*e.g.* knowledge of the inclination of the rotation axis of the star), it can be estimated that on average, for a sample of M0 to M5 dwarf stars, uniformly distributed by spectral type, 1 in 667 program stars will be inclined to present an observable transit consistent with the above mentioned orbital parameters. If we weight the program star list with physically larger M0 stars that ratio becomes more favorable: 1 in 209 stars for Mercury-sized orbits. Thus we might expect to observe as many as 17 planetary transits in a five year period.

COOL STAR SAMPLE

We are in the process of creating a new catalog of M dwarf stars located in a large area of the sky in the direction of the North Galactic Pole. This catalog is unusual in that it represents a faint, color-selected sample, rather than a potentially biased kinematic sample. This catalog will serve as a source of program stars for the planetary search project.

The source survey for selection of the M dwarf candidates is the Near Infrared Photographic Sky Survey (NIPSS) by Craine (1990). The NIPSS project consists of photographs of the northern celestial hemisphere in two spectral bands, V and I, to a limit of about 19 mag and 16.5 mag respectively. Each program field has a diameter of 4.5 deg and an image scale of 110 arcsec/mm. Exposure times were matched so that images of neutral color appear with (V-I) of about 2.5 mag, making the detection of very red objects obvious. Our experience is that the probability that a star so selected is an M dwarf exceeds 95%; the probability that it is an Me dwarf exceeds 50% (Craine *et al.* 1989). At present our catalog contains about 400 stars; we expect this number to increase to about 1,200 upon completion of the project.

We anticipate that the NIPSS data could be examined further to encompass a larger number of fields with the likelihood of yielding an even larger number of cool dwarfs. Although we initiated this program to search for relatively rare, very late type dwarfs, it is clear that the photometric search for extra-solar system planets can greatly benefit from the detection of large numbers of earlier M dwarfs. Indeed, infrared photometry of a portion of this sample (Craine et al 1991) clearly indicates that about 90% of the stars in this catalog have spectral types between M0 and M5.

ACKNOWLEDGEMENTS

This work has been supported in part by a NASA grant (NAS5-30923) and by observing and computer time at Kitt Peak National Observatory.

REFERENCES

Black, D.C., 1991,*Sci. Am.*, **264**, 76.

Craine, E.R., Tapia, S. and Culver, R.B. 1989, in Automatic Small Telescopes, ed. D.S. Hayes and R.M. Genet, Fairborn Press, Mesa, Arizona p.95

Craine, E.R. 1990, Log of the Near Infrared Photographic Sky Survey, Western Research Co., Tucson, AZ.

Craine, E.R., Culver R.B., Probst, R.G. and Engel, J.R. 1991, Cool Stars, Stellar Systems, and the Sun, Volume 7.

Crawford, D.L., Genet, R.M. and Hayes, D.S. 1989, in Automatic Small Telescopes, ed. D.S. Hayes and R.M. Genet, Fairborn Press, Mesa Arizona p.115.

Seventh Cambridge Workshop on Cool Stars, Stellar Systems, and the Sun
ASP Conference Series, Vol. 26, 1992
Mark S. Giampapa and Jay A. Bookbinder (eds.)

DO DUST GRAIN SIGNATURES VARY DURING THE LIGHT CYCLE OF A MIRA?

IRENE R. LITTLE-MARENIN[1,2] and ROBERT E. STENCEL[1]
[1]Center for Astrophysics and Space Astronomy, Campus Box 389, U. Colorado, Boulder, CO, 80309 and [2]Wellesley College, Wellesley, MA 02181

ABSTRACT Individual IRAS LRS spectra obtained at different phases of the light cycle of several Mira variables show that the contrast (strength) of the 10 μm silicate dust emission feature varies in phase with the optical light curve. The emission feature is stronger near maximum light and weaker near minimum light. The phase coverage of the IRAS data is too sparse to establish a phaselag between maximum strength and maximum optical light.

Keywords: circumstellar shells; dust grains; time variability

INTRODUCTION

Asymptotic giant branch stars are noted for their luminosity variability, enhanced mass loss and the development of circumstellar dust shells. As the temperature of the star varies during its pulsation, so will the radiative flux incident at the inner edge of the shell, which will determine the amount of dust that condenses and evaporates. Hence, the emission characteristics of dust grain features should also vary with time.

ANALYSIS and DISCUSSION

The spectral emission characteristics of amorphous silicate, aluminum oxide and silicon carbide (SiC) dust grains can be identified in the 8-22 μm spectra obtained with the Low Resolution Spectrometer (LRS) on IRAS (Little-Marenin and Little 1990; Little-Marenin *et al.* 1987). Since IRAS observed sources more than once, it noted variations in flux and assigned a variability index to the individual objects. However, only the averages of the individual broad band fluxes and LRS spectra are published. We extracted individual LRS spectral scans of several Miras from the extended LRS (XIRASX) database. Analyses of these spectra show that strengths of the 10 μm silicate emission feature and the 12 and 25 μm broad band fluxes vary with optical phase. Figure 1 shows three LRS scans for the Mira variable R LMi (M5e-9e; P=372 days) taken at phase 0.98, 0.00 and 0.47. The variations in flux and contrast (strength) of the silicate emission are clearly evident. The contrast varies from 1.51 to 1.36. Contrast is defined as peak feature flux divided by black body flux at peak wavelength; contrast = 1.0 indicates no spectral feature. Figure 2a compares for R LMi the variations of the 12, 25, 60 and 100 μm PSC2 broad band fluxes and the contrast to the visual light curve (obtained from the AAVSO). The contrast decreases by 10% and the 12 and 25 μm fluxes decrease by about 25% as the optical light decreases by 5 magnitudes. Analysis of three LRS spectra of the short period Mira R Cet (M4e-M9; P=166 days) taken at phase 0.67, 0.73 and 0.85 when the star was brightening shows much larger changes. The 12 and 25 μm fluxes almost double during this time period and the contrast increases from 2.3 to 3.1 (Fig. 2b). In general we find that the IR (12 and 25 μm) fluxes and strengths of the emission features track the visual light curve. We interpret this variation as implying that the 10 μm feature is formed in a relatively thin region which is subject to inhomogeneities possibly due to shocks. A phaselag between the IR fluxes and optical light curve could not be established since the phase coverage available in the IRAS LRS data are very sparse, typically being limited to less than three distinct phases since the IRAS data are taken hours, weeks and months apart. However, we anticipate finding additional cases of variability which will lead to establishing the composite behavior of Miras and to a test of this preliminary finding. Analysis of three carbon stars shows that the infrared flux also varies in phase; however, the contrast of the SiC emission feature appears to show much less variation with phase than the silicate feature.

ACKNOWLEDGEMENTS

This research was in part supported by NSF grant RII-9002960 and NASA grant NAG5- 1667.

REFERENCES

Little-Marenin, I.R. and Little, S.J. 1990, Astron.J., **99**, 1173.

Little-Marenin, I.R., Ramsay, M.E., Stephenson, C.B., Little, S.J. and Price, S.D. 1987, Astron.J., **93**, 663.

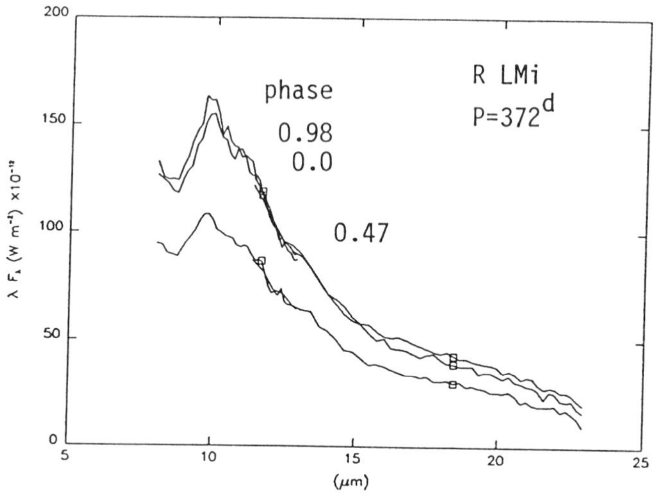

Fig. 1: Three LRS scans of the Mira R LMi are shown.

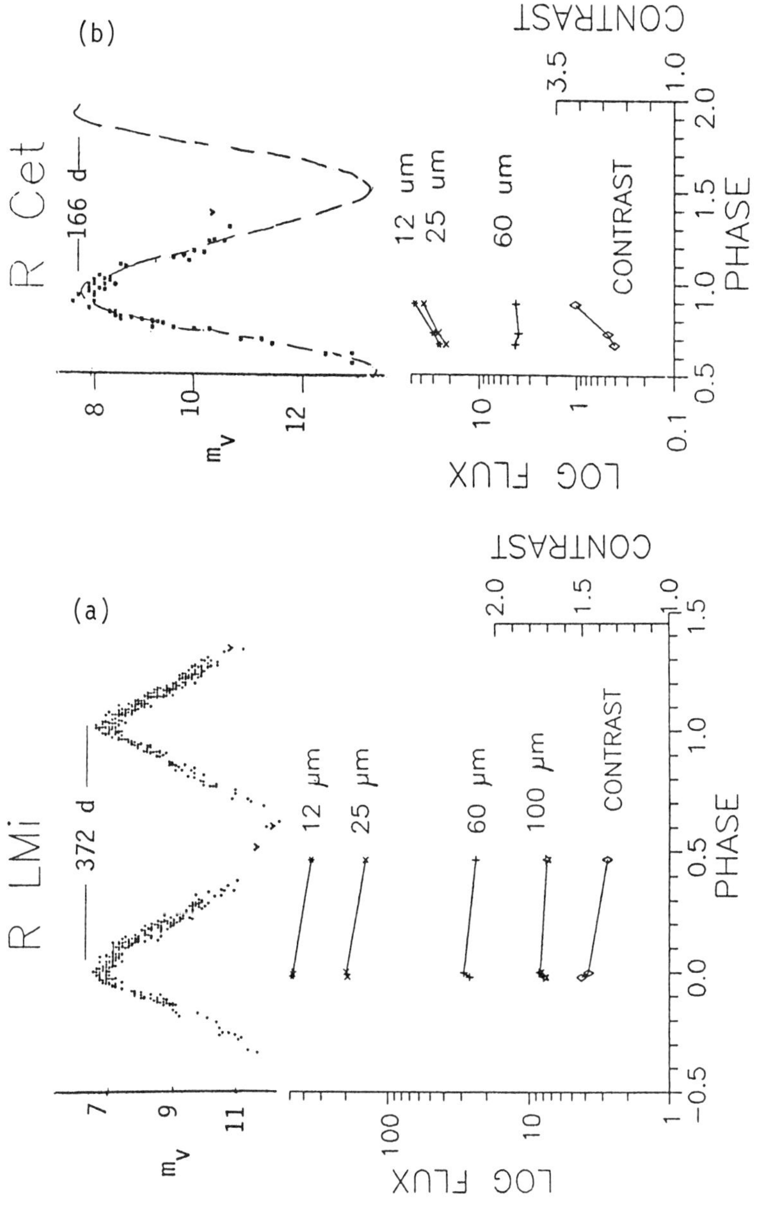

Fig.2: a) The 12, 25, 60 and 100 μm broad band fluxes (Jy) and contrast of R LMi obtained at three different phases are compared to its visual lightcurve. b) The broad band fluxes and contrast of the short period Mira R Cet are compared to its visual light curve. The decrease and increase in the 12 and 25 μm fluxes and contrast track the visual lightcurve.

Seventh Cambridge Workshop on Cool Stars, Stellar Systems, and the Sun
ASP Conference Series, Vol. 26, 1992
Mark S. Giampapa and Jay A. Bookbinder (eds.)

WATER MASER AND OPTICAL VARIABILITY IN THE SILICATE-CARBON STAR V778 CYGNI: 1987-1991

IRENE R. LITTLE-MARENIN[1,2] and PRICILLA J. BENSON[2]
[1]Center for Astrophysics and Space Astronomy, Campus Box 389, U. of Colorado, Boulder, CO 80309 and [2]Wellesley College, Wellesley, MA 02181

ROBERT R. CADMUS
Department of Physics, Grinnell College, Grinnell, IA 50112

ABSTRACT The optical light curve of the silicate-carbon V778 Cyg from 1987-1991 shows very regular, low amplitude variations ($\Delta m=0.5$ mag) with a period of 302 days. The 22 GHz water maser emission associated with V778 Cyg also shows large variations in flux that correlate with the optical light of the C star but are shifted in phase by 0.23 phase relative to the optical minimum.

Keywords: water maser; circumstellar shells; carbon stars; variability

INTRODUCTION

In 1987 water maser emission at 22 GHz was discovered to be associated with the carbon star V778 Cyg (Little-Marenin, Benson and Dickinson 1988). Since then we have monitored the 22,445,925 MHz water maser emission at approximately monthly intervals with the 37m telescope at Haystack Observatory. V778 Cyg is classified as an Lb (irregular) variable in the General Catalog of Variable Stars, but no light curve has been published. In order to define the relationship of the optical and water maser variability, we are monitoring V778 Cyg photoelectrically with the 0.61m telescope of the Grant O. Gale Observatory at Grinnell College since June 1989.

V778 Cyg is one of seven optically identified carbon stars that are known to be associated with oxygen-rich circumstellar shells: the silicate-carbon stars. The oxygen-rich nature of their circumstellar shells was first identified by their strong 10 and 18 μm silicate dust emission features in their IRAS low resolution spectra (Little-Marenin 1986; Willems and de Jong 1986). The nature of these stars has not yet been determined; (*i.e.* does the oxygen-rich dust surround the carbon star or is the silicate dust around an invisible companion to the carbon star?) Correlations of the water maser variability (associated with the shell) and the optical variability of the C star will show if the variations of the water masers are determined by flux variations of the C star.

ANALYSIS and DISCUSSION

The optical light curve of V778 Cyg from June 1989 through August 1991 clearly shows broad maxima and sharp minima with an amplitude of 0.5 magnitudes.

The period between successive minima is P=302 days as can be seen in Figure 1. The shape of the light curve is typical of many carbon star LPVs. Since the amplitude is smaller than 2.5 magnitude, the star will be classified as semi-regular (despite its very regular variations) and not as a Mira variable. The B-V color index varies by about 0.25 magnitudes and shows that the star gets cooler as it dims.

The water maser spectrum of V778 Cyg usually shows three features at -15, -17 and -20 km s^{-1}. The intensity of these features has varied by up to factors of 25 during the last four years. Figure 2 plots the peak antenna temperature (K) for the two stronger components and Figure 3 the integrated area (K km s^{-1}) from 1987-1991. In 1987 (around JD 6900-7000) the -15 km s^{-1} line was weak but has become the dominant line by 1991 whereas the line at -17 km s^{-1} has become the weaker one. The behavior of the -20 km s^{-1} feature is not shown. It tends to be weak and is currently not visible. Both lines show large random as well as periodic flux variations [1 K=10 Jy before the summer of 1990 (JD around 8100). Since then an upgrade of the antenna at Haystack has increased the sensitivity to about 1 K=8 Jy]. It is apparent from an overlay of the light curve and the peak flux plot that the maser flux varies in phase with the light curve; however, with a phase lag of about 70 days (0.23 phase) between optical and maser minima. The optical minima shifted by 0.23 phase are indicated in Figure 2 by vertical lines. The central velocities of the three features shows small, (on the order of about 0.8 km s^{-1}) possibly systematic variations between 1987-1991 (Fig. 4). The velocities of the features can be measured to an accuracy of better than 0.1 km s^{-1}. The optical velocity ($\pm$1.5 km s^{-1})(Barnbaum *et al.* 1991 (preprint)) falls within the range of the maser velocities.

Stellar water masers are located in expanding circumstellar shells of oxygen-rich LPVs, roughly at 10-100 AU from the central star. The silicate dust grain emission comes from approximately the same location in a circumstellar shell. For oxygen-rich Mira variables, we usually observe a phase lag between 0.2-0.4 phase between the optical and water maser emission. If we assume a similar mechanism for producing the water maser variability, then our phase lag data for V778 Cyg suggest that the variations in the water maser emission is driven by the optical variability of the C star. It suggests that the oxygen-rich shell is directly associated with the C star, however the data do not exclude the possibility that the oxygen-rich shell, located around a non-variable companion, is influenced by flux variations of the nearby carbon star.

ACKNOWLEDGEMENTS

This research is in part supported by NSF grant RII-9002960, AST-8913001 and AST- 8914917. Radio astronomy at Haystack Observatory of the Northeast Radio Observatory Corporation is supported by a grant from the National Science Foundation.

REFERENCES

Barnbaum, C., Morris, M., Likkel, L., and Kastner, J.H. 1991 (preprint)
Little-Marenin, I.R. 1986, *Ap. J. Letters*, **307**, L15.
Little-Marenin, I.R., Benson, P.J., Dickinson, D.F. 1988, *Ap. J.*, **330**, 828.
Willems, F.J. and de Jong, T. 1986, *Ap. J. Letters*, **309**, L39.

Fig. 1 (top left): Plots the light curve and the B-V color index from June 1989 through August 1991 for the C star V778 Cyg. A period of 302 days is clearly present.

Fig. 2 (top right): Plots the 22 GHz water maser flux of the two strongest lines from 1987-1991. The short vertical lines indicate the observed and estimated times of optical minimum of V778 Cyg shifted by 70 days (0.23 phase). The shifted optical minima coincide with minima in the maser flux.

Fig.3 (bottom right): The integrated area of the three maser lines also show the 302 day period.

Fig.4 (bottom left): The three water maser lines show small, possibly systematic velocity variations. The accuracy of the H_2O maser lines is better than 0.1 km s^{-1} . The optical velocity (± 1.5 km s^{-1}) of V778 Cyg lies in the range of the water maser emission.

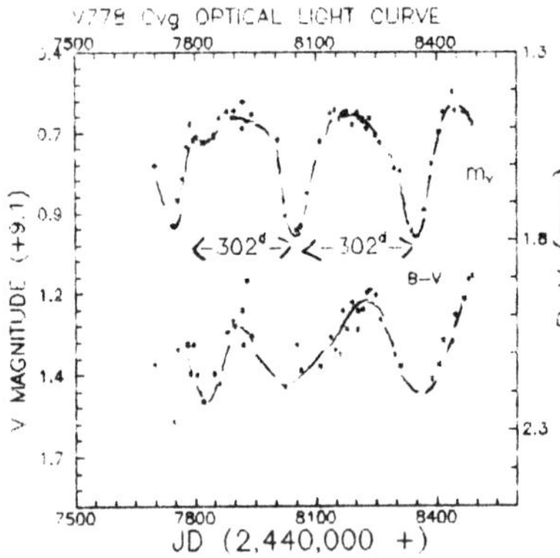

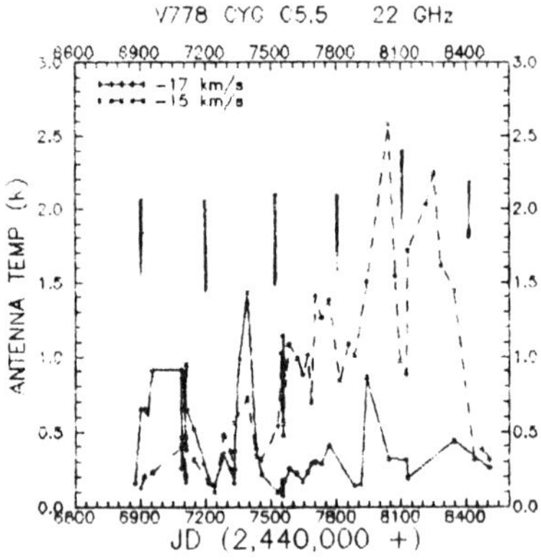

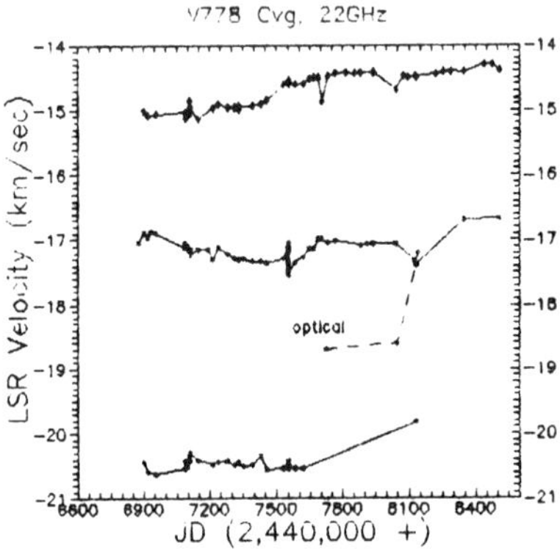

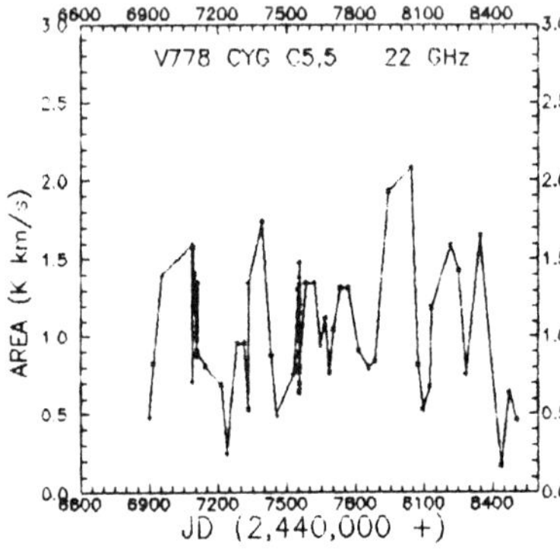

PART VIII

The Future

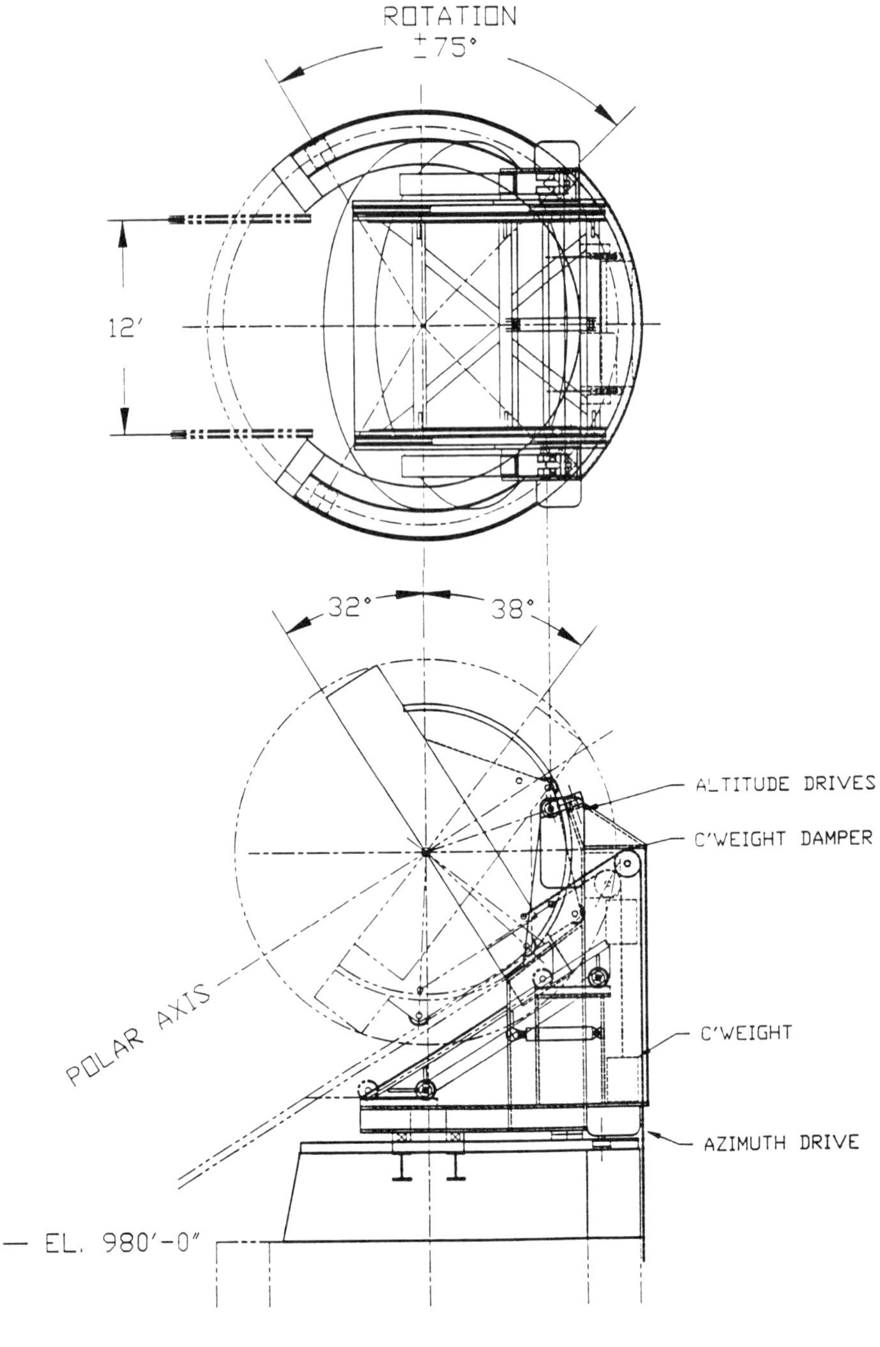

Figure 1 from "Upgrade to the McMath" by W. Livingston and L. Barr.

Seventh Cambridge Workshop on Cool Stars, Stellar Systems, and the Sun
ASP Conference Series, Vol. 26, 1992
Mark S. Giampapa and Jay A. Bookbinder (eds.)

ON SELF-LICKING ICE CREAM CONES

S. PETE WORDEN

Deputy for Technology
Strategic Defense Initiative Organization
Department of Defense

Based on my experience as a Washington bureaucrat for most of the past decade – both within the Pentagon and until a few weeks ago in the White House on the National Space Council – I've been asked to close this conference with a few words on politics. As we discuss a host of exciting new possibilities for space experiments and observatories, it is important to understand the obstacles standing in our way. And, if at all possible, we must devise strategies to work around these problems.

In this short discussion, I will start with some problems which are beyond our control and move to discuss some which we have some potential to address.

Keywords: Ice Cream Cones; Pork; NASA; Mafia; Congress

PROBLEMS, NATIONAL, THE BUDGET PROCESS

Some of the current difficulty with the civil space program stems from the budget process. First and foremost, the current budget cycle prevents fast action. To understand this, let's review the process.

Sometime in the spring of 1992, NASA's Comptroller will send out preliminary guidance for Fiscal Year (FY) 1994 (Starts in October 1993) to each Associate Administrator, for example the Associate Administrator for Space Science and Applications (OSSA). They, in turn, will task each directorate (such as Astrophysics) to prepare submissions. Over the summer, these submissions will be refined, ultimately by the NASA Administrator for submission to the Office of Management and Budget (OMB) in the White House in early September 1992. During the fall, OMB will try to balance the budget and consider how to (almost always) cut NASA's request. In early December, OMB will "passback" their budget guidance to NASA. About one week later, the NASA Administrator can appeal these decisions in a one-hour meeting with the President. This appeal rarely works, and almost never involves science missions.

The President's FY 1994 budget will be submitted to Congress in January 1993. After solemn declarations on the part of Congress that it is a ridiculous proposal, Congress will begin its deliberations in March 1993. These deliberations are divided into two parts – the authorization process and appropriations process. In principle, the authorization committees decide what

NASA will do and the appropriations committees decides how much money NASA will get. In practice the authorization committees are virtually powerless.

NASA's authorizing committees, particularly in the House of Representatives (The Committee on Science, Technology and Space), are very supportive of science programs. They usually pass on NASA's proposals with minimal cuts. Unfortunately, they usually take all summer to make their decisions. Once the Senate and House pass their independent authorizing bills, they are reconciled in conference and passed by the full Congress. In the last few years there has not been a NASA authorizing bill passed by both houses of Congress.

The real power lies in the appropriations committees. Here is where the problem is. The appropriations committee is divided into 13 subcommittees. Particularly in the House, where spending bills originate, the Chairmen of each subcommittee (the "Thirteen Cardinals") make all of the decisions. This is where the "pork" is larded onto the bill. For example, NASA's bill last year included a variety of parking garages and other foolishness in appropriations committee members' districts. A serious problem for NASA in the appropriations process is that it is handled by the HUD, Veterans Administration, and Independent Agencies (includes NASA, EPA, NSF, etc.) Subcommittee. With about $70 billion to spend and being required to compete, NASA versus these other, far better vote-getting programs, the possibility for increases in NASA budgets is small. About June 1992, after only one or two subcommittee hearings, the appropriations subcommittees will "mark up" the NASA program. In September they will conference with the other house of Congress and we'll have our FY 1994 NASA budget. Although this budget is supposed to include only a top-line number, it includes detailed management instructions which NASA slavishly adheres to in the form of "report language."

This process is almost impossible to manage. Appropriations bills are almost never vetoed by the President, so once the White House sends its budget to Congress there is little chance for the Administration to affect the outcome. The one time the authorizing committees challenged the appropriations committees on a NASA issue and won was last summer in the House of Representatives. Unfortunately, the success involved trading science programs for the Space Station.

This unfortunate train of events has resulted in a NASA which, more than any other agency, believes it works only for the appropriations committees. The senior staff of those committees, who have little interest in science or space, effectively run NASA. NASA senior officials' noses are usually found at waist level near these committee staffers.

While there is little that the astronomical community can do to affect this process, the best hope is to recognize where the power is and attempt to form one's own interaction with Congress, and in particular appropriations committee members and staff.

PROBLEMS, CLOSER TO HOME, NASA

"The Self-Licking Ice Cream Cone"

Since NASA effectively works for the most porkish part of Congress, it is not surprising that their programs are designed to maximize and perpetuate jobs programs in key Congressional districts. The Space Shuttle-Space Station is an outrageous example. Almost two-thirds of NASA's budget is tied up in this self-licking program. The Shuttle is an unbelievably costly way to get to space at $1 billion a pop. The Space Station is a silly design. Yet, this Station is designed so that it can only be built by the Shuttle and the Shuttle is the only way to construct the Station. Furthermore, the Shuttle has to be "improved" to support the Station with a new solid rocket motor which is to be built *you guessed it* in the district of the Chairman of the House Appropriations Committee. Since there are tens of thousands of jobs tied up in these programs and most of NASA's budget as well, there is not only no money to get out of this endless do-loop, there are positive political pressures to make sure that we don't get out. Witness the fact that not even $175 million could be found out of NASA's $14 billion budget to develop a new, cost-effective launch system. The appropriations committees effectively zeroed programs to develop this new launch system.

Lest we think the science programs are any better, consider the Hubble Space Telescope. It now absorbs almost half of the astrophysics budget ($400 million out of $900 million). Who would doubt that we could develop for this money, and in short time, a series of one or two meter-class instruments for placement at geo-synchronous altitudes. One could be launched every year. But there is now built-in resistance to this in the form of such institutions as the STScI, which just happens to reside in the district of the Chairman of the Senate appropriations subcommittee in charge of NASA.

"Inmates Running the Asylum"

NASA's vaunted "peer review" process is not a positive factor, but an example of the "pork" mentality within the scientific community. It results in needlessly complex programs whose primary objective is not putting instruments in orbit, but maximizing the number of constituencies and investigators, thereby maximizing the political invulnerability of the program.

NASA's proposed Earth Observing System (EOS) is a perfect example. Everyone agrees that the best way to conduct a scientific program – particularly one with a time-urgent objective such as finding out whether, and how bad, global warming may be – is with a series of rapid, small experiments in space. Yet NASA developed its mammoth EOS platforms, at almost 15,000 kilograms each, the largest, costliest and most complicated scientific instruments ever constructed by man. NASA's first step was to ensure that 500 principal investigators were chosen and that each understood his instrument would only fly if all flew together. Needless to say, this arrangement has made it almost impossible to move in a sensible direction.

"Mafia Tactics"

As with any monopoly, NASA is given to resorting to threats and intimidation when challenged. But, as programs grow less defensible on engineering and scientific grounds, NASA has carried these tactics to levels far above any other government agency. There is not an astronomer who hopes to get data from a NASA satellite, nor funding on a NASA program who does not feel this pressure.

The EOS is a case in point. About a year ago, encouraged by criticism from some quarters of Congress and in the press, some scientists and satellite contractors began proposing small, cheap, near-term alternatives to the EOS "battlestars." Senior NASA officials conducted, with impunity, an unbelievable campaign of threats against these critics. Members of White House advisory committees were told they would not get NASA funding if they continued to probe the program. Contractors were told they would lose other programs if they presented their ideas. Some scientists were actually denied funding on unrelated programs as a result of their public criticism of EOS.

SOLUTION, SELF-POLICING

"Faster, Cheaper, Better"

Technology exists, today, to do essentially all of NASA's scientific missions with "smallsats" systems weighing less then 1000 kilograms, costing less than $60 million, and taking no more than 3 years to construct. Even large satellites can be done on this time scale. In the late 1980s, my predecessor, now the NASA Associate Administrator for Exploration, developed, built and launched three medium-class satellites (4000 kg) in an average of 20 months from initial concept to launch.

The scientific community should insist that all NASA new starts be completed in three years. Those ongoing programs not meeting this requirement should be cancelled. In this way, a satellite program could be given – lock, stock and barrel – to a single scientific investigator and institution to conduct under their own full control. This would include contracting for the satellite bus and launch vehicle.

"Shoot the Sick Horses, and their Trainers"

It is outrageous that the Hubble disaster resulted in no repercussions. All we hear is that some un-named technician, no longer working for the contractor, made a mistake in the early 1980s. Even in the Defense Department, current officials would lose their jobs over allowing such an untested and expensive system to be launched.

It is time to start over on Hubble. It is not fixable. Its performance will not be corrected. The key instruments have failed and the satellite itself has far more problems than will be repaired in one, or even two Shuttle missions (at $1 billion each, I might add). Let's turn it off, start over, and do it right. Let's ask ourselves if the half-dozen magnificent spectra we have gotten (and will get no more of) are worth the literally tens of millions of dollars each they have cost.

Competition

The scientific community should insist on competition. New starts should involve at least two fundamentally different approaches. Success stories in American science and technology in the past have almost always involved several competing approaches being pursued simultaneously.

For its part, the Administration is trying, over NASA's strenuous objection, to develop competing space science programs. The Department of Energy is trying to establish its own civil space effort. This is an idea which should be warmly embraced by space scientists – to include encouraging Congress to support it. Further, there is no reason that NSF shouldn't do space missions. As mission costs descend into the few million dollar range, which technology shows they will, long-term university observatories could as easily be in space as they are on nearby mountains.

CONCLUSION

We can and must put our house in order. Despite national organizational problems, we can have an exciting space science and space astronomy program within even the limited resources we have. They key is to insist on fundamental changes in the way NASA does business.

Seventh Cambridge Workshop on Cool Stars, Stellar Systems, and the Sun
ASP Conference Series, Vol. 26, 1992
Mark S. Giampapa and Jay A. Bookbinder (eds.)

PROPOSED UPGRADE OF THE MCMATH SOLAR/STELLAR TELESCOPE TO A 4-METER APERTURE

WILLIAM LIVINGSTON National Solar Observatory, National Optical Astronomy Observatories*, Tucson, AZ 85726, USA

LARRY BARR Barr Engineering Arts, 1167 N. Winstel Blvd., Tucson, AZ 85716

ABSTRACT It appears perfectly feasible to increase the present McMath telescope from 1.6 to 4.0 m with minimal changes in the existing building structure and instruments. For solar physics such an upgrade would permit 0.8 arc-sec resolution in the IR at 12 μ where large Zeeman splitting and negligible mirror polarization would facilitate vector measurements of magnetic fields. For stellar physics a 4-m telescope could be scheduled for large blocks of time to explore stellar oscillations to V = 6-7 mag and anticipated results should promote the unification of solar-stellar seismology. A preliminary engineering study suggests that the light-feed be a 6-m all-aluminum actively supported mirror made of a mosaic of solid welded segments on an alt-az mount. The 4-m concave mirror would be of almost identical construction. Because it has a fixed orientation, its mount is simplified. The feasibility study is complete and the task of cost estimation has begun.

Keywords: McMath Telescope; Infrared Observing; Solar-stellar Observations

INTRODUCTION, WHY 4 METERS?

The McMath Telescope, with its relatively large 1.6 m aperture and all-reflecting unobscured optics, has long been favored for infrared observations of the resolved solar disk. Table I summarizes some of these IR projects in the past and the wavelengths at which they worked.

Study of the Sun's magnetism is the key to understanding solar activity. Until recently, the observations of surface fields have been mostly confined to visible wavelengths. Inadequate Zeeman splitting, inadequate spatial resolution, scattered light, and instrument polarization have all frustrated this work. By shifting to the IR domain these problems are largely avoided. This is because:

1) The ratio of Zeeman splitting to line Doppler width is proportional to wavelength. At 10 μ the relative magnetic broadening is 20 times that in the visible at 0.5 μ for a line of comparable Landé g.

* Operated by the Association of Universities for Research in Astronomy, Inc. under cooperative agreement with the National Science Foundation.

2) Linear polarization as measured by Stokes Q and U is linear with field strength B, just as for the circular polarization V. At 0.5 μ, Q and U are quadratic with B and difficult to determine.

3) Seeing improves by a factor of 2 at 10 μ, compared to identical conditions at 0.5 μ (Boyd 1978).

4) Umbral contrast decreases so that scattered light in sunspots becomes almost negligible. At 10 μ, Iu/Ip = 0.7 while at 0.5 μ, Iu/Ip = 0.1.

5) Telescope polarization, which is induced by oblique reflections, is reduced almost an order of magnitude at 10 μ compared to the value at 0.5 μ (Deming, 1991). This fact, combined with factor (2) above, means that Stokes polarimetry becomes straightforward at these long wavelengths.

ID	wavelength (μ)	g	U	P	use	ref
FeH FeH	1.006319 1.006270	0 ~1	x x		umbrae & pores umbrae, pores mag field	1
He	1.0830	1.2			high chromosphere	
HPβ	1.2818	~1			chromosphere	2
Fe	1.564852	3	x	x	plage, umbrae mag field	3,4,5,6
cont	1.63	-	x	x	opacity minimum	7
Na	2.208367	1.3	x	x	no pi component	8
Ti	2.227358	2.5	x		umbrae & pore mag field	8
CO	2.324	-	x	x	dT/dh	9
CO	4.665	-	x	x	oscillations, dT/dh	10
cont	10.0	-	x	x	limb dark, spots, seeing	11,12,13
OH	11.065	-		x	chromosphere oscillations	14
Mg	12.320	1.0		x	high photosphere mag field	15

where g = Landé factor, U = umbra, P = photosphere, dT/dh implies thermal structure
ref:

1 Wallace 1991
2 Livingston 1990
3 Harvey & Hall 1975
4 Solanki et al 1990
5 Stenflo et al 1987
6 Rabin et al 1990
7 Foukal et al 1990
8 Hall 1970
9 Ayres & Testerman 1981
10 Ayres & Brault 1990
11 Boyd 1978
12 Kopp & Livingston 1991
13 Turon & Lena 1970
14 Deming et al 1984
15 Deming et al 1991

TABLE I Infrared diagnostics of the solar atmosphere.

Two conditions speak against the use of the IR for magnetic field observations. First, the flux at 10 μ is smaller by a factor of 20 compared with 0.5 μ. Second, diffraction at the 1.6 m aperture limits resolution to about 2 arc-sec. Both of these constraints are relieved by scaling the telescope aperture to 4 m. The latter implies a flux gain of 7 and an improvement in spatial resolution to potentially 0.8 arc-sec. at 10 μ. At 5 μ the diffraction limit would be 0.4; at 2 μ it becomes 0.2 arc-sec. Telescope diffraction rather than seeing may often limit the resolution in practice. Thus it appears that the future of the observational study of the magnetic Sun lies in the use of IR techniques.

This conclusion finds support in the findings of the Solar Astronomy Panel for the "Bahcall Report" who propose a large-aperture IR telescope as a high priority project for the '90s (Nat. Acad. Sci., 1991).

As to the benefits for stellar astronomy, the advantages are practically self-evident: fainter objects (*i.e.*, more numerous stars and a greater variety of spectral types) for the same exposure time; higher time resolution for equivalent brightness. Observations of planets at small phase angles, now ruled out because of instrument danger from solar flux, would become possible. Assuming an NSO night-time program operating as it does now, it would mean that sizable blocks of time could be devoted to stellar seismology. Solar-type stars of V = 6-7 could be properly analyzed and an amalgamation of solar to stellar seismology achieved.

STRATEGY FOR THE UPGRADE

The existing McMath facility is a hallmark of conservative design and quality construction; the site is relatively dry except in summer; the elevation adequate to minimize telluric interference in spectral studies. For reasons of cost, we propose to preserve all these existing features, even the instrumentation (except for certain f/number modifications given below), and merely change out the optical train.

The present Cer-Vit concave mirror is a sphere of 1.61 m diameter and a focal length of 86.4 m. This results in an f/53.7 system. In the McMath building the diameter of the lower tunnel is about 5 m. We propose to replace this 1.61 m mirror with a 4-m one which will be actively supported and cooled to ambient temperature. Focal length would remain about 86 m, so the new optics would be nominally f/21.5.

More challenging is the replacement of the heliostat light feed. Here we envisage a 6 m flat actively supported on an altitude-azimuth mounting, see fig. 1.

Protection against wind would be via a translating wind-fence whose optimum position would be as close to the mirror as field-of-view permits, see fig. 2.

In bad weather, or for stowage, the 6-m will be turned to the auto-collimation position and the wind fence brought into actual contact with the support cell so as to seal the front surface with a louvered type panel.

Both the no. 2 mirror, a 4-m concave, and the no. 3 mirror, a 1.5-m flat, will be similar in construction to the 6-m. They too will be actively controlled and liquid cooled, but because their orientation is fixed, support is simplified.

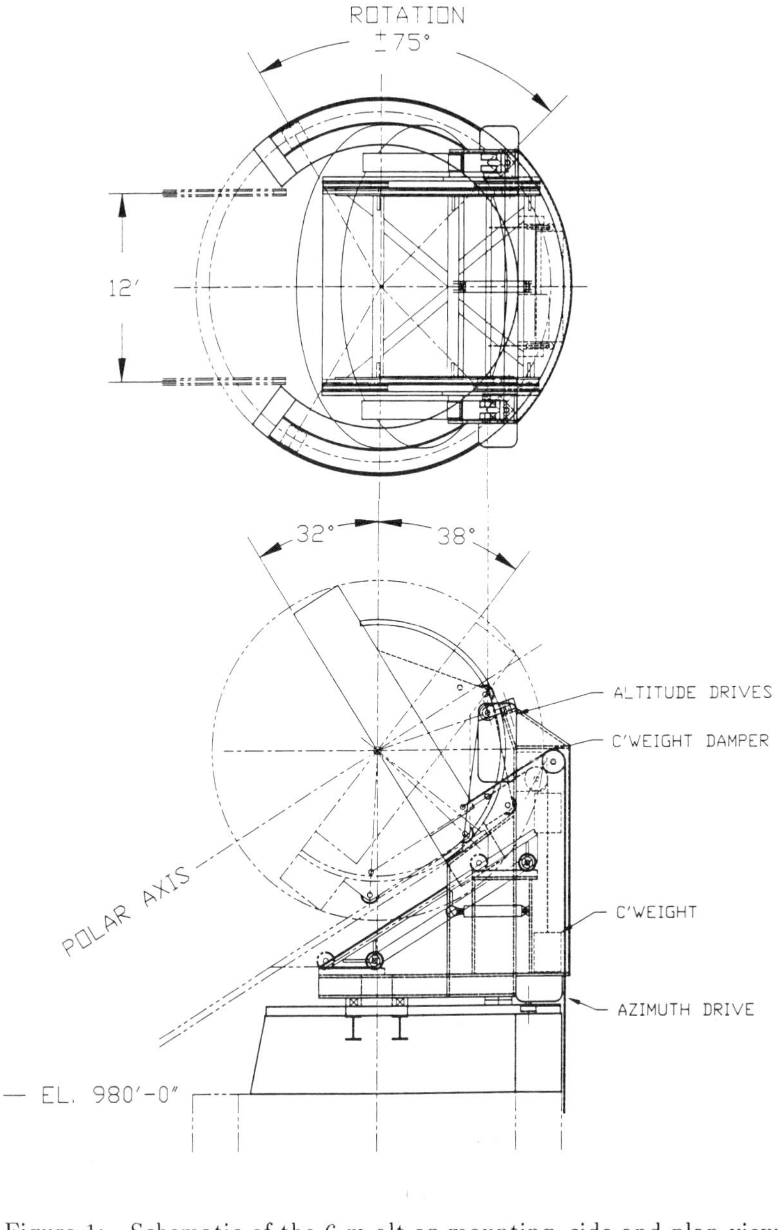

Figure 1: Schematic of the 6-m alt-az mounting, side and plan views.

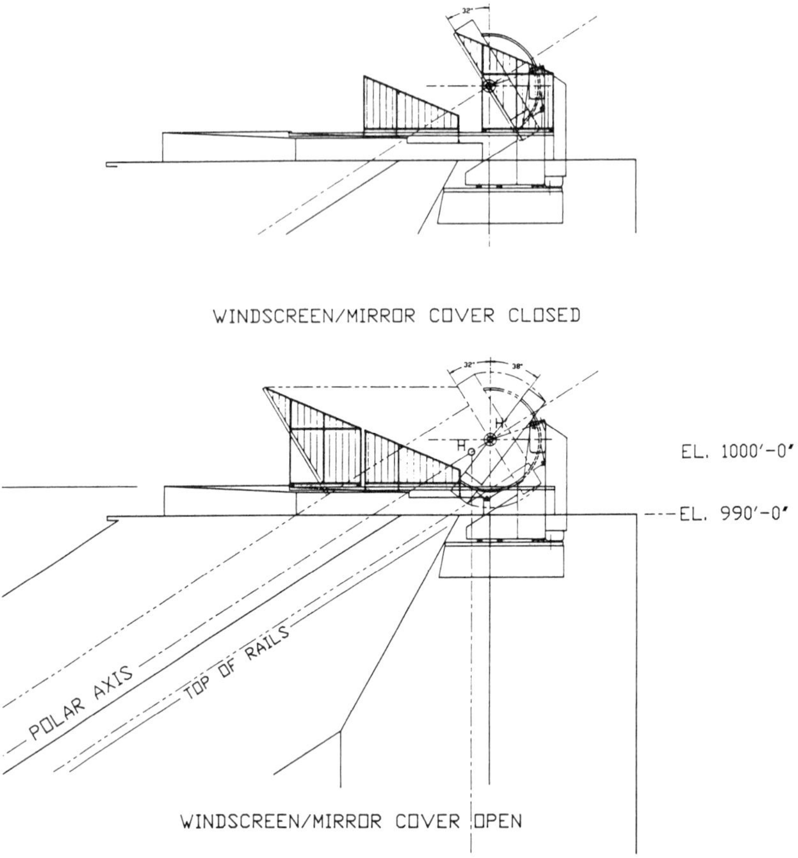

Figure 2: How the wind-screen translates to become a mirror cover. A louvered mirror cover (not shown) is built into the wind-fence and is closed to protect the mirror surface.

An unavoidable ancillary cost arises from the need for a tank adequate to coat the 6-m flat. To accommodate such a tank the present aluminizing room must be enlarged, although the existing mirror handling cranes are adequate.

We aim for an unvignetted field of 4 arc-min on the Sun. This is achieved with minor modifications to the building for feed to the solar and stellar spectrographs and to the 1-m FTS. Because the entire solar disk is not available in the focal plane, guiding will be done with an auxiliary optical system using light from the no. 1 alone.

Instrument modifications have not yet been laid out in detail. For the IR, the prime instrument would be the 13.5-m spectrograph, which features a choice of two gratings, one the world's largest IR echelle (Harrison 1973). It would need to have an enlarged collimating mirror capable of swinging into position as an alternative to the present collimator (which already overfills the visible light grating). Otherwise the spectrograph remains as is. Optionally, the 4-m concave may be diaphragmed to 1.6 m to allow the continuation of certain on-going synoptic visible-light programs. Changes to the stellar spectrograph are minimal: a transfer lens now converts the f/53.7 to a faster beam; only this lens needs to be altered. Similarly, the FTS collimator focal length must be modified.

ON MIRROR TECHNOLOGY

At this juncture, we believe that a solid aluminum mirror offers the lowest cost together with the important capability of significant cooling to avoid mirror-air interface seeing disturbance. Techniques exist to produce such a mirror blank at the 6 m size (Le Blanc and Rozelot, 1991). Mirror segments 15 cm thick are cast, forged, electron-beam welded together, machined to the desired perfection, heat stabilized, and finally nickel (Kanigen) coated on both sides with a 150 μ layer ready for polishing. As mentioned, active support schemes for all mirrors correct for predicted gravity and thermal aberrations via "look-up" tables. Additionally, the figure of the no. 2 will be controlled to minimize residual aberrations in the focused image. On-line information for this purpose will be intermittently analyzed by means of an appropriate imaging camera.

Polishing of the mirrors would follow conventional practice except for the no. 2. Its long focal length suggests that this concave be figured on the mountain at the telescope with the present 2-m heliostat serving as a folding flat to allow testing at center-of-curvature. This plan offers the additional advantage that optical tests can be performed with the mirror supported in its final attitude and by its final support mechanism. Testing will be scheduled to avoid interference with normal operation of the existing McMath telescope.

THERMAL CONSIDERATIONS

An aluminum coated mirror absorbs about 12% of incident normal solar radiation. Mirror-to-air temperature differences greater than about 0.5 C are found to produce noticeable adverse effects in the images. These worsen as the difference increases (Barr *et al.* 1990; Iye, 1991). The surface of an aluminized Cer-Vit mirror under typical conditions at the McMath was measured to rise at 3 C/hr in response to solar radiation. Such a rise can be prevented in a high conductivity solid aluminum mirror that is cooled from the back side. Under conditions of maximum heat load, it will be necessary to extract 3560 watts from the 6-m, 1445 watts from the 4-m, 1280 watts from the 1.5 m no. 3 mirror, and finally to absorb 9340 watts at the focal plane less the energy to the instrument. This being done, all surfaces will remain at or below ambient, and "mirror seeing" should not be present. If water is the coolant, a flow rate of 93 l/min is required at the 6-m, with about 1/3 this at the other mirrors. This assumes a temperature rise of 0.55 C in the water passing through the mirror.

At times the solar image will not be centered at the focus, of course, and part or even most of the light from the no. 1 will fall on the tunnel walls. As the panels of this structure are already cooled to maintain an isothermal environment, this off-axis energy should induce no difficulty.

MECHANICAL CONSIDERATIONS

A. The 6-m flat

A solid aluminum mirror 6 m in diameter and 0.15 m thick will weigh about 26,000 lbs. Add to this the support and cooling system plus the cell and its bearing journals, and we have a total moving weight of 71,000 lbs. Support for this will be by hydrostatic pads against retainer wheels and roller drives. As shown in fig. 1, the off-axis weight component will be partially compensated by a cabled counterweight.

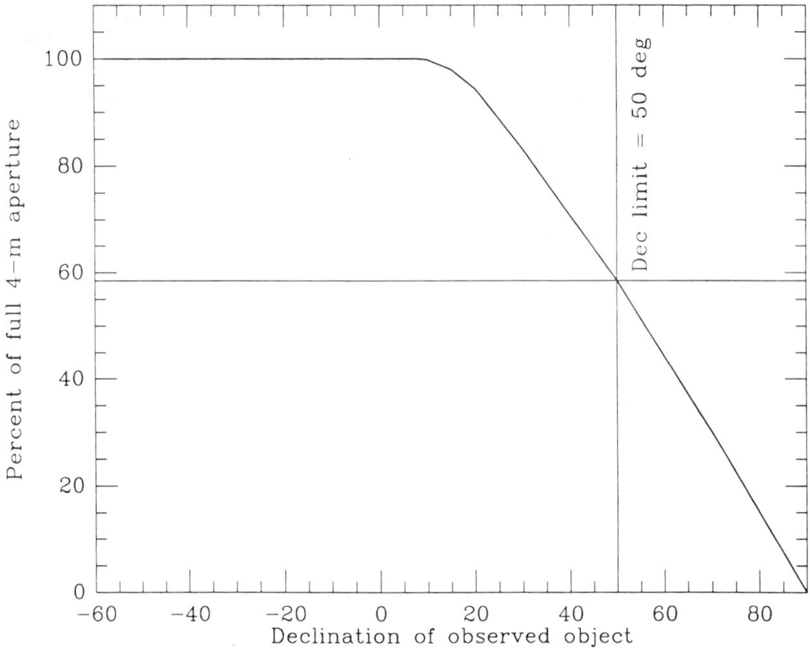

Fig 3: Vignetting of the no. 2 mirror as a function of declination.

Support for the 6-m flat will consist of approximately 150 axial and 75 lateral mechanisms. Axial pressure will be under computer control (look-up table) to compensate for temperature gradients, gravity, and optical imperfection. A partial vacuum will pull the mirror against the axial supports since, at times, the mirror partially inverted.

Procedure for re-aluminizing the 6-m mirror involves moving a translating set of rails outward to lift the mirror load off the journal bearings, disengaging the moving cell from its retaining structure, and rolling the cell down the tunnel rails on its transport trolley to the aluminizing gantry at ground level. The gantry crane will then lift the mirror from its cell and transport it to the coating chamber area.

This light-feed is capable of pointing to any celestial object at or above the horizon whose declination is less than +50 deg.

As shown in fig. 3, vignetting of the no. 2 begins at dec = +6 deg. 4 min. and increases to 50% of full mirror area at dec = +50 deg. Sky coverage is thus somewhat superior to the present McMath optics which begin to vignette at dec = −8 deg. 8 min.

B. The 4-m concave

The no. 2 mirror is a 4 m diameter 15 cm thick meniscus weighing 11,700 lbs. Active support will correct for temperature gradients and mirror figure. When observing the Sun, image corrections will be sensed sequentially at orthogonal solar limbs.

C. The 1.5-m flat

Support for this 1,700 lb. mirror will be a 6-point "whiffletree" and sling. Some mirror bending will be done for astigmatism correction.

PROJECT STATUS

A feasibility study of the project is now complete. Its conclusion is that it is reasonable, given today's mirror technology, to upgrade the McMath to a 4-m aperture. The next step will be a cost estimate based on this preliminary design.

ACKNOWLEDGEMENTS

The concept of a 4 m upgrade emerged from a conversation between R. Noyes and J. Leibacher. NSO staff: M. Giampapa, K. Pierce, D. Rabin, and J. Wagner contributed ideas. L. Daggert, E. Pearson, and observatory engineers likewise gave of their expertise. J. Hoey made the drawings. S. Wolff has encouraged and supported the project.

REFERENCES

Ayres, T.R. and Testerman, L. 1981, *Ap. J.*, **245**, 1124
Ayres, T.R. and Brault, J.W. 1990, *Ap. J.*, **363**, 705

Barr, L., Fox, J., Poczulp, G., and Roddier, C. 1990, SPIE Proc 1236, p. 492

Boyd, R. W. 1978, *J.O.S.A.* **68**, 877

Deming, D., Hewagama, T., Jennings, D.E., and Wiedemann, G. 1991, in Solar Polarimetry, ed. L.J. November (Sacramento Peak), p. 341

Deming, D., Hillman, J.J., Kostiuk, T., Mumma, M.J. and Zipoy, D.M. 1984, *Solar Phys.* **94**, 57

Foukal, P., Little, R., Graves, J., Rabin, D., and Lynch, D. 1990, *Ap. J.*, **353**, 712

Hall, D.N.B. 1970, Ph.D. Thesis, Harvard Univ., KPNO contr. no. 556

Harrison, G.R., 1973, *Appl. Optics* **12**, 2039

Harvey, J. and Hall, D.N.B. 1975, *Bull. A.A.S.*, , **7**, 459

Iye, M., Noguchi, T., Torii, Y., Mikami, Y., and Ando, H. 1991 *Pub. A.S.P.*, **103**, 712

Kopp, G. and Livingston, W. 1991 private communication

Leblanc, J-M. and Rozelot, J.-P. 1991 SPIE Proc. (in press)

Livingston, W. 1991, in Solar Polarimetry, ed. L.J. November (Sacramento Peak), p. 356

National Academy of Sciences 1991, in Working Papers Astronomy and Astrophysics Panel Reports, Nat. Acad. Press, Wash., p. IX -2

Rabin, D., Jaksha, D., Plymate, C., Wagner, J., and Iwata, K., 1991, in Solar Polarimetry, ed. L.J. November (Sacramento Peak), p. 361

Rozelot, J.P. and Leblanc, J-M. 1991, SPIE Proc. 1994, P. 481

Solanki, S., Biémont, E., and Mürset, U. 1990, *Astr. Ap. Suppl.*, **83**, 307

Stenflo, J.O., Solanki, S.K., and Harvey, J.W. 1987, *Astr. Ap.*, 173, 167

Turon, P.J. and Lena, P.J. 1970, *Solar Phys.* **14**, 112

Wallace, L. 1991 private communication

Seventh Cambridge Workshop on Cool Stars, Stellar Systems, and the Sun
ASP Conference Series, Vol. 26, 1992
Mark S. Giampapa and Jay A. Bookbinder (eds.)

THE EXTREME ULTRAVIOLET EXPLORER ALL-SKY SURVEY AND GUEST INVESTIGATOR SPECTROSCOPY MISSION

STUART BOWYER,[1] PATRICK JELINSKY, CAROL CHRISTIAN, AND ROGER F. MALINA   Center for EUV Astrophysics, 2150 Kittredge St., University of California, Berkeley, CA 94720

[1]Also with the Astronomy Department, University of California, Berkeley

ABSTRACT The *Extreme Ultraviolet Explorer* mission is described. For the first six months, an all-sky survey will be carried out covering 90 to 750 Å, or essentially the entire extreme ultraviolet (EUV) bandpass. This EUV survey will be made in four bands, or colors: $\lambda\lambda$ 90–150 Å, 170–250 Å, 400–600 Å, and 550–750 Å. A portion of the sky which is free from the normally intense 304 Å geocoronal helium background will be surveyed at greater sensitivity; the wavelength coverage of this band is from 90 to 400 Å. Following the sky survey portion of the mission, spectroscopy of individual sources will be carried out. Three spectrometers employing novel variable line-space gratings will provide spectra with ~1 Å resolution over the band from 70 to 760 Å. This spectroscopy will be carried out by guest observers chosen by NASA in a manner roughly analogous to the *International Ultraviolet Explorer* (*IUE*) guest observer program.

Keywords: *Extreme Ultraviolet Explorer*, EUV Astronomy, EUV Spectroscopy

INTRODUCTION

The *Extreme Ultraviolet Explorer* (*EUVE*) mission is the culmination of some 25 years of effort at the University of California at Berkeley to develop the field of extreme ultraviolet (EUV) astronomy (defined here as astronomy from roughly 100 to 1000 Å). A primary goal of the *EUVE* mission will be to carry out an all-sky survey over the entire EUV band. This survey will be conducted in four subbands, or colors. It was originally expected that this would be the first survey in this band, but in the passage of time the British Wide Field Camera, which was flown as part of the *ROSAT* mission, has carried out a sky survey at the shortest EUV wavelengths, which are similar to one of the *EUVE* bands. While it is certain that the majority of EUV sources will be observed at these shorter wavelengths, it is clear that the sky should be mapped over the entire EUV band at least once. In addition, our experience with X-ray astronomy, where at least five surveys of roughly equal sensitivity were conducted, has shown that multiple surveys bring out new and unexpected results.

The *EUVE* mission will also carry out a deeper survey over a limited portion of the sky along the ecliptic. These data will provide insights into the types of sources which would be discovered in a more sensitive all-sky survey. Specifically, would fainter sources be similar in character to the sources observed at higher intensities, or would an entirely new class of sources be evident?

Finally, *EUVE* will carry out spectroscopy of the brighter sources discovered in the survey phase of the mission. This spectroscopy will be carried out exclusively by guest observers, similar in manner to the *IUE* guest observer program.

In Figure 1 we show the sensitivity of the instrumentation on the *EUVE* satellite. For comparison we show the spectrum of the hot white dwarf HZ 43, the most intense EUV source now known. The all-sky/all-band surveys will be capable of detecting sources roughly one-hundredth as intense as HZ 43. The deep survey is much more sensitive but covers only part of the EUV band and only a part of the celestial sphere. The sensitivity of the spectrometer depends on the integration time and the character of the source; we show in Figure 1 the sensitivity for a continuum source and one day's integration.

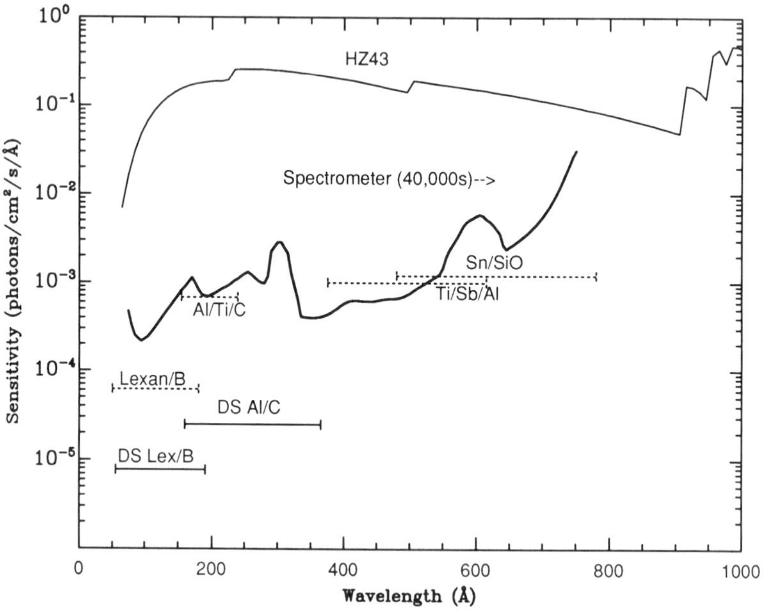

Fig. 1: The overall sensitivity of the *EUVE* instruments when combined with the anticipated sky integration times in the survey phase. The line that is marked "spectrometer" shows the sensitivity for a continuum source and about one calendar day's integration time.

THE EUVE INSTRUMENTATION

The mirror designs for *EUVE* are shown in Figure 2. The complement of mirrors used for the all-sky survey consists of two Wolter-Schwarzschild Type I mirrors used to carry out the short-wavelength bands of the survey and one Wolter-Schwarzschild Type II mirror for the longer wavelength bandpasses. The large graze angles in this design (Finley *et al.* 1988) prevent the (presumably more numerous) shorter wavelength sources from registering in the long-wavelength band and compromising this part of the survey. A standard Wolter-Schwarzschild Type II mirror is the collector for the three spectrometers and the deep survey instrument. The bandpass separation for the sky survey is provided by thin (1000–3000 Å thick) organic and metallic filters working in combination with the characteristics of the mirrors.

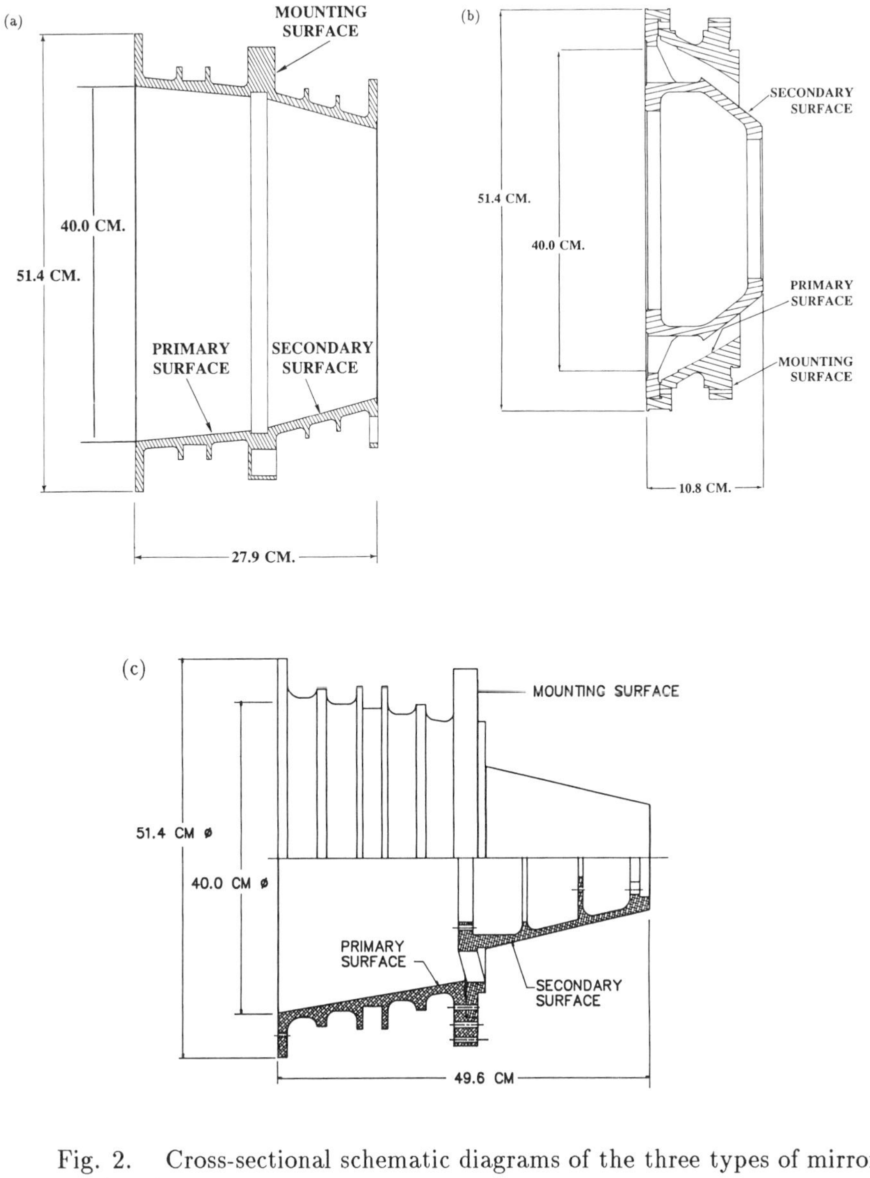

Fig. 2. Cross-sectional schematic diagrams of the three types of mirrors employed.

The detectors employed are photon-counting multichannel plate intensifiers with wedge-and-strip encoding. This type of detector was invented in our laboratory by Dr. Michael Lampton and co-workers (Martin *et al.*, 1981) and was extensively improved and space-qualified by Dr. Oswald Siegmund and co-workers (Siegmund *et al.*, 1986). These detectors possess a variety of qualities which commend them for use in space missions. The detectors in *EUVE* provide 1500 × 1500 rms resolution elements. They are stable, rugged, and solar-blind; they have low intrinsic background; they are linear to better than 0.5%; and they exhibit graceful degradation with high counting rate.

The spectrometer is an entirely new design invented at Berkeley (Hettrick and Bowyer 1983). This spectrometer uses variable line-space gratings, which provides a substantial number of advantages for spectroscopy with grazing incidence optics. First, it is highly efficient in that it requires a minimum number of grazing incidence reflections. Second, the image plane is nearly normal to the direction of the principal ray. Finally, it is readily adaptable to grazing incidence optics and provides the opportunity for optimizing parameters such as compactness, plate scale, and resolution. The concept of this spectrometer is illustrated in Figure 3.

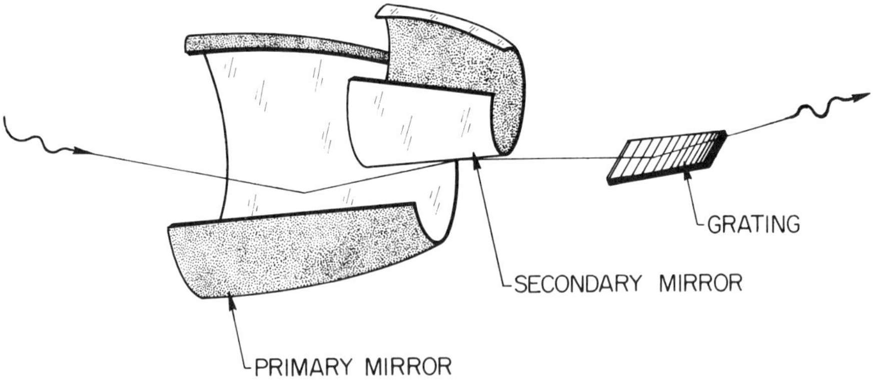

Fig. 3. A schematic illustration of the concept of the variable line-space grazing incidence spectrometer.

This concept has been implemented in the *EUVE* deep survey/spectrometer assembly (Martin *et al.*, 1985). This assembly is shown in Figure 4. Three spectrometers are employed so that the gratings will be reasonably on-blaze over the wavelength range from 70 to 760 Å. The characteristics of the gratings used in these spectrometers are listed in Table I.

TABLE I Grating Characteristics

	Short	Medium	Long
Bandpass (Å)	70–190	140–380	280–760
Line density (mm^{-1})	1675–3550	830–1750	415–875
Average plate scale (Å mm^{-1})	2.4	4.8	9.6
Blaze angle	3°	3°	3°
Coating	Rhodium	Platinum	Platinum
Filters	Lexan/B	Al/C and Lexan/B	Al and Al/C

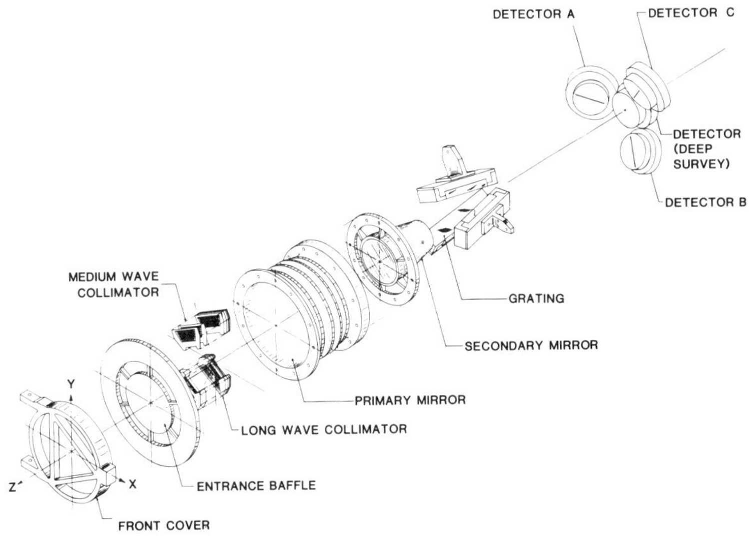

Fig. 4. An exploded schematic of the spectrometer on *EUVE*.

Figure 5 shows a plot of effective area versus wavelength for the three spectrometer bandpasses. The data points are calibration measurements, and curves are model fits. To determine the instrument effective areas during calibration, the aperture of each spectrometer channel was illuminated with a grid of monochromatic pencil beams. Throughputs were measured by comparing instrument counts with incident beam intensities. The effective areas were then derived by multiplying the measured throughputs by the geometric area of each spectrometer channel.

Figure 6 is a plot of the minimum detectable flux (MDF) for lines for a 3 σ detection in a simulated 40,000 second observation with the spectrometer. The MDF as a function of wavelength is produced by folding the projected instrument response together with expected background rates from diffuse airglow and detector noise. The MDF is then a threshold spectrum which source emissions

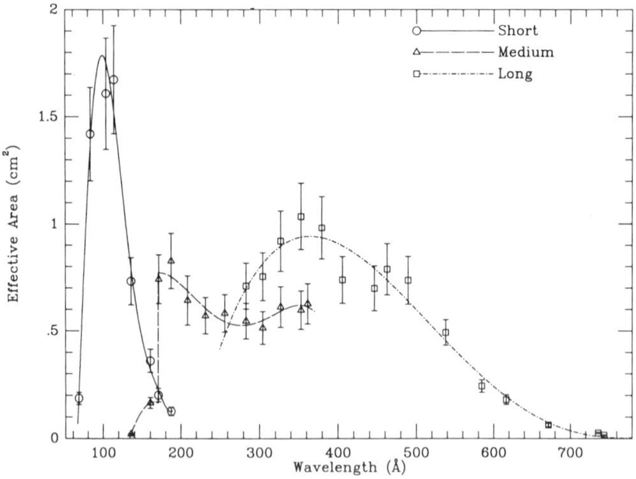

Fig. 5. Effective area versus wavelength for the three spectrometer bandpasses.

must exceed to be detected. This calibrated sensitivity of the instrument, in source flux units, can then be compared directly with model source spectra.

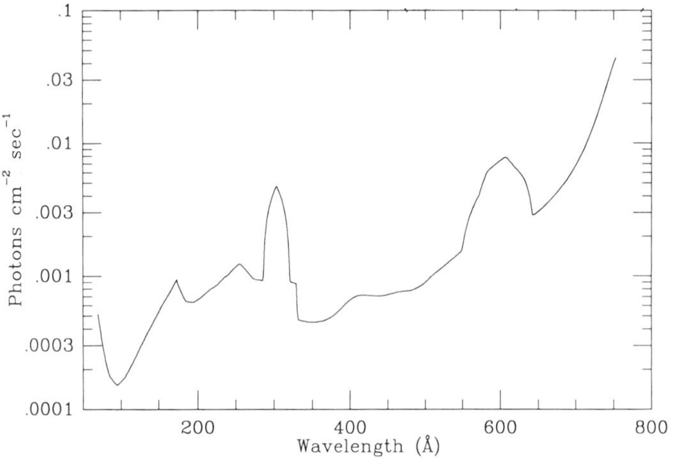

Fig. 6. Minimum detectable line flux in a simulated 40,000 second observation.

The spectral resolution and the positions of several wavelengths on the spectrometer detectors were measured using collimated beams of EUV light. Various calibration sources were used to produce complete spectra of magnesium, aluminum, argon, oxygen, and neon in the range 104–743 Å. Figure 7 shows the spectral resolution as a function of wavelength as derived from calibration data. The solid line is the best fit to the data. In Figure 8 we show the raw spectrum of a hypothetical chromospheric source observed by the spectrometer. The input spectrum employed was the solar EUV spectrum at the distance of Capella assuming the chromospheric emission is enhanced by the ratio of Capella's X-ray emission to the Sun. This spectrum was then absorbed by the known absorption to Capella and then folded through the measured spectrometer parameters.

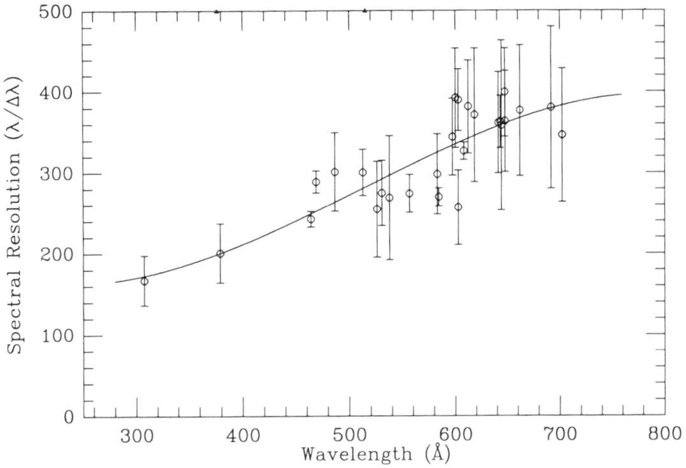

Fig. 7. Spectrometer spectral resolution as a function of wavelength.

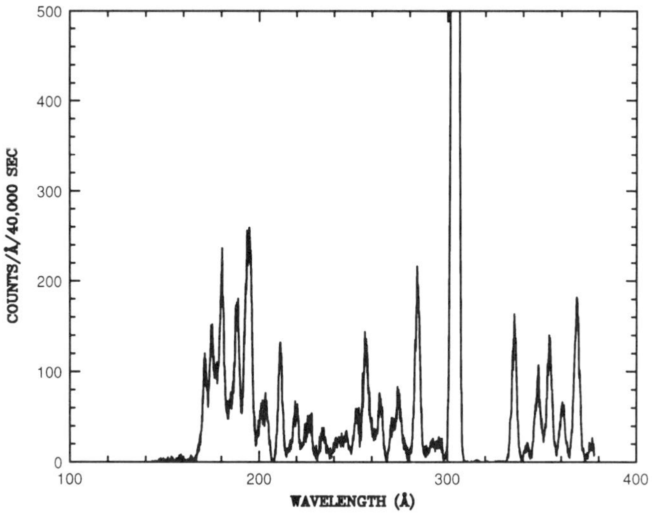

Fig. 8. Hypothetical spectrum of Capella as observed for 40,000 sec with the *EUVE* spectrometer.

THE NASA EUVE GUEST OBSERVER PROGRAM

NASA will conduct a Guest Observer program using the instruments on *EUVE* in a manner rather similar to the *IUE* Guest Observer program. The schedule for this program as of October 1991 is:

November 15, 1991	NASA Research Announcement release
January 12, 1992	*EUVE* launch date
February 15, 1992	Pointed phase proposals due
March 1992	*EUVE* sky survey begins
June 1992	NASA announces Allocation Awards
Summer 1992	Pointed spectroscopic observations begin

In the event that the launch date is revised, the relative offsets for the proposal due date and the awards announcements will be preserved.

Personnel at the Center for EUV Astrophysics will support scientists who have been chosen by NASA as guest observers during the pointed phase of the mission. Typical spectroscopic observations are expected to be 40,000 seconds in duration, observed with a 30% duty cycle, so approximately 150 guest observations per year are anticipated. The expected lifetime of the mission is at least 3.5 years, three of which would be devoted to pointed observations.

The *EUVE* Guest Observer support team at the Center for EUV Astrophysics will provide to NASA a facilities manual, or handbook, describing the instrumentation available for pointed observations. The document includes instrument performance data and sample technical feasibility calculations for proposers. This handbook will be available to respondents to the NASA Research Announcement. The team at Berkeley will provide assistance to proposers in order to clarify information contained in the handbook during the proposal preparation process.

The EUVE Guest Observer support team will monitor the acquisition of guest observer target data. This entails assisting operations in the scheduling of approved pointed observations and assisting in monitoring the quality of the incoming data. The data will then be provided to guest observers, and the *EUVE* Guest Observer support team will assist visiting scientists with data reductions. Guest observers will be encouraged to visit Berkeley at least once to gain familiarity with the data reduction programs being provided by the *EUVE* Guest Observer Center, who will write, document, and distribute the analysis software.

An analysis package for pointed observation data is being written to provide guest observers with routines which take advantage of the extensive design and calibration data available for the *EUVE* instrumentation. Scientists will be able to process data through reduction routines and, in the longer term, some instrument modeling packages will be provided. Guest observers will have access to data in a variety of forms: FITS tables, photon event files, and FITS images. The software modules are being written within the Image Reduction and Analysis Facility (IRAF), public domain software distributed by the National Optical Astronomical Observatories.

The motivation for supporting the spectrometer analysis software within the IRAF is to provide a set of packages that can be ported to the guest observers' home institutions for further use. The IRAF is one of the primary software environments available to astronomers for analyzing a variety of scientific data and for processing data. Facilities to be offered to visiting guest observers will be Unix workstations connected to the local network at the Center for EUV Astrophysics.

EUVE pointed spectrometer data become available in the public domain after a one year proprietary period. Information on the data will be accessible through the Astrophysics Data System. Tables containing instrument throughput information are available also through the Astrophysics Data System.

ACKNOWLEDGMENTS

The development of the *Extreme Ultraviolet Explorer* has been a cooperative effort involving a substantial number of people at NASA and UCB. We especially acknowledge the contributions of Michael Lampton, Oswald Siegmund, Herman Marshall, John Vallerga, David Finley, Peter Vedder, and Isabel Hawkins in the development of the instruments. This work has been supported by NASA contracts NAS5–29298 and NAS5–30180, which are administered by the Space Sciences Laboratory of the University of California at Berkeley.

REFERENCES

Finley, D.S., Jelinsky, P., Bowyer, S., and Malina, R.F. 1988, *Appl. Opt.*, **27**, 8, 1476.

Hettrick, M.C. and Bowyer, S. 1983, *Appl. Opt.*, **22**, 3921.

Martin, C., Jelinsky, P., Lampton, M., Malina, R.F., and Anger, H.O. 1981, *Rev. Sci. Instrum.*, **52**, 1967.

Martin, C., Mrowka, S., Bowyer, S., and Malina, R.F. 1985, *Proc. SPIE*, **597**, 284.

Siegmund, O.H.W., Lampton, M., Bixler, J., Chakrabarti, S., Vallerga, J., Bowyer, S., and Malina, R.F. 1986, *J. Opt. Soc. Am.-A*, Vol. **#3**, 2139–2145.

Seventh Cambridge Workshop on Cool Stars, Stellar Systems, and the Sun
ASP Conference Series, Vol. 26, 1992
Mark S. Giampapa and Jay A. Bookbinder (eds.)

FUSE – THE FAR ULTRAVIOLET SPECTROGRAPH EXPLORER

J. L. LINSKY[1] Joint Institute for Laboratory Astrophysics, NIST and Univ. of Colorado, Boulder, CO 80309-0440.

ABSTRACT I will summarize the past developments and present status of the FUSE Project. FUSE will observe a broad range of astrophysical sources, including cool stars, with high sensitivity and high spectral resolution in the far ultraviolet (912–1250Å). It will also study broad band spectra (100-1600Å) with high throughput and modest resolution.

Keywords: extreme ultraviolet spectroscopy; FUSE satellite; spectroscopy

THE ORIGIN AND PRESENT STATUS OF FUSE

I remember the morning of 17 January 1980 when the concept of The Far Ultraviolet Spectroscopic Explorer (FUSE) was first articulated. At an early meeting of the Ultraviolet, Optical, and Infrared Astronomy Panel of the Astronomy Survey Committee (the Field Committee), in San Francisco a few of us considered what might be a logical successor to the IUE spacecraft, which had already demonstrated its high scientific productivity. We had no inkling then that IUE would be an exceptionally productive mission for at least 13 years, but we did recognize that a successor should include many of the keys of IUE's success. These include broad simultaneous spectral coverage, a high orbit for long uninterrupted observations, a primarily guest observer program that involved many astronomers selected by peer review, flexible real time operations, and long mission life. At the same time the Copernicus satellite had provided a glimpse of the exciting science that could be accomplished with a spectrograph operating below 1200Å, and instrumental breakthroughs to increase the sensitivity over that of Copernicus by orders of magnitude were on the horizon. These considerations and the desire to complement the capabilities of HST (then expected to be launched in 1985) led the Astronomy Survey Committee to recommend that NASA study a "far-ultraviolet spectrograph in space" to observe in the 900–1200Å region.

NASA implemented this recommendation by establishing a Science Working Group for FUSE led by Al Boggess. In its April 1983 Final Report, the SWG laid out the scientific rationale for FUSE and a rough technical implementation plan for a high-orbit satellite with a grazing incidence telescope, and an assortment of possible spectrometers for high spectral resolution ($\lambda/\Delta\lambda \approx 30,000$) in the 900–1200Å band, and a grazing incidence spectrometer for the 100–900 Å spectral region.

[1] Staff member, Quantum Physics Division, National Institute of Standards and Technology.

NASA supported the FUSE concept by funding development on such basic items as optical designs, SiC optical coatings, and detectors that FUSE would require. Then in March 1986 NASA announced that it would evaluate proposals for Explorer concepts. Of the 45 proposals submitted in July 1986, NASA accepted four for Phase A study including a proposal entitled "LYMAN: The Far Ultraviolet Spectroscopic Explorer" submitted by a team headed by H. Warren Moos. This proposal developed the scientific objectives for FUSE in greater detail and described an optical arrangement consisting of an 80-cm diameter f/10 Wolter Type II glancing incidence telescope feeding a Rowland Spectrograph with three aspheric toroidal gratings to cover the 910–1250 Å spectral range in three exposures and a glancing incidence spectrograph for the 100-350Å range with a varied line-space straight groove grating.

Throughout this period there were meetings involving NASA, ESA, Australia, Canada, and the UK aimed at planning an international mission to accomplish the scientific objectives of FUSE. In particular, ESA supported several studies including a Phase A Study resulting in a Final Report in October 1988. The proposed ESA-NASA-Canada mission was not accepted for development by ESA. Subsequently, NASA accepted FUSE for Phase B development. At present FUSE is in Phase B as a joint NASA-Canada mission with possible UK collaboration, aiming for a launch in the 1999-2001 time frame, depending on the progress of missions before FUSE in the Explorer queue.

I have briefly summarized the history of FUSE to emphasize that the FUSE concept has survived many difficult challenges and extreme competition, because of the uniqueness and obvious importance of its scientific program, major technical accomplishments along the way, and strong leadership. I will emphasize the role of FUSE for studies of cool stars, but one should recognize that the prime driver for FUSE is cosmological and interstellar studies, in particular the determination of the primordial value of the deuterium to hydrogen ratio (D/H).

THE SPECTROSCOPIC RATIONALE FOR FUSE

The 912–1216Å spectral region contains a far richer selection of spectral lines formed over a broader range of temperatures than the 1200–3200Å region observed by IUE and HST. Cool stars emit emission line spectra with spectral lines formed at a few thousand K in the winds and circumstellar disks of M supergiants and pre-main sequence stars up to temperatures exceeding 10^7 K in stellar coronae. Table 1 summarizes some of the important spectral features that FUSE will detect in the spectra of cool stars. This list is compiled from tabulations of the observed solar spectrum (Dupree *et al.* 1973; Feldman and Doschek 1991) and the list of forbidden lines compiled by Feldman and Doschek (1991). This table is not a complete list, but it should include most of the bright lines that FUSE will observe in many cool stars. The temperatures of peak ion population are obtained from Shull and Van Steenberg (1982) for a homogeneous, low density plasma, but the density stratification and photoionizing radiation fields in real stars will modify these results, especially for the lower stages of ionization.

The Lyman alpha line is intrinsically one of the brightest emission lines emitted by a stellar chromosphere (Landsman and Simon 1991; Linsky *et al.* 1989), although interstellar absorption reduces the observed flux by a significant factor even for the nearest stars. The use of the higher members of the Lyman

Table 1. Important Spectral Features in the 912-1216Å Region

Species	log(T)	Important Lines (Å)
H_2, HD		hundreds of lines 912–1120
H I, D I		1216 (Lα), 1026 (Lβ), 973 (Lγ),950(Lδ),...
C I		1118m, 1140m, 1158m,1128m
N I		951, 964, 1133m, 1165m, 1200m
O I		989, 1152
He II		972, 1025, 1085
Si II	4.25	1190, 1194
C II	4.35	1010, 1036
N II	4.45	916m, 1084dm
S III	4.80	1012d, 1021d, 1077d, 1190
C III	4.85	977, 1175dm
N III	4.90	991dm
Si III	4.90	1109m, 1206
Si IV	5.05	1063d, 1073d, 1122, 1128
N IV	5.15	923
S IV	5.15	1062d, 1073d
S VI	5.30	933, 944
Ne V]	5.45	1136d, 1145d
O VI	5.50	1032, 1038
Ne VI]	5.60	999dm, 1005dm, 1010d
Si VII]	5.80	(1049)
Si VIII]	5.95	(944), (949)
Si IX]	6.10	(950)
Ca XIII]	6.40	(1106), 1133
Ca XIV]	6.50	(945)
Ca XV]	6.60	(1098), (1122)
Fe XVIIf	6.70	1153
Fe XVIIIf	6.80	975
Fe XIXf	6.85	1118

Notes:]=intersystem lines, f=forbidden lines, m=multiplet, d=member of a density-sensitive line ratio, ()=not observed in the solar spectrum.

series as chromospheric diagnostics will become possible for the first time with FUSE, since the effects of interstellar absorption decrease rapidly for higher members of the series. The bright C III and O VI resonance lines will be very useful in understanding the transition region plasma in cool stars, like the Si IV and C IV resonance lines in the IUE spectral range. There are two important differences, however; the O VI lines probe much hotter plasma (300,000 K) than C IV (100,000 K) and FUSE will be able to detect the O VI lines to a threshold 550 times fainter than the C IV lines that IUE can detect. In the 1020–1250Å spectral range FUSE will have an effective area of about 50 cm^2, and with low-background photon-counting detectors it should be able to detect lines with as few as 20 photons, corresponding to a flux of 2.7×10^{-16} erg cm^{-2} s^{-1}, in 8 hours. By comparison, the practical IUE limit for the detection of the C IV 1548-50Å resonance lines in high dispersion spectra is a flux of 1.5×10^{-13} erg cm^{-2} s^{-1}, the 3σ upper limit for Arcturus obtained by Ayres *et al.* (1986) with two exposures totaling 28.5 hours. FUSE will also have a "planetary" channel that will obtain spectra at lower sensitivity and resolution out to 1600Å to include lines of C II, Si IV, and C IV.

Shortward of about 650Å the solar EUV spectrum is dominated by permitted lines of highly-ionized species formed in the corona at temperatures above 1×10^6 K. In particular, the 100–200Å region contains many bright iron lines from Fe XII to Fe XXIV, which is formed at 1.8×10^7 K (Doschek and Cowan 1984). Since optical depth unity in the interstellar medium at 100Å corresponds to $N_H \approx 3 \times 10^{19}$ cm^{-2}, the EUV horizon extends out to ≈10 pc towards the Galactic Center and to ≥ 200 pc in the third quadrant of Galactic longitude (Paresce 1984). Thus a very large number of cool stars will be detectable by the EUV spectrograph on FUSE. Figure 1 demonstrates the rich spectrum of highly-ionized lines that FUSE will detect from Capella in only 1000 seconds. At somewhat longer wavelengths FUSE will also observe the resonance lines of He I and He II. Within the 100–350Å interval there are a number of density-sensitive line ratios that FUSE may be able to study (see Feldman *et al.* 1978).

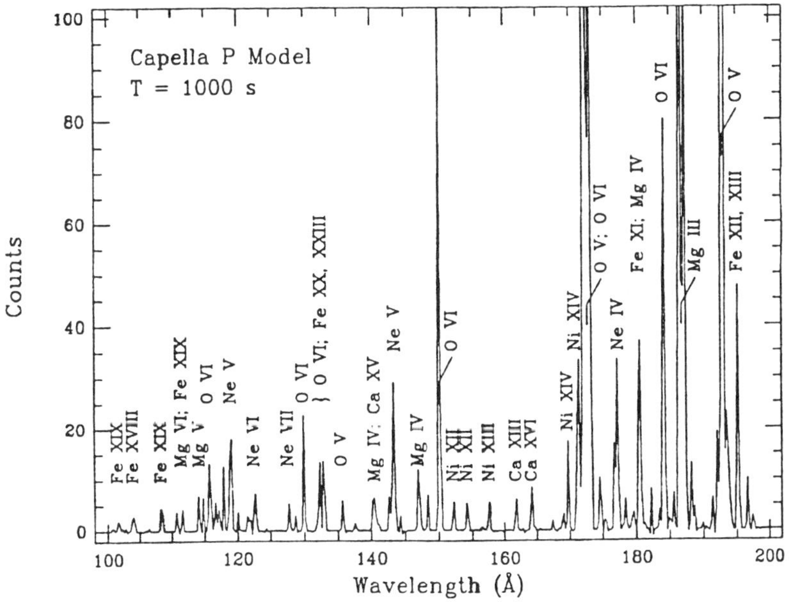

Fig. 1: The predicted spectrum of Capella (G9 III + G0 III) for a 1000 second exposure in the high resolution EUV channel of FUSE. From the LYMAN-FUSE Phase A Study Final Report.

SOME MAJOR QUESTIONS IN COOL STARS RESEARCH

The $\lambda/\Delta\lambda = 30,000$ and lower spectral resolutions, broad simultaneous spectral coverage, and high sensitivity of FUSE will allow observers to study many important problems concerning cool stars including the following:

(1) **The total radiative output from the outer layers of a stellar atmosphere.** FUSE will allow us to measure the radiative loss in the important Lyman lines, the middle transition region (from the O VI lines), together with the losses at lower and higher temperatures by determining the emission measure for $4.0 \leq log(T) \leq 7.2$ at one time.

(2) **Transition region dynamics and structure.** The C III 977Å and O VI 1032, 1038Å lines will be particularly useful in measuring transition region flows in active stars and for Doppler imaging of active regions in these stars. These results are needed to estimate the energy content of downflows during and outside of flares and for estimating the surface fluxes in active regions. The latter is needed for modeling the thermal structure and energy balance of active regions and quiet regions separately.

(3) **Flares and other time-dependent phenomena.** FUSE will time tag each photon to allow studies of the emission line fluxes and Doppler shifts during flares. The data may be binned flexibly in time and wavelength to optimize the signal/noise as the lines become bright and then faint. Observations with the planetary and EUV channels are best for studying line fluxes and the high resolution channels are best for studying flows.

(4) **Very young stars.** Since FUSE is about 550 times more sensitive than IUE, FUSE can easily study the emission lines of stars in young clusters like the Pleiades and pre-main sequence stars. These studies may tell us whether the lines are formed in a boundary layer between the circumstellar disk and the star or in solar-like chromospheres and transition regions.

(5) **Stellar winds.** FUSE may be able to measure the wind velocities of main sequence stars by measuring Doppler shifts in the O VI lines. These velocities refer to the beginning of the outflow. High-resolution spectra of lines of C I, N I, O I and H_2 should permit the determination of the wind speeds and mass loss rates for M giants and supergiants.

(6) **Do warm coronae exist?** Very high sensitivity in the resonance lines of C III and O VI will allow FUSE to measure small amounts of 100,000 to 300,000 K plasma, if present around stars with no evidence of hot coronae. These measurements will test the validity of the dividing line (*e.g.* Linsky and Haisch 1979) that appears to separate luminosity class III stars into groups with and without hot plasma.

A BRIEF OVERVIEW OF THE OPTICAL LAYOUT OF FUSE

Present plans are that the telescope will be a Wolter Type II with 64 cm aperture and f/8 focal ratio. It will be gold coated and have an optical effective area ≈1100 cm^2 at 1000Å. The telescope will feed a CCD fine error sensor and a number of different size slits arranged on a slit wheel or slit "farm". These slits direct the light to one of four gratings in the Prime Spectrograph or to the EUV glancing-incidence spectrograph. Table 2 summarizes the simultaneous spectral ranges, resolutions, grating coatings, effective areas, and paths to the detector for both spectrographs. Ray trace calculations verify that the spectrographs can achieve the desired resolutions. The Prime Spectrograph is a 1.63 m Rowland Circle mount with aspheric gratings to maintain good focus and greatly reduce astigmatism over the whole spectral range. The EUV spectrograph contains an ellipsoidal mirror and a variable line groove grating. The final selection of the two-dimensional detectors has not yet been made.

Table 2. Capabilities of the FUSE Spectrographs

Spectral Range (Å)	Spectral Resolution	Grating Coating	Effective Area (cm^2)	Path to the Detector
910–1030	30,000	SiC	40–50	Straight through
1020–1140	30,000	LiF	40–50	Periscope
1140–1250	30,000	MgF_2	40–50	Periscope
400–1600	800	SiC/LiF	40	Planetary channel
100–350	1,000–2,500	–	6–16	High resolution EUV
100–350	600–1,400	–	8–25	Low resolution EUV

Present plans call for a spatial resolving power of 2–3 arcseconds along the slit, and the planetary channel is so named because it will have a spatial field of view of 45 arcseconds to study extended solar-system objects. Three-axis stabilized pointing should have a 1σ accuracy of 0.2–0.25 arcseconds. FUSE is being designed with a mission life of 3 years and a goal of 5 years.

SOME UNANSWERED QUESTIONS

The successful operations of FUSE would clearly be an enormous stimulant for astrophysics as a whole, including studies of cool stars. While we are optimistic about the future of FUSE, I should inform you that some important problems are not yet solved. Here are a few of the major ones:

(1) **Schedule.** FUSE is in the Explorer queue behind EUVE, XTE, and ACE. The launch date of FUSE is therefore controlled by the funding of the Explorer line and possible cost overruns in the the previous missions in the queue. The present launch date of December 2000 is therefore optimistic.

(2) **Orbit.** The FUSE Science Team has requested a 24-hour highly eccentric orbit (1,000 x 70,000 km) with 28° inclination to provide high observing efficiency and about 17 hours of uninterrupted observing per day. This is now the baseline in NASA's planning, but is subject to change if fiscal or engineering realities so dictate.

(3) **Launch.** NASA has agreed to FUSE being launched by a Delta II rocket. The estimated launch weight, however, has only a 20% contingency, which is rather small for a mission this early in its development.

(4) **Radiation environment.** For the requested 24-hour orbit, FUSE would pass through the trapped proton belts around the Earth for up to 3 hours per day and pass through the outer Van Allen belts, which have MeV

electrons, for 4–7 hours per day. Cosmic rays and solar flare protons may also present problems at higher altitudes. This radiation environment may damage the electronics, degrade the CCD fine error sensor, and increase the detector backgrounds. There are ways of minimizing each of these problems, but more study is needed to assess the risks and to develop the most radiation-tolerant designs.

(5) **Detectors.** Although both the MAMA and the delay line detector technologies are advancing rapidly with funds primarily from other flight programs, full sized detectors meeting the FUSE specifications have not yet been demonstrated. The FUSE program is funding the construction of engineering models to demonstrate that they can meet specifications.

(6) **Optics.** Full sized gratings meeting FUSE specifications have also not yet been built, but programs are now funded to demonstrate these capabilities. Construction of the Wolter telescope will be challenging.

ACKNOWLEDGMENTS

I thank Webster Cash and Scott Friedman for their comments.

REFERENCES

Ayres, T.R., Judge, P., Jordan, C., Brown, A., and Linsky, J.L. 1986, *Ap. J.*, **311**, 947.
Doschek, G.A. and Cowan, R.D. 1984, *Ap. J. Suppl.*, **56**, 67.
Dupree, A.K., *et al.* . 1973, *Ap. J.*, **182**, 321.
Feldman, U. and Doschek, G.A. 1991, *Ap. J. Suppl.*, **75**, 925.
Feldman, U., Doschek, G.A., and Behring, W.E. 1978, *Space Sci. Rev.*, **22**, 191.
Landsman, W. and Simon, T. 1991, *Ap. J.*, **366**, L79.
Linsky, J.L. *et al.* . 1989, *Astr. Ap.*, 211173.
Linsky, J.L. and Haisch, B.M. 1979, *Ap. J.*, **229**, L27.
Paresce, F. 1984, *Astron. J.*, **89**, 1022.
Shull, M.J. and Van Steenberg, M. 1982, *Ap. J. Suppl.*, **48**, 95.

Seventh Cambridge Workshop on Cool Stars, Stellar Systems, and the Sun
ASP Conference Series, Vol. 26, 1992
Mark S. Giampapa and Jay A. Bookbinder (eds.)

NEXT GENERATION ULTRAVIOLET ASTRONOMY WITH THE DEEP ULTRAVIOLET EXPLORER OBSERVATORY

DONALD H. FERGUSON Ferguson Enterprises, 6802 Corte Nuevo, Pleasanton, CA 94566

MARK S. GIAMPAPA National Optical Astronomy Observatories, P. O. Box 26732, Tucson, AZ 85726

ABSTRACT Recent advances in control systems and sensors allow construction of an inexpensive yet high-performance orbiting observatory to collect data at ultraviolet wavelengths between 1150 Å and 3000 Å. The Deep Ultraviolet Explorer satellite (DUVE) will obtain all-sky imagery at various broad band wavelengths, high-resolution images, and spectra across the UV region. The DUVE program offers substantial performance advantages over current space-based observatories.

Keywords: DUVE; ultraviolet:images; ultraviolet:spectroscopy; sky survey

INTRODUCTION

The UV region is not observable from Earth and so is underexplored. Active galactic nuclei (AGNs), cataclysmic variable stars (CVs), hot white dwarfs, and of course young, hot main sequence stellar associations all emit a substantial component of their radiation in the UV. An all-sky survey will efficiently survey the sky down to a very faint limiting magnitude, $m_{bol} = 22$, leading to a number of statistically based studies of great importance. For example, DUVE will sample the early history of galaxy formation and evolution through identification of quasars and AGNs. Mass loss in the late stages of stellar evolution will be better constrained with the large sample of degenerate stars that will be observed. Uniformity of the diffuse UV background will be measured. A large number of unique individual objects will be identified for further study. Many additional observations will be obtained in the high-resolution mode. Hot stellar associations will be mapped in nearby galaxy clusters, yielding clues to star formation. A comprehensive search for cataclysmic variables in nearby globular clusters will be performed. Planets will be synoptically monitored at UV wavelengths and cometary jets can be observed with heretofore unavailable detail. The high resolution and excellent sensitivity of the spectrograph aboard DUVE will expand on the highly successful work of IUE. High-excitation phenomena, especially those identified with metal resonance lines, will be studied in a wide variety of objects: AGNs, CVs, active chromosphere stars, and many others. Additional UV features, such as low-z Lyman alpha absorption, various forbidden and intercombination lines, and mass loss phenomena can be studied in detail. These are only samples of the many discoveries DUVE is expected to provide. Given the quantum jump in performance and coverage over previous instruments, as shown in Table 1, experience suggests additional unanticipated, radically new discoveries will be made by DUVE (see Astronomy and Astrophysics Survey "Bahcall" Committee Report 1991).

THE DUVE CONCEPT

The DUVE program is designed to fit within the constraints of the NASA Medium Explorer program, as described in Figure 1. Use of existing space-qualified hardware designs is made to pack a relatively large aperture and state-of-the-art instrument package within the cost, envelope and weight restrictions. The DUVE payload is mounted atop a NASA standard spacecraft bus. DUVE features a particularly simple mechanical system – only two moving parts are required. The telescope and payload elements are modularized, further simplifying end item interfaces. Such simplicity allows the DUVE design team to reach design maturity typical of Critical Design Review before actually subcontracting out the end items, minimizing potential for expensive changes afterward. Such a procurement strategy, together with the the use of pre-existing hardware wherever possible, lends itself nicely to fixed price subcontracting for major hardware elements, further reducing costs. Additionally, the DUVE management effort is controlled by a close-knit group of astronomers and very experienced spacecraft designers – performance trades will be made early and efficiently. In this manner, the DUVE team expects to easily meet the $30 million cost target. Indeed, DUVE is cost-competitive with larger ground-based observatories.

DUVE DESIGN SYNOPSIS

DUVE is different than previous orbiting observatories in three ways. First, the aperture size is very large for the cost and volume envelope constraints. Second, the control system is based on active servo compensation, as opposed to assuming a stiff, monolithic structure. Third, new sensor technology allows use of large-format Charge Coupled Devices (CCDs) that are quite sensitive in the UV.

(i) Optics – DUVE's optical train is shown in Figure 2. The large 0.8-m Ritchey-Chretien telescope is the key to DUVE's excellent sensitivity. This telescope utilizes an 80% lightweighted primary and bipodal kinematical mounts for maximum stiffness relative to weight. Nevertheless, flexure is expected to be about one-half wave at 2000 Å – tests on an identical design used on an unrelated program will confirm this in early 1992. This limits best resolution to about 0.2 arcseconds, as will be discussed, but collects maximum light. Higher system resolution is not needed for the wide field survey or spectrograph instruments, and would seriously complicate the control system. The telescope body framework will consist of invar and the secondary will be hard-mounted to the spider. Focus over about three waves will be provided by inserting red-leak filters of various thicknesses near the sensor, as will be discussed, a factor of five more motion than is expected from preliminary thermal calculations. A fixed secondary results in no moving parts in the entire telescope subassembly; indeed, the only electrical interfaces will consist of temperature monitors and calibration lamp power supplies. MgF_2 overcoated aluminum will be used on the mirror surfaces with a scratch-dig specification suitable to reduce BRDF near the 2000 Å scattering maximum. The telescope will be well insulated to reduce the amplitudes of thermal gradients during long exposures.

The light cone extant from the telescope enters the carrousel assembly after reflecting off the fast steering mirror. While not at a pupil, the telescope focal length is sufficient that resolution degradation due to beam walk will not impact maximum system resolution of 0.2 arcseconds. A fixed field flattener at the entrance to the carrousel assembly improves image quality over the entire 2.8 degree field of view. The carrousel consists of two 12-position wheels arranged in series. The wheels are mounted on dual concentric shafts to avoid single-point bearing failure impacts. The entrance wheel is used to select various filtering and instrument options. The sensor wheel is used to control red-leaks and for telescope focusing. Each "hole" in the wheel is designed to cover the wide field CCD but not the visible track CCDs located alongside, as illustrated in the focal plane layout, Figure 2. Of course, non-red leak portions of the focus correction filters do overhand the visible CCDs. Figure 2 also describes the filter options available with DUVE. Two options are particularly noteworthy. First, a 10 × relay is included for high resolution work across the UV band. This allows high resolution work over the central 2/3 of the large-format CCD – the remainder is masked by mounting hardware and optics as the visible tracking CCDs must remain unobstructed. Second, a collimator-prism combination will allow for low resolution (~ 20 Å) classification-type spectra of every object within the 2.0 degree instantaneous field of view of the wide field imager.

(ii) Sensors – The sensor package is arranged around the wide field of view UV imager. A CCD chip of 4096 × 4096 format optimized for UV performance by thinning and coating will be used. Quantum efficiency exceeds 50% and readout noise is less than 10 electrons/pixel/frame total. Optics transmissions to the CCD is expected to average 12% in-band, for a total efficiency of 6%, including losses for particulate contaminations. This is considerably higher than current spaceborne UV systems and is largely the result of higher quantum efficiency. This chip and three 1024 × 512 chips optimized for visible tracking will be mounted on a monolithic, heat-conducting substrate cooled by redundant thermoelectric coolers. A slit alongside one edge of the large format CCD will serve as the entrance to the spectrograph, whose optics consist of the usual collimator, grating and camera mirror. The spectrograph detector will be a 2048 × 512 array of similar photodetection properties. The sensor package will be monolithically constructed from invar and thoroughly environmentally tested. There are no moving parts in the sensor subsystem – spectrograph focus is not critical in this application.

Many past UV-sensitive sensors have used red-insensitive photocathodes to reduce red-leak. With silicon-based CCDs optimized for UV performance, unfortunately, relative quantum efficiencies at UV and red wavelengths are similar. Additional filtering must be performed to reduce red-leak effects. This is accomplished in three steps. First, each bandpass filter located in the entrance carrousel wheel has about 10^{-2} rejection in the red relative to UV, as these are typically dichroics. Second, each filter in the sensor carrousel wheel contains a red cutoff filter of 10^{-4} relative transmission. Finally, a second red cut-off filter located on the CCD provides 10^{-4} additional protection, for a total of 10^{-10} relative transmission of red compared with UV light. This is adequate to give at least an order of magnitude greater signal in the UV for all but the coolest stars at the shortest wavelengths, near 1200Å, which are then of comparable signal.

(iii) Calibrations – Three lamp assemblies will be mounted on the secondary spider within the primary mirror pupil. Each assembly will consist of two lamps wired in parallel so that failure of a single lamp will not jeopardize the calibrations. Two of the lamp assemblies will respectively provide uniform diffuse illumination and a collimated wavelength comparison lamp for standard interpixel and spectroscopic comparison purposes. The fast steering mirror can be used to locate the wavelength comparison lamp either on the slit or in the wide field imager – the latter is useful to calibration of the low resolution classification spectra obtained on the large format CCD. The third lamp assembly is a collimated visible source designed to fall on one of the visible track CCDs and adjust the servo loop gain and track processor parameters.

(iv) Control Systems – The NASA spacecraft bus will point to within 2 arcseconds. This precision is adequate to preserve the wide field imagery resolution of 2.5 arcseconds, but will degrade spectroscopic throughput somewhat and completely blur the high resolution mode imagery. A more precise control scheme is needed. Use of a high bandwidth steering mirror to remove structural motions was successfully demonstrated by the Relay Mirror Experiment, an SDIO-sponsored spacecraft. The DUVE application uses stellar images on the visible CCDs alongside the wide field imager to supply track imagery to two weighted centroid track processors. These provide pointing control error signals to the fast steering mirror to null out the star positions and, hence, precisely maintain spacecraft pointing. Large offsets and roll corrections will be offloaded to the spacecraft bus pointing system at low rate. The carrousel and sensor packages are built structurally stiff, which is not difficult because of their relatively small sizes. Vibrations of up to 50 Hz above the fast steering mirror are removed by the servo system, as constrained by the CCD readout rate of 120 Hz. A simple Type 2 servo with large angle offload to the spacecraft bus is adequate due to the lack of significant acceleration during pointing. DUVE will use an analog servo loop, but with gain setable for the ground. The track processors feature adjustable track window sizes and threshold levels for differentiating amongst track stars. The tracking arrays are sized so that the likelihood of two stars of at least V=9 falling on two of the three CCD trackers is very high. The error signal is formed by blurring the visible image over several pixels, then using the relatively high signal per image per frame to average over the image size. In this way, subpixel tracking is obtained. Similar approaches have proven highly effective in various SDI pointing experiments.

(v) Mission Operations – The DUVE team would prefer the satellite placed in geosynchronous orbit. This would greatly simplify mission operations. For example, target identification would proceed in real-time, like with IUE, by downlinking the visible tracker data in RS-170 form, with reduced resolution. Then too, only one ground site would be required. This simplicity must, however, be balanced by the additional cost of the larger boost system needed to reach such an orbit. Nevertheless, with on-board storage and commands provided by an enhanced data processing system, operation in low earth orbit is achievable. Such a selection is also suitable for certain earth-oriented, non-astronomical missions whose sponsors may help to defray costs. In any event, coarse pointing will be provided by the NASA-supplied spacecraft bus, greatly simplifying command and telemetry requirements for the payload. The DUVE team anticipates that data will be available at the ground site within a few tens of minutes after any observation. All raw data will be cataloged and stored, and each image and spectrum will be reduced with the most current calibrations, documented and stored in pipeline fashion, after, of course, a quick overview by the observer. This is necessary due to the large amount of data expected from this satellite. Reduced data can then be called up separately as desired for further analysis.

SUMMARY AND CONCLUSIONS

The ultraviolet region between 1150 Å and 3200 Å has not been systematically surveyed and explored. IUE spectra and recent glimpses from the ASTRO mission indicate a great many valuable scientific discoveries await. DUVE is a space-based observatory providing all-sky imagery, high-resolution imagery, and high-resolution spectroscopy at wavelengths between 1150 Å and 3200 Å. Sky coverage, resolution, and sensitivity are each equal or better than available on existing spacecraft. Recent advances in controls, sensors, and telescope lightweighting coupled with space-qualified designs combine to make the DUVE project affordable within the guidelines of the Medium Explorer-class mission scenario. The DUVE team is experienced and has adopted a management approach that ensures success within resource constraints.

REFERENCES

Gillespie, B., 1991, 'Space Telescope Science Institute Memorandum', Hubble Space Telescope Cycle 2 Proposal Submissions–Updated Technical Information.

Harris, A. W. and Sonnenborn, G. 1987, in 'Scientific Accomplishments of the IUE', ed. Y. Kondo (Dordrecht: Reidel), pp. 729-749.

'The Decade of Discovery in Astronomy and Astrophysics', J. N. Bahcall, chair, Astronomy and Astrophysics Survey Committee, National Research Council (Washington, D. C.: National Academy Press), 1991.

Table 1

Comparison of DUVE w/Existing Programs

Program	Sensitivity FLAMBDA -- ergs/s/cm2/Å (20 min Integration)		Resolution	Field of View	Duty Cycle	Cost
Spacecraft	Imagery (SNR = 2.5)	Spectra (SNR = 10)	Imagery (arcsec)	Imagery Only (arcmin)	(%)	($M)
IUE	—	3E - 14	—	—	70	300
HST	1E - 17	1.5E - 14	0.2	2.6 (WFC)	30	1,500
DUVE	1E - 17	2E - 15	0.2	170 (Wide)	85	60

HST: (imagery) -- PC at 0.15 μm
Best Sectral Resolution w/non-echelle mode
(spectra) -- FOS/"Blue"/G19OH

Payload Data Sheet

Estimated science instrument characteristics

NASA Specifications	DUVE
Mass (150-300 kgs)	300 kgs
Width/Diameter (less than 200 cm, launch configuration)	120 cm
Height (under 300 cm, launch configuration)	240 cm
Power (up to 250 Watts)	50 W
Average Data Rate (less than 100 kbits/sec)	100 kbits/sec
Pointing accuracy (up to 1 arcsecond with input from science instrument, 30 arcseconds without)	≤ 2.0"
Lifetime needed to complete science investigation (1-3 years)	1 year, 3 years optional
Optimum orbit: - Apogee - Perigee - Inclination - Period	23,000 mi., circular Equatorial (0°) inclination 24 hr period
Time required for mission development (3-4 years)	3.5 years
Instrument payload cost (FY91 dollars)	$30M

Fig . 1

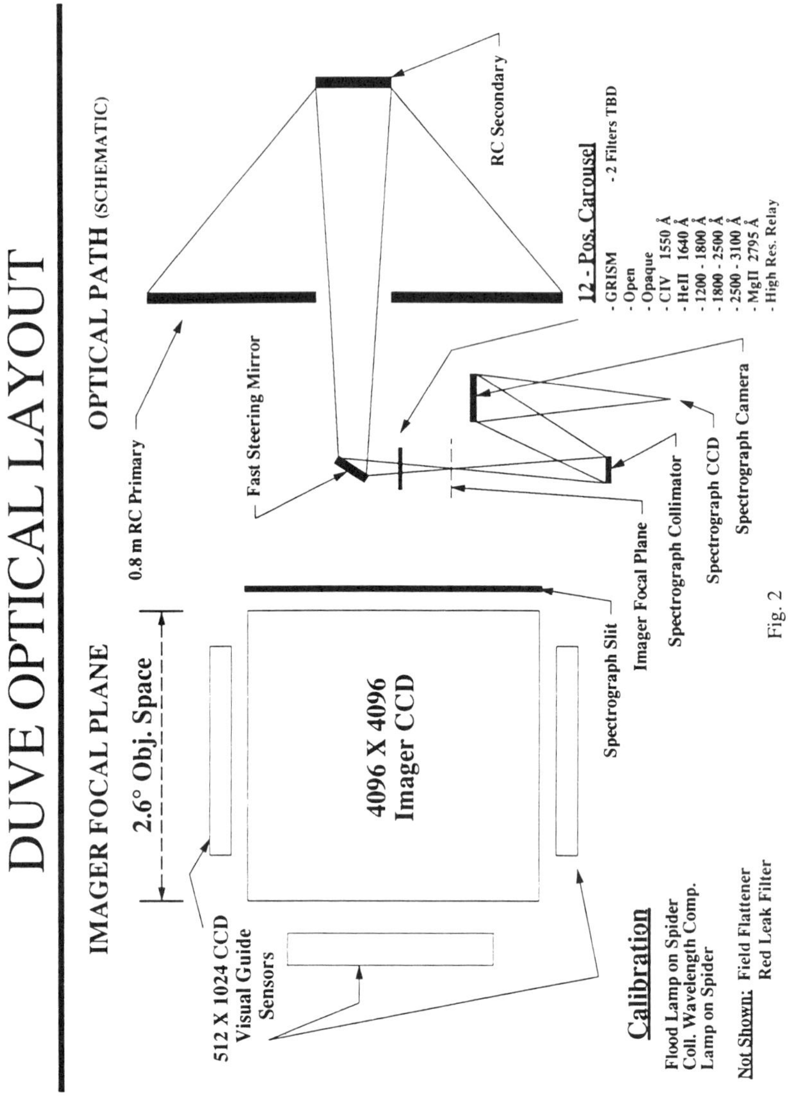
DUVE OPTICAL LAYOUT
IMAGER FOCAL PLANE
OPTICAL PATH (SCHEMATIC)
2.6° Obj. Space
0.8 m RC Primary
RC Secondary
Fast Steering Mirror
4096 X 4096
Imager CCD
512 X 1024 CCD
Visual Guide
Sensors
Spectrograph Slit
Imager Focal Plane
Spectrograph Collimator
Spectrograph CCD
Spectrograph Camera
12 - Pos. Carousel
- GRISM
- Open
- Opaque
- CIV 1550 Å
- HeII 1640 Å
- 1200 - 1800 Å
- 1800 - 2500 Å
- 2500 - 3100 Å
- MgII 2795 Å
- High Res. Relay
- 2 Filters TBD
Calibration
Flood Lamp on Spider
Coll. Wavelength Comp.
Lamp on Spider
Not Shown: Field Flattener
Red Leak Filter

Fig. 2

Seventh Cambridge Workshop on Cool Stars, Stellar Systems, and the Sun
ASP Conference Series, Vol. 26, 1992
Mark S. Giampapa and Jay A. Bookbinder (eds.)

THE MUSICOS NETWORK FOR MULTI-SITE CONTINUOUS SPECTROPHOTOMETRY

B.H.FOING Institut d'Astrophysique Spatiale (IAS), BP 10, 91371 Verrieres le Buisson, France

C. CATALA, J. BAUDRAND, T. BOEHM Paris-Meudon Observatory, 92195, Meudon, France

and the MUSICOS team

ABSTRACT Multi-site networks can be useful for the study of late-type stars and the solar/stellar connection (asteroseismology, stellar rotational modulation, surface structures, Doppler imaging, variable winds, coordinated multi-frequency observations with space satellites...) specially for programs requiring a continuous spectroscopic coverage around the clock during several days. MUSICOS, (for MUlti SIte COntinuous Spectroscopy) is a project for a multisite network of high resolution spectrometers around the world partly dedicated to continuous spectroscopy, under development in France, with European and extra-European collaborators. The stellar activity and asteroseismological goals drive the highest constraints on the spectrometer (high efficiency, high S/N, wavelength range over 500Å, spectral stability). Accordingly, a prototype of the MUSICOS spectrograph has been developed, tested in the laboratory and qualified at the 2m telescope at Pic du Midi.The instrument design and performances (low cost fiber-fed cross-dispersed echelle spectrograph with a resolution 38000) are described and the strategy for a multisite network of similar spectrographs is presented. The first MUSICOS campaign dedicated to 3 programmes in december 1989, involved sites and telescopes in Mauna Kea (CFH and 2.2m UH), Kitt Peak McMath, La Silla 1.4m CAT, France 1.5m OHP, Crimea 2.6m Shajn and China 2.16m Xinglong with two fiber-fed spectrographs transported for this campaign at Hawaii and Xinglong.This allowed the Doppler imaging of surface structures and the monitoring of two exceptional white-light flares on HR1099.

Keywords: MUSICOS, multi-wavelength observing

THE NEED FOR MULTI-SITE SPECTROSCOPY

We organised in June 1988 a national workshop in France about the scientific use of multi-site spectroscopy (see Catala and Foing 1988). The workshop was attended by 50 participants, from different fields of solar and stellar physics. The interest for continuous spectroscopy was recognised for various topics for the study of asteroseismology, stellar rotational modulation, surface inhomogeneities, for stellar activity short term phenomena, or Doppler imaging of surface structures. Also the study of wind variability, flare patrol, or eclipse imaging was stressed. Additionally, a joint network of high resolution spectroscopy and photometry would give a simultaneous support to continuous satellite observations. The need for multi-site continuous coverage (taking as example networks of solar seismology) was especially recognised for the asteroseismology programs e.g. of OB, Be, delta Scu, fast rotating B stars and solar type stars (see Dappen 1988, Catala and Foing 1988).

THE MUSICOS PROJECT

MUSICOS is an international project started in 1988 which aims at developing a multi-site network of high resolution spectrographs. This project has received support from many scientists, from European countries, America and Asia so far. The goal of Musicos is to facilitate multi-site spectroscopic observations, first by setting up an organisation helping the coordination of observations from existing instruments at different sites, then by designing, developing and installing similar spectrometers in well-chosen sites around the world, for which part of the time would be devoted to multi-site observations.

Given the positive impact of the first MUSICOS workshop and the interest shown by several other European teams , we have decided to get started on the practical and technical aspects of the project as soon as 1989. In a first phase, we have been organising multi-site campaigns using resident instruments on various telescopes around the world and a transportable fiber-fed mono-order spectrometer (ISIS, developed at Meudon Observatory and used until 1988 at the 1.93 m telescope at OHP). In a second phase, we continued in 1990-1991 the design and development of an cross echelle fiber-fed spectrometer, and perform tests in the laboratory and at telescope on this prototype to assess the scientific performances of this fiber-fed spectrograph. In a third phase, we shall develop an instrument that can be proposed for duplication to different foreign collaborators. This will allow at long term multi-site campaigns with identical instruments, and to reach the limits in quality, sensitivity and homogeneity of ground based continuous spectroscopy that are set by asteroseismology requirements.

REQUIREMENTS FOR ASTEROSEISMOLOGY AND STELLAR ACTIVITY

Asteroseismology requires high S/N, and the possibility of observing a large number of photospheric lines, together with the specific requirements of a high spectral spectral stability and very accurate velocity calibration. The need for pre-reduction of data at each site, in order to control in real time the instrumental parameters (quality of night, drifts) is also important.

For solar-type oscillations, the velocity signal is much reduced (of the order of 15 cm/s per stronger solar mode). An observing continuity better than 60% in order to decrease the effect of parasite sidelobes and aliasing in the power spectrum, and to increase the signal to noise and resolution on the oscillation modes was requested, which argues for multi-site network. Classical slit spectrographs using one or a few lines are not stable and efficient enough. Resonance cell techniques have a good stability but can apply only to 0 or 1st magnitude stars with 4m class telescopes, which are oversubscribed and make difficult a continuous observing from several sites. For having access to fainter stars from later-type closest stars or from other classes in the H-R diagram, the multi-line concept may be relevant in the future. For seismology velocity measurements, the MUSICOS spectrometer at 30000 resolution only with a 2m telescope would give on a 3 magnitude star a S/N of 100 in 1 mn, which coupled to the measurement of 400 lines relatively free of blending available on the spectral domain in the cross echelle mode would in principle allow to reach an oscillation detection of 3m/s per mn exposure, and thus a noise of 5cm/s for a continuous 72 hours observation.

Stellar activity requires also the observation of different lines simultaneously, for instance: photospheric lines to be used for magnetic measurements; different unblended lines for Doppler imaging of spots; chromospheric lines such as the Ca II H and K or infrared triplet allow to probe different chromospheric heights; in particular the H alpha line is a very sensitive indicator of extended active structures; Balmer lines give very important temperature, density and dynamic diagnostics for flare studies; both H Balmer line and strong resonance lines such as Na or Ca II can probe circumstellar absorbing material transiting over the disk; also high excitation lines such as He I allow to follow more energetic phenomena or variations of EUV or X radiation.

TECHNICAL DESIGN OF MUSICOS INSTRUMENT

From the scientific requirements after the MUSICOS workshop it was realised that a large range of scientific programmes require a spectral resolution 10000, 30000, and 80000. The main requirements for stellar activity studies and asteroseismology is the availability of large spectral domain in order to measure a number of photospheric lines in solar-type stars, and the stringent constraint for spectral stability down to the 10m/s range. The spectrograph can be fed by an optical fiber (Felenbok, 1988) which makes the instrument transportable and adaptable on another telescope.

The design of the instrument included the following specifications: a spectral resolution of 30000 minimum, a spectral domain 3900-8700 A that can be covered in two successive exposures (blue and red), an efficiency allowing a S/N better than 200 in 1 hour for a star of 6th magnitude with a 2m telescope and an astronomical CCD; a Thomson CCD 1024x1024 (19 μ pixel) proposed as a detector; a fiber feeding; the possibility to acquire 2 spectra simultaneously. A prototype of the MUSICOS spectrograph has been developed (Baudrand and Boehm, 1992). The selected fiber POLYMICRO FHA has a core of 50 μ, which corresponds to 2 arcsecs for an entrance aperture f/2.5. Its chromatic transmission is near 90 percent for a large domain in the visible and in the red, and remains better than 70 percent at 3900 A. Thus, an image slicer is not anymore necessary. Light is directed towards a 400 mm focal length collimator, and then on an echelle grating, a Milton Roy replica (31.6 gr/mm and 63.5 degrees blaze angle). The cross dispersion is given by two SCHOTT rectangle prism working in double pass at deviation minimum, covering the dispersion requirements of the two wavelength regions, 380-540 nm and 540-880 nm. The passage from blue to red needs 4 independent mechanical drives, motorized and simultaneously commanded, requiring less than 3 minutes for the change and yielding a positioning precision within 1 pixel. The whole is articulated around a CANON objective (from the market). A slight tilt of the grating, and a lateral mirror allow a comfortable mounting of the detector cryostat. This set up has the advantage of being very compact, luminous and inexpensive. We estimate that this spectrograph could be duplicated with a moderate budget of 40.000.- US Doll. and within a short month delay. Laboratory results indicated 17 throughput efficiency spectral resolution equivalent to 38000, a spectral coverage in two exposures from 380-880 nm complete until 700 nm using the Thomson 1024 x 1024 CCD, and a minimum separation between adjacent orders of 360 micron. An automatic procedure (MUSBIC) to reduce MUSICOS Echelle spectra was adapted from the ECHFIC programme developed by M. Spite. The extraction of the curved orders and auxiliary spectra (flatfield, Thorium), and wavelength calibration are described in Baudrand and Boehm, 1992. The output can be converted into standard data formats.

THE MUSICOS DECEMBER 1989 OBSERVING CAMPAIGN

For this first MUSICOS campaign at the end of 1989, three programs were chosen among those which really require multi-site spectroscopic observations:

1) Short-period spectroscopic variations in Be stars

2) Corotating stream structures in the winds of PMS Herbig Ae stars

3) Doppler Imaging and flare monitoring of RSCVn-type active stars

Initially, the sites, telescopes and observers to be involved in this campaign, were the following: Mauna Kea 2.2m UH, 3.6m CFHT, Kitt Peak McMath, La Silla 1.4m CAT, France 1.5m OHP, Crimea 2.6m Shajn, China 2.16m Xinglong.

After the general announcement sent by J. Butler through the Multi wavelength IAU Working group, and also the IAU circular for our multi-site continuous spectroscopy campaign from 8 to 17 december 1989 (MUSICOS 89 campaign), we received a large observing support (ground based spectroscopic or photometric, or satellite observations) and finally 17 telescopes , including IUE, were pointed towards our targets . In addition to resident spectrographs at OHP, Crimea, CFH, ESO, Lick, Mc Math , we used the two existing versions of the ISIS fiber-fed spectrograph in mono-order mode (one of them used until 1988 at OHP), and specially transported for the MUSICOS campaign on the 2.2m University of Hawaii telescope and on the 2.16m newly installed telescope at the Chinese XingLong station.

The first results from the MUSICOS 89 campaign have been presented in the 2nd MUSICOS workshop (Catala and Foing, Eds, 1990). On the RS CVn-type HR 1099, this includes a complete phase coverage allowing a Doppler imaging of photospheric spots. Quasi-simultaneously, we observed modulation of CaII K profile due to chromospheric plage regions. At least two exceptional white-light flares were detected photometrically and by their remarkable spectral signatures in H alpha. The 14 December flare starting at 13:00 UT was followed from China and Catania, and the 15 December flare starting at 2:00 UT was followed from ESO, Hawaii, China and Catania (Foing et al, 1990). We estimated the energy budget for these two events, the first of this importance detected on HR 1099. The interpretation of a filament ejection occurring over a magnetic arcade and the magnetic energy budget are discussed in Foing et al (1991). These observations bring a new light and a puzzle about energy transport mechanisms in stellar flares.

MUSICOS PROJECT ORGANIZATION AND COLLABORATIONS

There is in France a MUSICOS project group composed of C. Catala (Meudon) and B.H.Foing (ESA/IAS) as Principal Investigators, J. Baudrand (Meudon) is the instrument project manager. A group of Co-investigators and associated scientists is involved in the project. Also contacts were established with European groups in Italy (in Trieste and Catania), in Scandinavia (in Upssala, Helsinki, and Aarhus), in Spain, in United Kingdom (at Armagh). Associate countries such as US (in Hawaii, in Boulder, Goddard, and with the SYNOP group), in USSR (in Crimea) and in China (at Beijing Astronomical observatory) wish to participate to the multi-site project. The MUSICOS project will be open to any input and collaboration during the development phase , and through the observing proposals to the community during the operational phase. The philosophy of the MUSICOS project is to associate the groups interested also on the scientific return , and this is the reason why we start already collaborative campaigns to learn how making, reducing , analysing the results from multisite data. Asteroseismology will require continuous coverage and a good control on the stability and the quality of the measurements. Also, some programs require special observational strategy and might be interleaved with other programs or require service observing.

MUSICOS PLANNING AND PERSPECTIVES

Since 1982, with our collaborators, we have participated to Multi-Site Multi-Wavelength Observing campaign, giving the collaboration framework for operations and scientific analysis. At the end of 1989, we organised an observing campaigns including transport of ISIS and ISIS-bis instruments in complementary sites (Xing Long, Hawaii). In january-mars 1990, we worked on the preliminary reduction and analysis of the results from this campaign. A MUSICOS workshop organised at Meudon on 27-30 March allowed presentation of these results, further discussion and future organisation and realisation of the MUSICOS project (see Catala and Foing 1990). 1989-1990 was devoted also to the design and development of the spectrograph.

At the end of 1991, instrument tests and qualification of the MUSICOS prototype spectrograph took place at Pic du Midi Observatory, with the B. Lyot 2 m telescope. Spectra were recorded with a Pic du Midi CCD detector. The instrument speed resolving power, wavelength stability and scattered light were determined (Baudrand and Boehm, 1992). The MUSICOS spectrograph will be operated routinely on the 2 m B. Lyot telescope and will also be transported to remote sites for future multi-site MUSICOS campaigns, such as the one planned for December 1992. This campaign would include sites in Hawaii, ESO, OHP, Pic du Midi, Canarias, Crimea, Kitt Peak, AAT, China etc). 1991 has seen the final development of the MUSICOS spectrometer model to be duplicated by the participant countries. In 1992-1993, we plan the duplication and installation in remote sites (such as Hawaii, Xinglong, Canarias), and the start of the full network operations. When the network is operational, a Multi-Site Guest Observer program will be offered to the community.

Multi-site multi-frequency campaigns in coordination with satellite continuous observations will be also a driver for the MUSICOS operations. A multisite spectrophotometric network will complement several space missions such as EVRIS on Soviet Mars 94 probe, PRISMA currently under study at ESA for asteroseismology, or Hubble Space Telescope, ROSAT, and future UV and X-ray observatories for stellar activity or microvariability programmes. Also, moon-based astronomy (either from the future lunar base or its precursor missions) will provide very favourable viewing conditions for photometric asteroseismology and multi-frequency stellar activity studies, that can be complemented by global networks.

REFERENCES

Baudrand, J., Bohm, T, 1992 *Astr. Ap.*, in press.

Catala, C. and Foing , B.H., editors, 1988 Proc. 1st MUSICOS Workshop on MUlti SIte COntinuous Spectroscopy Meudon Observatory Pub.

Catala, C. and Foing , B.H., editors, 1990 Proc. 2nd MUSICOS Workshop on MUlti SIte COntinuous Spectroscopy, Meudon Observatory Pub.

Dappen, W., 1988 Proceedings 1st MUSICOS Workshop on MUlti SIte COntin -uous Spectroscopy eds.C.Catala, B.Foing, (Meudon Observatory Publications), p. 11.

Felenbok, P., 1988 Proceedings 1st MUSICOS Workshop on MUlti SIte COntin -uous Spectroscopy, eds.C.Catala, B.Foing, (Meudon Observatory Publications), p. 123.

Foing, B.H. *et al.*, 1990 Proceedings 2nd MUSICOS Workshop on MUlti SIte COntinuous Spectroscopy, eds.C.Catala, B.Foing, (Meudon Observatory Publications), p. 117.

Foing, B.H. *et al. Astr. Ap.*, submitted.

Seventh Cambridge Workshop on Cool Stars, Stellar Systems, and the Sun
ASP Conference Series, Vol. 26, 1992
Mark S. Giampapa and Jay A. Bookbinder (eds.)

PRISMA: A SPACE FACILITY FOR STUDYING THE ROTATION AND INTERIOR OF STARS

P. LEMAIRE I.A.S., BP 10, F-91371 Verrieres le Buisson Cedex, France

T. APPOURCHAUX, A. JONES ESTEC/SSD, P.O. Box 299, 2200 AG Noordwijk, The Netherlands

C. CATALA Observatoire de Paris-Meudon, F-92100 Meudon, France

S. CATALANO Instituto di Astronomia, I-95125 Catania, Italy

S. FRANDSEN Astronomisk Institut, DK-8000 Aarhus,Denmark

W. WEISS Institut fur Astronomie, A-1180 Wien, Austria

ABSTRACT The development of helioseismology has renewed the study of solar internal structure. The coupling between oscillations and activity is a key to understanding the solar dynamo and magnetic field generation. The PRISMA mission (Probing Rotation and Interior of Stars: Microvariability and Activity) is a space mission selected for phase A at ESA (European Space Agency). The combined measurement of frequencies, amplitudes and lifetimes of stellar oscillations with observations of emerging magnetic fields through activity indices is the method proposed to infer the internal structure and rotation rates of stars, and to follow their evolution with age. We present here the scientific objectives of the mission, the measurements to be made, the techniques to be used, and the main characteristics of the model payload.

Keywords: stellar structure, asteroseismology, stellar rotation, dynamo, stellar atmospheres, angular momentum

INTRODUCTION

Stars are the most numerous objects illuminating our visible universe. While being the basic constituent of galaxies and clusters the fundamental characteristics that drive their evolution are not very well known, mainly due to the difficulty of inferring their internal properties. Stars are point-like objects, thus most of our knowledge of their internal structure and external atmosphere follows from techniques that have already been applied to the Sun. Nevertheless the Sun only represents one set of stellar parameters. In order to constrain theories attempting to explain its structure and evolution, and to understand its past and future, it is necessary to look at other stars with different physical parameters (temperature, chemical composition, rotation rate, gravity, convection zone, magnetic field generation, angular momentum, etc...). Conversely, the evolution and behavior of galaxies and clusters cannot be understood without a clear description of the stars that constitute their basic ingredient. Improving models of stellar internal structure and evolution requires observational data other than just effective temperatures and luminosities. Additionally, observations of surface magnetic fields and of internal hydrodynamics (rotation, convection) must constrain theories of dynamo generation of magnetic fields.

Observations performed by HIPPARCOS are providing accurate parallaxes that are needed to determine precise luminosity measurements. From X–ray observations (EINSTEIN, EXOSAT, ROSAT), stellar coronae have been detected in many stars. Using the spectroscopic capability of IUE new methods of mapping the stellar surface structure have been developed. It is time now to combine these results with new observables in a mission dedicated to measure a comprehensive set of parameters required to constrain models of stellar interiors and stellar evolution.

In response to the ESA call for Medium Size Mission in 1989, the PRISMA (Probing Interior of Stars: Microvariability and Activity) mission has been proposed to fill this gap in observational data (Lemaire *et al.* 1989). The proposed mission being selected for an Assessment Study (Lemaire *et al.* 1991; Appourchaux *et al.* 1991) is now in Phase A. The mission objective is to sound stellar interiors and constrain stellar structure models by surface flux microvariability and surface magnetic activity observations using a sampling of stars spanning across the Hertzsprung-Russel diagram along and near the solar evolutionary path.

This paper describes the scientific objectives of the mission, the measurements to be made, the techniques to be used, and gives the main characteristics and performances of the proposed model payload.

SCIENTIFIC OBJECTIVES

The primary scientific objectives of the PRISMA mission (see Figure 1) are the study of structure, evolution, dynamics and dynamo mechanisms of stars through the determination of stellar parameters from the interior (core to surface) to corona. Two complementary tools will be used to attain those goals:
- asteroseismology, which is the measurement of frequencies, amplitudes and lifetimes of eigen-modes of oscillations.

Tests and improvements of stellar structural models through the determination of oscillation frequencies for many stars with different masses and ages will yield a reliable determination of the basic stellar parameters (mass, age, chemical composition), as well as a better characterization of physical processes (mixing length, equation of state, opacities) and mode excitation processes (related to convection).

The measurement of stellar internal rotation will give an observational basis to models of angular momentum loss and transport. - stellar activity monitoring, which is the observation of the surface magnetic field traced by UV and X–ray emission lines probing the various levels of the stellar atmosphere (chromosphere, transition region and corona). The resulting time-resolved 3-D picture of the atmosphere will provide the topology of the surface magnetic field and the structuration of the atmosphere.

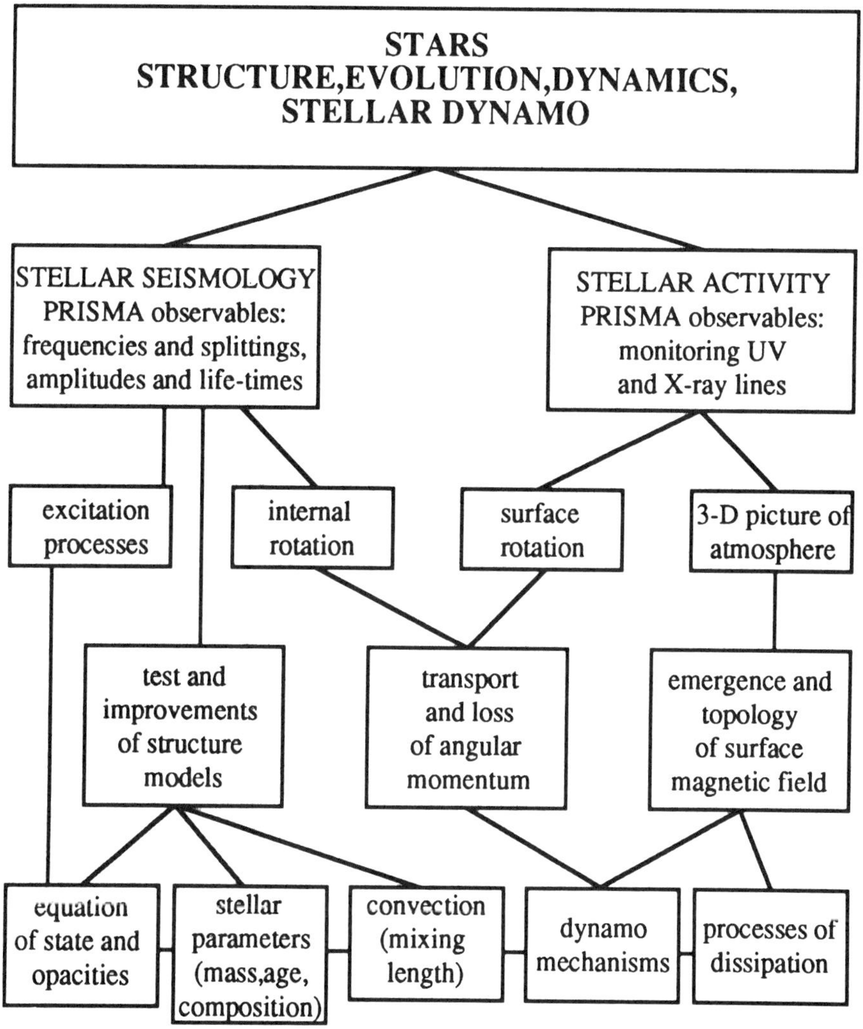

Figure 1: Summary of scientific objectives.

The study of the emergence of magnetic fields at the stellar surface provides tests of dynamo and convection models (when combined with the results of the asteroseismology experiment) and of the processes of energy deposition in atmospheric layers.

The precise determination of surface rotation rates will yield estimates of the radial differential rotation (when combined with results from asteroseismology) and will pose constraints on the depth of the dynamo action.

The mapping of the large scale magnetic structures (complemented with the asteroseismology experiment) will help to understanding the angular momentum loss process and efficiency.

MEASUREMENTS AND TECHNIQUES

Asteroseismology. Stellar oscillation studies require accurate determinations of frequencies, amplitudes, widths and lifetimes of eigen-modes. From the ground the solar Doppler shift measurements in photospheric lines (Gelly *et al.* 1988) have given valuable results to determine internal parameters. Efforts have been made to observe solar-type stars oscillations using Doppler measurements. Given the lack of photons, only the power above the noise has been detected for Procyon (Brown *et al.* 1991) and Arcturus (Belmonte *et al.* 1990). Up-to-now, the best resolved stellar flux oscillation spectrum has been obtained on the ground by the WET team (Winget *et al.* 1991) on a pre-white star from 264.1hr of nearly continuous time-series photometry, limited to 10-4 fractional amplitude by the residual atmospheric noise.

TABLE Ia Observational requirements: Photometric segment.

Observables	Range	Accuracy	Scientific keywords
Frequencies	0.02mHz-0.1Hz	0.1μHz	Stellar structure,g modes, He content, neutrino physics
Amplitudes	10^{-6} - 10^{-1}	10^{-7} - 10^{-5}	Excitation, stellar spots and plages, non linear pulsators?, convection
Line-widths	1-10μHz	0.05μHz	Excitation, convection, surface structure
Frequency separations	>0.3 μHz	0.1μHz	Basic stellar parameters, stellar structure, rotation, magnetic fields

TABLE Ib Observational requirements: Spectroscopic segment.

Observables	Range	Accuracy	Scientific keywords
UV flux	10^{-10} - 10^{-14} erg s^{-1} cm^{-2}	5%	3-D pictures
UV line profiles	120-280 nm	0.02 nm resolution	Doppler imaging
17 nm X-ray flux	10^{-11} - 10^{-14} erg s^{-1} cm^{-2}	20%	Corona, flares
Rotation	0.4-12μHz	0.4μHz	Angular momentum loss

From space, solar brightness fluctuation measurement (Woodard and Hudson 1983, Fröhlich *et al.* 1991) has proven to be a precise tool in competition with the best Doppler shift measurements. Applying this method to stars is the most promising technique to perform an extensive survey of the H-R diagram, being less limited by photon flux in using very wide spectral band and with no limitation given by the broadening of spectral lines produced by rotational velocity. To obtain significant results in understanding the internal structure of solar-type stars, photometric stability in the order of 0.1 to 1 ppm must be reached, combined with better than 1 5Hz frequency resolution to measure accurate line-width and line separation. To meet this goal it is necessary to performed continuous, uninterrupted observations for 10-30 days on the same star (see TABLE Ia). Activity monitoring. Once again the Sun is used as a reference. All activity phenomena such as spots, plages, coronal holes and flares are connected to magnetic field distribution and structure. Activity monitoring of stars takes advantage of star rotation which produces modulation of the line or continuum flux when an atmospheric structure transits across the visible part of the star. Stellar spots will be detected as dips in the photometric curves (*e.g.* smaller than 0.1as a star). Plages will enhance emission flux in UV lines (*e.g.* Mg II, He II, C IV and H I Ly-α lines) and in high temperature EUV coronal lines (*e.g.* Fe VIII-XI lines near 17 nm). Rapid transient events, such as flares, that can be directly associated with disruption-reconnection of magnetic field lines will be detected in UV-EUV spectral line intensities and sometimes in the visible photometry (white light flares). The Doppler imaging technique applied to strong lines, such as Mg II lines will provide the location and size of individual plages for the fast rotating stars (Neff *et al.* 1989). The estimation of the size and location of active regions at different atmospheric levels, leading to the determination of the topology and structure of the surface magnetic field will be derived from a combination of photometric and spectroscopic observations.

Nevertheless, activity is a source of noise for solar photometry. Stars may be more active than the Sun by several orders of magnitude, therefore a simultaneous registration of asteroseismology and activity measurements is necessary to correctly interpret the photometric data.

Stellar activity studies require high sensitivity and stability in photometry combined with multiwavelength coverage of activity tracers formed at different atmospheric temperatures. To be able to detect short term events and derive the size and location of magnetic structures through the signal modulation introduced by the star rotation, long series of observations (typically more than one period of star rotation) with sufficient time resolution (minute range) need to be performed on the same stars that are observed by the photometric segment. The requirements on the activity segments are listed in TABLE Ib.

STRATEGY OF OBSERVATION

Our main goal is the understanding of internal structure of solar-like stars to derive the basic parameters that drive the different steps of their evolution. So, we will concentrate along and above the main sequence near F, G, K spectral types and III, IV and V luminosity classes. In order to gain insights into the dynamics of stellar interiors and provide tests of dynamo theories, observation of stars with different rotation rates will be observed. Testing of the stellar internal structure models can be best achieved if one or several basic stellar parameters are known independently. Also, stars for which independent information is available have high priority in our strategy of observation, for example binary systems for which mass and sometimes radius are known, members of open clusters for which a good estimate of age is available,as well as stars for which we have a good estimate of the absolute luminosity through HIPPARCOS observations.

Within the constraints imposed on the mission during its definition, we have tentatively selected a 24 hour highly eccentric orbit with low inclination on the ecliptic plane for a nominal two-year mission. To obtain a significant sampling in the H-R diagram, *i.e.*, more than 100 stars with continuous time series of one month per star for accurate asteroseismology measurement, it is necessary to simultaneously observe several stars at the same time.

Using the model payload described in the following section, it appears that the choice of fields that can fulfill the scientific objectives, taking into account the technical constraints, is large. As an example, a possible selection of targets for the two-year mission includes:

- the Hyades and the Coma clusters that contain solar- type stars brighter than mv=8, each observed twice during the two-year mission
- 3 δ Scuti stars
- 2 rapidly pulsating CP2 stars (ro Ap)
- 2 binaries
- and 132 field stars offering a good sampling of the H-R diagram.

MODEL PAYLOAD

A model payload that meets the requirements of the mission has been proposed and is composed of 4 instruments (TABLE II).

TABLE II A model payload

Instrument	Characteristics	Field of View
Large Photometer 40 cm	10-6 magnitude in 1 month 8th magnitude multi-object	1.5° x 1.5°
UV Spectrometer 40 x 30 cm	$\Delta\lambda$ = 0.01-0.02 nm Cross dispersion 8th magnitude mono-objet	1.5° x 1.5°
XUV Telescope 15 cm	8th magnitude multi-object	1.5° x 1.5°
Small Photometer 15 cm	10-6 magnitude in 1 month 6th magnitude multi-object	3° x 3°

The first instrument is a large white light photometer (LP) with a 40 cm diameter equivalent effective collecting area. Its 1.5° x 1.5° field of view permits simultaneous observations of the broad band visible flux from several stars brighter than 8th magnitude and the detection of their micro-magnitude flux variations (see figure 2). The star images are recorded on a CCD detector, located slightly out of focus, to expand the blur over several pixels in order to provide good photometric accuracy on each individual measurement (Buffington *et al.* 1991). The radiation damage caused by the daily passage of the payload in the radiation belt can be lowered by several orders of magnitude by using multi-pinned- phase operation of the CCD and by proper shielding of the detectors. The field selection must be such that there is only one star (down to 10-12 magnitude) within the blurred image of each target. An off-axis three-mirror photometer with field stop and Lyot stop limits the out of field sky background contribution to a very small level. If required by the limited pointing accuracy of the three-axis pointing platform, a servo-loop system may be used to actuate a small secondary mirror to stabilize the stars inside one arcsecond during the observation. This photometer drives the pointing axis of the UV spectrometer and XUV telescope imager near the anti-Sun direction.

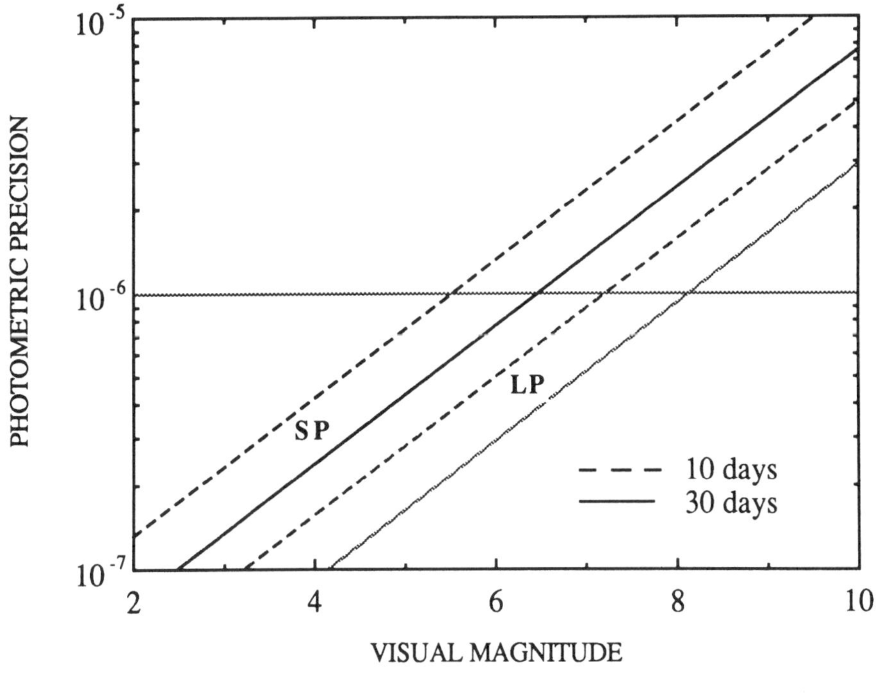

Figure 2: Photometer performances.

The second instrument is a small white light photometer (SP) with a 15 cm diameter equivalent effective collecting area. Its 3° x 3° field of view permits simultaneous observation of the broad band visible flux of several stars brighter than 6th magnitude and to detect their micro-magnitude flux variations (see figure 2). The detector choice may be a diode array that collects the out of focus blurred images of stars. The SP photometer optical scheme is very similar to that of the LP photometer. It includes an adjustable coelostat mirror to select the 3° x 3° field independently of the other instruments.

The third instrument is an UV spectrometer (UVS) with 40 x 30 cm^2 collecting area, providing a resolution between 0.01 to 0.02 nm and simultaneous coverage of the spectral region between 120 and 285 nm (cross-dispersion mounting, single object mode). The primary mirror of the telescope can rotate around its focus to select any star in the 1.5° x 1.5° field of the Large Photometer. The observation scheme within the field is performed in such a way that the period of rotation of each star is well covered by several sets of observation. The secondary mirror of the telescope is actuated in order to keep the star image inside the entrance hole of the spectrometer. Use of a photon counting detector will also greatly enhance the performances of this instrument which has characteristics very similar to the IUE high dispersion spectrometer.

Finally the fourth instrument is an eXtreme Ultra Violet Telescope (XUVT) with a bandpass of few nanometers near 17 nm. It images the 1.5° x 1.5° field of the Large Photometer at the high coronal temperature range sampled by the Fe VIII-XI lines. A multilayer coating deposited on the single mirror telescope used in normal incidence, combined with a broad band filter, allows the simultaneous field coverage and line selection (Delaboudiniere *et al.* 1989). The detector is a photon counting device.

CONCLUSION

Probing the interior of stars by measuring microvariability and activity is a challenging Phase A study in ESA. The mission objective is to sound stellar interior and to register surface activity using asteroseismology and UV spectrometry techniques, in order to derive the parameters required to constrain theoretical models of internal structure and stellar evolution. The scientific impact includes: convection and oscillation mode excitation, magnetic dynamo, angular momentum distribution and transport.

Long time series of continuous observations (typically one month) will provide the resolution needed to detect eigen-modes of stellar oscillations, and also to cover completely the phase of stellar rotation for activity measurements.

An appropriate model payload demonstrates the possibility to fulfill the scientific objective.

ACKNOWLEDGEMENTS

The Assessment Study in ESA has been performed with the active support of O. Pace, S. Volonté from the Scientific Directorate and with technical support from the Technical Directorate and the Directorate of Operations.

One of us (P.L) thanks the Conference Scientific Committee for the invitation to make this presentation.

REFERENCES

Appourchaux, T., Catala, C., Catalano, S., Frandsen, S., Jones, A., Lemaire, P., Pace, O., Volonté, S., Weiss, W. 1991, PRISMA Report on the Assessment Study, ESA-SC(91) 5.

Belmonte, J.A., Jones, A.R., Pallffi, P.L., and Roca Cortffls, T. 1990, *Ap. J.*, **358**, 595.

Brown, T.M., Gilliland, R.L., Noyes, R.W., and Ramsey, L.W. 1991, *Ap. J.*, **368**, 599.

Buffington, A. , Booth, C.H., and Hudson, H.S. 1991, *Pub. A.S.P.*, **103**, 685.

Delaboudiniere, J.-P., Gabriel, A.H., Artzner, G.E., Michels, D., Dere, K., Howard, R., Catura, R., Stern, R., Lemen, J., Neupert, W., Cugnon, P., Koeckelenbergh, A., Van Dessel, E.L., Jamar, C., Maucherat, A., Chauvineau, J.P., Marioge, J.P. 1989, in The SOHO Mission, ESA SP-1104, p. 43.

Fröhlich, C., Toutain, Th., and Schrijver C.J. 1991, *Adv. Space Res.*, **11(4)**, 69.

Gelly, B., Fossat, E., and Grec, G. 1988, in Seismology of the Sun and Sun-like Stars, ESA SP-286, p. 579.

Lemaire, P., Catala, C., Christensen-Dalsgaard, C., Hammer, R., Isaak, G., Mangeney, A., Praderie, F., Rodono, M., Roxburgh, I.W. 1989, PRISMA - Probing Rotation and Interior of Stars: Microvariability and Activity, Proposal to ESA, November 1989

Lemaire, P., Appourchaux, T., Catala, C., Catalano, S., Frandsen, S., Jones, A., Weiss, W. 1991,*Adv. Space Res.*, 11,(4), 141.

Neff, J.E., Walter, F.M., Rodono, M., Linsky, J.L. 1989, *Astr. Ap.*, **215**, 30

Winget, D.E., Nather, R.E., Clemens, J.C., Provencal, J., Kleinman, S.J., Bradley, P.A., Wood, M.A., Claver, C.F., Frueh, M.L., Grauer, A.D., Hine, B.P., Hansen, C.J., Fontaine, G., Achilleos, N., Wickramasinghe, D.T., Marar, T.M.K., Seetha, S., Ashoka, B.N., O'Donoghue, D., Warner, B., Kurtz, D.W., Buckley, D.A., Brickhill, J., Vauclair, G., Dolez, N., Chevreton, M., Barstow, M.A., Solheim, J.E., Kanaan, A., Kepler, S.O., Henry, G.W.,and Kawaler, S.D. 1991, *Ap. J.*, **378**, 326.

Woodard, M.F. and Hudson, H.S. 1983, **Solar Phys.**, **82**, 67.

Seventh Cambridge Workshop on Cool Stars, Stellar Systems, and the Sun
ASP Conference Series, Vol. 26, 1992
Mark S. Giampapa and Jay A. Bookbinder (eds.)

THE SPECTRUM-UV PROJECT FOR THE SPACE ASTROPHYSICAL OBSERVATORY

R.E.GERSHBERG
Crimean Astrophysical Observatory, Nauchny, Crimea, 334413, USSR.

ABSTRACT A preliminary description of the SPECTRUM-UV Project is given. The space astrophysical observatory is intended to consist of a main 1.7 m telescope for the 912 – 3400 Å range and a set of smaller telescopes for shorter wavelengths down to 100 Å. The main telescope will have spectrometers with resolving powers from 100 to 36000 or higher along with an imaging camera and a set of filters. The SPECTRUM-UV mission is planned as an international facility to be launched into a high apogee orbit in the middle of the 90s. Several scientific problems concerning cool stars that are to be included in the mission's program are briefly discussed.

Keywords: Instrumental Project; Cool Stars; SPECTRUM-UV

INTRODUCTION

In the USSR the Babakin Research Centre and the Lavochkin Science & Industry Corporation are now developing a new, unified spacecraft. It is intended for three main Soviet astrophysical space projects:

SPECTRUM-ROENTGEN-GAMMA: an orbital observatory for studies in high energy astrophysics;

RADIO-ASTRON: a space radio telescope with 10 m antenna; and

SPECTRUM-UV: a space astrophysical observatory with the 1.7 m main telescope and several smaller ones.

All these missions are being worked out in the framework of a broad international cooperation and are planned to be launched in the middle of the 90s. I discuss the SPECTRUM-UV Project only.

Acad. A.A. Boyarchuk is the chief scientist of the SPECTRUM-UV Project. The Crimean Astrophysical Observatory of the USSR Academy of Sciences, under the directorship of Prof. N.V. Steshenko, is the head Institute of the Project. This is due to the fact that the largest previous Soviet UV astrophysical mission, ASTRON, which was launched in 1983 and operated up to 1989, was also carried out under the scientific leadership of Acad. A.A. Boyarchuk. All the computations, manufacturing, verification and assembling of the optics of ASTRON were carried out under the guidance of Prof. N.V. Steshenko, while the staff of the Crimean Observatory was responsible for the operation of the UV part of the ASTRON mission. In that experiment, the Lavochkin SIC built the spacecraft and a 0.7 m telescope. French colleagues under the leadership of Prof G. Courtes, manufactured the spectrometer and contributed to the scientific programs. All of the results obtained with ASTRON are now collected and will be published in 1992 as a separate book (Boyarchuk 1992).

Short general descriptions of the SPECTRUM-UV Project have already been published (Boyarchuk 1989; Steshenko and Tanzi 1990; Rodonó *et al.* 1991). Italy, Germany and Canada are now participating in the Project, and the list of participating country will probably increase.

Here I shall repeat some of the published characteristics of the Project. However, I will mainly discuss those aspects of the Project that may be of interest to participants of this Workshop as well as those that are of interest to myself, both as a present contributor and a probable future user of the mission.

GENERAL CHARACTERISTICS OF THE PROJECT

The orbital SPECTRUM-UV mission is being designed as a versatile astrophysical observatory equipped with the 1.7 m reflector and a set of smaller telescopes-satellites. The main telescope of the mission, T-170, will have a primary mirror with an effective area some 4.5 times larger than either the COPERNICUS or ASTRON missions, and a factor of two smaller than the Hubble Space Telescope. T-170 is intended to obtain both direct images of celestial objects and their spectra within the wavelength range from 912 to 3400 Å. Two smaller T-50 telescopes with 50*cm* diameter primary mirrors will be installed for far UV imaging and spectral measurements in the range 400 – 1200 Å. The smallest four telescopes, T-20, will be used for recording images in the extreme UV region of 100 – 500 Å.

The SPECTRUM-UV spacecraft will be launched by the PROTON booster into a high apogee orbit with a period of about 7 days and with an initial inclination of 51°.

The on-board scientific payload will be controlled and operated through a local network with a central processor of high efficiency. The main operational mode of the spacecraft will be long-term observations up to 24 hours. The information obtained will be stored in the on-board memory and subsequently transmitted to the Earth. The ground segment of the mission will allow astronomers to control the operation of the space observatory, to monitor the scientific instruments, to obtain information from spacecraft, and to handle the data.

At any time, about 65% of the sky will be accessible for observations and during a year the whole sky can be observed. All systems of the satellite are guaranteed for operational life-times of not less than 3 years, and during this period several thousand celestial objects will be studied.

THE T-170 TELESCOPE

The T-170 telescope will be a reflector of the Ritchey-Chretien type with a primary mirror diameter of 1.7 m and an effective focal length of 17 m. The radius of curvature in the top of the primary mirror is 9.4444 m, its eccentricity-squared is 1.0539; corresponding values for the secondary mirror are 3.3846 m and 3.6829. The diameter of the primary mirror central hole is 0.68 m ($\epsilon = 0.4$). In the diffraction limit for $\lambda = 6328$ Å such a system has an Airy spot diameter of 0″.16 and 58.4% of the light is collected within the spot; 88.5% of the light is collected within the first diffraction circle with diameter 0″.37 and 90.3% within a 0″.5 circle. Since the telescope scale is 0.082 mm per arcsec, the system that collects 90% of the light into a square of $0.04 \times 0.04\,mm^2$ or 2×2 pixels may be regarded as rather close to the ideal one.

Fine guidance of T-170 will be carried out by tilting the secondary mirror along two axes around its neutral point.

The central region of the T-170 field of view is reserved for an imaging camera. Two sets of entrance diaphragms of spectrometers will be located in the focal plane in two opposite places at distances of about 5 cm from the telescope axis. Each diaphragm set will operate with a corresponding guidance system and will have at least 3 entrance diaphragms: 2 operational ones with sizes 1″ × 1″ and 0″.5 × 0″.5 and an auxiliary knee-shaped narrow slit for focusing the telescope using built-in photometers. According to ray-tracing computations, in the case of ideal optics and perfect operation of the guidance system, the 0″.5 × 0″.5 diaphragm will allow at least 91% of all the light collected by the telescope to pass into the spectrometer if the tilt of the secondary mirror compensates for a misalignment of a star of 1′; at least 51% of light if the compensation is made for 2′; and at least 35% for 3′. If one takes into account computations of the diffraction picture, for the above values of compensation for the star misalignments, one finds that it is possible to reach the usage of 82%, 46% and 31% of light with the 0″.5 × 0″.5 diaphragms and 95%, 95% and 77% with the 1″ × 1″ diaphragms.

To reduce the scattered light from the Sun, the Moon and the Earth, T-170 will have a rotatable shield as well as inner baffles; light shutters will be installed near the secondary mirror.

T-170 SCIENTIFIC PAYLOAD

Current ideas on the complement of the T-170 scientific payload are given in Table I. However, we are continuing to discuss actively these devices and therefore the Table should be regarded as preliminary.

Six spectrometers are now under consideration. One of the entrance diaphragm sets will be used for two High Resolution Spectrometers. According to one of the proposals, these HRSs are similar to the HRSs in IUE: the ultraviolet wavelength range is divided into two parts, from 1100 Å to 1900 Å and from 1900 Å to 3400 Å, and an echelle spectrometer is built for each part, spectrometers *A* and *B*, respectively. They will operate alternatively. In spectrometer *A* the light beam passing through the entrance diaphragm falls onto a collimator mirror, then a parallel beam comes to an echelle; then, diffracted beams come to a concave diffraction grating which splits spectral orders across the echelle dispersion and forms images of spectral strips on a detector. Flat mirrors insertable in front of each echelle will allow us to obtain spectra of rather faint objects with a significantly reduced spectral resolution. The configuration of spectrometer *B* is the same as for *A*, but just behind its entrance diaphragms there are two flat mirrors that shift the light beam coming into spectrometer *B*. This separates these two devices spatially.

Table I Scientific equipment of the telescope T-170

	wavelength range (Å)	resolving power ($\lambda/\delta\lambda$)
High resolution spectrometer *A*	1100 - 1900	36000 and 1300*
High resolution spectrometer *B*	1900 - 3400	36000 and 1300*
Lyman spectrometer	900 - 1200	10000
Moderate resolution spectrometer	1200 - 3000	3000
Low resolution spectrometer	1200 - 3400	500
Very low resolution spectrometer	1200 - 3400	100
High resolution UV field camera	1200-3400	100 - 30
Two FGS field cameras (guide sensors)	3500-7000	without filters

* the resolving powers are indicated for two operational modes: for high resolution mode with an echelle and a grating, and for moderate resolution mode with the grating only.

There will be two differences between the IUE spectrometers and ours. First, the light in a number of diffraction orders of the echelle, which are far removed from those that will be used to form echellegrams are provided for simultaneous photometric measurements. When one of the spectrometers operates in the moderate resolution mode, we shall use the zero order light from the concave grating. These built-in photometers can have rather high time resolution. It is possible that photometric measurements made behind an entrance diaphragm will be affected by errors of focusing and guidance; nevertheless, these measurements will give reliable color indices for variable objects during their spectroscopy where simultaneous ground observations are not always feasible. Second, instead of TV tube Secons installed on the IUE, we envisage using new two-dimensional systems that are to provide a higher accuracy for positional measurements in the spectrum as well as wider dynamic range with a compatible sensitivity.

Preliminary estimates show that the effective area of the system T-170 + spectrometer *A* is about 950 cm^2. The resolving power of 36000 can be achieved in this spectrometer with an entrance diaphragm of $1'' \times 1''$. The wavelength range from 1097 Å to 1900 Å can be overlapped by 70 spectral strips spaced at least 0.45 mm apart and covering an area of 29×54 mm^2. For spectrometer *B* we obtain an effective area of 2600 cm^2. The same resolving power for the same entrance diaphragm and the wavelength range from 1894 Å to 3430 Å can be overlapped by 71 spectral strips with the same mutual spacing and covering an area of 32×57 mm^2.

According to another proposal, spectrometers *A* and *B* are different from the IUE instruments. The spectrometer *A* has a resolving power of 57000 and, due to smaller spacing between spectral strips, it gives the whole echellegram on an area of 40×30 mm^2. Spectrometer *B* has a prism as a cross-dispersing element and a resolving power of 34000. It also gives an echellegram covering $40 \times 30\,mm^2$. A quantitative comparison of these proposals is under consideration.

The second entrance diaphragm set and its guiding system are planned to be used with several spectrometers of the Rowland type and with different resolving powers. Due to a decrease in the number of reflecting surfaces, the effective area of the system T-170 + spectrometer with a concave diffraction grating increases up to about $4600\,cm^2$.

To date, four instruments have been proposed for this box: a Lyman and a moderate resolution spectrometer, and a low and very-low resolution spectrometer. The Lyman spectrometer will cover the wavelength range from 912 Å to 1200 Å. The LRS and VLRS will be able to detect spectra of very faint stars and galaxies; these instruments will operate with different concave gratings but with the same detector. The MRS will record spectra in the 1200 – 3000 Å wavelength range. All concave diffraction gratings should have toroidal surfaces. All of the spectrometers must have built-in photometers for focusing the telescope and to check the guidance system operation. However, the final content of this instrument box of four spectrometers is still not determined.

SMALLER TELESCOPES IN THE FAR AND EXTREME ULTRAVIOLET

In order to work in the far and extreme UV, it was decided to mount two T-50 telescopes intended for spectroscopic and photometric observations in the range from 400 Å to 1200 Å, and four T-20 telescopes to obtain images of small regions on the sky in broad spectral ranges. One T-50 will be equipped with a grating spectrometer, while all other small telescopes will have the UV imaging detectors operating in a photon counting mode at the primary focus. The information concerning these telescopes is presented in Table II. The concept and content of the small telescopes' assembly are still under consideration.

Table II Characteristics of small telescopes

	$T-50S$	$T-50I$	T-20
Optical scheme	off-axis paraboloid		
Diameter of mirror (m)	0.5	0.5	0.2
Focal length (m)	2.5	2.5	2.5
Coating	Ir, Os, SiC		multilayer
Spectral band (Å)	400-1200	400-800	120, 240, 300, 430*
Spectral resolution (Å)	30	400	100
Angular resolution (arcsec)	20	10	10
Field of view 2β (arcmin)	4	4	6

* Each of four T-20 telescopes records radiation only in one of the bands, centered at the indicated wavelengths.

SPECTRUM-UV PROJECT AND COOL STARS

The SPECTRUM-UV mission is intended to cover the wavelength range from 100 Å to 3400 Å, and a broad range of stellar brightnesses down to 20^m - 21^m. Therefore, it will be a good tool to study cool stars. The mission will enable us to investigate the structure, chemical composition and kinematics of stellar atmospheres, and to study phenomena occurring in atmospheres and circumstellar volumes of various stationary and variable stars. In each of these areas, one may expect to obtain significant results. Let me describe briefly several scientific goals for which the mission is expected to be very valuable.

1. UV spectral observations are used to construct models of homogeneous upper layers of stellar atmospheres. A series of such observations allow one to estimate the degree of spottedness and the life-time of the spot features at different levels from the photosphere up to the chromosphere-corona transition zone. Such observations were initiated with the IUE and they should be continued with an instrument having better sensitivity and higher spectral resolution. Specifically, the plan is to observe emission and absorption lines of He II, C IV, Si IV, Mg II, C II, Si II, Fe II, Ti II in spectra of stars of different temperatures and luminosities, to estimate absolute temperatures and luminosities, to construct line profiles, and to determinate characteristic times of their variability. It should be noted here that the stellar spottedness is often regarded only as surface freckles on the face of a star that, in turn, have no contact with the stellar interior. However, as a manifestation of the general magnetism of a star, starspots can be in reality not freckles, but acupuncture points of the stellar body. Then, a new approach to stellar magnetism in the framework of the thermodynamics of open non-linear systems is required (Gershberg 1986). Starspot observations will give boundary conditions for such a model.

2. It may seem a paradox, but *Supernova* 1987*a* had the spectrum of a normal cool K-star (Boyarchuk *et al.* 1987). It should be remembered that due to spectral observations of SN 1987a by the ASTRON mission, the enrichment of its envelope by titanium was discovered 120 days after the explosion (Lyubimkov 1991). It was the first discovery of an enrichment of the ejected envelope by products of nucleosynthesis in this supernova explosion. This important result should be confirmed, and since 1-3 supernovae brighter than 12^m are recorded every year, there is a hope to do this during the SPECTRUM-UV operation in orbit.

3. The existence of large outflows of gas from many stars and different oscillation motions with various temporal and energy characteristics are well known. One of the important problems related to understanding such events which should be included in the research program of SPECTRUM-UV is a study of the kinematics of the T Tauri-type stellar atmospheres. For a half century of investigations of these young stars, different models of stellar winds, accretion and bipolar outflows with equatorial accretion have been considered in order to interpret their rich emission spectra. Now, a model of stochastic stellar winds is being actively developed. A report on such a model is given in this volume by Drs. V.P. Grinin and A. Mitskevich. In their model, the filling factor is a significant parameter in their spherically symmetric model of an outflow consisting of

discrete blobs. Recently, a series of spectral observations has led to conclusions on the existence of a forward-and-return motion of gaseous masses that are called "stellar prominences" (Petrov 1990). It is very desirable to study these motions through the whole depth of a stellar atmosphere, and this is possible only with observations in the UV.

4. Processes that occur in atmospheres and circumstellar volumes of non-stationary stars are very different with respect to physical causes, temporal and energy characteristics, and to understand them UV data can be key. Let us consider only two such processes that occur at the opposite ends of the stellar variability energy range: the *Supernova* 1987*a* explosion and short-lived, numerous flares on red dwarf stars.

As is well known, supernovae explosions determine to a significant extent the physical conditions and structure of interstellar gas: remnants of their envelopes form beautiful filamentary nebulae and a noticeable part of interstellar space is filled up by low density and coronal temperature plasma heated by supernova explosions. However, the process of a strong shock wave formation within the interstellar medium has not yet been observed directly. And we shall have such a possibility in 1995-98 when the matter ejected during *Supernova* 1987*a* comes into contact with the old envelope of the exploded star: the existence of this old envelope was discovered several months after the explosion due to the optical echo. During such an interaction we may expect the appearance of a gas with temperatures of $10^5 - 10^6 K$ and, consequently, strong line emission in the UV which can be recorded by SPECTRUM-UV.

5. In contrast to the grandiose explosions of supernovae which bring to an end the rather short life of the most massive stars, red dwarfs' flares occur sporadically on stars of low and moderate masses during many millions and billions of years of their slow evolution. From the physical point of view, these flares are close or identical to the solar ones and they form a common area of astrophysical problems related to processes in localized magnetic fields. The first UV spectra of such stellar flares have been obtained with IUE; they gave us a general idea about radiation from stellar atmospheres which are disturbed through the whole depth from the photosphere to the transition zone chromosphere-corona. Unfortunately, in most cases, these data have been averaged over all phases of the flare, since the time resolution of these observations was of order 10 minutes which is comparable to the total life-time of a typical stellar flare. However, during the flare on the red dwarf star EV Lac, which was observed with the ASTRON, a spike of about 1 second duration was recorded in the intensity of the C IV lines (Burnasheva *et al.* 1989).

Confirmation of this unique observation is of significant interest for understanding both stellar and solar flares. In fact, during UV observations in solar active regions, very fast and sporadic flashes of brightness in C IV and Si IV lines in very small areas have been discovered. On the other hand, very recently, the old pinch model of such flares has been revived (Hayrapetyan 1991). In this renewed scenario such thermal short-lived flashes of line emission correspond to the beginning of the pinch process that can develop into a normal flare with a flux of high energy particles along with X-ray, optical and radio emission, and high speed mass motions. Therefore, the confirmation of this unique observation by ASTRON, that has been interpreted as an analog of the solar point flashes (Katsova and Livshits 1989), and the accumulation of data on such stellar flares in chromospheric, coronal and transition region lines will allow us to advance our understanding of the origin and mechanism of these wide-spread, non-stationary processes in the realm of stars.

The problems listed here are inevitably connected with the traditions of astrophysical investigations at the Crimean Observatory. However, in the course of preparing SPECTRUM-UV as an international mission, the list of problems on cool stars will be significantly expanded by other participants of the Project.

ACKNOWLEDGEEMENTS

I am thankful to Acad. A.A. Boyarchuk for discussions, and to Dr D.J. Mullan for the improvement of my English in the original version of the paper.

REFERENCES

Boyarchuk, A.A. 1989, Space Research Institute of the USSR Academy of Sciences, preprint 1640, Moscow.

Boyarchuk, A.A. 1992, Astrophysical Investigations at the Space Station ASTRON, (Moscow: Nauka), in press.

Boyarchuk, A.A., Gershberg, R.E., Zvereva, A.M., Petrov, P.P., Severny, A.B., Terebizh, A.V., Hua, C.T. and Sheikhet, A.I. 1987, *Pis'ma Astron.Zh.*, **13**, 739.

Burnasheva, B.A., Gershberg, R.E., Zvereva, A.M., Ilyin, I.V., Shakhovskaya, N.I. and Sheikhet, A.I. 1989, *Astron.Zh.* **66**, 328.

Gershberg, R.E. 1986, in Upper Main Sequence Stars with Anomalous Abundances, eds. C.R.Cowley *et al.* , (Dordrecht:Reidel), p. 25.

Hayrapetyan, V.S. "Pinch Model of Primary Energy Release in Solar and Stellar Flares". *Solar Phys.* (in press).

Katsova, M.M. and Livshits, M.A. 1989 *Astron.Zh.*, **66**, 307.

Lyubimkov, L.S. 1991, in Supernovae, ed. S.E. Woosley, (New-York:Springer-Verlag), p. 118.

Petrov, P.P. 1990, *Astrophys.Space Sci.*, **169**, 61.

Rodonó M., Tanzi, E.G., Boyarchuk, A.A. and Steshenko, N.V. 1991, in Angular Momentum Evolution of Young Stars, eds. S. Catalano and J. Stauffer, (Dordrecht:Kluwer), p. 405.

Steshenko, N.V. and Tanzi, E.G. 1990, in Proc. IAU Coll. 123 Observatories in Earth Orbit and Beyond, Greenbelt, April 24 - 27, 1990.

Seventh Cambridge Workshop on Cool Stars, Stellar Systems, and the Sun
ASP Conference Series, Vol. 26, 1992
Mark S. Giampapa and Jay A. Bookbinder (eds.)

DEVELOPMENT OF REFLECTING CORONAGRAPHS

RAYMOND N. SMARTT and SERGE KOUTCHMY‡
National Solar Observatory/Sacramento Peak, National Optical Astronomy Observatories[1] , Sunspot, New Mexico 88349, USA

‡Institut d'Astrophysique de Paris/CNRS, Paris, France

ABSTRACT Lyot singlet-lens objective coronagraphs are used to observe the emission- and white-light components of the solar corona on a routine basis from high mountain sites. But such conventional coronagraphs cannot meet the observational requirements of many of the current outstanding problems in coronal physics. These problems require fully achromatic coronagraphic systems with high angular resolution and high photon flux. Limitations inherent to the design of Lyot coronagraphs can be substantially overcome. Mirror-objective coronagraphs have previously been proposed that now are a reality due to the recent advances in the development of superpolished mirror technology. Two prototype reflecting coronagraphs are described, of 5-cm and 15-cm aperture, based on off-axis reflection from the primary mirror to an annular field mirror (15-cm instrument) that functions as an inverse occulting disk, reflecting the coronal light to the detection system. The optical design of a third coronagraph of 55 *cm* aperture has been completed. Based on this design concept, a much larger (2 or more meters aperture) instrument is planned.

Keywords: solar/stellar reflecting coronagraphs, solar corona

1. INTRODUCTION

1.1 Observational Limitations of Existing Instrumentation

Observations of the solar corona in the wavelengths of the visible emission lines, and also the white-light (K) component, are carried out routinely at various high-altitude sites around the world (see, for example, Altrock 1988a). Coronal observations have provided a wealth of data about the short- and long-term morphological characteristics of the corona. Images in the emission lines reveal the typical complex loop structure associated with the photospheric magnetic field. But the structure usually has a rather different form associated with different lines that are characterized by different temperatures. Gross features observed in the K-corona are found also in emission-line images (Koutchmy 1988), which is to be expected if the structures are describing the magnetic-field distribution (Querfeld and Smartt 1984). But lack of high-angular-resolution observations precludes precise comparison and hence considerable

[1] Operated by the Association of Universities for Research in Astronomy, Inc., (AURA) under cooperative agreement with the National Science Foundation.

uncertainties remain. The long-term characteristics of the distribution of the corona around the limb as a function of the solar cycle can be well-studied with existing instrumentation (see, for example, Altrock, 1988b), by recording the brightness of spectra just above the limb, or by using extremely sensitive coronal photometers (Smartt 1982). Photometric measurements usually cannot be obtained beyond a height $\sim 0.5\ R_o$ above the limb, however. The relationship of hot coronal material with cooler prominence material, and with cool material in general in the coronal environment, has not been well-studied due to the limitations of instrumental sensitivity and angular resolution. Short-time-scale dynamical events such as rapid realignments (Dunn 1971) and oscillations and/or waves have been observed, but mostly such observations have provided only very limited data (Tsubaki 1988). With data from space-based white-light coronagraphs, much has been learned about coronal transients and mass ejections, and their relationship to surface events, although details of the mechanisms involved remain mostly quite speculative. Post-flare loop systems have been well-studied. However, only with improved-quality data has a phenomenon of loop interaction and partial magnetic reconnection become evident (Smartt and Zhang 1987). Extremely faint (relative to the average coronal field) features are occasionally observed at the limit of detection of existing instrumentation, but these data suggest that a whole regime of faint events that possibly occur frequently in the corona are simply never observed.

Spectral observations can provide some information about flow velocities, temperatures and, with white-light observations, densities. Flow velocities associated with coronal holes are of critical importance in solar wind studies. Such measurements are extremely difficult to obtain because of the low light flux and hence extremely noisy signals, requiring long integration times that result in low precision data (Smartt and Zirker 1981). Further, measurements of the magnitude of the magnetic field require much higher light flux (to measure the full spectral profile) and high angular resolution data.

In summary, progress in the field of coronal physics requires substantial improvements in the quality of the observational data, and hence, major advances in the instrumentation.

1.2 Observational Requirements

The above discussion highlights the lack of the availability of advanced instrumentation capable of meeting the current needs of coronal astronomy. Coronal measurements are complicated by the long integration paths inherent in optically thin visible lines and continuua. In particular, a much higher angular- and temporal-resolution capability for both images and spectra is required, with the possibility of obtaining simultaneous measurements in different spectral lines. Key to improved understanding of the coronal plasma and its dynamical manifestations is measurement of Doppler velocities and magnetic fields in small-scale structures. Doppler measurements in different temperature regimes are needed to analyze phenomena connected with oscillations, wave propagation, proper motions, etc. High-angular and spectral-resolution analysis should show line profiles composed of many components (multi-component analysis). These components should be followed precisely in time, in order to give a detailed diagnostic of the processes, their evolution and connection with other known phenomena. Up to now, the profile of a coronal line has been described by introducing a "turbulent" velocity, the appropriate value of which is very uncertain because of the lack of adequate angular resolution (Koutchmy and Smartt 1989). Advances in our understanding require precise determination of line profiles of individual features, in contrast to the smearing that results from overlapping structures. No direct measurements of the magnetic field have yet been obtained, but it is clear that the static magnetic field configuration determines the detailed morphology, while instabilities in the field principally determine the dynamical behavior.

2. EXISTING CORONAGRAPH INSTRUMENTATION

2.1 Solar Applications

Ground-based emission-line and white-light coronagraphs use a singlet, primary objective to minimize scattered light, consistent with the original Lyot design. Examples are the Sacramento Peak general purpose 40-cm-aperture emission-line coronagraph and the 20-cm-aperture patrol emission-line coronagraph (Smartt *et al.* 1981), the white-light Mk III K-Coronameter operated by the High Altitude Observatory in Hawaii (Fisher *et al.* 1981), and the eight 52-cm-aperture coronagraphs of the Soviet Union, Poland and Czechoslovakia. A primary limitation of these coronagraphs arises from the color curve of a singlet-objective lens. The secondary optical system can correct this color over a wide range of wavelengths and over a relatively large field. By use of a Mangin mirror, the total system can be made achromatic over the wavelength range transmitted by the glass. However, while the solar image is color-free, the image of the occulting disk has an enormous amount of color, with magnification a function of wavelength. This property precludes simultaneous observations at widely differing wavelengths, while observations close to the limb require a variable diameter occulting disk whose position along the optical axis is a function of wavelength. Overall, Lyot-type coronagraphs suffer from a) the chromatic properties of the primary objective; b) aperture limitations of a singlet lens, with a realistic limit probably $< 1\ m$, limiting resolution and light flux; c) long- and short-wavelength limits as determined by the glass spectral transmittance; d) internal seeing due to heat flux of the solar image. Despite such limitations, many outstanding observations have been obtained with these instruments. The use of high-resolution image intensifiers and CCD recording has greatly improved the capability of these coronagraphs (Smartt *et al.* 1987). Nevertheless, the principal requirements as outlined in Section 1.2 cannot be met with existing instrumentation.

2.2 Nighttime Applications

There is considerable interest in obtaining observations of faint emission associated with different types of astronomical objects. Such observations demand coronagraphic-quality instrumentation where the faint emission is angularly close to a much brighter source. Examples are the study of reflectance spectra of solar system objects such as planetary rings and outer satellites, white-dwarf companions of binary systems, searches for protoplanetary disks around stars, and the study of other faint emissions associated with some stellar, galactic and extragalactic objects. Coronagraph-like instruments have been designed for stellar studies. These are optical systems that incorporate the essential features of the Lyot coronagraph, that is an occulting mask at the primary image plane, and one or more apertures (equivalent to a Lyot stop) to block diffracted light from the secondary-mirror support and apertures of the telescope itself. Successful observations of Sirius B have been obtained with such a device operating on the DuPont 2.5-m telescope, Las Campanas, Chile (Vilas and Smith 1987). A similar device has been proposed for the Hubble Space Telescope (Breckinridge *et al.* 1984). Limitations result from the use of telescopes with pupil obscuration, diffracted light arising from different components, and hence different planes, in the telescope, and scattered light contributions from more than one optical surface.

2.3 Comparative Requirements

While the basic principles involved in designing coronagraphs for solar applications and for nighttime applications are the same, performance requirements for representative observations are somewhat different. The visible solar emission corona has a spectral brightness typically in the range of 10^{-4} (at some locations close to the solar limb around the period of solar maximum) to 10^{-6} (or less) of the photospheric spectral brightness of the extended solar surface. The average brightness of the brightest visible coronal line (5303Å; Fe XIV) at $1.4R_o$ from the disk center ($\sim$ 400 arcsec from the limb) is $\sim 10^{-6}\ B_o$, while the spectral brightness of the K-corona is little more than $10^{-6}\ B_o$ at the limb (Koutchmy *et al.* 1990). The observational requirement to observe the solar emission corona at a field angle of, say, 400 arcsec from the solar limb is very different from observations of, for example, the white dwarf companion of Sirius, which is 10^{-4} times as bright as the major star, with a separation of only a few arcsec (Breckinridge *et al.* 1984). In the above particular solar example, the requirement is to have an extraordinary low level of instrumentally-scattered light at a relatively large field angle, whereas the stellar example involves smaller differences in brightness, but requires minimization of small-angle scattering. Hence, in the stellar case, macroscopic smoothness of the objective surface is crucially important with regard to small angle scattering, whereas microscopic smoothness (the normal definition of a "superpolished" surface) is probably more important for typical solar coronal studies.

3. REFLECTING CORONAGRAPHS

A coronagraph based on a mirror objective avoids the problems inherent in a singlet objective design, since the primary image is achromatic, and for an all-mirror design, infrared and ultraviolet observations are limited only by the transmittance of the earth's atmosphere. Further, large apertures are, at least in principle, possible. The idea of using mirrors as coronagraph objectives has been proposed previously (Newkirk and Bohlin 1963; Zirin and Newkirk 1963) and successfully applied to small externally-occulted rocket- and balloon-borne coronagraphs (see, for example, Kohl *et al.* 1978). But it is only recently that superpolished mirrors of requisite quality have become available to allow the practical reality of internally-occulted, reflecting coronagraphs. The amount of scattered light from a polished surface depends on the residual micro-roughness of the surface, apart from that due to surface defects, dust particles and other contaminants. Now surfaces can be polished with an rms roughness of only a few Angstroms, and, for small surfaces, even at the sub-Angstrom level. Moreover, several methods have been developed for cleaning mirrors without damaging the surfaces. The problem is minimized by using special mountings for the objective and protective covers when the coronagraph is not in use.

For space-based coronagraphs, contamination of the primary-objective mirror surface is not likely to be a problem, but experience with other space optical systems suggests that gradual degrading of the reflectance of the objective could occur, especially for a solar coronagraph. A small reflecting coronagraph has been designed and constructed as one of three coronagraphs that together form the cluster, "LASCO", for the SOHO Mission. The objective is a superpolished uncoated, silicon mirror. An Earth-orbital true coronagraphic telescope of 1.5-m aperture has been proposed to allow direct detection of a Jupiter-size planet around a solar type star out to a distance of about 10 parsecs (Terrile *et al.* 1987).

3.1 Prototype Instruments

We review here briefly features of an ongoing program at NSO/SP to develop the technology of reflecting coronagraphs. The first prototype Mirror Advanced Coronagraph (MAC I) constructed in this program is based on a 5-cm-aperture superpolished silicon objective mirror with a 1-m focal length, operating in an off-axis configuration. The secondary optical system is a conventional design consisting of an occulting disk at the primary image plane to block the image of the solar disk, followed by a field lens, Lyot stop, filter and detection system (Koutchmy and Smartt 1989). As mentioned above, contaminants (such as dust) on the surface of the primary objective could severely impair the coronagraph performance. However, even though this design did not include special means to prevent dust accumulation, experience with this instrument has indicated that the problem is not severe. Only occasional cleaning was necessary, presumably a result of the basic form of the optical system of a reflecting coronagraph in which the objective is protected deep within the instrument. Images have been obtained with MAC I of the emission-corona (5303Å; Fe XIV), the first such images ever obtained with a ground-based coronagraph that uses a mirror as the primary objective. This coronagraph design serves as an excellent prominence monitor and is being converted for this purpose, with CCD detection.

A second prototype instrument (MAC II) is under development that uses a 15-cm aperture, spherical, superpolished objective mirror made of fused silica coated with aluminum, with a focal length of 2.25-m. This mirror, and its active mount, were constructed at Institut d'Astrophysique de Paris under a joint program between IAP and NSO/SP. The secondary optical system features a metal field mirror, annular-shaped, that has a concave radius of curvature. It reflects only the coronal field, while the solar image passes through the central hole to an inclined plane mirror that reflects this light to a second plane mirror and then out of the system. A Lyot stop is located at a pupil image to occult diffracted light from the primary mirror aperture, followed by a filter and detection system. While this is not an all-reflecting system, it offers the possibility of obtaining simultaneous coronal images at widely differing wavelengths, including the near-infrared emission lines. Both the MAC I and MAC II are mounted on the spar of the NSO/SP Hilltop Facility.

A research-quality reflecting coronagraph (MAC III) with an aperture of 55-cm and focal length of 4.5 m is in the design phase. The small field mirror reflects only a limited part of the coronal field (10 x 10 arcmin^2) and is rotated around the solar disk image to select different regions of the corona. This is an all-reflecting design (four mirrors) that achieves a diffraction-limited correction over the above field of view. The design allows the incorporation of a spectrograph as an alternative to direct imaging to the final focal plane. Experience with the two smaller prototype instruments has provided key ideas that will be incorporated in this more advanced instrument. It is anticipated that this coronagraph will be used also for low-scattered-light observations of the solar disk. This would be accomplished simply by tilting the primary mirror such that the field mirror reflects the solar disk image along the secondary optical axis. MAC III will be mounted on the spar of the NSO/SP John W. Evans Solar Facility. A much larger reflecting coronagraph of the same basic design, with an aperture of two or more meters, is also planned. Such a major instrument would be used for both solar and nighttime studies. For the reasons pointed out above, the surface of the primary objective would need to have a finish such that both low-angle and higher-angle scatter is minimized, appropriate for the dual purpose application of this telescope. Such a quality of polish is in principle possible, but considerable development and testing work would be required to establish a satisfactory polishing technology.

4. CONCLUSION

It has been established that advances in the field of coronal physics demand major improvements in the quality of the observations. Further, independent work points to the need for low-scattered-light coronagraphic-type telescopes for many planetary, stellar, galactic and extragalactic studies. Both of these requirements can be met by the development of reflecting coronagraphs. A miniature reflecting coronagraph (MAC I) developed at NSO/SP has been used to obtain images of the solar emission corona in the green-line, thus establishing the validity of this reflecting coronagraph technology for ground-based applications. The development of a second prototype instrument and the design of a third, research-quality instrument are viewed as prerequisites for the eventual development of a much larger instrument suitable for both solar and nighttime studies.

REFERENCES

Altrock, R. C. 1988a, in Solar and Stellar Coronal Structure and Dynamics, ed. R. C. Altrock (Sunspot, NM: NSO/SP), 288.

Altrock, R. C. 1988b, in Solar and Stellar Coronal Structure and Dynamics, ed. R. C. Altrock (Sunspot, NM: NSO/SP), 414.

Breckinridge, J. A., Kuper T. G., and Shack, R. V. 1984, *Opt. Eng.*, **23**, 816.

Dunn, R. B. 1971, in Physics of the Solar Corona, ed. Macris (Dordrecht-Holland: Reidel), 114.

Fisher, R. R., Lee, R. H., MacQueen, R. M. and Poland, A. I. 1981, *Appl. Opt.* **20**, 1094.

Kohl, J. L., Reeves, E. M., and Kirkham, B. 1978, in New Instrumentation for Space Astronomy, ed. K. A. van der Hucht and G. Vaiana (New York: Pergamon), 91.

Koutchmy, S. 1988, in Solar and Stellar Coronal Structure and Dynamics, ed. R. C. Altrock (Sunspot, NM: NSO/SP), 208.

Koutchmy, S., and Smartt, R. N. 1989, in High Spatial Resolution Solar Observations, ed. O. von der Lühe (Sunspot, NM: NSO/SP) 560.

Koutchmy, S., Colley, S., Smartt, R. N., Nitschelm, C., and Zimmermann, J. P. 1990, *SPIE*, **1235**, 849.

Newkirk, G. and Bohlin, D. 1963, *Appl. Opt.*, **2**, 131.

Querfeld, C. W. and Smartt, R. N. 1984, *Solar Phys.*, **91**, 299.

Smartt, R. N. 1982, *SPIE*, **331**, 442.

Smartt, R. N., Dunn, R. B. and Fisher, R. 1981, *SPIE* **288**, 395.

Smartt, R. N., Koutchmy, S., and Zirker, J. B. 1987, *BAAS*, **19**, 1122.

Smartt, R. N. and Zhang, Z. 1987, in Theoretical Problems in High-Resolution Solar Physics: Workshop Proceedings (2nd), ed. G. Athay (Boulder, Co: NASA), 129.

Smartt, R. N. and Zirker, J. B. 1981, in Solar Terrestrial Physics, Proceedings of Second Indo-US Workshop on Solar-Terrestrial Physics, ed. M. R. Kundu, B. Biswas, B. M. Reddy, S. Ramadurai (New Delhi: CSIR), 269.

Terrile, R. J., Ftaclas, C., Siebert, E. T. 1987, *BAAS*, **19**, 830.

Tsubaki, T. 1988, in Solar and Stellar Coronal Structure and Dynamics, ed. R. C. Altrock (Sunspot, NM: NSO/SP), 140.

Vilas, F. and Smith, B. A. 1987, *Appl. Opt.*, **26**, 664.

Zirin, H. and Newkirk, G. Jr. 1963, *Appl. Opt.*, **2**, 977.

Seventh Cambridge Workshop on Cool Stars, Stellar Systems, and the Sun
ASP Conference Series, Vol. 26, 1992
Mark S. Giampapa and Jay A. Bookbinder (eds.)

CATALOG INTEROPERABILITY AND NASA'S MASTER DIRECTORY AS A NEW FACILITY FOR RESEARCH SUPPORT

N. PAUL M. KUIN Hughes STX/NSSDC Project, 7601 Ora Glen Dr., Suite 300, Greenbelt, MD 20770, USA.

ABSTRACT The Catalog Interoperability and NASA Master Directory Project are placed in the context of other astronomical facilities, such as the Astrophysics Data System, the European Space Information System, and other on-line archives, inventories, and projects. The current status and near-future plans are described. Ways are indicated for the community to help direct the development of the interoperable catalog system.

Keywords: Catalog System, Directory System

INTRODUCTION

After an initial development and evaluation period the NASA Master Directory (NMD) became fully operational in May 1991. As the core of the Catalog Interoperability project the NMD provides general descriptions of (eventually) *nearly all* available earth and space sciences datasets, including astronomy, solar physics, planetary sciences, and space physics. The directory and interoperability effort is shared with the Earth Program Office (EPO) of the European Space Agency (ESA) and the National Space Development Agency of Japan (NASDA) Earth Observation Center, all of which maintain identical databases and run, currently, the same interface. Many other agencies are also involved in data information management in various disciplines. The main database is publicly accessible through electronic networks and dial-in, and a PC-based version is under development. The NMD provides summary description, attributes, contact information, data center information, and currently has 45 network links to remote catalogs, inventories or observing logs. Table 1 summarizes the links of astronomical and solar-heliospheric interest.

RELATIONSHIP TO THE ASTROPHYSICS DATA SYSTEM

The Astrophysics Data System (ADS) is a software product that will provide a uniform interface to many astronomical databases and software over the networks. Contained in the ADS are all inventories of data which are accessible by the ADS system. The NMD, on the other hand, attempts to have high level descriptions of *all* available datasets which will make it possible for the investigator to retrieve the information on *all* datasets available given certain search criteria. In this way the NMD attempts to be *complete*. This also poses a limitation on the NMD. For example, data sets of vastly different size will have one entry each; the IUE original data set, and a data set from a single observatory each get one NMD entry. Searching by object, and retrieving all the *data granules* (the elements of the larger data sets) is not possible with the current implementation of the NMD. One has to find which data sets are of interest, and then obtain the inventory information about the data granules by subsequently accessing all remote inventories/data systems through the network links. The ADS and ESIS will consolidate many of these inventories and make these kind of searches efficient. The NMD thus provides a point of entry for more general searches and more completeness.

TABLE 1 NMD network links to remote data/catalog systems Oct. 1991.

Catalog System	Name/Description
Astronomy:	
ADC @ NASA/NSSDC	Astronomical Data Center
EINSTEIN @ SAO	Einstein Data Center
EXOSAT @ ESA/ESTEC	Exosat Data Center
HEASARC @ NASA/GSFC	High Energy Astronomical Science Analysis and Research Center
IUE @ NASA/GSFC	International Ultraviolet Explorer
NED @ IPAC	NASA/IPAC Extra-galactic DB
NSSDC @ NASA/GSFC	National Space Science Data Center
VILSPA @ ESA	IUE archive at Villafranca, Spain
VLA @ NRAO	Very Large Array Inform. System
Solar Physics	
BATSE SOLAR DB @ NSSDC	Compton GRO observatory solar
BIG BEAR @ CALTECH	Big Bear solar observatory
SEL @ NOAA/SESC BOULDER	Space Environment Laboratory
Space Physics	
BRUNET	(ground magnetograms, ISEE, etc.)
GISS @ NASA	Goddard Institute for Space Science
OMNIFILE @ NASA/NSSDC	OMNI merged solar wind params.
SWRI	South West Research Institute
SWRIDE @ SWRI	Dynamic Explorer Data Center
Planetary Sciences	
PDS @ JPL	Planetary Data System
Multi-disciplinary	
ESIS @ ESA/ESRIN	European Space Data System

CURRENT STATUS AND FUTURE PLANS

The current directory allows the user to locate whole datasets, and attempts completeness in coverage by national and international cooperation with the various agencies and institutions. Because the directory will not actually distribute the data, but only metadata (information about the data), there are less constraints on the exchange of this information. It can help to locate other investigators in your field for collaborative efforts. To attain completeness of data set coverage, investigators and users of our system will have to submit descriptions of data they hold and will have to alert us of omissions; much depends on their interest. Especially in places where the use of electronic networks is still absent or new, this often leads to the omission of particular data.

There is a relatively rapid growth of astronomical and of solar-heliospheric data systems, inventories and digital observing logs, and rapid growth also in bringing this information on-line. In the next ten years we therefore see the possibility of accessing these remote systems in order to locate data from a particular object, and determining who to contact for those data soon after they have been taken. This would make it feasible to determine if someone else observed an object during certain time period.

The National Space Science Data Center (NSSDC) Data Archive and Distribution System (NDADS) will soon be one of the largest astronomical data archives available. It will include the ADC catalogs, IUE, HEAO, IRAS, GRO and many other datasets either on-line, or near-on-line. The ground based data from the Isaac Newton Group at the Observatorio del Roque de los Muchachos is available from the La Palma Data Archive at the Royal Greenwich Observatory. I see this as a trend, and hope that during this phase of rapid development we can agree on common formats on the exchange of the meta data, the meta data exchange, and information exchange protocols - such as the ANSI Z39.50 standard as used in the WAIS development (Kahle and Medlar 1991) - in order to achieve a truly world-wide facility to locate data of interest. In the short term we will continue to complete the coverage of datasets in the NMD, establish links to the available catalogs, and collaborate with the ADS and ESIS to enhance interoperability. We are open to suggestions of the scientific community and request its support in establishing the database and interconnectivity.

ACCESSING THE DIRECTORY AND FURTHER INFORMATION

In Europe the directory is part of the ESIS system which can be accessed by local telepad, or via SPAN/DECNET. It can also be accessed directly at SPAN/DECNET node EPOCAT (29628::), username ESAPID. In Asia the directory is accessible through SPAN/DECNET nodc 41950:: ,or via internet at rd.tksc.nasda.go.jp (133.56.11.11). In the Americas the directory is accessible via the NODIS menu at NSI/DECNET node NSSDCA, username NSSDC, or via internet nssdca.gsfc.nasa.gov (183.128.63.36). There is no password required. Some additional documentation is available with FTP from the anonymous account at nssdca.gsfc.nasa.gov, or by writing to Ms. Angelia Bland at the same address as the author.

ACKNOWLEDGEMENTS

The author would like to thank all those that have supported the NMD effort so far and David Irvine for suggested improvements in this paper.

REFERENCES

B. Kahle and A. Medlar, "An Information System for Corporate Users: Wide Area Information Servers", Thinking Machines Corporation, TMC-199, 1991.

Author Index

Object Index

Though the editors have tried to be concientious in generating a complete list of all objects referenced in the text, the following object index makes no pretense for completeness. We apologize for any omissions. In addition, the user should note that some objects may be referenced multiple times; for example, α Ori and Betelgeuse may both appear, each with its own page listings. Finally, this index only references discussions about the *exact* object in question. Thus, "RS CVn" refers only to the proto-type, and not to the class of objects. For such, please see the subject index under "RS CVn". Finally, Greek designations, such as "α Ori", will be found following English entries.

Subject Index

While the editors have tried to be concientious in generating a complete subject index, there are probably a number of items missing. Further, space limitations precluded us from printing a permuted index. Thus, while "RS CVn binaries" may appear as an entry, "binaries, RS CVn" probably doesn't. Also note that specific names such as "Algol" refer to the class of objects known as (in this example) Algol systems or Algol binaries rather than Algol itself. Elements may be listed either as spelled or as abbreviated. For example, boron might be listed as "B" or boron. Users of the index should be creative in searching for particular topics.

B

E

F

G

H

I

M

N

O

Q

S